Abbreviations for Units

A	ampere	H	henry	nm	
Å	angstrom (10^{-10} m)	h	hour	pt	pint
atm	atmosphere	Hz	hertz	qt	quart
Btu	British thermal unit	in	inch	rev	revolution
Bq	becquerel	J	joule	R	roentgen
C	coulomb	K	kelvin	Sv	seivert
°C	degree Celsius	kg	kilogram	s	second
cal	calorie	km	kilometer	T	tesla
Ci	curie	keV	kilo-electron volt	u	unified mass unit
cm	centimeter	lb	pound	V	volt
dyn	dyne	L	liter	W	watt
eV	electron volt	m	meter	Wb	weber
°F	degree Fahrenheit	MeV	mega-electron volt	y	year
fm	femtometer, fermi (10^{-15} m)	Mm	megameter (10^6 m)	yd	yard
ft	foot	mi	mile	μm	micrometer (10^{-6} m)
Gm	gigameter (10^9 m)	min	minute	μs	microsecond
G	gauss	mm	millimeter	μC	microcoulomb
Gy	gray	ms	millisecond	Ω	ohm
g	gram	N	newton		

Some Conversion Factors

Length

1 m = 39.37 in = 3.281 ft = 1.094 yd

1 m = 10^{15} fm = 10^{10} Å = 10^9 nm

1 km = 0.6215 mi

1 mi = 5280 ft = 1.609 km

1 lightyear = 1 $c \cdot y$ = 9.461 $\times$ 10^{15} m

1 in = 2.540 cm

Volume

1 L = 10^3 cm^3 = 10^{-3} m^3 = 1.057 qt

Time

1 h = 3600 s = 3.6 ks

1 y = 365.24 d = 3.156 $\times$ 10^7 s

Speed

1 km/h = 0.278 m/s = 0.6215 mi/h

1 ft/s = 0.3048 m/s = 0.6818 mi/h

Angle–angular speed

1 rev = 2π rad = 360°

1 rad = 57.30°

1 rev/min = 0.1047 rad/s

Force–pressure

1 N = 10^5 dyn = 0.2248 lb

1 lb = 4.448 N

1 atm = 101.3 kPa = 1.013 bar = 76.00 cmHg = 14.70 lb/in^2

Mass

1 u = [(10^{-3} mol^{-1})/N_A] kg = 1.661 $\times$ 10^{-27} kg

1 tonne = 10^3 kg = 1 Mg

1 slug = 14.59 kg

1 kg weighs about 2.205 lb

Energy–power

1 J = 10^7 erg = 0.7373 ft·lb = 9.869 $\times$ 10^{-3} L·atm

1 kW·h = 3.6 MJ

1 cal = 4.184 J = 4.129 $\times$ 10^{-2} L·atm

1 L·atm = 101.325 J = 24.22 cal

1 eV = 1.602 $\times$ 10^{-19} J

1 Btu = 778 ft·lb = 252 cal = 1054 J

1 horsepower = 550 ft·lb/s = 746 W

Thermal conductivity

1 W/(m·K) = 6.938 Btu·in/(h·ft^2·F°)

Magnetic field

1 T = 10^4 G

Viscosity

1 Pa·s = 10 poise

fifth edition

PHYSICS

FOR SCIENTISTS AND ENGINEERS

Volume 1A
Mechanics

W. H. Freeman and Company
New York

Publisher:	Susan Finnemore Brennan
Senior Development Editor:	Kathleen Civetta/Jennifer Van Hove
Assistant Editors:	Rebecca Pearce/Amanda McCorquodale/Eileen McGinnis
Marketing Manager:	Mark Santee
Project Editors:	Georgia L. Hadler/Cathy Townsend, PreMediaONE, A Black Dot Group Company
Cover and Text Designers:	Marcia Cohen/Blake Logan
Illustrations:	Network Graphics/PreMediaONE, A Black Dot Group Company
Photo Editors:	Patricia Marx/Dena Betz
Production Manager:	Julia DeRosa
Media and Supplements Editors:	Brian Donnellan
Composition:	PreMediaONE, A Black Dot Group Company
Manufacturing:	RR Donnelley & Sons Company

Cover image: Digital Vision

ISBN: 0-7167-0900-7 (EAN: 9780716709008)

Library of Congress Cataloging-in-Publication Data

Physics for Scientists and Engineers. - 5th ed.

   p. cm.

   By Paul A. Tipler and Gene Mosca

   Includes index.

   ISBN: 0-7167-0809-4 (Vol. 1 Hardback Ch. 1-20, R)

   ISBN: 0-7167-0900-7 (Vol. 1A Softcover Ch. 1-13, R)

   ISBN: 0-7167-0903-1 (Vol. 1B Softcover Ch. 14-20)

   ISBN: 0-7167-0810-8 (Vol. 2 Hardback Ch. 21-41)

   ISBN: 0-7167-0902-3 (Vol. 2A Softcover Ch. 21-25)

   ISBN: 0-7167-0901-5 (Vol. 2B Softcover Ch. 26-33)

   ISBN: 0-7167-0906-6 (Vol. 2C Softcover Ch. 34-41)

   ISBN: 0-7167-8339-8 (Standard Hardback Ch. 1-33, R)

   ISBN: 0-7167-4389-2 (Extended Hardback Ch. 1-41)

© 2004 by W. H. Freeman and Company

Printed in the United States of America

Second printing

PT: For Claudia

GM: For Vivian

CONTENTS IN BRIEF

CONTENTS

CHAPTER 10

CONSERVATION OF ANGULAR MOMENTUM / 309

CHAPTER R

SPECIAL RELATIVITY / R-1

CHAPTER 11

GRAVITY / 339

CHAPTER 12 *

STATIC EQUILIBRIUM AND ELASTICITY / 370

CHAPTER 20 *

THERMAL PROPERTIES AND PROCESSES / 628

VOLUME 2

PART IV   **ELECTRICITY AND MAGNETISM/651**

CHAPTER 21

THE ELECTRIC FIELD I: DISCRETE CHARGE DISTRIBUTIONS / 651

CHAPTER 22

THE ELECTRIC FIELD II: CONTINUOUS CHARGE DISTRIBUTIONS / 682

CHAPTER 30

MAXWELL'S EQUATIONS AND ELECTROMAGNETIC WAVES / 971

PART V   LIGHT/997

CHAPTER 31

PROPERTIES OF LIGHT / 997

CHAPTER 32

OPTICAL IMAGES / 1038

CHAPTER 33

INTERFERENCE AND DIFFRACTION / 1084

PART VI   MODERN PHYSICS: QUANTUM MECHANICS, RELATIVITY, AND THE STRUCTURE OF MATTER

CHAPTER 34

WAVE PARTICLE DUALITY AND QUANTUM PHYSICS

CHAPTER 35

APPLICATIONS OF THE SCHRÖDINGER EQUATION

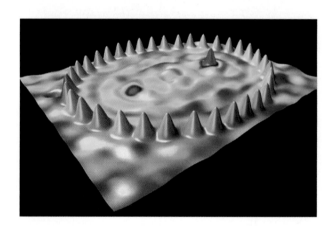

CHAPTER 36

ATOMS

CHAPTER 37

MOLECULES

CHAPTER 38

SOLIDS AND THE THEORY OF CONDUCTION

PREFACE

We are exceptionally pleased to present the fifth edition of *Physics for Scientists and Engineers*. Over the course of this revision, we have built upon the strengths of the fourth edition so that the new text is an even more reliable, engaging and motivating learning tool for the calculus-based introductory physics course. With the help of reviewers and the many users of the fourth edition we have carefully scrutinized and refined every aspect of the book, with an eye toward improving student comprehension and success. Our goals included helping students to increase their problem-solving ability, making the text more accessible and fun to read, and keeping the text flexible for the instructor.

Examples

One of the most important ways we've addressed our goals was to add some new features to the side-by-side worked examples that were introduced in the fourth edition. These examples juxtapose the problem-solving steps with the necessary equations so that it's easier for students to watch the problem unfold.

The side-by-side format for the worked examples came from a student suggestion; we've just added a few finishing touches:

- After each problem statement, students are asked to *Picture the Problem*. Here, the problem is analyzed both conceptually and visually, with students frequently directed to draw a free-body diagram. Each step of the solution is then presented with a written statement in the left-hand column and the corresponding mathematical equations in the right-hand column.

- *Remarks* at the end of the example point out the importance or relevance of the example, or suggest a different way to approach it.

- NEW *Plausibility Checks* remind students to check their results for mathematical accuracy, and for reasonableness as well.

- An *Exercise* often follows the solution of the example, allowing students to check their understanding by solving a similar problem without help. Answers are included with the Exercise to provide immediate feedback and alternative solutions.

- NEW *Master the Concept Exercises* appear at least once in each chapter and help build students' problem-solving skills online.

Every example has been scrutinized, with additional steps added wherever an assumption might have been made, new Remarks included, and new follow-up exercises, free-body diagrams added where appropriate. The answers are now boxed to make them easier to find. Our new features include the Plausibility Check, which offers quick tests that help students learn to evaluate their answers with logic. We've also added interactive Master the Concept exercises to help students work through key problems. The exercises follow examples in the textbook and are marked with a Master the Concept icon that directs students to our Web site. There, the exercise is set up with algorithmically generated variables and students work the problem with step-by-step guidance and immediate feedback.

This edition also includes two types of specialized examples that provide unique problem-solving opportunities for students. The Try it Yourself examples prompt students to take an active role in solving the problem, and the Put It in Context examples approximate the real life scenarios they might encounter as scientists.

Try It Yourself examples

Like the regular worked example, these use the side-by-side format, but here the Picture the Problem section is sometimes missing, and the descriptions in the left-hand column are more terse. These examples take students step-by-step through the solution without doing the math for them. Students find it helpful to cover the right-hand column and attempt to perform the calculations on their own before looking at the equations. In this way, students can think through the steps as they fill in the answers.

New Put It in Context examples

Each chapter now identifies at least one worked example as "context rich." These examples may include information not needed to solve the problem, or may require the student to find additional information in tables or to draw from experience or previously obtained information. Context-rich examples reflect the way that scientists and engineers solve problems in the real world. Laura McCullough of the University of Wisconsin, Stout, and Thomas Foster of Southern Illinois University, Edwardsville, initiated this feature and consulted with us in creating many of these examples.

Practice Problems

Care has been taken to improve the quality and clarity of the end-of-chapter problems. About twenty percent of the 4,500 problems are new, written by Charles Adler of St. Mary's College of Maryland. Conceptual problems have been grouped together at the beginning of each problem set, and a new category of Estimation and Approximation problems have been added to encourage students to think more like scientists or engineers. Answers to odd-numbered problems appear at the back of the text. Solutions to approximately twenty-five percent of the problems appear in the newly revised Student Solutions Manual. This was written by David Mills of the College of the Redwoods to provide detailed solutions and to mirror the popular side-by-side format of the textbook examples.

About 1,100 of the text's problems are included in the new iSOLVE homework service. These problems can be accessed at www.whfreeman.com/tipler5e. About a third of the iSOLVE problems are Checkpoint Problems, which ask students to note the key principles and equations they're using and indicate their confidence level.

Each problem is marked with:

• a series of one, two, or three bullets, which identify its level of difficulty

• a **SSM** icon if the answer is in the Student Solutions Manual

• an **iSOLVE** icon if the problem is part of the isolve homework service

and a **iSOLVE ✓** icon if the problem is a Checkpoint problem.

Features

This new edition of *Physics* has a number of textual features that make the book a valuable teaching tool. Key aspects of the last edition have been revised, and some new features have been added to make the book more engaging, inviting and up to date.

New chapter-opening pedagogy

• Each chapter now begins with a photograph and a question that is answered in a worked example within the chapter. These draw students into the material and provide motivation for problem solving.

• Chapter outlines list the major section headings, giving students a "road map" to the chapter.

• Chapter goal statements highlight the main ideas of the chapter.

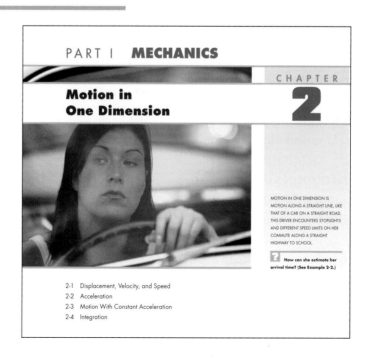

PART I **MECHANICS**

CHAPTER **2**

Motion in One Dimension

MOTION IN ONE DIMENSION IS MOTION ALONG A STRAIGHT LINE, LIKE THAT OF A CAR ON A STRAIGHT ROAD. THIS DRIVER ENCOUNTERS STOPLIGHTS AND DIFFERENT SPEED LIMITS ON HER COMMUTE ALONG A STRAIGHT HIGHWAY TO SCHOOL.

? How can she estimate her arrival time? (See Example 2-2.)

2-1 Displacement, Velocity, and Speed
2-2 Acceleration
2-3 Motion With Constant Acceleration
2-4 Integration

Content improvements

Chapter R, an optional "mini" chapter in Volume 1, brief enough to be covered in a lecture or two, allows instructors to include this popular modern topic early in the course. The chapter avoids the abstraction associated with the Lorentz transformations and focuses on the basic concepts of length contraction, time dilation, and simultaneity, using thought experiments involving meter sticks and light clocks. The relation between relativistic momentum and relativistic energy is also developed.

Quantum Theory: Chapters 17, "Wave-Particle Duality and Quantum Physics," and 27, "The Microscopic Theory of Electrical Conduction" of the fourth edition have been moved to their more traditional location in Volume II of the fifth edition as Chapters 37 and 38. Should instructors wish to include these chapters earlier in the course, both chapters are available on the web at www.whfreeman.com/tipler5e.

Changes in Approach: Dozens of smaller, yet significant improvements in content have been made throughout the book. For example:

- Motion-diagrams are introduced in Section 3-3 and used to estimate the direction of the acceleration vector using the definition of acceleration.

- In Section 4-4, frictional forces are now introduced qualitatively, allowing for free-body diagrams that include frictional forces. A quantitative treatment of frictional forces appears in Section 5-1.

- Section 4-7 introduces problems with two or more objects. Selecting a separate set of coordinate axes for each object is a robust problem-solving practice when using Newton's laws with systems consisting of two or more objects. The value of this practice is revealed in the example where Steve is sliding down the glacier while Paul has already fallen over its edge.

- In Section 8-8, "Systems With Variable Mass," the basic equation of motion for an object with continuously varying mass (the rocket equation) is developed using an object that is acquiring mass—like an open boxcar in the rain—rather than one that is losing mass—like a rocket spewing exhaust gasses. This approach facilitates both the development of the basic equation of motion and the application of it to certain situations.

- In Chapter 9, "Rotation," there is a new section that provides problem-solving guidelines for applying Newton's Second Law to rotation.

- In Section 13-3 the discussion of buoyancy now includes the buoyant force on objects supported by a submerged surface.

- In Chapter 18, work-energy relations are expressed in terms of the work done on the system. The first law of thermodynamics is now expressed in terms of the work done on the system also. (The Educational Testing Service has adopted the convention that the work term in first law of thermodynamics be the work done on the system. This will be adhered to on all Advanced Placement physics exams.)

More engineering and biological applications

Additional applications emphasize the relevance of physics to students' experiences, further studies, and future careers.

New focus on common pitfalls

Topics that commonly cause confusion are identified with a new ❶ icon where the difficulty is addressed. For example, in Section 3-4 the icon is used to identify the discussion pointing out that the horizontal and vertical motions are independent in projectile motion.

For instructor and student convenience, the fifth edition of *Physics for Scientists and Engineers* is available in five paperback volumes—

Vol. 1A Mechanics (Ch. 1-13, plus a mini-chapter on relativity, Ch. R) 0-7167-0900-7

Vol. 1B Oscillations & Waves; Thermodynamics (Ch. 14-20) 0-7167-0903-1

Vol. 2A Electricity (Ch. 21-25) 0-7167-0902-3

Vol. 2B Electrodynamics, Light (Ch. 26-33) 0-7167-0901-5

Vol. 2C Elementary Modern Physics (Ch. 34-41) 0-7167-0906-6

or in four hardcover versions—

Vol. 1 Mechanics, Oscillations and Waves; Thermodynamics (Ch. 1-20, R) 0-7167-0809-4

Vol. 2 Electricity, Magnetism, Light & Modern Physics (Ch. 21-41) 0-7167-0810-8

Standard Version (Vol 1A-2B) 0-7167-8339-8

Extended Version (Vol 1A-2C) 0-7167-4389-2

New design and improved illustrations

The book has a warmer, more colorful look. Each piece of art has been carefully considered and many have been revised to increase clarity. Approximately 245 new figures have been added, including many new free-body diagrams within the worked examples. New photos bring to life the many real-world applications of physics.

Optional sections

The book was designed to allow professors to be flexible by designating certain sections "optional." These sections are marked with an *, and professors who choose to skip this section can do so knowing that their students won't be missing any material they will need in later chapters.

Summary

End of chapter summaries are organized with important topics on the left and relevant remarks and equations on the right. Here the key equations from the chapter appear together for easy reference.

Exploring essays

Students are invited to examine interesting extensions of the chapter concepts in Exploring sections, which are now found on the Web. These short pieces relate the chapter concepts to everything from the weather to transducers.

Media and Print Supplements

The supplements package has been updated and improved in response to reviewer suggestions and those from users of the fourth edition.

For the Student:

Student Solutions Manual: *Vol. 1, 0-7167-8333-9; Vol. 2, 0-7167-8334-7.* The new manual prepared by David Mills of College of the Redwoods, Charles Adler of St. Mary's College of Maryland, Ed Whittaker of Stevens Institute of Technology, George Zober of Yough Senior High School and Patricia Zober of Ringgold High School provides solutions for about twenty-five percent of the problems in the textbook, using the same side-by-side format and level of detail as the textbook's worked examples.

Study Guide: *Vol. 1, 0-7167-8332-0; Vol. 2, 0-7167-8331-2.* Prepared by Gene Mosca of the United States Naval Academy and Todd Ruskell of Colorado School of Mines, the Study Guide describes the key ideas and potential pitfalls of each chapter, and also includes true and false questions that test essential definitions and relations, questions and answers that require qualitative reasoning, and problems and solutions.

Student Web Site: Robin Jordan of Florida Atlantic University has put together a site designed to make studying and testing easier for both students and professors. The Web site includes:

- **On-line quizzing:** Multiple choice quizzes are available for each chapter. Students will receive immediate feedback, and the quiz results are collected for the instructor in a grade book.

- **iSOLVE homework service**: *0-7167-5802-4.* About one-fourth of the book's end-of-chapter problems, 1,100 altogether, are available on-line in W.H. Freeman's iSOLVE homework service. This service will offer each student a different version of every problem similar to CAPA and WebAssign, and the iSOLVE problems will be marked with an icon in the textbook. Homework scores can be collected in a grade book. Students may purchase access to iSOLVE for three semesters at a time.

• **iSOLVE Checkpoint problems:** A third of our iSOLVE questions are Checkpoint problems, which prompt students to describe how they arrived at their answer and to indicate their confidence level. All student responses will be gathered and included in the instructor's grade book report. Rolf Enger of the U.S. Air Force Academy inspired the development of Checkpoints to help professors gauge their student's understanding of the material.

• **Master the Concept exercises:** For each chapter, one or more exercises from the book will be available on-line so students can practice working the problem with randomized variables and step-by-step guidance. The on-line exercise will walk the student slowly through the problem-solving process and use interactive animations, simulations, video, and other graphic aids to help students visualize the problem. Teachers can collect grade book information on their progress. These premium examples are called out in the book with a Master the Concept icon.

Homework services: In addition to the iSOLVE network, there are three other homework services that are compatible with this textbook. End of chapter problems are available in WebAssign as well as CAPA: A Computer-Assisted Personalized Approach. A list of all the fifth edition problems included in WebAssign and CAPA is posted on the instructor's section of the *Physics* Web site. Our text is also compatible with the University of Texas Interactive Homework Service.

The iSolve homework service is available at
www.whfreeman.com/tipler5e

For more information about WebAssign, CAPA or UTX homework services, find their Web sites at:
http://webassign.net/info
http://www.pa.msu.edu/educ/CAPA/
http://hw.utexas.edu/hw.html

For the Instructor:

Instructor's Resource CD-ROM: *0-7167-9839-5*. This multi-faceted resource will give instructors the tools to make their own Web sites and presentations. The CD contains illustrations from the text in .jpg format, Powerpoint Lecture Slides for each chapter of the book, Lab Demonstration Videos, and Applied Physics videos in QuickTime format, and Presentation Manager Pro v.2.0, as well as all of the solutions to the end-of-chapter problems in editable Microsoft Word format.

Instructor's Resource Manual: The updated IRM contains Classroom Demonstrations for each chapter, a film and video guide with suggestions for each chapter, links to valuable Web sites, and links to free sources for Physlets, animations, and other teaching tools. This manual will be available on the book's Web site at www.whfreeman.com/tipler5e.

Instructor's Solutions Manual: *Vol. 1, 0-7167-9640-6; Vol. 2, 0-7167-9639-2*. This guide contains fully worked solutions for all of the problems in the textbook, using the side-by-side format wherever possible. It is available in print and is also included in editable Word files on the Instructor's CD-ROM.

Test Bank: *In print, 0-7167-9652-X; CD-ROM, 0-7167-9653-8*. Prepared by Mark Riley of Florida State University and David Mills of College of the Redwoods, this set of more than 4,000 multiple choice questions is available both in print and on a CD-ROM for Windows and Macintosh users. All questions refer to specific sections in the book. The CD-ROM version of the Test Bank makes it easy to add, edit and re-sequence questions to suit your needs.

Transparencies: *0-7167-9664-3*. Approximately 150 full color acetates of figures and tables from the text are included, with type enlarged for projection.

Acknowledgments

We are grateful to the many instructors, students, colleagues, and friends who have contributed to this, and to earlier editions.

Charles Adler of St. Mary's College of Maryland authored the excellent new problems. David Mills of the College of the Redwoods extensively revised the solutions manual. Robin Jordan of Florida Atlantic University created the innovative Master the Concept exercises and iSOLVE Checkpoint problems. Laura McCullough of the University of Wisconsin, Stout, and Thomas Foster of Southern Illinois University, Edwardsville, drawing from their background in Physics Education Research, were instrumental in providing context-rich examples in every chapter as well as our new Estimation and Approximation problems. We received invaluable help in accuracy checking of text and problems from professors:

Karamjeet Arya,
San Jose State University

Michael Crivello,
San Diego Mesa College

David Faust,
Mt. Hood Community College

Jerome Licini,
Lehigh University

Dan Lucas,
University of Wisconsin, Madison

Jeannette Myers,
Clemson University

Marian Peters,
Appalachian State University

Paul Quinn,
Kutztown University

Michael G. Strauss,
University of Oklahoma

George Zober,
Yough Senior High School

Patricia Zober,
Ringgold High School

Many instructors and students have provided extensive and helpful reviews of one or more chapters. They have each made a fundamental contribution to the quality of this revision, and deserve our gratitude. We would like to thank the following reviewers:

Edward Adelson,
The Ohio State University

Todd Averett,
The College of William and Mary

Yildirim M. Aktas,
University of North Carolina at Charlotte

Karamjeet Arya,
San Jose State University

Alison Baski,
Virginia Commonwealth University

Gary Stephen Blanpied,
University of South Carolina

Ronald Brown,
California Polytechnic State University

Robert Coakley,
University of Southern Maine

Robert Coleman,
Emory University

Andrew Cornelius,
University of Nevada at Las Vegas

Peter P. Crooker,
University of Hawaii

N. John DiNardo,
Drexel University

William Ellis,
University of Technology - Sydney

John W. Farley,
University of Nevada at Las Vegas

David Flammer,
Colorado School of Mines

Tom Furtak,
Colorado School of Mines

Patrick C. Gibbons,
Washington University

John B. Gruber,
San Jose State University

Christopher Gould,
University of Southern California

Phuoc Ha,
Creighton University

Theresa Peggy Hartsell,
Clark College

James W. Johnson,
Tallahassee Community College

Thomas O. Krause,
Towson University

Donald C. Larson,
Drexel University

Paul L. Lee,
California State University, Northridge

Peter M. Levy,
New York University

Jerome Licini,
Lehigh University

Edward McCliment,
University of Iowa

Robert R. Marchini,
The University of Memphis

Pete E.C. Markowitz,
Florida International University

Fernando Medina,
Florida Atlantic University

Laura McCullough,
University of Wisconsin at Stout

John W. Norbury,
University of Wisconsin at Milwaukee

Melvyn Jay Oremland,
Pace University

Antonio Pagnamenta,
University of Illinois at Chicago

John Parsons,
Columbia University

Dinko Pocanic,
University of Virginia

Bernard G. Pope,
Michigan State University

Yong-Zhong Qian,
University of Minnesota

Ajit S. Rupaal,
Western Washington University

Todd G. Ruskell,
Colorado School of Mines

Mesgun Sebhatu,
Winthrop University

Marllin L. Simon,
Auburn University

Zbigniew M. Stadnik,
University of Ottawa

G. R. Stewart,
University of Florida

Michael G. Strauss,
University of Oklahoma

Chin-Che Tin,
Auburn University

Stephen Weppner,
Eckerd College

Suzanne E. Willis,
Northern Illinois University

Ron Zammit,
California Polytechnic State University

Problems/solutions reviewers

Lay Nam Chang,
Virginia Polytechnic Institute

Mark W. Coffey,
Colorado School of Mines

Brent A. Corbin,
UCLA

Alan Cresswell,
Shippensburg University

Ricardo S. Decca,
Indiana University-Purdue University

Michael Dubson,
University of Colorado at Boulder

David Faust,
Mount Hood Community College

Philip Fraundorf,
University of Missouri, Saint Louis

Clint Harper,
Moorpark College

Kristi R.G. Hendrickson,
University of Puget Sound

Michael Hildreth,
University of Notre Dame

David Ingram,
Ohio University

James J. Kolata,
University of Notre Dame

Eric Lane,
University of Tennessee, Chattanooga

Jerome Licini,
Lehigh University

Daniel Marlow,
Princeton University

Laura McCullough,
University of Wisconsin at Stout

Carl Mungan,
United States Naval Academy

Jeffry S. Olafsen,
University of Kansas

Robert Pompi,
The State University of New York
at Binghamton

R. J. Rollefson,
Wesleyan University

Andrew Scherbakov,
Georgia Institute of Technology

Bruce A. Schumm,
University of California, Santa Cruz

Dan Styer,
Oberlin College

Jeffrey Sundquist,
Palm Beach Community College - South

Cyrus Taylor,
Case Western Reserve University

Fulin Zuo,
University of Miami

Study Guide & Test Bank reviewers

Anthony J. Buffa,
California Polytechnic State University

Mirela S. Fetea,
University of Richmond

James Garner,
University of North Florida

Tina Harriott,
Mount Saint Vincent, Canada

Roger King,
City College of San Francisco

John A. McClelland,
University of Richmond

Chun Fu Su,
Mississippi State University

John A. Underwood,
Austin Community College

Media reviewers

Mick Arnett,
Kirkwood Community College

Colonel Rolf Enger,
U.S. Air Force Academy

John W. Farley,
The University of Nevada at Las Vegas

David Ingram,
Ohio University

Shawn Jackson,
The University of Tulsa

Dan MacIsaac,
Northern Arizona University

Peter E.C. Markowitz,
Florida International University

Dean Zollman,
Kansas State University

Media focus group participants

Edwin R. Jones,
University of South Carolina

William C. Kerr,
Wake Forest University

Taha Mzoughi,
Mississippi State University

Charles Niederriter,
Gustavus Adolphus College

Cindy Schwarz,
Vassar College

Dave Smith,
University of the Virgin Islands

D.J. Wagner,
Grove City College

George Watson,
University of Delaware

Frank Wolfs,
University of Rochester

We also remain indebted to the reviewers of past editions. We would therefore like to thank the following reviewers, who provided immeasurable support as we developed the fourth edition:

Michael Arnett,
Iowa State University

William Bassichis,
Texas A&M

Joel C. Berlinghieri,
The Citadel

Frank Blatt,
Michigan State University

John E. Byrne,
Gonzaga University

Wayne Carr,
Stevens Institute of Technology

George Cassidy,
University of Utah

I.V. Chivets,
Trinity College, University of Dublin

Harry T. Chu,
University of Akron

Jeff Culbert,
London, Ontario

Paul Debevec,
University of Illinois

Robert W. Detenbeck,
University of Vermont

Bruce Doak,
Arizona State University

John Elliott,
University of Manchester, England

James Garland,
Retired

Ian Gatland,
Georgia Institute of Technology

Ron Gautreau,
New Jersey Institute of Technology

David Gavenda,
University of Texas at Austin

Newton Greenberg,
SUNY Binghamton

Huidong Guo,
Columbia University

Richard Haracz,
Drexel University

Michael Harris,
University of Washington

Randy Harris,
University of California at Davis

Dieter Hartmann,
Clemson University

Robert Hollebeek,
University of Pennsylvania

Madya Jalil,
University of Malaya

Monwhea Jeng,
University of California – Santa Barbara

Ilon Joseph,
Columbia University

David Kaplan,
University of California – Santa Barbara

John Kidder,
Dartmouth College

Boris Korsunsky,
Northfield Mt. Hermon School

Andrew Lang (graduate student),
University of Missouri

David Lange,
University of California – Santa Barbara

Isaac Leichter,
Jerusalem College of Technology

William Lichten,
Yale University

Robert Lieberman,
Cornell University

Fred Lipschultz,
University of Connecticut

Graeme Luke,
Columbia University

Howard McAllister,
University of Hawaii

M. Howard Miles,
Washington State University

Matthew Moelter,
University of Puget Sound

Eugene Mosca,
United States Naval Academy

Aileen O'Donughue,
St. Lawrence University

Jack Ord,
University of Waterloo

Richard Packard,
University of California

George W. Parker,
North Carolina State University

Edward Pollack,
University of Connecticut

John M. Pratte,
Clayton College and State University

Brooke Pridmore,
Clayton State College

David Roberts,
Brandeis University

Lyle D. Roelofs,
Haverford College

Larry Rowan,
University of North Carolina
at Chapel Hill

Lewis H. Ryder,
University of Kent, Canterbury

Bernd Schuttler,
University of Georgia

Cindy Schwarz,
Vassar College

Murray Scureman,
Amdahl Corporation

Scott Sinawi,
Columbia University

Wesley H. Smith,
University of Wisconsin

Kevork Spartalian,
University of Vermont

Kaare Stegavik,
University of Trondheim, Norway

Jay D. Strieb,
Villanova University

Martin Tiersten,
City College of New York

Oscar Vilches,
University of Washington

Fred Watts,
College of Charleston

John Weinstein,
University of Mississippi

David Gordon,
Wilson, MIT

David Winter,
Columbia University

Frank L.H. Wolfe,
University of Rochester

Roy C. Wood,
New Mexico State University

Yuriy Zhestkov,
Columbia University

Of course, our work is never done. We hope to continue to receive comments and suggestions from our readers so that we can improve the text and correct any errors. If you believe you have found an error, or have any other comments, suggestions, or questions, send us a note at asktipler@whfreeman.com. We will incorporate corrections into the text during subsequent reprinting.

Finally, we would like to thank our friends at W. H. Freeman and Company for their help and encouragement. Susan Brennan, Kathleen Civetta, Georgia Lee Hadler, Julia DeRosa, Margaret Comaskey, Dena Betz, Rebecca Pearce, Brian Donnellan, Jennifer Van Hove, Patricia Marx, and Mark Santee were extremely generous with their creativity and hard work at every stage of the process. We are also grateful for the contributions of Cathy Townsend and Denise Kadlubowski at PreMediaONE and the help of our colleagues Larry Tankersley, John Ertel, Steve Montgomery, and Don Treacy.

Paul Tipler
Alameda, California

Gene Mosca
Annapolis, Maryland

ABOUT THE AUTHORS

PAUL A TIPLER

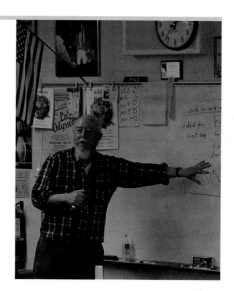

Paul Tipler was born in the small farming town of Antigo, Wisconsin, in 1933. He graduated from high school in Oshkosh, Wisconsin, where his father was superintendent of the Public Schools. He received his B.S. from Purdue University in 1955 and his Ph.D. at the University of Illinois in 1962, where he studied the structure of nuclei. He taught for one year at Wesleyan University in Connecticut while writing his thesis, then moved to Oakland University in Michigan, where he was one of the original members of the Physics department, playing a major role in developing the physics curriculum. During the next 20 years, he taught nearly all the physics courses and wrote the first and second editions of his widely used textbooks *Modern Physics* (1969, 1978) and *Physics* (1976, 1982). In 1982, he moved to Berkeley, California, where he now resides, and where he wrote *College Physics* (1987) and the third edition of *Physics* (1991). In addition to physics, his interests include music, hiking, and camping, and he is an accomplished jazz pianist and poker player.

GENE MOSCA

Gene Mosca was born in New York City and grew up on Shelter Island, New York. His undergraduate studies were at Villanova University and his graduate studies were at the University of Michigan and the University of Vermont, where he received his Ph.D. in 1974. He taught at Southampton High School, the University of South Dakota, and Emporia State University. Since 1986 Gene has been teaching at the U.S. Naval Academy. There he coordinated the core physics course for 16 semesters, and instituted numerous enhancements to both the laboratory and classroom. Proclaimed by Paul Tipler as, "the best reviewer I ever had," Mosca authored the popular Study Guide for the third and fourth editions of the text.

CHAPTER 1

Systems of Measurement

THE NUMBER OF GRAINS OF SAND ON A BEACH MAY BE TOO GREAT TO COUNT, BUT WE CAN ESTIMATE THE NUMBER BY USING REASONABLE ASSUMPTIONS AND SIMPLE CALCULATIONS.

How many grains of sand are on your favorite beach? (See Example 1-6.)

We have always been curious about the world around us. Since the beginnings of recorded thought, we have sought ways to impose order on the bewildering diversity of events that we observe. Science is a process of searching for fundamental and universal principles that govern causes and effects in the universe. The process of science is the building, testing, and connecting of falsifiable models in an effort to describe, explain, and predict reality. The process involves hypotheses, repeatable experiments and observations, and new hypotheses. The prime criteria for determining the value of a scientific model is its simplicity and its usefulness in correctly making predictions or explaining observations concerning a broad spectrum of phenomena.

Today we think of science as divided into separate fields, although this division occurred only in the last century or so. The separation of complex systems into smaller categories that can be more easily studied is one of the great successes of science in general. Biology, for example, is the study of living organisms. Chemistry deals with the interaction of elements and compounds. Geology is the study of the earth. Astronomy is the study of the solar system, the stars and galaxies, and the universe as a whole. Physics is the science of matter and energy, space and time. It includes the principles that govern the motion of particles and waves, the interactions of particles, and the properties of molecules, atoms, and atomic nuclei, as well as larger-scale systems such as gases, liquids, and solids.

Some consider physics the most fundamental science because its principles supply the foundation of the other scientific fields.

Physics is the science of the exotic and the science of everyday life. At the exotic extreme, black holes boggle the imagination. In everyday life, engineers, musicians, architects, chemists, biologists, doctors, and many others routinely command such subjects as heat transfer, fluid flow, sound waves, radioactivity, and stresses in buildings or bones to perform their daily work. Countless questions about our world can be answered with a basic knowledge of physics. Why must a helicopter have two rotors? Why do astronauts float in space? Why do moving clocks run slow? Why does sound travel around corners while light appears to travel in straight lines? Why does an oboe sound different from a flute? How do CD players work? Why is there no hydrogen in the atmosphere? Why do metal objects feel colder than wood objects at the same temperature? Why is copper an electrical conductor while wood is an insulator? Why is lithium, with its three electrons, extremely reactive, whereas helium, with two electrons, is chemically inert?

➤ **In this chapter, we will begin to prepare ourselves to answer some of these questions by examining the units and dimensions used. Any time a measurement is made, the accuracy of the measurement should be stated. If a fuel-gauge reading is 100 gallons, it does not mean that there are exactly 100 gallons of fuel left. So, what exactly does it mean, and how do I express it?**

Classical and Modern Physics

The earliest recorded efforts to systematically assemble knowledge concerning motion came from ancient Greece. In Aristotle's (384–322 B.C.) system of natural philosophy, explanations of physical phenomena were deduced from assumptions about the world, rather than derived from experimentation. For example, it was a fundamental assumption that every substance had a "natural place" in the universe. Motion was thought to be the result of a substance trying to reach its natural place. Because of the agreement between the deductions of Aristotelian physics and the motions observed throughout the physical universe, and the lack of experimentation that could overturn the ancient physical ideas, the Greek view was accepted for nearly two thousand years. It was the Italian scientist Galileo Galilei (1564–1642), whose brilliant experiments on motion established once and for all the absolute necessity of experimentation in physics, who initiated the disintegration of Aristotelian physics. Within a hundred years, Isaac Newton had generalized the results of Galileo's experiments into his three spectacularly successful laws of motion, and the reign of the natural philosophy of Aristotle was over.

Experimentation during the next two hundred years brought a flood of discoveries—and raised a flood of new questions. These inspired the development of new models to explain them. By the end of the nineteenth century, Newton's laws for the motions of mechanical systems had been joined by equally impressive laws from James Maxwell, James Joule, Sadi Carnot, and others to describe electromagnetism and thermodynamics. The subjects that occupied physical scientists through the end of the nineteenth century—mechanics, light, heat, sound, electricity and magnetism—are usually referred to as *classical physics*. Because classical physics is what we need to understand the macroscopic world we live in, it dominates Parts I through V of this text.

The remarkable success of classical physics led many scientists to believe that the description of the physical universe was complete. However, the discoveries of X rays by Wilhelm Röntgen in 1895, and of radioactivity by Antoine Becquerel and Marie and Pierre Curie in the next few years seemed to be outside the framework of classical physics. The theory of special relativity proposed by Albert Einstein in 1905 contradicted the ideas of space and time of Galileo and Newton. In the same year, Einstein suggested that light energy is quantized; that is, that light comes in discrete packets rather than being wavelike and continuous as was thought in classical physics. The generalization of this insight to the quantization of all types of energy is a central idea of quantum mechanics, one that has many amazing and important consequences. The application of special relativity, and particularly quantum theory, to such microscopic systems as atoms, molecules, and nuclei, which has led to a detailed understanding of solids, liquids, and gases, is often referred to as *modern physics*. Modern physics is the subject of Part VI of this text.

While classical physics is the main subject of this book, from time to time in the earlier parts of the text we will note the relationship between classical and modern physics. For example, when we discuss velocity in Chapter 2, we will take a moment to consider velocities near the speed of light and briefly cross over to the relativistic universe first imagined by Einstein. After discussing the conservation of energy in Chapter 7, we will discuss the quantization of energy and Einstein's famous relation between mass and energy, $E = mc^2$. Just a few chapters later, in Chapter R, we will study the nature of space and time as revealed by Einstein in 1903.

1-1 Units

We all know of things that cannot be measured—the beauty of a flower, or of a Bach fugue. As certain as our knowledge of these things may be, we readily admit that this knowledge is not science. The ability not only to define but also to measure is a requisite of science, and in physics, more than in any other field of knowledge, the precise definition of terms and the accurate measurement of quantities have led to great discoveries. We begin with some preliminary chores, such as establishing some basic definitions and learning about units and how to deal with them in equations. The fun comes later.

Measurement of any quantity involves comparison with some precisely defined unit value of the quantity. For example, to measure the distance between two points, we need a standard unit, such as a meter. The statement that a certain distance is 25 meters means that it is 25 times the length of the unit meter. That is, a standard meterstick fits into that distance 25 times. It is important to include the unit, in this case meters, along with the number 25 in expressing this distance, because there are other units of distance such as kilometers or miles that are in common use. To say that a distance is 25 is meaningless. The magnitude of any physical quantity must include both a number and a unit.

The International System of Units

A small number of fundamental units are sufficient to express all physical quantities. Many of the quantities that we shall be studying, such as velocity, force, momentum, work, energy, and power can be expressed in terms of three fundamental measures: length, time, and mass. The choice of standard units for these fundamental quantities determines a system of units. The system used universally in the scientific community is called SI (for *Système International*). The standard SI unit for length is the meter, the standard unit of time is the second, and the standard unit of mass is the kilogram. Complete definitions of the SI units are given in Appendix B.

Length The standard unit of length, the **meter** (abbreviated m), was originally defined by two scratches on a bar made of a platinum-iridium alloy kept at the International Bureau of Weights and Measures in Sèvres, France. This length was chosen so that the distance between the equator and the North Pole along the meridian through Paris would be 10 million meters (Figure 1-1). The meter is

FIGURE 1-1 The meter was originally chosen so that the distance from the equator to the North Pole along the meridian through Paris would be 10^7 m.

now defined in terms of the speed of light—the meter is the distance light travels through empty space in 1/299,729,458 second. (This makes the speed of light exactly 299,792,458 m/s.)

EXERCISE What is the circumference of the earth in meters? (*Answer* About 4×10^7 m)

Time The unit of time, the **second** (s), was originally defined in terms of the rotation of the earth and was equal to (1/60)(1/60)(1/24) of the mean solar day. The second is now defined in terms of a characteristic frequency associated with the cesium atom. All atoms, after absorbing energy, emit light with wavelengths and frequencies characteristic of the particular element. There is a set of wavelengths and frequencies for each element, with a particular frequency and wavelength associated with each energy transition within the atom. As far as we know, these frequencies remain constant. The second is defined so that the frequency of the light from a certain transition in cesium is exactly 9,192,631,770 cycles per second. With these definitions, the fundamental units of length and time are accessible to laboratories throughout the world.

Mass The unit of mass, the **kilogram** (kg), which equals 1000 grams (g), is defined to be the mass of a standard body, also kept at Sèvres. A duplicate of the standard 1-kg body is kept at the National Institute of Standards and Technology (NIST) in Gaithersburg, Maryland. We shall discuss the concept of mass in detail in Chapter 4, where we will see that the weight of an object at a given point on earth is proportional to its mass. Thus the masses of objects of ordinary size can be compared by weighing them.

In our study of thermodynamics and electricity, we shall need three more fundamental physical units: one for temperature, the kelvin (K) (formerly the degree Kelvin); one for the amount of a substance, the mole (mol); and one for electrical current, the ampere (A). There is another fundamental unit, the candela (cd) for luminous intensity, which we shall have no occasion to use in this book. These seven fundamental units, the meter (m), second (s), kilogram (kg), kelvin (K), ampere (A), mole (mol), and candela (cd), constitute the international system of units or SI units.

The unit of every physical quantity can be expressed in terms of the fundamental SI units. Some frequently used combinations are given special names. For example, the SI unit of force, kg·m/s² is called a newton (N). Similarly, the SI unit of power, 1 kg·m²/s³ = N·m/s, is called a watt (W). When a unit like the newton or the watt is someone's name, it is written starting with a lowercase letter. Abbreviations for such units start with uppercase letters.

Prefixes for common multiples and submultiples of SI units are listed in Table 1-1. These multiples are all powers of 10. Such a system is called a decimal system. The decimal system based on the meter is called the metric system. The prefixes can be applied to any SI unit; for example, 0.001 second is 1 millisecond (ms); 1,000,000 watts is 1 megawatt (MW).

(a)

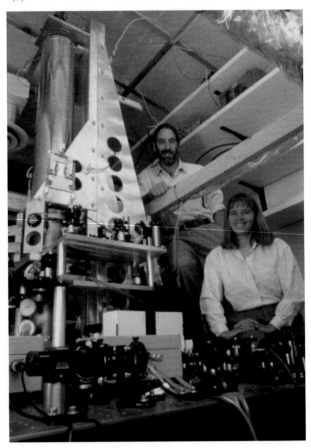

(b)

(*a*) Water clock used to measure time intervals in the thirteenth century. (*b*) Cesium fountain clock with developers Jefferts & Meekhof.

TABLE 1-1

Prefixes for Powers of 10†

Multiple	Prefix	Abbreviation
10^{18}	exa	E
10^{15}	peta	P
10^{12}	tera	T
10^{9}	giga	G
10^{6}	mega	M
10^{3}	kilo	k
10^{2}	hecto	h
10^{1}	deka	da
10^{-1}	deci	d
10^{-2}	centi	c
10^{-3}	milli	m
10^{-6}	micro	μ
10^{-9}	nano	n
10^{-12}	pico	p
10^{-15}	femto	f
10^{-18}	atto	a

† The prefixes hecto (h), deka (da), and deci (d) are not multiples of 10^{3} or 10^{-3} and are rarely used. The other prefix that is not a multiple of 10^{3} or 10^{-3} is centi (c). The prefixes frequently used in this book are printed in red. Note that all prefix abbreviations for multiples 10^{6} and higher are uppercase letters; all others are lowercase letters.

Other Systems of Units

Another decimal system still in use but gradually being replaced by SI units is the cgs system, based on the centimeter, gram, and second. The centimeter is defined as 0.01 m. The gram is now defined as 0.001 kg. Originally the gram was defined as the mass of one cubic centimeter (cm^{3}) of water at 4°C. The kilogram is then the mass of 1 liter (1000 cm^{3}) of water.

In another system of units, the U.S. customary system, a unit of force, the pound, is chosen to be a fundamental unit. In this system, the unit of mass is then defined in terms of the fundamental unit of force. The pound is defined in terms of the gravitational attraction of the earth at a particular place for a standard body. The fundamental unit of length in this system is the foot and the unit of time is the second, which has the same definition as the SI unit. The foot is defined as exactly one-third of a yard, which is now defined in terms of the meter:

$$1 \text{ yd} = 0.9144 \text{ m} \qquad\qquad 1\text{-}1$$

$$1 \text{ ft} = \tfrac{1}{3}\text{yd} = 0.3048 \text{ m} \qquad\qquad 1\text{-}2$$

making the inch exactly 2.54 cm. This scheme is not a decimal system. It is less convenient than the SI or other decimal systems because common multiples of the unit are not powers of 10. For example, 1 yd = 3 ft and 1 ft = 12 in. We will see in Chapter 4 that mass is a better choice for a fundamental unit than force because mass is an intrinsic property of an object, independent of its location. Relations between the U.S. customary system and SI units are given in Appendix A.

1-2 Conversion of Units

All physical quantities contain both a number and a unit. When such quantities are added, subtracted, multiplied, or divided in an algebraic equation, the unit can be treated like any other algebraic quantity. For example, suppose you want to find the distance traveled in 3 hours (h) by a car moving at a constant rate of 80 kilometers per hour (km/h). The distance is the product of the speed v and the time t:

$$x = vt = \frac{80 \text{ km}}{\cancel{\text{h}}} \times 3 \cancel{\text{h}} = 240 \text{ km}$$

We cancel the unit of time, the hours, just as we would any algebraic quantity to obtain the distance in the proper unit of length, the kilometer. This method of treating units makes it easy to convert from one unit of distance to another. Suppose we want to convert our answer of 240 km to miles (mi). Using the fact that 1 mi = 1.61 km, we divide each side of this equality by 1.61 km to obtain

$$\frac{1 \text{ mi}}{1.61 \text{ km}} = 1$$

Because any quantity can be multiplied by 1 without changing its value, we can now change 240 km to miles by multiplying by the factor (1 mi)/(1.61 km):

$$240 \text{ km} = 240 \cancel{\text{km}} \times \frac{1 \text{ mi}}{1.61 \cancel{\text{km}}} = 149 \text{ mi}$$

The factor (1 mi)/(1.61 km) is called a **conversion factor.** All conversion factors have a value of 1 and are used to convert a quantity expressed in one unit of measure into its equivalent in another unit of measure. By writing out the units explicitly and canceling them, we do not need to think about whether we multiply by 1.61 or divide by 1.61 to change kilometers to miles, because the units tell us whether we have chosen the correct or incorrect factor.

(*a*) Laser beam from the Macdonald Observatory used to measure the distance to the moon. The distance can be measured within a few centimeters by measuring the time required for the beam to go to the moon and back after reflecting off a mirror (*b*) placed on the moon by the Apollo 14 astronauts.

(*a*)

(*b*)

USING CONVERSION FACTORS   **EXAMPLE 1-1**

Your employer sends you on a trip to a foreign country where the road signs give distances in kilometers and the automobile speedometers are calibrated in kilometers per hour. If you drive 90 km/h, how fast are you going in meters per second and in miles per hour?

PICTURE THE PROBLEM We use the facts that 1000 m = 1 km, 60 s = 1 min, and 60 min = 1 h to convert to meters per second. The quantity 90 km/h is multiplied by a set of conversion factors each having the value 1, so the value of the speed is not changed. To convert to miles per hour, we use the conversion factor (1 mi)/(1.61 km) = 1.

1. Multiply 90 km/h by a set of conversion factors that convert km to m and h to s:

$$\frac{90 \ km}{h} \times \frac{1000 \ m}{1 \ km} \times \frac{1 \ h}{60 \ min} \times \frac{1 \ min}{60 \ s} = \boxed{25 \ m/s}$$

2. Multiply 90 km/h by 1 mi/1.61 km:

$$\frac{90 \ km}{h} \times \frac{1 \ mi}{1.61 \ km} = \boxed{55.9 \ mi/h}$$

EXERCISE What is the equivalent of 65 mi/h in m/s? (*Answer* 29.1 m/s)

1-3 Dimensions of Physical Quantities

The area of a surface is found by multiplying one length by another. For example, the area of a rectangle of sides 2 m and 3 m is $A = (2 \ m)(3 \ m) = 6 \ m^2$. The units of this area are square meters. Because area is the product of two lengths, it is said to have the dimensions of length times length, or length squared, often written L^2. The idea of dimensions is easily extended to other nongeometric quantities. For example, speed is said to have the dimensions of length divided by time, or L/T. The dimensions of other quantities such as force or energy are written in terms of the fundamental quantities of length, time, and mass. Adding two physical quantities makes sense only if the quantities have the same dimensions. For example, we cannot add an area to a speed to obtain a meaningful sum. If we have an equation like

$A = B + C$

the quantities A, B, and C must all have the same dimensions. Addition of B and C also requires that these quantities be in the same units. For example, if B is an area of 500 in.^{2} and C is 4 ft^2, we must either convert B into square feet or C into square inches in order to find the sum of the two areas.

We can often find mistakes in a calculation by checking the dimensions or units of the quantities in our result. Suppose, for example, that we mistakenly use the formula $A = 2\pi r$ for the area of a circle. We can see immediately that this cannot be correct because $2\pi r$ has dimensions of length whereas area must have dimensions of length squared. Dimensional consistency is a necessary but not a sufficient condition for an equation to be correct. An equation can have the correct dimensions in each term without describing any physical situation. Table 1-2 gives the dimensions of some quantities we will encounter in physics.

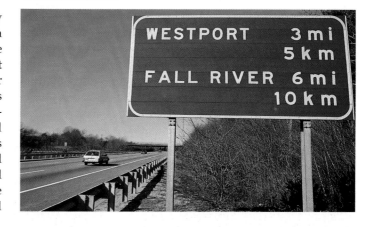

TABLE 1-2

Dimensions of Physical Quantities

Quantity	Symbol	Dimension
Area	A	L^2
Volume	V	L^3
Speed	v	L/T
Acceleration	a	L/T^2
Force	F	ML/T^2
Pressure (F/A)	p	M/LT^2
Density (M/V)	ρ	M/L^3
Energy	E	ML^2/T^2
Power (E/T)	P	ML^2/T^3

EXAMPLE 1-2

The pressure in a fluid in motion depends on its density ρ and its speed v. Find a simple combination of density and speed that gives the correct dimensions of pressure.

PICTURE THE PROBLEM We note from Table 1-2 that both pressure and density have units of mass in the numerator, whereas speed does not contain M. We therefore divide the units of pressure by those of density and inspect the result.

1. Divide the units of pressure by those of density:

$$\frac{[p]}{[\rho]} = \frac{M(L/T^2)}{M/L^3} = \frac{L^2}{T^2}$$

2. By inspection, we note that the result has dimensions of v^2. The dimensions of pressure are thus the same as the dimensions of density times speed squared:

$$[p] = [\rho][v^2] = \frac{M}{L^3}\left(\frac{L}{T}\right)^2 = \boxed{\frac{M}{LT^2}}$$

REMARKS When we study fluids in motion in Chapter 13, we will see from Bernoulli's law that for a fluid moving at a constant height, $p + \frac{1}{2}\rho v^2$ is constant where p is the pressure in the fluid.

1-4 Scientific Notation

Handling very large or very small numbers is simplified by using scientific notation. In this notation, the number is written as a product of a number between 1 and 10 and a power of 10, such as 10^2 ($= 100$) or 10^3 ($= 1000$). For example, the number 12,000,000 is written 1.2×10^7; the distance from the earth to the sun, about 150,000,000,000 m, is written 1.5×10^{11} m. The number 11 in 10^{11} is called the **exponent.** For numbers smaller than 1, the exponent is negative. For example, $0.1 = 10^{-1}$, and $0.000\,1 = 10^{-4}$. The diameter of a virus, which is about 0.00000001 m, is written 1×10^{-8} m.

When numbers in scientific notation are multiplied, the exponents are added; when divided, the exponents are subtracted. These rules can be seen from some simple examples:

$$10^2 \times 10^3 = 100 \times 1{,}000 = 100{,}000 = 10^5$$

Similarly,

$$\frac{10^2}{10^3} = \frac{100}{1000} = \frac{1}{10} = 10^{2-3} = 10^{-1}$$

In this notation, 10^0 is defined to be 1. To see why, suppose we divide 1000 by itself. We have

$$\frac{1000}{1000} = \frac{10^3}{10^3} = 10^{3-3} = 10^0 = 1$$

EXAMPLE 1-3

In 12 g of carbon, there are $N_A = 6.02 \times 10^{23}$ carbon atoms (Avogadro's number). If you could count 1 atom per second, how long would it take to count the atoms in 1 g of carbon? Express your answer in years.

PICTURE THE PROBLEM We need to find the total number of atoms to be counted, N, and then use the fact that the number counted equals the counting rate R multiplied by the time t.

1. The time is related to the total number of atoms N, and the rate of counting $R = 1$ atom/s:

$$t = \frac{N}{R}$$

2. Find the number of carbon atoms in 1 gram:

$$N = \frac{6.02 \times 10^{23} \text{ atoms}}{12 \text{ g}} \times 1 \text{ g} = 5.02 \times 10^{22} \text{ atoms}$$

3. Calculate the number of seconds it takes to count these at 1 per second:

$$t = \frac{N}{R} = \frac{5.02 \times 10^{22} \text{ atoms}}{1 \text{ atom/s}} = 5.02 \times 10^{22} \text{ s}$$

4. Calculate the number of seconds in a year:

$$n = \frac{365 \text{ d}}{1 \text{ y}} \times \frac{24 \text{ h}}{1 \text{ d}} \times \frac{3600 \text{ s}}{1 \text{ h}} = 3.15 \times 10^7 \text{ s/y}$$

5. Use the conversion factor 3.15×10^7 s/y (a handy quantity to remember) to convert the answer in step 3 to years:

$$t = 5.02 \times 10^{22} \text{ s} \times \frac{1 \text{ y}}{3.15 \times 10^7 \text{ s}}$$

$$= \frac{5.02}{3.15} \times 10^{22-7} \text{ y} = \boxed{1.59 \times 10^{15} \text{ y}}$$

REMARKS The time required is about 100,000 times the age of the universe.

EXERCISE If you divided the task so that each person counted different atoms, how long would it take for 5 billion (5×10^9) people to count the atoms in 1 g of carbon? (*Answer* 3.19×10^5 y)

How Much Water?   **EXAMPLE 1-4**

A liter (L) is the volume of a cube that is 10 cm by 10 cm by 10 cm. If you drink 1 liter of water, how much volume in cubic centimeters and in cubic meters would it occupy in your stomach?

PICTURE THE PROBLEM The volume of a cube of side ℓ is $V = \ell^3$. The volume in cubic centimeters is found directly from $\ell = 10$ cm. To find the volume in cubic meters, convert cm^3 to m^3 using the conversion factor 1 cm $= 10^{-2}$ m.

1. Calculate the volume in cm^3:

$$V = \ell^3 = (10 \text{ cm})^3 = 10^3 \text{ cm}^3$$

2. Convert to m^3:

$$10^3 \text{ cm}^3 = 10^3 \text{ cm}^3 \times \left(\frac{10^{-2} \text{ m}}{1 \text{ cm}}\right)^3 = 10^3 \text{ cm}^3 \times \left(\frac{10^{-6} \text{ m}^3}{1 \text{ cm}^3}\right) = \boxed{10^{-3} \text{ m}^3}$$

REMARKS Note that the conversion factor (which equals 1) can be raised to the third power without changing its value, enabling us to cancel units.

Care is required when adding or subtracting numbers written in scientific notation when their exponents don't match. Consider, for example,

$$(1.200 \times 10^2) + (8 \times 10^{-1}) = 120.0 + 0.8 = 120.8$$

To find the sum without converting both numbers into ordinary decimal form, it is sufficient to rewrite either of the numbers so that its power of 10 is the same as that of the other. For example, we can find the sum by writing $1.200 \times 10^2 = 1200 \times 10^{-1}$ and then adding:

$$(1200 \times 10^{-1}) + (8 \times 10^{-1}) = 1208 \times 10^{-1} = 120.8$$

When the exponents are very different, one of the numbers is much smaller than the other. The smaller number can often be neglected in addition or subtraction.

For example,

$$(2 \times 10^6) + (9 \times 10^{-3}) = 2{,}000{,}000 + 0.009$$

$$= 2{,}000{,}000.009 \approx 2 \times 10^6$$

where the symbol $\approx$ means "is approximately equal to."

When raising a power to another power, the exponents are multiplied. For example,

$$(10^2)^4 = 10^2 \times 10^2 \times 10^2 \times 10^2 = 10^8$$

1-5 Significant Figures and Order of Magnitude

Many of the numbers in science are the result of measurement and are therefore known only to within some degree of experimental uncertainty. The magnitude of the uncertainty depends on the skill of the experimenter and the apparatus used, and often can only be estimated. A rough indication of the uncertainty in a measurement is inferred by the number of digits used. For example, if we say that a table is 2.50 m long, we are often saying that its length is between 2.495 m and 2.505 m. That is, we know the length to about $\pm$ 0.005 m = $\pm$ 0.5 cm. If we used a meterstick with millimeter markings and measured the table length carefully, we might estimate that we could measure the length to $\pm$ 0.5 mm rather than $\pm$ 0.5 cm. We would indicate this precision when giving the length by using four digits, such as 2.503 m. A reliably known digit (other than a zero used to locate the decimal point) is called a **significant figure.** The number 2.50 has three significant figures; 2.503 m has four. The number 0.001 03 has three significant figures. (The first three zeroes are not significant figures but merely locate the decimal point.) In scientific notation, the number 0.001 03 is written 1.03×10^{-3}. A common student error is to carry more digits than the certainty of measurement warrants. Suppose, for example, that you measure the area of a circular playing field by pacing off the radius and using the formula for the area of a circle, $A = \pi r^2$. If you estimate the radius to be 8 m and use a 10-digit calculator to compute the area, you obtain $\pi(8 \text{ m})^2 = 201.0619298 \text{ m}^2$. The digits after the decimal point give a false indication of the accuracy with which you know the area. If you found the radius by pacing, you might expect that your measurement was accurate to only about 0.5 m. That is, the radius could be as great as 8.5 m or as small as 7.5 m. If the radius is 8.5 m, the area is $\pi(8.5 \text{ m})^2 = 226.9800692 \text{ m}^2$, whereas if it is 7.5 m, the area is $\pi(7.5 \text{ m})^2 = 176.714587 \text{ m}^2$. There is a general rule to guide you when combining several numbers in multiplication or division:

The number of significant figures in the result of multiplication or division is no greater than the least number of significant figures in any of the factors.

In the previous example, the radius is known to only one significant figure, so the area is also known only to one significant figure. It should be written as $2 \times 10^2 \text{ m}^2$, which says that the area is probably between 150 m² and 250 m².

The precision of the sum or difference of two measurements is only as good as the precision of the least precise of the two measurements. A general rule is:

The result of addition or subtraction of two numbers has no significant figures beyond the last decimal place where both of the original numbers had significant figures.

Benzene molecules of the order of 10^{-10} m in diameter as seen in a scanning electron microscope.

Chromosomes measuring on the order of 10^{-6} m across as seen in a scanning electron microscope.

SIGNIFICANT FIGURES

EXAMPLE 1-5

Find the sum of 1.040 and 0.21342.

PICTURE THE PROBLEM The first number, 1.040, has only three significant figures beyond the decimal point, whereas the second, 0.21342 has five. According to the rule stated above, the sum can have only three significant figures beyond the decimal point.

Sum the numbers, keeping only three digits beyond the decimal point:

$$1.040 + 0.21342 = \boxed{1.253}$$

EXERCISE Apply the appropriate rule for significant figures to calculate the following: (a) 1.58×0.03, (b) $1.4 + 2.53$, (c) $2.34 \times 10^2 + 4.93$ (Answer (a) 0.05, (b) 3.9, (c) 2.39×10^2)

Most examples and exercises in this book will be done with data to three (or sometimes four) significant figures, but occasionally we will say, for example, that a table top is 3 ft by 8 ft rather than taking the time and space to say it is 3.00 ft by 8.00 ft. Any data you see in an example or exercise can be assumed to be known to three significant figures unless otherwise indicated. The same assumption holds for the data in the end-of-chapter problems. In doing rough calculations or comparisons, we sometimes round off a number to the nearest power of 10. Such a number is called an **order of magnitude**. For example, the height of a small insect, say an ant, might be 8×10^{-4} m $\approx 10^{-3}$ m. We would say that the order of magnitude of the height of an ant is 10^{-3} m. Similarly, though the height of most people is about 2 m, we might round that off and say that the order of magnitude of the height of a person is 10^0 m. By this we do not mean to imply that a typical height is really 1 m but that it is closer to 1 m than to 10 m or to $10^{-1} = 0.1$ m. We might say that a typical human being is three orders of magnitude taller than a typical ant, meaning that the ratio of heights is about 1000 to 1. An order of magnitude does not provide any digits that are reliably known.

Distances familiar in our everyday world. The height of the woman is of the order of 10^0 m and that of the mountain is of the order of 10^4 m.

Earth, with a diameter of the order of 10^7 m, as seen from space.

The diameter of the Andromeda galaxy is of the order of 10^{21} m.

TABLE 1-3

The Universe by Orders of Magnitude

Size or Distance	(m)	Mass	(kg)	Time Interval	(s)
Proton	10^{-15}	Electron	10^{-30}	Time for light to cross nucleus	10^{-23}
Atom	10^{-10}	Proton	10^{-27}	Period of visible light radiation	10^{-15}
Virus	10^{-7}	Amino acid	10^{-25}	Period of microwaves	10^{-10}
Giant amoeba	10^{-4}	Hemoglobin	10^{-22}	Half-life of muon	10^{-6}
Walnut	10^{-2}	Flu virus	10^{-19}	Period of highest audible sound	10^{-4}
Human being	10^{0}	Giant amoeba	10^{-8}	Period of human heartbeat	10^{0}
Highest mountain	10^{4}	Raindrop	10^{-6}	Half-life of free neutron	10^{3}
Earth	10^{7}	Ant	10^{-4}	Period of earth's rotation	10^{5}
Sun	10^{9}	Human being	10^{2}	Period of earth's revolution	
Distance from earth		Saturn V rocket	10^{6}	around sun	10^{7}
to sun	10^{11}	Pyramid	10^{10}	Lifetime of human being	10^{9}
Solar system	10^{13}	Earth	10^{24}	Half-life of plutonium-239	10^{12}
Distance to nearest star	10^{16}	Sun	10^{30}	Lifetime of mountain range	10^{15}
Milky Way galaxy	10^{21}	Milky Way galaxy	10^{41}	Age of earth	10^{17}
Visible universe	10^{26}	Universe	10^{52}	Age of universe	10^{18}

It may be thought of as having no significant figures. Table 1-3 gives some typical order-of-magnitude values for a variety of sizes, masses, and time intervals encountered in physics.

In many cases the order of magnitude of a quantity can be estimated using reasonable assumptions and simple calculations. The physicist Enrico Fermi was a master at using cunning order-of-magnitude estimations to generate answers for questions that seemed impossible to calculate because of lack of information. Problems like these are often called **Fermi questions.** The following is an example of a Fermi question.

EXPLORING

How many piano tuners are there in Chicago? Find out this, and more, at www.whfreeman.com/tipler5e.

Burning Rubber **EXAMPLE 1 - 6**

What thickness of rubber tread is worn off the tire of an automobile as it travels 1 km (0.6 mi)?

PICTURE THE PROBLEM We assume the tread thickness of a new tire is 1 cm. This may be off by a factor of two or so, but 1 mm is certainly too small and 10 cm is too large. Since tires have to be replaced after about 60,000 km (about 37,000 mi), we will assume that the tread is completely worn off after 60,000 km. In other words, the rate of wear is 1 cm of tire per 60,000 km of travel.

Use 1 cm wear per 60,000 km travel to compute the thickness worn after 1 km of travel:

$$\frac{1 \text{ cm wear}}{60,000 \text{ km travel}} = \frac{1.7 \times 10^{-5} \text{ cm wear}}{1 \text{ km travel}}$$

$$\boxed{\approx 0.2 \ \mu\text{m wear per km of travel}}$$

EXERCISE How many grains of sand are on a 0.50 km stretch of beach that is 100 m wide? *Hint: Assume that the sand is 3 m deep. Estimate that the diameter of one grain of sand is 1.00 mm.* (*Answer* $\approx 2 \times 10^{14}$)

SUMMARY

The fundamental units in the SI system are the meter (m), the second (s), the kilogram (kg), the kelvin (K), the ampere (A), the mole (mol), and the candela (cd). The unit(s) of every physical quantity can be expressed in terms of these fundamental units.

Topic	Relevant Equations and Remarks
1. Units	The magnitude of physical quantities (for example, length, time, force, and energy) are expressed as a number times a unit.
Fundamental units	The fundamental units in the SI system (short for *Système International*) are the meter (m), the second (s), the kilogram (kg), the kelvin (K), the ampere (A), the mole (mol), and the candela (cd). The unit(s) of every physical quantity can be expressed in terms of these fundamental units.
Units in equations	Units in equations are treated just like any other algebraic quantity.
Conversion	Conversion factors, which are always equal to 1, provide a convenient method for converting from one kind of unit to another.
2. Dimensions	The two sides of an equation must have the same dimensions.
3. Scientific Notation	For convenience, very small and very large numbers are generally written as a factor times a power of 10.
4. Exponents	
Multiplication	When multiplying two numbers, the exponents are added.
Division	When dividing two numbers, the exponents are subtracted.
Raising to a power	When a number containing an exponent is itself raised to a power, the exponents are multiplied.
5. Significant Figures	
Multiplication and division	The number of significant figures in the result of multiplication or division is *no greater than* the least number of significant figures in any of the numbers.
Addition and subtraction	The result of addition or subtraction of two numbers has no significant figures beyond the last decimal place where both of the original numbers had significant figures.
6. Order of Magnitude	A number rounded to the nearest power of 10 is called an order of magnitude. The order of magnitude of a quantity can often be estimated using reasonable assumptions and simple calculations.

PROBLEMS

- Single-concept, single-step, relatively easy
- •• Intermediate-level, may require synthesis of concepts
- ••• Challenging
- SSM Solution is in the *Student Solutions Manual*
- iSOLVE Problems available on iSOLVE online homework service
- iSOLVE ✓ These "Checkpoint" online homework service problems ask students additional questions about their confidence level, and how they arrived at their answer

Conceptual Problems

1 • SSM iSOLVE Which of the following is *not* one of the fundamental physical quantities in the SI system? (a) Mass. (b) Length. (c) Force. (d) Time. (e) All of the above are fundamental physical quantities.

2 • iSOLVE In doing a calculation, you end up with m/s in the numerator and m/s^2 in the denominator. What are your final units? (a) m^2/s^3. (b) 1/s. (c) s^3/m^2. (d) s. (e) m/s.

3 • iSOLVE The prefix giga means (a) 10^3, (b) 10^6, (c) 10^9, (d) 10^{12}, (e) 10^{15}.

4 • iSOLVE The prefix mega means (a) 10^{-9}, (b) 10^{-6}, (c) 10^{-3}, (d) 10^6, (e) 10^9.

5 • SSM iSOLVE The prefix pico means (a) 10^{-12}, (b) 10^{-6}, (c) 10^{-3}, (d) 10^6, (e) 10^9.

6 • iSOLVE The number 0.0005130 has _____ significant figures. (a) one, (b) three, (c) four, (d) seven, (e) eight.

7 • iSOLVE The number 23.0040 has _____ significant figures. (a) two, (b) three, (c) four, (d) five, (e) six.

8 • What are the advantages and disadvantages of using the length of your arm for a standard length?

9 • True or false:

(a) Two quantities must have the same dimensions in order to be added.

(b) Two quantities must have the same dimensions in order to be multiplied.

(c) All conversion factors have the value 1.

Estimation and Approximation

10 •• SSM

The angle subtended by the moon's diameter at a point on the earth is about 0.524° (Figure 1-2). Use this and the fact that the moon is about 384 Mm away to find the

FIGURE 1-2 Problem 10

diameter of the moon. (The angle θ subtended by the moon is approximately D/r_m, where D is the diameter of the moon and r_m is the distance to the moon.)

11 •• SSM iSOLVE ✓ The sun has a mass of 1.99×10^{30} kg and is composed mostly of hydrogen, with only a small fraction being heavier elements. The hydrogen atom has a mass of 1.67×10^{-27} kg. Estimate the number of hydrogen atoms in the sun.

12 •• Most soft drinks are sold in aluminum cans. The mass of a typical can is about 0.018 kg. (a) Estimate the number of aluminum cans used in the United States in one year. (b) Estimate the total mass of aluminum in a year's consumption from these cans. (c) If aluminum returns $1/kg at a recycling center, how much is a year's accumulation of aluminum cans worth?

13 •• In his essay "There's plenty of room at the bottom," Richard Feynman proposed writing the entire *Encyclopaedia Brittanica* on the head of a pin. (a) Estimate the size of the letters needed if we assume that a pinhead is 1/16 in across (the value that Feynman used). (b) If the atomic spacing in a metal is about 0.5 nm (5×10^{-10} m), about how many atoms across is each letter?

14 •• SSM (a) Estimate the number of gallons of gasoline used per day by automobiles in the United States and the total amount of money spent on it. (b) If 19.4 gal of gasoline can be made from one barrel of crude oil, estimate the total number of barrels of oil imported into the United States per year to make gasoline. How many barrels per day is this?

15 •• There is an environmental debate over the use of cloth versus disposable diapers. (a) If we assume that between birth and 2.5 y of age, a child uses 3 diapers per day, estimate the total number of disposable diapers used in the United States per year. (b) Estimate the total landfill volume due to these diapers, assuming that 1000 kg of waste fills about 1 m^3 of landfill volume. (c) How many square miles of landfill area at an average height of 10 m is needed for the disposal of diapers each year?

16 ••• Each binary digit is termed a bit. A series of bits grouped together is called a word. An 8-bit word is called a byte. Suppose a computer hard disk has a capacity of 20 gigabytes. (a) How many bits can be stored on the disk? (b) Estimate the number of typical books that can be stored on the disk assuming each character requires one 8-bit word.

17 •• **SSM** Estimate the yearly toll revenue of the George Washington Bridge in New York. At last glance, the toll is $6 to go into New York from New Jersey; going from New York into New Jersey is free. There are a total of 14 lanes.

Units

18 • Express the following quantities using the prefixes listed in Table 1-1 and the abbreviations listed on page EP-1; for example, 10,000 meters = 10 km. (a) 1,000,000 watts, (b) 0.002 gram, (c) 3×10^{-6} meter, (d) 30,000 seconds.

19 • Write each of the following without using prefixes: (a) 40 μW, (b) 4 ns, (c) 3 MW, (d) 25 km.

20 • **SSM** Write the following (which are not SI units) without using abbreviations. For example, 10^3 meters = 1 kilometer: (a) 10^{-12} boo, (b) 10^9 low, (c) 10^{-6} phone, (d) 10^{-18} boy, (e) 10^6 phone, (f) 10^{-9} goat, (g) 10^{12} bull.

21 •• **SOLVE** In the following equations, the distance x is in meters, the time t is in seconds, and the velocity v is in meters per second. What are the SI units of the constants C_1 and C_2? (a) $x = C_1 + C_2 t$, (b) $x = \frac{1}{2} C_1 t^2$, (c) $v^2 = 2 C_1 x$, (d) $x = C_1 \cos C_2 t$, (e) $v^2 = 2 C_1 v - (C_2 x)^2$.

22 •• **SOLVE** If x is in feet, t is in seconds, and v is in feet per second, what are the units of the constants C_1 and C_2 in each part of Problem 21?

Conversion of Units

23 • From the original definition of the meter in terms of the distance from the equator to the North Pole, find in meters (a) the circumference of the earth and (b) the radius of the earth. (c) Convert your answers for (a) and (b) from meters into miles.

24 • **SOLVE** The speed of sound in air is 340 m/s. What is the speed of a supersonic plane that travels at twice the speed of sound? Give your answer in kilometers per hour and miles per hour.

25 • **SSM** **SOLVE** A basketball player is 6 ft $10\frac{1}{2}$ in. tall. What is his height in centimeters?

26 • Complete the following: (a) 100 km/h = _____ mi/h, (b) 60 cm = _____ in., (c) 100 yd = _____ m.

27 • The main span of the Golden Gate Bridge is 4200 ft. Express this distance in kilometers.

28 • **SSM** Find the conversion factor to convert from miles per hour into kilometers per hour.

29 • Complete the following: (a) 1.296×10^5 km/h^2 = _____ km/(h·s), (b) 1.296×10^5 km/h^2 = _____ m/s^2, (c) 60 mi/h = _____ ft/s, (d) 60 mi/h = _____ m/s.

30 • There are 1.057 quarts in a liter and 4 quarts in a gallon. (a) How many liters are there in a gallon? (b) A barrel equals 42 gallons. How many cubic meters are there in a barrel?

31 • **SOLVE** There are 640 acres in a square mile. How many square meters are there in one acre?

32 •• **SOLVE** A right circular cylinder has a diameter of 6.8 in and a height of 2 ft. What is the volume of the cylinder in (a) cubic feet, (b) cubic meters, (c) liters?

33 •• **SSM** In the following, x is in meters, t is in seconds, v is in meters per second, and the acceleration a is in meters per second squared. Find the SI units of each combination: (a) v^2/x, (b) $\sqrt{x/a}$, (c) $\frac{1}{2} a t^2$.

Dimensions of Physical Quantities

34 • What are the dimensions of the constants in each part of Problem 21?

35 •• The law of radioactive decay is $N(t) = N_0 e^{-\lambda t}$, where N_0 is the number of radioactive nuclei at $t = 0$, $N(t)$ is the number remaining at time t, and λ is a quantity known as the decay constant. What is the dimension of λ?

36 •• **SSM** The SI unit of force, the kilogram-meter per second squared (kg·m/s^2) is called the newton (N). Find the dimensions and the SI units of the constant G in Newton's law of gravitation $F = G m_1 m_2 / r^2$.

37 •• An object on the end of a string moves in a circle. The force exerted by the string has units of ML/T^2 and depends on the mass of the object, its speed, and the radius of the circle. What combination of these variables gives the correct dimensions?

38 •• **SOLVE** ✔ Show that the product of mass, acceleration, and speed has the dimensions of power.

39 •• The momentum of an object is the product of its velocity and mass. Show that momentum has the dimensions of force multiplied by time.

40 •• What combination of force and one other physical quantity has the dimensions of power?

41 •• **SSM** **SOLVE** ✔ When an object falls through air, there is a drag force that depends on the product of the surface area of the object and the square of its velocity, that is, $F_{air} = C A v^2$, where C is a constant. Determine the dimensions of C.

42 •• Kepler's third law relates the period of a planet to its orbital radius r, the constant G in Newton's law of gravitation ($F = G m_1 m_2 / r^2$), and the mass of the sun M_s. What combination of these factors gives the correct dimensions for the period of a planet?

Scientific Notation and Significant Figures

43 • **SSM** Express as a decimal number without using powers of 10 notation: (a) 3×10^4, (b) 6.2×10^{-3}, (c) 4×10^{-6}, (d) 2.17×10^5.

44 • Write the following in scientific notation: (a) 3.1 GW = _____ W, (b) 10 pm = _____ m, (c) 2.3 fs = _____ s, (d) 4 μs = _____ s.

45 • **SOLVE** Calculate the following, round off to the correct number of significant figures, and express your result in scientific notation: (a) $(1.14)(9.99 \times 10^4)$, (b) $(2.78 \times 10^{-8}) - (5.31 \times 10^{-9})$, (c) $12\pi/(4.56 \times 10^{-3})$, (d) $27.6 + (5.99 \times 10^2)$.

46 • Calculate the following, round off to the correct number of significant figures, and express your result in scientific notation: (a) $(200.9)(569.3)$, (b) $(0.000000513)(62.3 \times 10^7)$, (c) $28{,}401 + (5.78 \times 10^4)$, (d) $63.25/(4.17 \times 10^{-3})$.

47 • SSM ISOLVE A cell membrane has a thickness of about 7 nm. How many cell membranes would it take to make a stack 1 in high?

48 • Calculate the following, round off to the correct number of significant figures, and express your result in scientific notation: (a) $(2.00 \times 10^4)(6.10 \times 10^{-2})$, (b) $(3.141592)(4.00 \times 10^5)$, (c) $(2.32 \times 10^3)/(1.16 \times 10^8)$, (d) $(5.14 \times 10^3) + (2.78 \times 10^2)$, (e) $(1.99 \times 10^2) + (9.99 \times 10^{-5})$.

49 • SSM Perform the following calculations and round off the answers to the correct number of significant figures: (a) $3.141592654 \times (23.2)^2$, (b) $2 \times 3.141592654 \times 0.76$, (c) $4/(3\,\pi) \times (1.1)^3$, (d) $(2.0)^5/3.141592654$.

General Problems

50 • On many of the roads in Canada the speed limit is 100 km/h. What is the speed limit in miles per hour?

51 • SSM If you could count \$1 per second, how many years would it take to count 1 billion dollars (1 billion = 10^9)?

52 • Sometimes a conversion factor can be derived from the knowledge of a constant in two different systems. (a) The speed of light in vacuum is 186,000 mi/s = 3×10^8 m/s. Use this fact to find the number of kilometers in a mile. (b) The weight of 1 ft³ of water is 62.4 lb. Use this and the fact that 1 cm³ of water has a mass of 1 g to find the weight in pounds of a 1-kg mass.

53 •• ISOLVE The mass of one uranium atom is 4.0×10^{-26} kg. How many uranium atoms are there in 8 g of pure uranium?

54 •• ISOLVE ✓ During a thunderstorm, a total of 1.4 in. of rain falls. How much water falls on one acre of land? (1 mi² = 640 acres.)

55 •• An iron nucleus has a radius of 5.4×10^{-15} m and a mass of 9.3×10^{-26} kg. (a) What is its mass per unit volume in kg/m³? (b) If the earth had the same mass per unit volume, what would be its radius? (The mass of the earth is 5.98×10^{24} kg.)

56 •• Evaluate the following expressions: (a) $(5.6 \times 10^{-5})(0.0000075)/(2.4 \times 10^{-12})$, (b) $(14.2)(6.4 \times 10^7)(8.2 \times 10^{-9}) - 4.06$, (c) $(6.1 \times 10^{-6})^2(3.6 \times 10^4)^3/(3.6 \times 10^{-11})^{1/2}$, (d) $(0.000064)^{1/3}/[(12.8 \times 10^{-3})(490 \times 10^{-1})^{1/2}]$.

57 •• SSM The astronomical unit (AU) is defined in terms of the distance from the earth to the sun, namely 1.496×10^{11} m. The parsec is the radius of a circle for which a central angle of 1 s intercepts an arc of length 1 AU. The light-year is the distance that light travels in 1 y. (a) How many parsecs are there in one astronomical unit? (b) How many meters are in a parsec? (c) How many meters in a light-year? (d) How many astronomical units in a light-year? (e) How many light-years in a parsec?

58 •• If the average density of the universe is at least 6×10^{-27} kg/m³, then the universe will eventually stop expanding and begin contracting. (a) How many electrons are needed in a cubic meter to produce the critical density? (b) How many protons per cubic meter would produce the critical density? ($m_e = 9.11 \times 10^{-31}$ kg; $m_p = 1.67 \times 10^{-27}$ kg.)

59 •• SSM The Super-Kamiokande neutrino detector in Japan is a large transparent cylinder filled with ultrapure water. The height of the cylinder is 41.4 m and the diameter is 39.3 m. Calculate the mass of the water in the cylinder. Does this match the claim posted on the official Super-K Web site that the detector uses 50,000 tons of water? The density of water is 1,000 kg/m³.

60 ••• The table below gives experimental results for a measurement of the period of motion T of an object of mass m suspended on a spring versus the mass of the object. These data are consistent with a simple equation expressing T as a function of m of the form $T = Cm^n$, where C and n are constants and n is not necessarily an integer. (a) Find n and C. (There are several ways to do this. One is to guess the value of n and check by plotting T versus m^n on graph paper. If your guess is right, the plot will be a straight line. Another is to plot $\log T$ versus $\log m$. The slope of the straight line on this plot is n.) (b) Which data points deviate the most from a straight-line plot of T versus m^n?

Mass m, kg	0.10	0.20	0.40	0.50	0.75	1.00	1.50
Period T, s	0.56	0.83	1.05	1.28	1.55	1.75	2.22

61 ••• The table below gives the period T and orbit radius r for the motions of four satellites orbiting a dense, heavy asteroid. (a) These data can be fitted by the formula $T = Cr^n$. Find C and n. (b) A fifth satellite is discovered to have a period of 6.20 y. Find the radius for the orbit of this satellite, which fits the same formula.

Period T, y	0.44	1.61	3.88	7.89
Radius r, Gm	0.088	0.208	0.374	0.600

62 ••• SSM The period T of a simple pendulum depends on the length L of the pendulum and the acceleration of gravity g (dimensions L/T^2). (a) Find a simple combination of L and g that has the dimensions of time. (b) Check the dependence of the period T on the length L by measuring the period (time for a complete swing back and forth) of a pendulum for two different values of L. (c) The correct formula relating T to L and g involves a constant that is a multiple of π, and cannot be obtained by the dimensional analysis of Part (a). It can be found by experiment as in Part (b) if g is known. Using the value $g = 9.81$ m/s² and your experimental results from Part (b), find the formula relating T to L and g.

63 ••• ISOLVE ✓ The weight of the earth's atmosphere pushes down on the surface of the earth with a force of 14.7 lb for each square inch of the earth's surface. What is the weight in pounds of the earth's atmosphere? (The radius of the earth is about 6370 km.)

Motion in One Dimension

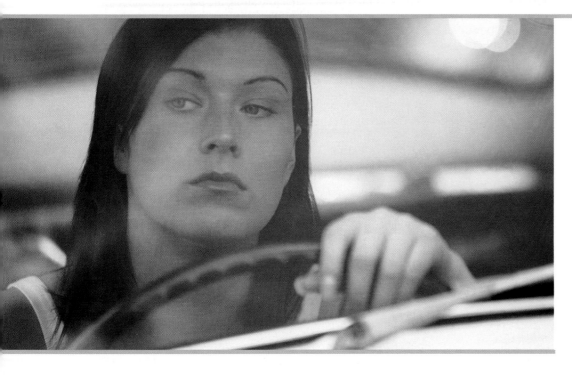

MOTION IN ONE DIMENSION IS MOTION ALONG A STRAIGHT LINE, LIKE THAT OF A CAR ON A STRAIGHT ROAD. THIS DRIVER ENCOUNTERS STOPLIGHTS AND DIFFERENT SPEED LIMITS ON HER COMMUTE ALONG A STRAIGHT HIGHWAY TO SCHOOL.

? **How can she estimate her arrival time? (See Example 2-2.)**

We begin our study of the physical universe by examining objects in motion. The study of motion, whose measurement, more than 400 years ago gave birth to physics, is called **kinematics.**

➤ **In this chapter, we start with the simplest case of kinematics, the motion of a particle along a straight line, like the motion of a car moving along a flat, straight, narrow road. A particle is an object whose position can be described by a single point. Anything can be considered to be a particle—a molecule, a person, or a galaxy—as long as we can reasonably ignore its internal structure.**

2-1 Displacement, Velocity, and Speed

Figure 2-1 shows a car at position x_i at time t_i and at position x_f at a later time t_f. The change in the car's position, called the **displacement,** is given by $x_f - x_i$. We use the Greek letter Δ (uppercase delta) to indicate the change in a quantity; thus, the change in x is written Δx:

$$\Delta x = x_f - x_i \qquad\qquad 2\text{-}1$$

DEFINITION—DISPLACEMENT

$$\Delta x = x_f - x_i$$

FIGURE 2-1 A car moving in a straight line. A coordinate axis consists of a line along the path of the car. A point on this line is chosen to be the origin O. Other points on it are assigned a number x, the value of x being proportional to its distance from O. Points to the right of O are positive as shown, and points to the left are negative. When the car travels from point x_i to point x_f, its displacement is $\Delta x = x_f - x_i$.

The notation Δx (read "delta x") stands for a single quantity, the change in x. It is not a product of Δ and x any more than $\cos \theta$ is a product of cos and θ. By convention, the change in a quantity is always its final value minus its initial value.

Velocity is the rate at which the position changes. The **average velocity** of the particle is defined as the ratio of the displacement Δx to the time interval $\Delta t = t_f - t_i$:

$$v_{av} = \frac{\Delta x}{\Delta t} = \frac{x_f - x_i}{t_f - t_i} \qquad\qquad 2\text{-}2$$

DEFINITION—AVERAGE VELOCITY

Displacement and average velocity may be positive or negative. A positive value indicates motion in the positive x direction. The SI unit of velocity is meters per second (m/s), and the U.S. customary unit is feet per second (ft/s).

DISPLACEMENT AND VELOCITY OF A COMET **EXAMPLE 2-1**

A comet moving toward the sun is first seen at $x_i = 3.0 \times 10^{12}$ m relative to the sun (see Figure 2-2). Exactly one year later, it is seen at $x_f = 2.1 \times 10^{12}$ m. Find its displacement and average velocity.

FIGURE 2-2

PICTURE THE PROBLEM Comets move in elliptical orbits around the sun. Here we consider just the distance from the sun as if the comet moved in one dimension. We are given x_i and x_f. If we choose $t_i = 0$, then $t_f = 1$ y $= 3.16 \times 10^7$ s. The average velocity is $\Delta x/\Delta t$.

1. The displacement is found from its definition:

$$\Delta x = x_f - x_i = (2.1 \times 10^{12}\text{ m}) - (3.0 \times 10^{12}\text{ m})$$
$$= -9 \times 10^{11}\text{ m}$$

2. The average velocity is the displacement divided by the time interval:

$$v_{av} = \frac{\Delta x}{\Delta t} = \frac{-9 \times 10^{11}\text{ m}}{3.16 \times 10^7\text{ s}}$$
$$= -2.85 \times 10^4\text{ m/s} = \boxed{-28.5\text{ km/s}}$$

REMARKS Both displacement and average velocity are negative, because the comet moved toward smaller values of x. Note that the units, m for Δx, and m/s or km/s for v_{av}, are essential parts of the answers. It is meaningless to say "the displacement is -9×10^{11}" or "the average velocity of a particle is -28.5."

EXERCISE A jet plane leaves the gate in Detroit at 2:15 P.M. Its average velocity is 500 km/h for the trip to Chicago, which is 483 km away. When does it arrive at the gate in Chicago? (*Answer* 3:13 P.M. Detroit time, which is actually 2:13 P.M. Chicago time)

DRIVING TO SCHOOL | **EXAMPLE 2-2** Put It in Context

It normally takes you 10 min to travel 5 mi to school along a straight road. You leave home 15 min before class begins. Delays caused by a broken traffic light slow down traffic to 20 mph for the first 2 mi of the trip. Will you be late for class?

PICTURE THE PROBLEM You need to find the total time that it will take you to travel to class. To do so, you must find the time $\Delta t_{2\,mi}$ that you will be driving at 20 mph, and the time $\Delta t_{3\,mi}$ for the remainder of the trip, during which you are driving at your usual speed.

1. The total time equals the time to travel the first 2 mi plus the time to travel the remaining 3 mi:

$$\Delta t_{tot} = \Delta t_{2\,mi} + \Delta t_{3\,mi}$$

2. Using $\Delta x = v_{av}\Delta t$, solve for the time taken to travel 2 mi at 20 mi/h:

$$\Delta t_{2\,mi} = \frac{\Delta x}{v_{av}} = \frac{2\,mi}{20\,mi/h} = 0.1\,h = 6\,min$$

3. Using $\Delta x = v_{av}\Delta t$, relate the time to travel the last 3 mi at the usual speed:

$$\Delta t_{3\,mi} = \frac{\Delta x}{v_{av}} = \frac{3\,mi}{v_{usual}}$$

4. Using $\Delta x = v_{av}\Delta t$, solve for the v_{usual}, the speed needed for you to travel the 5 mi in 10 min:

$$v_{usual} = \frac{\Delta x_{tot}}{\Delta t_{usual}} = \frac{5\,mi}{10\,min} = 0.5\,mi/min$$

5. Solve for $t_{3\,mi}$:

$$\Delta t_{3\,mi} = \frac{3\,mi}{0.5\,mi/min} = 6\,min$$

6. Solve for the total time:

$$\Delta t_{tot} = \Delta t_{2\,mi} + \Delta t_{3\,mi} = 12\,min$$

7. The trip takes 12 min with the delay, compared to the usual 10 minutes. Because you wisely allowed yourself 15 min for the trip, *you will not be late for class.*

The **average speed** of a particle is the ratio of the total distance traveled to the total time from start to finish:

$$\text{Average speed} = \frac{\text{total distance}}{\text{total time}} = \frac{s}{t} \qquad 2\text{-}3$$

Since the total distance and total time are both always positive, the average speed is always positive.

AVERAGE SPEED IN A FOOT RACE | **EXAMPLE 2-3**

You run 100 m in 12 s then turn around and jog 50 m back toward the starting point in 30 s (see Figure 2-3). Calculate (*a*) your average speed, and (*b*) your average velocity for the total trip.

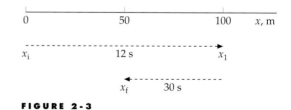

FIGURE 2-3

PICTURE THE PROBLEM We use the definitions of average speed and average velocity, noting that average *speed* is the total *distance* divided by Δt, whereas the average *velocity* is the *net displacement* divided by Δt:

(*a*) 1. Your average speed equals the total distance divided by the total time:

$$\text{Average speed} = \frac{s}{t}$$

2. Calculate the total distance traveled and the total time:

$$s = s_1 + s_2 = 100\,m + 50\,m = 150\,m$$

$$t = 12\,s + 30\,s = 42\,s$$

3. Use s and t to find your average speed:

$$\text{Average speed} = \frac{150 \text{ m}}{42 \text{ s}} = \boxed{3.57 \text{ m/s}}$$

(b) 1. Your average velocity is the ratio of the net displacement Δx to the time interval Δt:

$$v_{av} = \frac{\Delta x}{\Delta t}$$

2. Your net displacement is $x_f - x_i$, where $x_i = 0$ is your initial position and $x_f = 50$ m is your final position:

$$\Delta x = x_f - x_i = 50 \text{ m} - 0 = 50 \text{ m}$$

3. Use Δx and Δt to find your average velocity:

$$v_{av} = \frac{\Delta x}{\Delta t} = \frac{50 \text{ m}}{42 \text{ s}} = \boxed{1.19 \text{ m/s}}$$

PLAUSIBILITY CHECK The world record for a 100-m race is just under 10 s, so 10 m/s is about the maximum possible speed obtainable. The result of 3.57 m/s for the average speed in (a) is reasonable, given that you merely jogged for one-third of the distance. If you had obtained 35.7 m/s for the average speed, that would have been a clue that something was wrong with the calculation.

REMARKS Note that your average speed is greater than your average velocity because the total distance traveled is greater than the total displacement. Also, note that the net displacement is the sum of the individual displacements. That is, $\Delta x = \Delta x_1 + \Delta x_2 = (100 \text{ m}) + (-50 \text{ m}) = 50 \text{ m}$, which is the Part (b), step 2 result.

A TRAIN-HOPPING BIRD   **EXAMPLE 2-4**

Two trains 75 km apart approach each other on parallel tracks, each moving at 15 km/h. A bird flies back and forth between the trains at 20 km/h until the trains pass each other. How far does the bird fly?

PICTURE THE PROBLEM This problem seems difficult at first, but viewed in the right way it is actually quite simple. We approach it by first writing an equation for the quantity to be found, the total distance s flown by the bird.

1. The total distance traveled by the bird equals its speed times the time:

$$s = (\text{average speed})_{bird} \times t$$
$$= (\text{speed})_{av\ bird} \times t$$

2. The time that the bird is in the air is the time taken for the trains to meet. The sum of the distances traveled by the two trains is $D = 75$ km. Find the time it will take the two trains to travel a total distance D:

$$s_1 + s_2 = (\text{speed})_{av\ 1} \times t + (\text{speed})_{av\ 2} \times t = D$$

so

$$t = \frac{D}{(\text{speed})_{av\ 1} + (\text{speed})_{av\ 2}}$$

3. The total distance traveled by the bird is therefore:

$$s = (\text{speed})_{av\ bird} \times t$$
$$= (\text{speed})_{av\ bird} \frac{D}{(\text{speed})_{av\ 1} + (\text{speed})_{av\ 2}}$$
$$= 20 \text{ km/h} \frac{75 \text{ km}}{15 \text{ km/h} + 15 \text{ km/h}}$$
$$= \boxed{50 \text{ km}}$$

REMARKS Some try to solve this problem by finding and summing the distances flown by the bird each time it moves from one train to the other. This makes a relatively easy problem quite difficult. It is important to develop a thoughtful, systematic approach to solving problems. Begin by writing an equation for the unknown quantity in terms of other quantities. Then proceed by determining the values for each of the other quantities in the equation.

Figure 2-4 depicts average velocity graphically. A straight line connects points P_1 and P_2 and forms the hypotenuse of the triangle having sides Δx and Δt. The ratio $\Delta x / \Delta t$ is the line's **slope,** which gives us a geometric interpretation of average velocity:

> The average velocity is the slope of the straight line connecting the points (t_1, x_1) and (t_2, x_2).

GEOMETRIC INTERPRETATION OF AVERAGE VELOCITY

Generally, average velocity depends on the time interval on which it is based. In Figure 2-4, for example, the smaller time interval indicated by t_2' and P_2' gives a larger average velocity, as shown by the greater steepness of the line connecting points P_1 and P_2'.

Instantaneous Velocity

On first consideration, defining the velocity of a particle at a single instant seems impossible. At a given instant, a particle is at a single point. If it is at a single point, how can it be moving? If it is not moving, how can it have a velocity? This age-old paradox is resolved when we realize that observing and defining motion requires that we look at the position of the object at more than one time. Consider Figure 2-5. As we consider successively shorter time intervals beginning at t_1, the average velocity for the interval approaches the slope of the tangent at t_1. We define the slope of this tangent as the **instantaneous velocity** at t_1. This tangent is the limit of the ratio $\Delta x / \Delta t$ as Δt, and therefore Δx, approaches zero. So we can say:

> The instantaneous velocity is the limit of the ratio $\Delta x / \Delta t$ as Δt approaches zero.
>
> $$v(t) = \lim_{\Delta t \to 0} \frac{\Delta x}{\Delta t}$$
>
> = slope of the line tangent to the x-versus-t curve[†]   2-4

DEFINITION—INSTANTANEOUS VELOCITY

This limit is called the **derivative** of x with respect to t. In the usual calculus notation, the derivative is written dx/dt

$$v(t) = \lim_{\Delta t \to 0} \frac{\Delta x}{\Delta t} = \frac{dx}{dt} \qquad 2\text{-}5$$

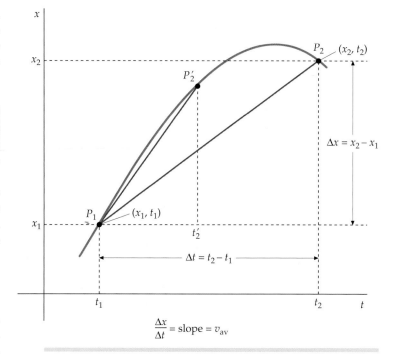

$$\frac{\Delta x}{\Delta t} = \text{slope} = v_{\text{av}}$$

FIGURE 2-4 Graph of x versus t for a particle moving in one dimension. Each point on the curve represents the location x at a particular time t. We have drawn a straight line between positions P_1 and P_2. The displacement $\Delta x = x_2 - x_1$ and the time interval $\Delta t = t_2 - t_1$ between these points are indicated. The straight line between P_1 and P_2 is the hypotenuse of the triangle having sides Δx and Δt, and the ratio $\Delta x/\Delta t$ is its slope. In geometric terms, the slope is a measure of the line's steepness.

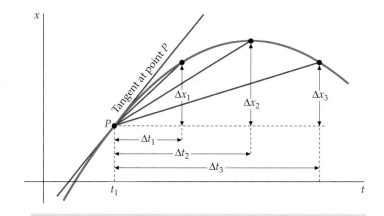

FIGURE 2-5 Graph of x versus t. Note the sequence of successively smaller time intervals, Δt_1, Δt_2, Δt_3, ... The average velocity of each interval is the slope of the straight line for that interval. As the time intervals become smaller, these slopes approach the slope of the tangent to the curve at point t_1. The slope of this line is defined as the instantaneous velocity at time t_1.

A line's slope may be positive, negative, or zero; consequently, instantaneous velocity (in one-dimensional motion) may be positive (x increasing), negative (x decreasing), or zero (no motion). The magnitude of the instantaneous velocity is the **instantaneous speed.**

[†] The slope of the line tangent to a curve is often referred to more simply as the "slope of the curve."

POSITION OF A PARTICLE AS A FUNCTION OF TIME   **EXAMPLE   2 - 5**   Try It Yourself

The position of a particle as a function of time is given by the curve shown in Figure 2-6. Find the instantaneous velocity at time $t = 2$ s. When is the velocity greatest? When is it zero? Is it ever negative?

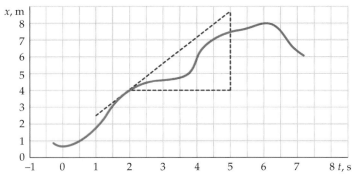

PICTURE THE PROBLEM In Figure 2-6, we have sketched the line tangent to the curve at $t = 2$ s. The tangent line's slope is the instantaneous velocity of the particle at the given time. You can use this figure to measure the slope of the tangent line.

FIGURE 2-6

Cover the column to the right and try these on your own before looking at the answers.

Steps

1. Find the values x_1 and x_2 on the tangent line at times $t_1 = 2$ s and $t_2 = 5$ s.

2. Compute the slope of the tangent line from these values. This slope equals the instantaneous velocity at $t = 2$ s.

3. From the figure, the slope (and therefore velocity) is greatest at about $t = 4$ s. The slope and velocity are zero at $t = 0$ and $t = 6$ s and are negative before 0 and after 6 s.

Answers

$x_1 \approx 4$ m, $x_2 \approx 8.5$ m

$v = \text{slope} \approx \dfrac{8.5 \text{ m} - 4 \text{ m}}{5 \text{ s} - 2 \text{ s}} = \boxed{1.5 \text{ m/s}}$

EXERCISE What is the average velocity of this particle between $t = 2$ s and $t = 5$ s? (*Answer*   1.17 m/s)

A STONE DROPPED FROM A CLIFF   **EXAMPLE   2 - 6**

The position of a stone dropped from a cliff is described approximately by $x = 5t^2$, where x is in meters measured downward from the original position at $t = 0$, and t is in seconds. Find the velocity at any time t. (We omit explicit indication of units to simplify the notation.)

PICTURE THE PROBLEM We can compute the velocity at some time t by computing the derivative dx/dt directly from the definition in Equation 2-4. The corresponding curve giving x versus t is shown in Figure 2-7. Tangent lines are drawn at times t_1, t_2, and t_3. The slopes of these tangent lines increase steadily, indicating that the instantaneous velocity increases steadily with time.

FIGURE 2-7

1. By definition the instantaneous velocity is:

$$v(t) = \lim_{\Delta t \to 0} \frac{\Delta x}{\Delta t} = \lim_{\Delta t \to 0} \frac{x(t + \Delta t) - x(t)}{\Delta t}$$

2. We compute the displacement Δx from the position function $x(t)$:

$$x(t) = 5t^2$$

3. At a later time $t + \Delta t$, the position is $x(t + \Delta t)$, given by:

$$x(t + \Delta t) = 5(t + \Delta t)^2 = 5[t^2 + 2t\Delta t + (\Delta t)^2]$$
$$= 5t^2 + 10t\Delta t + 5(\Delta t)^2$$

4. The displacement for this time interval is thus:

$$\Delta x = x(t + \Delta t) - x(t)$$
$$= \left[5t^2 + 10t\Delta t + 5(\Delta t)^2 \right] - 5t^2$$
$$= 10t\Delta t + 5(\Delta t)^2$$

5. Divide Δx by Δt to find the average velocity for this time interval:

$$v_{av} = \frac{\Delta x}{\Delta t} = \frac{10t\Delta t + 5(\Delta t)^2}{\Delta t} = 10t + 5\Delta t$$

6. As we consider shorter and shorter time intervals, Δt approaches zero and the second term $5\Delta t$ approaches zero, though the first term, $10t$, remains unchanged:

$$v(t) = \lim_{\Delta t \to 0} \frac{\Delta x}{\Delta t} = \lim_{\Delta t \to 0} (10t + 5\Delta t) = \boxed{10t}$$

REMARKS If we had set $\Delta t = 0$ in steps 4 and 5, the displacement would be $\Delta x = 0$, in which case the ratio $\Delta x / \Delta t$ would be undefined. Instead, we leave Δt as a variable until the final step, when the limit $\Delta t \to 0$ is well defined.

To find derivatives quickly, we use rules based on the limiting process above (see Appendix Table A-4). A particularly useful rule is

$$\text{If } x = Ct^n, \quad \text{then} \quad \frac{dx}{dt} = Cnt^{n-1} \qquad\qquad 2\text{-}6$$

where C and n are any constants. Using this rule in Example 2-6, we have $x = 5t^2$, and $v = dx/dt = 10t$, in agreement with our previous results.

Relative Velocity

If you are sitting in an airplane moving with a velocity of 500 mi/h toward the east, your velocity is also 500 mi/h toward the east. However, 500 mi/h toward the east might be your velocity relative to the surface of the earth, or it might be your velocity relative to the air outside the airplane. (If the plane is flying in a jet stream, these two velocities would be very different.) Furthermore, your velocity is zero relative to the airplane itself. To specify the velocity of a particle you must also specify the **frame of reference**. In this discussion three different frames of reference are specified, the surface of the earth, the air outside the airplane, and the airplane itself.

Midair refueling. Each plane is nearly at rest relative to the other, though both are moving with very large velocities relative to the earth.

A frame of reference is an extended object whose parts are at rest relative to each other.

DEFINITION—FRAME OF REFERENCE

To make position measurements we use coordinate axes that are attached to reference frames. For a horizontal coordinate axis attached to the plane your position remains constant, at least it does if you remain in your seat. However, for a horizontal coordinate axis attached to the surface of the earth, and for a horizontal coordinate axis attached to a balloon floating in the air outside the plane, your position keeps changing. If you have trouble imagining a coordinate axis attached to the air outside the plane, instead imagine a coordinate axis attached to a balloon that is suspended in the air. The air and the balloon are at rest relative to each other, so together they form a single reference frame.

If a particle moves with velocity v_{pA} relative to reference frame A, which is in turn moving with velocity v_{AB} relative to reference frame B, the velocity of the particle relative to B is

$$v_{pB} = v_{pA} + v_{AB} \qquad\qquad 2\text{-}7a$$

For example, if you swim in a river parallel to the direction of the flow, your velocity relative to the shore v_{ys} equals your velocity relative to the water v_{yw} plus the velocity of the water relative to the shore v_{ws}:

$$v_{ys} = v_{yw} + v_{ws}$$

If you are swimming upstream at 2 m/s relative to the water, and the water is moving at 1.2 m/s relative to the shore, then your velocity relative to the shore is $v_{ys} = -2$ m/s $+ 1.2$ m/s $= -0.8$ m/s, where we have chosen downstream to be the positive direction.

A great surprise of twentieth-century physics was the discovery that Equation 2-7a is only an approximation. When we study the theory of relativity in Chapter 39, we will see that the exact expression for relative velocities is

$$v_{pB} = \frac{v_{pA} + v_{AB}}{1 + v_{pA}v_{AB}/c^2} \qquad\qquad 2\text{-}7b$$

where $c = 3 \times 10^8$ m/s is the speed of light in a vacuum. In all everyday cases with macroscopic objects, v_{pA} and v_{AB} are both so much smaller than c that Equations 2-7a and b give essentially the same result, but for high-speed objects, such as electrons or distant galaxies, the difference between these two equations becomes significant. Equation 2-7b has the interesting property that if $v_{pA} = c$, then v_{pB} also equals c, which is a core tenet of relativity, namely that the speed of light is the same in all reference frames.

EXERCISE Using Equation 2-7b, substitute c for v_{pA} and solve for v_{pB}, verifying that Equation 2-7b is in agreement with the result "the speed of light is the same in all reference frames."

2-2 Acceleration

Acceleration is the rate of change of the instantaneous velocity. When you step on your car's accelerator, for example, you expect to change your velocity. The **average acceleration** for a particular time interval $\Delta t = t_2 - t_1$ is defined as the ratio $\Delta v/\Delta t$, where $\Delta v = v_2 - v_1$:

$$a_{av} = \frac{\Delta v}{\Delta t} \qquad\qquad 2\text{-}8$$

DEFINITION—AVERAGE ACCELERATION

Acceleration has dimensions of length divided by time squared. The SI unit is meters per second squared, m/s². (In Equation 2-8, if the numerator is in m/s and the denominator is in s, then $\Delta v/\Delta t$ is in units of (m/s)/s. Multiplying the

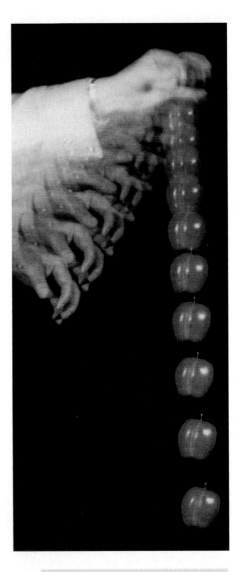

A falling apple captured by strobe photography at 60 flashes per second. The acceleration of the apple is indicated by the widening spaces between the images.

numerator and the denominator by 1 s, we find the units of $\Delta v/\Delta t$ to equal m/s^2.) We can write Equation 2-8 as $\Delta v = a_{av}\Delta t$. Thus, if a particle at rest accelerates at 5.1 m/s^2, its velocity after 1 s is 5.1 m/s, its velocity after 2 s is 10.2 m/s, and so on. **Instantaneous acceleration** is the limit of the ratio $\Delta v/\Delta t$ as Δt approaches zero. On a plot of velocity versus time, the instantaneous acceleration at time t is the slope of the line tangent to the curve at that time:

$$a = \lim_{\Delta t \to 0} \frac{\Delta v}{\Delta t}$$

$$= \text{slope of the line tangent to the } v\text{-versus-}t \text{ curve} \qquad \text{2-9}$$

DEFINITION—INSTANTANEOUS ACCELERATION

Thus, acceleration is the derivative of velocity with respect to time, dv/dt. Since velocity is the derivative of the position x with respect to t, acceleration is the second derivative of x with respect to t, d^2x/dt^2. We can see the reason for this notation when we write the acceleration as dv/dt and replace v with dx/dt:

$$a = \frac{dv}{dt} = \frac{d(dx/dt)}{dt} = \frac{d^2x}{dt^2} \qquad \text{2-10}$$

If acceleration remains zero, there is no change in velocity over time—velocity is constant. In this case, the curve of x versus t is a straight line. If acceleration is nonzero and constant, as in Example 2-9, then velocity varies linearly with time and x varies quadratically with time.

A Fast Cat	**EXAMPLE 2-7**

A cheetah can accelerate from 0 to 96 km/h (60 mi/h) in 2 s, whereas a Corvette requires 4.5 s. Compute the average accelerations for the cheetah and Corvette and compare them with the free-fall acceleration due to gravity, $g = 9.81 \text{ m/s}^2$.

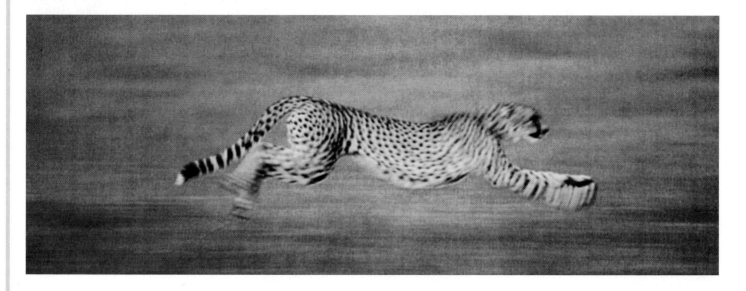

1. Find the average acceleration from the information given: cat $a_{av} = \dfrac{\Delta v}{\Delta t} = \dfrac{96 \text{ km/h} - 0}{2 \text{ s}}$

$$= 48 \text{ km/(h·s)}$$

car $a_{av} = \dfrac{\Delta v}{\Delta t} = \dfrac{96 \text{ km/h} - 0}{4.5 \text{ s}}$

$$= 21.3 \text{ km/(h·s)}$$

2. Convert to m/s² using 1 h = 3600 s = 3.6 ks: cat $\dfrac{48 \text{ km}}{\text{h·s}} \times \dfrac{1 \text{ h}}{3.6 \text{ ks}} = 13.3 \text{ m/s}^2$

car $\dfrac{21.3 \text{ km}}{\text{h·s}} \times \dfrac{1 \text{ h}}{3.6 \text{ ks}} = 5.92 \text{ m/s}^2$

3. To compare the result with the acceleration due to gravity, multiply each by the conversion factor 1g/9.81 m/s²: cat $13.3 \text{ m/s}^2 \times \dfrac{1g}{9.81 \text{ m/s}^2} = \boxed{1.36g}$

car $5.92 \text{ m/s}^2 \times \dfrac{1g}{9.81 \text{ m/s}^2} = \boxed{0.60g}$

REMARKS Note that by expressing the time in kiloseconds in step 2, the kilo prefixes in km and ks cancel.

EXERCISE A car is traveling at 45 km/h at time $t = 0$. It accelerates at a constant rate of 10 km/(h·s). (a) How fast is it traveling at $t = 2$ s? (b) At what time is the car traveling at 70 km/h?

(*Answer* (a) 65 km/h (b) 2.5 s)

EXERCISE IN DIMENSIONAL ANALYSIS If a car starts from rest at $x = 0$ with constant acceleration a, its velocity v depends on a and the distance traveled x. Which of the following equations has the correct dimensions and therefore could be a possible equation relating x, a, and v?

(a) $v = 2ax$ (c) $v = 2ax^2$

(b) $v^2 = 2a/x$ (d) $v^2 = 2ax$

(*Answer* Only (d) has the same dimensions on both sides of the equation. Although we cannot obtain the exact equation from dimensional analysis, we can often obtain the functional dependence.)

VELOCITY AND ACCELERATION AS FUNCTIONS OF TIME   **EXAMPLE 2 - 8**

The position of a particle is given by $x = Ct^3$, where C is a constant having units of m/s³. Find the velocity and acceleration as functions of time.

1. We find the velocity by applying $dx/dt = Cnt^{n-1}$ (Equation 2-6):

$$x = Ct^3$$

$$v = \frac{dx}{dt} = \boxed{3Ct^2}$$

2. The time derivative of velocity gives the acceleration:

$$a = \frac{dv}{dt} = \boxed{6Ct}$$

PLAUSIBILITY CHECK We can check the units of our answers. For velocity, $[v] = [C][t^2] = (\text{m/s}^3)(\text{s}^2) = \text{m/s}$. For acceleration, $[a] = [C][t] = (\text{m/s}^3)(\text{s}) = \text{m/s}^2$.

2-3 Motion With Constant Acceleration

The motion of a particle that has constant acceleration is common in nature. For example, near the earth's surface all unsupported objects fall vertically with constant acceleration (provided air resistance is negligible). If a particle has a constant acceleration a, it follows that the average acceleration for any time interval is also a. Thus,

$$a_{av} = \frac{\Delta v}{\Delta t} = a \qquad\qquad 2\text{-}11$$

If the velocity is v_0 at time $t = 0$, and v at some later time t, the corresponding acceleration is

$$a = \frac{\Delta v}{\Delta t} = \frac{v - v_0}{t - 0} = \frac{v - v_0}{t}$$

Rearranging yields v as a function of time:

$$v = v_0 + at \qquad\qquad 2\text{-}12$$

CONSTANT ACCELERATION, v VERSUS t

This is the equation for a straight line in a v-versus-t plot (Figure 2-8). The line's slope is the acceleration a and its v intercept is the initial velocity v_0.

The displacement $\Delta x = x - x_0$ in the time interval $\Delta t = t - 0$ is

$$\Delta x = v_{av}\Delta t = v_{av}t \qquad\qquad 2\text{-}13$$

For constant acceleration, the velocity varies linearly with time, and the average velocity is the mean value of the initial and final velocities. (This relation holds only if the acceleration is constant.) If v_0 is the initial velocity and v is the final velocity, the average velocity is

$$v_{av} = \tfrac{1}{2}(v_0 + v) \qquad\qquad 2\text{-}14$$

CONSTANT ACCELERATION, v_{av}

The displacement is then

$$\Delta x = x - x_0 = v_{av}t = \tfrac{1}{2}(v_0 + v)t \qquad\qquad 2\text{-}15$$

We can eliminate v by substituting $v = v_0 + at$ from Equation 2-12:

$$\Delta x = \tfrac{1}{2}(v_0 + v)t = \tfrac{1}{2}(v_0 + v_0 + at)t = v_0 t + \tfrac{1}{2}at^2$$

The displacement is thus

$$\Delta x = x - x_0 = v_0 t + \tfrac{1}{2}at^2 \qquad\qquad 2\text{-}16$$

CONSTANT ACCELERATION, $x(t)$

The first term on the right, $v_0 t$, is the displacement that would occur if a were zero, and the second term, $\tfrac{1}{2}at^2$, is the additional displacement due to the constant acceleration.

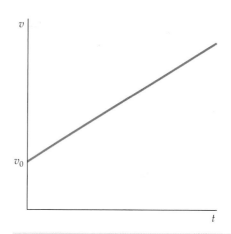

FIGURE 2-8 Graph of velocity versus time for constant acceleration.

"It goes from zero to 60 in about 3 seconds."
© Sydney Harris

Let's eliminate t from Equations 2-12 and 2-14 and find a relation between Δx, a, v, and v_0. From Equation 2-12, $t = (v - v_0)/a$. Substituting this into Equation 2-14,

$$\Delta x = v_{av}t = \frac{1}{2}(v_0 + v)t = \frac{1}{2}(v_0 + v)\left(\frac{v - v_0}{a}\right) = \frac{v^2 - v_0^2}{2a}$$

or

$$v^2 = v_0^2 + 2a\Delta x \qquad\qquad 2\text{-}17$$

CONSTANT ACCELERATION

Equation 2-17 is useful, for example, if we want to find the final velocity of a ball dropped from rest at some height x and we are not interested in the time the fall takes.

Problems with One Object

Many practical problems deal with objects in free-fall, that is, falling freely under the influence of gravity only. All objects in free-fall with the same initial velocity move identically. As shown in Figure 2-9, a feather and an apple, simultaneously released from rest in a large vacuum chamber, fall with identical motions. They have the same acceleration. The magnitude of this acceleration, designated by g, has the approximate value

$$g \approx 9.81 \text{ m/s}^2 = 32.2 \text{ ft/s}^2$$

Because g is the *magnitude* of the acceleration, it is always positive. If downward is the positive direction, then the free-fall acceleration is $a = g$; if upward is positive, then $a = -g$.

FIGURE 2-9

FIGURE 2-10

THE FLYING CAP   **EXAMPLE 2-9**

Upon graduation, a joyful physics student throws her cap straight upward with an initial speed of 14.7 m/s. Given that its acceleration is 9.81 m/s² downward (we neglect air resistance), (*a*) how long does it take to reach its highest point? (*b*) What is the distance to the highest point? (*c*) Assuming the cap is caught at the same height from which it was released, what is the total time the cap is in flight?

PICTURE THE PROBLEM When the cap is at its highest point, its instantaneous velocity is zero. Thus, we translate the phrase "at its highest point" into the mathematical condition $v = 0$.

(*a*) 1. Make a sketch of the cap in its initial position and again at its highest point. Include a coordinate axis and label the origin and the two positions of the cap.

2. The time is related to the velocity and acceleration: $\qquad v = v_0 + at$

3. To find the time at which the cap reaches its greatest height, set $v = 0$ and solve for t: $\qquad t = \dfrac{0 - v_0}{a} = \dfrac{-14.7 \text{ m/s}}{-9.81 \text{ m/s}^2} = \boxed{1.50 \text{ s}}$

(b) We can find the displacement from the time t and the average velocity:

$$\Delta y = v_{av}t = \tfrac{1}{2}(v_0 + v)t$$

$$= \tfrac{1}{2}(14.7 \text{ m/s} + 0)(1.50 \text{ s}) = \boxed{11.0 \text{ m}}$$

(c) 1. Set $\Delta y = 0$ in Equation 2-16 and solve for t:

$$\Delta y = v_0 t + \tfrac{1}{2}at^2$$

$$0 = (v_0 + \tfrac{1}{2}at)t$$

2. There are two solutions for t if $\Delta y = 0$. The first corresponds to the time at which the cap is released, the second to the time at which the cap is caught:

$$t = 0 \qquad \text{(first solution)}$$

$$t = -\frac{2v_0}{a} = -\frac{2(14.7 \text{ m/s})}{-9.81 \text{ m/s}^2} = \boxed{3 \text{ s}} \qquad \text{(second solution)}$$

REMARKS The $t = 3$ s solution also follows from a symmetry in the system—it takes the same time for the cap to fall from its greatest height as to rise to that height (see Figure 2-11). In reality the cap will not have a constant acceleration because air resistance has a significant effect on a light object like a cap. If air resistance is not negligible, the fall time will exceed the rise time.

EXERCISE Find $y_{max} - y_0$ using (a) Equation 2-15 and (b) Equation 2-16. (c) Find the velocity of the cap when it returns to its starting point. (*Answer* (a) and (b) $y_{max} - y_0 = 11.0$ m (c) -14.7 m/s; notice that the final speed is the same as the initial speed)

EXERCISE What is the velocity of the cap at the following points in time? (a) 0.1 s before it reaches its highest point; (b) 0.1 s after it reaches its highest point. (c) Compute $\Delta v/\Delta t$ for this 0.2-s-long time interval. (*Answer* (a) $+ 0.981$ m/s (b) $- 0.981$ m/s (c) $[(-0.981 \text{ m/s}) - (+0.981 \text{ m/s})]/(0.2 \text{ s}) = -9.81 \text{ m/s}^2$)

EXERCISE A car starting from rest gains speed at a constant rate of 8 m/s². (a) How fast is it going after 10 s? (b) How far has it gone after 10 s? (c) What is its average velocity for the interval $t = 0$ to $t = 10$ s? (*Answer* (a) 80 m/s (b) 400 m (c) 40 m/s)

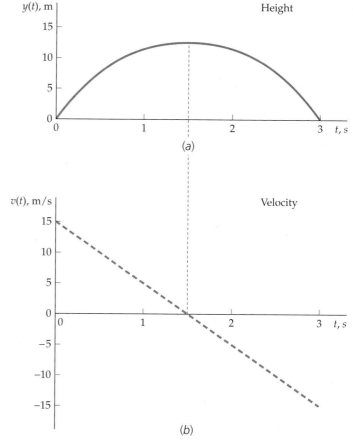

(a)

(b)

FIGURE 2-11

The next example concerns a car's **stopping distance**—how far it travels while coming to a halt.

A Car's Stopping Distance   **EXAMPLE 2-10**

On a highway at night you see a stalled vehicle and brake your car to a stop with an acceleration of magnitude 5 m/s². (An acceleration that reduces the speed is often called a deceleration.) What is the car's stopping distance if its initial speed is (a) 15 m/s (about 34 mi/h) or (b) 30 m/s?

PICTURE THE PROBLEM If we choose the direction of motion to be positive, the stopping distance and the initial velocity are positive, but the acceleration is negative. Thus, the initial velocity is $v_0 = 15$ m/s, the final velocity is $v = 0$, and the acceleration is $a = -5$ m/s^2. We seek the distance traveled, Δx. We do not need to know the time it takes for the car to stop, so Equation 2-17 is the most convenient formula to use.

(a) Using Equation 2-17 with $v = 0$, calculate the displacement Δx:

$$v^2 = v_0^2 + 2a\Delta x$$

so

$$\Delta x = \frac{v^2 - v_0^2}{2a} = \frac{0^2 - v_0^2}{2a} = -\frac{v_0^2}{2a}$$

$$= -\frac{(15 \text{ m/s})^2}{2(-5 \text{ m/s}^2)} = \boxed{22.5 \text{ m}}$$

(b) From the previous step we see that if $v = 0$, then $\Delta x = -v_0^2/(2a)$. Thus, Δx is proportional to the square of the initial speed. Using this observation and the Part (a) result, find the stopping distance for an initial speed equal to twice that in Part (a):

$$\Delta x = 2^2(22.5 \text{ m}) = \boxed{90 \text{ m}}$$

REMARKS The answer to (b) can also be gotten by directly substituting the initial speed of 30 m/s into the expression for Δx obtained in Part (a). Ninety meters is a considerable distance, roughly the length of a football field. Changing v_0 by a factor of 2 changes the stopping distance by a factor $2^2 = 4$ (see Figure 2-12). The practical implication of this squared dependence is that even modest increases in speed cause significant increases in stopping distance.

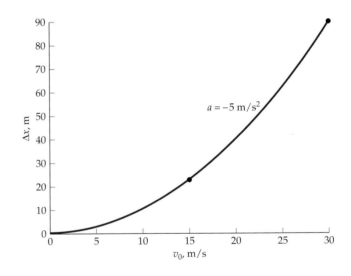

FIGURE 2-12 Stopping distance as a function of the initial velocity. The curve shows the case for Example 2-10, where the acceleration is $a = -5.0$ m/s^2; the points shown on the red curve are the solutions to parts (a) and (b).

STOPPING DISTANCE   **EXAMPLE 2-11** **Try It Yourself**

In Example 2-10, (a) how much time does it take for the car to stop if its initial velocity is 30 m/s? (b) How far does the car travel in the last second?

PICTURE THE PROBLEM (a) Except for the numbers, this is the same as Part (a) of Example 2-9. Use the same procedure shown in Example 2-10. (b) Since the speed decreases by 5 m/s each second, the velocity 1 s before the car stops must be 5 m/s. Find the average velocity during the last second and use that to find the distance traveled.

Cover the column to the right and try these on your own before looking at the answers.

Steps   Answers
(a) Find the total stopping time t. $t = 6$ s

(b) 1. Find the average velocity during the last second. $v_{av} = \boxed{2.5 \text{ m/s}}$

   2. Compute the distance traveled from $\Delta x = v_{av}\Delta t$. $\Delta x_1 = v_{av}\Delta t = \boxed{2.5 \text{ m}}$

REMARKS If Part (b) asked for the average velocity during the last 1.3 seconds (rather than during the last second), we can find the initial velocity v_1 during this interval using $\Delta v = a\Delta t$ (Equation 2-11).

Sometimes valuable insight can be gained about the motion of an object by assuming that the constant-acceleration formulas still apply even when the acceleration is not constant. This is the case in the following example.

THE CRASH TEST   **EXAMPLE 2 - 1 2**

In a crash test, a car traveling 100 km/h (about 62 mi/h) hits an immovable concrete wall. What is its acceleration?

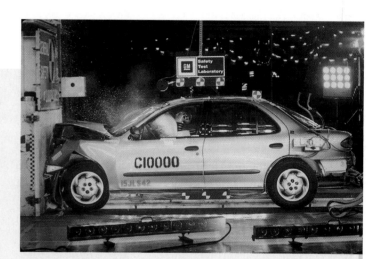

PICTURE THE PROBLEM In this example, it is not appropriate to treat the car as a particle because different parts of the vehicle will have different accelerations as the car crumples to a halt. Moreover, the accelerations are not constant. However, we *can* treat a bolt in the center of the car as a particle. We need additional information to solve this problem—either the stopping distance or the time to stop. We can estimate the stopping distance using common sense. Upon impact, the center of the car will certainly move forward less than half the length of the car. We'll choose 0.75 m as a reasonable estimate of the stopping distance. Since the problem neither asks for nor provides the time we will use the relation $v^2 = v_0^2 + 2a\Delta x$.

1. Using $v^2 = v_0^2 + 2a\Delta x$, solve for the acceleration:

$$v^2 = v_0^2 + 2a\Delta x$$

so

$$a = \frac{v^2 - v_0^2}{2\Delta x} = \frac{0^2 - (100 \text{ km/h})^2}{2(0.75 \text{ m})}$$

2. Convert the velocity from km/h to m/s. In one hour there are 60^2 s $= 3.6$ ks:

$$(100 \text{ km/h}) \times \left(\frac{1 \text{ h}}{3.6 \text{ ks}}\right) = 27.8 \text{ m/s}$$

3. Complete the calculation of the acceleration:

$$a = \frac{0^2 - (100 \text{ km/h})^2}{2(0.75 \text{ m})} = -\frac{(27.8 \text{ m/s})^2}{1.5 \text{ s}}$$

$$= -514 \text{ m/s}^2 \approx \boxed{-500 \text{ m/s}^2}$$

REMARKS Note that the magnitude of this acceleration is greater than $50g$. This estimate is what the magnitude of the acceleration would be both if the displacement of the center of the car were actually 0.75 m and if the acceleration were constant.

A TRAVELING ELECTRON   **EXAMPLE 2 - 1 3**   Try It Yourself

An electron in a cathode-ray tube accelerates from rest with a constant acceleration of 5.33×10^{12} m/s^2 for 0.15 μs (1 μs $= 10^{-6}$ s). The electron then drifts with constant velocity for 0.2 μs. Finally, it comes to rest with an acceleration of -2.67×10^{13} m/s^2. How far does the electron travel?

PICTURE THE PROBLEM The equations for constant acceleration do not apply to this problem directly because the acceleration of the electron varies with time. Divide the electron's motion into three intervals, each with a different constant acceleration and use the final position and velocity for one interval as the initial conditions for the next interval. Choose the origin to be at the electron's starting position, and the positive direction to be the direction of motion.

Cover the column to the right and try these on your own before looking at the answers.

Steps

1. Find the displacement and final velocity for the first 0.15-μs interval.

2. Use this final velocity as the constant velocity to find the displacement while it drifts at constant velocity.

3. Use this same velocity as the initial velocity and Equation 2-17 with $v = 0$ to find the displacement for the third interval, in which the electron slows down.

4. Add the displacements found in steps 1, 2, and 3 to find the total displacement.

Answers

$\Delta x_1 = 6.00$ cm, $v_1 = 8.00 \times 10^5$ m/s

$\Delta x_2 = 16$ cm

$\Delta x_3 = 1.20$ cm

$\Delta x = \Delta x_1 + \Delta x_2 + \Delta x_3$

$= 6.00$ cm $+ 16$ cm $+ 1.20$ cm $= \boxed{23.2 \text{ cm}}$

REMARKS In an X-ray machine electrons are accelerated from a hot wire to a metal target. They crash into it, coming abruptly to rest. As a result, the target emits X rays characteristic of the target metal.

(*left*) The two-mile-long linear accelerator at Stanford University, used to accelerate electrons and positrons in a straight line to nearly the speed of light. (*right*) Cross-section of the accelerator's electron beam as shown on a video monitor.

TOSSING BINOCULARS   **EXAMPLE 2 - 1 4**   Try It Yourself

John climbs a tree to get a better view of the speaker at an outdoor graduation ceremony. Unfortunately, he leaves his binoculars behind. Marsha throws them up to John, but her strength is greater than her accuracy. The binoculars pass John's outstretched hand after 0.69 s and again 1.68 s later. How high is John?

PICTURE THE PROBLEM There are two unknowns in this problem, John's height h and the initial velocity of the binoculars v_0. We know that $y = h$ at $t_1 = 0.69$ s and $y = h$ at $t_2 = 0.69$ s $+ 1.68$ s $= 2.37$ s. Expressing h as a function of time t gives us two equations from which the two unknowns can be determined.

Cover the column to the right and try these on your own before looking at the answers.

Steps

Answers

1. Using $\Delta y = v_0 t + \frac{1}{2}at^2$, equate y for times t_1 and t_2, noting that $y = h$, and $a = -g$ in each case.

$$h = v_0 t_1 - \tfrac{1}{2}g t_1^2 \quad \text{and} \quad h = v_0 t_2 - \tfrac{1}{2}g t_2^2$$

2. Eliminate v_0 from these two equations and solve for h in terms of the times t_1 and t_2. This can be done by solving the first equation for v_0 and substituting into the second equation.

$$h = \left(\frac{h + \frac{1}{2}g t_1^2}{t_1}\right)t_2 - \tfrac{1}{2}g t_2^2$$

so

$$h = \boxed{8.02 \text{ m}}$$

REMARKS We have two unknowns, h and v_0, but are given two times, t_1 and t_2, so we can write two equations and solve them for either or both of the two unknowns.

EXERCISE Find the initial velocity of the binoculars and the velocity of the binoculars as they pass John on the way down. (*Answer* $v_0 = 15.0$ m/s; $v_2 = -8.24$ m/s)

Problems with Two Objects

We now give some examples of problems involving two objects moving with constant acceleration.

CATCHING A SPEEDING CAR   **EXAMPLE 2-15**

A car is speeding at 25 m/s (~90 km/h; ~56 mi/h) in a school zone. A police car starts from rest just as the speeder passes and accelerates at a constant rate of 5 m/s². (*a*) When does the police car catch the speeding car? (*b*) How fast is the police car traveling when it catches up with the speeder?

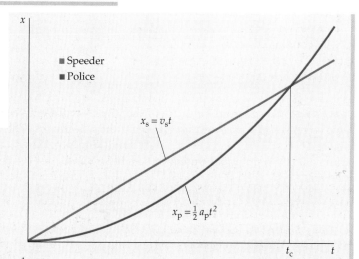

PICTURE THE PROBLEM To determine when the two cars will be at the same position, we write the positions of the speeder x_s and of the police car x_p as functions of time and solve for the time t when $x_s = x_p$.

(*a*) 1. Write the position functions for the speeder and the police car:

$$x_s = v_s t \quad \text{and} \quad x_p = \tfrac{1}{2}a_p t^2$$

2. Set $x_s = x_p$ and solve for the time t_c, for $t_c > 0$:

$$v_s t_c = \tfrac{1}{2}a_p t_c^2 \Rightarrow v_s = \tfrac{1}{2}a_p t_c \quad t_c \neq 0$$

$$t_c = \frac{2v_s}{a_p} = \frac{2(25 \text{ m/s})}{5 \text{ m/s}^2} = \boxed{10 \text{ s}}$$

FIGURE 2-13 The two curves depict the positions of the speeder and the police car. They have the same position at $t = 0$ and again at $t = t_c$.

(*b*) 1. The velocity of the police car is given by $v = v_0 + at_c$, with $v_0 = 0$:

$$v_p = a_p t_c = (5 \text{ m/s}^2)(10 \text{ s}) = \boxed{50 \text{ m/s}}$$

REMARKS Notice that the final speed of the police car in (*b*) is exactly twice that of the speeder. Since the two cars covered the same distance in the same time, they must have had the same average speed. The speeder's average speed, of course, is 25 m/s. For the police car to start from rest and have an average speed of 25 m/s, it must reach a final speed of 50 m/s.

EXERCISE How far have the cars traveled when the police car catches the speeder? (*Answer* 250 m)

THE POLICE CAR   **EXAMPLE 2-16**   **Try It Yourself**

How fast is the police car in Example 2-15 traveling when it is 25 m behind the speeding car?

PICTURE THE PROBLEM The speed is given by $v_p = at_1$, where t_1 is the time at which $D = x_s - x_p = 25$ m.

Cover the column to the right and try these on your own before looking at the answers.

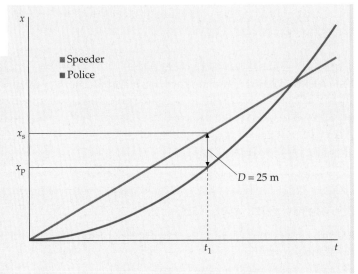

FIGURE 2-14

Steps	Answers
1. Sketch an *x*-versus-*t* curve showing the positions of the two cars at time t_1 (Figure 2-14).	
2. Using the equations for x_p and x_s from Example 2-15, solve for t_1 when $x_s - x_p = 25$ m. We expect two solutions, one shortly after the start time and one shortly before the speeder is caught.	$t_1 = (5 \pm \sqrt{15})$ s
3. Use $v_{p1} = a_p t_1$ to compute the speed of the police car at $t = t_1$.	$v_{p1} = $ 5.64 m/s and 44.4 m/s

REMARKS We see from Figure 2-14 that the distance between the cars starts at zero, increases to a maximum value, and then decreases. The separation at any time is $D = x_s - x_p = v_s t - \frac{1}{2} a_p t^2$. At maximum separation, $dD/dt = 0$, which occurs at $t = 5$ s. At equal time intervals before and after $t = 5$ s, the separations are equal.

A MOVING ELEVATOR   **EXAMPLE 2-17**

While standing in an elevator, you see a screw fall from the ceiling. The ceiling is 3 m above the floor. How long does it take for the screw to hit the floor if the elevator is moving upward and gaining speed at a constant rate of $a_f = 4.0$ m/s²?

PICTURE THE PROBLEM Write the position as a function of time for both the screw, y_s, and the floor, y_f. When the screw hits the floor $y_s = y_f$. Choose the origin to be the initial position of the floor, and designate "upward" as the positive direction.

1. Make a sketch of the elevator and the screw as shown in Figure 2-15. Include a coordinate axis indicating the positions of the screw and the floor:

2. Write the position function for the elevator floor and for the screw:

$$y_f - y_{0f} = v_{0f}t + \tfrac{1}{2}a_f t^2$$

$$y_s - y_{0s} = v_{0s}t + \tfrac{1}{2}a_s t^2$$

3. At $t = t_1$ the screw hits the floor. At this time these positions are equal:

$$y_s = y_f$$

$$y_{0s} + v_{0s}t_1 + \tfrac{1}{2}a_s t_1^2 = y_{0f} + v_{0f}t_1 + \tfrac{1}{2}a_f t_1^2$$

4. At $t = 0$, the floor and the screw have the same velocity. Use this to simplify the step 3 result:

$$v_{0f} = v_{0s}$$

so

$$y_{0s} + v_{0s}t_1 + \tfrac{1}{2}a_s t_1^2 = y_{0f} + v_{0s}t_1 + \tfrac{1}{2}a_f t_1^2$$

$$y_{0s} + \tfrac{1}{2}a_s t_1^2 = y_{0f} + \tfrac{1}{2}a_f t_1^2$$

5. Use the given information to further simplify:

$$y_{0f} = 0, \quad a_f = 4.0 \text{ m/s}^2$$

$$y_{0s} = h = 3 \text{ m}, \quad a_s = -g$$

so

$$h - \tfrac{1}{2}g t_1^2 = 0 + \tfrac{1}{2}a_f t_1^2$$

or

$$h = \tfrac{1}{2}(g + a_f)t_1^2$$

6. Solve for the time:

$$t_1 = \sqrt{\frac{2h}{g + a_f}} = \sqrt{\frac{2(3 \text{ m})}{9.81 \text{ m/s}^2 + 4.0 \text{ m/s}^2}}$$

$$= \boxed{0.659 \text{ s}}$$

FIGURE 2-15 The coordinate axis is fixed to the building.

The figure at right shows: $a_f = 4$ m/s^2, $a_s = -g$, with the elevator, screw, y_s, y_f, $h = 3$ m, and velocity v upward.

REMARKS The time of fall depends on the acceleration of the elevator, but not on its velocity. There is an "effective gravity" $g' = g + a_f$ in the frame of reference of the elevator. In the case (presumably hypothetical) in which the elevator itself is in free-fall, that is, $a_f = -g'$, the time of fall becomes infinite and the screw appears "weightless."

EXERCISE The speed of a good base runner is 9.5 m/s. The distance between bases is 26 m, and the pitcher is about 18.5 m from home plate. If a runner on first base edges 2 m off the base and then begins running at the instant the ball leaves the pitchers's hand, what is the likelihood that the runner will steal second base safely?

THE MOVING ELEVATOR   **EXAMPLE   2 - 1 8**   Try It Yourself

Consider the elevator and screw in Example 2-17. Assume the velocity of the elevator is 16 m/s upward when the screw separates from the ceiling. (*a*) How far does the elevator rise while the screw is falling? How far does the screw fall? (*b*) What is the velocity of the screw and the velocity of the elevator at impact? (*c*) What is the velocity of the screw relative to the floor at impact?

PICTURE THE PROBLEM The time of flight of the screw is obtained in the solution of Example 2-17. Use this time to solve parts (a) and (b). For part (c), the velocity of the screw relative to the building equals the velocity of the screw relative to the elevator floor plus the velocity of the elevator floor relative to the building.

Cover the column to the right and try these on your own before looking at the answers.

Steps

Answers

(a) 1. Using Equation 2-16, find the distance the floor rises in time t_1.

$\Delta y_f = v_{f0}t_1 + \frac{1}{2}a_f t_1^2 = \boxed{11.4 \text{ m}}$

2. The screw starts out 3 m above the floor.

$\Delta y_s = \boxed{8.42 \text{ m}}$

(b) Using Equation 2-12, find the impact velocity of the screw and of the floor at impact.

$v = v_0 + at$, so

$v_s = v_{s0} - gt_1 = \boxed{9.53 \text{ m/s}}$

$v_f = v_{f0} + a_f t_1 = \boxed{18.6 \text{ m/s}}$

(c) Using Equation 2-7a, find the velocity of the screw relative to the elevator floor.

$v_{sb} = v_{sf} + v_{fb}$

so

$v_{sf} = v_{sb} - v_{fb} = 9.53 \text{ m/s} - 18.6 \text{ m/s}$

$= \boxed{-9.10 \text{ m/s}}$

REMARKS The screw strikes the floor 8.4 m above its position when it leaves the ceiling. Relative to the building it is still rising when it strikes the floor. Note that at impact the velocity of the screw relative to the building is positive.

2-4 Integration

To find the velocity from a given acceleration, we note that the velocity is the function $v(t)$ whose derivative is the acceleration $a(t)$:

$$\frac{dv(t)}{dt} = a(t)$$

If the acceleration is constant, the velocity is that function of time which, when differentiated, equals this constant. One such function is

$v = at$, $a = $ constant

More generally, we can add any constant to at without changing the time derivative. Calling this constant c, we have

$v = at + c$

When $t = 0$, $v = c$. Thus, c is the initial velocity v_0.
    Similarly, the position function $x(t)$ is that function whose derivative is the velocity:

$$\frac{dx}{dt} = v = v_0 + at$$

We can treat each term separately. The function whose derivative is a constant v_0 is $v_0 t$ plus any constant. The function whose derivative is at is $\frac{1}{2}at^2$ plus any constant. Writing x_0 for the combined arbitrary constants, we have

$$x = x_0 + v_0 t + \tfrac{1}{2}at^2$$

When $t = 0$, $x = x_0$. Thus, x_0 is the initial position.

Whenever we find a function from its derivative, we must include an arbitrary constant in the general function. Since we go through the integration process twice to find $x(t)$ from the acceleration, two constants arise. These constants are usually determined from the velocity and position at some given time, which is usually chosen to be $t = 0$. They are therefore called the **initial conditions.** A common problem, called the **initial-value problem,** takes the form "given $a(t)$ and the initial values of x and v, find $x(t)$." This problem is particularly important in physics because the acceleration of a particle is determined by the forces acting on it. Thus, if we know the forces acting on a particle and the position and velocity of the particle at some particular time, we can find its position at all other times.

A function $F(t)$ whose derivative (with respect to t) equals the function $f(t)$ is called the **antiderivative** of $f(t)$. Finding the antiderivative of a function is related to the problem of finding the area under a curve. Consider motion with a constant velocity v_0. The change in position Δx during an interval Δt is

$$\Delta x = v_0 \Delta t$$

This is the area under the v-versus-t curve (Figure 2-16). If v_0 is negative, both the displacement and the area under the curve are negative. Normally we think of area as a quantity that cannot be negative. However, in this context that is not the case. In this case the "area under the curve" (the area between the curve and the time axis) is a negative quantity.

The geometric interpretation of the displacement as the area under the v-versus-t curve is true not only for constant velocity, but it is true in general, as illustrated in Figure 2-17. To show this we first divide the time interval into numerous small intervals, Δt_1, Δt_2, and so on. Then we draw a set of rectangles as shown. The area of the rectangle corresponding to the ith time interval Δt_i (shaded in the figure) is $v_i \Delta t_i$, which is approximately equal to the displacement Δx_i during the interval Δt_i. The sum of the rectangular areas is therefore approximately the sum of the displacements during the time intervals and is approximately equal to the total displacement from time t_1 to t_2. Mathematically, we write this as

$$\Delta x \approx \sum_i v_i \Delta t_i$$

where the Greek letter Σ (uppercase sigma) stands for sum. We can make the approximation as accurate as we wish by putting enough rectangles under the curve, each rectangle having a sufficiently small value for Δt. For the limit of smaller and smaller time intervals (and more and more rectangles), the resulting sum approaches the area under the curve, which in turn equals the displacement. The limit as Δt approaches zero (and the number of rectangles approaches infinity) is called an **integral** and is written

$$\Delta x = x(t_2) - x(t_1) = \lim_{\Delta t \to 0}\left(\sum_i v_i \Delta t_i\right) = \int_{t_1}^{t_2} v\,dt \qquad 2\text{-}18$$

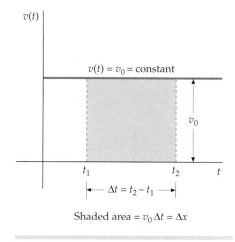

Shaded area $= v_0 \Delta t = \Delta x$

FIGURE 2-16 The displacement for the interval Δt equals the area under the velocity-versus-time curve for that interval. For $v(t) = v_0 = $ constant, the displacement equals the area of the shaded rectangle.

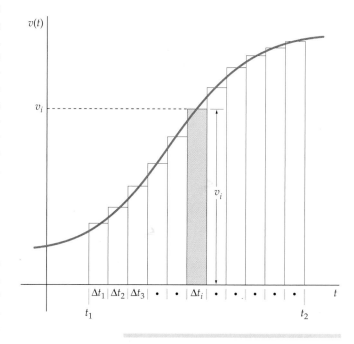

FIGURE 2-17 Graph of a general $v(t)$-versus-t curve. The total displacement from t_1 to t_2 is the area under the curve for this interval, which can be approximated by summing the areas of the rectangles.

It is helpful to think of the integral sign $\int$ as an elongated S indicating a sum. The limits t_1 and t_2 indicate the initial and final values of the variable t. The displacement is thus the area under the v-versus-t curve. Figure 2-18 demonstrates that the average velocity has a simple geometric interpretation in terms of the "area under a curve."

To illustrate that the displacement equals the area under the v-versus-t curve, consider what happens when you throw a golf ball straight up. The ball rises a meter or so, reverses direction, and then descends, gaining speed until you catch it. Assuming air resistance is negligible, the velocity of the ball is given by $v = v_0 + at$ (Equation 2-12), where the up direction is taken as positive and $a = -g$. Figure 2-19 is a plot of this velocity during the time that the ball is in free-fall. As shown, the velocity, which is initially positive, equals zero half way through the flight. As the ball descends, the velocity remains negative and, just before the ball is caught, reaches $-v_0$. During the rising portion of the motion, the area under the curve is positive, whereas during the descending portion it is negative. Thus, the total area under the curve for the entire flight is zero. It is easy to see that the displacement of the ball is also zero. Because the ball is thrown from the same place at which it is caught, the change in position is zero. Therefore, the displacement and the area under the v-versus-t curve are equal because they both are zero.

The process of computing an integral is called **integration**. In Equation 2-18, v is the derivative of x, and x is the antiderivative of v. This is an example of the fundamental theorem of calculus, whose formulation in the seventeenth century greatly accelerated the mathematical development of physics.

$$\text{If } f(t) = \frac{dF(t)}{dt}, \quad \text{then} \quad F(t_2) - F(t_1) = \int_{t_1}^{t_2} f(t)\, dt \qquad 2\text{-}19$$

FUNDAMENTAL THEOREM OF CALCULUS

The antiderivative of a function is also called the indefinite integral of the function and is written without limits on the integral sign, as in

$$x = \int v\, dt$$

Finding the function x from the derivative v (that is, finding the antiderivative) is also called integration. For example, if $v = v_0$, a constant, then

$$x = \int v_0\, dt = v_0 t + x_0$$

where x_0 is the arbitrary constant of integration. We can find a general rule for the integration of a power of t from Equation 2-6, which gives the general rule for the derivative of a power. The result is

$$\int t^n\, dt = \frac{t^{n+1}}{n+1} + C, \qquad n \neq -1 \qquad 2\text{-}20$$

where C is an arbitrary constant. This can be checked by differentiating the right side using the rule of Equation 2-6. (For the special case $n = -1$, $\int t^{-1}\, dt = \ln t + C$, where $\ln t$ is the natural logarithm of t.)

The change in velocity for some time interval can similarly be interpreted as the area under the a-versus-t curve for that interval. This is written

$$\Delta v = \lim_{\Delta t \to 0}\left(\sum_i a_i \Delta t_i\right) = \int_{t_1}^{t_2} a\, dt \qquad 2\text{-}21$$

We can now derive the constant-acceleration equations by computing the indefinite integrals of the acceleration and velocity. If a is constant, we have

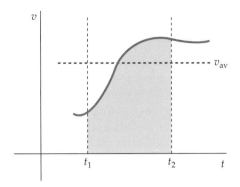

FIGURE 2-18 The displacement Δx during the time interval $\Delta t = t_2 - t_1$ is equal to the area of the shaded region. We know from the definition of average velocity that $\Delta x = v_{av}\, \Delta t$. This is just the area of a rectangle of height v_{av} and width Δt. Thus, the rectangular area $v_{av}\, \Delta t$ and the area under the v-versus-t curve must be equal.

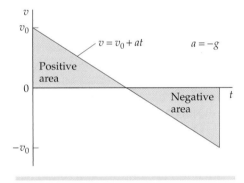

FIGURE 2-19 A v-versus-t curve for a golf ball thrown straight up that is caught by the thrower. The area under the curve is positive for the rising portion of the motion and negative for the descending portion. The area under the curve for the entire flight is zero.

$$v = \int a\, dt = a \int dt = v_0 + at \qquad\qquad 2\text{-}22$$

where we have expressed a times the constant of integration as v_0. Integrating again, and writing x_0 for the constant of integration, gives

$$x = \int (v_0 + at)dt = x_0 + v_0 t + \tfrac{1}{2}at^2 \qquad\qquad 2\text{-}23$$

It is instructive to derive Equations 2-22 and 2-23 using definite integrals instead of indefinite ones. For constant acceleration, Equation 2-21, with $t_1 = 0$, gives

$$v(t_2) - v(0) = a \int_0^{t_2} dt = a(t_2 - 0)$$

where the time t_2 is arbitrary. Because it is arbitrary, we can set $t_2 = t$ to obtain

$$v = v_0 + at$$

where $v = v(t)$ and $v_0 = v(0)$. To derive Equation 2-23, we substitute $v_0 + at$ for v in Equation 2-18 with $t_1 = 0$. This gives

$$x(t_2) - x(0) = \int_0^{t_2} (v_0 + at)dt$$

This integral is equal to the area under the v-versus-t curve (Figure 2-20). Evaluating the integral and solving for x gives

$$x(t_2) - x(0) = \int_0^{t_2} (v_0 + at)dt = v_0 t + \tfrac{1}{2}at^2 \Big|_0^{t_2} = v_0 t_2 + \tfrac{1}{2}at_2^2$$

where t_2 is arbitrary. Setting $t_2 = t$ we obtain

$$x = x_0 + v_0 t + \tfrac{1}{2}at^2$$

where $x = x(t)$ and $x_0 = x(0)$.

Having derived the constant-acceleration kinematic equations without any reference to average velocity, we can now show that the average velocity for the special case of constant acceleration is the mean value between the initial and final velocities as given by Equation 2-14. Let v_0 be the initial velocity at $t = 0$, and v be the final velocity at time t. According to the definition of average velocity, the displacement is

$$\Delta x = v_{av}\Delta t = v_{av}(t - 0) = v_{av}t \qquad\qquad 2\text{-}24$$

Also, from Equation 2-23, we have

$$\Delta x = v_0 t + \tfrac{1}{2}at^2$$

We eliminate the acceleration using $a = (v - v_0)/t$ from Equation 2-12. This gives

$$\Delta x = v_0 t + \frac{1}{2}\left(\frac{v - v_0}{t}\right)t^2 = v_0 t + \frac{1}{2}vt - \frac{1}{2}v_0 t = \frac{1}{2}(v + v_0)t \qquad\qquad 2\text{-}25$$

Comparing this with $\Delta x = v_{av}t$ (Equation 2-24), we have

$$v_{av} = \tfrac{1}{2}(v_0 + v_f)$$

which is Equation 2-14.

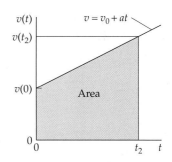

FIGURE 2-20 The area under the v-versus-t curve equals the displacement $\Delta x = x(t_2) - x(0)$.

The average velocity can be visualized using a v-versus-t curve (Figure 2-21). The displacement Δx equals the area under this curve. However, the average velocity is the area under the curve $v = v_{av}$ for the same time interval. Thus, the height of the $v = v_{av}$ curve is such that the areas under the two curves are equal. This implies that the areas of the two gray-shaded triangles are equal and that $v_{av} = \frac{1}{2}(v_1 + v_2)$.

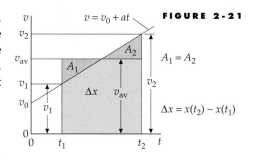

FIGURE 2-21

$$A_1 = A_2$$
$$\Delta x = x(t_2) - x(t_1)$$

A COASTING BOAT

EXAMPLE 2-19

A Shelter Island ferryboat moves with constant velocity $v_0 = 8$ m/s for 60 s. It then shuts off its engines and coasts. Its coasting velocity is given by $v = v_0 t_1^2/t^2$, where $t_1 = 60$ s. What is the displacement of the boat for the interval $0 < t < \infty$?

PICTURE THE PROBLEM This velocity function is shown in Figure 2-22. The total displacement is calculated as the sum of the displacement Δx_1 during the interval $0 < t < t_1 = 60$ s and the displacement Δx_2 during the interval $t_1 < t < \infty$.

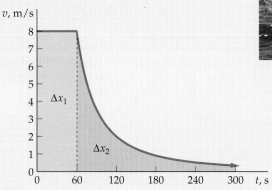

FIGURE 2-22

1. The velocity of the boat is constant during the first 60 s; thus the displacement is simply the velocity times the elapsed time:

$$\Delta x_1 = v_0 \Delta t = v_0 t_1 = (8 \text{ m/s})(60 \text{ s}) = 480 \text{ m}$$

2. The remaining displacement is given by the integral of the velocity from $t = t_1$ to $t = \infty$. We use Equation 2-18 to calculate the integral:

$$\Delta x_2 = \int_{t_1}^{\infty} v\, dt = \int_{t_1}^{\infty} \frac{v_0 t_1^2}{t^2}\, dt = v_0 t_1^2 \int_{t_1}^{\infty} t^{-2}\, dt = v_0 t_1^2 \left.\frac{t^{-1}}{-1}\right|_{t_1}^{\infty}$$

$$= -v_0 t_1^2 \left(\frac{1}{\infty} - \frac{1}{t_1}\right) = v_0 t_1 = (8 \text{ m/s})(60 \text{ s}) = 480 \text{ m}$$

3. The total displacement is the sum of the displacements found above:

$$\Delta x = \Delta x_1 + \Delta x_2 = 480 \text{ m} + 480 \text{ m} = \boxed{960 \text{ m}}$$

REMARKS Note that the area under the v-versus-t curve is finite. Thus, even though the boat never stops moving, it travels only a finite distance. A better representation of the velocity of a coasting boat might be the exponentially decreasing function $v = v_0 e^{-b(t-t_1)}$, where b is a positive constant. In that case, the boat would also coast a finite distance in the interval $t_1 \le t \le \infty$.

Displacement, velocity, and acceleration are important *defined* kinematics quantities.

Topic	Relevant Equations and Remarks	
1. Displacement	$\Delta x = x_2 - x_1$	2-1
Graphical interpretation	Displacement is the area under the v-versus-t curve.	
2. Velocity		
Average velocity	$v_{av} = \dfrac{\Delta x}{\Delta t}$	2-2
Instantaneous velocity	$v(t) = \lim\limits_{\Delta t \to 0} \dfrac{\Delta x}{\Delta t} = \dfrac{dx}{dt}$	2-5
Graphical interpretation	The instantaneous velocity is represented graphically as the slope of the x-versus-t curve.	
Relative velocity	If a particle moves with velocity v_{pA} relative to reference frame A, which is in turn moving with velocity v_{AB} relative to a second reference frame B, the velocity of the particle relative to B is	
	$v_{pB} = v_{pA} + v_{AB}$	2-7
3. Speed		
Average speed	$\text{Average speed} = \dfrac{\text{total distance}}{\text{total time}} = \dfrac{s}{t}$	2-3
Instantaneous speed	Instantaneous speed is the magnitude of the instantaneous velocity.	
4. Acceleration		
Average acceleration	$a_{av} = \dfrac{\Delta v}{\Delta t}$	2-8
Instantaneous acceleration	$a = \dfrac{dv}{dt} = \dfrac{d^2x}{dt^2}$	2-10
Graphical interpretation	The instantaneous acceleration is represented graphically as the slope of the v-versus-t curve.	
Acceleration due to gravity	The acceleration of an object near the surface of the earth in free-fall under the influence of gravity is directed downward and has the magnitude	
	$g = 9.81 \text{ m/s}^2 = 32.2 \text{ ft/s}^2$	
5. Displacement and velocity as integrals	Displacement is represented graphically as the area under the v-versus-t curve. This area is the integral of v over time from some initial time t_1 to some final time t_2 and is written	
	$\Delta x = \lim\limits_{\Delta t \to 0} \sum\limits_{i} v_i \Delta t_i = \int_{t_1}^{t_2} v \, dt$	2-18

Similarly, change in velocity is represented graphically as the area under the *a*-versus-*t* curve:

$$\Delta v = \lim_{\Delta t \to 0} \sum_i a_i \Delta t_i = \int_{t_1}^{t_2} a \, dt \qquad \qquad \textbf{2-21}$$

Velocity	$v = v_0 + at$	**2-12**
Displacement in terms of v_{av}	$\Delta x = x - x_0 = v_{av}t = \frac{1}{2}(v_0 + v)t$	**2-15**
Displacement in terms of a	$\Delta x = x - x_0 = v_0 t + \frac{1}{2}at^2$	**2-16**
v in terms of a and Δx	$v^2 = v_0^2 + 2a\,\Delta x$	**2-17**

PROBLEMS

- Single-concept, single-step, relatively easy
- ●● Intermediate-level, may require synthesis of concepts
- ●●● Challenging
- **SSM** Solution is in the *Student Solutions Manual*
- **iSOLVE** Problems available on iSOLVE online homework service
- **iSOLVE** ✓ These "Checkpoint" online homework service problems ask students additional questions about their confidence level, and how they arrived at their answer

In a few problems, you are given more data than you actually need; in a few other problems, you are required to supply data from your general knowledge, outside sources, or informed estimates.

For all problems, use $g = 9.81$ m/s² for the acceleration due to gravity and neglect friction and air resistance unless instructed to do otherwise.

Conceptual Problems

1 • What is the average velocity over the "round trip" of an object that is launched straight up from the ground and falls straight back down to the ground?

2 • **SSM** An object thrown straight up falls back to the ground. Its time of flight is T, its maximum height is H, and its height at release is negligible. Its average speed for the entire flight is (a) H/T, (b) 0, (c) $H/(2T)$, (d) $2H/T$.

3 • **iSOLVE** To avoid falling too fast during a landing, an airplane must maintain a minimum airspeed (the speed of the plane relative to the air). However, the slower the ground speed (speed relative to the ground) during a landing, the safer the landing. Is it safer for an airplane to land with the wind or against the wind?

4 • Give an example of one-dimensional motion where (a) the velocity is positive and acceleration is negative, and (b) the velocity is negative and the acceleration is positive.

5 • **SSM** Stand in the center of a large room. Call movement to your right "positive," and movement to your left "negative." Walk across the room along a straight line in such a way that, after getting started, your velocity is negative but your acceleration is positive. (a) Is your displacement initially positive or negative? Explain. (b) Describe how you vary your speed as you walk. (c) Sketch a graph of v versus t for your motion.

6 • True/false; explain: The displacement *always* equals the product of the average velocity and the time interval.

7 • True/false; explain:
(a) For the velocity to remain constant, the acceleration *must* remain zero.
(b) For the speed to remain constant, the acceleration *must* remain zero.

8 ●● **SSM** Draw careful graphs of the position and velocity and acceleration over the time period $0 \le t \le 25$ s for a cart that

(a) moves away from the origin at a slow and steady (constant) velocity for the first 5 s;
(b) moves away at a medium-fast, steady (constant) velocity for the next 5 s;
(c) stands still for the next 5 s;
(d) moves toward the origin at a slow and steady (constant) velocity for the next 5 s;
(e) stands still for the last 5 s.

9 • True/false; explain: The average velocity *always* equals one-half the sum of initial plus final velocities.

10 • Identical twin brothers standing on a horizontal bridge each throw a rock straight down into the water below. They throw rocks at exactly the same time, but one hits the water before the other. How can this occur if the rocks have the same starting time?

11 •• **SSM** Dr. Josiah S. Carberry stands at the top of the Sears Tower in Chicago. Wanting to emulate Galileo, and ignoring the safety of the pedestrians below, he drops a bowling ball from the top of the tower. One second later, he drops a second bowling ball. While the balls are in the air, does their separation (a) increase over time, (b) decrease, or (c) stay the same? Ignore any effects that air resistance may have.

12 •• Which of the position-versus-time curves in Figure 2-23 best shows the motion of an object with constant positive acceleration?

FIGURE 2-23 Problem 12

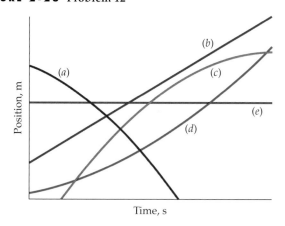

13 • **SSM** Which of the velocity-versus-time curves in Figure 2-24 best describes the motion of an object with constant positive acceleration?

FIGURE 2-24 Problem 13

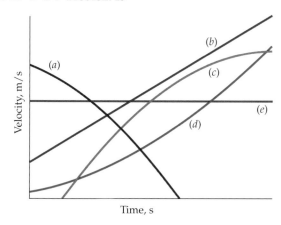

14 • Does the following statement make sense? "The average velocity of the car at 9 A.M. was 60 km/h."

15 • **SSM** Is it possible for the average velocity of an object to be zero during some interval, even though its average velocity for the first half of the interval is not zero? Explain.

16 • The diagram in Figure 2-25 tracks the path of an object moving in a straight line along the x axis. Assuming that the object is at the origin ($x_o = 0$) at $t_o = 0$, which point on the position-versus-time graph represents the instant the object is farthest from the origin? (a) A (b) B (c) C (d) D (e) E.

FIGURE 2-25 Problem 16

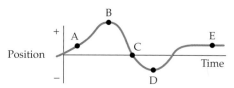

17 • **ISOLVE** If the instantaneous velocity does not change, will the average velocities for different intervals differ?

18 • If $v_{av} = 0$ for some time interval Δt, must the instantaneous velocity v be zero at some point in the interval? Support your answer by sketching a possible x-versus-t curve that has $\Delta x = 0$ for some interval Δt.

19 •• An object moves along a line as shown in Figure 2-26. At which point or points is its speed at a minimum? (a) A and E. (b) B, D, and E. (c) C only. (d) E only. (e) None of these is correct.

FIGURE 2-26 Problem 19

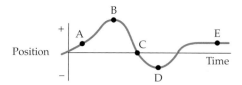

20 •• **SSM** **ISOLVE** For each of the four graphs of x versus t in Figure 2-27, answer the following questions. (a) Is the velocity at time t_2 greater than, less than, or equal to the velocity at time t_1? (b) Is the speed at time t_2 greater than, less than, or equal to the speed at time t_1?

FIGURE 2-27 Problem 20

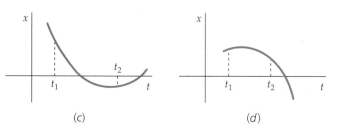

21 • True/false; explain:

(a) If the acceleration remains zero, the body cannot be moving.

(b) If the acceleration remains zero, the x-versus-t curve must be a straight line.

22 • Is it possible for a body to simultaneously have zero velocity and nonzero acceleration?

23 • ISOLVE A ball is thrown straight up. What is the velocity of the ball at the top of its flight? What is its acceleration at that point?

24 • Find the average speed over the "round trip" of an object that is launched straight up from the ground, reaches a height H, and falls straight back down to the ground, hitting it after T seconds have elapsed. Express this in terms of the initial launch speed v_0.

25 • A bowling ball is thrown upward. While it is in flight, its acceleration is (a) decreasing, (b) constant, (c) zero, (d) increasing.

26 • At $t = 0$, object A is dropped from the roof of a building. At the same instant, object B is dropped from a window 10 m below the roof. During their descent to the ground, the distance between the two objects (a) is proportional to t, (b) is proportional to t^2, (c) decreases, (d) remains 10 m throughout.

27 •• SSM Assume that the Porsche accelerates uniformly from 80.5 km/h (50 mi/h) at $t = 0$ to 113 km/h (70 mi/h) at $t = 9$ s. Which graph in Figure 2-28 best describes the motion of the car?

FIGURE 2-28 Problem 27

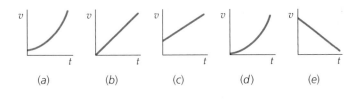

(a) (b) (c) (d) (e)

28 •• SSM An object is dropped from rest and falls a distance D in a given time. If the time during which it falls is doubled, the distance it falls will be (a) $4D$, (b) $2D$, (c) D, (d) $D/2$, (e) $D/4$.

29 •• A ball is thrown upward with an initial velocity v_0. Its velocity halfway to its highest point is (a) $0.25v_0$, (b) $0.5v_0$, (c) $0.707v_0$, (d) v_0, (e) cannot be determined from the information given.

30 • True or false:

(a) The equation $\Delta x = v_0 t + \frac{1}{2}at^2$ is valid for all particle motion in one dimension.
(b) If the velocity at a given instant is zero, the acceleration at that instant must also be zero.
(c) The equation $\Delta x = v_{av}\Delta t$ holds for all particle motion in one dimension.

31 • SSM If an object is moving at constant acceleration in a straight line, its instantaneous velocity halfway through any time interval is (a) greater than its average velocity, (b) less than its average velocity, (c) equal to its average velocity, (d) half its average velocity, (e) twice its average velocity.

32 • On a graph showing position on the vertical axis and time on the horizontal axis, a straight line with a negative slope represents motion with (a) constant positive acceleration, (b) constant negative acceleration, (c) zero velocity, (d) constant positive velocity, (e) constant negative velocity.

33 •• On a graph showing position on the vertical axis and time on the horizontal axis, a parabola that opens upward represents (a) a positive acceleration, (b) a negative acceleration, (c) no acceleration, (d) a positive followed by a negative acceleration, (e) a negative followed by a positive acceleration.

34 •• On a graph showing velocity on the vertical axis and time on the horizontal axis, a constant acceleration of zero is represented by (a) a straight line with positive slope, (b) a straight line with negative slope, (c) a straight line with zero slope, (d) either (a), (b), or (c), (e) none of the above.

35 •• On a graph showing velocity on the vertical axis and time on the horizontal axis, constant acceleration is represented by (a) a straight line with positive slope, (b) a straight line with negative slope, (c) a straight line with zero slope, (d) either (a), (b), or (c), (e) none of the above.

36 •• Which graph of v versus t in Figure 2-29 best describes the motion of a particle with positive velocity and negative acceleration?

FIGURE 2-29 Problem 36

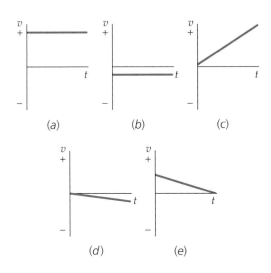

(a) (b) (c)

(d) (e)

37 •• ISOLVE ✓ Which graph of v versus t in Figure 2-29 best describes the motion of a particle with negative velocity and negative acceleration?

38 •• A graph of the motion of an object is plotted with the velocity on the vertical axis and time on the horizontal axis. The graph is a straight line. Which of these quantities *can* be determined from this graph? (a) The displacement from time $t = 0$ to any other time shown. (b) The initial velocity at $t = 0$. (c) The acceleration as a function of time. (d) The average velocity for any specified time interval. (e) All of the above.

39 •• **SSM** Figure 2-30 shows the position of a car plotted as a function of time. At which times t_0 to t_7 is the velocity (a) negative? (b) positive? (c) zero? At which times is the acceleration (a) negative? (b) positive? (c) zero?

FIGURE 2-30 Problem 39

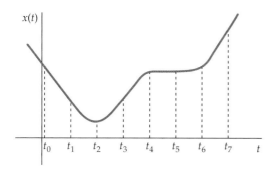

40 •• Sketch v-versus-t curves for each of the following conditions: (a) Acceleration is zero and constant while velocity is not zero. (b) Acceleration is constant but not zero. (c) Velocity and acceleration are both positive. (d) Velocity and acceleration are both negative. (e) Velocity is positive and acceleration is negative. (f) Velocity is negative and acceleration is positive. (g) Velocity is momentarily zero but the acceleration is not zero.

41 •• Figure 2-31 shows nine graphs of position, velocity, and acceleration for objects in motion along a straight line. Indicate the graphs that meet the following conditions: (a) velocity is constant, (b) velocity reverses its direction, (c) acceleration is constant, (d) acceleration is not constant. (e) Which graphs of position, velocity, and acceleration are mutually consistent?

FIGURE 2-31 Problem 41

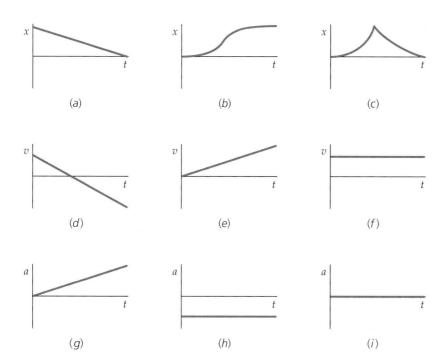

Estimation and Approximation

42 • Measure your own pulse rate (the number of heart beats per minute). Typical adult "resting rates" fall between 60 bpm (beats per min) and 80 bpm. (a) How many times will your heart beat during the time that it takes you to drive 1 mi at 60 mph? (b) How many times will your heart beat during your lifetime? (Assume a lifetime of 95 y.)

43 •• **SSM** **iSOLVE** Occasionally, people can survive after falling large distances if the surface they fall on is soft enough. During a traverse of the Eiger's infamous Nordvand, mountaineer Carlos Ragone's rock anchor pulled out and he plummeted 500 ft to land in snow. Amazingly, he suffered only a few bruises and a wrenched shoulder. (a) What final speed did he reach before impact? Ignore air resistance. (b) Assuming that his impact left a hole in the snow 4 ft deep, estimate his acceleration as he slowed to a stop. Assume that the acceleration was constant. Express this as a multiple of g (the magnitude of free-fall acceleration at the surface of the earth).

44 •• When we solve problems involving free-fall above the surface of the earth, it's important to remember that air resistance always exists; if we naively assume that objects always fall with constant acceleration, we may get answers that are wrong by orders of magnitude. How can we tell when it is valid to assume that a body is falling with (almost) constant acceleration? As a real body falls from rest through the air, as its speed increases, its acceleration downward decreases. The velocity will approach, but never quite reach, a *terminal velocity* that depends on the mass and cross-sectional area of the body; at the terminal velocity, the forces of gravity and air resistance exactly balance. For a "typical" skydiver falling through the air, a reasonable estimate for the terminal velocity is about 50 m/s (roughly 120 mph). At a speed of half the terminal velocity, the skydiver's acceleration will be $\frac{3}{4}g$. (a) Let's take half the terminal velocity as a reasonable "upper bound" beyond which we shouldn't use the constant acceleration formulas to calculate velocities and displacements. Roughly how far, and for how long, will the skydiver fall before we can't use these formulas anymore? (b) Repeat the analysis for a mouse, which has a terminal velocity of about 1 m/s.

45 •• On June 16, 1999 Maurice Greene of the United States set a new world's record for the 100-m dash with a time $t = 9.79$ s. Suppose that he started from rest at constant acceleration a and reached his maximum velocity in 3.00 s, which he then kept until the finish line. What was the acceleration a?

46 •• [SSM] The photograph in Figure 2-32 is a short-time exposure (1/30 s) of a juggler with two tennis balls in the air. The tennis ball near the top of its trajectory is less blurred than the lower one. Why is that? Can you estimate the speed of the lower ball from the picture?

FIGURE 2-32 Problem 46

47 •• Look up the speed at which a nerve impulse travels through the body. Estimate the time between stubbing your toe on a rock and feeling the pain due to this.

Speed, Displacement, and Velocity

48 • (a) An electron in a television tube travels the 16-cm distance from the grid to the screen at an average speed of 4×10^7 m/s. How long does the trip take? (b) An electron in a current-carrying wire travels at an average speed of 4×10^{-5} m/s. How long does it take to travel 16 cm?

49 • [SSM] A runner runs 2.5 km, in a straight line, in 9 min and then takes 30 min to walk back to the starting point. (a) What is the runner's average velocity for the first 9 min? (b) What is the average velocity for the time spent walking? (c) What is the average velocity for the whole trip? (d) What is the average speed for the whole trip?

50 • [SOLVE] A car travels in a straight line with an average velocity of 80 km/h for 2.5 h and then with an average velocity of 40 km/h for 1.5 h. (a) What is the total displacement for the 4-h trip? (b) What is the average velocity for the total trip?

51 • One busy air route across the Atlantic Ocean is about 5500 km. (a) How long does it take for a supersonic jet flying at 2 times the speed of sound to make the trip? Use 340 m/s for the speed of sound. (b) How long does it take a subsonic jet flying at 0.9 times the speed of sound to make the same trip? (c) Allowing 2 h at each end of the trip for ground travel, check-in, and baggage handling, what is your average speed, door to door, when traveling on the supersonic jet? (d) What is your average speed taking the subsonic jet?

52 • [SSM] The speed of light, c, is 3×10^8 m/s. (a) How long does it take for light to travel from the sun to the earth, a distance of 1.5×10^{11} m? (b) How long does it take light to travel from the moon to the earth, a distance of 3.84×10^8 m? (c) A light-year is a unit of distance equal to that traveled by light in 1 year. Convert 1 light-year into kilometers and miles.

53 • Proxima Centauri, a dim companion to Alpha Centauri, is 4.1×10^{13} km away. From the vicinity of this star, Gregor places an order at Tony's Pizza in Hoboken, New Jersey, communicating via light signals. Tony's fastest delivery craft travels at $10^{-4}c$ (see Problem 52). (a) How long does it take Gregor's order to reach Tony's Pizza? (b) How long does Gregor wait between sending the signal and receiving the pizza? If Tony's has a 1000-years-or-it's-free delivery policy, does Gregor have to pay for the pizza?

54 • A car making a 100-km journey travels 40 km/h for the first 50 km. How fast must it go during the second 50 km to average 50 km/h?

55 •• [SSM] An archer fires an arrow, which produces a muffled "thwok" as it hits a target. If the archer hears the "thwok" exactly 1 s after firing the arrow and the average speed of the arrow was 40 m/s, what was the distance separating the archer and the target? Use 340 m/s for the speed of sound.

56 •• John can run 6 m/s. Marcia can run 15% faster than John. (a) By what distance does Marcia beat John in a 100-m race? (b) By what time does Marcia beat John in a 100-m race?

57 • [SOLVE] Figure 2-33 shows the position of a particle as a function of time. Find the average velocities for the time intervals a, b, c, and d indicated in the figure.

FIGURE 2-33 Problem 57

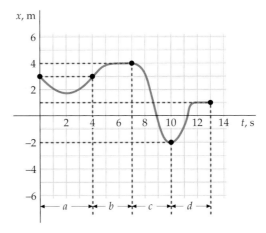

58 •• It has been found that galaxies are moving away from the earth at a speed that is proportional to their distance from the earth. This discovery is known as Hubble's law. The speed of a galaxy at a distance r from the earth is given by $v = Hr$, where H is the Hubble constant, equal to 1.58×10^{-18} s^{-1}. What is the speed of a galaxy (a) 5×10^{22} m from earth and (b) 2×10^{25} m from earth? (c) If each of these galaxies has traveled with constant speed, how long ago were they both located at the same place as the earth?

59 •• [SSM] [SOLVE] The cheetah can run as fast as $v_1 = 113$ km/h, the falcon can fly as fast as $v_2 = 161$ km/h, and the sailfish can swim as fast as $v_3 = 105$ km/h. The three of them run a relay with each covering a distance L at maximum speed. What is the average speed v of this relay team? Compare this with the average of the three speeds.

60 •• Two cars are traveling along a straight road. Car A maintains a constant speed of 80 km/h; car B maintains a constant speed of 110 km/h. At $t = 0$, car B is 45 km behind car A. How much farther will car A travel before it is overtaken by car B?

61 •• [SSM] A car traveling at a constant speed of 20 m/s passes an intersection at time $t = 0$, and 5 s later another car traveling at a constant speed of 30 m/s passes the same intersection in the same direction. (a) Sketch the position functions $x_1(t)$ and $x_2(t)$ for the two cars. (b) Determine when the second car will overtake the first. (c) How far from the intersection will the two cars be when they pull even? (d) Where is the first car when the second car passes the intersection?

62 • Joe and Sally tend to argue when they travel. Just as they reached the moving sidewalk at the airport, their tempers flared to a point where neither was talking to the other. Though they stepped on the moving belt at the same time, Joe chose to stand and ride, while Sally opted to keep walking. Sally reached the end in 1 min, while Joe took 2 min. How long would it have taken Sally if she had walked twice as fast relative to the moving belt?

63 •• Margaret has just enough gas in her speedboat to get to the marina, an upstream journey that takes 4.0 h. Finding it closed for the season, she spends the next 8.0 h floating back downstream (out of gas) to her shack. The entire trip took 12.0 h. How long would it have taken if she had bought gas at the marina? Assume that the effect of the wind is negligible.

Acceleration

64 • [SOLVE] A BMW-M3 sports car can accelerate in third gear from 48.3 km/h (30 mi/h) to 80.5 km/h (50 mi/h) in 3.7 s. (a) What is the average acceleration of this car in m/s²? (b) If the car continued at this acceleration for another second, how fast would it be moving?

65 • At $t = 5$ s, an object at $x = 3$ m has a velocity of $+5$ m/s. At $t = 8$ s, it is at $x = 9$ m and its velocity is -1 m/s. Find the average acceleration for this interval.

66 •• A particle moves with velocity $v = (8$ m/s²$)\, t - 7$ m/s. (a) Find the average acceleration for two 1-s intervals, one beginning at $t = 3$ s and the other beginning at $t = 4$ s. (b) Sketch v versus t. What is the instantaneous acceleration at any time?

67 •• [SOLVE]✔ The position of a certain particle depends on time according to the equation $x(t) = t^2 - 5t + 1$, where x is in meters if t is in seconds. (a) Find the displacement and average velocity for the interval $3\,\text{s} \le t \le 4\,\text{s}$. (b) Find the general formula for the displacement for the time interval from t to $t + \Delta t$. (c) Use the limiting process to obtain the instantaneous velocity for any time t.

68 •• [SSM] [SOLVE] The position of an object is related to time by $x = At^2 - Bt + C$, where $A = 8$ m/s², $B = 6$ m/s, and $C = 4$ m. Find the instantaneous velocity and acceleration as functions of time.

69 •• The one-dimensional motion of a particle is plotted in Figure 2-34. (a) What is the average acceleration in the intervals AB, BC, and CE? (b) How far is the particle from its starting point after 10 s? (c) Sketch the displacement of the particle as a function of time; label the instants A, B, C, D, and E on your figure. (d) At what time is the particle traveling most slowly?

FIGURE 2-34 Problem 69

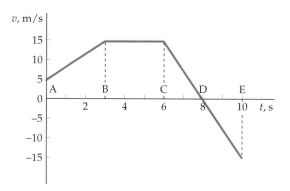

Constant Acceleration and Free-Fall

70 • [SSM] An object projected upward with initial velocity v_0 attains a height h. Another object projected up with initial velocity $2v_0$ will attain a height of (a) 4h, (b) 3h, (c) 2h, (d) h.

71 • A car starting at $x = 50$ m accelerates from rest at a constant rate of 8 m/s². (a) How fast is it going after 10 s? (b) How far has it gone after 10 s? (c) What is its average velocity for the interval $0 \le t \le 10$ s?

72 • An object with an initial velocity of 5 m/s has a constant acceleration of 2 m/s². When its speed is 15 m/s, how far has it traveled?

73 • [SSM] An object with constant acceleration has a velocity of 10 m/s when it is at $x = 6$ m and of 15 m/s when it is at $x = 10$ m. What is its acceleration?

74 • The speed of an object increases at a constant rate of 4 m/s each second. At $t = 0$, its velocity is 1 m/s and its position is $x = 7$ m. How fast is it moving when it is at $x = 8$ m and what does t equal then?

75 •• [SOLVE]✔ A ball is thrown upward with an initial velocity of 20 m/s. (a) How long is the ball in the air? (Neglect the height of the release point.) (b) What is the greatest height reached by the ball? (c) How long after release is the ball 15 m above the release point?

76 •• [SOLVE] ✓ In the Blackhawk landslide in California, a mass of rock and mud fell 460 m down a mountain and then traveled 8 km across a level plain. It has been theorized that the rock and mud moved on a cushion of water vapor. Assume that the mass dropped with the free-fall acceleration and then slid horizontally, losing speed at a constant rate. (a) How long did the mud take to drop the 460 m? (b) How fast was it traveling when it reached the bottom? (c) How long did the mud take to slide the 8 km horizontally?

77 •• [SSM] A load of bricks is being lifted by a crane at a steady velocity of 5 m/s when one brick falls off 6 m above the ground. (a) Sketch $x(t)$ to show the motion of the free brick. (b) What is the greatest height the brick reaches above the ground? (c) How long does it take to reach the ground? (d) What is its speed just before it hits the ground?

78 •• A bolt comes loose from underneath an elevator that is moving upward at a speed of 6 m/s. The bolt reaches the bottom of the elevator shaft in 3 s. (a) How high up is the elevator when the bolt comes loose? (b) What is the speed of the bolt when it hits the bottom of the shaft?

79 •• [SSM] An object is dropped from rest at a height of 120 m. Find the distance it falls during its final second in the air.

80 •• An object is released from rest at a height h. During the final second of its fall, it traverses a distance of 38 m. What is h?

81 • [SSM] A stone is thrown vertically from a 200-m-tall cliff. During the last half second of its flight the stone travels a distance of 45 m. Find the initial speed of the stone.

82 •• An object is released from rest at a height h. It travels 0.4h during the first second of its descent. Determine the average velocity of the object during its entire descent.

83 •• A bus accelerates at 1.5 m/s^2 from rest for 12 s. It then travels at constant velocity for 25 s, after which it slows to a stop with an acceleration of −1.5 m/s^2. (a) How far does the bus travel? (b) What is its average velocity?

84 •• [SSM] It is relatively easy to use a spreadsheet program such as Microsoft Excel to solve certain types of physics problems. For example, you probably solved Problem 75 using algebra. Let's solve Problem 75 in a different way, this time using a spreadsheet program. While we can solve this problem using algebra, there are many places in physics where we can't get an alternative solution so easily, and have to rely on numerical methods like the one shown here. (a) Using Microsoft Excel or some other spreadsheet program, generate a graph of the height versus time for the ball in Problem 75 (thrown upward with an initial velocity of 20 m/s). Determine the maximum height, the time it was in the air, and the time(s) when the ball is 15 m above the ground by inspection (i.e., look at the graph and find them.) (b) Now change the initial velocity to 10 m/s and find the maximum height the ball reaches and the time the ball spends in the air.

85 •• [SSM] [SOLVE] Al and Bert are jogging side-by-side on a trail in the woods at a speed of 0.75 m/s. Suddenly Al

sees the end of the trail 35 m ahead and decides to speed up to reach it. He accelerates at a constant rate of 0.5 m/s^2, leaving Bert behind, who continues on at a constant speed. (a) How long does it take Al to reach the end of the trail? (b) Once he reaches the end of the trail, he immediately turns around and heads back along the trail with a constant speed of 0.85 m/s. How long does it take him to reach Bert? (c) How far are they from the end of the trail when they meet?

86 •• Solve Problem 85, parts (b) and (c), using a spreadsheet program.

87 •• A rocket is fired vertically with an upward acceleration of 20 m/s^2. After 25 s, the engine shuts off and the rocket then continues rising (while in free-fall). The rocket eventually stops rising and then falls back to the ground. Calculate (a) the highest point the rocket reaches, (b) the total time the rocket is in the air, (c) the speed of the rocket just before it hits the ground.

88 •• [SOLVE] A flowerpot falls from the ledge of an apartment building. A person in an apartment below, coincidentally holding a stopwatch, notices that it takes 0.2 s for the pot to fall past his window, which is 4 m high. How far above the top of the window is the ledge from which the pot fell?

89 •• [SSM] In a classroom demonstration, a glider moves along an inclined air track with constant acceleration. It is projected from the start of the track with an initial velocity. After 8 s have elapsed, it is 100 cm from its starting point and is moving along the track at a velocity of −15 cm/s. Find the initial velocity and the acceleration.

90 •• A rock dropped from a cliff falls one-third of its total distance to the ground in the last second of its fall. How high is the cliff?

91 •• A typical automobile under hard braking loses speed at a rate of about 7 m/s^2; the typical reaction time to engage the brakes is 0.50 s. A school board sets the speed limit in a school zone to meet the condition that all cars should be able to stop in a distance of 4 m. (a) What maximum speed should be allowed for a typical automobile? (b) What fraction of the 4 m is due to the reaction time?

92 •• Two trains face each other on adjacent tracks. They are initially at rest, and their front ends are 40 m apart. The train on the left accelerates rightward at 1.4 m/s^2. The train on the right accelerates leftward at 2.2 m/s^2. How far does the train on the left travel before the front ends of the trains pass?

93 •• Two stones are dropped from the edge of a 60-m cliff, the second stone 1.6 s after the first. How far below the top of the cliff is the second stone when the separation between the two stones is 36 m?

94 •• [SSM] A motorcycle officer hidden at an intersection observes a car that ignores a stop sign, crosses the intersection, and continues on at constant speed. The police officer takes off in pursuit 2.0 s after the car has passed the stop sign, accelerates at 6.2 m/s^2 until her speed is 110 km/h, and then continues at this speed until she catches the car. At that instant, the car is 1.4 km from the intersection. How fast was the car traveling?

95 •• At $t = 0$, a stone is dropped from the top of a cliff above a lake. Another stone is thrown downward 1.6 s later from the same point with an initial speed of 32 m/s. Both stones hit the water at the same instant. Find the height of the cliff.

96 ••• A passenger train is traveling at 29 m/s when the engineer sees a freight train 360 m ahead of his train traveling in the same direction on the same track. The freight train is moving at a speed of 6 m/s. (a) If the reaction time of the engineer is 0.4 s, what is the minimum rate at which the passenger train must lose speed if a collision is to be avoided? (b) If the engineer's reaction time is 0.8 s and the train loses speed at the minimum rate described in part (a), what is the relative speed with which the two trains collide? How far will the passenger train have traveled in the time between the sighting of the freight train and the collision?

97 • Intent on studying the effects of gravity near the surface of the earth, a student launches a small projectile straight up with a speed of 300 m/s. How high will the projectile rise? (Ignore air resistance.)

98 • **SSM** At the end of *Charlie and the Chocolate Factory*, Willie Wonka presses a button that shoots the great glass elevator through the roof of his chocolate factory. (a) If the elevator reaches a maximum height of 10 km above the roof, what was its speed immediately after crashing through the roof? Ignore air resistance, even though in this case it makes little sense to ignore it. (b) Assume that the elevator's speed just after it crashes through the roof was half of what it was just before its impact with the roof. Assuming that it started from rest on the ground floor of the chocolate factory, and that the height of the roof is 150 m above the ground floor, what uniform acceleration is needed for it to reach this high speed?

99 •• The click beetle can project itself vertically with an acceleration of about $a = 400g$ (an order of magnitude more than a human could stand). The beetle jumps by "unfolding" its legs, which are about $d = 0.6$ cm long. How high can the click beetle jump? How long is the beetle in the air? (Assume constant acceleration while in contact with the ground, and neglect air resistance.)

100 • ✔ A test of the prototype of a new automobile shows that the minimum distance for a controlled stop from 98 km/h to zero is 50 m. Find the acceleration, assuming it to be constant, and express your answer as a fraction of the free-fall acceleration. How much time does the car take to stop?

101 •• **SSM** Consider the motion of a particle that experiences free-fall with a constant acceleration. Before the advent of computer-driven data-logging software, we used to do a free-fall experiment in which a coated tape was placed vertically next to the path of a dropped conducting puck. A high-voltage spark generator would cause an arc to jump between two vertical wires through the falling puck and through the tape, thereby marking the tape at fixed time intervals Δt. Show that the change in height in successive time intervals for an object falling from rest follows *Galileo's Rule of Odd Numbers*: $\Delta y_{21} = 3\Delta y_{10}$, $\Delta y_{32} = 5\Delta y_{10}, \ldots$, where Δy_{10} is the change in y during the first interval of duration Δt, Δy_{21} is the change in y during the second interval of duration Δt, etc.

102 •• A particle moves with a constant acceleration of 3 m/s². At a time of 4 s, it is at a position of 100 m with respect to some coordinate system; at a time of 6 s, it has a velocity of 15 m/s. Find its position at a time of 6 s.

103 •• **SSM** ✔ A plane landing on a small tropical island has just 70 m of runway on which to stop. If its initial speed is 60 m/s, (a) what is the minimum acceleration of the plane during landing, assuming it to be constant? (b) How long does it take for the plane to stop with this acceleration?

104 •• ✔ An automobile accelerates from rest at 2 m/s² for 20 s. The speed is then held constant for 20 s, after which there is an acceleration of −3 m/s² until the automobile stops. What is the total distance traveled?

105 •• **SSM** If it were possible for a spacecraft to maintain a constant acceleration indefinitely, trips to the planets of the Solar System could be undertaken in days or weeks, while voyages to the nearer stars would only take a few years. (a) Show that g, the magnitude of free-fall acceleration on earth, is approximately $1\ c \cdot y/y^2$. (See Problem 52 for the definition of a light-year.) (b) Using data from the tables at the back of the book, find the time it would take for a one-way trip from Earth to Mars (at Mars' closest approach to Earth). Assume that the spacecraft starts from rest, travels along a straight line, accelerates halfway at $1g$, and then flips around and decelerates at $1g$ for the rest of the trip.

106 • ✔ The Stratosphere Tower in Las Vegas is 1137 ft high. It takes 1 min, 20 s to ascend from the ground floor to the top of the tower using the high-speed elevator. Assuming that the elevator starts and ends at rest, and its acceleration has a constant *magnitude* when moving, find the acceleration of the elevator. Express it in terms of a multiple of the acceleration of gravity.

107 •• ✔ A train pulls away from a station with a constant acceleration of 0.4 m/s². A passenger arrives at the track 6.0 s after the end of the train has passed the very same point. What is the slowest constant speed at which she can run and catch the train? Sketch curves for the motion of the passenger and the train as functions of time.

108 ••• Ball A is dropped from the top of a building at the same instant that ball B is thrown vertically upward from the ground. When the balls collide, they are moving in opposite directions, and the speed of A is twice the speed of B. At what fraction of the height of the building does the collision occur?

109 ••• Solve Problem 108 if the collision occurs when the balls are moving in the same direction and the speed of A is 4 times that of B.

110 •• **SSM** Starting at one station, a subway train accelerates from rest at a constant rate of 1.0 m/s² for half the distance to the next station, then slows down at the same rate for the second half of the journey. The total distance between stations is 900 m. (a) Sketch a graph of the velocity v as a function of time over the full journey. (b) Sketch a graph of the distance covered as a function of time over the full journey. Place appropriate numerical values on both axes.

111 •• [SOLVE] A speeder traveling at a constant speed of 125 km/h races past a billboard. A patrol car pursues from rest with constant acceleration of (8 km/h)/s until it reaches its maximum speed of 190 km/h, which it maintains until it catches up with the speeder. (a) How long does it take the patrol car to catch the speeder if it starts moving just as the speeder passes? (b) How far does each car travel? (c) Sketch $x(t)$ for each car.

112 •• When the patrol car in Problem 111 (traveling at 190 km/h), pulls within 100 m behind the speeder (traveling at 125 km/h), the speeder sees the police car and slams on his brakes, locking the wheels. (a) Assuming that each car can brake at 6 m/s² and that the driver of the police car brakes instantly as she sees the brake lights of the speeder (reaction time = 0 s), show that the cars collide. (b) At what time after the speeder applies his brakes do the two cars collide? (c) Discuss how reaction time affects this problem.

113 •• Urgently needing the cash prize, Lou enters the Rest-to-Rest auto competition, in which each contestant's car begins and ends at rest, covering a distance L in as short a time as possible. The intention is to demonstrate mechanical and driving skills, and to consume the largest amount of fossil fuels in the shortest time possible. The course is designed so that maximum speeds of the cars are never reached. (a) If Lou's car has a maximum acceleration of a and a maximum deceleration of 2a, then at what fraction of L should Lou move his foot from the gas pedal to the brake? (b) What fraction of the time for the trip has elapsed at that point?

114 •• [SOLVE✓] A physics professor demonstrates her new "anti-gravity parachute" by exiting from a helicopter at an altitude of 575 m with zero initial velocity. For 8 s, she falls freely. Then she switches on the "parachute" and her rate of descent slows at a constant rate of 15 m/s² until her downward speed reaches 5 m/s, whereupon she adjusts her controls to maintain that speed until she reaches the ground. (a) On a single graph, sketch her acceleration and velocity as functions of time. (Take upward to be positive.) (b) What is her speed at the end of the first 8 s? (c) For how long is she losing speed? (d) How far does she travel while losing speed? (e) How much time is required for the entire trip from the helicopter to the ground? (f) What is her average velocity for the entire trip?

Integration of the Equations of Motion

115 • [SSM] The velocity of a particle is given by $v(t) = (6\ \text{m/s}^2)t + (3\ \text{m/s})$. (a) Sketch v versus t and find the area under the curve for the interval $t = 0$ to $t = 5$ s. (b) Find the position function $x(t)$. Use it to calculate the displacement during the interval $t = 0$ to $t = 5$ s.

116 • Figure 2-35 shows the velocity of a particle versus time. (a) What is the magnitude in meters represented by the area of the shaded box? (b) Estimate the displacement of the particle for the two 1-s intervals beginning at $t = 1$ s and at $t = 2$ s. (c) Estimate the average velocity for the interval

$1\ \text{s} \leq t \leq 3\ \text{s}$. (d) The equation of the curve is $v = (0.5\ \text{m/s}^3)t^2$. Find the displacement of the particle for the interval $1\ \text{s} \leq t \leq 3$ s by integration and compare this answer with your answer for part (b). Is the average velocity equal to the mean of the initial and final velocities for this case?

FIGURE 2-35 Problem 116

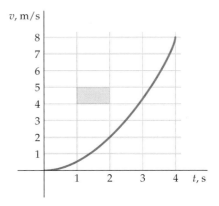

117 •• [SSM] The velocity of a particle is given by $v = (7\ \text{m/s}^3)t^2 - 5\ \text{m/s}$, where t is in seconds and v is in meters per second. If the particle starts from the origin, $x_o = 0$, at $t_o = 0$, find the general position function $x(t)$.

118 •• Consider the velocity graph in Figure 2-36. Assuming $x = 0$ at $t = 0$, write correct algebraic expressions for $x(t)$, $v(t)$, and $a(t)$ with appropriate numerical values inserted for all constants.

FIGURE 2-36 Problem 118

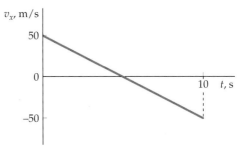

119 ••• Figure 2-37 shows the acceleration of a particle versus time. (a) What is the magnitude, in m/s, of the area of the shaded box? (b) The particle starts from rest at $t = 0$. Estimate the velocity at $t = 1$ s, 2 s, and 3 s by counting the boxes under the curve. (c) Sketch the curve v versus t from your results for part (b), and using it, estimate how far the particle travels in the interval $t = 0$ to $t = 3$ s.

FIGURE 2-37 Problem 119

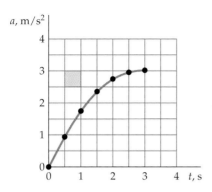

120 •• Figure 2-38 is a graph of v versus t for a particle moving along a straight line. The position of the particle at time $t_0 = 0$ is $x_0 = 5$ m. (a) Find x for various times t by counting boxes, and sketch x as a function of t. (b) Sketch a graph of the acceleration a as a function of the time t.

FIGURE 2-38 Problem 120

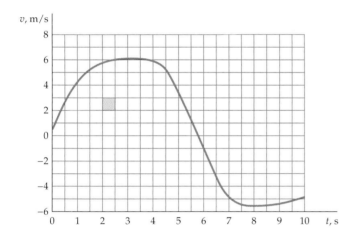

121 •• SSM Figure 2-39 shows a plot of x versus t for a body moving along a straight line. Sketch rough graphs of v as a function of t and a as a function of t for this motion.

FIGURE 2-39 Problem 121

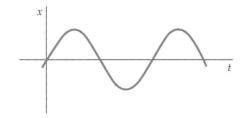

122 •• The acceleration of a certain rocket is given by $a = bt$, where b is a positive constant. (a) Find the general position function $x(t)$. (b) Find the position and velocity at $t = 5$ s if $x = 0$ and $v = 0$ at $t = 0$ and $b = 3$ m/s³.

123 •• ISOLVE In the time interval from 0.0 s to 10.0 s, the acceleration of a particle is given by $a = (0.20 \text{ m/s}^3)t$ for one-dimensional motion. If a particle starts from rest at the origin, (a) calculate first its *instantaneous velocity* at any time during the interval, then (b) calculate its *average velocity* over the time interval from 2.0 s to 7.0 s.

124 • Consider the motion of a particle that experiences a nonconstant acceleration a given by $a = a_0 + bt$, where a_0 and b are constant. (a) Find the instantaneous velocity as a function of time. (b) Find the position as a function of time. (c) Find the average velocity over the time interval with an initial time of zero and arbitrary final time t.

General Problems

125 ••• ISOLVE The following apparatus is used in a science class to determine the free-fall acceleration. Two photogates are set up, one at the edge of a table exactly 1.0 m high, the second one directly below it at a height of 0.5 m. A marble is dropped through the photogates from the height of 1 m above the floor (i.e., exactly the same as the height of the table), starting a timer when it enters the first photogate and stopping it when it enters the second one. The magnitude of free-fall acceleration is determined by $g_{exp} = (1 \text{ m})/(\Delta t)^2$, where Δt is the time measured by the timer. A careless student sets up the first photogate 0.5 cm below the edge of the table. (Assume that the second photogate is properly placed at a height of 0.5 m.) What value of g_{exp} will he determine? What percentage difference does this represent from the accepted value at sea level?

126 ••• SSM The position of a body oscillating on a spring is given by $x = A \sin \omega t$, where A and ω are constants with values $A = 5$ cm and $\omega = 0.175$ s⁻¹. (a) Plot x as a function of t for $0 \le t \le 36$ s. (b) Measure the slope of your graph at $t = 0$ to find the velocity at this time. (c) Calculate the average velocity for a series of intervals beginning at $t = 0$ and ending at $t = 6, 3, 2, 1, 0.5$, and 0.25 s. (d) Compute dx/dt and find the velocity at time $t = 0$. (e) Compare your results in parts (c) and (d).

127 ••• ISOLVE Consider an object that is attached to a driving motor so that the object moves with a velocity given by $v = v_{max} \sin(\omega t)$, where ω is in radians/s. (a) What is the acceleration of the object and is it constant? (b) At $t = 0$, the position is x_0. What is the position as a function of time?

128 ••• Suppose the acceleration of a particle is a function of x, where $a(x) = (2 \text{ s}^{-2}) x$. (a) If the velocity at $x = 1$ m is zero, what is the speed of the particle at $x = 3$ m? (b) How long does it take the particle to travel from $x = 1$ m to $x = 3$ m?

129 ••• Suppose that a particle moves in a straight line such that, at each instant of time, its position and velocity have the same numerical value if expressed in SI units. (*a*) Express the position x as a function of time t. (*b*) Show that at each instant of time the acceleration has the same numerical value as the position and velocity.

130 ••• A small rock sinking through water experiences an exponentially decreasing acceleration as a function of time given by $a(t) = ge^{-bt}$, where b is a positive constant that depends on the shape and size of the rock and the physical properties of the water. Based upon this result, derive an expression for the position of the rock as a function of time. Assume that its initial velocity is 0.

131 ••• [SSM] In Problem 130, a rock falls through water with a continuously decreasing acceleration of the form $a(t) = ge^{-bt}$, where b is a positive constant. In physics, we are not often given acceleration directly as a function of time, but usually either as a function of position or of velocity. Assume that the rock's acceleration as a function of *velocity* has the form $a = g - bv$ where g is the magnitude of free-fall acceleration and v is the rock's speed. Prove that if the rock has an initial velocity $v = 0$ at time $t = 0$, it will have the dependence on *time* given above.

132 ••• The acceleration of a skydiver jumping from an airplane is given by the formula $a = g - cv^2$, where c is a constant depending on the skydiver's cross-sectional area and the density of the surrounding atmosphere she is diving through. (*a*) If her initial speed is 0 when jumping from the plane, show that her speed as a function of time is given by the formula $v(t) = v_t \tanh(t/T)$, where v_T is the terminal velocity $(v_T = \sqrt{g/c})$ and $T = v_T/g$ is a time scale determining very roughly the time it takes for her speed to approach v_T. (*b*) Use a spreadsheet program such as Microsoft Excel to graph $v(t)$ as a function of time, using a terminal velocity of 56 m/s (use this to calculate c and T). Does the resulting curve make sense?

Motion in Two and Three Dimensions

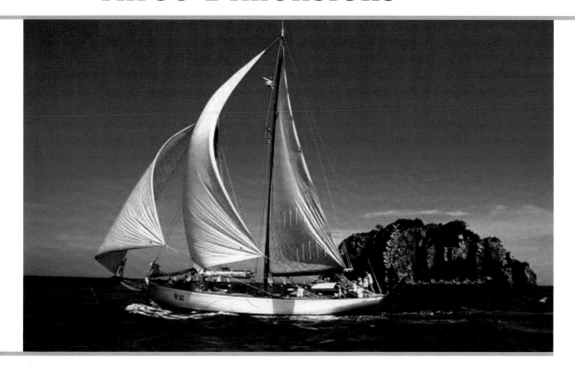

SAILBOATS DON'T TRAVEL IN STRAIGHT LINES TO THEIR DESTINATIONS, BUT INSTEAD MUST "TACK" BACK AND FORTH ACROSS THE WIND. THIS BOAT MUST SAIL EAST, THEN SOUTH, AND THEN EAST AGAIN, IN ITS JOURNEY TO A SOUTHEASTERN PORT.

 How can we calculate its displacement and its average velocity? (See Example 3-3.)

Here we extend the ideas of Chapter 2 to two and three dimensions. To do this, we introduce vectors and show how they are used to analyze and describe motion.

➤ **Our major goal in this chapter is to develop the concept of the acceleration vector, a concept that is central to the development of Newton's laws in Chapters 4 and 5.**

3-1 The Displacement Vector

When motion occurs, the displacement of a particle has a direction in space as well as a magnitude. The quantity that gives the straight-line distance between two points in space as well as the direction is a line segment called the **displacement vector.** It is represented graphically by an arrow whose direction is the same as that of the displacement vector and whose length is proportional to the magnitude of the displacement vector. We denote vectors by italic boldface letters with an overhead arrow, $\vec{A}$. The magnitude of $\vec{A}$ is written $|\vec{A}|$, $\|\vec{A}\|$, or simply A, and has dimensions of length. The magnitude of a vector is never negative.

Addition of Displacement Vectors

Figure 3-1 shows the path of a particle that moves from point P_1 to a second point P_2 and then to a third point P_3. The vector $\vec{A}$ represents the displacement from P_1 to P_2, while $\vec{B}$ represents the displacement from P_2 to P_3. Note that the displacement vectors depend only on the end points and not on the actual path of the particle. The *resultant* displacement from P_1 to P_3, labeled $\vec{C}$, is called the sum of the two successive displacements $\vec{A}$ and $\vec{B}$:

$$\vec{C} = \vec{A} + \vec{B} \qquad\qquad 3\text{-}1$$

Two displacement vectors are added graphically by placing the tail of one at the head of the other (Figure 3-2). The resultant vector extends from the tail of the first to the head of the second. Note that C does not equal $A + B$ unless $\vec{A}$ and $\vec{B}$ are in the same direction. That is, $\vec{C} = \vec{A} + \vec{B}$ does not imply that $C = A + B$.

FIGURE 3-1

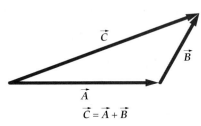

FIGURE 3-2
Head-to-tail method of vector addition.

$\vec{C} = \vec{A} + \vec{B}$

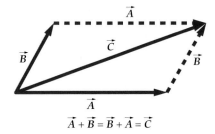

FIGURE 3-3
Parallelogram method for adding vectors.

$\vec{A} + \vec{B} = \vec{B} + \vec{A} = \vec{C}$

An equivalent way of adding vectors, called the **parallelogram method,** is to move $\vec{B}$ so that it is tail-to-tail with $\vec{A}$. The diagonal of the parallelogram formed by $\vec{A}$ and $\vec{B}$ then equals $\vec{C}$. From Figure 3-3, we can see that it makes no difference in which order we add two vectors; that is, $\vec{A} + \vec{B} = \vec{B} + \vec{A}$.

YOUR DISPLACEMENT   **EXAMPLE 3 - 1**

You walk 3 km east and then 4 km north. What is your resultant displacement?

PICTURE THE PROBLEM Draw and label the two displacements $\vec{A}$ and $\vec{B}$ as well as the resultant displacement $\vec{C}$ (Figure 3-4). Include axes indicating the directions north and east. $\vec{A}$ and $\vec{B}$ are at right angles to each other, and $\vec{C} = \vec{A} + \vec{B}$ is the hypotenuse of the corresponding right triangle. The magnitude of $\vec{C}$ can be obtained using the Pythagorean theorem, and the direction of $\vec{C}$ is found using trigonometry.

FIGURE 3-4

1. The magnitude of the resultant displacement is related to the magnitudes of the two sequential displacements by the Pythagorean theorem:

$$C^2 = A^2 + B^2$$
$$= (3 \text{ km})^2 + (4 \text{ km})^2$$
$$= 25 \text{ km}^2$$

so

$$C = \sqrt{25 \text{ km}^2} = \boxed{5 \text{ km}}$$

2. Let θ be the angle from the east axis to the resultant displacement $\vec{C}$. From the figure we find $\tan \theta$; using a calculator with trigonometric functions yields θ:

$$\tan \theta = \frac{B}{A}$$

so

$$\theta = \tan^{-1} \frac{B}{A} = \tan^{-1} \frac{4 \text{ km}}{3 \text{ km}} = \boxed{53.1°}$$

REMARKS A vector is described by its magnitude and its direction. Your resultant displacement is a vector of length 5 km in a direction 53.1° north of east.

3-2 General Properties of Vectors

Many quantities in physics have magnitude and direction, and add like displacements. Examples include velocity, acceleration, momentum, and force. Such quantities are called **vectors.** Quantities with magnitude but no associated direction—for example, distance and speed—are called **scalars.**

Vectors are quantities with magnitude and direction that add like displacements.

DEFINITION—VECTORS

A vector is represented graphically by an arrow drawn in the same direction as that of the vector, and with a length that is proportional to the magnitude of the vector. When the magnitude of a vector is given, its units must also be given. The magnitude of a velocity vector, for example, requires units such as m/s. Two vectors are defined to be equal if they have the same magnitude and the same direction. Graphically, this means that they have the same length and are parallel to each other. A consequence of this definition is that moving a vector so that it remains parallel to itself does not change it. Thus, all the vectors in Figure 3-5 are equal. If we translate or rotate the coordinate system, all the vectors in Figure 3-5 remain equal. Vectors do not depend on the coordinate system used to represent them (except for position vectors, which are introduced in Section 3.3).

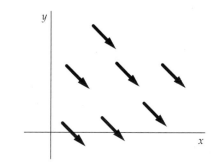

FIGURE 3-5 Vectors are equal if their magnitudes and directions are the same. All vectors in this figure are equal.

Multiplying a Vector by a Scalar

A vector $\vec{A}$ multiplied by a scalar s is the vector $\vec{B} = s\vec{A}$, which has magnitude $|s|A$ and is parallel to $\vec{A}$ if s is positive and antiparallel if s is negative. The vector $-\vec{A}$ is the vector you add to $\vec{A}$ to get zero. Thus, $-\vec{A}$ has the same magnitude as $\vec{A}$ but points in the opposite direction. The dimensions of $s\vec{A}$ are those of s multiplied by those of A.

Subtracting Vectors

We subtract vector $\vec{B}$ from vector $\vec{A}$ by adding $-\vec{B}$ to $\vec{A}$. The result is $\vec{C} = \vec{A} + (-\vec{B}) = \vec{A} - \vec{B}$ (Figure 3-6a). An equivalent way of subtracting $\vec{B}$ from $\vec{A}$ is to draw them tail-to-tail (Figure 3-6b) and noting that $\vec{C}$ is the vector that must be added to $\vec{B}$ to obtain the resultant vector $\vec{A}$. The rules for adding or subtracting any two vectors, such as two velocity vectors or two acceleration vectors, are the same as for adding or subtracting displacement vectors.

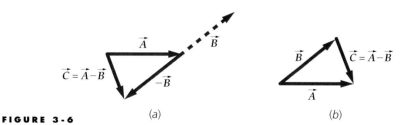

FIGURE 3-6

(a)   (b)

Components of Vectors

The component of a vector in a given direction is a number whose absolute value is the length of the projection of the vector onto an axis (directed line) in that direction. It is found by dropping a perpendicular from the head of the vector to the axis, as shown in Figure 3-7. The sign of the component is positive if the projection of the head of the vector is in the positive direction relative to the tail of the vector. Thus, A_S is positive and B_S is negative. The components of a vector along the x, y, and z directions, illustrated in Figure 3-8 for a vector in the xy plane, are called rectangular components. Note that the components of a vector *do* depend on the coordinate system used, although the vector itself does not.

Rectangular components are useful for the addition or subtraction of vectors. If θ is the angle between $\vec{A}$ and the x axis, then

$$A_x = A \cos \theta \qquad \text{3-2}$$

X COMPONENT OF A VECTOR

and

$$A_y = A \sin \theta \qquad \text{3-3}$$

Y COMPONENT OF A VECTOR

where A is the magnitude of $\vec{A}$.

If we know A_x and A_y, we can find the angle θ from

$$\tan \theta = \frac{A_y}{A_x}, \qquad \theta = \tan^{-1} \frac{A_y}{A_x} \qquad \text{3-4}$$

and the magnitude A from the Pythagorean theorem:

$$A = \sqrt{A_x^2 + A_y^2} \qquad \text{3-5}a$$

In three dimensions,

$$A = \sqrt{A_x^2 + A_y^2 + A_z^2} \qquad \text{3-5}b$$

Components can be positive or negative. For example, if $\vec{A}$ points in the negative x direction, then A_x is negative.

Consider two vectors $\vec{A}$ and $\vec{B}$ that lie in the xy plane. The rectangular components of each vector and those of the sum $\vec{C} = \vec{A} + \vec{B}$ are shown in Figure 3-9. We see that the vector equation $\vec{C} = \vec{A} + \vec{B}$ is equivalent to the two component equations

$$C_x = A_x + B_x \qquad \text{3-6}a$$

and

$$C_y = A_y + B_y \qquad \text{3-6}b$$

EXERCISE A car travels 20 km in a direction 30° north of west. Let the x axis point east and the y axis point north as in Figure 3-10. Find the x and y components of the displacement vector of the car. (*Answer* $A_x = -17.3$ km, $A_y = +10$ km)

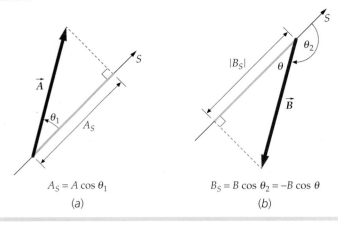

$$A_S = A \cos \theta_1 \qquad\qquad B_S = B \cos \theta_2 = -B \cos \theta$$

(a)   (b)

FIGURE 3-7 Definition of the component of a vector. The component of the vector $\vec{A}$ in the positive S direction is A_S, and A_S is positive. The component of the vector $\vec{B}$ in the positive S direction is B_S, and B_S is negative.

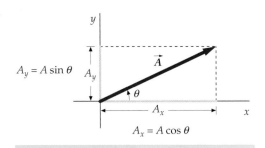

FIGURE 3-8 The rectangular components of a vector. $A_x = A \cos \theta$, $A_y = A \sin \theta$.

FIGURE 3-9

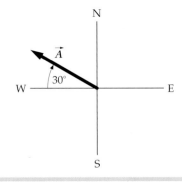

FIGURE 3-10

A TREASURE MAP   **EXAMPLE 3-2**   **Put It in Context**

You are working as a counselor at a tropical resort. You've been given a map and told to follow its directions in order to bury the "treasure" at a specific location. You don't want to waste time walking around the island, because you want to finish early and go surfing. The directions are to walk 3 km headed due west and then 4 km headed 60° north of east. Where should you head and how far must you walk to get the job done quickly? Find your answer (*a*) graphically and (*b*) using vector components.

PICTURE THE PROBLEM   You need to find your resultant displacement, which is $\vec{C}$ in Figure 3-11. The triangle formed by the three vectors is not a right triangle, so the magnitudes of the vectors are not related by the Pythagorean theorem. We find the resultant graphically by drawing each of the displacements to scale and measuring the resultant displacement.

FIGURE 3-11

(*a*) If you draw the first displacement vector $\vec{A}$ 3 cm long and the second one $\vec{B}$ 4 cm long, you find the resultant vector $\vec{C}$ to be about 3.5 cm long. Thus, the magnitude of the resultant displacement is $\boxed{3.5 \text{ km}}$. The angle θ made between the resultant displacement and the west direction can then be measured with a protractor. It is about 75°. Therefore, you should walk 3.5 km headed $\boxed{75° \text{ north of west}}$.

(*b*) 1. To solve using vector components, let $\vec{A}$ be the first displacement and choose the direction of increasing *x* to be east and the direction of increasing *y* to be north. Compute A_x and A_y from Equations 3-2 and 3-3:

$A_x = -3 \text{ km}$ and $A_y = 0$

2. Similarly, compute the components of the second displacement $\vec{B}$:

$B_x = (4 \text{ km}) \cos 60° = 2 \text{ km}$

$B_y = (4 \text{ km}) \sin 60° = 2\sqrt{3} \text{ km} = 3.46 \text{ km}$

3. The components of the resultant displacement $\vec{C} = \vec{A} + \vec{B}$ are found by addition:

$C_x = A_x + B_x = -3 \text{ km} + 2 \text{ km} = -1 \text{ km}$

$C_y = A_y + B_y = 0 + 2\sqrt{3} \text{ km} = 3.46 \text{ km}$

4. The Pythagorean theorem gives the magnitude of $\vec{C}$:

$C^2 = C_x^2 + C_y^2$

$= (-1 \text{ km})^2 + (2\sqrt{3} \text{ km})^2 = 13 \text{ km}^2$

$C = \sqrt{13} \text{ km} = \boxed{3.61 \text{ km}}$

5. The ratio of C_y to $|C_x|$ equals the tangent of the angle θ between $\vec{C}$ and the negative *x* direction:

$\tan \theta = \dfrac{C_y}{|C_x|}$

so

$\theta = \tan^{-1} \dfrac{2\sqrt{3} \text{ km}}{1 \text{ km}} = \tan^{-1} (2\sqrt{3})$

$= \boxed{74° \text{ north of west}}$

REMARKS   Since the displacement (which is a vector) was asked for, the answer must include either both the magnitude *and* direction, or both components. In (*b*) we could have stopped at step 3 because the *x* and *y* components completely define the displacement vector. We converted to the magnitude and direction to compare with the answer to Part (*a*). Note that step 5 of Part (*b*) gives the angle as 74°. This agrees with the results in (*a*) within the accuracy of our measurement.

Unit Vectors

A **unit vector** is a *dimensionless* vector with unit magnitude. The vector $A^{-1}\vec{A}$ is an example of a unit vector that points in the direction of $\vec{A}$. Unit vectors are often written with an overhead caret as in $\hat{A} = A^{-1}\vec{A}$. Unit vectors that point in the positive x, y, and z directions are convenient for expressing vectors in terms of their rectangular components. They are usually written $\hat{i}$, $\hat{j}$, and $\hat{k}$, respectively. For example, the vector $A_x\hat{i}$ has a magnitude $|A_x|$ and points in the positive x direction if A_x is positive (or the negative x direction if A_x is negative). A general vector $\vec{A}$ can be written as the sum of three vectors, each of which is parallel to a coordinate axis (Figure 3-12):

$$\vec{A} = A_x\hat{i} + A_y\hat{j} + A_z\hat{k} \qquad 3\text{-}7$$

The addition of two vectors $\vec{A}$ and $\vec{B}$ can be written in terms of unit vectors as

$$\vec{A} + \vec{B} = (A_x\hat{i} + A_y\hat{j} + A_z\hat{k}) + (B_x\hat{i} + B_y\hat{j} + B_z\hat{k})$$
$$= (A_x + B_x)\hat{i} + (A_y + B_y)\hat{j} + (A_z + B_z)\hat{k} \qquad 3\text{-}8$$

The general properties of vectors are summarized in Table 3-1.

(a)

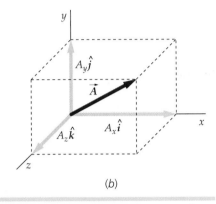

(b)

FIGURE 3-12 (*a*) The unit vectors $\hat{i}$, $\hat{j}$, and $\hat{k}$ in a rectangular coordinate system. (*b*) The vector $\vec{A}$ in terms of unit vectors: $\vec{A} = A_x\hat{i} + A_y\hat{j} + A_z\hat{k}$.

TABLE 3-1

Properties of Vectors

Property	Explanation	Figure	Component representation
Equality	$\vec{A} = \vec{B}$ if $|\vec{A}| = |\vec{B}|$ and their directions are the same		$A_x = B_x$ $A_y = B_y$ $A_z = B_z$
Addition	$\vec{C} = \vec{A} + \vec{B}$		$C_x = A_x + B_x$ $C_y = A_y + B_y$ $C_z = A_z + B_z$
Negative of a vector	$\vec{A} = -\vec{B}$ if $|\vec{B}| = |\vec{A}|$ and their directions are opposite		$A_x = -B_x$ $A_y = -B_y$ $A_z = -B_z$
Subtraction	$\vec{C} = \vec{A} - \vec{B}$		$C_x = A_x - B_x$ $C_y = A_y - B_y$ $C_z = A_z - B_z$
Multiplication by a scalar	$\vec{B} = s\vec{A}$ has magnitude $|\vec{B}| = |s||\vec{A}|$ and has the same direction as $\vec{A}$ if s is positive or as $-\vec{A}$ if s is negative		$B_x = sA_x$ $B_y = sA_y$ $B_z = sA_z$

EXERCISE Given two vectors

$$\vec{A} = (4\text{ m})\hat{i} + (3\text{ m})\hat{j} \qquad \text{and} \qquad \vec{B} = (2\text{ m})\hat{i} - (3\text{ m})\hat{j}$$

find (a) A, (b) B, (c) $\vec{A} + \vec{B}$, and (d) $\vec{A} - \vec{B}$. (*Answers* (a) $A = 5$ m, (b) $B = 3.61$ m, (c) $\vec{A} + \vec{B} = (6\text{ m})\hat{i}$, (d) $\vec{A} - \vec{B} = (2\text{ m})\hat{i} + (6\text{ m})\hat{j}$)

3-3 Position, Velocity, and Acceleration

Position and Velocity Vectors

The **position vector** of a particle is a vector drawn from the origin of a co-ordinate system to the position of the particle. For a particle at the point (x, y), the position vector $\vec{r}$ is

$$\vec{r} = x\hat{i} + y\hat{j} \tag{3-9}$$

DEFINITION—POSITION VECTOR

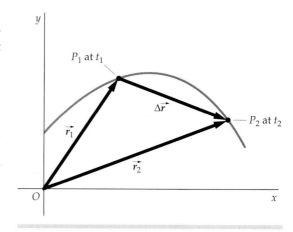

FIGURE 3-13 The displacement vector $\Delta\vec{r}$ is the difference in the position vectors, $\Delta\vec{r} = \vec{r}_2 - \vec{r}_1$. Equivalently, $\Delta\vec{r}$ is the vector that, when added to $\vec{r}_1$, yields the new position vector $\vec{r}_2$.

Figure 3-13 shows the actual path or trajectory of the particle. (Don't confuse the trajectory with the *x*-versus-*t* plots of Chapter 2.) At time t_1, the particle is at P_1, with position vector $\vec{r}_1$; by t_2, the particle has moved to P_2, with position vector $\vec{r}_2$. The particle's change in position is the displacement vector $\Delta\vec{r}$:

$$\Delta\vec{r} = \vec{r}_2 - \vec{r}_1 \tag{3-10}$$

DEFINITION—DISPLACEMENT VECTOR

The ratio of the displacement vector to the time interval $\Delta t = t_2 - t_1$ is the **average-velocity vector:**

$$\vec{v}_{av} = \frac{\Delta\vec{r}}{\Delta t} \tag{3-11}$$

DEFINITION—AVERAGE-VELOCITY VECTOR

This vector points in the direction of the displacement.

The magnitude of the displacement vector is less than the distance traveled along the curve unless the particle moves in a straight line. However, if we consider smaller and smaller time intervals (Figure 3-14), the magnitude of the displacement approaches the distance along the curve, and the direction of $\Delta\vec{r}$ approaches the tangent to the curve at the beginning of the interval. We define the **instantaneous-velocity vector** as the limit of the average-velocity vector as Δt approaches zero:

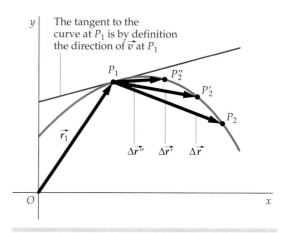

FIGURE 3-14 As the time interval is made smaller, the direction of the displacement approaches the tangent to the curve.

$$\vec{v} = \lim_{\Delta t \to 0}\frac{\Delta\vec{r}}{\Delta t} = \frac{d\vec{r}}{dt} \tag{3-12}$$

DEFINITION—INSTANTANEOUS-VELOCITY VECTOR

The instantaneous-velocity vector is the derivative of the position vector with respect to time. Its magnitude is the speed and its direction is the direction of motion of the particle along the line tangent to the curve.

To calculate the derivative in Equation 3-12, we write the position vectors in terms of their components:

$$\Delta \vec{r} = \vec{r}_2 - \vec{r}_1 = (x_2 - x_1)\hat{i} + (y_2 - y_1)\hat{j} = \Delta x\hat{i} + \Delta y\hat{j}$$

Then

$$\vec{v} = \lim_{\Delta t \to 0}\frac{\Delta \vec{r}}{\Delta t} = \lim_{\Delta t \to 0}\frac{\Delta x\hat{i} + \Delta y\hat{j}}{\Delta t} = \lim_{\Delta t \to 0}\left(\frac{\Delta x}{\Delta t}\right)\hat{i} + \lim_{\Delta t \to 0}\left(\frac{\Delta y}{\Delta t}\right)\hat{j}$$

or

$$\vec{v} = \frac{dx}{dt}\hat{i} + \frac{dy}{dt}\hat{j} = v_x\hat{i} + v_y\hat{j} \qquad\qquad 3\text{-}13$$

THE VELOCITY OF A SAILBOAT   **EXAMPLE   3-3**

A sailboat has coordinates $(x_1, y_1) = (110\text{ m}, 218\text{ m})$ at $t_1 = 60$ s. Two minutes later, at time t_2, it has coordinates $(x_2, y_2) = (130\text{ m}, 205\text{ m})$. (*a*) Find the average velocity for this 2-min interval. Express $\vec{v}_{av}$ in terms of its rectangular components. (*b*) Find the magnitude and direction of this average velocity. (*c*) For $t \geq 20$ s, the position of the sailboat as a function of time is $x(t) = b_1 + b_2 t$ and $y(t) = c_1 + c_2/t$, where $b_1 = 100$ m, $b_2 = \frac{1}{6}$ m/s, $c_1 = 200$ m, and $c_2 = 1080$ m·s. Find the instantaneous velocity at a general time $t \geq 20$ s.

PICTURE THE PROBLEM The initial and final positions of the sailboat are given. (*a*) The average-velocity vector points from the initial toward the final position. (*b*) The instantaneous-velocity components are calculated from Equation 3-13: $v_x = dx/dt$ and $v_y = dy/dt$.

FIGURE 3-15

(*a*) 1. Draw a coordinate system (Figure 3-15) showing the displacement of the sailboat. The average-velocity vector and the displacement are in the same direction:

    2. The x and y components of the average velocity $\vec{v}_{av}$ are calculated directly from their definitions:

$$\vec{v}_{av} = v_{x,av}\,\hat{i} + v_{y,av}\,\hat{j}$$

where

$$v_{x,av} = \frac{\Delta x}{\Delta t} = \frac{130\text{ m} - 110\text{ m}}{120\text{ s}} = 0.167\text{ m/s}$$

$$v_{y,av} = \frac{\Delta y}{\Delta t} = \frac{205\text{ m} - 218\text{ m}}{120\text{ s}} = -0.108\text{ m/s}$$

so

$$\boxed{\vec{v}_{av} = (0.167\text{ m/s})\hat{i} - (0.108\text{ m/s})\hat{j}}$$

(*b*) 1. The magnitude of $\vec{v}_{av}$ is found from the Pythagorean theorem:

$$v_{av} = \sqrt{(v_{x,av})^2 + (v_{y,av})^2} = \boxed{0.199\text{ m/s}}$$

2. The ratio of $v_{y,\,\text{av}}$ to $v_{x,\,\text{av}}$ gives the tangent of the angle θ between $\vec{v}_{\text{av}}$ and the x axis:

$$\tan \theta = \frac{v_{y,\text{av}}}{v_{x,\text{av}}}$$

so

$$\theta = \tan^{-1}\frac{v_{y,\text{av}}}{v_{x,\text{av}}} = \tan^{-1}\frac{-0.108 \text{ m/s}}{0.167 \text{ m/s}} = \boxed{-33.0°}$$

(c) We find the instantaneous velocity $\vec{v}$ by calculating dx/dt and dy/dt:

$$\vec{v} = \frac{dx}{dt}\hat{i} + \frac{dy}{dt}\hat{j} = b_2\hat{i} - c_2 t^{-2}\hat{j} = \boxed{\left(\frac{1}{6}\text{ m/s}\right)\hat{i} - \frac{1080 \text{ m}\cdot\text{s}}{t^2}\hat{j}}$$

REMARKS The magnitude of $\vec{v}$ can be found from $v = \sqrt{v_x^2 + v_y^2}$ and its direction can be found from $\tan\theta = v_y/v_x$.

EXERCISE Find the x and y components and the magnitude and direction of the instantaneous velocity of the sailboat at time $t_1 = 60$ s. (*Answer* $\vec{v}_1 = (\frac{1}{6}\text{ m/s})\hat{i} - (0.30 \text{ m/s})\hat{j}, v_1 = 0.34 \text{ m/s}, \theta_1 = -60.9°$)

Relative Velocity

Relative velocities in two and three dimensions combine just as they do in one dimension, except that the velocity vectors are not necessarily along the same line. If a particle moves with velocity $\vec{v}_{\text{pA}}$ relative to reference frame A, which is in turn moving with velocity $\vec{v}_{\text{AB}}$ relative to reference frame B, the velocity of the particle relative to reference frame B is

$$\vec{v}_{\text{pB}} = \vec{v}_{\text{pA}} + \vec{v}_{\text{AB}} \qquad\qquad 3\text{-}14$$

RELATIVE VELOCITY

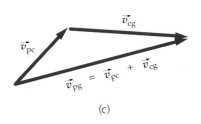

For example, if you are on a railroad flatcar that is moving with velocity $\vec{v}_{\text{cg}}$ relative to the ground (Figure 3-16a), and you are walking with velocity $\vec{v}_{\text{pc}}$ (Figure 3-16b) relative to the car, then your velocity relative to the ground is the vector sum of these two velocities: $\vec{v}_{\text{pg}} = \vec{v}_{\text{pc}} + \vec{v}_{\text{cg}}$ (Figure 3-16c).

The velocity of object A relative to object B is equal in magnitude and opposite in direction to the velocity of object B relative to object A. For example, $\vec{v}_{\text{pc}} = -\vec{v}_{\text{cp}}$ where $\vec{v}_{\text{cp}}$ is the velocity of the car relative to the person. The addition of relative velocities is done in the same way as the addition of displacements, either graphically, using either the head-to-tail or the parallelogram method, or analytically, using vector components.

FIGURE 3-16

A FLYING PLANE   **E X A M P L E   3 - 4**

A pilot wants to fly a plane due north. The speed of the plane relative to the air is 200 km/h and the wind is blowing from west to east at 90 km/h. (*a*) In which direction should the plane head? (*b*) How fast does the plane travel relative to the ground?

PICTURE THE PROBLEM Because the wind is blowing toward the east, a plane headed due north will drift off course toward the east. To compensate for the crosswind, the plane must head west of due north. The velocity of the plane relative to the ground $\vec{v}_{pg}$ will be the sum of the velocity of the plane relative to the air $\vec{v}_{pa}$ and the velocity of the air relative to the ground $\vec{v}_{ag}$:

FIGURE 3-17

(a) 1. The velocity of the plane relative to the ground is given by Equation 3-14:

$$\vec{v}_{pg} = \vec{v}_{pa} + \vec{v}_{ag}$$

2. Make a velocity addition diagram showing the addition of the vectors in step 1. Include direction axes as shown in Figure 3-17.

3. The sine of the angle θ between the velocity of the plane and north equals the ratio of v_{ag} and v_{pa}:

$$\sin \theta = \frac{v_{ag}}{v_{pa}}$$

so

$$\theta = \sin^{-1} \frac{v_{ag}}{v_{pa}} = \sin^{-1} \frac{90 \text{ km/h}}{200 \text{ km/h}} = \boxed{26.7°}$$

(b) Because $\vec{v}_{ag}$ and $\vec{v}_{pg}$ are perpendicular, we can use the Pythagorean theorem to find the magnitude of $\vec{v}_{pg}$:

$$v_{pa}^2 = v_{pg}^2 + v_{ag}^2$$

so

$$v_{pg} = \sqrt{v_{pa}^2 - v_{ag}^2}$$

$$= \sqrt{(200 \text{ km/h})^2 - (90 \text{ km/h})^2} = \boxed{179 \text{ km/h}}$$

The Acceleration Vector

The **average-acceleration** is the ratio of the change in the instantaneous-velocity vector $\Delta \vec{v}$ to the time interval Δt:

$$\vec{a}_{av} = \frac{\Delta \vec{v}}{\Delta t}$$

3-15

DEFINITION—AVERAGE-ACCELERATION VECTOR

The **instantaneous-acceleration** is the limit of this ratio as Δt approaches zero; in other words, it is the derivative of the velocity vector with respect to time:

$$\vec{a} = \lim_{\Delta t \to 0} \frac{\Delta \vec{v}}{\Delta t} = \frac{d\vec{v}}{dt}$$

3-16

DEFINITION—INSTANTANEOUS-ACCELERATION VECTOR

To calculate the instantaneous acceleration, we express $\vec{v}$ in rectangular coordinates:

$$\vec{v} = v_x \hat{i} + v_y \hat{j} + v_z \hat{k} = \frac{dx}{dt} \hat{i} + \frac{dy}{dt} \hat{j} + \frac{dz}{dt} \hat{k}$$

Then

$$\vec{a} = \frac{dv_x}{dt} \hat{i} + \frac{dv_y}{dt} \hat{j} + \frac{dv_z}{dt} \hat{k} = \frac{d^2x}{dt^2} \hat{i} + \frac{d^2y}{dt^2} \hat{j} + \frac{d^2z}{dt^2} \hat{k}$$

$$= a_x \hat{i} + a_y \hat{j} + a_z \hat{k}$$

3-17

A THROWN BASEBALL   **EXAMPLE 3-5**

The position of a thrown baseball is given by $\vec{r} = 1.5$ m $\hat{i}$ + (12 m/s $\hat{i}$ + 16 m/s $\hat{j}$)t − 4.9 m/s² $\hat{j}\,t^2$. Find its velocity and acceleration.

PICTURE THE PROBLEM Because $\vec{r} = x\hat{i} + y\hat{j}$, we have $x = 1.5$ m + (12 m/s)t and $y = (16$ m/s)$t - (4.9$ m/s²)t^2. We can find the x and y components of the velocity and acceleration by differentiating x and y.

1. The x and y components of the velocity are found by differentiating x and y:

$$v_x = \frac{dx}{dt} = \frac{d}{dt}\left[1.5\text{ m} + (12\text{ m/s})t\right] = 12\text{ m/s}$$

$$v_y = \frac{dy}{dt} = \frac{d}{dt}\left[(16\text{ m/s})t - (4.9\text{ m/s}^2)t^2\right]$$

$$= 16\text{ m/s} - (9.8\text{ m/s}^2)t$$

2. We differentiate again to obtain the components of the acceleration:

$$a_x = \frac{dv_x}{dt} = 0$$

$$a_y = \frac{dv_y}{dt} = -9.8\text{ m/s}^2$$

3. In vector notation, the velocity and acceleration are:

$$\vec{v} = \boxed{(12\text{ m/s})\,\hat{i} + \left[16\text{ m/s} - (9.8\text{ m/s}^2)t\right]\hat{j}}$$

$$\vec{a} = \boxed{(-9.8\text{ m/s}^2)\hat{j}}$$

REMARKS This is an example of projectile motion, a topic we will study in Section 3-4.

For a vector to be constant, both its magnitude and direction must remain constant. If either changes, the vector changes. Thus, if a car rounds a curve in the road at constant speed, it is accelerating because the velocity is changing due to the change in direction of the velocity vector.

FIGURE 3-18

CORNERING A TURN   **EXAMPLE 3-6**

A car is traveling east at 60 km/h. It rounds a curve, and 5 s later it is traveling north at 60 km/h. Find the average acceleration of the car.

PICTURE THE PROBLEM The initial and final velocity vectors are shown in Figure 3-18. We choose the unit vector $\hat{i}$ to be east and $\hat{j}$ to be north, and calculate the average acceleration from its definition, $\vec{a}_{av} = \Delta\vec{v}/\Delta t$. Note that $\Delta\vec{v}$ is the vector that, when added to $\vec{v}_i$, results in $\vec{v}_f$.

1. The average acceleration is the ratio of the velocity change to the time interval:

$$\vec{a}_{av} = \frac{\Delta\vec{v}}{\Delta t}$$

2. The change in velocity is related to the initial and final velocities:

$$\Delta\vec{v} = \vec{v}_f - \vec{v}_i$$

3. Express the initial and final velocities as vectors:

$$\vec{v}_i = (60\text{ km/h})\hat{i}, \qquad \vec{v}_f = (60\text{ km/h})\hat{j}$$

4. Substitute these results to find the average acceleration:

$$\vec{a}_{av} = \frac{\vec{v}_f - \vec{v}_i}{\Delta t} = \frac{(60\text{ km/h})\hat{j} - (60\text{ km/h})\hat{i}}{5\text{ s}}$$

$$= \boxed{-\left[(12\text{ km/h})/\text{s}\right]\hat{i} + \left[(12\text{ km/h})/\text{s}\right]\hat{j}}$$

(a)

(b)

REMARKS Note that the car accelerates, even though its speed does not change.

EXERCISE Find the magnitude and direction of the average acceleration vector. (*Answer* $a_{av} = (17.0 \text{ km/h})/\text{s}$ at $45°$ west of north)

The motion of an object traveling in a circle is a common example of motion in which the velocity of an object changes even though its speed remains constant.

The Direction of the Acceleration Vector In the next few chapters we will want to determine the direction of the acceleration vector from a description of the motion. For example, consider a bungee jumper as she slows down prior to reversing direction at the lowest point of her jump. To find the direction of her acceleration as she loses speed we make a series of dots representing her position at successive ticks of a clock, as shown in Figure 3-19a. The faster she moves, the greater the distance she travels between ticks, and the greater the space between the dots in the diagram. Next we number the dots, starting with zero and increasing in the direction of her motion. At time t_0 she is at dot 0, at time t_1 she is at dot 1, and so forth. To determine the direction of the acceleration at time t_3 we draw vectors representing the jumper's velocities at times t_2 and t_4. The average acceleration during the interval from t_2 to t_4 equals $\Delta\vec{v}/\Delta t$, where $\Delta\vec{v} = \vec{v}_4 - \vec{v}_2$ and $\Delta t = t_4 - t_2$. We use this as an estimate of her acceleration at time t_3. That is, $\vec{a}_3 \approx \Delta\vec{v}/\Delta t$. Since $\vec{a}_3$ and $\Delta\vec{v}$ are in the same direction, by finding the direction of $\Delta\vec{v}$ we also find the direction of $\vec{a}_3$. The direction of $\Delta\vec{v}$ is obtained by using the relation $\vec{v}_2 + \Delta\vec{v} = \vec{v}_4$ and drawing the corresponding vector addition diagram (Figure 3-19b). Because the jumper is moving faster (the dots are farther apart) at t_2 than at t_4, we draw $\vec{v}_2$ longer than $\vec{v}_4$. From this figure we get the direction of $\Delta\vec{v}$, and thus the direction of $\vec{a}_3$.

EXERCISE Figure 3-20 is a motion diagram of the bungee jumper before, during, and after time t_6, when she momentarily comes to rest at the lowest point in her descent. During the part of her ascent shown she is moving upward with increasing speed. Use this diagram to determine the direction of the jumper's acceleration (a) at time t_6 and (b) at time t_9. (*Answer* (a) upward, (b) upward)

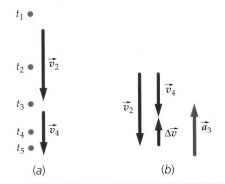

FIGURE 3-19 (*a*) A motion diagram of a bungee jumper losing speed as she descends. The dots are drawn at successive ticks of a clock. (*b*) We draw vectors $\vec{v}_2$ and $\vec{v}_4$ starting from the same point. Then we draw $\Delta\vec{v}$ from the head of $\vec{v}_2$ to the head of $\vec{v}_4$ to obtain the graphical expression of the relation $\vec{v}_2 + \Delta\vec{v} = \vec{v}_4$. The acceleration $\vec{a}_3$ is in the same direction as $\Delta\vec{v}$.

FIGURE 3-20 The dots for the bungee jumper's ascent are drawn to the right of those for her descent so that they do not overlap each other. Her motion, however, is straight down and then straight up.

ANOTHER FLYING CAP   **EXAMPLE 3-7**

A graduating physics student throws his cap into the air with an initial angle of 60° above the horizontal. Using a motion diagram, find the direction of the acceleration of the cap during the ascending portion of its flight.

PICTURE THE PROBLEM As the cap rises it both loses speed and changes direction. We should draw a motion diagram and then make a sketch of the relation $\vec{v}_i + \Delta\vec{v} = \vec{v}_f$ to find the direction of $\Delta\vec{v}$ and thus the direction of the acceleration.

1. Make a motion diagram (Figure 3-21a) of the cap's motion for the ascending portion of its flight. Because it slows as it ascends, the spacing between dots should decrease as it rises.

2. Pick a dot on the motion diagram and draw a velocity vector on the diagram for both the preceding and the following dots. These vectors should be drawn tangent to the path of the cap.

FIGURE 3-21

3. Draw the graphical expression of the relation $\vec{v}_i + \Delta\vec{v} = \Delta\vec{v}_f$. Begin by drawing the two velocity vectors from the same point. These vectors should have the same magnitude and direction as the vectors drawn for step 2. Then draw the $\Delta\vec{v}$ vector connecting their heads.

4. Draw the acceleration vector in the same direction as $\Delta\vec{v}$, but not necessarily the same length since $\vec{a} = \Delta\vec{v}/\Delta t$.

(c)   (d)

FIGURE 3-21 *continued*

REMARKS The process of finding the direction of the acceleration using a motion diagram is not precise. Therefore the result is an estimate of the direction of the acceleration, as opposed to a precise determination.

EXERCISE Use a motion diagram to find the direction of the acceleration of the cap in Example 3-7 during the descending portion of its flight. (*Answer*   directly downward)

3-4 Special Case 1: Projectile Motion

Figure 3-22 shows a particle launched with initial speed v_0 at angle θ_0 with the horizontal axis. Let the launch point be at (x_0, y_0); y is positive upward and x is positive to the right.

The initial velocity then has components

$$v_{0x} = v_0 \cos\theta_0 \qquad\qquad\qquad 3\text{-}18a$$

$$v_{0y} = v_0 \sin\theta_0 \qquad\qquad\qquad 3\text{-}18b$$

In the absence of air resistance, the acceleration is free-fall acceleration, vertically downward:

$$a_x = 0 \qquad\qquad\qquad\qquad\qquad 3\text{-}19a$$

and

$$a_y = -g \qquad\qquad\qquad\qquad\qquad 3\text{-}19b$$

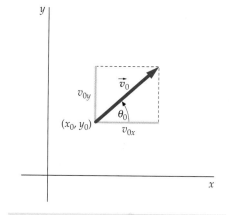

FIGURE 3-22

Since the acceleration is constant, we can use the kinematics equations discussed in Chapter 2. The x component of the velocity is constant because there is no horizontal acceleration:

$$v_x = v_{0x} \qquad\qquad\qquad\qquad\qquad 3\text{-}20a$$

The y component of the velocity varies with time according to Equation 2-12, with $a = -g$:

$$v_y = v_{0y} - gt \qquad\qquad\qquad\qquad 3\text{-}20b$$

❶ Notice that v_x does not depend on v_y and vice versa: *The horizontal and vertical components of projectile motion are independent.* (Dropping a ball from a desktop and projecting a second ball horizontally at the same time can demonstrate this. The two balls strike the floor simultaneously.) The displacements x and y are given by (see Equation 2-16)

$$x(t) = x_0 + v_{0x}t \qquad\qquad 3\text{-}21a$$

$$y(t) = y_0 + v_{0y}t - \tfrac{1}{2}gt^2 \qquad\qquad 3\text{-}21b$$

EQUATIONS OF MOTION FOR A PROJECTILE

The notation $x(t)$ and $y(t)$ simply emphasizes that x and y are functions of time. If the y component of the initial velocity is known, the time t for which the particle is at height y can be found from Equation 3-21b. The horizontal position at that time can then be found using Equation 3-21a. The total horizontal distance a projectile travels is called its **range.** (The vector forms of equations 3-19 to 3-21 are given on page 71.)

A CAP IN THE AIR

EXAMPLE 3 - 8

A delighted physics graduate throws her cap into the air with an initial velocity of 24.5 m/s at 36.9° above the horizontal. It is later caught by another student. Find (*a*) the total time the cap is in the air and (*b*) the total horizontal distance traveled.

PICTURE THE PROBLEM We choose the origin to be the initial position of the cap so that $x_0 = y_0 = 0$. We assume it is caught at the same height. The total time the cap is in the air is found by setting $y = 0$ in Equation 3-21b. We can then use this result in Equation 3-21a to find the total horizontal distance traveled.

(*a*) 1. Set $y = 0$ in Equation 3-21b and solve for t:

$$y = v_{0y}t - \tfrac{1}{2}gt^2 = t(v_{0y} - \tfrac{1}{2}gt) = 0$$

2. There are two solutions for t:

$$t = 0 \qquad \text{(initial conditions)}$$

$$t = \frac{2v_{0y}}{g}$$

3. Compute the vertical component of the initial velocity vector:

$$v_{0y} = v_0 \sin\theta_0 = (24.5\text{ m/s})\sin 36.9° = 14.7\text{ m/s}$$

4. Substitute for v_{0y} in the Step 2 result to find the total time t:

$$t = \frac{2v_{0y}}{g} = \frac{2v_0 \sin\theta_0}{g}$$

$$= \frac{2(24.5\text{ m/s})\sin 36.9°}{9.81\text{ m/s}^2} = \boxed{3.00\text{ s}}$$

(*b*) Use this value for the time to calculate the total horizontal distance traveled:

$$x = v_{0x}t = (v_0 \cos\theta_0)t$$

$$= (24.5\text{ m/s})\cos 36.9°(3.00\text{ s}) = \boxed{58.8\text{ m}}$$

REMARKS The vertical component of the initial velocity of the cap is 14.7 m/s, the same as that of the cap in Example 2-9 (Chapter 2), where the cap was thrown straight up with $v_0 = 14.7$ m/s. The time the cap is in the air is also the same as in Example 2-9. Figure 3-23 shows the height y versus t for the cap. This curve is identical to Figure 2-11 (Example 2-9) because the caps each have the same vertical acceleration and vertical velocity. Figure 3-23 can be reinterpreted as a graph of y versus x if its time scale is converted to a distance scale, as shown in the figure. This can be done by multiplying the time values by 19.6 m/s, because the cap moves at (24.5 m/s) cos 36.9° = 19.6 m/s horizontally. The curve y versus x is a parabola.

FIGURE 3-23

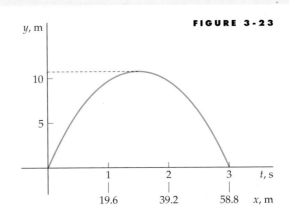

Figure 3-24 shows graphs of the vertical heights versus the horizontal distances for projectiles with an initial speed of 24.5 m/s and several different initial angles. The angles drawn are 45°, which has the maximum range, and pairs of angles at equal amounts above and below 45°. Notice that the paired angles have the same range. The green curve has an initial angle of 36.9° (0.64 rad), as in this example.

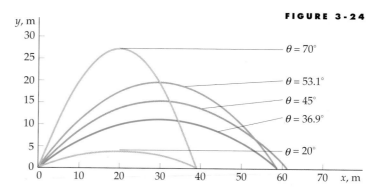

FIGURE 3-24

The general equation for the path $y(x)$ of a projectile can be obtained from Equations 3-21 by eliminating the variable t. Choosing $x_0 = 0$ and $y_0 = 0$, we obtain $t = x/v_{0x}$ from Equation 3-21a. Substituting this into Equation 3-21b gives

$$y(x) = v_{0y}\left(\frac{x}{v_{0x}}\right) - \frac{1}{2}g\left(\frac{x}{v_{0x}}\right)^2 = \left(\frac{v_{0y}}{v_{0x}}\right)x - \left(\frac{g}{2v_{0x}^2}\right)x^2$$

Writing out the velocity components yields

$$y(x) = (\tan\theta_0)x - \left(\frac{g}{2v_0^2\cos^2\theta_0}\right)x^2 \qquad\qquad 3\text{-}22$$

PATH OF A PROJECTILE

for the projectile's path. This is of the form $y = ax + bx^2$, the equation for a parabola passing through the origin. Figure 3-25 shows the path of a projectile with its velocity vector and components at several points. The path is for a projectile that impacts the ground at P. The horizontal distance between launch and impact is the range R.

If the initial and final elevations are equal, the range of a projectile can be written in terms of its initial speed and the angle of projection. As in the preceding examples, we find the range by multiplying the x component of the velocity by the total time that the projectile is in the air. The total flight time T is obtained by setting $y = 0$ in Equation 3-21b.

$$y(T) = v_{0y}T - \tfrac{1}{2}gT^2 = 0, \qquad T > 0$$

Dividing through by T gives

$$v_{0y} - \tfrac{1}{2}gT = 0$$

The flight time of the projectile is thus

$$T = \frac{2v_{0y}}{g} = \frac{2v_0}{g}\sin\theta_0$$

and the range is

$$R = v_{0x}T = (v_0\cos\theta_0)\left(\frac{2v_0}{g}\sin\theta_0\right) = \frac{2v_0^2}{g}\sin\theta_0\cos\theta_0$$

This can be further simplified by using the following trigonometric identity:

$$\sin 2\theta = 2\sin\theta\cos\theta$$

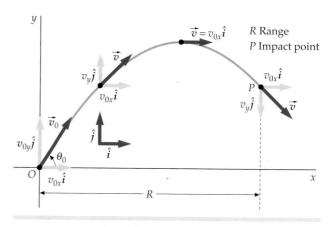

FIGURE 3-25 Path of a projectile showing velocity vectors.

Thus,

$$R = \frac{v_0^2}{g} \sin 2\theta_0 \qquad\qquad\qquad 3\text{-}23$$

RANGE OF A PROJECTILE FOR EQUAL INITIAL AND FINAL ELEVATIONS

EXERCISE Use Equation 3-22 for the path to derive Equation 3-23. (*Answer* Set $y(x) = 0$ and solve for x)

Equation 3-23 is useful if you want to find the range for many projectiles with equal initial and final elevations. More importantly, this equation shows how the range depends on θ. Since the maximum value of $\sin 2\theta$ is 1 when $2\theta = 90°$ or $\theta = 45°$, the range is greatest when $\theta = 45°$. In many practical applications, the initial and final elevations may not be equal, and other considerations are important. For example, in the shot put, the ball ends its flight when it hits the ground, but it is projected from an initial height of about 2 m above the ground. This causes the range to be maximum at an angle somewhat lower than 45°, as shown in Figure 3-26. Studies of the best shot-putters show that maximum range occurs at an initial angle of about 42°.

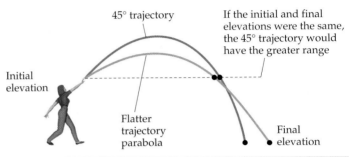

FIGURE 3-26 If a projectile lands at an elevation lower than the elevation of projection, the maximum range is achieved when the projection angle is somewhat lower than 45°.

A Supply Drop   **EXAMPLE 3-9**

A helicopter drops a supply package to flood victims on a raft on a swollen lake. When the package is released, the helicopter is 100 m above the raft and flying at 25 m/s at an angle $\theta_0 = 36.9°$ above the horizontal. (*a*) How long is the package in the air? (*b*) How far from the raft does the package land? (*c*) If the helicopter flies at constant velocity, where is the helicopter when the package lands?

FIGURE 3-27

PICTURE THE PROBLEM The time in the air depends only on the vertical aspects of the motion. Using Equation 3-21*b*, solve for the time. Choose the origin to be at the location of the package when it is released. The initial velocity of the package is the initial velocity of the helicopter. The horizontal distance traveled by the package is given by Equation 3-21*a*, where *t* is the time the package is in the air.

(*a*) 1. Sketch the trajectory of the package during the time it is in the air. Include coordinate axes as shown in Figure 3-27.

2. To find the time of flight, write $y(t)$:

$$y(t) = y_0 + v_{0y}t - \tfrac{1}{2}gt^2$$

3. Set $y_0 = 0$ and apply the quadratic formula to solve for *t*:

$$y(t) = v_{0y}t - \tfrac{1}{2}gt^2 \quad \text{or} \quad 0 = \tfrac{1}{2}gt^2 - v_{0y}t + y(t)$$

so

$$t = \frac{v_{0y} \pm \sqrt{v_{0y}^2 - 2gy(t)}}{g}$$

4. Solve for the time when $y(t) = -100$ m. First solve for v_{0y}:

$$v_{0y} = v_0 \sin \theta_0 = (25\text{ m/s})\sin 36.9° = 15\text{ m/s}$$

so

$$t = \frac{15 \text{ m/s} \pm \sqrt{(15 \text{ m/s})^2 - 2(9.81 \text{ m/s}^2)(-100 \text{ m})}}{9.81 \text{ m/s}^2}$$

Because the package is released at $t = 0$, the time of impact cannot be negative. Hence:

so

$t = -3.24 \text{ s}$ or $t = 6.30 \text{ s}$

$t = \boxed{6.30 \text{ s}}$

(b) At impact the package has traveled a horizontal distance x, where x is the horizontal velocity times the time of flight. First solve for the horizontal velocity:

$v_{0x} = v_0 \cos \theta_0 = (25 \text{ m/s}) \cos 36.9° = 20 \text{ m/s}$

so

$x = v_{0x}t = (20 \text{ m/s})(6.3 \text{ s}) = \boxed{126 \text{ m}}$

(c) The coordinates of the helicopter at the time of impact are:

$x_h = v_{0x}t = (20 \text{ m/s})(6.3 \text{ s}) = \boxed{126 \text{ m}}$

$y_h = y_{h0} + v_{h0}t = 0 + (15 \text{ m/s})(6.30 \text{ s}) = \boxed{94.5 \text{ m}}$

The height of the helicopter is 194.5 m directly above the package.

REMARKS The positive time is appropriate because it corresponds to a time after the package is dropped (which occurs at $t = 0$). The negative time is when the package would have been at $y = -100$ m if its motion had started earlier as shown in Figure 3-28. Note that the helicopter is directly above the package when the package hits the water (and at all other times before then). Figure 3-29 shows a graph of y versus x for various initial angles and an initial speed of 25 m/s. The curve with an initial angle of 36.9° is the one given in this example. Note that the maximum range occurs at an angle less than 45°.

FIGURE 3-29

FIGURE 3-28

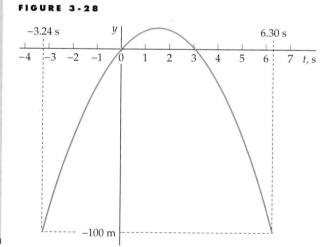

TO CATCH A THIEF   **EXAMPLE   3-10**   Try It Yourself

A police officer chases a master jewel thief across city rooftops. They are both running when they come to a gap between buildings that is 4 m wide and has a drop of 3 m (Figure 3-30). The thief, having studied a little physics, leaps at 5 m/s and at 45° and clears the gap easily. The police officer did not study physics and thinks he should maximize his horizontal velocity, so he leaps at 5 m/s horizontally. (a) Does he clear the gap? (b) By how much does the thief clear the gap?

FIGURE 3-30

PICTURE THE PROBLEM Assuming they both clear the gap, the total time in the air depends only on the vertical aspects of the motion. Choose the origin at the launch point, with upward positive so that Equations 3-21 apply. Use Equation 3-21b for $y(t)$ and solve for the time when $y = -3$ m for $\theta_0 = 0°$ and again for $\theta_0 = 45°$. The horizontal distances traveled are the values of x at these times.

Cover the column to the right and try these on your own before looking at the answers.

Steps	Answers
(a) 1. Write $y(t)$ for the police officer and solve for t when $y = -3$ m.	$t = 0.782$ s
2. Substitute this into the equation for $x(t)$ and find the horizontal distance traveled during this time.	$x = \boxed{3.91 \text{ m}}$
	Since this is less than 4 m, it appears the police officer fails to make it across the gap between buildings.
(b) 1. Write $y(t)$ for the thief and solve for t when $y = -3$ m. $y(t)$ is a quadratic equation with two solutions, but only one of its solutions is acceptable.	$t = -0.5$ s or $t = 1.22$ s She must land after she leaps, so $t = 1.22$ s
2. Find the horizontal distance covered for the positive value of t.	$x = v_{0x}t = 4.31$ m
3. Subtract 4.0 m from this distance.	$4.31 \text{ m} - 4.0 \text{ m} = \boxed{0.31 \text{ m}}$

REMARKS The thief probably knew that she should jump at slightly less than 45°, but she didn't have time to solve the problem exactly. The police officer actually did make it across by tightening his abdominal muscles before impact. This raised his feet more than the 9 cm needed for him to complete the leap.

DROPPING SUPPLIES   **EXAMPLE   3 - 1 1   Try It Yourself**

In Example 3-9, find (a) the time t_1 for the package to reach its greatest height h above the water, (b) its greatest height h, and (c) the time t_2 for the package to fall to the water from its greatest height.

PICTURE THE PROBLEM The time t_1 is the time at which the vertical component of the velocity is zero. Using Equation 3-20b solve for t_1.

Cover the column to the right and try these on your own before looking at the answers.

Steps	Answers
(a) 1. Write $v_y(t)$ for the package.	$v_y(t) = v_{0y} - gt$
2. Set $v_y(t_1) = 0$ and solve for t_1.	$t_1 = \boxed{1.53 \text{ s}}$
(b) 1. Find $v_{y \text{ av}}$ during the time the package is moving up.	$v_{y \text{ av}} = 7.5$ m/s
2. Use $v_{y \text{ av}}$ to find the distance traveled up. Then find h.	$\Delta y = 11.5$ m, $h = \boxed{111.5 \text{ m}}$
(c) Find the time for the package to fall a distance h.	$t_1 = \boxed{4.77 \text{ s}}$

REMARKS Note that $t_1 + t_2 = 6.3$ s, in agreement with Example 3-9.

EXERCISE Solve Part (b) of Example 3-11 using $y(t)$ (Equation 3-21b) instead of finding $v_{y\,av}$.

Equations 3-19a and b can be expressed in vector form. Multiplying through Equation 3-19a by $\hat{i}$, multiplying through Equation 3-19b by $\hat{j}$, and then adding the two equations gives $a_x\hat{i} + a_y\hat{j} = -g\hat{j}$, or

$$\vec{a} = \vec{g} \qquad\qquad 3\text{-}19c$$

where $\vec{g}$ is the free-fall acceleration vector. At the surface of the earth the magnitude of $\vec{g}$ is $g = 9.81$ m/s². Equations 3-20a and b can also be expressed in vector form. Multiplying through Equation 3-20a by $\hat{i}$, multiplying through Equation 3-20b by $\hat{j}$, and then adding the two equations gives $(v_x\hat{i} + v_y\hat{j}) = (v_{0x}\hat{i} + v_{0y}\hat{j}) - gt\hat{j}$ or

$$\vec{v} = \vec{v}_0 + \vec{g}t \qquad \text{or} \qquad \Delta\vec{v} = \vec{g}t \qquad\qquad 3\text{-}20c$$

where $\vec{v} = v_x\hat{i} + v_y\hat{j}$, $\vec{v}_0 = v_{0x}\hat{i} + v_{0y}\hat{j}$, and $\vec{g} = -g\hat{j}$. Repeating the process for Equations 3-21a and b we obtain

$$\vec{r} = \vec{r}_0 + \vec{v}_0 t + \tfrac{1}{2}\vec{g}t^2 \qquad \text{or} \qquad \Delta\vec{r} = \vec{v}_0 t + \tfrac{1}{2}\vec{g}t^2 \qquad\qquad 3\text{-}21c$$

where $\vec{r} = x\hat{i} + y\hat{j}$ and $\vec{r}_0 = x_0\hat{i} + y_0\hat{j}$. For a number of problems the vector forms of the kinematic equations (Equations 3-20c and 3-21c) are more suitable. This is the case in the following example.

THE RANGER AND THE MONKEY   **E X A M P L E   3 - 1 2**

A park ranger with a tranquilizer dart gun intends to shoot a monkey hanging from a branch. The ranger aims directly at the monkey, not realizing that the dart will follow a parabolic path that will pass below the present position of the creature. The monkey, seeing the gun discharge, lets go of the branch and drops out of the tree, expecting to avoid the dart. (a) Show that the monkey will be hit regardless of the initial speed of the dart so long as it is great enough for the dart to travel the horizontal distance to the tree. Assume the reaction time of the monkey is negligible. (b) Let $\vec{v}_{d0}$ be the velocity of the dart as it leaves the gun. Find the velocity of the dart relative to the monkey at an arbitrary time t during the dart's flight.

PICTURE THE PROBLEM We apply Equation 3-21c to both the monkey and the gun.

(a) 1. Apply Equation 3-21c to the monkey at time t:

$$\Delta\vec{r}_m = \tfrac{1}{2}\vec{g}t^2$$

where the initial velocity of the monkey is zero.

FIGURE 3-31

2. Apply Equation 3-21c to the dart at time t:

$$\Delta\vec{r}_d = \vec{v}_{d0}t + \tfrac{1}{2}\vec{g}t^2$$

where $\vec{v}_{d0}$ is the velocity of the dart as it leaves the gun.

3. Make a sketch of the monkey, the dart, and the gun, as shown in Figure 3-31. Show the dart and the monkey at their initial locations and at their locations a time t later. On the figure draw a vector representing each term in the step 1 and step 2 results. Note that at time t the dart and the monkey both are a distance $\tfrac{1}{2}gt^2$ below the line of sight of the gun. The dart will strike the monkey when it reaches the monkey's line of fall:

(b) 1. The velocity of the dart relative to the monkey equals the velocity of the dart relative to the gun plus the velocity of the gun relative to the monkey:

$$\vec{v}_{dm} = \vec{v}_{dg} + \vec{v}_{gm}$$

2. The velocity of the gun relative to the monkey is the negative of the velocity of the monkey relative to the gun:

$$\vec{v}_{dm} = \vec{v}_{dg} - \vec{v}_{mg}$$

3. Using Equation 3-20c, express the velocity of the dart relative to the gun and the velocity of the monkey relative to the gun:

$$\vec{v}_{dg} = \vec{v}_{d0} + \vec{g}t$$
$$\vec{v}_{mg} = \vec{g}t$$

4. Substitute these expressions into the Part (b) step 2 result:

$$\vec{v}_{dm} = (\vec{v}_{d0} + \vec{g}t) - (\vec{g}t) = \boxed{\vec{v}_{d0}}$$

REMARKS Relative to the monkey, the dart moves with constant speed v_{d0} in a straight line. The dart strikes the monkey at time $t = L/v_{d0}$, where L is the distance from the muzzle of the gun to the initial position of the monkey.

REMARKS In a familiar lecture demonstration, a target is suspended by an electromagnet. When the dart leaves the gun, the circuit to the magnet is broken and the target falls. The initial velocity of the dart is varied so that for large v_{d0} the target is hit very near its original height and for small v_{d0} it is hit just before it reaches the floor.

EXERCISE A hockey puck at ice level is struck such that it misses the net and clears the top of the Plexiglass wall of height $h = 2.80$ m. The flight time at the moment the puck clears the wall is $t_1 = 0.650$ s, and the horizontal distance is $x_1 = 12.0$ m. (a) Find the initial speed and direction of the puck. (b) When does the puck reach its maximum height? (c) What is the maximum height of the puck? (*Answer* (a) $\vec{v} = 20.0$ m/s, $\theta_0 = 22.0°$, (b) $t = 0.764$ s, (c) $v_{y \, av} \, t = 2.86$ m)

FIGURE 3-32

3-5 Special Case 2: Circular Motion

Figure 3-32 shows a pendulum bob swinging back and forth in a vertical plane. The path of the bob is a segment of a circular path. Motion along a circular path, or a segment of a circular path, is called **circular motion.**

FIGURE 3-33

(a)

(b)

(c)

A Swinging Pendulum   **EXAMPLE 3 - 1 3**

Consider the motion of the pendulum bob shown in Figure 3-32. Using a motion diagram (Figure 3-33), find the direction of the acceleration vector when the bob is swinging from left to right and (a) on the descending portion of the path, (b) passing through the lowest point on the path, and (c) on the ascending portion of the path.

PICTURE THE PROBLEM As the bob descends it both gains speed and changes direction. The acceleration is related to the change in velocity by $\vec{a} \approx \Delta\vec{v}/\Delta t$. The direction of the acceleration at a point can be estimated by constructing a vector addition diagram for the relation $\vec{v}_i + \Delta\vec{v} = \vec{v}_f$ to find the direction of $\Delta\vec{v}$, and thus the direction of the acceleration vector.

(a) 1. Make a motion diagram for a full left-to-right swing of the bob. The spacing between dots is greatest at the lowest point where the speed is greatest.

2. Pick a dot on the descending portion of the motion diagram and draw a velocity vector on the diagram for both the preceding and the following dot. The velocity vectors should be drawn tangent to the path and with lengths proportional to the speed.

3. Draw the graphical expression of the relation $\vec{v}_i + \Delta\vec{v} = \vec{v}_f$. On this figure draw the acceleration vector. Since $\vec{a} \approx \Delta\vec{v}/\Delta t$, $\vec{a}$ is in the same direction as $\Delta\vec{v}$.

(b) Repeat steps 2 and 3 for the lowest point on the path.

(c) Repeat steps 2 and 3 for a point on the ascending portion of the path.

FIGURE 3-33 *continued*

(d)

(e)

REMARKS At the lowest point the acceleration vector is directed straight upward (Figure 3-34)—toward point P at the center of the circle. Where the speed is increasing (on the descending portion), the acceleration has a component in the forward direction as well as toward P. Where the speed is decreasing, the acceleration has a rearward component as well as a component toward P.

If a particle moves along a circular arc, the direction from the particle toward the center of the circle is called the **centripetal direction.** In Example 3-13 the acceleration at the lowest point of the pendulum bob's path was found to be in the centripetal direction. At other points the acceleration was found to have a tangential component as well as a centripetal component.

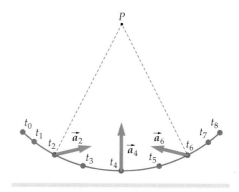

FIGURE 3-34

Uniform Circular Motion

Motion in a circle at constant speed is called **uniform circular motion.** To find an expression for the acceleration of a particle moving in a circle at constant speed we will extend the method used in Example 3-13 to find the direction of the acceleration. The position and velocity vectors for a particle moving in a circle at constant speed are shown in Figure 3-35. The angle $\Delta\theta$ between $\vec{v}(t)$ and $\vec{v}(t + \Delta t)$ is the same as between $\vec{r}(t)$ and $\vec{r}(t + \Delta t)$ because the position and velocity vectors must both rotate through equal angles to remain mutually perpendicular. An isosceles triangle is formed by the two velocity vectors and $\Delta\vec{v}$, and a second isosceles triangle is formed by the two position vectors and $\Delta\vec{r}$. To find the direction of the acceleration vector we examine the triangle formed by the two velocity vectors and $\Delta\vec{v}$. The sum of the angles of any triangle is 180° and the base angles of any isosceles triangle are equal. In the limit that Δt approaches zero, $\Delta\theta$ also approaches zero, so in this limit the two base angles must each approach 90°. This means $\Delta\vec{v}$ is perpendicular to the velocity. If $\Delta\vec{v}$ is drawn from the position of the particle then it points in the centripetal direction.

The two triangles are similar, so $|\Delta\vec{v}|/v = |\Delta\vec{r}|/r$ (corresponding lengths of similar shapes are proportional). Dividing both sides by Δt and rearranging gives

$$\frac{|\Delta\vec{v}|}{\Delta t} = \frac{v|\Delta\vec{r}|}{r\ \Delta t}$$

In the limit as Δt approaches zero, the term $|\Delta\vec{v}|/\Delta t$ approaches a, the magnitude of the instantaneous acceleration, and the term $|\Delta\vec{r}|/\Delta t$ approaches v, the magnitude of the instantaneous velocity (the speed). With these substitutions we have

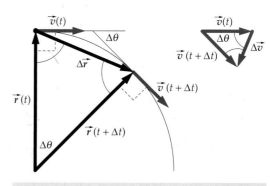

FIGURE 3-35 Position and velocity vectors for a particle moving in a circle at constant speed.

$$a = a_c = \frac{v^2}{r}$$

3-24

CENTRIPETAL ACCELERATION

The motion of a particle moving in a circle with constant speed is often described in terms of the time T required for one complete revolution, called the **period**. During one period, the particle travels a distance of $2\pi r$ (where r is the radius of the circle), so its speed is related to r and T by

$$v = \frac{2\pi r}{T}$$

3-25

A SATELLITE'S MOTION

EXAMPLE 3-14

A satellite moves at constant speed in a circular orbit about the center of the earth and near the surface of the earth. If its acceleration is $g = 9.81$ m/s², find (a) its speed and (b) the time for one complete revolution.

PICTURE THE PROBLEM Since the satellite orbits near the surface of the earth, we take the radius of the orbit to be the radius of the earth, $r = 6370$ km.

(a) Set the centripetal acceleration v^2/r equal to g and solve for the speed v:

$$a = \frac{v^2}{r} = g \quad \text{or}$$

$$v = \sqrt{rg} = \sqrt{(6370 \text{ km})(9.81 \text{ m/s}^2)}$$

$$= \boxed{7.91 \text{ km/s} = 17{,}700 \text{ mi/h}}$$

(b) Use Equation 3-25 to get the period T:

$$T = \frac{2\pi r}{v} = \frac{2\pi (6370 \text{ km})}{7.91 \text{ km/s}} = \boxed{5060 \text{ s} = 84.3 \text{ min}}$$

REMARKS For actual satellites in orbit a few hundred kilometers above the earth's surface, the orbital radius r is slightly greater than 6370 km. As a result, the centripetal acceleration is slightly less than 9.81 m/s² because of the decrease in the gravitational force with distance from the center of the earth. Many satellites are launched into such orbits, and their periods are roughly 90 min.

EXERCISE A car rounds a curve of radius 40 m at 48 km/h. What is its centripetal acceleration? (*Answer* 4.44 m/s²)

A particle moving in a circle with *varying speed* has a component of acceleration tangent to the circle, dv/dt, as well as the radially inward centripetal acceleration, v^2/r. For general motion along a curve, we can treat a portion of the curve as an arc of a circle (Figure 3-36). The particle then has acceleration v^2/r toward the center of curvature, and if the speed v is changing, it has tangential acceleration

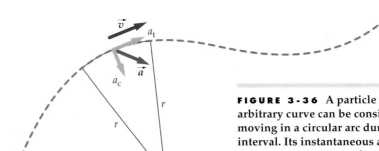

FIGURE 3-36 A particle moving along an arbitrary curve can be considered to be moving in a circular arc during a small time interval. Its instantaneous acceleration vector has a component $a_c = v^2/r$ toward the center of curvature of the arc and a component $a_t = dv/dt$ that is tangential to the curve.

$$a_t = \frac{dv}{dt}$$

3-26

TANGENTIAL ACCELERATION

Topic	Relevant Equations and Remarks
1. Vectors	
Definition	Vectors are quantities that have both magnitude and direction. Vectors add like consecutive displacements.
Components	The component of a vector in a direction in space is its projection on an axis in that direction. If $\vec{A}$ makes an angle θ with the positive x direction, its x and y components are $$A_x = A \cos \theta \qquad \text{3-2}$$ $$A_y = A \sin \theta \qquad \text{3-3}$$
Magnitude	$$A = \sqrt{A_x^2 + A_y^2} \qquad \text{3-5}a$$
Adding vectors graphically	Any two vectors whose magnitudes have the same units may be added graphically by placing the tail of one arrow at the head of the other.
Adding vectors using components	If $\vec{C} = \vec{A} + \vec{B}$, then $$C_x = A_x + B_x \qquad \text{3-6}a$$ and $$C_y = A_y + B_y \qquad \text{3-6}b$$
Unit vectors	A vector $\vec{A}$ can be written in terms of unit vectors $\hat{i}$, $\hat{j}$, and $\hat{k}$, which have unit magnitude and lie along the x, y, and z axes, respectively $$\vec{A} = A_x\hat{i} + A_y\hat{j} + A_z\hat{k} \qquad \text{3-7}$$
Position vector	The position vector $\vec{r}$ points from the origin of the coordinate system to the particle's position.
Instantaneous-velocity vector	The velocity vector $\vec{v}$ is the rate of change of the position vector. Its magnitude is the speed and it points in the direction of motion. $$\vec{v} = \lim_{\Delta t \to 0}\frac{\Delta \vec{r}}{\Delta t} = \frac{d\vec{r}}{dt} \qquad \text{3-12}$$
Instantaneous-acceleration vector	$$\vec{a} = \lim_{\Delta t \to 0}\frac{\Delta \vec{v}}{\Delta t} = \frac{d\vec{v}}{dt} \qquad \text{3-16}$$
2. Relative Velocity	If a particle moves with velocity $\vec{v}_{pA}$ relative to reference frame A, which is in turn moving with velocity $\vec{v}_{AB}$ relative to reference frame B, the velocity of the particle relative to B is $$\vec{v}_{pB} = \vec{v}_{pA} + \vec{v}_{AB} \qquad \text{3-14}$$
3. Projectile Motion	The positive x direction is horizontal and the positive y direction is upward for the equations in this section.
Independence of motion	In projectile motion, the horizontal and vertical motions are independent. Thus, $$a_x = 0 \qquad \text{and} \qquad a_y = -g.$$

Dependence on time	$v_x(t) = v_{0x} + a_x t, \qquad v_y(t) = v_{0y} + a_y t$	2-14
	$x(t) = x_0 + v_{0x}t + \frac{1}{2}a_x t^2, \qquad y(t) = y_0 + v_{0y}t + \frac{1}{2}a_y t^2$	2-16
	where $a_x = 0$, $a_y = -g$, $v_{0x} = v_0 \cos\theta_0$, and $v_{0y} = v_0 \sin\theta_0$. Alternatively,	
	$\Delta\vec{v} = \vec{g}t, \qquad \Delta\vec{r} = \vec{v}_0 t + \frac{1}{2}\vec{g}t^2$	3-20c, 3-21c
	where $\vec{g} = -g\hat{j}$	
Range	The range is found by multiplying v_x by the total time the projectile is in the air.	

3. Circular Motion

Centripetal acceleration	$a_c = \dfrac{v^2}{r}$	3-24
Tangential acceleration	$a_t = \dfrac{dv}{dt}$	
	where v is the speed.	3-26
Period	$v = \dfrac{2\pi r}{T}$	3-25

PROBLEMS

- Single-concept, single-step, relatively easy
- Intermediate-level, may require synthesis of concepts
- Challenging
- **SSM** Solution is in the *Student Solutions Manual*
- **iSOLVE** Problems available on iSOLVE online homework service
- **iSOLVE✓** These "Checkpoint" online homework service problems ask students additional questions about their confidence level, and how they arrived at their answer

In a few problems, you are given more data than you actually need; in a few other problems, you are required to supply data from your general knowledge, outside sources, or informed estimates.

Conceptual Problems

1 • **SSM** Can the magnitude of the displacement of a particle be less than the distance traveled by the particle along its path? Can its magnitude be more than the distance traveled? Explain.

2 • Give an example in which the distance traveled is a significant amount, yet the corresponding displacement is zero.

3 • What is the approximate average velocity of the race cars during the Indianapolis 500?

4 • **iSOLVE** True or false: The magnitude of the sum of two vectors *must be* greater than the magnitude of either vector.

5 • Can a component of a vector have a magnitude greater than the magnitude of the vector? Under what circumstances can a component of a vector have a magnitude equal to the magnitude of the vector?

6 • **SSM** Can a vector be equal to zero and still have one or more components not equal to zero?

7 • **iSOLVE** Are the components of $\vec{C}$, where $\vec{C} = \vec{A} + \vec{B}$, necessarily larger than the corresponding components of either $\vec{A}$ or $\vec{B}$?

8 • **SSM** True or false: The instantaneous-acceleration vector is *always* in the direction of motion.

9 • If an object is moving toward the west at some instant, in what direction is its acceleration? (a) North (b) East (c) West (d) South (e) May be any direction.

10 • **iSOLVE** A golfer drives the ball from the tee down the fairway in a high arcing shot. When the ball is at the highest point of its flight, (a) its velocity and acceleration are both zero, (b) its velocity is zero but its acceleration is nonzero, (c) its velocity is nonzero but its acceleration is zero, (d) its velocity and acceleration are both nonzero, (e) insufficient information is given to answer correctly.

11 • The velocity of a particle is in the eastward direction while the acceleration is directed toward the northwest as shown in Figure 3-37. The particle is (a) speeding up and turning toward the north, (b) speeding up and turning toward the south, (c) slowing down and turning toward the north, (d) slowing down and turning toward the south, (e) maintaining constant speed and turning toward the south.

FIGURE 3-37
Problem 11

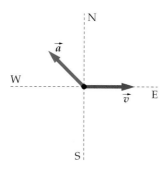

12 • **SSM** Assuming constant acceleration, if you know the position vectors of a particle at two points on its path and also know the time it took to move from one point to the other, you can then compute (a) the particle's average velocity, (b) the particle's average acceleration, (c) the particle's instantaneous velocity, (d) the particle's instantaneous acceleration, (e) insufficient information is given to describe the particle's motion.

13 •• Consider the path of a particle as it moves in space. (a) How is the velocity vector related geometrically to the path of the particle? (b) Sketch a curved path and draw the velocity vector for the particle for several positions along the path.

14 • The acceleration of a car is zero when it is (a) turning right at a constant speed, (b) driving up a long straight incline at constant speed, (c) topping the crest of a hill at constant speed, (d) bottoming out at the lowest point of a valley at constant speed, (e) speeding up as it descends a long straight decline.

15 • **SSM** Give examples of motion in which the directions of the velocity and acceleration vectors are (a) opposite, (b) the same, and (c) mutually perpendicular.

16 • **iSOLVE** How is it possible for a particle moving at constant speed to be accelerating? Can a particle with constant velocity be accelerating at the same time?

17 •• Imagine throwing a dart straight upward so that it sticks into the ceiling. After it leaves your hand, it steadily slows down as it rises before it sticks. (a) Draw the dart's velocity vector at times t_1 and t_2, where $\Delta t = t_2 - t_1$ is small. From your drawing find the direction of the change in velocity $\Delta \vec{v} = \vec{v}_2 - \vec{v}_1$, and thus the direction of the acceleration vector. (b) After it has stuck in the ceiling for a few seconds, the dart falls down to the floor. As it falls it speeds up, of course, until it hits the floor. Repeat part (a) to find the direction of its acceleration vector as it falls. (c) Now imagine tossing the dart horizontally. What is the direction of its acceleration vector after it leaves your hand, but before it strikes the floor?

18 •• **SSM** As a bungee jumper approaches the lowest point in her descent, the rubber band holding her stretches and she loses speed as she continues to move downward. Assuming that she is dropping straight down, make a motion diagram to find the direction of her acceleration vector as she slows down by drawing her velocity vectors at times t_1 and t_2, where $\Delta t = t_2 - t_1$ is small. From your drawing find the direction of the change in velocity $\Delta \vec{v} = \vec{v}_2 - \vec{v}_1$, and thus the direction of the acceleration vector.

19 •• After reaching the lowest point in her jump at time t_{low}, the bungee jumper in Problem 18 moves upward, gaining speed for a short time until gravity again dominates her motion. Draw her velocity vectors at times t_1 and t_2, where $\Delta t = t_2 - t_1$ is small and $t_1 < t_{\text{low}} < t_2$. From your drawing find the direction of the change in velocity $\Delta \vec{v} = \vec{v}_2 - \vec{v}_1$, and thus the direction of the acceleration vector.

20 • A river is 0.76 km wide. The banks are straight and parallel (Figure 3-38). The current is 4.0 km/h and is parallel to the banks. A boat has a maximum speed of 4 km/h in still water. The pilot of the boat wishes to go on a straight line from A to B, where AB is perpendicular to the banks. The pilot should (a) head directly across the river, (b) head 53° upstream from the line AB, (c) head 37° upstream from the line AB, (d) give up—the trip from A to B is not possible with a boat of this limited speed, (e) do none of the above.

FIGURE 3-38
Problem 20

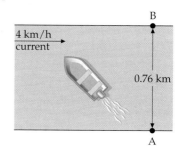

21 • **SSM** True or false: When a projectile is fired horizontally, it takes the same amount of time to reach the ground as an identical projectile dropped from rest from the same height. Ignore the effects of air resistance.

22 • **iSOLVE** A projectile is fired at 35° above the horizontal. At the highest point in its trajectory, its speed is 200 m/s. The initial velocity had a horizontal component of (a) 0; (b) (200 m/s) cos 35°; (c) (200 m/s) sin 35°; (d) (200 m/s)/cos 35°; (e) 200 m/s. Neglect the effects of air resistance.

23 • Figure 3-39 represents the parabolic trajectory of a ball going from A to E. What is the direction of the acceleration at point B? (a) Up and to the right (b) Down and to the left (c) Straight up (d) Straight down (e) The acceleration of the ball is zero.

FIGURE 3-39
Problems 23 and 24

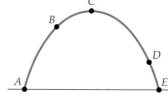

24 • Referring to the motion described in Problem 23, (a) at which point(s) is the speed the greatest? (b) At which point(s) is the speed the lowest? (c) At which two points is the speed the same? Is the velocity the same at those points?

25 • ▐SOLVE▐ True or false:

(a) If the speed is constant, the acceleration must be zero.
(b) If the acceleration is zero, the speed must be constant.

26 • The initial and final velocities of an object are as shown in Figure 3-40. Indicate the direction of the average acceleration.

FIGURE 3-40 Problem 26

27 • The velocities of objects A and B are shown in Figure 3-41. Draw a vector that represents the velocity of B relative to A.

FIGURE 3-41 Problem 27

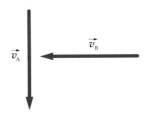

28 •• **SSM** A vector $\vec{A}(t)$ has a constant magnitude but is changing direction. (a) Find $d\vec{A}/dt$ in the following manner: Draw the vectors $\vec{A}(t + \Delta t)$ and $\vec{A}(t)$ for a small time interval Δt, and find the difference $\Delta \vec{A} = \vec{A}(t + \Delta t) - \vec{A}(t)$ graphically. How is the direction of $\Delta \vec{A}$ related to $\vec{A}$ for small time intervals? (b) Interpret this result for the special cases where $\vec{A}$ represents the position of a particle with respect to some coordinate system. (c) Could $\vec{A}$ represent a velocity vector? Explain.

29 •• The automobile path shown in Figure 3-42 is made up of straight lines and arcs of circles. The automobile starts from rest at point A. After it reaches point B, it travels at constant speed until it reaches point E. It comes to rest at point F. (a) At the middle of each segment (AB, BC, CD, DE, and EF), what is the direction of the velocity vector? (b) At which of these points does the automobile have an acceleration? In those cases, what is the direction of the acceleration? (c) How do the magnitudes of the acceleration compare for segments BC and DE?

FIGURE 3-42
Problem 29

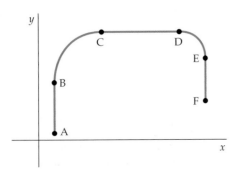

30 •• **SSM** Two cannons are pointed directly toward each other as shown in Figure 3-43. When fired, the cannonballs will follow the trajectories shown—P is the point where the trajectories cross each other. If we want the cannonballs to hit each other, should the gun crews fire cannon A first, cannon B first, or should they fire simultaneously? Ignore the effects of air resistance.

FIGURE 3-43 Problem 30

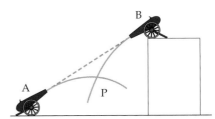

31 •• Galileo wrote the following in his *Dialogue concerning the two world systems:* "Shut yourself up . . . in the main cabin below decks on some large ship, and . . . hang up a bottle that empties drop by drop into a wide vessel beneath it. When you have observed [this] carefully . . . have the ship proceed with any speed you like, so long as the motion is uniform and not fluctuating this way and that. . . . The droplets will fall as before into the vessel beneath without dropping towards the stern, although while the drops are in the air the ship runs many spans." Explain this quotation.

32 • A man swings a stone attached to a rope in a horizontal circle at constant speed. Figure 3-44 represents the path of the rock looking down from above. (a) Which of the vectors $\vec{A}$ to $\vec{E}$ could represent the velocity of the stone? (b) Which could represent the acceleration?

FIGURE 3-44 Problem 32

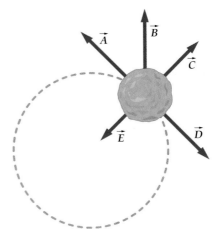

33 • True or false: An object cannot move in a circle unless it is accelerating.

34 •• Using a motion diagram, find the direction of the acceleration of the bob of a pendulum when the bob is at a point where it is just reversing its direction.

35 • The speed of a batted baseball immediately after being struck can reach 110 mph. Let's say that the baseball has a "launch angle" of 35°, which is fairly typical for the sport. Naively using the range equation (Equation 3-23) to calculate the distance the ball will travel, we find a range of 760 ft (232 m)! In reality, it will only travel about 400 ft. Can you give a reason why the range equation breaks down so badly here? Be specific: If you can, look up the *terminal speed* for a baseball.

Estimation and Approximation

36 •• SSM Estimate how far you can throw a ball if you throw it (*a*) horizontally while standing on level ground, (*b*) at $\theta = 45°$ while standing on level ground, (*c*) horizontally from the top of a building 12 m high, (*d*) at $\theta = 45°$ from the top of a building 12 m high.

37 •• SOLVE In 1978, Geoff Capes of Great Britain threw a heavy brick a horizontal distance of 44.5 m. Find the approximate velocity of the brick at the highest point of its flight, neglecting the effects of air resistance.

Vectors, Vector Addition, and Coordinate Systems

38 • A wall clock has a minute hand that has a length of 0.5 m and an hour hand with a length of 0.25 m. Taking the center of the clock as the origin, and choosing an appropriate coordinate system, write the position of the hour and minute hands as vectors when the time reads (*a*) 12:00, (*b*) 3:30, (*c*) 6:30, (*d*) 7:15. (*e*) Call the position of the tip of the minute hand $\vec{A}$ and the position of the tip of the hour hand $\vec{B}$. Find $\vec{A} - \vec{B}$ for the times given in (*a*)–(*d*) above.

39 • SSM A bear walks northeast for 12 m and then east for 12 m. Show each displacement graphically and find the resultant displacement vector graphically, as in Example 3-2(*a*).

40 • SOLVE ✔ A circular arc is centered at $x = 0$, $y = 0$. (*a*) A student walks along the circular arc from the position $x = 5$ m, $y = 0$ to a final position $x = 0$, $y = 5$ m. What is her displacement? (*b*) A second student walks from the same initial position along the x axis to the origin and then along the y axis to $y = 5$ m and $x = 0$. What is his displacement?

41 • SSM SOLVE ✔ For the two vectors $\vec{A}$ and $\vec{B}$ of Figure 3-45, find the following graphically as in Example 3-2(*a*): (*a*) $\vec{A} + \vec{B}$, (*b*) $\vec{A} - \vec{B}$, (*c*) $2\vec{A} + \vec{B}$, (*d*) $\vec{B} - \vec{A}$, (*e*) $2\vec{B} - \vec{A}$.

FIGURE 3-45 Problem 41

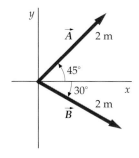

42 • A Scout walks 2.4 km due east from camp, then turns left and walks 2.4 km along the arc of a circle centered at the campsite, and finally walks 1.5 km directly toward camp. (*a*) How far is the Scout from camp at the end of his walk? (*b*) In what direction is the Scout's position relative to the campsite? (*c*) What is the ratio of the final magnitude of the displacement to the total distance walked?

43 • A velocity vector has an x component of $+5.5$ m/s and a y component of -3.5 m/s. Which diagram in Figure 3-46 shows the direction of the vector correctly?

FIGURE 3-46 Problem 43

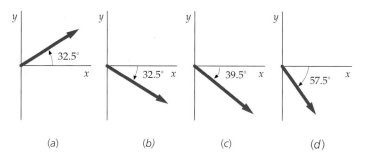

(a)   (b)   (c)   (d)

(e) None of the above.

44 • SOLVE Three vectors $\vec{A}$, $\vec{B}$, and $\vec{C}$ have the following x and y components: $A_x = 6$, $A_y = -3$; $B_x = -3$, $B_y = 4$; $C_x = 2$, $C_y = 5$. The magnitude of $\vec{A} + \vec{B} + \vec{C}$ is (*a*) 3.3, (*b*) 5.0, (*c*) 11, (*d*) 7.8, (*e*) 14.

45 • Find the rectangular components of the following vectors $\vec{A}$ which lie in the xy plane and make an angle θ with the x axis (Figure 3-47) if (*a*) $A = 10$ m, $\theta = 30°$, (*b*) $A = 5$ m, $\theta = 45°$, (*c*) $A = 7$ km, $\theta = 60°$, (*d*) $A = 5$ km, $\theta = 90°$, (*e*) $A = 15$ km/s, $\theta = 150°$, (*f*) $A = 10$ m/s, $\theta = 240°$, (*g*) $A = 8$ m/s², $\theta = 270°$.

FIGURE 3-47
Problem 45

46 • SSM Vector $\vec{A}$ has a magnitude of 8 m at an angle of 37° with the positive x axis; vector $\vec{B} = (3$ m$)\hat{i} - (5$ m$)\hat{j}$; vector $\vec{C} = (-6$ m$)\hat{i} + (3$ m$)\hat{j}$. Find the following vectors: (*a*) $\vec{D} = \vec{A} + \vec{C}$, (*b*) $\vec{E} = \vec{B} - \vec{A}$, (*c*) $\vec{F} = \vec{A} - 2\vec{B} + 3\vec{C}$, (*d*) a vector $\vec{G}$ such that $\vec{G} - \vec{B} = \vec{A} + 2\vec{C} + 3\vec{G}$.

47 •• Find the magnitude and direction of the following vectors: (*a*) $\vec{A} = 5\hat{i} + 3\hat{j}$, (*b*) $\vec{B} = 10\hat{i} - 7\hat{j}$, (*c*) $\vec{C} = -2\hat{i} - 3\hat{j} + 4\hat{k}$.

48 • Find the magnitude and direction of $\vec{A}$, $\vec{B}$, and $\vec{C} = \vec{A} + \vec{B}$ for (*a*) $\vec{A} = -4\hat{i} - 7\hat{j}$, $\vec{B} = 3\hat{i} - 2\hat{j}$, and (*b*) $\vec{A} = 1\hat{i} - 4\hat{j}$, $\vec{B} = 2\hat{i} + 6\hat{j}$.

49 • Describe the following vectors using the unit vectors $\hat{i}$ and $\hat{j}$: (*a*) a velocity of 10 m/s at an angle of elevation of 60°, (*b*) a vector $\vec{A}$ of magnitude $A = 5$ m and $\theta = 225°$, (*c*) a displacement from the origin to the point $x = 14$ m, $y = -6$ m.

50 • For the vector $\vec{A} = 3\hat{i} + 4\hat{j}$, find any three other vectors $\vec{B}$ that also lie in the xy plane and have the property that $A = B$ but $\vec{A} \neq \vec{B}$. Write these vectors in terms of their components and show them graphically.

51 •• [SSM] The faces of a cube with 3-m-long edges are parallel to the coordinate planes. The cube has one corner at the origin. A fly begins at the origin and walks along three edges until it is at the far corner. Write the displacement vector of the fly using the unit vectors $\hat{i}$, $\hat{j}$, and $\hat{k}$, and find the magnitude of this displacement.

52 • [SSM] A ship at sea receives radio signals from two transmitters A and B, which are 100 km apart, one due south of the other. The direction finder shows that transmitter A is $\theta = 30°$ south of east, while transmitter B is due east. Calculate the distance between the ship and transmitter B.

Velocity and Acceleration Vectors

53 • A stationary radar operator determines that a ship is 10 km south of him. An hour later the same ship is 20 km southeast. If the ship moved at constant speed and always in the same direction, what was its velocity during this time?

54 • A particle's position coordinates (x, y) are (2 m, 3 m) at $t = 0$; (6 m, 7 m) at $t = 2$ s; and (13 m, 14 m) at $t = 5$ s. (a) Find the average velocity v_{av} from $t = 0$ to $t = 2$ s. (b) Find v_{av} from $t = 0$ to $t = 5$ s.

55 • [SSM] [SOLVE✓] A particle moving at a velocity of 4.0 m/s in the positive x direction is given an acceleration of 3.0 m/s² in the positive y direction for 2.0 s. The final speed of the particle is (a) -2.0 m/s, (b) 7.2 m/s, (c) 6.0 m/s, (d) 10 m/s, (e) none of the above.

56 • Initially, a particle is moving due west with a speed of 40 m/s; 5 s later it is moving due north with a speed of 30 m/s. (a) What was the change in the magnitude of the particle's velocity during this time? (b) What was the change in the direction of the velocity? (c) What are the magnitude and direction of $\Delta\vec{v}$ for this interval? (d) What are the magnitude and direction of $\vec{a}_{av}$ for this interval?

57 • At $t = 0$, a particle located at the origin has a velocity of 40 m/s at $\theta = 45°$. At $t = 3$ s, the particle is at $x = 100$ m and $y = 80$ m with a velocity of 30 m/s at $\theta = 50°$. Calculate (a) the average velocity and (b) the average acceleration of the particle during this interval.

58 •• [SSM] [SOLVE✓] A particle moves in the xy plane with constant acceleration. At time zero, the particle is at $x = 4$ m, $y = 3$ m and has velocity $\vec{v} = (2 \text{ m/s})\hat{i} + (-9 \text{ m/s})\hat{j}$. The acceleration is given by $\vec{a} = (4 \text{ m/s}^2)\hat{i} + (3 \text{ m/s}^2)\hat{j}$. (a) Find the velocity at $t = 2$ s. (b) Find the position at $t = 4$ s. Give the magnitude and direction of the position vector.

59 •• [SOLVE] A particle has a position vector given by by $\vec{r} = (30t)\hat{i} + (40t - 5t^2)\hat{j}$, where r is in meters and t is in seconds. Find the instantaneous-velocity and instantaneous-acceleration vectors as functions of time t.

60 •• A particle has a constant acceleration of $\vec{a} = (6 \text{ m/s}^2)\hat{i} + (4 \text{ m/s}^2)\hat{j}$. At time $t = 0$, the velocity is zero and the position vector is $\vec{r}_0 = (10 \text{ m})\hat{i}$. (a) Find the velocity and position vectors at any time t. (b) Find the equation of the particle's path in the xy plane and sketch the path.

61 ••• [SOLVE] Starting from rest at a dock, a motor boat heads *north* while accelerating at a constant 3 m/s² for 20 s. The boat then turns *west* at the speed that it had at 20 s and travels west at this constant speed for 10 s. (a) What was the average velocity of the boat during the 30-s trip? (b) What was the average acceleration of the boat during the 30-s trip? (c) What is the displacement of the boat from the dock at the end of the 30-s trip?

62 ••• [SSM] Mary and Robert decide to rendezvous on Lake Michigan. Mary departs in her boat from Petoskey at 9:00 A.M. and travels due north at 8 mi/h. Robert leaves from his home on the shore of Beaver Island, 26 mi, 30° west of north of Petoskey, at 10:00 A.M. and travels at a constant speed of 6 mi/h. In what direction should Robert be heading to intercept Mary, and where and when will they meet?

Relative Velocity

63 •• A plane flies at an airspeed of 250 km/h. There is a wind blowing at 80 km/h in the northeast direction at exactly 45° to the east of north. (a) In what direction should the plane head in order to fly due north? (b) What is the speed of the plane relative to the ground?

64 •• [SOLVE✓] A swimmer heads directly across a river, swimming at 1.6 m/s relative to the water. She arrives at a point 40 m downstream from the point directly across the river, which is 80 m wide. (a) What is the speed of the river current? (b) What is the swimmer's speed relative to the shore? (c) In what direction should the swimmer head to arrive at the point directly opposite her starting point?

65 •• [SSM] A small plane departs from point A heading for an airport 520 km due north at point B. The airspeed of the plane is 240 km/h and there is a steady wind of 50 km/h blowing northwest to southeast. Determine the proper heading for the plane and the time of flight.

66 •• [SOLVE✓] Two boat landings are 2.0 km apart on the same bank of a stream that flows at 1.4 km/h. A motorboat makes the round trip between the two landings in 50 min. What is the speed of the boat relative to the water?

67 •• In radio-controlled model airplane competition each plane must fly from the center of a 1-km-radius circle to any point on the circle and back to the center. The winner is the plane with the shortest round-trip time. The contestants are free to fly their planes along any route so long as the plane begins at the center, travels to the circle, and then returns to the center. On the day of the race, a steady wind blows out of the north at 5 m/s. Your plane can maintain an airspeed of 15 m/s. Strategy is paramount. Should you fly your plane upwind on the first leg and downwind on the trip back, or across the wind flying east and then west? Optimize your chances by calculating the round-trip time for both routes.

68 • [SOLVE] The pilot of a small plane maintains an air speed of 150 kts (knots, or nautical miles per hour) and wants to fly due north (000°) with respect to the earth. If a wind of 30 kts is blowing *from* the east (090°), calculate the heading (azimuth) the pilot must take.

69 •• [SSM] Car A is traveling east at 20 m/s toward an intersection. As car A crosses the intersection, car B starts from rest 40 m north of the intersection and moves south with a constant acceleration of 2 m/s². Six seconds after A crosses the intersection find (a) the position of B relative to A, (b) the velocity of B relative to A, (c) the acceleration of B relative to A.

70 ••• **SSM** A tennis racket is held horizontally and a tennis ball is held above the racket. When the ball is dropped from rest, and bounces off the strings of the racket, the ball always rebounds to 64% of its initial height. (a) Express the speed of the tennis ball just after it bounces as some fraction of the speed of the ball just before the bounce. (b) The tennis ball is now thrown up into the air and served using the same racket. Assuming the ball's pre-impact speed is zero, and that the racket's speed through impact is 25 m/s, with what speed does the tennis ball come off the racket strings? *Hint: Using the results of part (a), solve for the post-impact speed of the ball in the reference frame of the racket, and then calculate the ball's speed in the reference frame of the earth.* (c) From some well-established laws of physics, we never see a ball bounce higher than the point from which it was released. From this, can you give an upper bound on the speed of a served tennis ball in relation to the speed of the racket, no matter how well the racket is designed? (We will see later that these results can be explained in a different context: the idea of *conservation of momentum*.)

Circular Motion and Centripetal Acceleration

71 • What is the acceleration of the extreme tip of the minute hand of the clock in Problem 38? Express it as a fraction of the magnitude of free-fall acceleration g.

72 • A centrifuge spins at a rate of 15,000 rev/min. (a) Calculate the centripetal acceleration of a test-tube sample held in the centrifuge arm 15 cm from the rotation axis. (b) It takes 1 min, 15 s for the centrifuge to spin up to its maximum rate of revolution from rest. Calculate the *magnitude* of the tangential acceleration of the centrifuge while it is spinning up, assuming that the tangential acceleration is constant.

73 • An object resting on the equator has an acceleration toward the center of the earth due to the earth's rotational motion about its axis, and an acceleration toward the sun due to the earth's orbital motion. Calculate the magnitudes of both of these accelerations, and express them as a fraction of the magnitude of free-fall acceleration g. Use values from the physical-data table in the textbook.

74 •• **SSM** Determine the acceleration of the moon toward the earth, using values for its mean distance and orbital period from the physical-data table in the textbook. Assume a circular orbit. Express the acceleration as a fraction of the magnitude of free-fall acceleration g.

75 • A boy whirls a ball on a string in a horizontal circle of radius 0.8 m. How many revolutions per minute does the ball make if the magnitude of its centripetal acceleration is g (the magnitude of the free-fall acceleration)?

Projectile Motion and Projectile Range

76 • A pitcher throws a fastball at 140 km/h (about 87 mi/h) toward home plate, which is 18.4 m away. Neglecting air resistance (not a good idea if you are the batter), how far does the ball drop because of gravity by the time it reaches home plate?

77 • A projectile is launched with speed v_0 at an angle of θ_0 with the horizontal. Find an expression for the maximum height it reaches above its starting point in terms of v_0, θ_0, and g.

78 •• **SSM** A cannonball is fired with initial speed v_0 at an angle 30° above the horizontal from a height of 40 m above the ground. The projectile strikes the ground with a speed of $1.2v_0$. Find v_0.

79 •• In Figure 3-48, if x is 50 m and $h = 10$ m, what is the minimum initial speed of the dart if it is to hit the monkey before hitting the ground, which is 11.2 m below the initial position of the monkey?

FIGURE 3-48 Problem 79

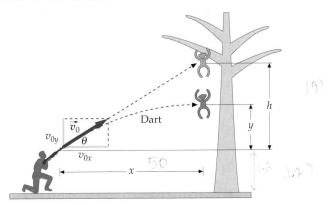

80 •• A projectile is fired with an initial speed of 53 m/s. Find the angle of projection such that the maximum height of the projectile is equal to its horizontal range.

81 •• A ball thrown into the air lands 40 m away 2.44 s later. Find the direction and magnitude of the initial velocity.

82 •• **SSM** **SOLVE** Consider a ball that is thrown with initial speed v_0 at an angle θ above the horizontal. If we consider its speed v at some height h above the ground, show that $v(h)$ is independent of θ.

83 •• At $\frac{1}{2}$ of its maximum height, the speed of a projectile is $\frac{3}{4}$ of its initial speed. What was its launch angle?

84 • **SOLVE** ✓ A cargo plane is flying horizontally at an altitude of 12 km with a speed of 900 km/h when a large crate falls out of the rear loading ramp. (a) How long does it take the crate to hit the ground? (b) How far horizontally is the crate from the point where it fell off when it hits the ground? (c) How far is the crate from the aircraft when the crate hits the ground, assuming that the plane continues to fly with constant velocity?

85 •• **SSM** **SOLVE** Wile E. Coyote (*Carnivorous hungribilous*) is chasing the Roadrunner (*Speedibus cantcatchmi*) yet again. While running down the road, they come to a deep gorge, 15 m straight across and 100 m deep. The Roadrunner launches itself across the gorge at a launch angle of 15° above the horizontal, and lands with 1.5 m to spare. (a) What was the Roadrunner's launch speed? Ignore air resistance. (b) Wiley Coyote launches himself across the gorge with the same initial speed, but at a different launch angle. To his horror, he is short the other lip by 0.5 m. What was his launch angle? (Assume that it was lower than 15°.)

86 • A cannon is elevated at an angle of 45°. It fires a ball with a speed of 300 m/s. (*a*) What height does the ball reach? (*b*) How long is the ball in the air? (*c*) What is the horizontal range of the cannon?

87 •• A stone thrown horizontally from the top of a 24-m tower hits the ground at a point 18 m from the base of the tower. (*a*) Find the speed with which the stone was thrown. (*b*) Find the speed of the stone just before it hits the ground.

88 •• [SOLVE] A projectile is fired into the air from the top of a 200-m cliff above a valley (Figure 3-49). Its initial velocity is 60 m/s at 60° above the horizontal. Where does the projectile land?

FIGURE 3-49 Problem 88

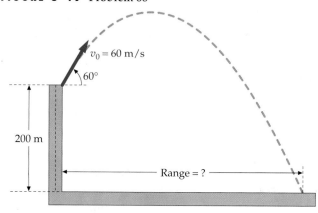

89 •• The range of a cannonball fired horizontally from a cliff is equal to the height of the cliff. What is the direction of the velocity vector when the projectile strikes the ground?

90 • A projectile is fired at an angle of 60° above the horizontal with an initial speed of 300 m/s. Calculate (*a*) the horizontal distance traveled and (*b*) the vertical height attained in the first 6 s.

91 •• A cannonball is fired with an initial speed of 42.2 m/s at an angle of 30° above the horizontal from an initial height of 40 m. Find the range of the cannonball.

92 •• [SSM] [SOLVE] The speed of an arrow fired from a compound bow is about 45 m/s. (*a*) A Tartar archer sits astride his horse and launches an arrow into the air, elevating the bow at an angle of 10° above the horizontal. If the bow is 2.25 m above the ground, what is the arrow's range? Assume that the ground is level, and ignore air resistance. (*b*) Now assume that his horse is at full gallop, moving in the same direction as he will fire the arrow, and that he elevates the bow in the same way as in part (*a*) and fires. If the horse's speed is 12 m/s, what is the arrow's range now?

93 • [SOLVE]✓ Using a potato cannon, Chuck launches a spud-plug horizontally with an initial velocity of 50.0 m/s (about 112 mi/h). (*a*) If Chuck holds the potato cannon so that it is 1.00 m above ground, how long is the spud-plug in the air? (*b*) How far does the plug travel before hitting the ground? (You can find more about potato cannons on the Internet.)

94 •• Compute $dR/d\theta$ from $R = (v_0^2/g)\sin(2\theta_0)$ and show that setting $dR/d\theta = 0$ gives $\theta = 45°$ for the maximum range.

95 • [SSM] In a science fiction short story written in the 1970s, Ben Bova described a conflict between two hypothetical colonies on the moon—one founded by the United States and the other by the USSR. In the story, colonists from each side started firing bullets at each other, only to find to their horror that their rifles had a high enough muzzle velocity that the bullets went into orbit. (*a*) If the magnitude of free-fall acceleration on the moon is 1.67 m/s², what is the maximum range of a rifle bullet with a muzzle velocity of 900 m/s? (Assume the curvature at the surface of the moon is negligible.) (*b*) What would the muzzle velocity have to be to send the bullet into a circular orbit just above the surface of the moon? (You will first need to look up the radius of the moon.)

96 ••• In the text, we calculated the range for a projectile that lands at the same elevation from which it is fired as $R = (v_0^2/g)\sin 2\theta_0$. Show that the change in the range for a small change in free-fall acceleration is given by $\Delta R/R = -\Delta g/g$.

97 ••• In the text, we calculated the range for a projectile that lands at the same elevation from which it is fired as $R = (v_0^2/g)\sin 2\theta_0$. Show that the change in the range for a small change in launch velocity is given by $\Delta R/R = 2\Delta v_0/v_0$.

98 ••• In the text, we calculated the range for a projectile that lands at the same elevation from which it is fired as $R = (v_0^2/g)\sin 2\theta_0$. Show that the range for the more general problem (Figure 3-50) where $\Delta y \neq 0$ is given by

FIGURE 3-50 Problem 98

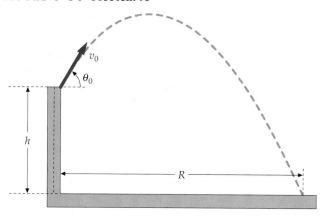

$$R = \left(1 + \sqrt{1 + \frac{2gh}{v_0^2 \sin^2\theta_0}}\right) \frac{v_0^2}{2g} \sin 2\theta_0$$

99 •• [SSM] A projectile is launched over level ground at an elevation angle of θ. An observer standing at the launch site sights the projectile at the point of its highest elevation, and measures the angle ϕ shown in Figure 3-51. Show that $\tan \phi = \frac{1}{2} \tan \theta$.

FIGURE 3-51 Problem 99

100 • A projectile, fired with unknown initial velocity, lands 20 s later on the side of a hill, 3000 m away horizontally and 450 m vertically above its starting point. (*a*) What is the vertical component of its initial velocity? (*b*) What is the horizontal component of its initial velocity?

101 •• **SSM** A stone is thrown horizontally from the top of an incline that makes an angle θ with the horizontal. If the stone's initial speed is v_0, how far down the incline will it land?

102 ••• A toy cannon is placed on a ramp that has a slope of angle ϕ. (*a*) If the cannonball is projected up the hill at an angle of θ_0 above the horizontal (Figure 3-52) and has a muzzle speed of v_0, show that the range R of the cannonball (as measured along the ramp) is given by

$$R = \frac{2v_0^2 \cos^2 \theta_0 \left(\tan \theta_0 - \tan \phi\right)}{g \cos \phi}$$

Ignore air resistance.

FIGURE 3-52 Problem 102

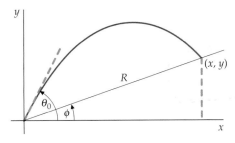

103 •• A rock is thrown from the top of a 20-m building at an angle of 53° above the horizontal. If the horizontal range of the throw is equal to the height of the building, with what speed was the rock thrown? What is the velocity of the rock just before it strikes the ground?

104 •• A girl throws a ball at a vertical wall 4 m away (Figure 3-53). The ball is 2 m above ground when it leaves the girl's hand with an initial velocity of $\vec{v}_0 = (10 \text{ m/s})(\hat{i} + \hat{j})$ or $10\sqrt{2}$ m/s at 45°. When the ball hits the wall, the horizontal component of its velocity is reversed; the vertical component remains unchanged. Where does the ball hit the ground? *Hint: The wall can be thought of as a mirror. Determine the range, neglecting the wall, and then consider the mirror-like reflection.*

FIGURE 3-53 Problem 104

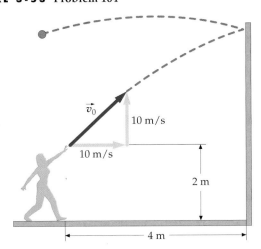

Hitting Targets and Related Problems

105 • A boy uses a slingshot to project a pebble at a shoulder-height target 40 m away. He finds that to hit the target he must aim 4.85 m above the target. Determine the velocity of the pebble on leaving the slingshot and the time of flight of the pebble.

106 •• **SSM** The distance from the pitcher's mound to home plate is 18.4 m. The mound is 0.2 m above the level of the field. A pitcher throws a fastball with an initial speed of 37.5 m/s. At the moment the ball leaves the pitcher's hand, it is 2.3 m above the mound. What should the angle between $\vec{v}_0$ and the horizontal be so that the ball crosses the plate 0.7 m above ground? (Neglect interaction with air.)

107 •• Suppose that a hockey puck is struck in such a way that, when it is at its highest point, it just clears a Plexiglass wall of height $h = 2.80$ m. Find v_{0y}, the time t to reach the wall, and v_{0x}, v_0, and θ_0 for this case. Assume that the horizontal distance is $x_1 = 12.0$ m.

108 •• Carlos is on his trail bike, approaching a creek bed that is 7 m wide. A ramp with an incline of 10° has been built for daring people who try to jump the creek. Carlos is traveling at his bike's maximum speed, 40 km/h. (*a*) Should Carlos attempt the jump or emphatically hit the brakes? (*b*) What is the minimum speed a bike must have to make this jump? Assume equal elevations on either side of the creek.

109 •• If a bullet that leaves the muzzle of a gun at 250 m/s is to hit a target 100 m away at the level of the muzzle, the gun must be aimed at a point above the target. How far above the target is that point?

General Problems

110 • The displacement vectors $\vec{A}$ and $\vec{B}$ in Figure 3-54 both have a magnitude of 1 m. (*a*) Find their x and y components. (*b*) Find the components, magnitude, and direction of the sum $\vec{S} = \vec{A} + \vec{B}$. (*c*) Find the components, magnitude, and direction of the difference $\vec{D} = \vec{A} - \vec{B}$.

FIGURE 3-54 Problem 110

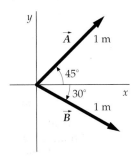

111 • **SSM** A plane is inclined at an angle of 30° from the horizontal. Choose the x axis pointing down the slope of the plane and the y axis perpendicular to the plane. Find the x and y components of the acceleration of gravity, which has the magnitude 9.81 m/s² and points vertically down.

112 • Two vectors $\vec{A}$ and $\vec{B}$ lie in the xy plane. Under what conditions does the ratio A/B equal A_x/B_x?

113 • The position vector of a particle is given by $\vec{r} = (5\,\text{m/s})t\hat{i} + (10\,\text{m/s})t\hat{j}$, where t is in seconds and $\vec{r}$ is in meters. (a) Draw the path of the particle in the xy plane. (b) Find $\vec{v}$ in component form and then find its magnitude.

114 •• [SOLVE] A worker on the roof of a house drops her hammer, which slides down the roof at constant speed of 4 m/s. The roof makes an angle of 30° with the horizontal, and its lowest point is 10 m from the ground. What is the horizontal distance traveled by the hammer between the time it leaves the roof of the house and the time it hits the ground?

115 •• In 1940, Emanuel Zacchini flew about 53 m as a human cannonball, a record that remains unbroken. His initial velocity was 24.2 m/s at an angle θ. Find θ and the maximum height h Emanuel achieved during the record flight. Ignore the effects of air resistance.

116 •• A particle moves in the xy plane with constant acceleration. At $t = 0$ the particle is at $\vec{r}_1 = (4\,\text{m})\hat{i} + (3\,\text{m})\hat{j}$, with velocity $\vec{v}_1$. At $t = 2$ s the particle has moved to $\vec{r}_2 = (10\,\text{m})\hat{i} - (2\,\text{m})\hat{j}$ and its velocity has changed to $\vec{v}_2 = (5\,\text{m/s})\hat{i} - (6\,\text{m/s})\hat{j}$. (a) Find $\vec{v}_1$. (b) What is the acceleration of the particle? (c) What is the velocity of the particle as a function of time? (d) What is the position vector of the particle as a function of time?

117 •• [SSM] A small steel ball is projected horizontally off the top landing of a long, rectangular staircase. The initial speed of the ball is 3 m/s. Each step is 0.18 m high and 0.3 m wide. Which step does the ball strike first?

118 •• Suppose you can throw a ball a distance x_0 when standing on level ground. How far can you throw it from a building of height $h = x_0$ if you throw it at (a) 0°? (b) 30°? (c) 45°?

119 ••• Darlene is a stunt motorcyclist in a traveling circus. For the climax of her show, she takes off from the ramp at angle θ, clears a fiery ditch of width x, and lands on an elevated platform (height h) on the other side (Figure 3-55). (a) For a given height h, find the minimum necessary takeoff speed $v_{\min}$ needed to make the jump successfully. (b) What is $v_{\min}$ for a launch angle $\theta = 30°$ with a pit width of 8 m and a platform height $h = 4$ m? (c) Show that no matter what her takeoff speed is, the maximum height of the platform is $h_{\max} < x \tan \theta$. Interpret this result physically. (Neglect the effects of air resistance and treat the bike as if it were a particle.)

FIGURE 3-55 Problem 119

120 ••• A small boat is headed for a harbor 32 km northwest of its current position when it is suddenly engulfed in heavy fog. The captain maintains a compass bearing of northwest and a speed of 10 km/h relative to the water. The fog lifts 3 h later and the captain notes that he is now exactly 4.0 km south of the harbor. (a) What was the average velocity of the current during those 3 h? (b) In what direction should the boat have been heading to reach its destination along a straight course? (c) What would its travel time have been if it had followed a straight course?

121 •• [SSM] Galileo showed that, if air resistance is neglected, the ranges for projectiles whose angles of projection exceed or fall short of 45° by the same amount are equal. Prove Galileo's result.

122 •• Two balls are thrown with equal speeds from the top of a cliff of height h. One ball is thrown at an angle of α above the horizontal. The other ball is thrown at an angle of β below the horizontal. Show that each ball strikes the ground with the same speed, and find that speed in terms of h and the initial speed v_0.

Newton's Laws

THIS AIRPLANE IS ACCELERATING AS IT HEADS DOWN THE RUNWAY BEFORE TAKEOFF. NEWTON'S LAWS RELATE AN OBJECT'S ACCELERATION TO ITS MASS AND THE FORCES ACTING ON IT.

? **If you were a passenger on this plane, how might you use Newton's laws to determine the plane's acceleration? (See Example 4-9.)**

N ow that we have studied how objects move in one, two, and three dimensions, we can ask the question, "why do objects start to move?" What causes a moving object to gain speed or change direction?

Classical mechanics relates the forces objects exert on each other, and relates changes in the motion of an object to the forces that act on it. It describes phenomena using Newton's three laws of motion. While we may already have an intuitive idea of a force as a push or a pull, like that exerted by our muscles or by stretched rubber bands and springs, Newton's laws allow us to refine our understanding of forces.

➤ **In this chapter, we describe Newton's three laws of motion and begin using them to solve problems involving objects in motion and at rest.**

A modern wording of Newton's laws is:

First law. An object at rest stays at rest *unless* acted on by an external force. An object in motion continues to travel with constant velocity *unless* acted on by an external force.

Second law. The direction of the acceleration of an object is in the direction of the net external force acting on it. The acceleration is proportional to the net external force $\vec{F}_{net}$, in accordance with $\vec{F}_{net} = m\vec{a}$, where m is the mass of the object. The net force acting on an object, also called the resultant force, is the vector sum of all the forces acting on it: $\vec{F}_{net} = \Sigma\vec{F}$. Thus,

$$\Sigma\vec{F} = m\vec{a}$$

4-1

NEWTON'S SECOND LAW

Third law. Forces always occur in equal and opposite pairs. If object A exerts a force $\vec{F}_{A,B}$ on object B, an equal but opposite force $\vec{F}_{B,A}$ is exerted by object B on object A. Thus,

$$\vec{F}_{B,A} = -\vec{F}_{A,B}$$

4-2

NEWTON'S THIRD LAW

4-1 Newton's First Law: The Law of Inertia

Friction is greatly reduced by a cushion of air that supports the hovercraft.

Push a piece of ice on a counter top: It slides, then stops. If the counter is wet, the ice will travel farther before stopping. A piece of dry ice (frozen carbon dioxide) riding on a cushion of carbon dioxide vapor slides quite far with little change in velocity. Before Galileo it was thought that a force, such as a push or pull, was always needed to keep an object moving with constant velocity. But Galileo, and later Newton, recognized that the slowing of objects in everyday experience is due to the force of friction. If friction is reduced, the rate of slowing is reduced. A water slick or a cushion of gas is especially effective at reducing friction, allowing the object to slide a great distance with little change in velocity. Galileo reasoned that, if we could remove from an object all external forces including friction, then the velocity of the object would never change—a property of matter he described as **inertia**. This conclusion, restated by Newton as his first law, is also called the **law of inertia.**

Inertial Reference Frames

Newton's first law makes no distinction between an object at rest and an object moving with constant (nonzero) velocity. Whether an object remains at rest or remains moving with constant velocity depends on the reference frame in which the object is observed. Suppose you are a passenger on an airplane that is flying along a straight path at constant altitude and you carefully place a small ball on your seat tray (which is horizontal). Relative to the plane, the ball will remain at rest as long as the plane continues to fly at constant velocity relative to the ground. Relative to the ground, the ball remains moving with the same velocity as the plane.

Now, suppose that the pilot opens the throttle and the plane suddenly accelerates forward (relative to the ground). You will then observe that the ball on your tray suddenly starts to roll backward, accelerating (relative to the plane) even though there is no horizontal force acting on it.

A reference frame accelerating relative to an inertial reference frame is not an inertial reference frame. *Newton's first law thus gives us the criterion for determining if a reference frame is an inertial frame.* In fact, it is useful to think of Newton's first law as a statement that defines inertial reference frames.

> If no forces act on an object, any reference frame with respect to which the acceleration of the object remains zero is an **inertial reference frame.**

DEFINITION—INERTIAL REFERENCE FRAME

Both the plane, when cruising at constant velocity, and the ground are, to a good approximation, inertial reference frames. Any reference frame moving with constant velocity relative to an inertial reference frame is also an inertial reference frame.

A reference frame attached to the surface of the earth is not quite an inertial reference frame because of the small acceleration of the surface of the earth due to the rotation of the earth and the small acceleration of the earth itself due to its revolution around the sun. However, these accelerations are of the order of 0.01 m/s^2 or less, so to a good approximation, a reference frame attached to the surface of the earth is an inertial reference frame.

The concept of inertial reference frame is of central importance because *Newton's first, second, and third law statements are valid only in inertial reference frames.*

4-2   Force, Mass, and Newton's Second Law

Newton's first and second laws allow us to define force. A **force** is an external influence on an object that causes it to accelerate relative to an inertial reference frame. (We assume there are no other forces acting.) The direction of the force is the direction of the acceleration it causes. The magnitude of the force is the product of the mass of the object and the magnitude of its acceleration. This definition is given in Equation 4-1.

Forces can be compared by stretching identical rubber bands. For example, if two identical rubber bands are stretched by the same amount, then they exert forces of equal magnitudes.

Objects intrinsically resist being accelerated. Imagine kicking a soccer ball or a bowling ball. The bowling ball resists being accelerated much more than does the soccer ball, as would be evidenced by your sore toes. This intrinsic property is called the object's **mass.** It is a measure of the object's inertia. The ratio of two masses is defined quantitatively by applying the same force to each and comparing their accelerations. If a force F produces acceleration a_1 when applied to an object of mass m_1, and an equal force produces acceleration a_2 when applied to an object of mass m_2, then the ratio of the two masses is defined by

$$\frac{m_2}{m_1} = \frac{a_1}{a_2}$$

4-3

DEFINITION—MASS

This definition agrees with our intuitive idea of mass. If a force is applied to an object and a force of equal magnitude is applied to a second object, then the object with more mass will accelerate less. The ratio a_1/a_2 produced by forces of equal magnitude acting on two objects is found experimentally to be independent of the magnitude, direction, or type of force used. Mass is an intrinsic property of an object that does not depend on its location—it remains the same whether the object is on the earth, on the moon, or in outer space.

If a direct comparison shows that $m_2/m_1 = 2$ and $m_3/m_1 = 4$, then m_3 will be twice m_2 when objects 2 and 3 are compared with each other. We can therefore establish a mass scale by choosing a standard object and assigning it a mass of 1 unit. As we noted in Chapter 1, the object chosen as the international standard for mass is a platinum-iridium alloy cylinder carefully preserved at the International Bureau of Weights and Measures at Sèvres, France. The mass of the standard object is 1 **kilogram** (kg), the SI unit of mass. The force required to produce an acceleration of 1 m/s² on the standard object is defined to be 1 **newton** (N). The force that produces an acceleration of 2 m/s² on the standard object is 2 N, and so on.

A SLIDING ICE CREAM CARTON   **EXAMPLE 4 - 1**

A given force produces an acceleration of 5 m/s² on the standard object of mass m_1. When an equal force is applied to a carton of ice cream of mass m_2, it produces an acceleration of 11 m/s². (a) What is the mass of the carton of ice cream? (b) What is the magnitude of the force?

PICTURE THE PROBLEM Apply $\Sigma\vec{F} = m\vec{a}$ to each object and solve for the mass of the ice-cream carton and the magnitude of the force.

(a) 1. Apply $\Sigma\vec{F} = m\vec{a}$ to each object. There is only one force and we only need to consider magnitudes of the vector quantities:

$$F_1 = m_1 a_1 \quad \text{and} \quad F_2 = m_2 a_2$$

2. The ratio of the masses varies inversely as the ratio of the accelerations under applied forces of equal magnitude:

$$F_1 = F_2 = F, \quad \text{so} \quad m_1 a_1 = m_2 a_2$$

and

$$\frac{m_2}{m_1} = \frac{a_1}{a_2} = \frac{5 \text{ m/s}^2}{11 \text{ m/s}^2}$$

3. Solve for m_2 in terms of m_1, which is 1 kg:

$$m_2 = \frac{5}{11} m_1 = \frac{5}{11}(1 \text{ kg}) = \boxed{0.45 \text{ kg}}$$

(b) The magnitude F is found by using the mass and acceleration of either object:

$$F = m_1 a_1 = (1 \text{ kg})(5 \text{ m/s}^2) = \boxed{5 \text{ N}}$$

EXERCISE A force of 3 N produces an acceleration of 2 m/s² on an object of unknown mass. (a) What is the mass of the object? (b) If the force is increased to 4 N, what is the acceleration? (*Answer* (a) 1.5 kg (b) 2.67 m/s²)

It is found experimentally that two or more forces acting on an object accelerate it as if the object were acted upon by a single force equal to the vector sum of the individual forces. That is, forces combine as vectors. Newton's second law is thus

$$\Sigma\vec{F} = \vec{F}_{\text{net}} = m\vec{a}$$

A WALK IN SPACE   **EXAMPLE 4 - 2**

You're stranded in space away from your spaceship. Fortunately, you have a propulsion unit that provides a constant force $\vec{F}$ for 3 s. After 3 s you have moved 2.25 m. If your mass is 68 kg, find $\vec{F}$.

The propulsion unit (not shown) is pushing the astronaut to the right.

PICTURE THE PROBLEM The force acting on you is constant, so your acceleration $\vec{a}$ is also constant. We can use the kinematic equations of Chapter 2 to find
$\vec{a}$, and then obtain the force from $\Sigma\vec{F} = m\vec{a}$. Choose $\vec{F}$ to be along the x axis, so
that $\vec{F} = F_x\hat{i}$ (Figure 4-1). The component of Newton's second law along the x axis
is then $F_x = ma_x$.

FIGURE 4-1

1. Apply $\Sigma\vec{F} = m\vec{a}$ to relate the net force to the mass and $F_x = ma_x$
   the acceleration:

2. To find the acceleration, we use Equation 2-15 with $v_0 = 0$: $\Delta x = v_0 t + \frac{1}{2}a_x t^2 = \frac{1}{2}a_x t^2$

$$a_x = \frac{2\Delta x}{t^2} = \frac{2(2.25 \text{ m})}{(3 \text{ s})^2} = 0.500 \text{ m/s}^2$$

3. Substitute $a_x = 0.500$ m/s^2 and $m = 68$ kg to find the $F_x = ma_x = (68 \text{ kg})(0.500 \text{ m/s}^2) = \boxed{34.0 \text{ N}}$
   force:

A PARTICLE SUBJECTED TO FORCES   **E X A M P L E   4 - 3**   **Try It Yourself**

**A particle of mass 0.4 kg is subjected simultaneously to two forces $\vec{F}_1 =$
$-2 \text{ N }\hat{i} - 4 \text{ N }\hat{j}$ and $\vec{F}_2 = -2.6 \text{ N }\hat{i} + 5 \text{ N }\hat{j}$. If the particle is at the origin and
starts from rest at $t = 0$, find (a) its position vector $\vec{r}$ and (b) its velocity $\vec{v}$ at
$t = 1.6$ s.**

PICTURE THE PROBLEM Since $\vec{F}_1$ and $\vec{F}_2$ are constant, the acceleration of the
particle is constant. Hence, we use the kinematic equations of Chapter 2 to determine the particle's position and velocity as functions of time.

Cover the column to the right and try these on your own before looking at the answers.

Steps   **Answers**

(a) 1. Write the general equation for the position vector $\vec{r}$ $\vec{r} = \vec{r}_0 + \vec{v}_0 t + \frac{1}{2}\vec{a}t^2 = \frac{1}{2}\vec{a}t^2$
   as a function of time t for constant acceleration $\vec{a}$ in
   terms of $\vec{r}_0$, $\vec{v}_0$, and $\vec{a}$, and substitute $\vec{r}_0 = \vec{v}_0 = 0$.

2. Use $\Sigma\vec{F} = m\vec{a}$ to write the acceleration $\vec{a}$ in terms of $\vec{a} = \dfrac{\Sigma\vec{F}}{m}$
   the resultant force $\Sigma\vec{F}$ and the mass m.

3. Compute $\Sigma\vec{F}$ from the given forces. $\Sigma\vec{F} = \vec{F}_1 + \vec{F}_2 = -4.6 \text{ N }\hat{i} + 1.0 \text{ N }\hat{j}$

4. Find the acceleration vector $\vec{a}$. $\vec{a} = \dfrac{\Sigma\vec{F}}{m} = -11.5 \text{ m/s}^2\,\hat{i} + 2.5 \text{ m/s}^2\,\hat{j}$

5. Find the position vector $\vec{r}$ for a general time t. $\vec{r} = \frac{1}{2}\vec{a}t^2 = \frac{1}{2}a_x t^2\hat{i} + \frac{1}{2}a_y t^2\hat{j}$

$= -5.75 \text{ m/s}^2\,t^2\hat{i} + 1.25 \text{ m/s}^2\,t^2\hat{j}$

6. Find $\vec{r}$ at $t = 1.6$ s. $\vec{r} = \boxed{-14.7 \text{ m }\hat{i} + 3.20 \text{ m }\hat{j}}$

(b) Write the velocity vector $\vec{v}$ in terms of the accelera- $\vec{v} = \vec{a}t = (-11.5 \text{ m/s}^2\,\hat{i} + 2.5 \text{ m/s}^2\,\hat{j})t$
   tion and time and compute its components for the time
   $t = 1.6$ s. $= \boxed{-18.4 \text{ m/s }\hat{i} + 4.00 \text{ m/s }\hat{j}}$

4-3 The Force Due to Gravity: Weight

If we drop an object near the earth's surface, it accelerates toward the earth. If we neglect air resistance, all objects have the same acceleration, called the free-fall acceleration $\vec{g}$, at any location. The force causing this acceleration is the gravitational force on the object, called its weight.[†] If its weight $\vec{w}$ is the *only* force acting on an object, the object is said to be in **free-fall.** If its mass is m, Newton's second law ($\Sigma\vec{F} = m\vec{a}$) defines the weight $\vec{w}$:

$$\vec{w} = m\vec{g} \qquad\qquad\qquad 4\text{-}4$$

WEIGHT

Since $\vec{g}$ is the same for all objects, it follows that the weight of an object is proportional to its mass. The vector $\vec{g}$ is the force per unit mass exerted by the earth on any object and is called the **gravitational field** of the earth. It is equal to the free-fall acceleration.[‡] Near the surface of the earth, g has the value

$$g = 9.81 \text{ N/kg} = 9.81 \text{ m/s}^2$$

Careful measurements show that $\vec{g}$ varies with location. In particular, at points above the surface of the earth, $\vec{g}$ points toward the center of the earth and varies inversely with the square of the distance to the center of the earth. Thus an object weighs slightly less at very high altitudes than it does at sea level. The gravitational field also varies slightly with latitude because the earth is not exactly spherical but is slightly flattened at the poles. **Thus weight, unlike mass, is not an intrinsic property of an object.** Although the weight of an object varies from place to place because of changes in g, this variation is too small to be noticed in most practical applications on or near the surface of the earth.

An example should help clarify the difference between mass and weight. Consider a bowling ball near the moon. Its weight is the gravitational force exerted on it by the moon, but that force is a mere sixth of the force exerted on the bowling ball when it is similarly positioned on earth. The ball weighs about one-sixth as much on the moon, and lifting the ball on the moon requires one-sixth the force. However, because the mass of the ball is the same on the moon as on the earth, throwing the ball with some horizontal acceleration requires the same force on the moon as on the earth or in free space.

Although the weight of an object may vary from one place to another, at any particular location the weight of the object is proportional to its mass. Thus we can conveniently compare the masses of two objects at a given location by comparing their weights.

Our sensation of our own weight comes from other forces that balance it. When you sit on a chair, you feel a force exerted by the chair that balances your weight and prevents you from falling to the floor. When you stand on a spring scale, your feet feel the force exerted by the scale. The scale is calibrated to read the force it must exert (by the compression of its springs) to balance your weight. This force is called your **apparent weight.** If there is no force to balance your weight, as in free-fall, your apparent weight is zero. This condition, called **weightlessness,** is experienced by astronauts in orbiting satellites. A satellite in a circular orbit near the surface of the earth is accelerating toward the earth. The only force acting on the satellite is that of gravity (its weight), so it is in free-fall.

---

† Referring to the gravitation force as "its weight" is unfortunate because it appears to imply that "its weight" is a property of the object rather than an external force acting on it. To avoid buying into the apparent implication of the wording, every time you read "its weight," mentally translate it to read "the gravitational force acting on it."
‡ $\vec{g}$ refers to the free-fall acceleration, which is the acceleration (of an object in free-fall) relative to the ground. It is not entirely correct to attribute it to the gravitational attraction of the earth. This distinction is discussed further in Chapter 11.

Astronauts in the satellite are also in free-fall. The only force on them is their weight, which produces the acceleration g. Because there is no force balancing the force of gravity, the astronauts have zero apparent weight.

Units of Force and Mass

Like the second and the meter, the SI unit of mass, the kilogram, is a fundamental unit. The unit of force, the newton, and the units for other quantities that we will study such as momentum and energy are derived from the three fundamental units s, m, and kg.

As noted in Section 4-2, the newton is defined as the force that produces an acceleration of 1 m/s² when it acts on an object with a mass of 1 kg. Then Newton's second law gives

$$1\,N = (1\,kg)(1\,m/s^2) = 1\,kg{\cdot}m/s^2 \qquad 4\text{-}5$$

A convenient standard unit for mass in atomic and nuclear physics is the **unified mass unit** (u), which is defined as one-twelfth of the mass of the neutral carbon-12 (^{12}C) atom. The unified mass unit is related to the kilogram by

$$1\,u = 1.660\,540 \times 10^{-27}\,kg \qquad 4\text{-}6$$

The mass of a hydrogen atom is approximately 1 u.

Although we will generally use SI units in this book, we need to know another scheme, the U.S. customary system, still used in the United States, which is based on the second, the foot, and the pound. The U.S. customary system differs from SI in that a unit of force, the pound, has been chosen as a fundamental unit rather than a unit of mass. The **pound** was originally defined as the weight of a particular standard object at a particular location. It is now defined as 4.448222 N. Rounding to three places, we have 1 lb ≈ 4.45 N. Since 1 kg weighs 9.81 N, its weight in pounds is

$$9.81\,N = 2.20\,lb \qquad 4\text{-}7$$

WEIGHT OF 1 KG

The unit of mass in the U.S. customary system is the rarely encountered slug, defined as the mass of an object that weighs 32.2 lb. When working problems in the U.S. customary system, we substitute w/g for mass m, where w is the weight in pounds and g is the acceleration due to gravity in feet per second per second:

$$g = 32.2\,ft/s^2 \qquad 4\text{-}8$$

AN ACCELERATING STUDENT   **E X A M P L E   4 - 4**

The net force acting on a 130-lb student is 25 lb. What is her acceleration?

PICTURE THE PROBLEM Apply $\Sigma F = ma$ and solve for the acceleration. The mass can be found from the student's weight.

According to Newton's second law, the student's acceleration is the force divided by her mass:

$$a = \frac{F}{m} = \frac{F}{w/g} = \frac{25\,lb}{(130\,lb)/(32.2\,ft/s^2)}$$

$$= \boxed{6.19\,ft/s^2}$$

EXERCISE What force is needed to give an acceleration of 3 ft/s² to a 5-lb block? (*Answer* 0.466 lb)

4-4 Forces in Nature

The full power of Newton's second law emerges when it is combined with the force laws that describe the interactions of objects. For example, Newton's law for gravitation, which we study in Chapter 11, gives the gravitational force exerted by one object on another in terms of the distance between the objects and the masses of each. This, combined with Newton's second law, enables us to calculate the orbits of planets around the sun, the motion of the moon, and variations with altitude of g, the gravitational field strength.

(a)

The Fundamental Forces

All the different forces observed in nature can be explained in terms of four basic interactions that occur between elementary particles (see Figure 4.2):

1. The gravitational force—the force of mutual attraction between objects
2. The electromagnetic force—the force between electric charges
3. The strong nuclear force—the force between subatomic particles
4. The weak nuclear force—the force between subatomic particles during certain radioactive decay processes

The everyday forces that we observe between macroscopic objects are due to either the gravitational force or the electromagnetic force.

(b)

Action at a Distance

The fundamental forces of gravity and electromagnetism act between particles that are separated in space. This creates a philosophical problem referred to as **action at a distance.** Newton perceived action at

(c)

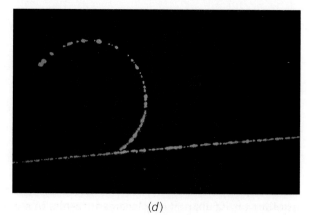

(d)

FIGURE 4-2 (*a*) The gravitational force between the earth and an object near the earth's surface is the weight of the object. The gravitational force exerted by the sun on the earth and the other planets is responsible for keeping the planets in their orbits around the sun. Similarly, the gravitational force exerted by the earth on the moon keeps the moon in its nearly circular orbit around the earth. The gravitational forces exerted by the moon and the sun on the oceans of the earth are responsible for the tides. Mont-Saint-Michel, France, shown in the photo, is an island when the tide is in. (*b*) The electromagnetic force includes both the electric and the magnetic forces. A familiar example of the electric force is the attraction between small bits of paper and a comb that is electrified after being run through hair. The lightning bolts above the Kitt Peak National Observatory, shown in the photo, are the result of the electromagnetic force. (*c*) The strong nuclear force occurs between elementary particles called hadrons, which include protons and neutrons, the constituents of atomic nuclei. This force results from the interaction of quarks, which are the building blocks of hadrons, and is responsible for holding nuclei together. The hydrogen bomb explosion shown here illustrates the strong nuclear force. (*d*) The weak nuclear force occurs between leptons (which include electrons and muons) and between hadrons (which include protons and neutrons). This false-color cloud chamber photograph illustrates the weak interaction between a cosmic ray muon (green) and an electron (red) knocked out of an atom.

a distance as a flaw in his theory of gravitation but avoided giving any other hypothesis. Today the problem is avoided by introducing the concept of a field, which acts as an intermediary agent. For example, we consider the attraction of the earth by the sun in two steps. The sun creates a condition in space that we call the gravitational field. This field then exerts a force on the earth. Similarly, the earth produces a gravitational field that exerts a force on the sun. Your weight is the force exerted by the gravitational field of the earth on you. When we study electricity and magnetism (Chapters 21–30) we will study electric fields, which are produced by electrical charges, and magnetic fields, which are produced by electrical charges in motion.

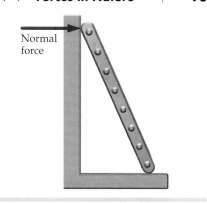

FIGURE 4-3 The wall supports the ladder by pushing on the ladder with a force normal to the wall.

Contact Forces

Many forces we encounter are exerted by objects in direct contact—they touch. These forces are electromagnetic in origin. They are exerted between the surface molecules of the objects in contact.

Solids If a surface is pushed against, it pushes back. Consider the ladder leaning against a wall shown in Figure 4-3. At the region of contact, the ladder pushes against the wall with a horizontal force, compressing the molecules in the surface of the wall. Like mattress springs, the compressed molecules in the wall push back on the ladder with a horizontal force. Such a force, *perpendicular* to the contacting surfaces, is called a **normal force** (the word *normal* means perpendicular). The wall bends slightly in response to a load, though this is rarely noticeable to the naked eye.

FIGURE 4-4 The frictional force exerted by the floor on the block opposes its sliding motion or its tendency to slide.

Normal forces can vary over a wide range of magnitudes. A table, for instance, will exert an upward normal force on any object resting on it. As long as the table doesn't break, this normal force will balance the downward weight force on the object. Furthermore, if you press down on the object, the upward normal force exerted by the table will increase, countering the extra force, thus preventing the object from accelerating downward.

Surfaces in contact can also exert forces on each other that are *parallel* to the contacting surfaces. Consider the large block on the floor shown in Figure 4-4. If the block is pushed sideways with a gentle enough force, it will not slide. The surface of the floor exerts a force back on the block, opposing its tendency to slide in the direction of the push. However, if the block is pushed sideways with a sufficiently hard force, it will start to slide. To keep it sliding it is necessary to continue to push it. If the sideways push is not sustained, the contact force will slow the motion of the box until it stops. A component of a contact force that opposes sliding, or the tendency to slide, is called a **frictional force;** it acts parallel to the contacting surfaces.

Although the frictional and normal forces are shown in diagrams as if they act at a single point, they are, in reality, distributed over the entire region of contact. Frictional forces are treated in more depth in Chapter 5.

Springs When a spring is compressed or extended by a small amount Δx, the force it exerts is found experimentally to be

$$F_x = -k\,\Delta x \qquad\qquad 4\text{-}9$$

HOOKE'S LAW

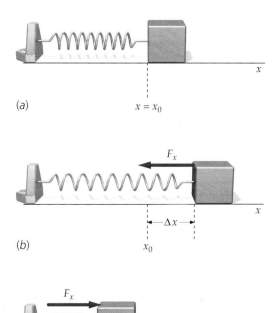

FIGURE 4-5 A horizontal spring. (*a*) When the spring is unstretched, it exerts no force on the block. (*b*) When the spring is stretched so that Δx is positive, it exerts a force of magnitude $k\,\Delta x$ in the negative x direction. (*c*) When the spring is compressed so that Δx is negative, the spring exerts a force of magnitude $k\,|\Delta x|$ in the positive direction.

where k is the force constant, a measure of the stiffness of the spring (Figure 4-5). The negative sign in Equation 4-9 signifies that when the spring is stretched or compressed, the force it exerts back is in the opposite direction. This relation, known as Hooke's law, turns out to be quite important. An object at rest under the influence of forces that balance is said to be in static equilibrium. If a small displacement results in a net restoring force toward the equilibrium position, the equilibrium is called stable equilibrium. For small displacements, nearly all restoring forces obey Hooke's law.

The molecular force of attraction between atoms in a molecule or solid varies approximately linearly with the change in separation (for small changes); the force varies much like that of a spring. We can therefore use two masses on a spring to model a diatomic molecule, or a set of masses connected by springs to model a solid as shown in Figure 4-6.

(a)   (b)

FIGURE 4-6 (a) Model of a solid consisting of atoms connected to each other by springs. The springs are very stiff (large force constant) so that when a weight is placed on the solid its deformation is not visible. However, compression such as that produced by the clamp on the plastic block in (b) leads to stress patterns that are visible when viewed with polarized light.

THE SLAM DUNK   **EXAMPLE 4 - 5**

A 110-kg basketball player hangs on the rim following a slam dunk (Figure 4-7). Prior to dropping to the floor, he hangs motionless with the front of the rim deflected down a distance of 15 cm. Assume the rim can be approximated by a spring and calculate the force constant k.

PICTURE THE PROBLEM Since the acceleration of the player is zero, the net force exerted on him must also be zero. The upward force exerted by the rim balances his weight. Let $y = 0$ be the original position of the rim and choose down to be the positive y direction. Then Δy is positive, the weight mg is positive, and the force exerted by the rim, $-k\,\Delta y$, is negative.

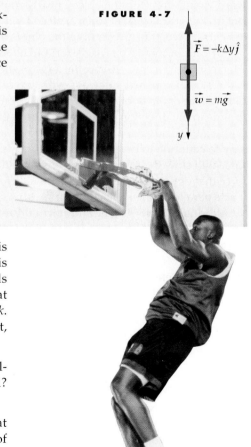

FIGURE 4-7

$\vec{F} = -k\Delta y\hat{j}$

$\vec{w} = m\vec{g}$

y

Apply $\Sigma\vec{F} = m\vec{a}$ to the player, and solve for k:

$$\Sigma F_y = w_y + F_y = ma_y$$
$$mg + (-k\Delta y) = 0$$
$$k = \frac{mg}{\Delta y} = \frac{(110\ \text{kg})(9.81\ \text{N/kg})}{0.15\ \text{m}}$$
$$= \boxed{7.19 \times 10^3\ \text{N/m}}$$

REMARKS Although a basketball rim doesn't look much like a spring, the rim is suspended by a hinge with a spring that is distorted when the front of the rim is pulled down. As a result, the upward force the rim exerts on the player's hands is proportional to the rim front's displacement and oppositely directed. Note that we used N/kg for the units of g so that kg cancels, giving N/m for the units of k. We can use either 9.81 N/kg or 9.81 m/s^2 for g, whichever is more convenient, because 1 N/kg = 1 m/s^2.

EXERCISE A 4-kg bunch of bananas is suspended motionless from a spring balance whose force constant is $k = 300$ N/m. By how much is the spring stretched? (*Answer* 13.1 cm)

EXERCISE A spring of force constant 400 N/m is attached to a 3-kg block that rests on a horizontal air track that renders friction negligible. What extension of the spring is needed to give the block an acceleration of 4 m/s^2 upon release? (*Answer* 3.0 cm)

EXERCISE IN DIMENSIONAL ANALYSIS An object of mass m oscillates at the end of an ideal spring of force constant k. The time for one complete oscillation is the period T. Assuming that T depends on m and k, use dimensional analysis to find the form of the relationship $T = f(m, k)$, ignoring numerical constants. This is most easily found by looking at the units. Note that the units of k are $N/m = (kg \cdot m/s^2)/m = kg/s^2$, and the units of m are kg. (*Answer* $T = C\sqrt{m/k}$ where C is some dimensionless constant. The correct expression for the period, as we will see in Chapter 14, is $T = 2\pi\sqrt{m/k}$.

Strings Strings (ropes) are used to pull things. We can think of a string as a spring with such a large force constant that the extension of the string is negligible. Strings are flexible, however, so unlike springs, they cannot push things. Instead they flex or bend. The magnitude of the force that one segment of a string exerts on an adjacent segment is called **tension.** It follows that if a string pulls on an object, the magnitude of the force equals the tension. The concept of tension in a string or rope is further developed in Section 4-7.

Constraints A railroad car moves along a track. A wooden pony on a merry-go-round moves in a circle. A sled sliding along the surface of a frozen pond moves in a horizontal plane. Such conditions on the motion of objects are called **constraints.**

4-5 Problem Solving: Free-Body Diagrams

Imagine a dogsled being pulled across icy ground. The dog in front pulls on a rope attached to the sled (Figure 4-8a) with a horizontal force causing the sled to gain speed. We can think of the sled and rope together as a single object. What forces act on the sled-rope object? Both the dog and the ice touch the object, so we know that the dog and the ice exert contact forces on it. We also know that the earth exerts a gravitational force on the sled-rope (the object's weight). Thus, a total of three forces act on the object (assuming that friction is negligible):

1. The weight of the sled-rope $\vec{w}$.
2. The contact force $\vec{F}_n$ exerted by the ice (without friction, the contact force is directed normal to the ice.)
3. The contact force $\vec{F}$ exerted by the dog.

A diagram that shows schematically all the forces acting on a system, such as Figure 4-8b, is called a **free-body diagram.** It is called a free-body diagram because the body (object) is drawn free from (without) its surroundings. Drawing the force vectors on a free-body diagram to scale requires that we first determine the direction of the acceleration vector using kinematic methods. We know the object is moving to the right with increasing speed. It follows from kinematics that its acceleration vector is in the direction of its motion—to the right. Note that $\vec{F}_n$ and $\vec{w}$ in the diagram have equal magnitudes. The magnitudes must be equal because the vertical component of the acceleration is zero. As a check on the correctness of our free-body diagram, we draw a vector-addition diagram (Figure 4-9) verifying that the sum of the force vectors is in the direction of the acceleration vector.

The x component of Newton's second law gives

$$\Sigma F_x = F_{n,x} + w_x + F_x = ma_x$$
$$0 + 0 + F = ma_x$$

or

$$a_x = \frac{F}{m}$$

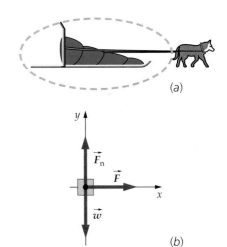

(a)

(b)

FIGURE 4-8 (*a*) A dog pulling a sled. The first step in problem solving is to isolate the system to be analyzed. In this case, the closed dashed curve represents the boundary between the sled-rope object and its surroundings. (*b*) The forces acting on the sled in Figure 4.8*a*.

FIGURE 4-9

The y component of Newton's second law gives

$$\Sigma F_y = F_{n,y} + w_y + F_y = ma_y$$
$$F_n - w + 0 = 0$$

or

$$F_n = w$$

In this simple example, we found two things: the horizontal acceleration ($a_x = F/m$), and the vertical force $\vec{F}_n$ exerted by the ice ($F_n = w$).

A Dogsled Race   **EXAMPLE 4-6**

During your winter break, you enter a dogsled race in which students replace the dogs. Wearing cleats for traction, you begin the race by pulling on a rope attached to the sled with a force of 150 N at 25° with the horizontal. The mass of the sled-passenger-rope object is 80 kg and there is negligible friction between the sled runners and the ice. Find (a) the acceleration of the sled and (b) the normal force $\vec{F}_n$ exerted by the surface on the sled.

(a)

PICTURE THE PROBLEM Three forces act on the object: its weight $\vec{w}$, which acts downward; the normal force $\vec{F}_n$, which acts upward; and the force with which you pull the rope $\vec{F}$, directed 25° above the horizontal. Since the forces are not all parallel to a single line, we study the system by applying Newton's second law to the x and y directions separately.

FIGURE 4-10   (b)

(a) 1. Sketch a free-body diagram (Figure 4-10b) of the sled. Include a coordinate system with one of the coordinate axes in the direction of the sled's acceleration. The object moves to the right with increasing speed, so we know the acceleration is in that direction:

2. *Note:* Add the force vectors on the free-body diagram (Figure 4-11) to verify that their sum can be in the direction of the acceleration:

3. Apply Newton's second law to the object. Write out the equation in both vector and component form:

$$\vec{F}_n + \vec{w} + \vec{F} = m\vec{a}$$

or

FIGURE 4-11

$$F_{n,x} + w_x + F_x = ma_x$$
$$F_{n,y} + w_y + F_y = ma_y$$

4. Express the x components of $\vec{F}_n$, $\vec{w}$, and $\vec{F}$:

$$F_{n,x} = 0, \quad w_x = 0, \quad \text{and} \quad F_x = F \cos \theta$$

5. Substitute the step 4 results into the x component equation in step 3. Then solve for the acceleration a_x:

$$0 + 0 + F \cos \theta = ma_x$$

$$a_x = \frac{F \cos \theta}{m} = \frac{(150 \text{ N}) \cos 25°}{80 \text{ kg}} = \boxed{1.70 \text{ m/s}^2}$$

(b) 1. Express the y component of $\vec{a}$:

$$a_y = 0$$

2. Express the y components of $\vec{F}_n$, $\vec{w}$, and $\vec{F}$:

$$F_{n,y} = F_n, \quad w_y = -mg, \quad \text{and} \quad F_y = F \sin \theta$$

3. Substitute the step b1 and b2 results into the y component equation in step a3. Then solve for F_n:

$$\Sigma F_y = F_n - mg + F \sin \theta = 0$$

$$F_n = mg - F \sin \theta$$

$$= (80 \text{ kg})(9.81 \text{ N/kg}) - (150 \text{ N}) \sin 25° = \boxed{721 \text{ N}}$$

REMARKS Note that only the x component of $\vec{F}$, which is $F\cos\theta$, causes the object to accelerate. Also note that the ice supports less than the full weight of the object since part of the weight, $F\sin\theta$, is supported by the rope.

◐ PLAUSIBILITY CHECK If $\theta = 0$, the object is accelerated by a force F and the ice supports all the object's weight. Our results agree, giving $a_x = F/m$ and $F_n = mg$.

EXERCISE If $\theta = 25°$, what is the maximum force F that can be applied to the rope without lifting the sled off the surface? (*Answer $F = 1.86$ kN*)

Example 4-6 illustrates a general method for solving problems using Newton's laws:

1. Draw a neat diagram that includes the important features of the problem.

2. Isolate the object (particle) of interest, and draw a free-body diagram showing each external force that acts on it. If there is more than one object of interest in the problem, draw a separate free-body diagram for each. Choose a convenient coordinate system for each object and include it on that object's free-body diagram. If the direction of the acceleration is known, choose a coordinate axis that is parallel to it. For objects sliding along a surface, choose one coordinate axis parallel to the surface and the other perpendicular to it.

3. Apply Newton's second law, $\Sigma\vec{F} = m\vec{a}$, usually in component form.

4. For problems involving two or more objects, make use of Newton's third law, $\vec{F}_{A,B} = -\vec{F}_{B,A}$, and any constraints to simplify the equations obtained from applying $\Sigma\vec{F} = m\vec{a}$.

5. Solve the resulting equations for the unknowns.

6. Check to see whether your results have the correct units and seem plausible. Substituting extreme values into your solution is a good way to check your work for errors.

SOLVING PROBLEMS USING NEWTON'S LAWS

·

UNLOADING A TRUCK   **EXAMPLE 4 - 7**   **Put It in Context**

You are working for a big delivery company, and must unload a large, fragile package from your truck, using a delivery ramp (Figure 4-12). If the downward component of the velocity of the package when it reaches the bottom of the ramp is greater than 2.5 m/s (the speed an object would have if it were dropped from a height of about 1 ft), the package will break. What is the largest angle at which you can safely unload? The ramp is 1-m high, has rollers (i.e., the ramp is approximately frictionless), and is inclined at an angle θ to the horizontal.

PICTURE THE PROBLEM Two forces act on the box, the weight $\vec{w}$ and the normal force $\vec{F}_n$. Since these forces are not parallel to a single line, they cannot sum to zero, hence there is a net force on the box causing it to accelerate. The ramp constrains the box to move parallel to its surface, so we choose down the incline as the positive x direction. To determine the acceleration we apply Newton's second law to the box. Once the acceleration is known, we can use kinematics to determine the largest safe angle.

FIGURE 4-12

1. Relate the downward component of the velocity of the box to its speed v along the ramp:

$$v_d = v \sin \theta$$

2. The speed v is related to the displacement Δx along the ramp by the kinematic equation:

$$v^2 = v_0^2 + 2a_x \Delta x$$

3. To find a_x we apply Newton's second law ($\Sigma F_x = ma_x$) to the package. First we draw a free-body diagram (Figure 4-13). Two forces act on the package, the weight force and the normal force. We choose the direction of the acceleration, down the ramp, as the $+x$ direction.

Note: The angle between $\vec{w}$ and the negative y axis is the same as the angle between the incline and the horizontal as we see from the free-body diagram. We can also see that $w_x = w \sin \theta$.

FIGURE 4-13

4. Applying Newton's second law gives:

$$F_{n,x} + w_x = ma_x$$
where

Note: F_n is perpendicular to the x axis and $w = mg$.

$$F_{n,x} = 0 \quad \text{and} \quad w_x = w \sin \theta = mg \sin \theta$$

5. Substituting and solving for the acceleration gives:

$$0 + mg \sin \theta = ma_x$$

so

$$a_x = g \sin \theta$$

6. Substituting for a_x in the kinematic equation (step 2) and setting v_0 to zero gives:

$$v^2 = 2g \sin \theta \, \Delta x$$

7. From Figure 4-12 we can see that when Δx equals the length of the ramp, $\Delta x \sin \theta = h$, where h is the height of the ramp:

$$v^2 = 2gh$$

8. Using $v_d = v \sin \theta$, solve for v_d:

$$v_d = \sqrt{2gh} \sin \theta$$

9. Solve for the maximum angle:

$$2.5 \text{ m/s} = \sqrt{2(9.81 \text{ m/s}^2)(1.0 \text{ m})} \sin \theta_{max}$$

$$\theta_{max} = \boxed{34.4°}$$

REMARKS The acceleration down the incline is constant and equal to $g \sin \theta$. Also, the speed v at the bottom ($v = \sqrt{2gh}$) does not depend on the angle θ.

EXERCISE Apply $\Sigma F_y = ma_y$ to the package and show that $F_n = mg \cos \theta$.

PICTURE HANGING   **E X A M P L E   4 - 8**   Try It Yourself

A picture weighing 8 N is supported by two wires with tensions T_1 and T_2, as shown in Figure 4-14. Find each tension.

PICTURE THE PROBLEM Because the picture does not accelerate, the net force acting on it must be zero. The three forces acting on it (its weight $\vec{w}$, the tension forces $\vec{T}_1$ in one wire and $\vec{T}_2$ in the other wire) must therefore sum to zero.

FIGURE 4-14

Cover the column to the right and try these on your own before looking at the answers.

Steps: Answers:

T_2

T_1

T_{2y} 60°

30° T_{1y}

T_{2x} T_{1x}

w

FIGURE 4-15

1. Draw a free-body diagram for the picture (Figure 4-15). On your diagram show the x and y components of the two tension forces.

2. Apply $\Sigma \vec{F} = m\vec{a}$ in vector form to the picture.

$$\vec{T}_1 + \vec{T}_2 + \vec{w} = m\vec{a}$$

3. Resolve each force into its x and y components. This gives you two equations for the two unknowns T_1 and T_2.

$$T_{1,x} + T_{2,x} + w_x = 0$$
$$T_{1,y} + T_{2,y} + w_y = 0$$
$$T_1 \cos 30° - T_2 \cos 60° + 0 = 0$$
$$T_1 \sin 30° + T_2 \sin 60° - w = 0$$

4. Solve the x component equation for T_2 in terms of T_1.

$$T_2 = T_1 \frac{\cos 30°}{\cos 60°} = T_1 \sqrt{3}$$

5. Substitute your result for T_2 (from step 4) into the y component equation and solve for T_1.

$$T_1 \sin 30° + (T_1 \sqrt{3}) \sin 60° - w = 0$$
$$T_1 = \tfrac{1}{2}w = \boxed{4\text{N}}$$

6. Use your result for T_1 to find T_2.

$$T_2 = T_1 \sqrt{3} = \boxed{6.93 \text{ N}}$$

REMARKS Note that the more vertical of the two wires supports the greater share of the load, as you might expect. Also, we see that $T_1 + T_2 > 8$ N. The "extra" force is due to the wires pulling to the right and left.

AN ACCELERATING JET PLANE   **EXAMPLE   4 - 9**

As your jet plane speeds down the runway on takeoff, you decide to determine its acceleration, so you take out your yo-yo and note that when you suspend it, the string makes an angle of 22° with the vertical (Figure 4-16a). (a) What is the acceleration of the plane? (b) If the mass of the yo-yo is 40 g, what is the tension in the string?

$\vec{a}$

θ

(a)

PICTURE THE PROBLEM The yo-yo and plane have the same acceleration to the right. The net force on the yo-yo is in the direction of its acceleration. This force is supplied by the horizontal component of the tension force $\vec{T}$. The vertical component of $\vec{T}$ balances the weight of the yo-yo. We choose a coordinate system in which the x direction is parallel to the acceleration vector $\vec{a}$ and the y direction is vertical. Writing Newton's second law for both the x and y directions gives two equations to determine the two unknowns, a and T.

(a) 1. Draw a free-body diagram for the yo-yo (Figure 4-16b). Choose the postive x direction to be the direction of the acceleration.

y

θ

$\vec{T}$

x

$\vec{w}$

2. Apply $\Sigma F_x = ma_x$ to the yo-yo. Then simplify using trigonometry:

$$T_x + w_x = ma_x$$
$$T \sin \theta + 0 = ma_x$$

or

(b)

$$T \sin \theta = ma_x$$

FIGURE 4-16

3. Apply $\Sigma F_y = ma_y$ to the yo-yo. Then simplify using trigonometry (Figure 4-16c) and $w = mg$. Since the acceleration is in the positive x direction, $a_y = 0$:

$$T_y + w_y = ma_y$$

$$T \cos \theta - mg = 0$$

or

$$T \cos \theta = mg$$

(c)

FIGURE 4-16

4. Divide the step 2 result by the step 3 result and solve for the acceleration. Since the acceleration vector is in the positive x direction, $a = a_x$:

$$\frac{T \sin \theta}{T \cos \theta} = \frac{ma_x}{mg}, \quad \text{so} \quad \tan \theta = \frac{a_x}{g}$$

and

$$a = g \tan \theta = (9.81 \text{ m/s}^2) \tan 22° = \boxed{3.96 \text{ m/s}^2}$$

(b) Using the step 3 result, solve for the tension:

$$T = \frac{mg}{\cos \theta} = \frac{(0.04 \text{ kg})(9.81 \text{ m/s}^2)}{\cos 22°} = \boxed{0.423 \text{ N}}$$

REMARKS Notice that T is greater than the weight of the yo-yo ($mg = 0.392$ N) because the cord not only keeps the yo-yo from falling but also accelerates it in the horizontal direction. Here we use the units m/s² for g because we are calculating acceleration.

PLAUSIBILITY CHECK At $\theta = 0$, we find that $T = mg$ and $a = 0$.

EXERCISE For what acceleration magnitude a would the tension in the string be equal to $3mg$? What is θ in this case? (*Answer* $a = 27.8$ m/s², $\theta = 70.5°$)

Our next example is the application of Newton's second law to objects that are at rest relative to a reference frame that is itself accelerating.

YOUR WEIGHT IN AN ELEVATOR

EXAMPLE 4-10

Suppose that your mass is 80 kg, and you are standing on a scale fastened to the floor of an elevator. The scale measures force and is calibrated in newtons. What does the scale read when (*a*) the elevator is rising with upward acceleration of magnitude *a*; (*b*) the elevator is descending with downward acceleration of magnitude *a'*; (*c*) the elevator is rising at 20 m/s and its speed is decreasing at a rate of 8 m/s²?

PICTURE THE PROBLEM The scale reading is the magnitude of the normal force F_n exerted by the scale on you (Figure 4-17). Because you are at rest relative to the elevator, you and the elevator have the same acceleration. Two forces act on you: the downward force of gravity, mg, and the upward normal force from the scale, F_n. The sum of these forces gives you the observed acceleration. We choose upward to be the positive direction.

(*a*) 1. Draw a free-body diagram of yourself (Figure 4-18):

(a) (b)

FIGURE 4-17

y

$\vec{F}_n$

$\vec{w}$

FIGURE 4-18

2. Apply $\Sigma\vec{F} = m\vec{a}$ in the y direction:

$$F_{n,y} + w_y = ma_y$$
$$F_n - mg = ma$$

3. Solve for F_n. This is the reading on the scale (your apparent weight):

$$F_n = mg + ma = \boxed{m(g + a)}$$

(b) 1. Apply $\Sigma\vec{F} = m\vec{a}$ in the y direction for the case in which the elevator accelerates downward with magnitude a':

$$F_{n,y} + w_y = ma_y$$
$$F_n - mg = m(-a')$$

2. Solve for F_n:

$$F_n = mg - ma' = \boxed{m(g - a')}$$

(c) 1. Apply $\Sigma\vec{F} = m\vec{a}$ in the y direction. Note that the acceleration is downward. (Why is that?) It follows that a_y is negative:

$$F_{n,y} + w_y = ma_y$$

2. Solve for F_n:

$$F_n - mg = ma_y$$
$$F_n = m(g + a_y) = (80\ \text{kg})(9.81\ \text{m/s}^2 - 8.00\ \text{m/s}^2) = \boxed{145\ \text{N}}$$

REMARKS Whether the elevator is ascending or descending, if it accelerates upward, then your apparent weight is greater than mg by ma. For you, it is as if gravity were increased from g to $g + a$. If it accelerates downward, then your apparent weight is less than mg by the amount ma'. You feel lighter, as if gravity were $g - a'$. If $a' = g$, the elevator is in free-fall, and you experience weightlessness.

EXERCISE An elevator descending comes to a stop with an acceleration of magnitude 4 m/s². If your mass is 70 kg and you are standing on a scale in the elevator, what does the scale read as the elevator is stopping? (*Answer*   967 N)

EXERCISE A man stands on a scale in an elevator that has an upward acceleration a. The scale reads 960 N. When he picks up a 20 kg box, the scale reads 1200 N. Find the mass of the man, his weight, and the acceleration a.

4-6 Newton's Third Law

When two objects interact, they exert forces on each other. Newton's third law states that these forces are equal in magnitude and opposite in direction. That is, if object A exerts a force on object B, then object B exerts a force on A that is equal in magnitude and opposite in direction. Thus forces always occur in pairs. It is common to refer to one force in the pair as an action and the other as a reaction. This terminology is unfortunate because it sounds like one force "reacts" to the other, which is not the case. The two forces occur simultaneously. **Either can be called the action and the other the reaction.** If we refer to an external force acting on a particular object as an action force, then the corresponding reaction force must act on a different object. Thus no two external forces acting on a single object can ever constitute an action–reaction pair.

In Figure 4-19, a block rests on a table. The force acting downward on the block is the weight $\vec{w}$ due to the attraction of the earth. An equal and opposite force $\vec{w}' = -\vec{w}$ is exerted by the block on the earth. These forces form an action–reaction pair. If they were the only forces present, the block would accelerate downward because it would have only a single force acting on it (and the earth would accelerate upward toward the block). However, the table exerts an

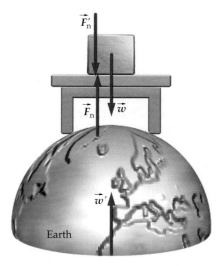

FIGURE 4-19

upward force $\vec{F}_n$ on the block that balances the block's weight. The block also exerts a force $\vec{F}_n' = -\vec{F}_n$ downward on the table. The forces $\vec{F}_n$ and $\vec{F}_n'$ also form an action–reaction pair.

EXERCISE Do the forces $\vec{w}$ and $\vec{F}_n$ in Figure 4-19 form an action–reaction pair? (*Answer* No, they do not. These forces are both external forces and they both act on the same object, the block. Thus, they cannot constitute an action–reaction pair.)

THE HORSE BEFORE THE CART   **EXAMPLE 4-11**

A horse refuses to pull a cart (Figure 4-20a). The horse reasons, "according to Newton's third law, whatever force I exert on the cart, the cart will exert an equal and opposite force on me, so the net force will be zero and I will have no chance of accelerating the cart." What is wrong with this reasoning?

PICTURE THE PROBLEM Because we are interested in the motion of the cart, we draw a simple diagram for it (Figure 4-20b). The force exerted by the horse on the harness is labeled $\vec{F}$. (The harness is attached to the cart, and so we can consider it to be part of the cart.) Other forces acting on the cart are its weight $\vec{w}$, the vertical support force of the ground $\vec{F}_n$, and the horizontal force exerted by the pavement, labeled $\vec{f}$ (for friction).

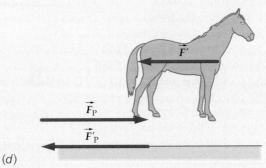

(a)

1. Draw a free-body diagram for the cart (see Figure 4-20c). Because the cart does not accelerate vertically, the verticle forces must sum to zero. The horizontal forces are $\vec{F}$ to the right and $\vec{f}$ to the left. The cart will accelerate to the right if $\vec{F}$ is greater than $\vec{f}$.

2. Note that the reaction force to $\vec{F}$, which we call $\vec{F}'$, is exerted on the horse, not on the cart (Figure 4-20d). It has no effect on the motion of the cart, but it does affect the motion of the horse. If the horse is to accelerate to the right, there must be a force $\vec{F}_P$ (to the right) exerted by the pavement on the horse's hooves that is greater than $\vec{F}'$.

(b)

(c)

REMARKS This example illustrates the importance of drawing a simple diagram when solving mechanics problems. Had the horse done so, he would have seen that he need only push back hard against the pavement so that the pavement will push him forward.

EXERCISE As you stand facing a friend, place your palms against your friend's palms and push. Can your friend exert a force on you if you do not exert a force back? Try it.

EXERCISE True or false: The force exerted by the cart on the horse is equal and opposite to the force exerted by the horse on the cart, but only when the horse and cart are not accelerating. (*Answer* False! An action–reaction pair of forces describes the interaction of two objects. One force cannot exist without the other. They are *always* equal and opposite.)

(d)

FIGURE 4-20

4-7 Problems With Two or More Objects

In some problems, two or more objects are in contact or are connected by a string or spring. Such problems are solved by drawing
a separate free-body diagram for each object and then applying
Newton's second law to each object. The resultant equations, together with any equations describing interactions and constraints,
are solved simultaneously for the unknown quantities. If the
objects are in direct contact, the forces they exert on each other
must be equal and opposite, as stated in Newton's third law. For
two objects, each moving in a straight line, that are connected by a
taut nonstretching string, the acceleration components parallel to
the string are the same for both objects. This is so because, for each
object, its motion parallel to the string is identical with that of the
other object. If the string passes over a pulley or peg, the phrase
"parallel to the string" means parallel with that segment of the
string attached to the object.

Consider the motion of Steve and Paul in Figure 4-21. The rate
at which Paul descends equals the rate at which Steve slides along
the glacier. That is, Paul's velocity component parallel with
the segment of the rope attached to him equals Steve's velocity
component parallel with the segment attached to him. These two
velocity components must remain equal. If Steve and Paul are
changing speed, they do so at the same rate. Their acceleration
components parallel with the rope must be equal.

FIGURE 4-21

The tension in a string or rope is the magnitude of the force that one segment
of the rope exerts on a neighboring segment. The tension can vary throughout
the rope. For a rope dangling from a girder at the ceiling of a gymnasium, the
tension is greatest near the top because the segment at the top has to support
the weight of all the rope below it. However, for the problems in this book, the
masses of strings and ropes are normally assumed to be so small that variations
in tension due to the weight of a string or rope can be neglected. Conveniently,
this also means that variations in the tension due to any acceleration of the
rope can also be neglected. To see that this is so, consider the free-body diagram
of a segment of the rope attached to Steve, where Δm_s is the segment's mass
(Figure 4-22).

FIGURE 4-22

Applying Newton's second law to the segment gives $T - T' = \Delta m_s a_x$. If the
mass of the segment is negligible, then $T = T'$. No net force is needed to give
the segment an acceleration. (That is, only a negligible difference in tension is
needed to give a rope segment of negligible mass any finite acceleration.)

Next we consider the entire rope connecting Steve and Paul. Neglecting gravity, there are three forces acting on the rope. Steve and Paul each exert a force, as
does the ice at the edge of the glacier. Neglecting any friction between the ice
and the rope means that the force exerted by the ice is always a normal force
(Figure 4-23). A normal force has no component along the rope, so it cannot produce a change in the tension. Thus the tension is the same throughout the entire
length of the rope. To summarize, if a taut rope of negligible mass changes direction by passing over a frictionless surface, the tension is the same throughout
the rope.

FIGURE 4-23

EXERCISE Suppose that instead of passing over the edge of a glacier, the rope
passed around a pulley with frictionless bearings as shown in Figure 4-24.
Would the tension then be the same throughout the length of the rope? (*Answer*
No. Doing away with friction in the bearing is one thing, but the pulley still
has inertia (mass). A difference in tension is needed to change the rate of rotation of the pulley.)

FIGURE 4-24

THE ROCK CLIMBERS   **EXAMPLE 4-12**

Paul (mass m_P) accidentally falls off the edge of a glacier as shown in Figure 4-21. Fortunately he is tied by a long rope to Steve (mass m_S), who has a climbing ax. Before Steve sets his ax to stop them, he slides without friction along the ice, attached by the rope to Paul. Assume there is no friction between the rope and the glacier. Find the acceleration of each person and the tension in the rope.

PICTURE THE PROBLEM The tension forces $\vec{T}_1$ and $\vec{T}_2$ have equal magnitudes because the rope is assumed to be massless and the glacier ice is assumed to be frictionless. The rope does not stretch or become slack, so Paul and Steve have equal speeds at all times. Their accelerations $\vec{a}_S$ and $\vec{a}_P$ must therefore be equal in magnitude, but not in direction. Steve accelerates down the face of the glacier whereas Paul accelerates vertically downward.

FIGURE 4-25

Newton's second law relates each person's acceleration to the forces acting on him. Apply $\Sigma\vec{F} = m\vec{a}$ to each, and solve for the accelerations and the tension.

1. Draw separate free-body diagrams for Steve and Paul (Figure 4-25). Put axes x and y on Steve's diagram, choosing the direction of Steve's acceleration as the positive x direction. Choose the direction of Paul's acceleration as the positive x' direction.

2. Apply $\Sigma\vec{F} = m\vec{a}$ in the x direction to Steve:

$$F_{n,x} + T_{1,x} + m_S g_x = m_S a_{S,x}$$

3. Apply $\Sigma\vec{F} = m\vec{a}$ in the x' direction to Paul:

$$T_{2,x'} + m_P g_{x'} = m_P a_{P,x'}$$

4. Because they are each moving in a straight line and are connected by a taut length of rope that does not stretch, the accelerations of Paul and Steve are related. Express this relation:

$$a_{P,x'} = a_{S,x} = a_x$$

5. Because the rope is of negligible mass and slides over the ice with negligible friction, the forces $\vec{T}_1$ and $\vec{T}_2$ are related. Express this relation:

$$T_2 = T_1 = T$$

6. Substitute the steps 4 and 5 results into the step 2 and step 3 equations:

$$T + m_S g \sin\theta = m_S a_x$$
$$-T + m_P g = m_P a_x$$

7. Solve the step 6 equations for the acceleration by eliminating T and solving for a_x:

$$\boxed{a_x = \frac{m_S \sin\theta + m_P}{m_S + m_P} g}$$

8. Substitute the step 7 result into either step 6 equation and solve for T:

$$\boxed{T = \frac{m_S m_P}{m_S + m_P}(1 - \sin\theta)g}$$

REMARKS In Step 3 we chose downward to be positive to keep the solution as simple as possible. With this choice, when Steve moves in the positive x direction (down the glacier), Paul moves in the positive x' direction (downward).

❶ **PLAUSIBILITY CHECK** If m_P is very much greater than m_S, we expect the acceleration to be approximately g and the tension to be approximately zero. Substituting $m_S = 0$ does indeed give $a = g$ and $T = 0$ for this case. If m_P is much less than m_S, we expect the acceleration to be approximately $g \sin\theta$ (see Example 4-8) and the tension to be zero. Substituting $m_P = 0$ in steps 7 and 8, we indeed obtain $a_x = g \sin\theta$ and $T = 0$. At the extreme value of the inclination ($\theta = 90°$) we check our answers. Substituting $\theta = 90°$ in steps 7 and 8, we obtain $a_x = g$ and $T = 0$. This seems right since Steve and Paul would be in free-fall for $\theta = 90°$.

EXERCISE (*a*) Find the acceleration if $\theta = 15°$ and if the masses are $m_S = 78$ kg and $m_P = 92$ kg. (*b*) Find the acceleration if these two masses are interchanged. (*Answer* (*a*) $a_x = 0.660g$ (*b*) $a_x = 0.599g$)

BUILDING A SPACE STATION | **EXAMPLE 4-13** | Try It Yourself

You are an astronaut constructing a space station, and you push on a box of mass m_1 with a force of $\vec{F}_A$. The box is in direct contact with a second box of mass m_2 (Figure 4-26). (*a*) What is the acceleration of the boxes? (*b*) What is the magnitude of the force exerted by one box on the other?

PICTURE THE PROBLEM Let $\vec{F}_{2,1}$ be the force exerted by box 2 on box 1, and $\vec{F}_{1,2}$ be the force exerted by box 1 on box 2. In accord with Newton's third law, these forces are equal and opposite ($\vec{F}_{2,1} = -\vec{F}_{1,2}$), so $F_{2,1} = F_{1,2}$. Apply Newton's second law to each box separately. The motions of the two boxes are identical, so the accelerations $\vec{a}_1$ and $\vec{a}_2$ are equal.

FIGURE 4-26

Cover the column to the right and try these on your own before looking at the answers.

Steps

(*a*) 1. Draw free-body diagrams for the two boxes (Figure 4-27).

2. Apply $\Sigma\vec{F} = m\vec{a}$ to box 1.

3. Apply $\Sigma\vec{F} = m\vec{a}$ to box 2.

4. Express both the relation between the two accelerations and the relation between the magnitudes of the forces the blocks exert on each other.

5. Substitute these back into the step 2 and step 3 results and solve for a_x.

(*b*) Substitute your expression for a_x into either the step 2 or the step 3 result and solve for F.

Answers

$F_A - F_{2,1} = m_1 a_{1,x}$

$F_{1,2} = m_2 a_{2,x}$

$a_{2,x} = a_{1,x} = a_x$

$F_{2,1} = F_{1,2} = F$

$a_x = \dfrac{F_A}{m_1 + m_2}$

$F = \dfrac{m_2}{m_1 + m_2} F_A$

FIGURE 4-27

REMARKS Note that the result in step 5 is the same as if the force $\vec{F}_A$ had acted on a single mass equal to the sum of the masses of the two boxes. In fact, since the two boxes have the same acceleration, we can consider them to be a single object with mass $m_1 + m_2$.

EXERCISE (*a*) Find the acceleration and the contact force if $m_1 = 2$ kg, $m_2 = 3$ kg, and $F_A = 12$ N. (*b*) Find the contact force if the two boxes are interchanged so that the first block has a mass of 3 kg and the second block has a mass of 2 kg. (*Answer* (*a*) $a_x = 2.4$ m/s^2, $F = 7.2$ N (*b*) $F = 4.8$ N)

SUMMARY

1. Newton's laws of motion are fundamental laws of nature that serve as the basis for our understanding of mechanics.

2. Mass is an *intrinsic* property of an object.

3. Force is an important *derived* dynamic quantity.

Topic	Relevant Equations and Remarks
1. Newton's Laws	
First law	An object at rest stays at rest unless acted on by an external force. An object in motion continues to travel with constant velocity unless acted on by an external force. (Reference frames in which this occurs are called inertial reference frames.)
Second law	The magnitude of the acceleration is proportional to the magnitude of the net external force $\vec{F}_{net}$, in accordance with $\vec{F}_{net} = m\vec{a}$, where m is the mass of the object. The net force acting on an object, also called the resultant force, is the vector sum of all the forces acting on it: $\vec{F}_{net} = \Sigma\vec{F}$. Thus $$\Sigma\vec{F} = m\vec{a} \qquad\qquad 4\text{-}1$$
Third law	Forces always occur in equal and opposite pairs. If object A exerts a force on object B, an equal but opposite force is exerted by object B on object A: $$\vec{F}_{A,B} = -\vec{F}_{B,A} \qquad\qquad 4\text{-}2$$
2. Inertial Reference Frames	Our statements of Newton's laws are valid only in an inertial reference frame—a reference frame for which an object at rest remains at rest if no force acts on the object. Any reference frame that is moving with constant velocity relative to an inertial reference frame is itself an inertial reference frame, and any reference frame that is accelerating relative to an inertial frame is not an inertial reference frame. The earth's surface is, to a good approximation, an inertial reference frame.
3. Force, Mass, and Weight	
Force	Force is defined in terms of the acceleration it produces on a given object. A force of 1 newton (N) is that force which produces an acceleration of 1 m/s^2 on a mass of 1 kilogram (kg).
Mass	Mass is the intrinsic property of an object that measures its inertial resistance to acceleration. Mass does not depend on the location of the object. Applying identical forces to each of two objects and measuring their respective accelerations allows the masses of two objects to be compared. The ratio of the masses of the objects is equal to the inverse ratio of the accelerations produced: $$\frac{m_2}{m_1} = \frac{a_1}{a_2} \qquad\qquad 4\text{-}3$$
Weight	The weight $\vec{w}$ of an object is the force of gravitational attraction exerted by the earth on the object. It is proportional to the mass m of the object and the gravitational field $\vec{g}$, which equals the free-fall acceleration: $$\vec{w} = m\vec{g} \qquad\qquad 4\text{-}4$$ Weight is not an intrinsic property of an object; it depends on the location of the object.
4. Fundamental Forces	All the forces observed in nature can be explained in terms of four basic interactions: 1. The gravitational force 2. The electromagnetic force 3. The strong nuclear force (also called the hadronic force) 4. The weak nuclear force
5. Contact Forces	Contact forces of support and friction and those exerted by springs and strings are due to molecular forces that arise from the basic electromagnetic force.
Hooke's law	When a relaxed spring is compressed or extended by a small amount Δx, the force it exerts is proportional to Δx: $$F_x = -k\,\Delta x \qquad\qquad 4\text{-}9$$

PROBLEMS

- Single-concept, single-step, relatively easy
- •• Intermediate-level, may require synthesis of concepts
- ••• Challenging
- SSM Solution is in the *Student Solutions Manual*
- iSOLVE Problems available on iSOLVE online homework service
- iSOLVE ✓ These "Checkpoint" online homework service problems ask students additional questions about their confidence level, and how they arrived at their answer

In a few problems, you are given more data than you actually need; in a few other problems, you are required to supply data from your general knowledge, outside sources, or informed estimates.

For all problems, use $g = 9.81$ m/s² for the free-fall acceleration and neglect friction and air resistance unless instructed to do otherwise.

Conceptual Problems

1 •• SSM How can you tell if a particular reference frame is an inertial reference frame?

2 •• Suppose you observe an object from a reference frame and find that it has an acceleration $\vec{a}$ when there are no forces acting on it. How can you use this information to find an inertial frame?

3 • If an object has no acceleration when observed from an inertial reference frame, can you conclude that no forces are acting on it?

4 • SSM If only a single nonzero force acts on an object, must the object have an acceleration relative to any inertial reference frame? Can it ever have zero velocity?

5 • If an object is acted upon by a single known force, can you tell in which direction the object will move, using no other information?

6 • An object is observed to be moving at constant velocity in an inertial reference frame. It follows that (a) no forces act on the object, (b) a constant force acts on the object in the direction of motion, (c) the net force acting on the object is zero, (d) the net force acting on the object is equal and opposite to its weight.

7 • Suppose an object was sent far out in space, away from galaxies, stars, or other bodies. How would its mass change? Its weight?

8 • SSM How would an astronaut in apparent weightlessness be aware of her mass?

9 • SSM Under what circumstances would your apparent weight be greater than your true weight?

10 •• It is often said that Newton's first and second laws imply that it is impossible to use the laws of mechanics to tell if you are standing still or moving with a constant velocity. Explain.

11 • Suppose a block of mass m_1 rests on a block of mass m_2 and the combination rests on a table as shown in Figure 4-28. Find the force exerted (a) by m_1 on m_2, (b) by m_2 on m_1, (c) by m_2 on the table, (d) by the table on m_2.

FIGURE 4-28 Problem 11

12 • SSM True or false:
(a) If two external forces that are both equal in magnitude and opposite in direction act on the same object, the two forces can never be an action-reaction force pair.
(b) Action equals reaction only if the objects are not accelerating.

13 • An 80-kg man on ice skates pushes his 40-kg son, also on skates, with a force of 100 N. The force exerted by the boy on his father is (a) 200 N, (b) 100 N, (c) 50 N, (d) 40 N.

14 • A girl holds a bird in her hand. The reaction force to the weight of the bird is (a) the gravitational force of the earth on the bird, (b) the gravitational force of the bird on the earth, (c) the contact force of the hand on the bird, (d) the contact force of the bird on the hand, (e) the gravitational force of the earth on the hand.

15 • A baseball player hits a ball with a bat. If the force with which the bat hits the ball is considered the action force, what is the reaction force? (a) The force the bat exerts on the batter's hands. (b) The force on the ball exerted by the glove of the person who catches it. (c) The force the ball exerts on the bat. (d) The force the pitcher exerts on the ball while throwing it. (e) Friction, as the ball rolls to a stop.

16 • Consider any situation in which an external force, say a push, is applied to an object. If Newton's third law requires that *for every action there is an equal and opposite reaction*, why doesn't the reaction force always cancel out the applied force, leaving no acceleration at all?

17 • SSM A 2.5-kg object hangs at rest from a string attached to the ceiling. (a) Draw a diagram showing all the forces on the object and indicate the reaction force to each. (b) Do the same for each force acting on the string.

18 • Which of the free-body diagrams in Figure 4-29 represents a block sliding down a frictionless inclined surface?

FIGURE 4-29 Problem 18

(a)   (b)   (c)   (d)

19 • For an observer in an inertial reference frame, identify which (if any) of the following statements are true and which (if any) are false.

(a) If there are no forces acting on an object, it will not accelerate.
(b) If an object is not accelerating, there must be no forces acting on it.
(c) The motion of an object is always in the direction of the resultant force.
(d) The mass of an object depends on its location.

20 • A sky diver of weight w is descending near the surface of the earth. What is the magnitude of the force exerted by her body *on the earth*? (a) w. (b) Greater than w. (c) Less than w. (d) $9.8w$. (e) 0. (f) It depends on the air resistance.

21 • SSM The net force on a moving object is suddenly reduced to zero and remains zero. As a consequence, the object (a) stops abruptly, (b) stops during a short time interval, (c) changes direction, (d) continues at constant velocity, (e) changes velocity in an unknown manner.

22 • A clothesline is stretched taut between two poles. Then a wet towel is hung at the center of the line. Can the line remain horizontal? Explain.

23 • What effect does the velocity of an elevator have on the apparent weight of a person in the elevator?

Estimation and Approximation

24 •• A car traveling 90 km/h crashes into the rear end of an unoccupied stalled vehicle. Fortunately, the driver is wearing a seat belt. Using reasonable values for the mass of the driver and the stopping distance, estimate the force (assuming it to be constant) exerted on the driver by the seat belt.

25 ••• SSM Making any necessary assumptions, find the normal force and the tangential force exerted by the road on the wheels of your bicycle (a) as you climb an 8% grade at constant speed, and (b) as you descend the 8% grade at constant speed. (An 8% grade means that the angle of inclination θ is given by $\tan \theta = 0.08$.)

Newton's First and Second Laws: Mass, Inertia, and Force

26 • A particle of mass m is traveling at an initial speed $v_0 = 25.0$ m/s. When a net force of 15.0 N acts on it, it comes to a stop in a distance of 62.5 m. What is m? (a) 37.5 kg. (b) 3.00 kg. (c) 1.50 kg. (d) 6.00 kg. (e) 3.75 kg.

27 • (a) An object has an acceleration of 3 m/s² when the only force acting on it is F_0. What is its acceleration when this force is doubled? (b) A second object has an acceleration of 9 m/s² under the influence of the force F_0. What is the ratio of the masses of the two objects? (c) If the two objects are glued together, what acceleration will the force F_0 produce?

28 • ISOLVE ✓ A tugboat tows a ship with a constant force F_1. The increase in the ship's speed in a 10-s interval is 4 km/h. When a second tugboat applies an additional constant force F_2 in the same direction, the speed increases by 16 km/h in a 10-s interval. How do the magnitudes of the two forces compare? (Neglect water resistance.)

29 •• SSM ISOLVE A bullet of mass 1.8×10^{-3} kg moving at 500 m/s impacts a large fixed block of wood and travels 6 cm before coming to rest. Assuming that the acceleration of the bullet is constant, find the force exerted by the wood on the bullet.

30 •• SSM A cart on a horizontal, linear track has a fan attached to it. The cart is positioned at one end of the track, and the fan is turned on. Starting from rest, the cart takes 4.55 s to travel a distance of 1.5 m. The mass of the cart plus fan is 355 g. Assume that the cart travels with constant acceleration. (a) What is the net force exerted on the cart? (b) Weights are added to the cart until its mass is 722 g, and the experiment is repeated. How long does it take for the cart to travel 1.5 m now? Ignore the effects of friction.

31 • A horizontal force F_0 causes an acceleration of 3 m/s² when it acts on an object of mass m sliding on a frictionless surface. Find the acceleration of the same object in the circumstances shown in Figure 4-30a and b.

FIGURE 4-30 Problem 31

(a)   (b)

32 • ISOLVE A force $\vec{F} = (6 \text{ N})\hat{i} - (3 \text{ N})\hat{j}$ acts on an object of mass 1.5 kg. Find the acceleration $\vec{a}$. What is the magnitude a?

33 • A single force of 12 N acts on a particle of mass m. The particle starts from rest and travels in a straight line a distance of 18 m in 6 s. Find m.

34 • SSM Al and Bert stand in the middle of a large frozen lake. Al pushes on Bert with a force of 20 N for a period of 1.5 s. Bert's mass is 100 kg. Assume that both are at rest before Al pushes Bert. (a) What is the speed that Bert reaches as he is pushed away from Al? Treat the ice as frictionless. (b) What speed does Al reach if his mass is 80 kg?

35 • If I push a block whose mass is m_1 across a frictionless floor with a force of a given magnitude, the block has acceleration 12 m/s². If I push on a different block whose mass is m_2 with a force of the same magnitude, its acceleration is 3 m/s². (Both forces are applied horizontally.) (a) What acceleration will this force give to a block with mass $m_2 - m_1$? The force is still applied horizontally. (b) What acceleration will this force give to a block with mass $m_2 + m_1$?

36 • To drag a 75-kg log along the ground at constant velocity, you have to pull on it with a horizontal force of 250 N. (*a*) What is the resistive force exerted by the ground? (*b*) What horizontal force must you exert if you want to give the log an acceleration of 2 m/s²?

37 • ISOLVE ✓ A 4-kg object is subjected to two forces, $\vec{F}_1 = (2\,\text{N})\hat{i} + (-3\,\text{N})\hat{j}$ and $\vec{F}_2 = (4\,\text{N})\hat{i} - (11\,\text{N})\hat{j}$. The object is at rest at the origin at time $t = 0$. (*a*) What is the object's acceleration? (*b*) What is its velocity at time $t = 3$ s? (*c*) Where is the object at time $t = 3$ s?

Mass and Weight

38 • SSM On the moon, the acceleration due to gravity is only about 1/6 of that on earth. An astronaut whose weight on earth is 600 N travels to the lunar surface. His mass as measured on the moon will be (*a*) 600 kg, (*b*) 100 kg, (*c*) 61.2 kg, (*d*) 9.81 kg, (*e*) 360 kg.

39 • ISOLVE Find the weight of a 54-kg student in (*a*) newtons and (*b*) pounds.

40 • Find the mass of a 165-lb engineer in kilograms.

Contact Forces

41 • SSM ISOLVE ✓ A vertical spring of force constant 600 N/m has one end attached to the ceiling and the other to a 12-kg block resting on a horizontal surface so that the spring exerts an upward force on the block. The spring stretches by 10 cm. (*a*) What force does the spring exert on the block? (*b*) What is the force that the surface exerts on the block?

42 • A 6-kg box on a frictionless horizontal surface is attached to a horizontal spring with a force constant of 800 N/m. If the spring is stretched 4 cm from its equilibrium length, what is the acceleration of the box?

Free-Body Diagrams: Static Equilibrium

43 • A traffic light is supported by two wires as in Figure 4-31. Is the tension in the wire that is more nearly vertical greater than or less than the tension in the other wire?

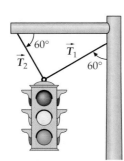

FIGURE 4-31 Problem 43

44 • ISOLVE A lamp with mass $m = 42.6$ kg is hanging from wires as shown in Figure 4-32. The ring has negligible mass. The tension T_1 in the vertical wire is (*a*) 209 N, (*b*) 418 N, (*c*) 570 N, (*d*) 360 N, (*e*) 730 N.

FIGURE 4-32 Problem 44

45 •• SSM ISOLVE ✓ In Figure 4-33*a*, a 0.500-kg block is suspended from a 1.25-m-long string. The ends of the string are attached to the ceiling at points separated by 1.00 m. (*a*) What angle does the string make with the ceiling? (*b*) What is the tension in the string? (*c*) The 0.500-kg block is removed and two 0.250-kg blocks are attached to the string such that the lengths of the three string segments are equal (Figure 4-33*b*). What is the tension in each segment of the string?

FIGURE 4-33 Problem 45

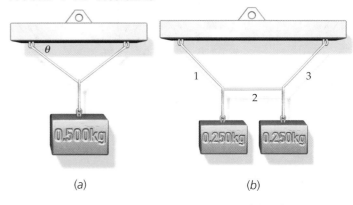

(a)   (b)

46 • A 100-N body is shown suspended from a system of cords. What is the tension in the horizontal cord?

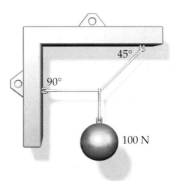

FIGURE 4-34 Problem 46

47 • ISOLVE A 10-kg object on a frictionless table is subjected to two horizontal forces, $\vec{F}_1$ and $\vec{F}_2$, with magnitudes $F_1 = 20$ N and $F_2 = 30$ N, as shown in Figure 4-35. (*a*) Find the acceleration $\vec{a}$ of the object. (*b*) A third force $\vec{F}_3$ is applied so that the object is in static equilibrium. Find $\vec{F}_3$.

FIGURE 4-35 Problem 47

48 • SSM ISOLVE ✓ A vertical force $\vec{T}$ is exerted on a 5-kg object near the surface of the earth, as shown in Figure 4-36. Find the acceleration of the object if (*a*) $T = 5$ N, (*b*) $T = 10$ N, and (*c*) $T = 100$ N.

FIGURE 4-36 Problem 48

49 •• A 2-kg picture is hung by two wires of equal length. Each makes an angle θ with the horizontal, as shown in Figure 4-37. (a) Find the general equation for the tension T, given θ and the weight w for the picture. For what angle θ is T the least? The greatest? (b) If $\theta = 30°$, what is the tension in the wires?

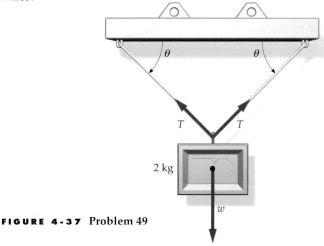

FIGURE 4-37 Problem 49

50 ••• [SSM] Balloon arches are often seen at festivals or celebrations; they are made by attaching helium-filled balloons to a rope that is fixed to the ground at each end. The lift from the balloons raises the structure into the arch shape. Figure 4-38a shows the geometry of such a structure: N balloons are attached at equally spaced intervals along a massless rope of length L, which is attached to two supports. Each balloon provides a lift force F. The horizontal and vertical coordinates of the point on the rope where the ith balloon is attached are x_i and y_i, and T_i is the tension in the ith segment (with segment 0 being the segment between the point of attachment and the first balloon, and segment N being the segment between the last balloon and the other point of attachment).

FIGURE 4-38 Problem 50

(a)

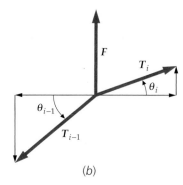

(b)

(a) Figure 4-38b shows a free-body diagram for the ith balloon. From this diagram, show that the horizontal component of the force T_i (call it T_H) is the same for all the balloons, and that by considering the vertical component of the force, one can derive the following equation relating the tension in the ith and $(i - 1)$th segments:

$$T_{i-1} \sin \theta_{i-1} - T_i \sin \theta_i = F$$

(b) Show that $\tan \theta_0 = -\tan \theta_{N+1} = NF/2T_H$.
(c) From the diagram and the two expressions above, show that

$$\tan \theta_i = (N - 2i)F/2T_H$$

and that

$$x_i = \frac{L}{N+1} \sum_{j=0}^{i-1} \cos \theta_j, \qquad y_i = \frac{L}{N+1} \sum_{j=0}^{i-1} \sin \theta_j$$

(d) Write a **spreadsheet** program to make a graph of the shape of a balloon arch with the following parameters: $N = 10$ balloons giving a lift force $F = 1$ N each attached to a rope length $L = 10$ m, with a horizontal component of tension $T_H = 10$ N. How far apart are the two points of attachment? How high is the arch at its highest point?
(e) Note that we haven't specified the spacing between the supports—it is determined by the other parameters. Vary T_H while keeping the other parameters the same until you create an arch that has a spacing of 8 m between the supports. What is T_H then? As you increase T_H, the arch should get flatter and more spread out. Does your spreadsheet model show this?

51 •• A 1000-kg load is being moved by a crane. Find the tension in the cable that supports the load as (a) it moves upward with a speed increasing by 2 m/s each second, (b) it is lifted at constant speed, and (c) it moves upward with speed decreasing by 2 m/s each second.

52 •• [ISOLVE] For the systems in equilibrium in Figures 4-39a, 4-39b, and 4-39c, find the unknown tensions and masses.

FIGURE 4-39 Problem 52

(a)

(b)

(c)

53 •• [ISOLVE] ✓ Your car is stuck in a mud hole. You are alone, but you have a long, strong rope. Having studied physics, you tie the rope tautly to a telephone pole and pull on it sideways, as shown in Figure 4-40. (a) Find the force exerted by the rope on the car when the angle θ is 3° and you are pulling with a force of 400 N but the car does not move. (b) How strong must the rope be if it takes a force of 600 N to move the car when θ is 4°?

FIGURE 4-40 Problem 53

Free-Body Diagrams: Inclined Planes and the Normal Force

54 • [SSM] [ISOLVE] A large box whose mass is 20 kg rests on a frictionless floor. A mover pushes on the box with a force of 250 N at an angle 35° below the horizontal. What is the acceleration of the box across the floor?

55 • [ISOLVE] The box from Problem 54 now rests on a frictionless ramp with a 15° slope. The mover pulls up on a rope attached to the box to pull it up the incline (see Figure 4-41). If the rope makes an angle of 40° with the horizontal, what is the smallest force F the mover will have to exert to move the box up the ramp?

FIGURE 4-41 Problem 55

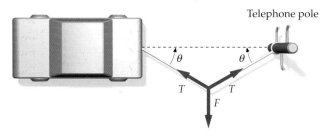

56 • A box slides down a frictionless inclined plane. Draw a diagram showing the forces acting on the box. For each force in your diagram, indicate the reaction force.

57 • The system shown in Figure 4-42 is in equilibrium and the incline is frictionless. It follows that the mass m is (a) 3.5 kg, (b) 3.5 sin 40° kg, (c) 3.5 tan 40° kg, (d) none of these answers.

FIGURE 4-42 Problem 57

58 • [SSM] In Figure 4-43, the objects are attached to spring balances calibrated in newtons. Give the reading of the balance(s) in each case, assuming that the strings are massless.

FIGURE 4-43 Problem 58

(a)   (b)

(c)   (d)

59 •• A box is held in position by a cable along a frictionless incline (Figure 4-44). (*a*) If $\theta = 60°$ and $m = 50$ kg, find the tension in the cable and the normal force exerted by the incline. (*b*) Find the tension as a function of θ and m, and check your result for $\theta = 0°$ and $\theta = 90°$.

FIGURE 4-44 Problem 59

60 •• A horizontal force of 100 N pushes a 12-kg block up a frictionless incline that makes an angle of 25° with the horizontal. (*a*) What is the normal force that the incline exerts on the block? (*b*) What is the acceleration of the block?

61 •• **SSM** **iSOLVE** A 65-kg student weighs himself by standing on a scale mounted on a skateboard that is rolling down an incline, as shown in Figure 4-45. Assume there is no friction so that the force exerted by the incline on the skateboard is normal to the incline. What is the reading on the scale if $\theta = 30°$?

FIGURE 4-45 Problem 61

62 •• A block of mass m slides across a frictionless floor and then up a frictionless ramp (see Figure 4-46). The angle of the ramp is θ and the speed of the block before it starts up the ramp is v_0. The block will slide up to some maximum height h above the floor before starting to slide back down. Show that h is independent of θ.

FIGURE 4-46 Problem 62

Free-Body Diagrams: Elevators

63 • An object is suspended from the ceiling of an elevator that is descending at a constant speed of 9.81 m/s. The tension in the string holding the object is (*a*) equal to the weight of the object, (*b*) less than the weight of the object, (*c*) greater than the weight of the object, (*d*) zero.

64 • Suppose you are standing on a force scale in a descending elevator as it comes to a stop on the ground floor. Will the scale's report of your weight be high, low, or correct as the elevator slows down?

65 • **SSM** A person of weight w is in an elevator going up when the cable suddenly breaks. What is the person's apparent weight immediately after the cable breaks? (*a*) w. (*b*) Greater than w. (*c*) Less than w. (*d*) $9.8w$. (*e*) Zero.

66 • **iSOLVE** A person in an elevator is holding a 10-kg block by a cord rated to withstand a tension of 150 N. When the elevator starts up, the cord breaks. What was the minimum acceleration of the elevator?

67 •• A 2-kg block hangs from a spring scale calibrated in newtons that is attached to the ceiling of an elevator (Figure 4-47). What does the scale read when (*a*) the elevator is moving up with a constant velocity of 30 m/s, (*b*) the elevator is moving down with a constant velocity of 30 m/s, (*c*) the elevator is ascending at 20 m/s and gaining speed at a rate of 3 m/s²? (*d*) From $t = 0$ to $t = 5$ s, the elevator moves up at 10 m/s. Its velocity is then reduced uniformly to zero in the next 4 s, so that it is at rest at $t = 9$ s. Describe the reading of the scale during the interval $0 < t < 9$ s.

FIGURE 4-47 Problem 67

Free-Body Diagrams: Ropes, Tension, and Newton's Third Law

68 • Two boxes of mass m_1 and m_2 connected by a massless string are accelerated uniformly on a frictionless surface, as shown in Figure 4-48. The ratio of the tensions T_1/T_2 is given by (a) m_1/m_2, (b) m_2/m_1, (c) $(m_1 + m_2)/m_2$, (d) $m_1/(m_1 + m_2)$, (e) $m_2/(m_1 + m_2)$.

FIGURE 4-48 Problem 68

69 •• A box of mass $m_2 = 3.5$ kg rests on a frictionless horizontal shelf and is attached by strings to boxes of masses $m_1 = 1.5$ kg and $m_3 = 2.5$ kg, which hang freely, as shown in Figure 4-49. Both pulleys are frictionless and massless. The system is initially held at rest. After it is released, find (a) the acceleration of each of the boxes and (b) the tension in each string.

FIGURE 4-49 Problem 69

70 •• **SSM** Two blocks are in contact on a frictionless horizontal surface. The blocks are accelerated by a horizontal force $\vec{F}$ applied to one of them (Figure 4-50). Find the acceleration and the contact force for (a) general values of F, m_1, and m_2, and (b) for $F = 3.2$ N, $m_1 = 2$ kg, and $m_2 = 6$ kg.

FIGURE 4-50 Problems 70 and 71

71 •• Repeat Problem 70, but with the two blocks interchanged.

72 •• **ISOLVE** ✓ Two 100-kg boxes are dragged along a frictionless surface with a constant acceleration of 1.0 m/s², as shown in Figure 4-51. Each rope has a mass of 1 kg. Find the force F and the tension in the ropes at points A, B, and C.

FIGURE 4-51 Problem 72

73 •• A block of mass m is being lifted vertically by a rope of mass M and length L. The rope is being held at its top end, and the rope and block are accelerating upward with acceleration a. The distribution of mass in the rope is uniform. Show that the tension in the rope at a distance x ($< L$) above the block is $(a + g)[m + (x/L)M]$.

74 •• **SSM** **ISOLVE** A chain consists of 5 links, each having a mass of 0.1 kg. The chain is lifted vertically with an upward acceleration of 2.5 m/s². The chain is held at the top link; no point of the chain touches the floor. Find (a) the force F exerted on the top of the chain, (b) the net force on each link, and (c) the force each link exerts on the link below it.

75 • A 40.0-kg object supported by a vertical rope is initially at rest. The object is then accelerated upward. The tension in the rope needed to give the object an upward speed of 3.50 m/s in 0.700 s is (a) 590 N, (b) 390 N, (c) 200 N, (d) 980 N, (e) 720 N.

76 • **ISOLVE** ✓ A hovering helicopter of mass m_h is lowering a truck of mass m_t. If the truck's downward speed is increasing at the rate of 0.1g, what is the tension in the supporting cable? (a) 1.1$m_t g$. (b) $m_t g$. (c) 0.9$m_t g$. (d) 1.1($m_h + m_t$)g. (e) 0.9($m_h + m_t$)g.

77 •• Two objects are connected by a massless string, as shown in Figure 4-52. The incline and pulley are frictionless. Find the acceleration of the objects and the tension in the string for (a) general values of θ, m_1, and m_2, and (b) $\theta = 30°$ and $m_1 = m_2 = 5$ kg.

FIGURE 4-52 Problem 77

78 • **ISOLVE** In a stage production of Peter Pan, the 50-kg actress playing Peter has to fly in vertically, and to be in time with the music, she must be lowered, starting from rest, a distance of 3.2 m in 2.2 s. Backstage, a smooth surface sloped at 50° supports a counterweight of mass m, as shown in Figure 4-53. Show the calculations that the stage manager must perform to find (a) the mass of the counterweight that must be used and (b) the tension in the wire.

FIGURE 4-53 Problem 78

79 •• **ISOLVE** An 8-kg block and a 10-kg block connected by a rope that passes over a frictionless peg slide on frictionless inclines, as shown in Figure 4-54. (*a*) Find the acceleration of the blocks and the tension in the rope. (*b*) The two blocks are replaced by two others of masses m_1 and m_2 such that there is no acceleration. Find whatever information you can about the masses of these two new blocks.

FIGURE 4-54 Problem 79

80 •• A heavy rope of length 5 m and mass 4 kg lies on a frictionless horizontal table. One end is attached to a 6-kg block. At the other end of the rope, a constant horizontal force of 100 N is applied. (*a*) What is the acceleration of the system? (*b*) Give the tension in the rope as a function of position along the rope.

81 •• **SSM** A 60-kg house-painter stands on a 15-kg aluminum platform. The platform is attached to a rope that passes through an overhead pulley, which allows the painter to raise herself and the platform (Figure 4-55). (*a*) To accelerate herself and the platform at a rate of 0.8 m/s², with what force F must she pull on the rope? (*b*) When her speed reaches 1 m/s, she pulls in such a way that she and the platform go up at a constant speed. What force is she exerting on the rope? (Ignore the mass of the rope.)

FIGURE 4-55 Problem 81

82 ••• Figure 4-56 shows a 20-kg block sliding on a 10-kg block. All surfaces are frictionless. Find the acceleration of each block and the tension in the string that connects the blocks.

FIGURE 4-56 Problem 82

83 ••• **ISOLVE** ✓ A 20-kg block with a pulley attached slides along a frictionless ledge. It is connected by a massless string to a 5-kg block via the arrangement shown in Figure 4-57. (*a*) Find the horizontal distance the 20-kg block moves when the 5-kg block descends a distance of 10 cm. (*b*) Find the acceleration of each block and the tension in the connecting string.

FIGURE 4-57 Problem 83

Free-Body Diagrams: The Atwood's Machine

84 •• **SSM** The apparatus in Figure 4-58 is called an *Atwood's machine* and is used to measure the free-fall acceleration g by measuring the acceleration of the two blocks. Assuming a massless, frictionless pulley and a massless string, show that the magnitude of the acceleration of either body and the tension in the string are

$$a = \frac{m_1 - m_2}{m_1 + m_2}g \quad \text{and} \quad T = \frac{2m_1 m_2 g}{m_1 + m_2}$$

FIGURE 4-58
Problems 84–87

85 •• **ISOLVE** If one of the masses of the Atwood's machine in Figure 4-58 is 1.2 kg, what should be the other mass so that the displacement of either mass during the first second following release is 0.3 m?

86 •• A very small pebble of mass m rests on the block of mass m_2 of the Atwood's machine in Figure 4-58. Find the force exerted by the pebble on m_2.

87 •• Find the force exerted by the Atwood's machine on the hanger to which the pulley is attached, as shown in Figure 4-58, while the blocks accelerate. Neglect the mass of the pulley. Check your answer by considering limiting values for m_1 and/or m_2 for which you can determine the answer by qualitative reasoning.

88 ••• The acceleration of gravity g can be determined by measuring the time t it takes for a mass m_2 in an Atwood's machine to fall a distance L, starting from rest. (*a*) Find an expression for g in terms of m_1, m_2, L, and t. (*b*) Show that if there is a small error in the time measurement dt, it will lead to an error in the determination of g by an amount dg given by $dg/g = -2dt/t$. (*c*) If $L = 3$ m and m_1 is 1 kg, find the value of m_2 such that g can be measured with an accuracy of ±5% with a time measurement that is accurate to 0.1 s. Assume that the only significant uncertainty in the measurement is the time of fall.

89 •• [SSM] You are given an Atwood's machine and a set of weights whose total mass is M. You are told to attach some of the weights to one side of the machine, and the rest to the other side. If m_1 represents the mass attached to the left side and m_2 is the mass attached to the right side, the tension in the rope is given by the expression

$$T = \frac{2m_1m_2}{m_1 + m_2} g$$

as was shown in Problem 85. Show that the tension will be greatest when $m_1 = m_2 = M/2$.

90 ••• An Atwood's machine has a fixed mass m_1 attached on one side and variable mass m_2 ($> m_1$) on the other side. (a) Show that the largest possible magnitude of the tension in the rope is $2m_1g$. (b) Interpret this result physically, without the use of calculus.

General Problems

91 • A redheaded woodpecker hits the bark of a tree extremely hard—the speed of its head reaches approximately $v = 3.5$ m/s before impact. If the mass of the bird's head is 0.060 kg, and the average force acting on the head during impact is $F = 6.0$ N, find (a) the acceleration of its head (assuming it is constant), (b) the depth of penetration into the bark, and (c) the time t it takes the woodpecker's head to stop.

92 •• [SSM] A simple accelerometer can be made by suspending a small object from a string attached to a fixed point on an accelerating object. Suppose such an accelerometer is attached to the ceiling of an automobile traveling on a large flat surface. When there is acceleration, the object will deflect and the string will make some angle with the vertical. (a) How is the direction in which the suspended object is deflected related to the direction of the acceleration? (b) Show that the acceleration a is related to the angle θ that the string makes by $a = g \tan \theta$. (c) Suppose the automobile brakes to rest from 50 km/h in a distance of 60 m. What angle will the accelerometer make? Will the object swing forward or backward?

93 •• [SOLVE] The mast of a sailboat is supported at bow and stern by stainless steel wires, the forestay and backstay, anchored 10 m apart (Figure 4-59). The 12-m long mast weighs 800 N and stands vertically on the deck of the boat. The mast is positioned 3.6 m behind where the forestay is attached. The tension in the forestay is 500 N. Find the tension in the backstay and the force that the mast exerts on the deck.

FIGURE 4-59 Problem 93

94 •• [SOLVE] A large uniform chain is hanging from the ceiling, supporting a block of mass 50 kg. The mass of the chain itself is 20 kg, and the length of the chain is 1.5 m. Determine the tension in the chain (a) at the point where the chain is supporting the block, (b) midway up the chain, and (c) at the top of the chain where it is attached to the ceiling.

95 ••• [SSM] [SOLVE] A man pushes a 24-kg box across a frictionless floor. The box begins moving from rest. He initially pushes on the box gently, but gradually increases his force so that the force he exerts on the box varies in time as $F = (8$ N/s$)t$. After 3 s, he stops pushing the box. The force is always exerted in the same direction. (a) What is the velocity of the box after 3 s? (b) How far has the man pushed the box in 3 s? (c) What is the average velocity of the box between 0 s and 3 s? (d) What is the average force that the man exerts on the box while he is pushing it?

96 •• Suppose that a frictionless surface is inclined at an angle of 30° to the horizontal. The 270-g block is attached to a 75-g hanging weight using a pulley, as shown in Figure 4.60. (a) Draw two free-body diagrams, one for the block and the other for the hanging weight. (b) Find the tension in the string and the acceleration of the block. (c) The block is released from rest. How long does it take for it to slide a distance of 1.00 m down the surface?

FIGURE 4-60 Problem 96

97 •• A box of mass m_1 is pulled along a smooth horizontal surface by a force F exerted at the end of a rope that has a much smaller mass m_2, as shown in Figure 4-61. (a) Find the acceleration of the rope and block, assuming them to be one object. (b) What is the net force acting on the rope? (c) Find the tension in the rope at the point where it is attached to the block. (d) The diagram, with the rope perfectly horizontal along its length, is not quite accurate. Correct the diagram and state how this correction affects your solution.

FIGURE 4-61 Problem 97

98 •• [SSM] [SOLVE] A 2-kg block rests on a frictionless wedge that has an inclination of 60° and an acceleration a to the right such that the mass remains stationary relative to the wedge (Figure 4-62). (a) Find a. (b) What would happen if the wedge were given a greater acceleration?

FIGURE 4-62 Problem 98

99 •• **iSOLVE** ✓ The masses attached to each side of an Atwood's machine consist of a stack of five washers, each of mass m, as shown in Figure 4-63. The tension in the string is T_0. When one of the washers is removed from the left side, the remaining washers accelerate and the tension decreases by 0.3 N. (a) Find m. (b) Find the new tension and the acceleration of each mass when a second washer is removed from the left side.

FIGURE 4-63 Problems 99 and 100

100 •• Consider the Atwood's machine in Figure 4-63. When N washers are transferred from the left side to the right side, the right side drops 47.1 cm in 0.40 s. Find N.

101 •• Blocks of mass m and $2m$ are connected by a string (Figure 4-64). (a) If the forces are constant, find the tension in the connecting string. (b) If the forces vary with time as $F_1 = Ct$ and $F_2 = 2Ct$, where C is a constant and t is time, find the time t_0 at which the tension in the string is T_0.

FIGURE 4-64 Problem 101

102 ••• **SSM** **iSOLVE** The pulley in an Atwood's machine is given an upward acceleration a, as shown in Figure 4-65. Find the acceleration of each mass and the tension in the string that connects them. *Hint: a constant upward acceleration has the same effect as an increase in the acceleration due to gravity.*

FIGURE 4-65 Problem 102

Applications of Newton's Laws

 **What factors determine
how fast the car can go around
the corner without skidding? (See
Example 5-10.)**

I n Chapter 4 we introduced Newton's laws and applied them to situations where action was restricted to straight-line motion, and the quantitative effects of friction were excluded.

➤ **In this chapter we will extend the application of Newton's laws to motion along curved paths, and we will include the quantitative effects of friction.**

5-1 Friction

Without friction our ground-based transportation system, from walking to automobiles, could not function. To start walking on a horizontal surface requires friction, and once you are already walking, friction is required if you are to change either speed or direction. Friction is required to hold a nut on a screw or a nail in wood. As important as friction is, it is often not desirable. Lubricants, such as motor oil in an automobile engine, or synovial fluid in our joints, can reduce friction.

Static Friction

When you apply a small horizontal force $\vec{F}$ (see Figure 5-1) to a large box resting on the floor, the box may not move because the force of **static friction** $\vec{f}_s$ exerted by the floor on the box, balances the force you are applying. The force of static friction, which opposes the applied force on the box, can vary from zero to some maximum force $f_{s,max}$, depending on how hard you push. Data show that $f_{s,max}$ is proportional to the normal force exerted by one surface on the other:

FIGURE 5-1

$$f_{s,max} = \mu_s F_n \qquad \text{5-1}$$

DEFINITION—COEFFICIENT OF STATIC FRICTION

where the proportionality constant μ_s, called the **coefficient of static friction,** depends on the nature of the surfaces in contact. If you exert a horizontal force smaller than $f_{s,max}$ on the box, the frictional force will just balance this horizontal force. In general, we can write

$$f_s \leq \mu_s F_n \qquad \text{5-2}$$

Kinetic Friction

If you push the box in Figure 5-1 hard enough, it will slide across the floor. As it slides, the floor exerts a force of **kinetic friction** $\vec{f}_k$ (also called sliding friction) that opposes the motion. To keep the box sliding with constant velocity, you must exert a force on the box that is equal in magnitude and opposite in direction to the force of kinetic friction exerted by the floor.

The **coefficient of kinetic friction** μ_k is the ratio of the magnitudes of the kinetic frictional force f_k and the normal force F_n:

$$f_k = \mu_k F_n \qquad \text{5-3}$$

DEFINITION—COEFFICIENT OF KINETIC FRICTION

where μ_k depends on the nature of the surfaces in contact. Experimentally, it is found that μ_k is less than μ_s, and is approximately constant for speeds ranging from about 1 cm/s to several meters per second—the only situations we will consider.

Rolling Friction

When an ideal, rigid wheel rolls *at constant speed* along an ideal, rigid horizontal road without slipping, no frictional force slows its motion. However, because real tires and roads continually deform and because the tread and the road are continually peeled apart, the road exerts a force of **rolling friction** $\vec{f}_r$ that opposes the motion. To keep the wheel rolling with constant velocity, you must exert a force on the wheel that is equal in magnitude and opposite in direction to the force of rolling friction exerted by the road.

The **coefficient of rolling friction** μ_r is the ratio of the magnitudes of the rolling frictional force f_r and the normal force F_n:

$$f_r = \mu_r F_n \qquad \text{5-4}$$

DEFINITION—COEFFICIENT OF ROLLING FRICTION

where μ_r depends on the nature of the surfaces in contact and the composition of the wheel and road. Typical values of μ_r are 0.01 to 0.02 for rubber tires on concrete and 0.001 to 0.002 for steel wheels on steel rails.

Friction Explained

Friction is a complex, incompletely understood phenomenon that arises from the attraction of molecules between two surfaces that are in close contact. The nature of this attraction is electromagnetic—the same as the molecular bonding that holds an object together. This attractive force is short ranged and becomes negligible at distances of only a few atomic diameters.

As shown in Figure 5-2, ordinary objects that look smooth and feel smooth are rough and bumpy at the microscopic (atomic) scale. This is the case even if the surfaces are highly polished. When surfaces come into contact, they touch only at widely spaced prominences, called asperities, shown in Figure 5-2. The normal force exerted by a surface is exerted at the tips of these asperities where the force per unit area is very large, large enough to flatten the tips of the asperities. As the normal force increases, so does this flattening, resulting in a larger microscopic contact area. Under a wide range of conditions the microscopic area of contact is proportional to the normal force. The frictional force is proportional to the microscopic contact area, so it is also proportional to the normal force.

FIGURE 5-2 The microscopic area of contact between box and floor is only a small fraction of the macroscopic area of the box's bottom surface. This fraction is proportional to the normal force exerted between the surfaces. If the box rests on its side, the macroscopic area is increased, but the force per unit area is decreased so the microscopic area of contact is unchanged. Whether the box is upright or on its side, the same horizontal applied force F is required to keep it sliding at constant speed.

Magnified section of a polished steel surface showing surface irregularities. The irregularities are about 5×10^{-5} cm high, corresponding to several thousand atomic diameters.

The computer graphic shows gold atoms (bottom) adhering to the fine point of a nickel probe (top) that has been in contact with the gold surface.

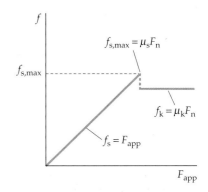

Figure 5-3 shows a plot of the frictional force exerted on the box by the floor as a function of the applied force. The force of friction balances the applied force until the applied force equals $\mu_s F_n$, at which point the box begins to slide. Then the frictional force is constant and equal to $\mu_k F_n$. Table 5-1 lists some approximate values of μ_s and μ_k for various surfaces.

FIGURE 5-3

TABLE 5-1

Approximate Values of Frictional Coefficients

Materials	μ_S	μ_k
Steel on steel	0.7	0.6
Brass on steel	0.5	0.4
Copper on cast iron	1.1	0.3
Glass on glass	0.9	0.4
Teflon on Teflon	0.04	0.04
Teflon on steel	0.04	0.04
Rubber on concrete (dry)	1.0	0.80
Rubber on concrete (wet)	0.30	0.25
Waxed ski on snow (0°C)	0.10	0.05

A GAME OF SHUFFLEBOARD　　　　　　　　　　**EXAMPLE 5-1**

A cruise-ship passenger uses a shuffleboard cue to push a shuffleboard disk of mass 0.40 kg horizontally along the deck so that the disk leaves the cue with a speed of 5.5 m/s. The disk then slides a distance of 8 m before coming to rest. Find the coefficient of kinetic friction between the disk and the deck.

PICTURE THE PROBLEM The force of kinetic friction is the only horizontal force acting on the disk after it separates from the cue. Since the frictional force is constant, the acceleration is constant. We can find the acceleration using the constant-acceleration equations of Chapter 2 and relate it to μ_k using $\Sigma F_x = ma_x$.

FIGURE 5-4

1. Draw a free-body diagram for the disk after it leaves the cue (Figure 5-4). The arrows on the x and y axes indicate the positive x and y directions.

2. The coefficient of friction relates the frictional and normal forces:

$$f_k = \mu_k F_n$$

3. Apply $\Sigma F_y = ma_y$ to the disk. Solve for the normal force. Then, using the step 2 result, solve for the frictional force:

$$\Sigma F_y = ma_y \Rightarrow F_n - mg = 0$$

so

$$F_n = mg \quad \text{and} \quad f_k = \mu_k mg$$

4. Apply $\Sigma F_x = ma_x$ to the disk. Using the step 3 result, solve for the acceleration:

$$\Sigma F_x = ma_x \Rightarrow -f_k = ma_x$$

so

$$-\mu_k mg = ma_x \quad \text{and} \quad a_x = -\mu_k g$$

5. The acceleration is constant. Relate it to the total distance traveled and the initial velocity using Equation 2-15. Using the step 4 result, solve for μ_k:

$$v^2 = v_0^2 + 2a_x \Delta x \Rightarrow 0 = v_0^2 - 2\mu_k g \, \Delta x$$

so

$$\mu_k = \frac{v_0^2}{2g \, \Delta x} = \frac{(5.5 \text{ m/s})^2}{2(9.81 \text{ m/s}^2)(8 \text{ m})} = \boxed{0.193}$$

REMARKS Note that the mass m of the disk cancels. The greater the mass, the harder it is to stop the disk, but the greater mass is also accompanied by greater friction. The net result is that mass has no effect.

Example 5-1 illustrates guidelines for solving problems involving friction. They are as follows:

1. Choose the y axis normal to the contacting surfaces. Choose the x axis parallel with the surface and either parallel or antiparallel with the frictional force.

2. Apply $\Sigma F_y = ma_y$ and solve for the normal force F_n.
   • If the friction is *kinetic*, solve for the frictional force using $f_k = \mu_k F_n$.
   • If the friction is *static*, relate the maximum frictional force to the normal force using $f_{s,max} = \mu_s F_n$.
   • If the friction is *rolling*, solve for the frictional forces using $f_r = \mu_r F_n$.

3. Apply $\Sigma F_x = ma_x$ to the object and solve for the desired quantity.

SOLVING PROBLEMS INVOLVING FRICTION

A SLIDING COIN   **E X A M P L E   5 - 2**

A hardcover book is resting on a table top with its front cover facing upward. You place a coin on this cover and very slowly open the book until the coin starts to slide. The angle θ_{max} is the angle the front cover makes with the horizontal just as the coin starts to slide. Find the coefficient of static friction μ_s between the book cover and the coin in terms of θ_{max}.

PICTURE THE PROBLEM The forces acting on the coin are its weight mg, the normal force F_n exerted by the plane, and the force of friction f. We should follow the guidelines for solving problems with static friction.

FIGURE 5-5

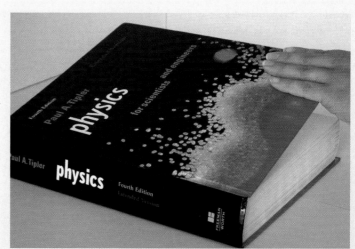

1. Draw a free-body diagram for the coin with the book cover inclined at angle θ, where $\theta \le \theta_{max}$ (Figure 5-6):

FIGURE 5-6

2. Apply $\Sigma F_y = ma_y$ to the coin and solve for the normal force. Then determine the maximum static frictional force:

$$\Sigma F_y = ma_y \Rightarrow F_n - mg \cos \theta = 0$$

so

$$F_n = mg \cos \theta$$

$$f_{s,max} = \mu_s F_n \quad \text{so} \quad f_{s,max} = \mu_s mg \cos \theta_{max}$$

3. Apply $\Sigma F_x = ma_x$ to the coin and solve for the frictional force. Then substitute into the step 2 result:

$$\Sigma F_x = ma_x \Rightarrow -f_{s,max} + mg \sin \theta_{max} = 0$$

so

$$f_{s,max} = mg \sin \theta_{max} \quad \text{and} \quad mg \sin \theta_{max} = \mu_s mg \cos \theta_{max}$$

4. Solve the step 3 result for μ_s:

$$\mu_s = \frac{mg \sin \theta_{max}}{mg \cos \theta_{max}} = \boxed{\tan \theta_{max}}$$

EXERCISE The coefficient of static friction between a car's tires and the road on a particular day is 0.7. What is the steepest angle of inclination of the road for which the car can be parked with its wheels locked and not slide down the hill? (*Answer* 35°)

From Example 5-2 we see that the coefficient of static friction is related to the angle θ_{max} at which an object begins to slip by

$$\mu_s = \tan \theta_{max} \qquad\qquad 5\text{-}5$$

ANGLE OF REPOSE

and θ_{max} is called the angle of repose.

PULLING A SLED   **EXAMPLE 5-3**

Two children sitting on a sled at rest in the snow ask you to pull them. You oblige by pulling on the sled's rope, which makes an angle of 40° with the horizontal (Figure 5-7). The children have a combined mass of 45 kg and the sled has a mass of 5 kg. The coefficients of static and kinetic friction are $\mu_s = 0.2$ and $\mu_k = 0.15$. Find the frictional force exerted by the ground on the sled and the acceleration of the children and sled, starting from rest, if the tension in the rope is (*a*) 100 N and (*b*) 140 N.

PICTURE THE PROBLEM First we need to find out whether the frictional force is static or kinetic. To do this we solve for the maximum tension in the rope without the sled sliding.

FIGURE 5-7

(*a*) 1. Draw a free-body diagram for the sled (Figure 5-8):

FIGURE 5-8

2. Apply $\Sigma F_y = ma_y$ to the sled and solve for the normal force. Then solve for the maximum static frictional force:

$$\Sigma F_y = ma_y \Rightarrow F_n + T \sin \theta - mg = 0$$

so

$$F_n = mg - T \sin \theta$$

$$f_{s,max} = \mu_s F_n \quad \text{so} \quad f_{s,max} = \mu_s(mg - T_{max} \sin \theta)$$

3. Apply $\Sigma F_x = ma_x$ to the sled and solve for the frictional force. Then substitute into the step 2 result:

$$\Sigma F_x = ma_x \Rightarrow -f_{s,max} + T_{max} \cos \theta = 0$$

so

$$f_{s,max} = T_{max} \cos \theta \quad \text{and} \quad T_{max} \cos \theta = \mu_s(mg - T_{max} \sin \theta)$$

$$T_{max} \cos \theta = \mu_s(mg - T_{max} \sin \theta)$$

4. Solve the step 3 result for the maximum tension without slipping:

$$T_{max}(\mu_s \sin \theta + \cos \theta) = \mu_s mg$$

so

$$T_{max} = \frac{\mu_s mg}{\mu_s \sin \theta + \cos \theta}$$

$$= \frac{0.2(50 \text{ kg})(9.81 \text{ m/s}^2)}{0.2 \sin 40° + \cos 40°} = 110 \text{ N}$$

5. The tension is 100 N, which is less than 110 N. The sled is *not* sliding. To find the frictional force, use the step 3 expression for f_s:

$$a_x = \boxed{0}$$

$$f_s = T \cos \theta = (100 \text{ N})\cos 40° = \boxed{76.6 \text{ N}}$$

(b) 1. The tension is 140 N, which is greater than $T_{max} = 110 \text{ N}$, so the sled is sliding. Relate the kinetic frictional force f_k to the normal force:

$$f_k = \mu_k F_n$$

2. In step (a)2 we applied $\Sigma F_y = ma_y$ to the sled and found $F_n = mg - T \sin \theta$. Use this result along with the step (b)1 result to solve for the kinetic frictional force:

$$f_k = \mu_k(mg - T \sin \theta)$$

$$= 0.15\left[(50 \text{ kg})(9.81 \text{ N/kg}) - (140 \text{ N}) \sin 40°\right]$$

$$= \boxed{60.1 \text{ N}}$$

3. Apply $\Sigma F_x = ma_x$ to the sled and solve for the frictional force. Then substitute the step 2 result for f_k and solve for the acceleration:

$$\Sigma F_x = ma_x \Rightarrow -f_k + T \cos \theta = ma_x$$

so

$$a_x = \frac{-f_k + T \cos \theta}{m}$$

$$= \frac{(-60.1 \text{ N}) + (140 \text{ N})\cos 40°}{50 \text{ kg}}$$

$$= \boxed{0.943 \text{ m/s}^2}$$

REMARKS There are two important points to note about this example: (1) the normal force is not equal to the weight of the children and the sled. That is because the vertical component of the tension helps lift the sled off the ground. (2) In Part (a), the force of static friction is less than $\mu_s F_n$.

A SLIDING BLOCK **EXAMPLE 5-4** Try It Yourself

The block of mass m_2 in Figure 5-9 has been adjusted so that the block of mass m_1 is on the verge of sliding. (a) If $m_1 = 7$ kg and $m_2 = 5$ kg, what is the coefficient of static friction between the table and the block? (b) With a slight nudge, the blocks move with acceleration a. Find a if the coefficient of kinetic friction between the table and the block is $\mu_k = 0.54$.

FIGURE 5-9

PICTURE THE PROBLEM Apply Newton's second law to each block. By neglecting the masses of both the rope and the pulley, and by neglecting friction in the pulley bearing, the tension has the same magnitude throughout the rope, so $T_1 = T_2 = T$, and, because the rope remains taut but does not stretch, the accelerations have the same magnitude, so $a_1 = a_2 = a$.

To find the coefficient of static friction μ_s, as required in Part (a), set the force of static friction on m_1 equal to its maximum value $f_{max} = \mu_s F_n$ and set the acceleration equal to zero.

Cover the column to the right and try these on your own before looking at the answers.

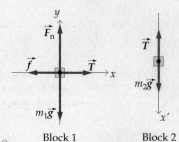

FIGURE 5-10

Block 1   Block 2

Steps   **Answers**

(a) 1. Draw a free-body diagram for each block (Figure 5-10). Choose the positive x and x' directions to be the same as the directions of the accelerations once the blocks are moving.

2. Apply $\Sigma F_y = ma_y$ to block 1 and solve for the normal force. Then solve for the static frictional force.

$$\Sigma F_y = m_1 a_{1y} \Rightarrow F_n - m_1 g = 0$$

so

$$F_n = m_1 g$$

$$f_{s,max} = \mu_s F_n \quad \text{so} \quad f_{s,max} = \mu_s m_1 g$$

3. Apply $\Sigma F_x = ma_x$ to block 1 and solve for the frictional force. Then substitute into the step 2 result.

$$\Sigma F_x = m_1 a_{1x} \Rightarrow T - f_{s,max} = 0$$

so

$$f_{s,max} = T \quad \text{and} \quad T = \mu_s m_1 g$$

4. Apply $\Sigma F_x = ma_x$ to block 2 and solve for the tension. Then substitute into the step 3 result.

$$\Sigma F_{x'} = m_2 a_{2x'} \Rightarrow m_2 g - T = 0$$

so

$$T = m_2 g \quad \text{and} \quad m_2 g = \mu_s m_1 g$$

5. Solve the step 4 result for μ_s

$$\mu_s = \frac{m_2}{m_1} = \frac{5 \text{ kg}}{7 \text{ kg}} = \boxed{0.714}$$

(b) 1. During sliding the frictional force is kinetic. Relate the kinetic frictional force f_k to the normal force. The normal force was found in step 2 of Part (a).

$$f_k = \mu_k F_n$$

so

$$f_k = \mu_k m_1 g$$

2. Apply $\Sigma F_x = ma_x$ to block 1. Then substitute for the frictional force using the result from step 1 of Part (b).

$$\Sigma F_x = m_1 a_{1x} \Rightarrow T - f_k = m_1 a$$

so

$$T - \mu_k m_1 g = m_1 a$$

3. Apply $\Sigma F_{x'} = ma_{x'}$ to block 2.

$$\Sigma F_{x'} = m_2 a_{2x'} \Rightarrow m_2 g - T = m_2 a$$

4. Add the equations in steps 2 and 3 of Part (b) and solve for a.

$$a = \frac{m_2 - \mu_k m_1}{m_1 + m_2} g = \boxed{0.997 \text{ m/s}^2}$$

PLAUSIBILITY CHECK Note that $\mu_k = 0$ gives the expression for the acceleration derived in Example 4-12 with $\theta = 0$.

EXERCISE What is the tension in the rope when the blocks are sliding? (*Answer* $T = m_2(g - a) = 44.1$ N)

THE RUNAWAY BUGGY   **EXAMPLE 5-5**

A runaway baby buggy is sliding without friction across a frozen pond toward
a hole in the ice (Figure 5-11). You race after the buggy on skates. As you grab
it, you and the buggy are moving toward the hole at speed v_0. The coefficient
of friction between your skates and the ice as you turn out the blades to brake
is μ_k. D is the distance to the hole when you reach the buggy, M is the total
mass of the buggy, and m is your mass. (*a*) What is the lowest value of D such
that you stop the buggy before it reaches the hole in the ice? (*b*) What force do
you exert on the buggy?

PICTURE THE PROBLEM Initially, you and the
buggy are moving toward the hole with speed v_0,
which we take to be in the positive x direction. If you
exert a force $\vec{F} = -F\hat{i}$ on the buggy, the buggy, in accord with Newton's third law, exerts a force $\vec{F'} = F\hat{i}$
on you. Apply Newton's second law to determine the
acceleration. After finding the acceleration, find the
distance D the buggy travels while slowing to a stop.
The lowest value of D is that for which your speed
reaches zero just as the buggy reaches the hole.

FIGURE 5-11

(*a*) 1. Draw separate free-body diagrams for yourself
and the buggy.

Yourself   Buggy   **FIGURE 5-12**

2. Apply $\Sigma F_y = ma_y$ to yourself and solve first for the
normal force and then for the frictional force:

$$\Sigma F_y = ma_y \Rightarrow F_n - mg = 0$$

and

$$f_k = \mu_k F_n \quad \text{so} \quad f_k = \mu_k mg$$

3. Apply $\Sigma F_x = ma_x$ to yourself. Then substitute in the
step 2 result:

$$\Sigma F_x = ma_x \Rightarrow F - f_k = ma_x$$

so

$$F - \mu_k mg = ma_x$$

4. Apply $\Sigma F_x = ma_x$ to the buggy. Then substitute for F
in the step 3 result:

$$\Sigma F_x = Ma_x \Rightarrow -F = Ma_x$$

so

$$-Ma_x - \mu_k mg = ma_x$$

5. Solve the step 4 result for a_x:

$$a_x = -\frac{\mu_k}{1 + M/m}g$$

(The acceleration is negative, as expected.)

6. Substitute the step 5 result into a kinematic equation
and solve for D:

$$v_x^2 = v_{0x}^2 + 2a_x\Delta x \Rightarrow 0 = v_0^2 + 2a_x D$$

so

$$D = -\frac{v_0^2}{2a_x} = \boxed{\left(1 + \frac{M}{m}\right)\frac{v_0^2}{2\mu_k g}}$$

(*b*) F can be found from Newton's second law applied to
the buggy:

$$F = -Ma_x \Rightarrow F = \boxed{\frac{\mu_k M}{1 + M/m}g}$$

REMARKS The minimum value of D is proportional to v_0^2 and inversely proportional to μ_k. Figure 5-13 shows the stopping distance D versus initial velocity squared for values of M/m equal to 0.1, 0.3, and 1.0, with $\mu_k = 0.5$. Note that the larger the mass ratio M/m, the greater the distance D needed to stop for a given initial velocity. This is akin to stopping a car that is pulling a trailer that does not have its own brakes. The mass of the trailer increases the stopping distance for a given speed.

FIGURE 5-13

PULLING A CHILD ON A TOBOGGAN   **EXAMPLE 5-6**   **Try It Yourself**

A child of mass m_c sits on a toboggan of mass m_t, which in turn sits on a frozen pond assumed to be frictionless (Figure 5-14). The toboggan is pulled with a horizontal force as shown. The coefficients of static and sliding friction between the child and toboggan are μ_s and μ_k. (a) Find the maximum value of F for which the child will not slide relative to the toboggan. (b) Find the acceleration of the toboggan and child when F is greater than this value.

FIGURE 5-14

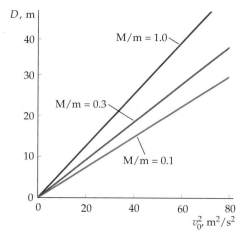

PICTURE THE PROBLEM The only force accelerating the child forward is the frictional force exerted by the toboggan on the child. In Part (a) the challenge is to find F when this frictional force is static and maximum. To do this, apply $\Sigma \vec{F} = m\vec{a}$ to the child and solve for the acceleration when static frictional force is maximum. Then apply $\Sigma \vec{F} = m\vec{a}$ to the toboggan and solve for F. In Part (b), we follow a parallel procedure. However, in this part F is given and we solve for the acceleration of the toboggan.

Cover the column to the right and try these on your own before looking at the answers.

Steps

Answers

(a) 1. Draw a free-body diagram for each object (Figure 5-15).

Two alternative styles for the free-body diagram are shown for the toboggan. In the first, the two downward forces are displaced slightly so they can be seen. In the second, they are drawn head to tail.

FIGURE 5-15

2. Equate the magnitudes of the forces in each action–reaction force pair appearing in the two free-body diagrams. Express the relation between the accelerations due to the nonslipping constraint.

$F'_n = F_n$ and $f'_{s,max} = f_{s,max}$

and

$a_{cx} = a_{tx} = a_x$

3. Apply $\Sigma F_y = ma_y$ to the child. Solve first for the normal force, then the frictional force.

$\Sigma F_{cy} = m_c a_y \Rightarrow F_n - m_c g = 0$

$F_n = m_c g$ and

$f_{s,max} = \mu_s F_n$ so $f_{s,max} = \mu_s m_c g$

4. Apply $\Sigma F_x = ma_x$ to the child and solve for the acceleration.

$\Sigma F_{cx} = m_c a_x \Rightarrow f_{s,max} = m_c a_x$

and

$\mu_s m_c g = m_c a_x$, so $a_x = \mu_s g$

5. Apply $\Sigma F_x = ma_x$ to the toboggan and, using the acceleration relations from step 2 and the result from step 3, solve for F.

$$\Sigma F_{tx} = m_t a_x \Rightarrow F - f_{s,max} = m_t a_x$$

and

$$F - \mu_s m_c g = m_t \mu_s g \quad \text{so} \quad \boxed{F = (m_c + m_t)\mu_s g}$$

(b) 1. Equate the magnitudes of each action–reaction force pair and express the change in the relation between the accelerations due to the removal of the nonslipping constraint.

$$F'_n = F_n \quad \text{and} \quad f'_k = f_k$$

but

$$a_{cx} < a_{tx}$$

2. Solve for the kinetic frictional force using the result from step 3 of Part (a) for the normal force.

$$f_k = \mu_k F_n \quad \text{so} \quad f_k = \mu_k m_c g$$

3. Apply $\Sigma F_x = ma_x$ to the child and solve for *her acceleration*.

$$\Sigma F_{cx} = m_c a_{cx} \Rightarrow f_k = m_c a_{cx}$$

and

$$\mu_k m_c g = m_c a_{cx} \quad \text{so} \quad a_{cx} = \mu_k g$$

4. Apply $\Sigma F_x = ma_x$ to the toboggan. Using the result from step 2 of Part (b), solve for *its acceleration*.

$$\Sigma F_{tx} = m_t a_{tx} \Rightarrow F - f_k = m_t a_{tx}$$

and

$$F - \mu_k m_c g = m_t a_{tx} \quad \text{so} \quad a_{tx} = \boxed{\dfrac{F - \mu_k m_c g}{m_t}}$$

REMARKS Friction is a force between two surfaces in contact, and it is not correct that friction always opposes motion, or the tendency to motion. In this example, friction does not oppose the motion of the child, it causes it. It is correct that friction always opposes motion, or the tendency to motion, of each surface *relative to the other surface.* For example, even though the child moves forward *relative to the ice,* she moves, or tends to move, backward (leftward) *relative to the toboggan.* Friction opposes this relative motion or tendency to motion.

Figure 5-16 shows the forces acting on a front-wheel-drive car that is just starting to move from rest on a horizontal road. The weight of the car is balanced by the normal force F_n exerted on the tires. To start the car moving, the engine delivers power to the axle that makes the wheels rotate (we discuss power in Chapter 6). If the road were perfectly frictionless, the wheels would merely spin. When friction is present, the frictional force exerted by the road on the tires is in the forward direction, opposing the tendency of the tire surface to slip backward. This frictional force provides the acceleration needed for the car to start moving forward. If the power delivered by the engine is small enough so that the frictional force exerted by the tire surface on the road surface is sufficiently small, then the two surfaces do not slip. The wheels roll without slipping and the tire tread touching the road is at rest relative to the road. The friction between the road and the tire tread is then static friction. The largest frictional force that the tire can exert on the road (and that the road can exert on the tire) is $\mu_s F_n$.

For a car moving in a straight line with speed v relative to the road, the center of each of its wheels also moves with speed v, as shown in Figure 5-17. If a wheel is rolling without slipping, its top is moving faster than v whereas its bottom is moving slower than v. However, *relative to the car,* each point on the perimeter of the wheel moves in a circle with the same speed v. Moreover, the speed of the point on the tire momentarily in contact with the ground is zero *relative to the ground.* (Otherwise, the tire would be skidding along the road.)

If the engine power is great enough, the tire will slip and the wheels will spin. Then the force that accelerates the car is the force of kinetic friction, which is less than the force of static friction. If we are stuck on ice or snow, our chances of getting free are better if we use a light touch on the accelerator pedal. Similarly,

FIGURE 5-16 Forces acting on a car (front-wheel drive). The normal forces $\vec{F}_n$ are usually larger on the front tires because typically the engine of the car is mounted at the front of the car.

FIGURE 5-17 In this figure, dashed lines represent velocities relative to the body of the car, solid lines represent velocities relative to the ground.

when braking a car to a stop, the force exerted by the road on the tires may be either static friction or kinetic friction, depending on how the brakes are applied. If the brakes are applied so hard that the wheels lock, the tires will skid along the road and the stopping force will be that of kinetic friction. If the brakes are applied less forcefully, so that no slipping occurs between the tires and the road, the stopping force will be that of static friction. Antilock braking systems in cars have wheel-speed sensors. If the control unit senses that a wheel is about to lock, the module signals the brake pressure modulator to drop, hold, and then restore the pressure to that wheel up to 15 times per second. This varying of pressure is much like "pumping" the brake, but with the ABS system, the wheel that is locking is the only one being pumped. This is called threshold braking. With threshold braking, maximum friction for stopping is maintained.

When wheels do lock and tires skid, two undesirable things happen. The minimum stopping distance is increased *and* the ability to control the direction of the car's motion is greatly diminished. Obviously, this loss of directional control can have dire consequences.

THE EFFECT OF ANTILOCK BRAKES   **E X A M P L E   5 - 7**

A car is traveling at 30 m/s along a horizontal road. The coefficients of friction between the road and the tires are $\mu_s = 0.5$ and $\mu_k = 0.3$. How far does the car travel before stopping if (*a*) the car is braked with an antilock braking system so that threshold braking is sustained, and (*b*) the car is braked hard with no antilock braking system so that the wheels lock?

PICTURE THE PROBLEM The force that stops a car when it brakes without skidding is the force of static friction exerted by the road on the tires (Figure 5-18). We use Newton's second law to solve for the frictional force and the car's acceleration. Kinematics is then used to find the stopping distance.

FIGURE 5-18

(*a*) 1. Draw a free-body diagram for the car (Figure 5-19). Treat all four wheels as if they were a single point of contact with the ground. Assume further that the brakes are applied to all four wheels. Let $\vec{f} = \vec{f}' + \vec{f}''$.

FIGURE 5-19

2. Assuming that the acceleration is constant, we use Equation 2-15 to relate the stopping distance Δx to the initial speed v_0. The coefficients of friction change with temperature. Since skidding heats up the tires, these coefficients can be expected to change. Such changes are neglected here:

$$v^2 = v_0^2 + 2a_x\Delta x$$

When $v = 0$,

$$\Delta x = -\frac{v_0^2}{2a_x}$$

3. Apply $\Sigma F_y = ma_y$ to the car. Solve first for the normal force, then for the frictional force:

$$\Sigma F_y = ma_y \Rightarrow F_n - mg = 0, \quad \text{so}$$

$$F_n = mg \quad \text{and}$$

$$f_{s,max} = \mu_s F_n, \quad \text{so} \quad f_{s,max} = \mu_s mg$$

4. Apply $\Sigma F_x = ma_x$ to the car and solve for the acceleration:

$$\Sigma F_x = ma_x \Rightarrow -f_{s,max} = ma_x$$

and

$$-\mu_s mg = ma_x \quad \text{so} \quad a_x = -\mu_s g$$

5. Substituting these results in the equation for Δx in step 2 gives the stopping distance:

$$\Delta x = -\frac{v_0^2}{2a_x} = \frac{v_0^2}{2\mu_s g}$$

$$= \frac{(30 \text{ m/s})^2}{2(0.5)(9.81 \text{ m/s}^2)} = \boxed{91.8 \text{ m}}$$

(b) 1. When the wheels lock, the force exerted by the road on the car is that of kinetic friction. Using reasoning similar to that in Part (a), we obtain for the acceleration:

$$a_x = -\mu_k g$$

2. The stopping distance is then:

$$\Delta x = -\frac{v_0^2}{2a_x} = \frac{v_0^2}{2\mu_k g}$$

$$= \frac{(30 \text{ m/s})^2}{2(0.3)(9.81 \text{ m/s}^2)} = \boxed{153 \text{ m}}$$

REMARKS Notice that the stopping distance is more than 50% greater when the wheels are locked. Also note that the stopping distance is independent of the car's mass—the stopping distance is the same for a subcompact car as for a large truck—provided the coefficients of friction are the same.

EXERCISE What must be the coefficient of static friction between the road and the tires of a four-wheel-drive car if the car is to accelerate from rest to 25 m/s in 8 s? (*Answer* 0.319)

EXPLORING

Newton's laws are not valid in noninertial reference frames. Explore noninertial reference frames, pseudoforces, and cyclones at www.whfreeman.com/tipler5e.

5-2 Motion Along a Curved Path

In Chapter 3 we established that a particle moving with speed v along a curved path with a radius of curvature r has an acceleration component $a_c = v^2/r$ in the centripetal direction (toward the center of curvature), and an acceleration component $a_t = dv/dt$ in the tangential direction.

As with any acceleration, the net force is in the direction of the acceleration. The component of the net force in the centripetal direction is called the **centripetal force**. The centripetal force is not a new force, or a new kind of force. It is merely the name for the net-force component perpendicular to the direction of motion. It may be due to a string, spring, or other contact force such as a normal or frictional force; it may be an action-at-a-distance type of force such as a gravitational force; or it may be any combination of these. It is always directed inward—toward the center of curvature of the path.

SWINGING A PAIL   **EXAMPLE 5-8**

You swing a pail of water in a vertical circle of radius r (Figure 5-20). If its speed is v_t at the top of the circle, find (*a*) the force exerted on the water by the pail at the top of the circle, and (*b*) the minimum value of v_t for the water to remain in the pail. (*c*) What is the force exerted by the pail on the water at the bottom of the circle, where the pail's speed is v_b?

PICTURE THE PROBLEM We apply Newton's second law to find the force exerted by the pail on the water. Since the water moves in a circular path it has an acceleration component v^2/r toward the center of the circle.

(*a*)1. Draw free-body diagrams for the water at the top and bottom of the circle (Figure 5-21). Choose the positive axis direction to be toward the center of the circle in each case.

FIGURE 5-20

FIGURE 5-21

2. Apply $\Sigma F_y = ma_y$ to the water as it passes through the top of the circle with speed v_t. Solve for the force F_P exerted on the water by the pail:

$$\Sigma F_y = ma_y \Rightarrow F_P + mg = m\frac{v_t^2}{r}$$

so

$$\boxed{F_P = m\left(\frac{v_t^2}{r} - g\right)}$$

(b) The pail can push on the water, but not pull on it. The minimum force it can exert is zero. Set $F_P = 0$ and solve for the minimum speed:

$$0 = m\left(\frac{v_{t,min}^2}{r} - g\right) \Rightarrow v_{t,min} = \boxed{\sqrt{rg}}$$

(c) Apply $\Sigma F_{y'} = ma_{y'}$ to the water as it passes through the bottom of its path with speed v_b. Solve for F_P:

$$\Sigma F_{y'} = ma_{y'} \Rightarrow F_P - mg = \frac{mv_b^2}{r}$$

so

$$\boxed{F_P = m\left(\frac{v_b^2}{r} + g\right)}$$

REMARKS Note that there is no arrow for centripetal force in the free-body diagrams. Centripetal force is not a kind of force exerted by some agent; it is just the name for the component of the resultant force in the centripetal direction. When a whirling bucket is at the top of its circle, both gravity and the contact force of the pail contribute to the force in the centripetal direction acting on the water. When the water is moving at the minimum speed at the top of the circle it is in free-fall with acceleration $\vec{g}$. The only force acting on it at this point is its weight, $m\vec{g}$. At the bottom of the circle, F_P must be greater than the weight mg to provide a net force in the centripetal direction (upward).

PLAUSIBILITY CHECK When $v = 0$ at the bottom, $F_P = mg$.

EXERCISE Estimate (a) the minimum speed at the top of the circle and (b) the maximum period of revolution that will keep you from getting wet if you swing a pail of water in a vertical circle at constant speed. (*Answer* (a) Assuming $r \approx 1$ m, we find $v_{t,min} \approx 3$ m/s (b) $T = 2\pi r/v \approx 2$ s)

TETHER BALL | **EXAMPLE 5-9** Try It Yourself

A tether ball of mass m is suspended from a length of rope and travels at constant speed v in a horizontal circle of radius r as shown. The rope makes an angle θ with the vertical. Find (a) the direction of the acceleration, (b) the tension in the rope, and (c) the speed of the ball.

FIGURE 5-22

PICTURE THE PROBLEM Two forces act on the ball; its weight and the tension in the rope (Figure 5-22). The vector sum of these forces is in the direction of the acceleration.

Cover the column to the right and try these on your own before looking at the answers.

Steps | Answers

(a) The ball is moving in a horizontal circle at constant speed. The acceleration is in the centripetal direction. | The acceleration is horizontal and directed from the ball toward the center of the circle it is moving in.

(b) 1. Draw a free-body diagram for the ball. Choose the positive *x* direction as the direction of the ball's acceleration.

FIGURE 5-23

2. Apply $\Sigma F_y = ma_y$ to the ball and solve for the tension T.

$$\Sigma F_y = ma_y \Rightarrow T\cos\theta - mg = 0$$

so

$$\boxed{T = \frac{mg}{\cos\theta}}$$

(c) 1. Apply $\Sigma F_x = ma_x$ to the ball.

$$\Sigma F_x = ma_x \Rightarrow T\sin\theta = m\frac{v^2}{r}$$

2. Substitute $mg/\cos\theta$ for T and solve for v.

$$\frac{mg}{\cos\theta}\sin\theta = m\frac{v^2}{r} \Rightarrow \tan\theta = \frac{v^2}{rg}$$

so

$$v = \boxed{\sqrt{rg\tan\theta}}$$

REMARKS An object attached to a string and moving in a horizontal circle so that the string makes an angle θ with the vertical is called a *conical pendulum*.

When a car rounds a curve on a horizontal road, both the centripetal force and the forward forces are provided by the force of static friction exerted by the road on the tires of the car. If the car is traveling at constant speed, then the forward component of the frictional force is balanced by the rearward-directed forces of air drag and rolling friction. If air drag is negligible, then the forward component of the frictional force is zero.

A ROAD TEST   **EXAMPLE 5-10**   **Put It in Context**

You have a summer job as part of an automobile tire design team. You are testing new prototype tires to see whether or not the tires perform as well as predicted. In a skid test, a new BMW 530i was able to travel at constant speed in a circle of radius 45.7 m in 15.2 s without skidding. (*a*) What was its speed *v*? (*b*) What is the acceleration? (*c*) Assuming air drag and rolling friction to be negligible, what is the minimum value for the coefficient of static friction between the tires and the road?

PICTURE THE PROBLEM Figure 5-24 shows the forces acting on the car. The normal force $\vec{F}_n$ balances the downward force due to gravity $m\vec{g}$. The horizontal force is the force of static friction, which provides the centripetal acceleration. The faster the car travels, the greater the centripetal acceleration. The speed can be found from the circumference of the circle and the period T. This speed puts a lower limit on the maximum value of the coefficient of static friction.

FIGURE 5-24

(a) 1. Draw a free-body diagram for the car (Figure 5-25). The positive *r* direction is away from the center of curvature.

FIGURE 5-25

2. Use "speed equals distance divided by the time" to determine the speed v:

$$v = \frac{2\pi r}{T} = \frac{2\pi(45.7 \text{ m})}{15.2 \text{ s}} = \boxed{18.9 \text{ m/s}}$$

(b) Use v to calculate the acceleration:

$$a_c = \frac{v^2}{r} = \frac{(18.9 \text{ m/s})^2}{(45.7 \text{ m})} = \boxed{7.81 \text{ m/s}^2}$$

$$a_t = \frac{dv}{dt} = \boxed{0}$$

The acceleration is 7.81 m/s² in the centripetal direction.

(c) 1. Apply $\Sigma F_y = ma_y$ to the car. Solve for the normal and maximum frictional force:

$$\Sigma F_y = ma_y \Rightarrow F_n - mg = 0$$

so

$$F_n = mg \quad \text{and} \quad f_{s,max} = \mu_s mg$$

2. Apply $\Sigma F_r = ma_r$ to the car. Substituting from step (c), solve 1, for μ_s:

$$\Sigma F_r = ma_r \Rightarrow -f_{s,max} = m\left(-\frac{v^2}{r}\right)$$

so

$$\mu_s mg = m\frac{v^2}{r} \quad \text{and} \quad \mu_s = \frac{v^2}{rg}$$

$$\mu_s = \frac{(18.9 \text{ m/s})^2}{(45.7 \text{ m})(9.81 \text{ m/s}^2)} = \boxed{0.796}$$

REMARKS A good practice when calculating values to three figures is to calculate all intermediate values to at least four figures. For example, if you use the values shown in Part (b) you obtain $a_c = 7.816$ m/s². This should not be rounded to 7.82 m/s² because in step 2 of Part (a), substituting the exact values gives, to four figures, $v = 18.89$ m/s. Calculating a_c using $v = 18.89$ m/s (rather than 18.9 m/s) gives $a_c = 7.808$ m/s², which rounds up to $a_c = 7.81$ m/s². Storing intermediate values in the memory of your calculator and recalling them for later calculations facilitates this process.

❶ PLAUSIBILITY CHECK If μ_s were equal to 1, the inward force would be equal to mg and the centripetal acceleration would be g. Here μ_s is about 0.8 and the centripetal acceleration is about 0.8 g.

*Banked Curves

If a curved road is not horizontal but banked, the normal force of the road will have a component directed inward toward the center of the circle that will contribute to the centripetal force. The banking angle can be chosen so that, for a given speed, no friction is needed for a car to complete the curve.

As a car rounds the curve, the tire is distorted by the frictional force exerted by the road.

Large centrifuge used for research at Sandia National Laboratories.

FIGURE 5-26

ROUNDING A BANKED CURVE   **EXAMPLE 5 - 1 1**

A curve of radius 30 m is banked at an angle θ. Find θ for which a car can round the curve at 40 km/h even if the road is covered with ice so that friction is negligible.

PICTURE THE PROBLEM In this case only two forces act on the car: gravity and the normal force. Since the car is traveling in a circle at constant speed, the acceleration is in the centripetal direction. The vector sum of the two forces is in the direction of the acceleration.

1. In Figure 5-26a the forces exerted by the road on the car are represented by $\vec{F}_{n1}$ and $\vec{F}_{n2}$. These forces are combined into $\vec{F}_n$ in Figure 5-26b. The angle between the normal force $\vec{F}_n$ and the vertical is θ, the same as the banking angle. Draw a free-body diagram for the car.

FIGURE 5-27

2. Apply $\Sigma F_y = ma_y$ to the car:

$$\Sigma F_y = ma_y \Rightarrow F_n \cos \theta - mg = 0$$

so

$$F_n = \frac{mg}{\cos \theta}$$

3. Apply $\Sigma F_x = ma_x$ to the car. Substitute for F_n using the step 2 result, then solve for θ:

$$\Sigma F_x = ma_x \Rightarrow F_n \sin \theta = m\frac{v^2}{r}$$

and

$$\frac{mg}{\cos \theta} \sin \theta = m\frac{v^2}{r} \quad so \quad \theta = \tan^{-1}\frac{v^2}{rg}$$

$$\theta = \tan^{-1}\frac{[(40{,}000 \text{ m})/(3600 \text{ s})]^2}{(30 \text{ m})(9.81 \text{ m/s}^2)} = \boxed{22.8°}$$

REMARKS The banking angle θ depends on v and r, but not the mass m; θ increases with increasing v, and decreases with increasing r. When the banking angle, speed, and radius satisfy $\tan \theta = v^2/rg$, the car rounds the curve smoothly, with no tendency to slide either inward or outward. If the car speed is greater than $\sqrt{rg \tan \theta}$, the road will exert a frictional force down the incline. This force has an inward horizontal component, which provides the additional centripetal force needed to keep the car from moving outward (sliding up the incline). If the car speed is less than this amount, the road must exert a frictional force up the incline.

ALTERNATIVE SOLUTION In the preceding solution we followed the guideline to choose one of the coordinate axis directions to be the direction of the acceleration vector, the centripetal direction. However, the solution is no more difficult if we choose one of the axis directions to be down the incline. This choice is taken in the following solution.

FIGURE 5-28

1. Draw a free-body diagram for the car (Figure 5-28). The x axis direction is down the incline and the y axis direction is the normal direction.

2. Apply $\Sigma F_x = ma_x$ to the car:

$$\Sigma F_x = ma_x \Rightarrow mg \sin \theta = ma_x$$

3. Draw a sketch and use trigonometry to obtain an expression for a_x in terms of a and θ (Figure 5-29):

$$a_x = a \cos \theta$$

FIGURE 5-29

4. Substitute the step 3 result into the step 2 result. Then substitute v^2/r for a and solve for θ:

$$mg \sin \theta = ma \cos \theta$$

$$g \sin \theta = \frac{v^2}{r} \cos \theta$$

$$\tan \theta = \frac{v^2}{rg} \Rightarrow \theta = \tan^{-1} \frac{v^2}{rg}$$

EXERCISE Find the component of the acceleration normal to the road surface. (*Answer* 1.60 m/s^2)

*5-3 Drag Forces

When an object moves through a fluid such as air or water, the fluid exerts a **drag force** or retarding force that opposes the motion of the object. The drag force depends on the shape of the object, the properties of the fluid, and the speed of the object relative to the fluid. Unlike ordinary friction, the drag force increases as the speed of the object increases. At low speeds, the drag force is approximately proportional to the speed of the object; at higher speeds, it is more nearly proportional to the square of the speed.

Consider an object dropped from rest and falling under the influence of the force of gravity, which we assume to be constant. Now add a drag force of magnitude bv^n, where b and n are constants. We then have a constant downward force mg and an upward force bv^n (Figure 5-30).

If we take the downward direction to be positive, we obtain from Newton's second law

$$\Sigma F_y = mg - bv^n = ma_y \qquad 5\text{-}6$$

FIGURE 5-30 Free-body diagram showing forces on an object falling with air resistance.

At $t = 0$, the instant when the object is dropped, the speed is zero, so the drag force is zero and the acceleration is g downward. As the speed of the object increases, the drag force increases and the acceleration becomes less than g. Eventually, the speed is great enough for the magnitude of the drag force bv^n to approach the force of gravity mg. As this happens, the acceleration approaches zero and the speed approaches the **terminal speed** v_t. At terminal speed the drag force balances the weight force and the acceleration is zero. Setting the acceleration a_y in Equation 5-6 equal to zero, we obtain

$$bv_t^n = mg$$

Solving for the terminal speed, we get

$$v_t = \left(\frac{mg}{b}\right)^{1/n} \qquad 5\text{-}7$$

The larger the constant b, the smaller the terminal speed. A parachute is designed to maximize b so that the terminal speed will be small. On the other hand, cars are designed to minimize b to reduce the effect of wind resistance.

The terminal speed of a sky diver before release of the parachute is about 200 km/h, which is about 60 m/s. When the parachute is opened, the drag force rapidly increases, becoming greater than the force of gravity, and the sky diver experiences an upward acceleration while falling; that is, the speed of the descending sky diver decreases. As the speed of the sky diver decreases, the drag force decreases and the speed approaches a new terminal speed of about 20 km/h.

TERMINAL SPEED

EXAMPLE 5-12

A sky diver of mass 64 kg falls with a terminal speed of 180 km/h with her arms and legs outspread. (*a*) **What is the magnitude of the upward drag force F_d on the sky diver?** (*b*) **If the drag force is equal to bv^2, what is the value of b?**

(*a*) 1. Draw a free-body diagram.

FIGURE 5-31

2. Apply $\Sigma F_y = ma_y$. Since the sky diver is moving with constant velocity, the acceleration is zero:

$$\Sigma F_y = ma_y \Rightarrow F_d - mg = 0$$

so

$$F_d = mg = (64 \text{ kg})(9.81 \text{ N/kg}) = \boxed{628 \text{ N}}$$

(b) 1. To find b we set $F_d = bv^2$:

$$F_d = mg = bv^2$$

so

$$b = \frac{mg}{v^2}$$

2. Find the speed in m/s, then calculate b:

$$180 \text{ km/h} = \frac{180 \text{ km}}{1 \text{ h}} \left(\frac{1000 \text{ m}}{1 \text{ km}}\right)\left(\frac{1 \text{ h}}{3600 \text{ s}}\right)$$

$$= \frac{180 \text{ km}}{1 \text{ h}} \left(\frac{1 \text{ m/s}}{3.6 \text{ km/h}}\right) = 50 \text{ m/s}$$

$$b = \frac{(64 \text{ kg})(9.81 \text{ N/kg})}{(50 \text{ m/s})^2} = \boxed{0.251 \text{ kg/m}}$$

*5-4 Numerical Integration: Euler's Method

If a particle moves under the influence of a *constant* force, its acceleration is constant and we can find its velocity and position from the constant-acceleration kinematic formulas in Chapter 2. But consider a particle moving through space where the force on it, and therefore its acceleration, depends on its position and its velocity. The position, velocity, and acceleration of the particle at one instant determine the position and velocity the next instant, which then determine its acceleration at that instant. The actual position, velocity, and acceleration of an object all change continuously with time. We can approximate this by replacing the continuous time variations with small time steps of duration Δt. The simplest approximation is to assume constant acceleration during each step. This approximation is called **Euler's method.** If the time interval is sufficiently short, the change in acceleration during the interval will be small and can be neglected.

Let x_0, v_0, and a_0 be the known position, velocity, and acceleration of a particle at some initial time t_0. If we assume constant acceleration during Δt, the velocity at time $t_1 = t_0 + \Delta t$ is given by

$$v_1 = v_0 + a_0 \Delta t$$

Similarly, if we neglect any change in velocity during the time interval, the new position is given by

$$x_1 = x_0 + v_0 \Delta t$$

We can use the values v_1 and x_1 to compute the new acceleration a_1 using Newton's second law, and then use x_1, v_1, and a_1 to compute x_2 and v_2.

$$x_2 = x_1 + v_1 \Delta t \qquad\qquad 5\text{-}8$$

$$v_2 = v_1 + a_1 \Delta t \qquad\qquad 5\text{-}9$$

The connection between the position and velocity at time t_n and time $t_{n+1} = t_n + \Delta t$ is given by

$$x_{n+1} = x_n + v_n \Delta t \tag{5-10}$$

and

$$v_{n+1} = v_n + a_n \Delta t \tag{5-11}$$

To find the velocity and position at some time t, we therefore divide the time interval $t - t_0$ into a large number of smaller intervals Δt and apply Equations 5-10 and 5-11, beginning at the initial time t_0. This involves a large number of simple, repetitive calculations that are most easily done on a computer. The technique of breaking the time interval into small steps and computing the acceleration, velocity, and position at each step using the values from the previous step is called numerical integration.

Drag Forces   To illustrate the use of numerical integration, let us consider a problem in which a sky diver is dropped from rest at some height under the influences of both gravity and a drag force that is proportional to the square of the speed. We will find the velocity v and the distance traveled x as functions of time.

The equation describing the motion of an object of mass m dropped from rest is Equation 5-6 with $n = 2$:

$$mg - bv^2 = ma$$

where down is the positive direction. The acceleration is thus

$$a = g - \frac{b}{m} v^2 \tag{5-12}$$

It is convenient to write the constant b/m in terms of the terminal speed v_t. Setting $a = 0$ in Equation 5-12 we obtain

$$0 = g - \frac{b}{m} v_t^2$$

$$\frac{b}{m} = \frac{g}{v_t^2}$$

Substituting g/v_t^2 for b/m in Equation 5-12 gives

$$a = g \left(1 - \frac{v^2}{v_t^2} \right) \tag{5-13}$$

The acceleration at time t_n is calculated using the values x_n and v_n.

To solve Equation 5-13 numerically, we need to use numerical values for g and v_t. A reasonable terminal speed for a sky diver is 60 m/s. If we choose $x_0 = 0$ for the initial position, the initial values are $x_0 = 0$, $v_0 = 0$, and $a_0 = g = 9.81$ m/s². To find the velocity v and position x after some time, say $t = 20$ s, we divide the time interval $0 < t < 20$ s into many small intervals Δt and apply Equations 5-10, 5-11, and 5-13. We do this by writing a computer program or by using a computer spreadsheet, shown in Figure 5-32. This spreadsheet has $\Delta t = 0.5$ s and at $t = 20$ s, the computed values are $x = 59.89$ m and $v = 939.9$ m/s.

FIGURE 5-32 (*a*) Spreadsheet to compute the position and speed of a sky diver with air drag proportional to v^2. (*b*) The same Excel spreadsheet displaying the formulas rather than the values.

	A	B	C	D
1	Δt =	0.5	s	
2	x0 =	0	m	
3	v0 =	0	m/s	
4	a0 =	9.81	m/s^2	
5	vt =	60	m/s	
6				
7	t	x	v	a
8	(s)	(m)	(m/s)	(m/s^2)
9	0.00	0.0	0.00	9.81
10	0.50	0.0	4.91	9.74
11	1.00	2.5	9.78	9.55
12	1.50	7.3	14.55	9.23
13	2.00	14.6	19.17	8.81
14	2.50	24.2	23.57	8.30
15	3.00	36.0	27.72	7.72
41	16.00	701.0	59.55	0.15
42	16.50	730.7	59.62	0.16
43	17.00	760.6	59.68	0.10
44	17.50	790.4	59.74	0.09
45	18.00	820.3	59.78	0.07
46	18.50	850.2	59.82	0.06
47	19.00	880.1	59.85	0.05
48	19.50	910.0	59.87	0.04
49	20.00	939.9	59.89	0.04
50				

(*a*)

	A	B	C	D
1	Δt =	0.5	s	
2	x0 =	0	m	
3	v0 =	0	m/s	
4	a0 =	9.81	m/s^2	
5	vt =	60	m/s	
6				
7	t	x	v	a
8	(s)	(m)	(m/s)	(m/s^2)
9	0	=B2	=B3	=$B$4*(1−C9^2/$B$5^2)
10	=A9+$B$1	=B9+C9*$B$1	=C9+D9*$B$1	=$B$4*(1−C10^2/$B$5^2)
11	=A10+$B$1	=B10+C10*$B$1	=C10+D10*$B$1	=$B$4*(1−C11^2/$B$5^2)
12	=A11+$B$1	=B11+C11*$B$1	=C11+D11*$B$1	=$B$4*(1−C12^2/$B$5^2)
13	=A12+$B$1	=B12+C12*$B$1	=C12+D12*$B$1	=$B$4*(1−C13^2/$B$5^2)
14	=A13+$B$1	=B13+C13*$B$1	=C13+D13*$B$1	=$B$4*(1−C14^2/$B$5^2)
15	=A14+$B$1	=B14+C14*$B$1	=C14+D14*$B$1	=$B$4*(1−C15^2/$B$5^2)
41	=A40+$B$1	=B40+C40*$B$1	=C40+D40*$B$1	=$B$4*(1−C41^2/$B$5^2)
42	=A41+$B$1	=B41+C41*$B$1	=C41+D41*$B$1	=$B$4*(1−C42^2/$B$5^2)
43	=A42+$B$1	=B42+C42*$B$1	=C42+D42*$B$1	=$B$4*(1−C43^2/$B$5^2)
44	=A43+$B$1	=B43+C43*$B$1	=C43+D43*$B$1	=$B$4*(1−C44^2/$B$5^2)
45	=A44+$B$1	=B44+C44*$B$1	=C44+D44*$B$1	=$B$4*(1−C45^2/$B$5^2)
46	=A45+$B$1	=B45+C45*$B$1	=C45+D45*$B$1	=$B$4*(1−C46^2/$B$5^2)
47	=A46+$B$1	=B46+C46*$B$1	=C46+D46*$B$1	=$B$4*(1−C47^2/$B$5^2)
48	=A47+$B$1	=B47+C47*$B$1	=C47+D47*$B$1	=$B$4*(1−C48^2/$B$5^2)
49	=A48+$B$1	=B48+C48*$B$1	=C48+D48*$B$1	=$B$4*(1+C49^2/$B$5^2)
50				

(*b*)

Figure 5-33 shows graphs of v versus t and x versus t plotted from this data.

But how accurate are our computations? We can estimate the accuracy by running the program again using a smaller time interval. If we use $\Delta t = 0.25$ s, one-half of the value we originally used, we obtain $v = 59.86$ m/s and $x = 943.1$ m at $t = 20$ s. The difference in v is about 0.05 percent and that in x is about 0.4 percent. These are our estimates of the accuracy of the original computations.

Since the difference between the value of a_{av} for some time interval Δt and the value of a at the beginning of the interval becomes smaller as the time interval becomes smaller, we might expect that it would be better to use very small time intervals, say $\Delta t = 0.000000001$ s. But there are two reasons for not using very small time intervals. First, the smaller the time interval, the larger the number of calculations that are required and the longer the program takes to run. Second, the computer keeps only a fixed number of digits at each step of the calculation, so that at each step there is a round-off error. These round-off errors add up. The larger the number of calculations, the more significant the total round-off errors become. When we first decrease the time interval, the accuracy improves because a_i more nearly approximates a_{av} for the interval. However, as the time interval is decreased further, the round-off errors build up and the accuracy of the computation decreases. A good rule of thumb to follow is to use no more than about 10^4 or 10^5 time intervals for the typical numerical integration.

(a)

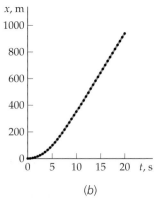

(b)

FIGURE 5-33 (*a*) Graph of v versus t for a sky diver, found by numerical integration using $\Delta t = 0.5$ s. The horizontal dashed line is the terminal speed $v_t = 60$ m/s. (*b*) Graph of x versus t using $\Delta t = 0.5$ s.

Friction and drag forces are complex phenomena empirically approximated by simple equations.

Topic	Relevant Equations and Remarks
1. Friction	Two objects in contact exert frictional forces on each other. These forces are parallel to the surfaces of the objects at the points of contact and directed opposite to the direction of sliding or tendency to slide.
Static friction	$$f_s \leq \mu_s F_n \qquad \text{5-2}$$ where F_n is the normal force of contact and μ_s is the coefficient of static friction.
Kinetic friction	$$f_k = \mu_k F_n \qquad \text{5-3}$$ where μ_k is the coefficient of kinetic friction. The coefficient of kinetic friction is slightly less than the coefficient of static friction.
2. Motion Along a Curved Path	A particle moving along an arbitrary curve can be considered to be moving along a circular arc during a small time interval. Its instantaneous acceleration vector has a component $a_c = v^2/r$ toward the center of curvature of the arc and a component $a_t = dv/dt$ that is tangential to the curve. If the particle is moving along a circular path of radius r at constant speed v, $a_t = 0$ and the speed, radius, and period T are related by $2\pi r = vT$.
3. *Drag Forces	When an object moves through a fluid it experiences a drag force that opposes its motion. The drag force increases with increasing speed. If the body is dropped from rest, its speed increases. As it does, the magnitude of the drag force comes closer and closer to the magnitude of the force of gravity, so the net force, and thus the acceleration, approaches zero. As the acceleration approaches zero, the speed approaches a constant value called its terminal speed. The terminal speed depends on the shape of the body and on the medium through which it falls.
4. *Numerical Integration: Euler's Method	To estimate the position x and velocity v at some time t, we first divide t into a large number of small intervals, each of length Δt. The initial acceleration a_0 is then calculated from the initial position x_0 and velocity v_0. The position x_1 and velocity v_1 a time Δt later are estimated using the relations $$x_{n+1} = x_n + v_n \Delta t \qquad \text{5-10}$$ and $$v_{n+1} = v_n + a_n \Delta t \qquad \text{5-11}$$ with $n = 0$. The acceleration a_{n+1} is calculated using the values for x_{n+1} and v_{n+1} and the process is repeated. This continues until estimations for the position and velocity at time t are calculated.

PROBLEMS

- Single-concept, single-step, relatively easy
- •• Intermediate-level, may require synthesis of concepts
- ••• Challenging
- **SSM** Solution is in the *Student Solutions Manual*
- **iSOLVE** Problems available on iSOLVE online homework service
- **iSOLVE** ✓ These "Checkpoint" online homework service problems ask students additional questions about their confidence level, and how they arrived at their answer

In a few problems, you are given more data than you actually need; in a few other problems, you are required to supply data from your general knowledge, outside sources, or informed estimates.

Conceptual Problems

1 • Various objects lie on the floor of a truck moving along a straight horizontal road. If the truck gradually speeds up, what force acts on the objects to cause them to speed up too?

2 • **SSM** Any object resting on the floor of a truck will slide if the truck's acceleration is too great. How does the critical acceleration at which a light object slips compare with that at which a much heavier object slips?

3 • A block of mass m rests on a plane inclined at an angle θ with the horizontal. It follows that the coefficient of static friction between the block and plane is (a) $\mu_s \geq g$, (b) $\mu_s = \tan \theta$, (c) $\mu_s \leq \tan \theta$, (d) $\mu_s \geq \tan \theta$.

4 • **SSM** A block of mass m is at rest on a plane inclined at an angle of 30° with the horizontal, as shown in Figure 5-34. Which of the following statements about the force of static friction is necessarily true? (a) $f_s > mg$. (b) $f_s > mg \cos 30°$. (c) $f_s = mg \cos 30°$. (d) $f_s = mg \sin 30°$. (e) None of these statements are true.

FIGURE 5-34 Problem 4

5 •• On an icy winter day, the coefficient of friction between the tires of a car and a roadway is reduced to one-half of its value on a dry day. As a result, the maximum speed at which a curve of radius R can be safely negotiated is (a) the same as on a dry day, (b) reduced to 71% of its value on a dry day, (c) reduced to 50% of its value on a dry day, (d) reduced to 25% of its value on a dry day, (e) reduced by an unknown amount depending on the car's mass.

6 •• **SSM** Show with a force diagram how a motorcycle can travel in a circle on the inside vertical wall of a hollow cylinder. Assume reasonable parameters (coefficient of friction, radius of the circle, mass of the motorcycle, or whatever is required), and calculate the minimum speed needed.

7 •• This is an interesting experiment that you can perform at home: take a wooden block and rest it on the floor or some other flat surface. Attach a spring (a slinky will work well) to the block and pull gently and steadily on the spring in the horizontal direction. At some point, the block will start moving, but it will not move smoothly. Instead, it will start moving, stop again, start moving again, stop again, etc. Explain why the motion of the block (called "stick–slip" motion) behaves this way.

8 • True or false: Viewed from an inertial reference frame, an object cannot move in a circle unless there is a net force acting on it.

9 •• A particle is traveling in a vertical circle at constant speed. One can conclude that the _____ is constant.
(a) velocity, (b) acceleration, (c) net force, (d) apparent weight, (e) none of the properties listed.

10 • **SSM** You place a lightweight piece of iron on a table and hold a small kitchen magnet above the iron at a distance of 1 cm. You find that the magnet cannot lift the iron, even though there is obviously a force between the iron and the magnet. Next, you again hold the piece of iron and the magnet 1 cm apart with the magnet above the iron, but this time you drop them from arm's length. As they fall, the magnet and the piece of iron are pulled together before hitting the floor. (a) Draw free-body diagrams illustrating all of the forces on the magnet and the iron for each demonstration. (b) Explain why the magnet and iron are pulled together when they are dropped, even though the magnet cannot pull up the piece of iron when it is sitting on the table.

11 ••• **SSM** The following question is an excellent "braintwister" invented by Boris Korsunsky[†]: Two identical blocks are attached by a massless string running over a pulley as shown in Figure 5-35. The rope initially runs over the pulley at its (the rope's) midpoint, and the surface that block 1 rests on is frictionless. Blocks 1 and 2 are initially at rest when block 2 is released with the string taut and horizontal. Will block 1 hit the pulley before or after block 2 hits the wall? (Assume that the initial distance from block 1 to the pulley is the same as the initial distance from block 2 to the wall.) There is a very simple solution.

---

† Boris Korsunsky, "Braintwisters for Physics Students," *The Physics Teacher*, **33**, 550 (1995).

FIGURE 5-35
Problem 11

12 • True or false: The terminal speed of a falling object depends on its shape.

13 • [SSM] As a sky diver falls through the air, her terminal speed (a) depends on her mass, (b) depends on her orientation as she falls, (c) depends on the density of the air, (d) depends on all of the above.

14 •• You are sitting in the passenger seat in a car driving around a circular, horizontal, flat racetrack at a high speed. As you sit there, you "feel" a "force" pushing you toward the outside of the track. What is the true direction of the force acting on you, and where does it come from? (Assume that you don't slide across the seat.)

15 • [SSM] The mass of the moon is about 1% of that of the earth. The centripetal force that keeps the moon in its orbit around the earth (a) is much smaller than the gravitational force exerted on the moon by the earth, (b) depends on the phase of the moon, (c) is much greater than the gravitational force exerted on the moon by the earth, (d) is the gravitational force exerted on the moon by the earth, (e) cannot be answered; we haven't studied Newton's law of gravity yet.

16 • A block is sliding on a frictionless surface along a loop-the-loop as in Figure 5-36. The block is moving fast enough that it never loses contact with the track. Match the points along the track to the appropriate free-body diagrams in Figure 5-37.

Problem 16 **FIGURE 5-36**

FIGURE 5-37

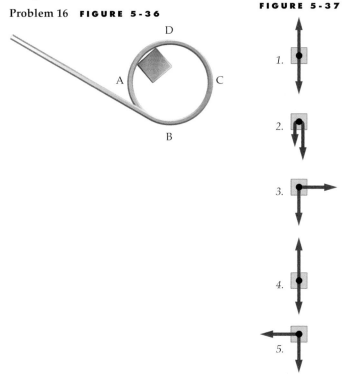

17 •• If you hold a rock and a feather at the same height above the ground and drop them, the rock will hit the ground first. From this you might conclude that the average aerodynamic drag force on the feather as it falls is larger than that on the rock, but in fact the opposite is true. Explain in detail why the rock hits the ground first. Assume that the drag force is given by the Newtonian formula $F_D = (1/2) CA\rho v^2$ (see Problem 18(c) for an explanation of this formula).

Estimation and Approximation

18 • [SSM] To determine the aerodynamic drag on a car, the "coast-down" method is often used. The car is driven on a long, flat road at some convenient speed (60 mph is typical), shifted into neutral, and allowed to coast to a stop. The time that it takes for the speed to drop by successive 5-mph intervals is measured and used to compute the net force slowing the car down. (a) It was found that a Toyota Tercel with a mass of 1020 kg coasted down from 60 mph to 55 mph in 3.92 s. What is the average force slowing the car down? (b) If the coefficient of rolling friction for the car is 0.02, what is the force of rolling friction that is acting to slow the car down? If we assume that the only two forces acting on the car are rolling friction and aerodynamic drag, what is the average drag force acting on the car? (c) The drag force will have the form $\frac{1}{2}CA\rho v^2$, where A is the cross-sectional area of the car facing the wind, v is the car's speed, ρ is the density of air, and C is a dimensionless constant of order 1. If the cross-sectional area of the car is 1.91 m², determine C from the data given. (The density of air is 1.21 kg/m³; use 55 mph for the speed of the car in this computation.)

19 • Using dimensional analysis, determine the units and dimensions of the constant b in the retarding force bv^n if (a) $n = 1$ and (b) $n = 2$. (c) Newton showed that the air resistance of a falling object with a circular cross section should be approximately $\frac{1}{2}\rho\pi r^2 v^2$, where $\rho = 1.2$ kg/m³, the density of air. Show that this is consistent with your dimensional analysis for part (b). (d) Find the terminal speed for a 56-kg sky diver; approximate his or her cross-sectional area as a disk of radius 0.30 m. The density of air near the surface of the earth is 1.20 kg/m³. (e) The density of the atmosphere decreases with height above the surface of the earth; at a height of 8 km, the density is only 0.514 kg/m³. What is the terminal velocity at this height?

20 •• While hailstones the size of golf balls are a bit unusual, hailstones are generally substantially larger than raindrops. Estimate the terminal velocity of a raindrop and a golf-ball sized hailstone.

Friction

21 • [SSM] A block of mass m slides at constant speed down a plane inclined at an angle of θ with the horizontal. It follows that (a) $\mu_k = mg \sin \theta$, (b) $\mu_k = \tan \theta$, (c) $\mu_k = 1 - \cos \theta$, (d) $\mu_k = \cos \theta - \sin \theta$.

22 • A block of wood is pulled by a horizontal string across a horizontal surface at a constant velocity with a force of 20 N. The coefficient of kinetic friction between the surfaces is 0.3. The force of friction is (a) impossible to determine without knowing the mass of the block, (b) impossible to determine without knowing the speed of the block, (c) 0.3 N, (d) 6 N, (e) 20 N.

23 • [SSM] [iSOLVE ✓] A 20-N block rests on a horizontal surface. The coefficients of static and kinetic friction between the surface and the block are $\mu_s = 0.8$ and $\mu_k = 0.6$. A horizontal string is attached to the block and a constant tension T is maintained in the string. What is the force of friction acting on the block if (a) $T = 15$ N or (b) $T = 20$ N.

24 • A block of mass m is pulled at a constant velocity across a horizontal surface by a string as shown in Figure 5-38. The magnitude of the frictional force is (a) $\mu_k \, mg$, (b) $T \cos \theta$, (c) $\mu_k (T - mg)$, (d) $\mu_k T \sin \theta$, (e) $\mu_k (mg - T \sin \theta)$.

FIGURE 5-38 Problem 24

25 • [iSOLVE] A tired worker pushes with a horizontal force of 500 N on a 100-kg crate initially resting on a thick pile carpet. The coefficients of static and kinetic friction are 0.6 and 0.4, respectively. Find the frictional force exerted by the carpet on the crate.

26 • [iSOLVE] A box weighing 600 N is pushed along a horizontal floor at constant velocity with a force of 250 N parallel to the floor. What is the coefficient of kinetic friction between the box and the floor?

27 • [iSOLVE ✓] The coefficient of static friction between the tires of a car and a horizontal road is $\mu_s = 0.6$. If the net force on the car is the force of static friction exerted by the road, (a) what is the magnitude of the maximum acceleration of the car when it is braked? (b) What is the shortest distance in which the car can stop if it is initially traveling at 30 m/s?

28 • [SSM] The force that accelerates a car along a flat road is the frictional force exerted by the road on the car's tires. (a) Explain why the acceleration can be greater when the wheels do not slip. (b) If a car is to accelerate from 0 to 90 km/h in 12 s, what is the minimum coefficient of friction needed between the road and tires? Assume that half the weight of the car is supported by the drive wheels.

29 • A 5-kg block is held at rest against a vertical wall by a horizontal force of 100 N. (a) What is the frictional force exerted by the wall on the block? (b) What is the minimum horizontal force needed to prevent the block from falling if the coefficient of friction between the wall and the block is $\mu_s = 0.40$?

30 • A tired and overloaded student is attempting to hold a large physics textbook wedged under his arm as shown in Figure 5-39. The textbook has a mass of 10.2 kg, while the coefficient of static friction of the textbook against the student's underarm is 0.32 and the coefficient of static friction of the book against the student's shirt is 0.16. (a) What is the minimum horizontal force that the student must apply to the textbook to prevent it from falling? (b) If the student can only exert a force of 195 N, what is the acceleration of the textbook as it slides from under his arm? The coefficient of kinetic friction of arm against textbook is 0.20, while that of shirt against textbook is 0.09.

FIGURE 5-39 Problem 30

31 • On a snowy day with the temperature near the freezing point, the coefficient of static friction between a car's tires and an icy road is 0.08. What is the maximum incline that a vehicle with four-wheel drive can climb at constant speed?

32 • [SSM] A 50-kg box that is resting on a level floor must be moved. The coefficient of static friction between the box and the floor is 0.6. One way to move the box is to push down on the box at an angle θ with the horizontal. Another method is to pull up on the box at an angle θ with the horizontal. (a) Explain why one method is better than the other. (b) Calculate the force necessary to move the box by each method if $\theta = 30°$ and compare the answer with the results when $\theta = 0°$.

33 • [iSOLVE ✓] A 3-kg block rests on a horizontal table, attached to a 2-kg block by a light string as shown in Figure 5-40. (a) What is the minimum coefficient of static friction such that the objects remain at rest? (b) If the coefficient of static friction is less than that found in part (a), and if the coefficient of kinetic friction between the block and the table is 0.3, find the time it takes for the 2-kg mass to fall 2 m to the floor if the system starts from rest.

FIGURE 5-40 Problem 33

34 •• A block on a horizontal plane is given an initial speed v. Traveling in a straight line, it comes to rest after sliding a distance d. Show that the coefficient of kinetic friction is $\mu_k = v^2/2gd$.

35 •• SSM A block of mass $m_1 = 250$ g is at rest on a plane that makes an angle of $\theta = 30°$ above the horizontal. The coefficient of kinetic friction between the block and the plane is $\mu_k = 0.100$. The block is attached to a second block of mass $m_2 = 200$ g that hangs freely by a string that passes over a frictionless, massless pulley (Figure 5-41). When the second block has fallen 30.0 cm, its speed is (a) 83 cm/s, (b) 48 cm/s, (c) 160 cm/s, (d) 59 cm/s, (e) 72 cm/s.

FIGURE 5-41 Problems 35–37

36 •• Returning to Figure 5-41, this time $m_1 = 4$ kg. The coefficient of static friction between the block and the incline is 0.4. (a) Find the range of possible values for m_2 for which the system will be in static equilibrium. (b) What is the frictional force on the 4-kg block if $m_1 = 1$ kg?

37 •• Returning once again to Figure 5-41, this time $m_1 = 4$ kg, $m_2 = 5$ kg, and the coefficient of kinetic friction between the inclined plane and the 4-kg block is $\mu_k = 0.24$. Find the acceleration of the masses and the tension in the cord.

38 •• SSM ISOLVE✓ The coefficient of static friction between the bed of a truck and a box resting on it is 0.30. The truck is traveling at 80 km/h along a horizontal road. What is the shortest distance in which the truck can stop if the box is not to slide?

39 •• A 4.5-kg block is given an initial velocity of 14 m/s up an incline that makes an angle of 37° with the horizontal. When its displacement is 8.0 m, its velocity up the incline has diminished to 5.2 m/s. Find (a) the coefficient of kinetic friction between the block and plane, (b) the displacement of the block from its starting point at the time when it momentarily comes to rest, and (c) the speed of the block when it again reaches its starting point.

40 •• An automobile is going up a grade of 15° at a speed of 30 m/s. The coefficient of static friction between the tires and the road is 0.7. (a) What minimum distance does it take to stop the car? (b) What minimum distance would it take if the car were going down the grade?

41 •• A rear-wheel-drive car supports 40 percent of its weight on its two drive wheels and has a coefficient of static friction of 0.7 with a horizontal straight road. (a) Find the vehicle's maximum acceleration. (b) What is the shortest possible time in which this car can achieve a speed of 100 km/h? (Assume the engine has unlimited power.)

42 •• SSM Lou bets an innocent stranger that he can place a 2-kg box against the side of a cart, as in Figure 5-42, and that the box will not fall to the ground, even though Lou will use no hooks, ropes, fasteners, magnets, glue, or adhesives of any kind. When the stranger accepts the bet, Lou begins to push the cart in the direction shown. The coefficient of static friction between the box and the cart is 0.6. (a) Find the minimum acceleration for which Lou will win the bet. (b) What is the magnitude of the frictional force in this case? (c) Find the force of friction on the box if the acceleration is twice the minimum needed for the box not to fall. (d) Show that, for a box of any mass, the box will not fall if the acceleration is $a \geq g/\mu_s$, where μ_s is the coefficient of static friction.

FIGURE 5-42 Problem 42

43 •• ISOLVE Two blocks attached by a string (Figure 5-43) slide down a 10° incline. The lower block has a mass of $m_2 = 0.25$ kg and a coefficient of kinetic friction $\mu_k = 0.2$. For the upper block, $m_1 = 0.8$ kg and $\mu_k = 0.3$. Find (a) the magnitude and direction of the acceleration of the blocks and (b) the tension in the string.

44 •• SSM As in Problem 43, two blocks of masses m_1 and m_2 are sliding down an incline as shown in Figure 5-43. They are connected by a massless *rod* this time; the rod behaves in exactly the same way as a string, except that the force can be compressive as well as tensile. The coefficient of kinetic friction for block 1 is μ_1 and the coefficient of kinetic friction for block 2 is μ_2. (a) Determine the acceleration of the two blocks. (b) Determine the force that the rod exerts on the two blocks. Show that the force is 0 when $\mu_1 = \mu_2$ and give a simple, nonmathematical argument why this is true.

FIGURE 5-43 Problems 43 and 44

45 •• Two blocks attached by a string are at rest on an inclined surface. The lower block has a mass of $m_1 = 0.2$ kg and a coefficient of static friction $\mu_s = 0.4$. The upper block has a mass $m_2 = 0.1$ kg and $\mu_s = 0.6$. The angle θ is slowly increased. (a) At what angle θ_c do the blocks begin to slide? (b) What is the tension in the string just before sliding begins?

46 •• Two blocks connected by a massless, rigid rod slide on a surface inclined at an angle of 20°. The lower block has a mass $m_1 = 1.2$ kg, and the upper block's mass is $m_2 = 0.75$ kg. (a) If the coefficient of kinetic friction is $\mu_k = 0.3$ for the lower block and $\mu_k = 0.2$ for the upper block, what is the acceleration of the blocks? (b) Determine the force exerted by the rod on either block.

47 •• [SSM] A block of mass m rests on a horizontal table (Figure 5-44). The block is pulled by a massless rope with a force $\vec{F}$ at an angle θ. The coefficient of static friction is 0.6. The minimum value of the force needed to move the block depends on the angle θ. (*a*) Discuss qualitatively how you would expect this force to depend on θ. (*b*) Compute the force for the angles $\theta = 0°$, $10°$, $20°$, $30°$, $40°$, $50°$, and $60°$, and make a plot of F versus θ for $mg = 400$ N. From your plot, at what angle is it most efficient to apply the force to move the block?

FIGURE 5-44 Problem 47

48 •• For the block in Problem 47, show that, in the general case for a block of mass m resting on a horizontal surface whose coefficient of static friction is μ_s: (*a*) If we want to apply the minimum possible force to move the block, it should be applied with the force pulling upward at an angle $\theta_{min} = \tan^{-1} \mu_s$, and (*b*) the minimum force necessary to start the block moving is $F_{min} = \left(\mu_s/\sqrt{1 + \mu_s^2}\right) mg$. (*c*) Once you start the block moving, if you want to apply the least possible force to *keep* it moving, should you keep the angle at which you are pulling the same, increase it, or decrease it?

49 •• Answer the questions in Problem 48, but for force $\vec{F}$ that pushes down on the block at an angle θ below the horizontal (Figure 5-45).

FIGURE 5-45 Problem 49

50 •• A 100-kg mass is pulled along a frictionless surface by a horizontal force $\vec{F}$ such that its acceleration is $a_1 = 6$ m/s² (Figure 5-46). A 20-kg mass slides along the top of the 100-kg mass and has an acceleration of $a_2 = 4$ m/s². (It thus slides backward relative to the 100-kg mass.) (*a*) What is the frictional force exerted by the 100-kg mass on the 20-kg mass? (*b*) What is the net force acting on the 100-kg mass? What is the force F? (*c*) After the 20-kg mass falls off the 100-kg mass, what is the acceleration of the 100-kg mass? (Assume that the force F does not change.)

FIGURE 5-46 Problem 50

51 •• [SOLVE] A 60-kg block slides along the top of a 100-kg block. The 60-kg block has an acceleration of 3 m/s² when a horizontal force $\vec{F}$ of 320 N is applied, as in Figure 5-47. There is no friction between the 100-kg block and a horizontal frictionless surface, but there is friction between the two blocks. (*a*) Find the coefficient of kinetic friction between the blocks. (*b*) Find the acceleration of the 100-kg block during the time that the 60-kg block remains in contact.

FIGURE 5-47 Problem 51

52 •• [SSM] [SOLVE] ✓ The coefficient of static friction between a rubber tire and the road surface is 0.85. What is the maximum acceleration of a 1000-kg four-wheel-drive truck if the road makes an angle of 12° with the horizontal and the truck is (*a*) climbing and (*b*) descending?

53 •• A 2-kg block sits on a 4-kg block that is on a frictionless table (Figure 5-48). The coefficients of friction between the blocks are $\mu_s = 0.3$ and $\mu_k = 0.2$. (*a*) What is the maximum horizontal force F that can be applied to the 4-kg block if the 2-kg block is not to slip? (*b*) If F has half this value, find the acceleration of each block and the force of friction acting on each block. (*c*) If F is twice the value found in (*a*), find the acceleration of each block.

FIGURE 5-48 Problem 53

54 •• In Figure 5-49, the mass $m_2 = 10$ kg slides on a frictionless table. The coefficients of static and kinetic friction between m_2 and $m_1 = 5$ kg are $\mu_s = 0.6$ and $\mu_k = 0.4$. (*a*) What is the maximum acceleration of m_1? (*b*) What is the maximum value of m_3 if m_1 moves with m_2 without slipping? (*c*) If $m_3 = 30$ kg, find the acceleration of each body and the tension in the string.

FIGURE 5-49 Problem 54

55 • A massive sandstone block is being dragged up a ramp by a counterweight suspended over a pulley, as shown in Figure 5-50. The mass of the block is 1600 kg, while the mass of the counterweight is 550 kg. The coefficient of kinetic friction of the block against the ramp is 0.15, and the ramp is sloped at an angle of 10°. (*a*) What is the acceleration of the block up the ramp? (*b*) Three seconds after the block begins moving up the ramp, the rope holding the counterweight breaks. How much farther will the block continue to slide up the ramp before stopping? (*c*) Unfortunately, the block begins to slide back down the ramp once it reaches its highest point. What is the acceleration of the block as it slides down the ramp?

FIGURE 5-50 Problem 55

56 ••• 【SOLVE】 A 10-kg block rests on a 5-kg bracket shown in Figure 5-51. The 5-kg bracket sits on a frictionless surface. The coefficients of friction between the 10-kg block and the bracket on which it rests are $\mu_s = 0.40$ and $\mu_k = 0.30$. (*a*) What is the maximum force *F* that can be applied if the 10-kg block is not to slide on the bracket? (*b*) What is the corresponding acceleration of the 5-kg bracket?

FIGURE 5-51 Problem 56

57 •• SSM On planet Vulcan, an introductory physics class performs several experiments involving friction. In one of these experiments the acceleration of a block is measured both when it is sliding up an incline and when it is sliding down the same incline. You copy the following data and diagram (Figure 5-52) out of one of the lab notebooks, but can't find any translations into metric units (Negative sign indicates that the acceleration is pointing down the incline.):

Acceleration of block
Going up inclined plane   −1.73 glapp/plip²
Going down plane   −1.42 glapp/plip²

FIGURE 5-52 Problem 57

0.73 glapp

3.82 glapp

Inclined plane

From these data, determine the acceleration of gravity on Vulcan (in glapps/plip²) and the coefficient of kinetic friction between the block and the incline.

58 •• SSM A 100-kg block on an inclined plane is attached to another block of mass *m* via a string, as in Figure 5-53. The coefficients of static and kinetic friction of the block and the incline are $\mu_s = 0.4$ and $\mu_k = 0.2$. The angle of the incline is 18° above horizontal. (*a*) Determine the range of values for *m*, the mass of the hanging block, for which the block on the incline will not move unless disturbed, but if nudged, will slide *down* the incline. (*b*) Determine a range of values for *m* for which the block on the incline will not move unless nudged, but if nudged will slide *up* the incline.

FIGURE 5-53 Problem 58

59 ••• 【SOLVE✔】 A block of mass 0.5 kg rests on the inclined surface of a wedge of mass 2 kg, as in Figure 5-54. The wedge is acted on by a horizontal force $\vec{F}$ and slides on a frictionless surface. (*a*) If the coefficient of static friction between the wedge and the block is $\mu_s = 0.8$ and the angle of the incline is 35°, find the maximum and minimum values of *F* for which the block does not slip. (*b*) Repeat part (*a*) with $\mu_s = 0.4$.

FIGURE 5-54 Problem 59

60 • The coefficient of kinetic friction between two surfaces is not truly independent of the relative velocity of the two objects in contact. In fact, it tends to decrease slightly with increasing velocity. For example, one series of experiments found that for wood sliding on wood, the coefficient of friction could be described by the equation

$$\mu_k = 0.11/(1 + 2.3\times10^{-4}\, v^2)^2$$

where v is measured in m/s. From this expression, find the force of kinetic friction acting on a 100-kg block of wood moving across a horizontal wooden surface at (a) 10 m/s and (b) 20 m/s.

61 •• In some experiments, it has been found that the coefficient of kinetic friction is not independent of the normal force acting on the object. Vaclav Konecny[†] found, in a set of experiments for wood sliding over wood, that the force of static friction between the two wood surfaces was given by the expression $f_k = 0.4\, F_n^{0.91}$, where f_k is the force of friction and F_n is the normal force acting on the object (both forces measured in newtons). Because of this, the acceleration of a block down an inclined plane will not be independent of the mass of the block. Using this formula, calculate the acceleration and stopping distance for (a) a 10-kg block and (b) a 100-kg block sliding along a horizontal surface with an initial speed of 10 m/s.

62 ••• SSM A block of wood with a mass of 10 kg is pushed, starting from rest, with a constant horizontal force of 70 N across a wooden floor. Assuming that the coefficient of kinetic friction varies with particle speed as $\mu_k = 0.11/(1 + 2.3\times10^{-4}\, v^2)^2$ (see Problem 60), write a spreadsheet program using Euler's method to calculate and graph the speed of the block and its displacement as a function of time from 0 to 10 s. Compare this to the case where the coefficient of kinetic friction is equal to 0.11, independent of v.

63 •• To determine the coefficient of kinetic friction of a block of wood on a horizontal table surface, you are given the following assignment: Take the block of wood and shove it across the surface of the table. Using a stopwatch, measure the time it takes for the block to come to a stop (Δt) and the total distance that the block travels after the push (Δx). (a) Show that from these measurements, $\mu_k = 2\Delta x/[g(\Delta t)^2]$. If the block slides a distance of 1.37 m in 0.97 s, calculate μ_k. (b) What was the initial speed of the block?

64 •• SSM The following data show the acceleration of a block down an inclined plane as a function of the angle of incline θ[‡]:

θ (degrees)	Acceleration (m/s²)
25	1.6909
27	2.1043
29	2.4064
31	2.8883
33	3.1750
35	3.4886
37	3.7812
39	4.1486
41	4.3257
43	4.7178
45	5.1056

(a) Show that, for a block sliding down an incline, graphing $a/\cos\theta$ versus $\tan\theta$ should give a straight line with slope g and y-intercept $-\mu_k g$. (b) Using a spreadsheet program, graph these data and fit a straight line to them to determine μ_k and g. What is the percentage error in g from the commonly accepted value of 9.81 m/s²?

Motion Along a Curved Path

65 •• A stone with a mass $m = 95$ g is being whirled in a horizontal circle on the end of a string that is 85 cm long. The length of time required for the stone to make one complete revolution is 1.22 s. The angle that the string makes with the horizontal is (a) 52°, (b) 46°, (c) 26°, (d) 23°, (e) 3°.

66 •• A 0.20-kg stone attached to a 0.8-m string is rotated in the horizontal plane. The string makes an angle of 20° with the horizontal. Determine the speed of the stone.

67 •• A 0.75-kg stone attached to a string is whirled in a horizontal circle of radius 35 cm as in the conical pendulum of Example 5-9. The string makes an angle of 30° with the vertical. (a) Find the speed of the stone. (b) Find the tension in the string.

68 •• SSM SOLVE A 50-kg pilot comes out of a vertical dive in a circular arc such that at the bottom of the arc her upward acceleration is 8.5g. (a) What is the magnitude of the force exerted by the airplane seat on the pilot at the bottom of the arc? (b) If the speed of the plane is 345 km/h, what is the radius of the circular arc?

69 •• SOLVE A 65-kg airplane pilot pulls out of a dive by following, at constant speed, the arc of a circle whose radius is 300 m. At the bottom of the circle, where her speed is 180 km/h, (a) what are the direction and magnitude of her acceleration? (b) What is the net force acting on her at the bottom of the circle? (c) What is the force exerted on the pilot by the airplane seat?

70 •• SOLVE Mass m_1 moves in a circular path of radius r on a frictionless horizontal table (Figure 5-55). It is attached to a string that passes through a frictionless hole in the center of the table. A second mass m_2 is attached to the other end of the string. Derive an expression for r in terms of m_1, m_2, and the time T for one revolution.

FIGURE 5-55 Problem 70

† Vaclav Konecny, "On the first law of friction," *American Journal of Physics,* **41,** 588 (1973).

‡ Data taken from Dennis W. Phillips, "Science Friction Adventure—Part II," *The Physics Teacher,* 553 (Nov. 1990).

71 •• [SSM] A block of mass m_1 is attached to a cord of length L_1, which is fixed at one end. The block moves in a horizontal circle on a frictionless table. A second block of mass m_2 is attached to the first by a cord of length L_2 and also moves in a circle, as shown in Figure 5-56. If the period of the motion is T, find the tension in each cord.

FIGURE 5-56 Problem 71

72 •• [SSM] A particle moves with constant speed in a circle of radius 4 cm. It takes 8 s to complete each revolution. Draw the path of the particle to scale, and indicate the particle's position at 1-s intervals. Draw displacement vectors for each interval. These vectors also indicate the directions for the average-velocity vectors for each interval. Find graphically the magnitude of the change in the average velocity $|\Delta\vec{v}|$ for two consecutive 1-s intervals. Compare $|\Delta\vec{v}|/\Delta t$, measured in this way, with the instantaneous acceleration computed from $a = v^2/r$.

73 •• A man swings his child in a circle of radius 0.75 m as shown in Figure 5-57. If the mass of the child is 25 kg and the child makes one revolution in 1.5 s, what is the magnitude and direction of the force that must be exerted by the man on the child? (Assume the child to be a point particle.)

FIGURE 5-57 Problem 73

74 •• The string of a conical pendulum is 50 cm long and the mass of the bob is 0.25 kg. Find the angle between the string and the horizontal when the tension in the string is six times the weight of the bob. Under those conditions, what is the period of the pendulum?

75 •• [iSOLVE ✔] A 100-g coin sits on a horizontally rotating turntable. The turntable makes one revolution each second. The coin is located 10 cm from the axis of rotation of the turntable. (a) What is the frictional force acting on the coin?

(b) The coin will slide off the turntable if it is located more than 16 cm from the axis of rotation. What is the coefficient of static friction?

76 •• A 0.25-kg tether ball is attached to a vertical pole by a 1.2-m cord. Assume that the cord is attached to the center of the ball. If the ball moves in a horizontal circle with the cord making an angle of 20° with the vertical, (a) what is the tension in the cord? (b) What is the speed of the ball?

77 •• [SSM] An object on the equator has both an acceleration toward the center of the earth because of the earth's rotation, and an acceleration toward the sun because of the earth's motion along its orbit. Calculate the magnitudes of both accelerations, and express them as fractions of the free-fall acceleration g at the earth's surface.

78 • (a) Using data available from the textbook, calculate the net force acting on the earth that holds it in orbit around the sun. Assume that the orbit is circular. (b) Using data available from the textbook, calculate the force acting on the moon that holds it in orbit around the earth. Assume that the orbit is circular.

79 •• [iSOLVE ✔] A small bead with a mass of 100 g slides without friction along a semicircular wire with a radius of 10 cm that rotates about a vertical axis at a rate of 2 revolutions per second, as in Figure 5-58. Find the value of θ for which the bead will remain stationary relative to the rotating wire.

FIGURE 5-58 Problem 79

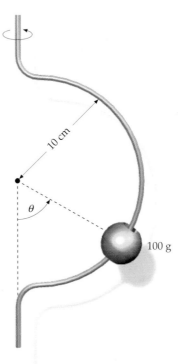

80 ••• Consider a bead of mass m that is free to move on a thin, circular wire of radius r. The bead is given an initial speed v_0, and there is a coefficient of kinetic friction μ_k. The experiment is performed in a spacecraft drifting in space. Find the speed of the bead at any subsequent time t.

81 ••• Revisiting Problem 80, (a) find the centripetal acceleration of the bead. (b) Find the tangential acceleration of the bead. (c) What is the magnitude of the resultant acceleration?

Concepts of Centripetal Force

82 • SSM A ride at an amusement park carries people in a vertical circle at constant speed such that the normal forces exerted by the seats are always inward—toward the center of the circle. At the top, the normal force exerted by a seat equals the person's weight, mg. At the bottom of the loop, the force exerted by the seat will be (a) 0, (b) mg, (c) $2mg$, (d) $3mg$, (e) greater than mg, but it cannot be calculated from the information given.

83 • The radius of curvature of a loop-the-loop roller coaster is 12.0 m. At the top of the loop, the force that the seat exerts on a passenger of mass m is $0.4mg$. Find the speed of the roller coaster at the top of the loop.

84 • A car speeds along the curved exit ramp of a freeway. The radius of the curve is 80 m. A 70-kg passenger holds the arm rest of the car door with a 220-N force in order to keep from sliding across the front seat of the car. (Assume the exit ramp is not banked and ignore friction with the car seat.) What is the car's speed? (a) 16 m/s. (b) 57 m/s. (c) 18 m/s. (d) 50 m/s. (e) 28 m/s.

85 ••• SSM Suppose you ride a bicycle on a horizontal surface in a circle with a radius of 20 m. The resultant force exerted by the road on the bicycle (normal force plus frictional force) makes an angle of 15° with the vertical. (a) What is your speed? (b) If the frictional force is half its maximum possible value, what is the coefficient of static friction?

86 •• An airplane is flying in a horizontal circle at a speed of 480 km/hr. Banked for this turn, the wings of the plane are tilted at an angle of 40° from the horizontal (Figure 5-59). Assume that a lift force acting perpendicular to the wings holds the aircraft in the sky. What is the radius of the circle in which the plane is flying?

FIGURE 5-59
Problem 86

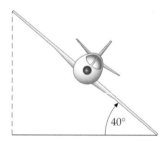

40°

87 • ISOLVE A 750-kg car travels around a curve of 160-m radius at 90 km/h. What should the banking angle of the curve be so that the force of the pavement on the tires of the car is in the normal direction?

88 •• SSM A curve of radius 150 m is banked at an angle of 10°. An 800-kg car negotiates the curve at 85 km/h without skidding. Find (a) the normal force exerted by the pavement on the tires, (b) the frictional force exerted by the pavement on the tires, (c) the minimum coefficient of static friction between the pavement and the tires.

89 •• On another occasion, the car in Problem 88 negotiates the curve at 38 km/h. Find (a) the normal force exerted on the tires by the pavement, and (b) the frictional force exerted on the tires by the pavement.

90 ••• SSM ISOLVE A civil engineer is asked to design a curved section of roadway that meets the following conditions: With ice on the road, when the coefficient of static friction between the road and rubber is 0.08, a car at rest must not slide into the ditch and a car traveling less than 60 km/h must not skid to the outside of the curve. What is the minimum radius of curvature of the curve and at what angle should the road be banked?

91 ••• ISOLVE A curve of radius 30 m is banked so that a 950-kg car traveling at 40 km/h can round it even if the road is so icy that the coefficient of static friction is approximately zero. Find the range of speeds at which a car can travel around this curve without skidding if the coefficient of static friction between the road and the tires is 0.3.

Drag Forces

92 • A small pollution particle settles toward the earth in still air with a terminal speed of 0.3 mm/s. The particle has a mass of 10^{-10} g and a retarding force of the form bv. What is the value of b?

93 • A Ping-Pong ball has a mass of 2.3 g and a terminal speed of 9 m/s. The retarding force is of the form bv^2. What is the value of b?

94 • SSM A sky diver of mass 60 kg can slow herself to a constant speed of 90 km/h by adjusting her form. (a) What is the magnitude of the upward drag force on the sky diver? (b) If the drag force is equal to bv^2, what is the value of b?

95 •• An 800-kg car rolls down a very long 6° grade. The drag force for motion of the car has the form $F_d = 100$ N $+ (1.2$ N·s^2/m$^2)v^2$. Neglect rolling friction. What is the terminal velocity for the car rolling down this grade?

96 ••• Small spherical particles experience a viscous drag force given by Stokes' law: $F_d = 6\pi\eta rv$, where r is the radius of the particle, v is its speed, and η is the viscosity of the fluid medium. (a) Estimate the terminal speed of a spherical pollution particle of radius 10^{-5} m and density of 2000 kg/m^3. (b) Assuming that the air is still and that η is 1.8×10^{-5} N·s/m^2, estimate the time it takes for such a particle to fall from a height of 100 m.

97 ••• SSM A sample of air containing pollution particles of the size and density given in Problem 96 is captured in a test tube 8.0 cm long. The test tube is then placed in a centrifuge with the midpoint of the test tube 12 cm from the center of the centrifuge. The centrifuge spins at 800 revolutions per minute. Estimate the time required for nearly all of the pollution particles to settle at the end of the test tube and compare this to the time required for a pollution particle to fall 8.0 cm under the action of gravity and subject to the viscous drag of air.

Euler's Method

98 •• You are riding in a hot air balloon when you throw a baseball straight down with an initial speed of 35 km/h. A baseball falls with a terminal speed of 150 km/h. Assuming air drag is proportional to the speed squared, use Euler's method to estimate the speed of the ball after 10 s. What is the uncertainty in this estimate? You drop a second baseball, this one released from rest. How long does it take for it to reach 99% of its terminal speed? How far does it fall during this time?

99 •• SSM You throw a baseball straight up with an initial speed of 150 km/h. Its terminal speed when falling is also 150 km/h. Use Euler's method to estimate its height 3.5 s after release. What is the maximum height it reaches? How long after release does it reach its maximum height? How much later does it return to the ground? Is the time the ball spends on the way up less than, the same as, or greater than the time it spends on the way down?

100 •• A 0.80-kg block on a horizontal frictionless surface is held against a massless spring, compressing it 30 cm. The force constant of the spring is 50 N/m. The block is released and the spring pushes it 30 cm. Use Euler's method with $\Delta t = 0.005$ s to estimate the time it takes for the spring to push the block the 30 cm. How fast is the block moving at this time? What is the uncertainty in this speed?

General Problems

101 • SOLVE A 4.5-kg block slides down an inclined plane that makes an angle of 28° with the horizontal. Starting from rest, the block slides a distance of 2.4 m in 5.2 s. Find the coefficient of kinetic friction between the block and plane.

102 • SOLVE A model airplane of mass 0.4 kg is attached to a horizontal string and flies in a horizontal circle of radius 5.7 m. (The weight of the plane is balanced by the upward "lift" force of the air on the wings of the plane.) The plane makes 1.2 revolutions every 4 s. (a) Find the speed v of the plane. (b) Find the tension in the string.

103 •• SSM An 800-N box rests on an incline making a 30° angle with the horizontal. A physics student finds that she can prevent the box from sliding if she pushes on it with a force of at least 200 N parallel to the incline. (a) What is the coefficient of static friction between the box and the incline? (b) What is the greatest force that can be applied to the box parallel to the incline before the box slides up the incline?

104 •• The position of a particle is given by the vector $\vec{r} = -10$ m $\cos \omega t \hat{i} + 10$ m $\sin \omega t \hat{j}$, where $\omega = 2$ s^{-1}. (a) Show that the path of the particle is a circle. (b) What is the radius of the circle? (c) Does the particle move clockwise or counterclockwise around the circle? (d) What is the speed of the particle? (e) What is the time for one complete revolution?

105 •• SOLVE A crate of books is to be put on a truck with the help of some planks sloping up at 30°. The mass of the crate is 100 kg, and the coefficient of sliding friction between it and the planks is 0.5. You and your friends push *horizontally* with a force $\vec{F}$. Once the crate has started to move, how large must F be in order to keep the crate moving at constant speed?

106 •• An object with a mass of 5.5 kg is allowed to slide from rest down an inclined plane. The plane makes an angle of 30° with the horizontal and is 72 m long. The coefficient of kinetic friction between the plane and the object is 0.35. The speed of the object at the bottom of the plane is (a) 5.3 m/s, (b) 15 m/s, (c) 24 m/s, (d) 17 m/s, (e) 11 m/s.

107 •• SSM A brick slides down an inclined plank at constant speed when the plank is inclined at an angle θ_0. If the angle is increased to θ_1, the block accelerates down the plank with acceleration a. The coefficient of kinetic friction is the same in both cases. Given θ_0 and θ_1, calculate a.

108 •• Three forces act on an object, shown in Figure 5-60, that is in static equilibrium. (a) If F_1, F_2, and F_3 represent the magnitudes of the forces acting on the object, show that $F_1/\sin \theta_{23} = F_2/\sin \theta_{31} = F_3/\sin \theta_{12}$. (b) Show that $F_1^2 = F_2^2 + F_3^2 + 2F_2F_3 \cos \theta_{23}$.

FIGURE 5-60 Problem 108

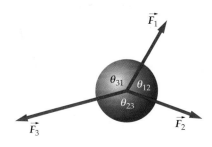

109 •• In a carnival ride, the passenger sits on a seat in a compartment that rotates with constant speed in a vertical circle of radius $r = 5$ m. The heads of the seated passengers always point toward the center of the circle. (a) If the carnival ride completes one full circle in 2 s, find the acceleration of the passenger. (b) Find the slowest rate of rotation (in other words, the longest time T_m to complete one full circle) if the seat belt is to exert no force on the passenger at the top of the ride.

110 •• SSM A flat-topped toy cart moves on frictionless wheels, pulled by a rope under tension T. The mass of the cart is m_1. A load of mass m_2 rests on top of the cart with a coefficient of static friction μ_s. The cart is pulled up a ramp that is inclined at angle θ above the horizontal. The rope is parallel to the ramp. What is the maximum tension T that can be applied without making the load slip?

111 •• A sled weighing 200 N rests on a 15° incline, held in place by static friction (Figure 5-61). The coefficient of static friction is 0.5. (*a*) What is the magnitude of the normal force on the sled? (*b*) What is the magnitude of the static frictional force on the sled? (*c*) The sled is now pulled up the incline at constant speed by a child. The child weighs 500 N and pulls on the rope with a constant force of 100 N. The rope makes an angle of 30° with the incline and has negligible mass. What is the magnitude of the kinetic frictional force on the sled? (*d*) What is the coefficient of kinetic friction between the sled and the incline? (*e*) What is the magnitude of the force exerted on the child by the incline?

FIGURE 5-61 Problem 111

112 • In 1976, Gerard O'Neill proposed the building of large space stations for human habitation in orbit around the earth and the moon. Because prolonged free-fall has adverse medical effects, he proposed making the stations in the form of long cylinders and spinning them around the cylinder axis to provide the inhabitants inside with the sensation of gravity. This idea has found its way into mainstream science fiction; for one example, the TV show *Babylon 5* was set on an O'Neill colony 5 mi long, with a diameter of 0.6 mi. Because of the rotation, someone on the inside of the colony would experience a sense of "gravity," because he or she would be in an accelerated frame of reference. (*a*) Show that the "acceleration of gravity" experienced by someone in an O'Neill colony is equal to his or her centripetal acceleration. *Hint: consider someone "looking in" from outside the colony.* (*b*) If we assume that the space station is composed of several decks which are at varying distances from the axis of rotation, show that the "acceleration of gravity" becomes weaker the closer one gets to the axis. (*c*) How many revolutions per minute would Babylon 5 have to make to give an "acceleration of gravity" of 9.8 m/s² at the outermost edge of the station?

113 •• A child slides down a slide inclined at 30° in time t_1. The coefficient of kinetic friction between her and the slide is μ_k. She finds that if she sits on a small cart with frictionless wheels, she slides down the same slide in time $\frac{1}{2}t_1$. Find μ_k.

114 •• **SSM** The position of a particle of mass $m = 0.8$ kg as a function of time is

$$\vec{r} = x\hat{i} + y\hat{j} = R \sin \omega t \, \hat{i} + R \cos \omega t \, \hat{j}$$

where $R = 4.0$ m and $\omega = 2\pi$ s⁻¹. (*a*) Show that the path of this particle is a circle of radius R with its center at the origin. (*b*) Compute the velocity vector. Show that $v_x/v_y = -y/x$. (*c*) Compute the acceleration vector and show that it is in the radial direction and has the magnitude v^2/r. (*d*) Find the magnitude and direction of the net force acting on the particle.

115 •• In an amusement-park ride, riders stand with their backs against the wall of a spinning vertical cylinder. The floor falls away and the riders are held up by friction. If the

radius of the cylinder is 4 m, find the minimum number of revolutions per minute necessary when the coefficient of static friction between a rider and the wall is 0.4.

116 •• **iSOLVE** A mass m_1 on a horizontal table is attached by a thin string that passes over a frictionless, massless pulley to a 2.5-kg mass m_2 that hangs over the side of the table 1.5 m above the ground (Figure 5-62). The system is released from rest at $t = 0$ and the 2.5-kg mass strikes the ground at $t = 0.82$ s. The system is now placed in its initial position and a 1.2-kg mass is placed on top of the block of mass m_1. Released from rest, the 2.5-kg mass now strikes the ground 1.3 s later. Determine the mass m_1 and the coefficient of kinetic friction between m_1 and the table.

FIGURE 5-62 Problem 116

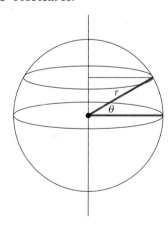

117 ••• **SSM** (*a*) Show that a point on the surface of the earth at latitude θ, shown in Figure 5-63, has an acceleration of magnitude (3.37 cm/s²)cos θ relative to a reference frame not rotating with the earth. What is the direction of this acceleration? (*b*) Discuss the effect of this acceleration on the apparent weight of an object near the surface of the earth. (*c*) The free-fall acceleration of an object at sea level measured *relative to the earth's surface* is 9.78 m/s² at the equator and 9.81 m/s² at latitude $\theta = 45°$. What are the values of the gravitational field g at these points?

FIGURE 5-63 Problem 117

118 ••• **SSM** A small block of mass 0.01 kg is at rest atop a smooth (frictionless) sphere of radius 0.8 m. The block is given a tiny nudge and starts to slide down the sphere. The mass loses contact with the sphere when the angle between the vertical and the line from the center of the sphere to the position of the mass is θ. Find the angle θ.

Work and Energy

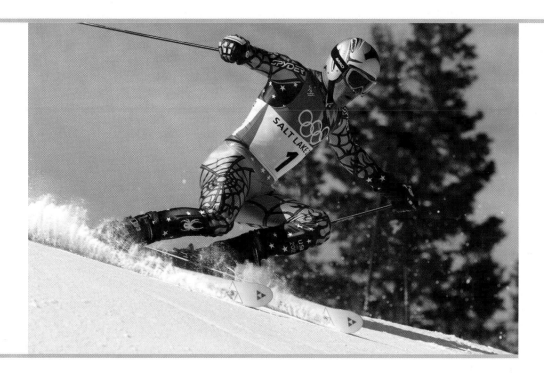

DOWNHILL RACING CAN LOOK LIKE A LOT OF FUN, OR A LOT OF WORK, DEPENDING UPON YOUR PERSPECTIVE.

? **Did you know that, in fact, work is being done on the skier? (See Example 6-12.)**

Work and energy are important concepts in physics as well as in our everyday lives. In physics, a force does **work** if its point of application moves through a distance and there is a component of the force in the direction of the velocity of the force's point of application. For a constant force in one dimension, the work done equals the force component in the direction of the displacement times the displacement. (This differs somewhat from the everyday use of the word work. When you study hard for an exam, the only work you do according to the use of the word in physics, is in pushing your pencil on the paper, or turning the pages of your book.)

Energy is closely associated with work. When work is done by one system on another, energy is transferred between the two systems. For example, when you do work pushing a swing, chemical energy of your body is transferred to the swing and appears as kinetic energy of motion or as gravitational potential energy of the earth–swing system. There are many forms of energy. Kinetic energy is associated with the motion of an object. Potential energy is associated with the configuration of a system, such as the separation distance between two objects that attract each other. Thermal energy is associated with the random motion of the molecules within a system and is closely connected with the temperature of the system.

➤ **In this chapter we study the concepts of work, kinetic energy, and potential energy. Because these concepts arise from Newton's laws, the concepts introduced in this chapter are a continuation of those introduced in previous chapters.**

These new concepts provide powerful methods of solving a wide class of problems. Many of the end-of-chapter problems for this chapter can be solved using the concepts and methods of the previous chapters. However, it is important that you resist any temptation to solve them in that way. The concepts and methods developed in this chapter are developed further in Chapter 7.

6-1 Work and Kinetic Energy

Motion in One Dimension With Constant Forces

The work W done by a constant force $\vec{F}$ moving through a displacement $\Delta x \hat{i}$ is given by

$$W = F_x \Delta x = F \cos \theta \, \Delta x \qquad\qquad 6\text{-}1$$

WORK BY A CONSTANT FORCE

where θ is the angle between the directions of $\vec{F}$ and $\hat{i}$, and $\Delta x \hat{i}$ is the displacement of the point of application of the force as shown in Figure 6-1.

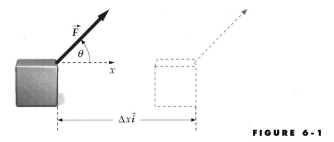

FIGURE 6-1

Work is a scalar quantity that is positive if Δx and F_x have the same signs and negative if they have opposite signs. The dimensions of work are those of force times distance. The SI unit of work and energy is the **joule** (J), which equals the product of a newton and a meter:

$$1 \, \text{J} = 1 \, \text{N} \cdot \text{m} \qquad\qquad 6\text{-}2$$

(In the U.S. customary system, the unit of work is the foot-pound: 1 ft·lb = 1.356 J.) A convenient unit of work and energy in atomic and nuclear physics is the electron volt (eV):

$$1 \, \text{eV} = 1.6 \times 10^{-19} \, \text{J} \qquad\qquad 6\text{-}3$$

Commonly used multiples are keV (10^3 eV) and MeV (10^6 eV). The work required to remove an electron from an atom is of the order of a few eV, whereas the work needed to remove a proton or a neutron from an atomic nucleus is of the order of several MeV.

EXERCISE A force of 12 N is exerted on a box at an angle of $\theta = 20°$, as in Figure 6-1. How much work is done by the force on the box as the box moves along the table a distance of 3 m? (*Answer* 33.8 J)

If there are several forces that do work, the total work is found by computing the work done by each force and summing.

$$W_{\text{total}} = F_{1x} \Delta x_1 + F_{2x} \Delta x_2 + F_{3x} \Delta x_3 + \dots$$

A **particle** is any object that moves so that all of its parts undergo identical displacements during any interval of time. That is, an object can be modeled as a particle as long as it remains perfectly rigid and moves without rotating.

When several forces do work on a *particle*, the displacements of the points of application of each these forces are equal. Let the displacement of the point of application of any one of the forces be Δx. Then

$$W_{\text{total}} = F_{1x}\Delta x + F_{2x}\Delta x + \cdots = (F_{1x} + F_{2x} + \cdots)\Delta x = F_{\text{net}\,x}\Delta x \qquad \text{6-4}$$

For a particle constrained to move along the x axis, the net force has only an x component. That is, $\vec{F}_{\text{net}} = F_{\text{net}\,x}\hat{\imath}$. Thus, for a particle the total work can be found by first finding the net force, and then multiplying the net force by the displacement.

The Work–Kinetic Energy Theorem

There is an important relation between the total work done on a particle and the initial and final speeds of the particle. If $F_{\text{net}\,x}$ is the net force acting on a particle, Newton's second law gives

$$F_{\text{net}\,x} = ma_x$$

For a constant force, the acceleration is constant, and we can relate the displacement to the initial speed v_i and final speed v_f by using the constant-acceleration formula $v_f^2 = v_i^2 + 2a_x\Delta x$ (Equation 2-16). Solving this for a_x gives

$$a_x = \frac{1}{2\Delta x}(v_f^2 - v_i^2)$$

Substituting for a_x in $F_{\text{net}\,x} = ma_x$ and then multiplying both sides by Δx gives

$$F_{\text{net}\,x}\Delta x = \tfrac{1}{2}mv_f^2 - \tfrac{1}{2}mv_i^2$$

We recognize the term on the left as the total work done on the particle. Thus

$$W_{\text{total}} = \tfrac{1}{2}mv_f^2 - \tfrac{1}{2}mv_i^2 \qquad \text{6-5}$$

The quantity $\tfrac{1}{2}mv^2$ is a scalar quantity called the **kinetic energy** K of the particle:

$$K = \tfrac{1}{2}mv^2 \qquad \text{6-6}$$

DEFINITION—KINETIC ENERGY

The quantity on the right side of Equation 6-5 is the change in the kinetic energy of the particle. Thus,

The total work done on a particle is equal to the change in its kinetic energy:

$$W_{\text{total}} = \Delta K \qquad \text{6-7}$$

WORK–KINETIC ENERGY THEOREM

This result is known as the **work–kinetic energy theorem.** The derivation presented here is valid only if the net force remains constant. However, as we will see later in this chapter, this theorem is valid even when the net force varies and the motion is not along a straight line.

EXERCISE A woman of mass 50 kg is running at 3.5 m/s. What is her kinetic energy? (*Answer* 306 J)

LOADING WITH A CRANE

EXAMPLE 6-1

A truck of mass 3000 kg is to be loaded onto a ship by a crane that exerts an upward force of 31 kN on the truck. This force, which is strong enough to overcome the gravitational force and get the truck started upward, is applied over a distance of 2 m. Find (*a*) the work done by the crane, (*b*) the work done by gravity, and (*c*) the upward speed of the truck after the 2 m.

PICTURE THE PROBLEM Sketch the truck at its initial and final positions and choose the positive y direction to be the direction of the displacement (Figure 6-2). Use the work–kinetic energy theorem to find the truck's final kinetic energy. The final speed of the truck can be obtained from the final kinetic energy. The total work is the sum of the results for (*a*) and (*b*).

FIGURE 6-2

(*a*) Calculate the work done by the applied force:

$$W_{app} = F_{app\,y}\,\Delta y = F_{app}\cos 0°\,\Delta y$$
$$= (31\text{ kN})(1)(2\text{ m}) = \boxed{62.0\text{ kJ}}$$

(*b*) Calculate the work done by the force of gravity:

$$W_g = mg_y\,\Delta y = mg\cos 180°\,\Delta y$$
$$= (3000\text{ kg})(9.81\text{ N/kg})(-1)(2\text{ m})$$
$$= \boxed{-58.9\text{ kJ}}$$

(*c*) Apply the work–kinetic energy theorem and solve for v_f:

$$W_{total} = \Delta K$$
$$W_{app} + W_g = K_f - K_i$$
$$= \tfrac{1}{2}mv_f^2 - \tfrac{1}{2}mv_i^2$$

$$v_f^2 = v_i^2 + \frac{2(W_{app} + W_g)}{m}$$

$$= 0 + \frac{2(62{,}000\text{ J} - 58{,}900\text{ J})}{3000\text{ kg}}$$

$$= 2.09\text{ m}^2/\text{s}^2$$

$$v_f = \boxed{1.45\text{ m/s}}$$

REMARKS Notice that we treat each force separately when calculating the work done. We could also find the total work by first adding the forces to obtain the net force, then applying $W_{total} = F_{net\,y}\,\Delta y$. In either case, the work–kinetic energy theorem applies only to the total work.

EXERCISE Find the final speed of the truck if the same upward force were applied for 2 m after it was already moving upward at 1 m/s. (*Answer* 1.73 m/s. Note that the answer is *not* 1.45 m/s + 1.00 m/s. Why not?)

FORCE ON AN ELECTRON

EXAMPLE 6-2

In a television tube, an electron is accelerated from rest to a kinetic energy of 2.5 keV over a distance of 80 cm. (The force that accelerates the electron is an electric force due to the electric field in the tube.) Find the force on the electron, assuming it to be constant and in the direction of motion.

PICTURE THE PROBLEM Because the electron starts from rest, the work done equals the final kinetic energy.

Set the work done to be equal to the change in kinetic energy and solve for the force. The initial and final kinetic energies are both given:

$$W_{total} = \Delta K$$

$$F_x \, \Delta x = K_f - K_i$$

$$F_x = \frac{K_f - K_i}{\Delta x}$$

$$= \frac{2500 \text{ eV} - 0}{0.8 \text{ m}} \times \frac{1.6 \times 10^{-19} \text{ J}}{1 \text{ eV}}$$

$$= \boxed{5.0 \times 10^{-16} \text{ N}}$$

REMARKS When we discuss electricity we will see that the work done per unit charge is called the potential difference and is measured in volts. Thus, 1 eV is the energy acquired or lost by a particle of charge e (an electron or proton, for example) when its potential difference changes by 1 V.

A DOGSLED RACE	**EXAMPLE 6-3**

During your winter break you enter a dogsled race across a frozen lake. To get started you pull the sled (total mass 80 kg) with a force of 180 N at 20° above the horizontal. Find (*a*) the work you do and (*b*) the final speed of the sled after it moves $\Delta x = 5$ m, assuming that it starts from rest and there is no friction.

FIGURE 6-3

PICTURE THE PROBLEM The work done by you is $F_x \, \Delta x$, where we choose the direction of the displacement as the positive x direction. This is also the *total* work done on the sled because the other forces, $m\vec{g}$ and $\vec{F}_n$, have no x components (Figure 6-3). The final speed of the sled is found by applying the work–kinetic energy theorem to the sled.

FIGURE 6-4

(*a*) 1. Sketch the sled both in its initial position and in its position after moving the 5 m. Draw the x axis in the direction of motion (Figure 6-4).

  2. The work done by you on the sled is $F_x \, \Delta x$. (This is also the total work done on the sled since the other forces act perpendicular to the x direction):

$$W = F_x \, \Delta x = F \cos \theta \, \Delta x$$

$$= (180 \text{ N})(\cos 20°)(5 \text{ m}) = \boxed{846 \text{ J}}$$

(*b*) Apply the work–kinetic energy theorem to the sled and solve for the final speed:

$$W_{total} = \tfrac{1}{2} mv_f^2 - \tfrac{1}{2} mv_i^2$$

$$v_f^2 = v_i^2 + \frac{2W_{total}}{m}$$

$$= 0 + \frac{2(846 \text{ J})}{80 \text{ kg}}$$

$$= 21.1 \text{ m}^2/\text{s}^2$$

$$v_f = \boxed{4.60 \text{ m/s}}$$

REMARKS We do not need to work out the units. If we have a correct equation, and all quantities are in SI units, the result will be in the correct SI units. However, as a check on the equation, we can show that 1 J/kg = 1 m²/s². We have 1 J/kg = 1 N·m/kg = (1 kg·m/s²)·m/kg = 1 m²/s².

EXERCISE What is the magnitude of the force you exert if the sled starts with a speed of 2 m/s and its final speed is 4.5 m/s after you pull it through a distance of 5 m? (*Answer* 138 N)

What if you hold a massive object in a fixed position? You are expending energy, but are you doing work? According to the definition of work, you are not doing work *on the object* because the object does not move in the direction of the force you exert (Figure 6-5). But your muscles are continually contracting and relaxing as you hold the weight. In this process internal chemical energy in your body is converted to thermal energy (Figure 6-6).

Work Done by a Variable Force

In Figure 6-7 we plot a constant force F_x as a function of position x. The work done on a particle whose displacement is Δx is represented by the area under the force-versus-position curve, indicated by the shading in Figure 6-7.

Many forces vary with position. For example, a stretched spring exerts a force proportional to the distance it is stretched. And the gravitational force the earth exerts

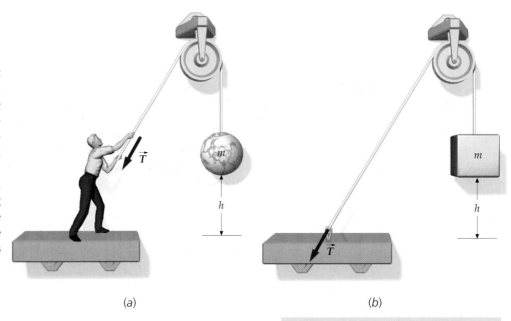

(a)   (b)

FIGURE 6-5 (*a*) The man standing on a ledge does not do work on the weight when holding it at a fixed position. (*b*) The same task could be accomplished by tying the rope to a fixed point.

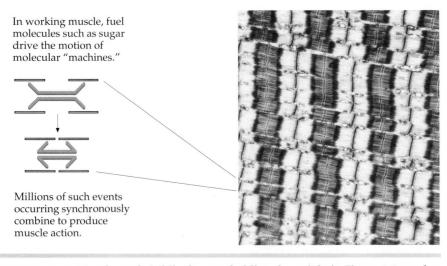

In working muscle, fuel molecules such as sugar drive the motion of molecular "machines."

Millions of such events occurring synchronously combine to produce muscle action.

FIGURE 6-6 Muscle work. While the man holding the weight in Figure 6-5 may be doing no work on the weight, his body is expending energy on the molecular level, as structures within the muscle slide over each other during muscular extension and contraction.

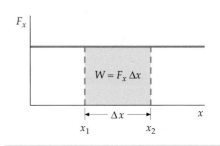

$$W = F_x \, \Delta x$$

FIGURE 6-7 The work done by a constant force is represented graphically as the area under the F_x-versus-x curve.

on a spaceship varies inversely with the square of the distance between the two bodies. We can approximate a variable force by a series of constant forces (Figure 6-8). The work done by a variable force is then

$$W = \lim_{\Delta x_i \to 0} \sum_i F_x \, \Delta x_i = \text{area under the } F_x\text{-versus-}x \text{ curve} \qquad 6\text{-}8$$

This limit is the integral of F_x over x. So the work done by a variable force F_x acting on a particle as it moves from x_1 to x_2 is

$$W = \int_{x_1}^{x_2} F_x \, dx = \text{area under the } F_x\text{-versus-}x \text{ curve} \qquad 6\text{-}9$$

WORK BY A VARIABLE FORCE

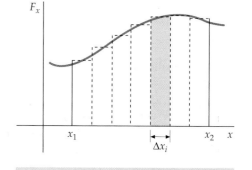

FIGURE 6-8 A variable force can be approximated by a series of constant forces over small intervals. The work done by the constant force in each interval is the area of the rectangle beneath the force curve. The sum of these rectangular areas is the sum of the work done by the set of constant forces that approximates the varying force. In the limit of infinitesimally small Δx_i, the sum of the areas of the rectangles equals the area under the complete force curve.

For each displacement interval Δx_i, the force is essentially constant. Therefore the work done equals the area of the rectangle of height $F_{x\,i}$ and width Δx_i. As was shown earlier in Section 6-1, this work equals the change in kinetic energy for this displacement interval (if the force is the net force). The total work done is the sum of the areas over all displacement intervals. It follows that the total work equals the change in kinetic energy for the entire displacement. Thus, $W_{\text{total}} = \Delta K$ holds for variable forces as well as for constant forces.

EXERCISE IN DIMENSIONAL ANALYSIS A spring is characterized by its force constant k, which has dimensions[†] N/m. How does the work required to stretch a spring by an amount x_0 depend on k and x_0? (*Answer* Because work has dimensions of N·m, the work must depend on k and x_0 in the combination kx_0^2. We will see in Example 6-5 that the actual expression is $W = \frac{1}{2} kx_0^2$. The factor $\frac{1}{2}$ arises because the force varies from 0 to a maximum value of kx_0, and has the average[‡] value $\frac{1}{2} kx_0$.)

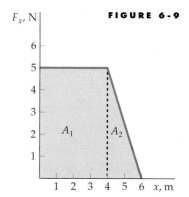

FIGURE 6-9

WORK DONE ON A PARTICLE   **EXAMPLE 6-4**

A force F_x varies with x as shown in Figure 6-9. Find the work done by the force on a particle as the particle moves from $x = 0$ to $x = 6$ m.

PICTURE THE PROBLEM The work is the area under the curve. Because the curve consists of straight-line segments, the easiest approach is to use the geometric formulas for area. (The alternative approach is to set up and execute an integration, which is done in Example 6-5.)

1. We find the work done by calculating the area under the F_x-versus-x curve: $W = A_{\text{total}}$

2. This area is the sum of the two areas shown. The area of a triangle is one half the altitude times the base:

$$W = A_{\text{total}} = A_1 + A_2$$
$$= (5\,\text{N})(4\,\text{m}) + \tfrac{1}{2}(5\,\text{N})(2\,\text{m})$$
$$= 20\,\text{J} + 5\,\text{J} = \boxed{25\,\text{J}}$$

EXERCISE The force shown is the only force that acts on a particle of mass 3 kg. If the particle starts from rest at $x = 0$, how fast is it moving when it reaches $x = 6$ m? (*Answer* 4.08 m/s)

---

† The SI units of k are N/m, but the dimensions of k are $[M][T]^{-2}$.
‡ Here average refers not to an average over time but to an average over distance.

WORK DONE ON A BLOCK BY A SPRING   **EXAMPLE 6-5**   **FIGURE 6-10**

A 4-kg block on a frictionless table is attached to a horizontal spring that obeys Hooke's law and exerts a force $\vec{F} = -kx\hat{i}$, where $k = 400$ N/m and x is measured from the equilibrium position of the block. The spring is originally compressed with the block at $x_1 = -5$ cm (Figure 6-10). Find (a) the work done by the spring on the block as the block moves from $x_1 = -5$ cm to its equilibrium position $x_2 = 0$ and (b) the speed of the block at $x_2 = 0$.

$x_1 = -5$ cm   0

PICTURE THE PROBLEM Make a graph of F_x versus x. The work done on the block as it moves from x_1 to $x_2 = 0$ equals the area under the F_x-versus-x curve between these limits, shaded in Figure 6-11, which can be calculated by integrating the force over the distance. The work done equals the change in kinetic energy, which is simply its final kinetic energy because the initial kinetic energy is zero. The speed of the block at $x = 0$ is found from the kinetic energy of the block.

FIGURE 6-11

(a) The work W done by the spring on the block is the integral of $F_x \, dx$ from $x_1 = -5$ cm to $x_2 = 0$:

$$W = \int_{x_1}^{x_2} F_x \, dx = \int_{x_1}^{x_2} -kx \, dx = -k \int_{x_1}^{x_2} x \, dx$$

$$= -\tfrac{1}{2} kx^2 \Big|_{x_1}^{x_2} = -\tfrac{1}{2} k(x_2^2 - x_1^2)$$

$$= -\tfrac{1}{2}(400 \text{ N/m})\big[0 - (0.05 \text{ m})^2\big] = \boxed{0.500 \text{ J}}$$

(b) Apply the work–kinetic energy theorem and solve for v_2:

$$W_{\text{total}} = \tfrac{1}{2} mv_2^2 - \tfrac{1}{2} mv_1^2$$

$$v_2^2 = v_1^2 + \frac{2W_{\text{total}}}{m} = 0 + \frac{2(0.500 \text{ J})}{4 \text{ kg}}$$

$$= 0.250 \text{ m}^2/\text{s}^2$$

$$v_2 = \boxed{0.500 \text{ m/s}}$$

REMARKS Besides the spring force, two other forces act on the block; the force of gravity, $m\vec{g}$, and the normal force of the table, $\vec{F}_n$. These forces do no work because they have no component in the direction of the displacement. Only the spring does work on the block. (The force it exerts does have a component in the direction of the displacement.)

EXERCISE Find the speed of the block when it reaches $x = 3$ cm if it starts from $x = 0$ with velocity $v_x = 0.5$ m/s. (*Answer* 0.4 m/s)

Note that we could *not* have solved Example 6-5 by first applying Newton's second law to find the acceleration, and then using the constant-acceleration kinematic equations. Because the force exerted by the spring on the block, $F_x = -kx$, varies with position, the acceleration also varies, and thus the constant-acceleration condition is not met.

6-2 The Dot Product

The component F_s in Figure 6-12 is related to the angle ϕ between the directions of $\vec{F}$ and $d\vec{s}$ by $F_s = F \cos \phi$, so the work done by $\vec{F}$ for the displacement $d\vec{s}$ is

$$W = F_s ds = F \cos \phi \, ds$$

This combination of two vectors and the cosine of the angle between their directions is called the **dot product** (or **scalar product**) of the vectors. The dot product of two general vectors $\vec{A}$ and $\vec{B}$ is written $\vec{A} \cdot \vec{B}$ and defined by

$$\vec{A} \cdot \vec{B} = AB \cos \phi \qquad \text{6-10}$$

DEFINITION—DOT PRODUCT

where ϕ is the angle between $\vec{A}$ and $\vec{B}$. (The angle between two vectors is defined as the angle between their directions in space.) The dot product $\vec{A} \cdot \vec{B}$ can be thought of either as A times the component of $\vec{B}$ in the direction of $\vec{A}$ ($A[B \cos \phi]$), or as B times the component of $\vec{A}$ in the direction of $\vec{B}$ ($B[A \cos \phi]$) (see Figure 6-13). Properties of the dot product are summarized in Table 6-1.

TABLE 6-1

Properties of Dot Products

If	then
$\vec{A}$ and $\vec{B}$ are perpendicular,	$\vec{A} \cdot \vec{B} = 0$ (since $\phi = 90°$, $\cos 90° = 0$)
$\vec{A}$ and $\vec{B}$ are parallel,	$\vec{A} \cdot \vec{B} = AB$ (since $\phi = 0°$, $\cos 0° = 1$)
$\vec{A} \cdot \vec{B} = 0$,	Either $\vec{A} = 0$ or $\vec{B} = 0$ or $\vec{A}$ and $\vec{B}$ are perpendicular
Furthermore,	
$\vec{A} \cdot \vec{A} = A^2$	Since $\vec{A}$ is parallel to itself
$\vec{A} \cdot \vec{B} = \vec{B} \cdot \vec{A}$	Commutative rule of multiplication
$(\vec{A} + \vec{B}) \cdot \vec{C} = \vec{A} \cdot \vec{C} + \vec{B} \cdot \vec{C}$	Distributive rule of multiplication

We can use unit vectors to write the dot product in terms of the rectangular components of the two vectors:

$$\vec{A} \cdot \vec{B} = (A_x \hat{i} + A_y \hat{j} + A_z \hat{k}) \cdot (B_x \hat{i} + B_y \hat{j} + B_z \hat{k})$$

The dot product of any unit vector with itself, like $\hat{i} \cdot \hat{i}$, is 1, so a term like $A_x \hat{i} \cdot B_x \hat{i}$ equals $A_x B_x$. Also, because the unit vectors $\hat{i}$, $\hat{j}$, and $\hat{k}$ are mutually perpendicular, the dot product of one of them with any other one, like $\hat{i} \cdot \hat{j}$, is zero. Thus, terms like $A_x \hat{i} \cdot B_y \hat{j}$ (called cross terms) each equal zero. The result is

$$\vec{A} \cdot \vec{B} = A_x B_x + A_y B_y + A_z B_z \qquad \text{6-11}$$

FIGURE 6-12 (*a*) A particle moving along an arbitrary curve in space. (*b*) The perpendicular component of the force $F_\perp$ changes the direction of the particle's motion but not its speed. The tangential component F_s changes the particle's speed but not its direction. F_s equals the mass m times the tangential acceleration dv/dt. Only this component does work.

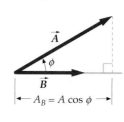

(*a*)

FIGURE 6-13 (*a*) The dot product $\vec{A} \cdot \vec{B}$ is the product of A and the projection of $\vec{B}$ on $\vec{A}$ or the product of B and the projection of $\vec{A}$ on $\vec{B}$. That is, $\vec{A} \cdot \vec{B} = AB \cos \phi = AB_A = BA_B$.

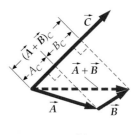

(*b*)

(*b*) $(\vec{A} + \vec{B}) \cdot \vec{C}$ equals $(\vec{A} + \vec{B})_C C$ (the projection of $\vec{A} + \vec{B}$ in the direction of $\vec{C}$ times C). However, $(\vec{A} + \vec{B})_C = A_C + B_C$, so $(\vec{A} + \vec{B}) \cdot \vec{C} = (A_C + B_C)C = A_C C + B_C C = \vec{A} \cdot \vec{C} + \vec{B} \cdot \vec{C}$. That is, for the dot product multiplication is distributive over addition.

The component of a vector in some direction can be written as the dot product of the vector and the unit vector in that direction. For example, the component A_x is found from

$$\vec{A} \cdot \hat{i} = (A_x\hat{i} + A_y\hat{j} + A_z\hat{k}) \cdot \hat{i} = A_x \qquad \text{6-12}$$

This suggests an algebraic procedure for obtaining a component equation, given a vector equation. That is, multiplying both sides of the vector equation $\vec{A} + \vec{B} = \vec{C}$ by $\hat{i}$ gives $\vec{A} \cdot \hat{i} + \vec{B} \cdot \hat{i} = \vec{C} \cdot \hat{i}$, which in turn gives $A_x + B_x = C_x$.

To establish the product rule for differentiating dot products we differentiate both sides of Equation 6-11. For brevity we do this for vectors with two dimensions.

$$\frac{d}{dt}(\vec{A} \cdot \vec{B}) = \frac{d}{dt}(A_xB_x + A_yB_y)$$

$$= \frac{dA_x}{dt}B_x + A_x\frac{dB_x}{dt} + \frac{dA_y}{dt}B_y + A_y\frac{dB_y}{dt}$$

Rearranging gives

$$\frac{d}{dt}(\vec{A} \cdot \vec{B}) = \frac{dA_x}{dt}B_x + \frac{dA_y}{dt}B_y + A_x\frac{dB_x}{dt} + A_y\frac{dB_y}{dt}$$

$$= \left(\frac{dA_x}{dt}\hat{i} + \frac{dA_y}{dt}\hat{j}\right) \cdot (B_x\hat{i} + B_y\hat{j}) + (A_x\hat{i} + A_y\hat{j}) \cdot \left(\frac{dB_x}{dt}\hat{i} + \frac{dB_y}{dt}\hat{j}\right)$$

so

$$\frac{d}{dt}(\vec{A} \cdot \vec{B}) = \frac{d\vec{A}}{dt} \cdot \vec{B} + \vec{A} \cdot \frac{d\vec{B}}{dt} \qquad \text{6-13}$$

USING THE DOT PRODUCT   **E X A M P L E   6 - 6**

(a) Find the angle between the vectors $\vec{A} = 3\text{ m}\hat{i} + 2\text{ m}\hat{j}$ and $\vec{B} = 4\text{ m}\hat{i} - 3\text{ m}\hat{j}$ (Figure 6-14). (b) Find the component of $\vec{A}$ in the direction of $\vec{B}$.

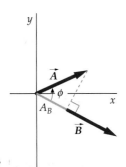

FIGURE 6-14

PICTURE THE PROBLEM We find the angle ϕ from the definition of the dot product. The component of $\vec{A}$ in the direction of $\vec{B}$ is found from the dot product $\vec{A} \cdot \hat{B}$, where $\hat{B} = \vec{B}/B$.

(a) 1. Write the dot product of $\vec{A}$ and $\vec{B}$ in terms of A, B, and $\cos \phi$ and solve for $\cos \phi$:

$$\vec{A} \cdot \vec{B} = AB\cos\phi, \quad \text{so}$$

$$\cos\phi = \frac{\vec{A} \cdot \vec{B}}{AB}$$

2. Find $\vec{A} \cdot \vec{B}$ from their components:

$$\vec{A} \cdot \vec{B} = A_xB_x + A_yB_y$$

$$= (3\text{ m})(4\text{ m}) + (2\text{ m})(-3\text{ m})$$

$$= 12\text{ m}^2 - 6\text{ m}^2 = 6\text{ m}^2$$

3. The magnitudes of the vectors are obtained from the dot product of the vector with itself:

$$\vec{A} \cdot \vec{A} = A^2 = A_x^2 + A_y^2$$

$$= (3 \text{ m})^2 + (2 \text{ m})^2 = 13 \text{ m}^2, \quad \text{so}$$

$$A = \sqrt{13} \text{ m}$$

and

$$\vec{B} \cdot \vec{B} = B^2 = B_x^2 + B_y^2$$

$$= (4 \text{ m})^2 + (-3 \text{ m})^2 = 25 \text{ m}^2, \quad \text{so}$$

$$B = 5 \text{ m}$$

4. Substitute these values into the equation in step 1 for $\cos \phi$ to find ϕ:

$$\cos \phi = \frac{\vec{A} \cdot \vec{B}}{AB} = \frac{6 \text{ m}^2}{(\sqrt{13} \text{ m})(5 \text{ m})} = 0.333$$

$$\phi = \boxed{70.6°}$$

(b) The component of $\vec{A}$ in the direction of $\vec{B}$ is the dot product of $\vec{A}$ with the unit vector $\hat{B} = \vec{B}/B$:

$$A_B = \vec{A} \cdot \hat{B} = \vec{A} \cdot \frac{\vec{B}}{B} = \frac{\vec{A} \cdot \vec{B}}{B}$$

$$= \frac{6 \text{ m}^2}{5 \text{ m}} = \boxed{1.2 \text{ m}}$$

❶ PLAUSIBILITY CHECK The component of A along B is $A \cos \phi = (\sqrt{13} \text{ m})$ $\cos 70.6° = 1.2 \text{ m}$.

EXERCISE (a) Find $\vec{A} \cdot \vec{B}$ for $\vec{A} = 3 \text{ m}\hat{i} + 4 \text{ m}\hat{j}$ and $\vec{B} = 2 \text{ m}\hat{i} + 8 \text{ m}\hat{j}$. (b) Find A, B, and the angle between $\vec{A}$ and $\vec{B}$ for these vectors. (*Answer* (a) 38 m² (b) $A = 5$ m, $B = 8.25$ m, $\phi = 23°$)

In dot-product notation, the work dW done by a force $\vec{F}$ on a particle over a displacement $d\vec{s}$ is

$$dW = F \cos \phi \, ds = \vec{F} \cdot d\vec{s} \qquad\qquad 6\text{-}14$$

INCREMENTAL WORK

where $ds = |d\vec{s}|$ (the magnitude of $d\vec{s}$). The work done on the particle as it moves from point 1 to point 2 is

$$W = \int_{s_1}^{s_2} \vec{F} \cdot d\vec{s} \qquad\qquad 6\text{-}15$$

THE DEFINITION OF WORK

(If the force remains constant, the work can be expressed $W = \vec{F} \cdot \vec{s}$, where $\vec{s}$ is the net displacement.)

When several forces $\vec{F}_i$ act on a particle whose displacement is $d\vec{s}$, the total work done on it is

$$dW_{\text{total}} = \vec{F}_1 \cdot d\vec{s} + \vec{F}_2 \cdot d\vec{s} + \cdots = (\vec{F}_1 + \vec{F}_2 + \cdots) \cdot d\vec{s} = (\Sigma \vec{F}_i) \cdot d\vec{s} \qquad 6\text{-}16$$

PUSHING A BOX   **E X A M P L E   6 - 7**

You push a box up a ramp using a horizontal 100-N force $\vec{F}$. For each 5 m of distance along the ramp the box gains 3 m of height. Find the work done by $\vec{F}$ for each 5 m it moves along the ramp (*a*) by directly computing the dot product from the components of $\vec{F}$ and the displacement $\vec{s}$, (*b*) by multiplying the product of the magnitudes of $\vec{F}$ and $\vec{s}$ with the cosine of the angle between their directions, (*c*) by finding F_s (the component of the force in the direction of the displacement) and multiplying it by the magnitude of the displacement, and (*d*) by finding the component of the displacement in the direction of the force and multiplying it by the magnitude of the force.

FIGURE 6-15

PICTURE THE PROBLEM Draw a sketch of the box in its initial and final positions. Place coordinate axes on the sketch with the x axis horizontal. Express the force and displacement vectors in component form and take the dot product. Then find the component of the force in the direction of the displacement, and vice versa.

(*a*) Express $\vec{F}$ and $\vec{s}$ in component form and take the dot product:

$$\vec{F} = 100 \text{ N}\hat{i} + 0\hat{j}$$

$$\vec{s} = 4 \text{ m}\hat{i} + 3 \text{ m}\hat{j}$$

$$W = \vec{F} \cdot \vec{s} = F_x\Delta x + F_y\Delta y$$

$$= (100 \text{ N})(4 \text{ m}) + 0(3 \text{ m}) = \boxed{400 \text{ J}}$$

(*b*) Calculate $Fs \cos \phi$, where ϕ is the angle between the directions of the two vectors as shown. Equate this expression with the Part (*a*) result and solve for $\cos \phi$; then solve for the work:

$$\vec{F} \cdot \vec{s} = Fs \cos \phi \quad \text{and} \quad \vec{F} \cdot \vec{s} = F_x\Delta x + F_y\Delta y$$

so

$$\cos \phi = \frac{F_x\Delta x + F_y\Delta y}{Fs} = \frac{(100 \text{ N})(4 \text{ m}) + 0}{(100 \text{ N})(5 \text{ m})} = 0.8$$

and

$$W = Fs \cos \phi$$

$$= (100 \text{ N})(5 \text{ m})0.8 = \boxed{400 \text{ J}}$$

(*c*) Find F_s and multiply it by s:

$$F_s = F \cos \phi = (100 \text{ N})0.8 = 80 \text{ N}$$

$$W = F_s s = (80 \text{ N})(5 \text{ m}) = \boxed{400 \text{ J}}$$

(*d*) Multiply F and s_F, where s_F is the component of $\vec{s}$ in the direction of $\vec{F}$. First calculate s_F:

$$s_F = s \cos \phi = (5 \text{ m})0.8 = 4\text{m}$$

$$W = Fs_F = (100 \text{ N})(4 \text{ m}) = \boxed{400 \text{ J}}$$

REMARKS For this problem, computing the work is easiest using the procedure in Part (*d*). For other problems, the procedure in Part (*a*), Part (*b*), or Part (*c*) may be the easiest. You should be comfortable with all four procedures. That way, for a given problem, you can choose the easiest procedure.

A DISPLACED PARTICLE   **E X A M P L E   6 - 8   Try It Yourself**

A particle is given a displacement $\vec{s} = 2 \text{ m}\hat{i} - 5 \text{ m}\hat{j}$ along a straight line. During the displacement, a constant force $\vec{F} = 3 \text{ N}\hat{i} + 4 \text{ N}\hat{j}$ acts on the particle. Find (*a*) the work done by the force and (*b*) the component of the force in the direction of the displacement.

FIGURE 6-16

PICTURE THE PROBLEM The work W is found by computing $W = \vec{F} \cdot \vec{s} = F_x \Delta x + F_y \Delta y$. Combining this with the relation $\vec{F} \cdot \vec{s} = F_s s$, we can find the component of $\vec{F}$ in the direction of the displacement. Make a sketch showing $\vec{F}, \vec{s}$, and $|F_s|$ (Figure 6-16).

Cover the column to the right and try these on your own before looking at the answers.

Steps

(a) Compute the work done W.

(b) 1. Compute $\vec{s} \cdot \vec{s}$ and use your result to find the distance $|\vec{s}|$.

2. Using $\vec{F} \cdot \vec{s} = F_s s$, solve for F_s.

Answers

$W = \vec{F} \cdot \vec{s} = \boxed{-14 \, \text{N} \cdot \text{m}}$

$|\vec{s}|^2 = \vec{s} \cdot \vec{s} = 29 \, \text{m}^2$, so

$|\vec{s}| = \sqrt{29} \, \text{m}$

$F_s = \boxed{-2.60 \, \text{N}}$

REMARKS The component of the force in the direction of the displacement is negative, so the work done is negative.

EXERCISE Find the magnitude of $\vec{F}$ and the angle ϕ between $\vec{F}$ and $\vec{s}$. (*Answer* $F = 5 \, \text{N}, \phi = 121°$)

DIFFERENTIATING A DOT PRODUCT   **EXAMPLE 6-9**

Show that $d(v^2)/dt = 2\vec{a} \cdot \vec{v}$, **where** v **is the speed,** $\vec{v}$ **the velocity, and** $\vec{a}$ **the acceleration.**

PICTURE THE PROBLEM Note that $v^2 = \vec{v} \cdot \vec{v}$, so the rule for differentiating dot products can be used here.

Apply the rule for differentiating dot products to the dot product $\vec{v} \cdot \vec{v}$:

$$\frac{d}{dt}(v^2) = \frac{d}{dt}(\vec{v} \cdot \vec{v})$$

$$= \frac{d\vec{v}}{dt} \cdot \vec{v} + \vec{v} \cdot \frac{d\vec{v}}{dt} = 2\frac{d\vec{v}}{dt} \cdot \vec{v}$$

so

$$\boxed{\frac{d}{dt}(v^2) = 2\vec{a} \cdot \vec{v}}$$

REMARKS This example involves only kinematic parameters, so the resulting relation is a strictly kinematic relation. Also, the result is unconditionally valid because the computation involved only the definition of acceleration and the rule for differentiating a dot product.

Power

The **power** P supplied by a force is the rate at which the force does work. Consider a particle moving with instantaneous velocity $\vec{v}$. In a short time interval dt, the particle has a displacement $d\vec{s} = \vec{v} \, dt$. The work done by a force $\vec{F}$ acting on the particle during this time interval is

$$dW = \vec{F} \cdot d\vec{s} = \vec{F} \cdot \vec{v} \, dt$$

The power delivered to the particle is then

$$P = \frac{dW}{dt} = \vec{F} \cdot \vec{v}$$

6-17

DEFINITION—POWER

The SI unit of power, one joule per second, is called a watt (W):

$$1 \text{ W} = 1 \text{ J/s}$$

Note the difference between power and work. Two motors that lift a given load a given distance expend the same amount of energy, but the one that does it in the least time is more powerful. Gas and electric companies charge for energy, not power, usually by the kilowatt-hour (kW·h). A kilowatt-hour of energy is

$$1 \text{ kW·h} = (10^3 \text{ W})(3600 \text{ s}) = 3.6 \times 10^6 \text{ W·s} = 3.6 \text{ MJ}$$

In the U.S. customary system, the unit of energy is the foot-pound and the unit of power is the foot-pound per second. A commonly used multiple of this unit, called a horsepower (hp) is defined as

$$1 \text{ hp} = 550 \text{ ft·lb/s} = 746 \text{ W}$$

FIGURE 6-17

THE POWER OF A MOTOR **EXAMPLE 6-10**

A small motor is used to operate a lift that raises a load of bricks weighing 800 N to a height of 10 m in 20 s (Figure 6-17). What is the minimum power the motor must deliver?

PICTURE THE PROBLEM To find the *minimum* power we assume that the bricks are lifted at constant speed. Because the acceleration is zero, the magnitude of the upward force $\vec{F}$ exerted by the motor is equal to the weight of the bricks, 800 N. The power delivered by the motor is the power supplied by $\vec{F}$.

The power is given by $\vec{F} \cdot \vec{v}$:

$$P = \vec{F} \cdot \vec{v} = Fv \cos \theta = Fv \cos 0$$

$$= (800 \text{ N})\frac{10 \text{ m}}{20 \text{ s}}(1) = \boxed{400 \text{ W}}$$

REMARKS This minimum power output of 400 W is slightly more than $\frac{1}{2}$ hp.

EXERCISE (*a*) Find the total work done by the force. (*b*) Calculate the power by dividing the total work by the total time. (*Answer* (*a*) 8000 J (*b*) 400 W)

POWER AND KINETIC ENERGY **EXAMPLE 6-11**

Show that the power delivered to a particle by the net force acting on it equals the rate at which the kinetic energy of the particle is changing.

PICTURE THE PROBLEM The power delivered by the net force equals $\vec{F}_{net} \cdot \vec{v}$. Show that $\vec{F}_{net} \cdot \vec{v} = dK/dt$, where $K = \frac{1}{2} mv^2$.

1. Apply the result of Example 6-9 together with $\vec{a} = \vec{F}_{net}/m$ and solve for $\vec{F}_{net} \cdot \vec{v}$:

$$2\vec{a} \cdot \vec{v} = \frac{d}{dt}(v^2)$$

$$2\frac{\vec{F}_{net}}{m} \cdot \vec{v} = \frac{d}{dt}(v^2)$$

$$\vec{F}_{net} \cdot \vec{v} = \frac{m}{2}\frac{d}{dt}(v^2)$$

2. The mass is constant so it can be moved inside the argument of the derivative:

$$\vec{F}_{net} \cdot \vec{v} = \frac{d}{dt}\left(\frac{1}{2}mv^2\right)$$

$$P_{net} = \frac{d}{dt}\left(\frac{1}{2}mv^2\right) = \boxed{\frac{dK}{dt}}$$

REMARKS In the next section we use the results of this example to obtain the work–kinetic energy theorem in three dimensions.

In Example 6-10, the power delivered to the bricks by the lower end of the rope was calculated. In that example the rate of change in kinetic energy of the rope is negligible, so the power delivered to the rope by the motor is equal to the power the rope delivers to the bricks.

6-3 Work and Energy in Three Dimensions

From Example 6-11 we have $\vec{F}_{net} \cdot \vec{v} = dK/dt$, where $K = \frac{1}{2} mv^2$. The work–kinetic energy theorem for three dimensions can be established by integrating both sides of this equation with respect to time. Integrating both sides gives

$$\int_{t_1}^{t_2} \vec{F}_{net} \cdot \vec{v}\, dt = \int_{t_1}^{t_2} \frac{dK}{dt}\, dt \qquad\qquad 6\text{-}18$$

Because $d\vec{s} = \vec{v}\, dt$, where $d\vec{s}$ is the displacement during time dt, and because $(dK/dt)\, dt = dK$, Equation 6-18 can be expressed

$$\int_1^2 \vec{F}_{net} \cdot d\vec{s} = \int_1^2 dK$$

where the integral on the left is the total work W_{total} done on the particle. (In Chapter 7 work–energy relations for objects that cannot be modeled as a particle are presented.) The integral on the right can be integrated giving

$$W_{total} = \int_1^2 \vec{F}_{net} \cdot d\vec{s} = K_2 - K_1 = \Delta K \qquad\qquad 6\text{-}19$$

WORK–KINETIC ENERGY EQUATION IN THREE DIMENSIONS

Equation 6-19 follows directly from Newton's second law of motion.

WORK DONE ON A SKIER   **EXAMPLE 6-12** **Put It in Context**

You and your friend are at ski resort with two ski runs, a beginner's run and an expert's run. Both runs begin at the top of the ski lift and end at a finish line at the bottom of the same lift. Let *h* be the vertical descent for both runs. The beginner's run is longer and less steep than the expert's run. You and your friend, who is a much better skier than you, are testing some experimental frictionless skis. To make things interesting, you offer a wager that if she takes the expert's run and you take the beginner's run, her speed at the finish line will not be greater than your speed at the finish line. She accepts the bet (momentarily forgetting that you are taking a physics course). The conditions are that you both start from rest at the top of the lift and both of you coast for the entire trip. Who wins the bet? (Assume airdrag is negligible.)

PICTURE THE PROBLEM Because you and your friend are coasting on the skis, you both can be modeled as particles. Two forces act on each of you, gravity $m\vec{g}$ and the normal force $\vec{F}_n$. Make a sketch of yourself and draw the two force vectors on the sketch (Figure 6-18*a*). Also include coordinate axes. The work–kinetic energy theorem, with $v_i = 0$, relates the final speed v_f to the total work. (The work–kinetic energy theorem works only for particles.)

FIGURE 6-18

(a)

1. The final speed is related to the final kinetic energy, which in turn is related to the total work by the work–kinetic energy theorem:

$$W_{total} = \tfrac{1}{2}mv_f^2 - \tfrac{1}{2}mv_i^2$$

2. For each of you, the total work is the work done by the normal force plus the work done by the gravitational force:

$$W_{total} = W_n + W_g$$

3. The force $m\vec{g}$ on you is constant, but the force $\vec{F}_n$ is not constant. First we calculate the work done by $\vec{F}_n$. Calculate the work dW_n done on you by $\vec{F}_n$ for an infinitesimal displacement $d\vec{s}$ (Figure 6-18*b*) at an arbitrary location along the run:

$$dW_n = \vec{F}_n \cdot d\vec{s} = F_n \cos\phi\, ds$$

4. Find the angle ϕ between the directions of $\vec{F}_n$ and $d\vec{s}$. The displacement $d\vec{s}$ is parallel to the slope:

$$\phi = 90°$$

(b)

5. Calculate the work done by $\vec{F}_n$ for the entire run:

$$dW_n = F_n \cos 90°\, ds = 0, \quad \text{so}$$

$$W_n = \int dW_n = 0$$

6. The force of gravity is constant, so the work done by gravity is $W_g = m\vec{g} \cdot \vec{s}$, where $\vec{s}$ (Figure 6-19) is the net displacement from the top to the bottom of the lift:

$$W_g = m\vec{g} \cdot \vec{s} = -mg\,\hat{j} \cdot (\Delta x\hat{i} + \Delta y\hat{j})$$
$$= -mg\,\Delta y$$

FIGURE 6-19

7. The skier is descending the hill, so Δy is negative. From Figure 6-18a, we see that $\Delta y = -h$:

$$\Delta y = -h$$

8. Substituting gives:

$$W_g = mgh$$

9. Apply the work–kinetic energy theorem to find v_f:

$$W_n + W_g = \Delta K$$

10. The final speed depends only on h, which is the same for both runs. Both of you will have the same final speeds.

$$0 + mgh = \tfrac{1}{2} mv_f^2 - 0$$

so

$$v_f = \sqrt{2gh}$$

YOU WIN! (The bet was that she would not be going faster than you.)

REMARKS Your friend on the steeper trail will cross the finish line in less time, but that was not the bet. What was shown here is that the work done by the gravitational force equals mgh. It does not depend upon the shape of the hill or upon the length of the path taken. It depends only upon the vertical drop h between the starting point and the finishing point.

In Example 6-12 we found that the work done by the gravitational force is independent of the path taken. This leads us to the concept of potential energy, which is the topic of the next section.

6-4 Potential Energy

The total work done on a particle equals the change in its kinetic energy. But we are often interested in the work done on a *system* consisting of two or more particles.[†] Often, the work done by external forces on a *system* does not increase the total kinetic energy *of the system*, but instead is stored as **potential energy**— energy associated with the configuration of the system.

Consider lifting a barbell of mass m to a height h. The work done by the gravitational force on the barbell is $-mgh$. The barbell starts at rest and ends at rest. Because the kinetic energy of the barbell does not change, we know the total work done on the barbell is zero. That means the work done by the force of your hands on the barbell is $+mgh$. Now consider the barbell and the earth to be a *system* of two particles. (You are not part of this system.) The external forces *on the earth–barbell system* are the gravitational attraction you exert *on the earth*, the force your feet exert *on the earth*, and the force mg exerted by your hands *on the barbell* (Figure 6-20). (We can neglect the gravitational force you exert on the barbell.) The barbell moves, but the motion of the earth is negligible, so only the force exerted on the barbell by your hands does work on the system. The total work done on the earth–barbell system by all forces *external* to the system is mgh. This work is stored as potential energy, which is energy associated with the position of the barbell relative to the earth. That is, it is energy associated with the configuration of the earth–barbell system. This kind of energy is called gravitational potential energy.

Another system that stores energy associated with its configuration is a spring. If you stretch or compress a spring, energy associated with the length of the spring is stored as potential energy. Consider the spring shown in Figure 6-21 as the system. You compress the spring, pushing it with equal and opposite forces $\vec{F}_1$ and $\vec{F}_2$. These forces sum to zero, so the net force on the spring remains

---

† Systems of particles are discussed more thoroughly in Chapter 8.

FIGURE 6-20 A system consisting of a barbell plus the earth, but not the person holding the barbell. When you lift the barbell, you do work on this system.

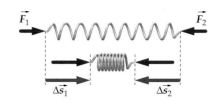

FIGURE 6-21 The spring is compressed by external forces $\vec{F}_1$ and $\vec{F}_2$. Both forces do positive work on the spring as they compress it, so the elastic potential energy of the spring increases as it is compressed.

zero. Thus, there is no change in the kinetic energy of the spring. The work you do on this system is stored not as kinetic energy, but as elastic potential energy. The configuration of this system has been changed, as evidenced by the change in the length of the spring. The total work done on the spring is positive because both $\vec{F}_1$ and $\vec{F}_2$ do positive work. (The work done by $\vec{F}_1$ is positive because both $\vec{F}_1$ and the displacement $\Delta \vec{s}_1$ of its point of application are in the same direction. The same can be said for $\vec{F}_2$ and $\Delta \vec{s}_2$.)

Conservative Forces

When you ride a ski lift to the top of a hill of height h, the work done by gravity on you is $-mgh$ and the work done by the lift on you is $+mgh$. When you ski down the hill to the bottom, the work done by gravity is $+mgh$ independent of the shape of the hill (as we saw in Example 6-12). The total work done by gravity on you during the round trip up and down the hill is zero, independent of the path you take. The force of gravity, exerted by the earth on you, is a **conservative force.**

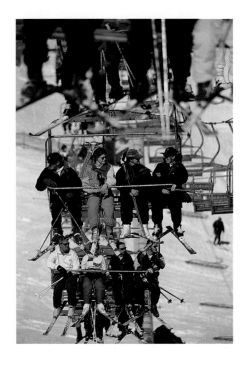

> A force is conservative if the total work it does on a particle is zero when the particle moves around *any* closed path, returning to its initial position.

DEFINITION—CONSERVATIVE FORCE

From Figure 6-22 we see that this definition implies that:

> The work done by a conservative force on a particle is independent of the path taken as the particle moves from one point to another.

ALTERNATIVE DEFINITION—CONSERVATIVE FORCE

Now consider yourself and the earth to be a *two-particle system.* (The ski lift is not part of this system.) When a ski lift raises you to the top of the hill, it does work mgh on the you–earth system. This work is stored as the gravitational potential energy of the system. When you ski down the hill, this potential energy is converted to the kinetic energy of your motion.

Potential-Energy Functions

Because the work done by a conservative force on a particle does not depend on the path, it can depend only on the endpoints 1 and 2. We can use this property to define the **potential-energy function** U that is associated with a conservative force. Note that when you ski down the hill, the work done by gravity *decreases* the potential energy of the system. We define the potential-energy function such that the work done by a conservative force equals the decrease in the potential-energy function:

$$W = \int_1^2 \vec{F} \cdot d\vec{s} = -\Delta U$$

or

$$\Delta U = U_2 - U_1 = -\int_1^2 \vec{F} \cdot d\vec{s} \qquad \text{6-20}a$$

DEFINITION—POTENTIAL-ENERGY FUNCTION

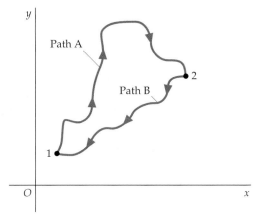

FIGURE 6-22 Two paths in space connecting the points 1 and 2. If the work done by a conservative force along path A from 1 to 2 is W, the work done on the return trip along path B must be $-W$ because the roundtrip work is zero. When traversing path B from 1 to 2, the force is the same at each point, but the displacement is opposite to that when going from 2 to 1. Then the work done along path B from 1 to 2 must also be W. It follows that the work done as a particle goes from point 1 to 2 is the same along any path connecting the two points.

For an infinitesimal displacement, we have

$$dU = -\vec{F} \cdot d\vec{s} \qquad\qquad 6\text{-}20b$$

We can calculate the potential-energy function associated with the gravitational force near the surface of the earth from Equation 6-20b. For the force $\vec{F} = -mg\hat{j}$, we have

$$dU = -\vec{F} \cdot d\vec{s} = -(-mg\hat{j}) \cdot (dx\hat{i} + dy\hat{j} + dz\hat{k}) = +mg\,dy$$

Integrating, we obtain

$$U = \int mg\,dy = mgy + U_0$$

$$U = U_0 + mgy \qquad\qquad 6\text{-}21$$

GRAVITATIONAL POTENTIAL ENERGY NEAR THE EARTH'S SURFACE

where U_0, the arbitrary constant of integration, is the value of the potential energy at $y = 0$. Because only a change in the potential energy is defined, the actual value of U is not important. We are free to choose U to be zero at any convenient reference point. For example, if the gravitational potential energy of the earth–skier system is chosen to be zero when the skier is at the bottom of the hill, its value when the skier is at a height h above that level is mgh. Or we could choose the potential energy to be zero when the skier is at point P half way down the ski slope, in which case its value at any other point would be mgy, where y is the height of the skier above point p. On the lower half of the slope the potential energy would then be negative.

EXERCISE A 55-kg window washer stands on a platform 8 m above the ground. What is the potential energy U of the window-washer–earth system if (a) U is chosen to be zero on the ground, (b) U is chosen to be zero 4 m above the ground, and (c) U is chosen to be zero 10 m above the ground? (*Answer* (a) 4.32 kJ (b) 2.16 kJ (c) −1.08 kJ)

A FALLING BOTTLE   **EXAMPLE 6-13**

A bottle of mass 0.350 kg falls from rest from a shelf that is 1.75 m above the floor. Find the potential energy of the bottle–earth system when the bottle is on the shelf and just before impact with the floor. Find the kinetic energy of the bottle just before impact.

PICTURE THE PROBLEM The work done by the earth on the bottle as it falls equals the negative of the change in the potential energy of the bottle–earth system. Knowing the work, we can use the work–kinetic energy theorem to find the kinetic energy.

1. Make a sketch showing the bottle on the shelf and again when it is about to impact the floor (Figure 6-23). Choose the potential energy of the bottle–earth system to be zero when the bottle is on the floor, and place a y axis on the sketch with the origin at the floor:

FIGURE 6-23

2. The only force doing work on the falling bottle is the force of gravity, so $W_{total} = W_g$. Apply the work–kinetic energy theorem to the falling bottle:

$$W_{total} = W_g = \Delta K$$

3. The force exerted by the earth on the falling bottle is internal to the bottle–earth system. It is also a conservative force, so the work done by it equals the change in the potential energy of the system:

$$W_g = -\Delta U = -(U_f - U_i) = -(mgy_f - mgy_i)$$
$$= mg(y_i - y_f) = mg(h - 0) = mgh$$

4. Substitute the step 3 result into the step 2 result and solve for the final kinetic energy. The original kinetic energy is zero:

$$mgh = \Delta K$$
$$mgh = K_f - K_i$$
$$K_f = K_i + mgh$$
$$= 0 + (0.350 \text{ kg})(9.81 \text{ N/kg})(1.75 \text{ m})$$
$$= \boxed{6.01 \text{ J}}$$

REMARKS In this example, the potential energy lost by the bottle–earth system is converted entirely to kinetic energy of the bottle as it falls. Note in step 4 that we have used the definition $1 \text{ J} = 1 \text{ N·m}$.

Potential energy is associated with the configuration of a *system of particles*, but we sometimes have systems such as the bottle–earth system, in which only one particle moves (the earth's motion is negligible). For brevity, then, we sometimes refer to the potential energy of the bottle–earth system as simply the potential energy of the bottle.

Potential Energy of a Spring   Another example of a conservative force is that of a stretched (or compressed) spring. Suppose we pull a block attached to a spring from a position $x = 0$ (equilibrium) to x_1 (Figure 6-24). The spring does negative work because the force it exerts on the block is opposite to the block's displacement. If we then release the block, the force of the spring does positive work on the block as it accelerates toward its initial position. The total work done by the spring on the block as it moves from its original position to $x = x_1$, and then back, is zero. This result is independent of the size of x_1 (as long as the stretching is not so great as to exceed the elastic limit of the spring). The force exerted by the spring is therefore a conservative force. We can calculate the potential-energy function associated with this force from Equation 6-20b:

$$dU = -\vec{F} \cdot d\vec{s} = -F_x \, dx = -(-kx) \, dx = +kx \, dx$$

Then

FIGURE 6-24 The applied force F_{app} pulls the block to the right, stretching the spring by x_1.

$$U = \int kx \, dx = \tfrac{1}{2}kx^2 + U_0$$

where U_0 is the potential energy when $x = 0$, that is, when the spring is unstretched. Choosing U_0 to be zero gives

$$U = \tfrac{1}{2}kx^2 \qquad\qquad\qquad 6\text{-}22$$

POTENTIAL ENERGY OF A SPRING

When we pull the block from $x = 0$ to $x = x_1$, we must exert an applied force to the spring. If the block starts from rest at $x = 0$ and ends at rest at $x = x_1$, the change in its kinetic energy is zero. The work–energy theorem then implies that the total work done on the block is zero. That is, $W_{app} + W_{spring} = 0$, or

$$W_{app} = -W_{spring} = \Delta U_{spring} = \tfrac{1}{2}kx_1^2 - 0 = \tfrac{1}{2}kx_1^2$$

This work is stored as potential energy in the spring–block system.

POTENTIAL ENERGY OF A BASKETBALL PLAYER

E X A M P L E   6 - 1 4

A system consists of a basketball player, the rim of a basket-ball hoop, and the earth. Assume that the potential energy of this system is zero when the player is standing on the floor and the rim is horizontal. Find the total potential energy of this system when the player is hanging on the rim (as in Figure 6-25). Also assume that the player can be described as a point mass of 110 kg at 0.8 m above the floor when standing and at 1.3 m above the floor when hanging. The force constant of the rim is 7.2 kN/m and the front of the rim is displaced a distance $s = 15$ cm.

FIGURE 6-25

$s = 15$ cm $= 0.15$ m

PICTURE THE PROBLEM In the player's change in position from standing on the floor to hanging on the rim, the total change in potential energy consists of gravitational potential energy, $U_g = mgy$, and energy stored in the deformed rim, whose potential energy can be measured just as if it were a spring: $U_s = \tfrac{1}{2}ks^2$. Choose $y = 0$ at 0.8 m above the floor for the gravitational potential-energy reference point.

The total potential energy is the sum of gravitational potential energy and the elastic potential energy of the rim (see Figure 6-26):

$$U = U_g + U_s = mgy + \tfrac{1}{2}ks^2$$
$$= (110 \text{ kg})(9.81 \text{ N/kg})(0.5 \text{ m}) + \tfrac{1}{2}(7.2 \text{ kN/m})(0.15 \text{ m})^2$$
$$= 540 \text{ J} + 81 \text{ J} = \boxed{621 \text{ J}}$$

FIGURE 6-26 The graph shows the total potential energy $U_g + U_s$ as a function of the downward extension of the rim.

REMARKS Most of the potential energy is gravitational in this case, because of the choice of the potential-energy reference point.

EXERCISE A 3-kg block is hung vertically from a spring with a force constant of 600 N/m. (a) By how much is the spring stretched? (b) How much potential energy is stored in the spring? (*Answer* (a) 4.9 cm (b) 0.72 J)

Nonconservative Forces

Not all forces are conservative. Suppose that you push a box across a table along a straight line from point A to point B and back, so that the box ends up at its original position. Friction opposes the block's motion, so the force you push with is in the direction of motion and does positive work on both legs of the round trip. The total work done by the push does not equal zero. Thus, the push is an example of a **nonconservative force** and no potential-energy function can be defined for it.

Sometimes we can show that a given force is not conservative by computing the work done by the force around some chosen closed curve and showing that it is not zero. Consider the force $\vec{F} = F_0\hat{\phi}$, where $\hat{\phi}$ is a unit vector directed tangent to a circle of radius r. The work done by this force as we move around a circle of radius r is $+F_0 2\pi r$ if we move in the direction of the force (and $-F_0 2\pi r$ if we move opposite to the force). Since this work is not zero, we conclude that the force is not conservative. This method is of limited use in investigating whether a given force is conservative or not. If the work done around *any* particular closed path is not zero, we may conclude that the force is not conservative. However, if the force is conservative, the work must be zero around *all* possible closed paths. Since there are infinitely many possible closed paths, it is impossible to calculate the work done for each one. In more advanced physics courses, more sophisticated mathematical methods for testing forces are discussed.

Potential Energy and Equilibrium

For a general conservative force in one dimension, $\vec{F} = F_x\hat{i}$, Equation 6-20b is

$$dU = -\vec{F} \cdot d\vec{s} = -F_x\, dx$$

The force is therefore the negative derivative of the potential-energy function:

$$F_x = -\frac{dU}{dx} \qquad\qquad 6\text{-}23$$

We can illustrate this general relation for a block–spring system by differentiating the function $U = \frac{1}{2}kx^2$. We obtain

$$F_x = -\frac{dU}{dx} = -\frac{d}{dx}\left(\frac{1}{2}kx^2\right) = -kx$$

Figure 6-27 shows a plot of $U = \frac{1}{2}kx^2$ versus x for a block and spring. The derivative of this function is represented graphically as the slope of the tangent line to the curve. The force is thus equal to the negative of the slope of the curve. At $x = 0$, the force $F_x = -dU/dx$ is zero and the block is in equilibrium.

A particle is in equilibrium if the net force acting on it is zero.

CONDITION FOR EQUILIBRIUM

When x is positive in Figure 6-27, the slope is positive and the force F_x is negative. When x is negative, the slope is negative and the force F_x is positive. In either case, the force is in the direction that will accelerate the block toward lower potential energy. If the block is displaced slightly from $x = 0$, the force is directed back toward $x = 0$. The equilibrium at $x = 0$ is thus **stable equilibrium** because a small displacement results in a restoring force that accelerates the particle back toward its equilibrium position.

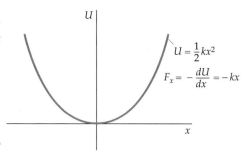

FIGURE 6-27 Plot of the potential-energy function U versus x for an object on a spring. A minimum in a potential-energy curve is a point of stable equilibrium. Displacement in either direction results in a force directed toward the equilibrium position.

In stable equilibrium, a small displacement in any direction results in a restoring force that accelerates the particle back toward its equilibrium position.

Figure 6-28 shows a potential-energy curve with a maximum rather than a minimum at the equilibrium point $x = 0$. Such a curve could represent the potential energy of a skier at the top of a hill. For this curve, when x is positive, the slope is negative and the force F_x is positive, and when x is negative, the slope is positive and the force F_x is negative. Again, the force is in the direction that will accelerate the particle toward lower potential energy, but this time the force is away from the equilibrium position. The maximum at $x = 0$ in Figure 6-28 is a point of **unstable equilibrium** because a small displacement results in a force that accelerates the particle away from its equilibrium position.

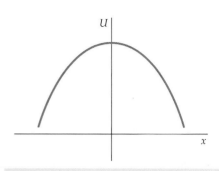

FIGURE 6-28 A particle at $x = 0$ on this potential-energy curve will be in unstable equilibrium because a displacement in either direction results in a force directed away from the equilibrium position.

In unstable equilibrium, a small displacement results in a force that accelerates the particle away from its equilibrium position.

Figure 6-29 shows a potential-energy curve that is flat in the region near $x = 0$. No force acts on a particle at $x = 0$, and hence the particle is at equilibrium; furthermore, there will be no resulting force if the particle is displaced slightly in either direction. This is an example of **neutral equilibrium.**

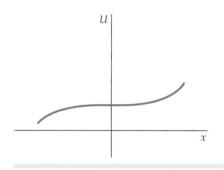

In neutral equilibrium, a small displacement results in zero force and the particle remains in equilibrium.

FIGURE 6-29 Neutral equilibrium. The force $F_x = -dU/dx$ is zero at $x = 0$ and at neighboring points, so displacement away from $x = 0$ results in no force, and the system remains in equilibrium.

FORCE AND THE POTENTIAL-ENERGY FUNCTION   **E X A M P L E   6 - 1 5**

In the region $-a < x < a$ the force on a particle is represented by the potential energy function

$$U = -b\left(\frac{1}{a + x} + \frac{1}{a - x}\right)$$

Try It Yourself

where a and b are positive constants. (*a*) Find the force F_x in the region $-a < x < a$. (*b*) At what value of x is the force zero? (*c*) At the location where the force equals zero, is the equilibrium stable or unstable?

PICTURE THE PROBLEM The force is the negative of the derivative of the potential-energy function. The equilibrium is stable where the potential-energy function is a minimum and it is unstable where the potential-energy function is a maximum.

Cover the column to the right and try these on your own before looking at the answers.

Steps

(*a*) Compute $F_x = -dU/dx$.

Answers

$$F_x = -\frac{d}{dx}\left[-b\left(\frac{1}{a + x} + \frac{1}{a - x}\right)\right]$$

$$= -b\left(\frac{1}{(x + a)^2} - \frac{1}{(x - a)^2}\right)$$

(b) Set F_x equal to zero and solve for x.

$F_x = 0$ at $\boxed{x = 0}$

(c) Compute d^2U/dx^2. If it is positive at the equilibrium position, then U is a minimum and the equilibrium is stable. If it is negative, then U is a maximum and the equilibrium is unstable.

At $x = 0$, $\dfrac{d^2U}{dx^2} = \dfrac{-4b}{a^3}$

Thus, $\boxed{\text{unstable}}$ equilibrium.

REMARKS This potential-energy function is that for a particle under the influence of the gravitational forces exerted by two identical fixed masses, one at $x = -a$ and the other at $x = +a$. The particle is located on the line joining the masses. Midway between the two masses the net force on the particle is zero. Otherwise, it is toward the closest mass.

SUMMARY

1. Work, kinetic energy, potential energy, and power are important derived dynamic quantities.
2. The work–kinetic energy theorem is an important relation derived from Newton's laws applied to a particle. (In this context, a particle is a perfectly rigid object that moves without rotating.)
3. The dot product of vectors is a mathematical definition that is useful throughout physics.

Topic	Relevant Equations and Remarks	
1. **Work (definition)**	$W = \displaystyle\int_1^2 \vec{F} \cdot d\vec{s}$	6-15
Constant force	$W = \vec{F} \cdot \vec{s}$	
In one dimension		
Constant force	$W = F_x \Delta x = F \cos\theta\, \Delta x$	6-1
Variable force	$W = \displaystyle\int_{x_1}^{x_2} F_x\, dx =$ area under the F_x-versus-x curve	6-9
2. **Kinetic Energy (definition)**	$K = \frac{1}{2}mv^2$	6-6
3. **Work–Kinetic Energy Theorem**	$W_{\text{total}} = \Delta K = \frac{1}{2}mv_f^2 - \frac{1}{2}mv_i^2$	6-7
4. **Dot Product (definition)**	$\vec{A} \cdot \vec{B} = AB \cos\phi$	6-10
In terms of components	$\vec{A} \cdot \vec{B} = A_xB_x + A_yB_y + A_zB_z$	6-11
Component of vector	$\vec{A} \cdot \hat{i} = A_x$	6-12
Derivative	$\dfrac{d}{dt}(\vec{A} \cdot \vec{B}) = \dfrac{d\vec{A}}{dt} \cdot \vec{B} + \vec{A} \cdot \dfrac{d\vec{B}}{dt}$	6-13
5. **Power**	$P = \dfrac{dW}{dt} = \vec{F} \cdot \vec{v}$	6-17

6. Conservative Force

A force is conservative if the total work it does on a particle is zero when the particle moves along any path that returns it to its initial position. Alternatively, the work done by a conservative force on a particle is independent of the path taken by the particle as it moves from one point to another.

7. Potential Energy

The potential energy of a system is the energy associated with the configuration of the system. The change in the potential energy of a system is defined as the negative of the work done by all internal conservative forces acting on the system.

Definition	$\Delta U = U_2 - U_1 = -\int_1^2 \vec{F} \cdot d\vec{s}$	6-20
	$dU = -\vec{F} \cdot d\vec{s}$	6-20
Gravitational	$U = U_0 + mgy$	6-21
Elastic (spring)	$U = \frac{1}{2}kx^2$	6-22
Conservative force	$F_x = -\dfrac{dU}{dx}$	6-23

| Potential-energy curve | At a minimum on the curve of the potential-energy function versus the displacement, the force is zero and the system is in stable equilibrium. At a maximum, the force is zero and the system is in unstable equilibrium. A conservative force always tends to accelerate a particle toward a position of lower potential energy. |

PROBLEMS

- Single-concept, single-step, relatively easy
- •• Intermediate-level, may require synthesis of concepts
- ••• Challenging
- SSM Solution is in the *Student Solutions Manual*
- iSOLVE Problems available on iSOLVE online homework service
- iSOLVE ✓ These "Checkpoint" online homework service problems ask students additional questions about their confidence level, and how they arrived at their answer

In a few problems, you are given more data than you actually need; in a few other problems, you are required to supply data from your general knowledge, outside sources, or informed estimates.

Take g = 9.81 N/kg = 9.81 m/s² and neglect friction in all problems unless otherwise stated.

Conceptual Problems

1 • SSM True or false:

(a) Only the net force acting on an object can do work.
(b) No work is done on a particle that remains at rest.
(c) A force that is always perpendicular to the velocity of a particle never does work on the particle.

2 • You are to move a heavy box from the top of one table to the top of another table of the same height on the other side of the room. What is the minimum amount of work you must do on the box to accomplish the move? Explain.

3 • True or False: A person on a Ferris wheel is moving in a circle at constant speed. Thus, no force is doing work on the person.

4 • SSM By what factor does the kinetic energy of a car change when its speed is doubled?

5 • A particle moves in a circle at constant speed. Only one of the forces acting on the particle is in the centripetal direction. Does the net force on the particle do work on it? Explain.

6 • An object initially has kinetic energy K. The object then moves in the opposite direction with three times its initial speed. What is the kinetic energy now? (a) K. (b) 3K. (c) −3K. (d) 9K. (e) −9K.

7 • SSM How does the work required to stretch a spring 2 cm from its natural length compare with that required to stretch it 1 cm from its natural length?

8 • Suppose there is a net force acting on a particle but it does no work. Can the particle be moving in a straight line?

9 • The dimension of power is (a) $[M][L]^2[T]^2$, (b) $[M][L]^2/[T]$, (c) $[M][L]^2/[T]^2$, (d) $[M][L]^2/[T]^3$.

10 • Two knowledge seekers decide to ascend a mountain. Sal chooses a short, steep trail, while Joe, who weighs the same as Sal, goes up via a long, gently sloped trail. At the top, they get into an argument about who gained more potential energy. Which of the following is true:

(a) Sal gains more gravitational potential energy than Joe.
(b) Sal gains less gravitational potential energy than Joe.
(c) Sal gains the same gravitational potential energy as Joe.
(d) To compare energies, we must know the height of the mountain.
(e) To compare the energies, we must know the length of the two trails.

11 • True or false:

(a) Only conservative forces can do work.
(b) If only conservative forces act, the kinetic energy of a particle does not change.
(c) The work done by a conservative force equals the decrease in the potential energy associated with that force.

12 •• SSM Figure 6-30 shows the plot of a potential-energy function U versus x. (a) At each point indicated, state whether the force F_x is positive, negative, or zero. (b) At which point does the force have the greatest magnitude? (c) Identify any equilibrium points, and state whether the equilibrium is stable, unstable, or neutral.

FIGURE 6-30 Problem 12

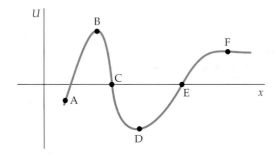

13 • True or false:

(a) The gravitational force cannot do work because it acts at a distance.
(b) Work is the area under the force-versus-time curve.

14 • Negative work means (a) the kinetic energy of the object increases, (b) the applied force is variable, (c) the angle between applied force is perpendicular to the displacement, (d) the applied force and the displacement is greater than 90°, (e) nothing; there is no such thing as negative work.

Estimation and Approximation

15 •• SSM A tightrope walker whose mass is 50 kg walks across a tightrope held between two supports 10 m apart; the tension in the rope is 5000 N. The height of the rope is 10 m above the ground. Estimate: (a) the sag in the tightrope when she stands in the exact center and (b) the change in her gravitational potential energy from just before stepping onto the tightrope to when she stands at its dead center.

16 • Estimate (a) the change in your potential energy on taking an elevator from the ground floor to the top of the Empire State building, (b) the average force acting on you by the elevator to bring you to the top, and (c) the average power due to that force. The building is 102 stories high.

17 • The nearest stars, apart from the sun, are at light-years away from the earth. (One light-year is the distance that light travels during a year: 9.47×10^{15} m.) If we ever want to send spacecraft to investigate the stars, they will have to have velocities that are an appreciable fraction of the speed of light. Calculate the kinetic energy of a 10,000-kg spacecraft traveling at a speed of 10 percent of the speed of light, and compare that to the amount of energy that the United States uses in a year (about 5×10^{20} J). NOTE At velocities approaching the speed of light, the theory of relativity tells us that the formula $\frac{1}{2}mv^2$ for kinetic energy is not correct. However, the correction is only about 1 percent for this particular speed ($0.1c$).

18 •• SSM The mass of the Space Shuttle orbiter is about 8×10^4 kg and the period of its orbit is 90 min. Estimate (a) the kinetic energy of the orbiter and (b) the change in its potential energy between resting on the surface of the earth and in its orbit, 200 mi above the surface of the earth. (c) Why is the change in potential energy much smaller than the shuttle's kinetic energy? Shouldn't they be equal?

19 • Ten inches of snow have fallen during the night, and you must shovel out your 50-ft-long driveway (Figure 6-31). Estimate how much work you do on the snow by completing this task. Make a plausible guess of any value(s) needed (the width of the driveway, for example), and state the basis for each guess.

FIGURE 6-31 Problem 19

50 ft

Work and Kinetic Energy

20 • SSM A 15-g bullet has a speed of 1.2 km/s. (a) What is its kinetic energy in joules? (b) What is its kinetic energy if its speed is halved? (c) What is its kinetic energy if its speed is doubled?

21 • Find the kinetic energy in joules of (a) a 0.145-kg baseball moving with a speed of 45 m/s and (b) a 60-kg jogger running at a steady pace of 9 min/mi.

22 • A 6-kg box is raised a distance of 3 m from rest by a vertical applied force of 80 N. Find (*a*) the work done by the force, (*b*) the work done by gravity, and (*c*) the final kinetic energy of the box.

23 • A constant force of 80 N acts on a box of mass 5.0 kg that is moving in the direction of the force with a speed of 20 m/s. A few seconds later the box is moving with a speed of 68 m/s. Determine the work done by this force.

24 •• **SSM** You run a race with a friend. At first you each have the same kinetic energy, but you find that she is beating you. When you increase your speed by 25 percent, you are running at the same speed she is. If your mass is 85 kg, what is her mass?

Work Done by a Variable Force

25 •• A 3-kg particle is moving along the *x* axis with a velocity of 2 m/s as it passes through $x = 0$. It is subjected to a single force F_x that varies with position as shown in Figure 6-32. (*a*) What is the kinetic energy of the particle as it passes through $x = 0$? (*b*) How much work is done by the force as the particle moves from $x = 0$ to $x = 4$ m? (*c*) What is the speed of the particle when it is at $x = 4$ m?

FIGURE 6-32 Problem 25

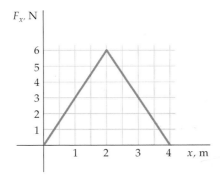

26 •• **SSM** ✔ A force F_x acts on a particle. The force is related to the position of the particle by the formula $F_x = Cx^3$, where *C* is a constant. Find the work done by this force on the particle as the particle moves from $x = 1.5$ m to $x = 3$ m.

27 •• Lou's latest invention, aimed at urban dog owners, is the X-R-Leash. It is made of a rubber-like material that exerts a force $F_x = -kx - ax^2$ when it is stretched a distance *x*, where *k* and *a* are constants. The ad claims, "You'll never go back to your old dog leash after you've had the thrill of an X-R-Leash experience. And you'll see a new look of respect in the eyes of your proud pooch." Find the work done on a dog by the leash if the person remains stationary, and the dog bounds off, stretching the X-R-Leash from $x = 0$ to $x = x_1$.

28 •• A 3-kg object moving along the *x* axis has a velocity of 2.40 m/s as it passes through the origin. It is acted on by a single force F_x that varies with *x* as shown in Figure 6-33. (*a*) Find the work done by the force from $x = 0$ to $x = 2$ m. (*b*) What is the kinetic energy of the object at $x = 2$ m? (*c*) What is the speed of the object at $x = 2$ m? (*d*) Find the work done on the object from $x = 0$ to $x = 4$ m. (*e*) What is the speed of the object at $x = 4$ m?

FIGURE 6-33 Problem 28

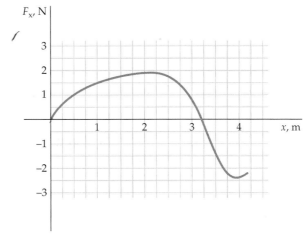

29 •• **SSM** Near Margaret's cabin is a 20-m water tower that attracts many birds during the summer months. During a hot spell, the tower went dry, and Margaret had to have her water hauled in. She was lonesome without the birds visiting, so she decided to carry some water up the tower to attract them back. Her bucket has a mass of 10 kg and holds 30 kg of water when it is full. However, the bucket has a hole, and as Margaret climbed at a constant speed, water leaked out at a constant rate. When she reached the top, only 10 kg of water remained for the birdbath. (*a*) Write an expression for the mass of the bucket plus water as a function of the height (*y*) climbed. (*b*) Find the work done by Margaret on the bucket.

Work, Energy, and Simple Machines

30 • ✔ A 6-kg block slides down a frictionless incline making an angle of 60° with the horizontal. (*a*) List all the forces acting on the block, and find the work done by each force when the block slides 2 m (measured along the incline). (*b*) What is the total work done on the block? (*c*) What is the speed of the block after it has slid 1.5 m if it starts from rest? (*d*) What is its speed after 1.5 m if it starts with an initial speed of 2 m/s?

31 • A 2-kg object attached to a horizontal string moves with a speed of 2.5 m/s in a circle of radius 3 m on a frictionless horizontal surface. (*a*) Find the tension in the string. (*b*) List the forces acting on the object, and find the work done by each force during one revolution.

32 • **SSM** *Simple machines* are used for reducing the amount of force that must be supplied to perform a task such as lifting a heavy weight. Such machines include the screw, block-and-tackle systems, and levers, but the simplest of the simple machines is the inclined plane. In Figure 6-34 you are raising the heavy box to the height of the truck bed by pushing it up an inclined plane (a ramp). (*a*) We define the *mechanical advantage M* of the inclined plane as the ratio of the force it would take to lift the block into the truck directly from the ground (at constant speed) to the force it takes to push it up the ramp (at constant speed). If the plane is frictionless, show that $M = 1/\sin \theta = L/H$, where *H* is the height of the truck bed and *L* is the length of the ramp. (*b*) Show that the work you do by moving the block into the truck is the same whether you lift it directly into the truck or push it up the frictionless ramp.

FIGURE 6-34 Problem 32

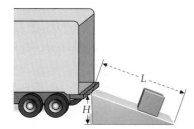

33 •• The screw is a form of the inclined plane. Figure 6-35 shows a type of car jack used for changing flat tires. The screw on the jack has a pitch (distance between threads) of p and a handle of length R. As the handle is turned through a full circle, the jack's height increases, raising the weight w by p. Assuming no friction, the work done turning the jack handle is the same as the increase in the potential energy of whatever the jack is lifting. Show that the jack has a mechanical advantage (defined in Problem 32) of $2\pi R/p$.

FIGURE 6-35 Problem 33

34 • Figure 6-36 shows two pulleys arranged to help lift a heavy load: a rope runs around two massless, frictionless pulleys and the weight $\vec{w}$ hangs from one pulley. You exert a force of magnitude F on the free end of the cord. (a) If the weight is to move up a distance h, through what distance must the force move? (b) How much work is done by the rope on the weight? (c) How much work do you do? (d) What is the mechanical advantage (defined in Problem 32) of this system?

FIGURE 6-36 Problem 34

Dot Products

35 • **SSM** What is the angle between the vectors $\vec{A}$ and $\vec{B}$ if $\vec{A} \cdot \vec{B} = -AB$?

36 • Two vectors $\vec{A}$ and $\vec{B}$ each have magnitudes of 6 m and make an angle of 60° with each other. Find $\vec{A} \cdot \vec{B}$.

37 • Find $\vec{A} \cdot \vec{B}$ for the following vectors: (a) $\vec{A} = 3\hat{i} - 6\hat{j}$, $\vec{B} = -4\hat{i} + 2\hat{j}$; (b) $\vec{A} = 5\hat{i} + 5\hat{j}$, $\vec{B} = 2\hat{i} - 4\hat{j}$; and (c) $\vec{A} = 6\hat{i} + 4\hat{j}$, $\vec{B} = 4\hat{i} - 6\hat{j}$.

38 • Find the angles between the vectors $\vec{A}$ and $\vec{B}$ in Problem 37.

39 • **iSOLVE ✓** A 2-kg object is given a displacement $\Delta\vec{s} = 3\,\text{m}\hat{i} + 3\,\text{m}\hat{j} - 2\,\text{m}\hat{k}$ along a straight line. During the displacement, a constant force $\vec{F} = 2\,\text{N}\hat{i} - 1\,\text{N}\hat{j} + 1\,\text{N}\hat{k}$ acts on the object. (a) Find the work done by $\vec{F}$ for this displacement. (b) Find the component of $\vec{F}$ in the direction of the displacement.

40 •• (a) Find the unit vector that is parallel to vector $\vec{A} = A_x\hat{i} + A_y\hat{j} + A_z\hat{k}$. (b) Find the component of the vector $\vec{A} = 2\hat{i} - \hat{j} - \hat{k}$ in the direction of the vector $\vec{B} = 3\hat{i} + 4\hat{j}$.

41 •• **SSM** Given two vectors $\vec{A}$ and $\vec{B}$, show that if $|\vec{A} + \vec{B}| = |\vec{A} - \vec{B}|$, then $\vec{A} \perp \vec{B}$.

42 •• $\hat{A}$ and $\hat{B}$ are two unit vectors in the xy plane. They make angles of θ_1 and θ_2 with the positive x-axis, respectively. (a) Find the x and y components of the two vectors. (b) By considering the dot product of $\hat{A}$ and $\hat{B}$, show that $\cos(\theta_1 - \theta_2) = \cos\theta_1 \cos\theta_2 + \sin\theta_1 \sin\theta_2$.

43 • If $\vec{A} \cdot \vec{B} = \vec{A} \cdot \vec{C}$, must $\vec{B} = \vec{C}$? If no, give a counterexample, if yes, explain why.

44 •• (a) Let $\vec{A}$ be a constant vector in the xy plane with its tail at the origin. Let $\vec{r} = x\hat{i} + y\hat{j}$ be a vector in the xy plane that satisfies the relation $\vec{A} \cdot \vec{r} = 1$. Show that the points (x,y) lie on a straight line. (b) If $\vec{A} = 2\hat{i} - 3\hat{j}$, find the slope and y intercept of the line. (c) If we now let $\vec{A}$ and $\vec{r}$ be vectors in three-dimensional space, show that the relation $\vec{A} \cdot \vec{r} = 1$ specifies a plane.

45 •• **SSM** When a particle moves in a circle centered at the origin with constant speed, the magnitudes of its position vector and velocity vectors are constant. (a) Differentiate $\vec{r} \cdot \vec{r} = r^2 =$ constant with respect to time to show that $\vec{v} \cdot \vec{r} = 0$ and therefore $\vec{v} \perp \vec{r}$. (b) Differentiate $\vec{v} \cdot \vec{v} = v^2 =$ constant with respect to time to show that $\vec{a} \cdot \vec{v} = 0$ and therefore $\vec{a} \perp \vec{v}$. What do the results of (a) and (b) imply about the direction of $\vec{a}$? (c) Differentiate $\vec{v} \cdot \vec{r} = 0$ with respect to time and show that $\vec{a} \cdot \vec{r} + v^2 = 0$ and therefore $a_r = -v^2/r$.

Power

46 •• Force A does 5 J of work in 10 s. Force B does 3 J of work in 5 s. Which force delivers greater power?

47 • A 5-kg box is being lifted upward at a constant velocity of 2 m/s by a force equal to the weight of the box. (a) What is the power input of the force? (b) How much work is done by the force in 4 s?

48 • Fluffy has just caught a mouse and decides to bring it to the bedroom so that his human roommate can admire it when she wakes up. A constant horizontal force of 3 N is enough to drag the mouse across the rug at a constant speed v. If Fluffy's force does work at the rate of 6 W, (a) what is his speed v? (b) How much work does Fluffy do in 4 s?

49 • A single force of 5 N acts in the x direction on an 8-kg object. (a) If the object starts from rest at $x = 0$ at time $t = 0$, find its velocity v as a function of time. (b) Write an expression for the power input as a function of time. (c) What is the power input of the force at time $t = 3$ s?

50 • [SOLVE] Find the power input of a force $\vec{F}$ acting on a particle that moves with a velocity $\vec{v}$ for (a) $\vec{F} = 4\,\mathrm{N}\hat{\imath} + 3\,\mathrm{N}\hat{k}$, $\vec{v} = 6\,\mathrm{m/s}\hat{\imath}$; (b) $\vec{F} = 6\,\mathrm{N}\hat{\imath} - 5\,\mathrm{N}\hat{\jmath}$, $\vec{v} = -5\,\mathrm{m/s}\hat{\imath} + 4\,\mathrm{m/s}\hat{\jmath}$; and (c) $\vec{F} = 3\,\mathrm{N}\hat{\imath} + 6\,\mathrm{N}\hat{\jmath}$, $\vec{v} = 2\,\mathrm{m/s}\hat{\imath} + 3\,\mathrm{m/s}\hat{\jmath}$.

51 • [SSM] A small food service elevator (dumbwaiter) in a cafeteria is connected over a pulley system to a motor as shown in Figure 6-37; the motor raises and lowers the dumbwaiter. The mass of the dumbwaiter is 35 kg. In operation, it moves at a speed of 0.35 m/s upward, without accelerating (except for a brief initial period just after the motor is turned on). If the output power from the motor is 27 percent of its input power, what is the input power to the motor? Assume that the pulleys are frictionless.

FIGURE 6-37 Problem 51

Motor

52 •• A sky diver falls though the air toward the ground at a constant speed of 120 mph, her terminal velocity, before opening her parachute. (a) If her mass is 55 kg, calculate the magnitude of the power due to the drag force. (b) After she opens her parachute, her speed slows to 15 mph. What is the magnitude of the power due to the drag force now?

53 •• [SSM] A cannon placed at the top of a cliff of height H fires a cannonball into the air with an initial speed v_0, shooting directly upward. The cannonball rises, falls back down, missing the cannon by a little bit, and lands at the foot of the cliff. Neglecting air resistance, calculate the velocity $\vec{v}(t)$ for all times while the cannonball is in the air, and show explicitly that the integral of $\vec{F} \cdot \vec{v}$ over the time that the cannonball spends in the air is equal to the change in the kinetic energy of the cannonball.

54 •• A particle of mass m moves from rest at $t = 0$ under the influence of a single constant force $\vec{F}$. Show that the power delivered by the force at time t is $P = F^2 t/m$.

Potential Energy

55 • [SOLVE]✓ An 80-kg man climbs up a 6-m high flight of stairs. What is the increase in the gravitational potential energy of the man–earth system?

56 • [SOLVE] Water flows over Victoria Falls, which is 128 m high, at an average rate of 1.4×10^6 kg/s. If half the potential energy of this water were converted into electric energy, how much power would be produced by these falls?

57 • [SOLVE] A 2-kg box slides down a long, frictionless incline of angle 30°. It starts from rest at time $t = 0$ at the top of the incline at a height of 20 m above the ground. (a) What is the original potential energy of the box relative to the ground? (b) From Newton's laws, find the distance the box travels during the interval $0 < t < 1$ s and its speed at $t = 1$ s. (c) Find the potential energy and the kinetic energy of the box at $t = 1$ s. (d) Find the kinetic energy and the speed of the box just as it reaches the bottom of the incline.

58 • A force $F_x = 6$ N is constant. (a) Find the potential-energy function $U(x)$ associated with this force for an arbitrary reference position x_0 at which $U = 0$. (b) Find $U(x)$ such that $U = 0$ at $x = 4$ m. (c) Find $U(x)$ such that $U = 14$ J at $x = 6$ m.

59 • [SOLVE]✓ A spring has a force constant $k = 10^4$ N/m. How far must it be stretched for its potential energy to be (a) 50 J and (b) 100 J?

60 •• [SSM] [SOLVE] A simple Atwood's machine (Figure 6-38) uses two masses m_1 and m_2. Starting from rest, the speed of the two masses is 4.0 m/s at the end of 3.0 s. At that time, the kinetic energy of the system is 80 J and each mass has moved a distance of 6.0 m. Determine the values of m_1 and m_2.

FIGURE 6-38 Problem 60

61 •• A straight rod of negligible mass is mounted on a frictionless pivot as shown in Figure 6-39. Masses m_1 and m_2 are attached to the rod at distances ℓ_1 and ℓ_2. (a) Write an expression for the gravitational potential energy of the masses as a function of the angle θ made by the rod and the horizontal. (b) For what angle θ is the potential energy a minimum? Is the statement "systems tend to move toward a configuration of minimum potential energy" consistent with your result? (c) Show that if $m_1\ell_1 = m_2\ell_2$, the potential energy is the same for all values of θ. (When this holds, the system will balance at any angle θ. This result is known as *Archimedes' law of the lever*.)

FIGURE 6-39 Problem 61

Force, Potential Energy, and Equilibrium

62 • (*a*) Find the force F_x associated with the potential-energy function $U = Ax^4$, where A is a constant. (*b*) At what point(s) is the force zero?

63 •• A potential-energy function is given by $U = C/x$, where C is a positive constant. (*a*) Find the force F_x as a function of x. (*b*) Is this force directed toward the origin or away from it? (*c*) Does the potential energy increase or decrease as x increases? (*d*) Answer parts (*b*) and (*c*) where C is a negative constant.

64 •• [SSM] On the potential-energy curve for U versus y shown in Figure 6-40, the segments AB and CD are straight lines. Sketch a plot of the force F_y versus y.

FIGURE 6-40 Problem 64

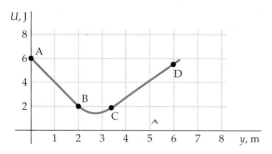

65 •• The force acting on an object is given by $F_x = a/x^2$. Determine the potential energy of the object as a function of x.

66 •• The potential energy of an object is given by $U(x) = 3x^2 - 2x^3$, where U is in joules and x is in meters. (*a*) Determine the force acting on this object. (*b*) At what positions is this object in equilibrium? (*c*) Which of these equilibrium positions are stable and which are unstable?

67 •• The potential energy of an object is given by $U(x) = 8x^2 - x^4$, where U is in joules and x is in meters. (*a*) Determine the force acting on this object. (*b*) At what positions is this object in equilibrium? (*c*) Which of these equilibrium positions are stable and which are unstable?

68 •• The force acting on an object is given by $F(x) = x^3 - 4x$. Locate the positions of unstable and stable equilibrium and show that at these points $U(x)$ is a local maximum or minimum, respectively.

69 •• The potential energy of a 4-kg object is given by $U = 3x^2 - x^3$ for $x \le 3$ m and $U = 0$ for $x \ge 3$ m, where U is in joules and x is in meters. (*a*) At what positions is this object in equilibrium? (*b*) Sketch a plot of U versus x. (*c*) Discuss the stability of the equilibrium for the values of x found in (*a*). (*d*) If the total energy of the particle is 12 J, what is its speed at $x = 2$ m?

70 •• A force is given by $F_x = Ax^{-3}$, where $A = 8$ N·m³. (*a*) For positive values of x, does the potential energy associated with this force increase or decrease with increasing x? (You can determine the answer to this question by imagining what happens to a particle that is placed at rest at some point x and is then released.) (*b*) Find the potential-energy function U associated with this force such that U approaches zero as x approaches infinity. (*c*) Sketch U versus x.

71 ••• [SSM] A novelty desk clock is shown in Figure 6-41: The clock (which has mass m) is supported by two light cables running over the two pulleys, which are attached to counterweights that each have mass M. (*a*) Find the potential energy of the system as a function of the distance y. (*b*) Find the value of y for which the potential energy of the system is a minimum. (*c*) If the potential energy is a minimum, then the system is in equilibrium. Apply Newton's second law to the clock and show that it is in equilibrium (the forces on it sum to zero) for the value of y obtained for part (*b*). Is this a point of stable or unstable equilibrium?

FIGURE 6-41 Problem 71

General Problems

72 • [SSM] In February 2002, a total of 60.7 billion kW·h of electrical energy was generated by nuclear power plants in the United States. At that time, the population of the United States was about 287 million people. If the average American has a mass of 60 kg, and if the entire energy output of all nuclear power plants was diverted to supplying energy for a single giant elevator, estimate the height h to which the entire population of the country could be lifted by the elevator. In your calculations, assume that 25 percent of the energy goes into lifting the people; assume also that g is constant over the entire height h.

73 • ✓ One of the most powerful cranes in the world, operating in Switzerland, can slowly raise a load of $M = 6000$ tonne to a height of $h = 12.0$ m (1 tonne = 1000 kg). (a) How much work is done by the crane? (b) If it takes 1.00 min to lift the load at constant velocity to this height, find the power developed by the crane.

74 • In Austria, there once was a ski lift of length 5.6 km. It took about 60 min for a gondola to travel all the way up. If there were 12 gondolas going up, each with a cargo of mass 550 kg, and if there were 12 empty gondolas going down, and the angle of ascent was $30°$, estimate the power P of the engine needed to operate the ski lift.

75 • ✓ A 2.4-kg object attached to a horizontal string moves with constant speed in a circle of radius R on a frictionless horizontal surface. The kinetic energy of the object is 90 J and the tension in the string is 360 N. Find R.

76 • SSM The movie crew arrives in the Badlands ready to shoot a scene. The script calls for a car to crash into a vertical rock face at 100 km/h. Unfortunately, the car won't start, and there is no mechanic is sight. The crew are about to skulk back to the studio to face the producer's wrath when the cameraman gets an idea. They use a crane to lift the car by its rear end and then drop it vertically, filming at an angle that makes the car appear to be traveling horizontally. How high should the 800-kg car be lifted so that it reaches a speed of 100 km/h in the fall?

77 ••• The four strings pass over the bridge of a violin as shown in Figure 6-42. The angle that the strings make with the normal to the plane of the instrument is $72°$ on either side. The total normal force pressing the bridge into the violin is 103 N. The length of the strings from the bridge to the peg to which each is attached is 32.6 cm. (a) Determine the tension in the violin strings, assuming that the tension is the same for each string. (b) One of the strings is plucked out a distance of 4 mm, as shown in the figure. Make a free-body diagram showing all of the forces acting on the string at that point, and determine the force pulling the string back to its equilibrium position. Assume that the tension in the string remains constant. (c) Determine the work done on the string in plucking it out that distance. Remember that the net force pulling the string back to its equilibrium position is changing as the string is being pulled out, but assume that the magnitude of the tension in the string remains constant.

78 •• The force acting on a particle that is moving along the x axis is given by $F_x = -ax^2$, where a is a constant. Calculate the potential-energy function U relative to $U = 0$ at $x = 0$, and sketch a graph of U versus x.

79 •• SSM A horizontal force acts on a cart of mass m such that the speed v of the cart increases with distance x as $v = Cx$, where C is a constant. (a) Find the force acting on the cart as a function of position. (b) What is the work done by the force in moving the cart from $x = 0$ to $x = x_1$?

80 •• ✓ A force $\vec{F} = (2 \text{ N/m}^2)x^2\hat{i}$ is applied to a particle. Find the work done on the particle as it moves a total distance of 5 m (a) parallel to the y axis from point (2 m, 2 m) to point (2 m, 7 m) and (b) in a straight line from (2 m, 2 m) to (5 m, 6 m).

81 •• A particle of mass m moves along the x axis. Its position varies with time according to $x = 2t^3 - 4t^2$, where x is in meters and t is in seconds. Find (a) the velocity and acceleration of the particle at any time t, (b) the power delivered to the particle at any time t, and (c) the work done by the force from $t = 0$ to $t = t_1$.

82 •• A 3-kg particle starts from rest at $x = 0$ and moves under the influence of a single force $F_x = 6 + 4x - 3x^2$, where F_x is in newtons and x is in meters. (a) Find the work done by the force as the particle moves from $x = 0$ to $x = 3$ m. (b) Find the power delivered to the particle when it is at $x = 3$ m.

83 •• SSM ✓ The initial kinetic energy imparted to a 0.020-kg bullet is 1200 J. Neglecting air resistance, find the range of this projectile when it is fired at an angle such that the range equals the maximum height attained.

84 •• ✓ A force F_x acting on a particle is shown as a function of x in Figure 6-43. (a) From the graph, calculate the work done by the force when the particle moves from $x = 0$ to the following values of x: $-4, -3, -2, -1, 0, 1, 2, 3$, and 4 m. (b) Plot the potential energy U versus x for the range of values of x from -4 m to $+4$ m, assuming that $U = 0$ at $x = 0$.

FIGURE 6-43 Problem 84

FIGURE 6-42 Problem 77

85 •• Repeat Problem 84 for the force F_x shown in Figure 6-44.

FIGURE 6-44 Problem 85

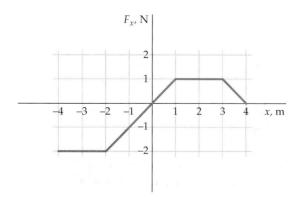

86 •• [SOLVE] A box of mass M is at rest at the bottom of a frictionless inclined plane (Figure 6-45). The box is attached to a string that pulls with a constant tension T. (a) Find the work done by the tension T as the box moves through a distance x along the plane. (b) Find the speed of the box as a function of x and θ. (c) Determine the power produced by the tension in the string as a function of x and θ.

FIGURE 6-45 Problem 86

87 ••• A force in the xy plane is given by $\vec{F} = (F_0/r)(y\hat{i} - x\hat{j})$, where F_0 is a constant and $r = \sqrt{x^2 + y^2}$. (a) Show that the magnitude of this force is F_0 and that its direction is perpendicular to $\vec{r} = x\hat{i} + y\hat{j}$. (b) Find the work done by this force on a particle that moves once around a circle of radius 5 m centered at the origin. Is this force conservative?

88 ••• [SSM] A force in the xy plane is given by:

$F = -(b/r^3)(x\hat{i} + y\hat{j})$, where b is a positive constant and $r = \sqrt{x^2 + y^2}$. (a) Show that the magnitude of the force varies as the inverse of the square of the distance to the origin, and that its direction is antiparallel (opposite) to the radius vector $\vec{r} = x\hat{i} + y\hat{j}$. (b) If $b = 3\ \text{N·m}^2$, find the work done by this force on a particle moving along a straight-line path between an initial position $x = 2$ m, $y = 0$ m and a final position $x = 5$ m, $y = 0$ m. (c) Find the work done by this force on a particle moving once around a circle of radius $r = 7$ m centered around the origin. (d) If this force is the only force acting on the particle, what is the particle's speed as it moves along this circular path? Assume that the particle's mass is $m = 2$ kg.

89 ••• In the words of Richard Feynman, one of the greatest physicists of the twentieth century, "If, in some cataclysm, all of scientific knowledge were to be destroyed, and only one sentence passed on to the next generation, . . . what statement would contain the most information in the fewest words? I believe it is . . . that all things are made of atoms—little particles . . . attracting each other when they are a little distance apart, but repelling upon being squeezed into each other."[†] When looking at the forces between atoms, modern physicists and chemists often model their interactions by the so-called "6–12" potential, where the potential energy function between two atoms is given by the functional form

$$U(r) = \frac{a}{r^{12}} - \frac{b}{r^6}$$

where r is the distance between atomic nuclei and a and b are constants that can be determined spectroscopically. Because they do not form atomic bonds, noble gas atoms have potential-energy functions that can be modeled well by a 6–12 potential and measured reasonably accurately. For argon, these parameters are $a = 1.09 \times 10^{-7}$ and $b = 6.84 \times 10^{-5}$, where r is measured in nm (1 nm $= 10^{-9}$ m) and U is measured in eV (1 eV $= 1.6 \times 10^{-19}$ J.) (a) Using a spreadsheet program, make a graph of the interatomic potential energy function for two argon atoms as a function of separation, for values of r between 0.3 nm and 0.7 nm. Does the shape of the potential energy function support Feynman's claim? Explain. (b) What is the minimum value of the potential energy (compared to where the atoms are separated by a very large distance)? At what separation does this occur? Is this minimum a point of stable or unstable equilibrium? (c) From either the graph or the formula given above, estimate the force of attraction between two argon atoms separated by a distance of 5 Å and the force of repulsion for a separation of 3.5 Å. Make sure that you convert to MKS units!

90 ••• [SSM] A theoretical formula for the potential energy associated with the nuclear force between two protons, two neutrons, or a neutron and a proton is the Yukawa potential $U(r) = -U_0(a/r)e^{-r/a}$, where U_0 and a are constants, and r is the separation between the two nucleons. (a) Using a spreadsheet program such as Microsoft Excel™, make a graph of U versus r, using $U_0 = 4$ pJ (a picojoule, pJ, is 1×10^{-12} J) and $a = 2.5$ fm (a femtometer, fm, is 1×10^{-15} m). (b) Find the force $F(r)$ as a function of the separation of the two nucleons. (b) Find the force $F(r)$ as a function of the separation of the two nucleons. (c) Compare the magnitude of the force at the separation $r = 2a$ to that at $r = a$. (d) Compare the magnitude of the force at the separation $r = 5a$ to that at $r = a$.

---

† Richard P. Feynman, Matthew L. Sands, and Robert B. Leighton, *The Feynman Lectures on Physics*, Vol. 1, p. 1.1. Boston: Addison-Wesley (1970).

Conservation of Energy

ENERGY CAN BE CONVERTED FROM ONE FORM INTO ANOTHER. THIS ROLLER COASTER CONVERTS ELECTRICAL ENERGY PURCHASED FROM THE POWER COMPANY INTO GRAVITATIONAL POTENTIAL ENERGY OF THE PASSENGERS AND THEIR VEHICLES. THEN GRAVITATIONAL POTENTIAL ENERGY IS CONVERTED INTO KINETIC ENERGY, SOME OF WHICH IS CONVERTED BACK TO POTENTIAL ENERGY. EVENTUALLY FRICTION TRANSFORMS THE KINETIC AND POTENTIAL ENERGY INTO THERMAL ENERGY. CHEMICAL ENERGY STORED IN THE MUSCLES OF THE PASSENGERS IS CONVERTED INTO SOUND ENERGY EACH TIME SOMEONE SCREAMS.

? **How does conservation of mechanical energy allow us to determine how high up the cars must be when they start their descent, in order to complete the loop-the-loop? (See Example 7-5.)**

A system is a collection of particles. External forces are any forces exerted by particles not in the system on particles that are in the system, and internal forces are those exerted by particles in the system on other particles in the system. The many ways there are to change the total energy of the system can be divided into just two categories, work and heat. Total energy can change if external forces do work on the system or if energy is transferred because of a temperature difference between the system and its surroundings. (Energy transferred due to a temperature difference is called heat.) Because we do not analyze systems where heat plays a significant role until Chapter 18, here we will consider the change in energy of the system to be equal to the total work done on it by external forces. We define the potential energy of a system so that the change in the potential energy of the system equals the negative of the total work done by all internal conservative forces. If no external forces do work on the system, and if the internal conservative forces are the only internal forces that do work, then the work they do equals the change in the system's kinetic energy. Because the change in the system's potential energy equals the negative of the change in its kinetic energy, the sum of the potential and kinetic energies does not change. This is known as the conservation of mechanical energy. It follows from Newton's laws and is a useful alternative to them for solving many problems in mechanics.

The usefulness of mechanical energy conservation is limited by the presence of nonconservative forces such as friction. When kinetic friction is present in a system, the mechanical energy of the system does not stay the same but instead decreases.[†] Because mechanical energy is often not conserved, the importance of energy conservation was not realized until the nineteenth century, when it was discovered that the disappearance of macroscopic mechanical energy is always accompanied by the appearance of some other kind of energy, often thermal energy, which is usually indicated by an increase in temperature or a change in phase (like the melting of ice). We now know that, on the microscopic scale, this thermal energy is associated with the kinetic and potential energies at the molecular level.

There are a number of different forms of energy, such as chemical energy, sound energy, electromagnetic energy, and nuclear energy. Whenever the energy of a system changes, we can account for the change by the appearance or disappearance of energy somewhere else. This experimental observation is the **law of conservation of energy,** one of the most fundamental and important laws in all of science. Although energy changes from one form to another, it is never created or destroyed.

➤ **In this chapter, we continue the study of energy begun in Chapter 6 by describing and applying the law of conservation of energy and examining different types of energy, including thermal energy. We also discuss Einstein's famous relation between mass and energy, and discover that energy changes for a system are not continuous, but occur in discrete "bundles" or "lumps" called quanta. Although in a macroscopic system the quantum of energy typically is so small that it goes unnoticed, its presence has profound consequences for microscopic systems such as atoms and molecules.**

7-1 The Conservation of Mechanical Energy

The total work done on each particle in a system equals the change in the kinetic energy ΔK_i of that particle, so the total work done by all the forces W_{total} equals the change in the total kinetic energy of the system ΔK_{sys}:

$$W_{total} = \sum \Delta K_i = \Delta K_{sys} \qquad 7\text{-}1$$

Each internal force is either conservative or nonconservative. The negative of the total work done by all the conservative internal forces $-W_c$ equals the change in the potential energy of the system ΔU_{sys}:

$$-W_c = \Delta U_{sys} \qquad 7\text{-}2$$

The total work done by all forces equals the work done by all external forces W_{ext}, plus the work done by all internal nonconservative forces W_{nc}, plus that done by all internal conservative forces W_c:

$$W_{total} = W_{ext} + W_{nc} + W_c$$

Rearranging gives $W_{ext} + W_{nc} = W_{total} - W_c$. Substituting from Equations 7-1 and 7-2 we have

$$W_{ext} + W_{nc} = \Delta K_{sys} + \Delta U_{sys} \qquad 7\text{-}3$$

The right side of this equation can be simplified as

$$\Delta K_{sys} + \Delta U_{sys} = \Delta(K_{sys} + U_{sys}) \qquad 7\text{-}4$$

The sum of the kinetic energy K_{sys} and the potential energy U_{sys} is called the **total mechanical energy** E_{mech}:

---

† Other factors in the system, such as a battery that runs a motor, can cause the mechanical energy to increase, so the presence of kinetic friction does not necessarily mean that the mechanical energy of the system will decrease.

$$E_{mech} = K_{sys} + U_{sys} \qquad\qquad 7\text{-}5$$

DEFINITION—TOTAL MECHANICAL ENERGY

Combining Equations 7-4 and 7-5, and then substituting into Equation 7-3 gives

$$W_{ext} = \Delta E_{mech} - W_{nc}$$

The mechanical energy of a system is conserved (E_{mech} = constant) if the total work done by all external forces and by all internal nonconservative forces is zero.

$$E_{mech} = K_{sys} + U_{sys} = \text{constant} \qquad\qquad 7\text{-}6$$

CONSERVATION OF MECHANICAL ENERGY

This is **conservation of mechanical energy** and is the origin of the expression "conservative force."

If $E_{mech\,i} = K_i + U_i$ is the initial mechanical energy of the system and $E_{mech\,f} = K_f + U_f$ is the final mechanical energy of the system, conservation of mechanical energy implies that

$$E_{mech\,f} = E_{mech\,i} \qquad (\text{or } K_f + U_f = K_i + U_i) \qquad 7\text{-}7$$

CONSERVATION OF MECHANICAL ENERGY

Many mechanics problems can be solved by setting the final mechanical energy of a system equal to its initial mechanical energy.

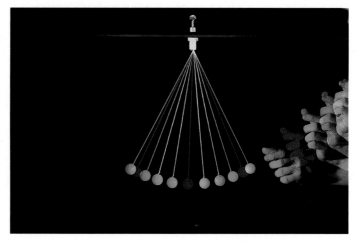

Multiflash photograph of a simple pendulum. As the bob descends, gravitational potential energy is converted into kinetic energy, and the speed increases as indicated by the increased spacing of the recorded positions. The speed decreases as the bob moves up, and the kinetic energy is changed into potential energy.

Applications

Suppose that you are wearing skis and, starting at rest from a height h_0 above the bottom of a hill, you coast down the hill. Assuming that the skis are frictionless, what is your speed as you pass through a gate at height h above the bottom of the hill? The mechanical energy of the earth–skier system is conserved because the only force doing work is the internal, conservative force of gravity. If we choose $U = 0$ at the bottom of the hill, the initial potential energy is mgh_0. This is also the total mechanical energy because the initial kinetic energy is zero. Thus,

$$E_{mech\,i} = K_i + U_i = 0 + mgh_0$$

At height h, the potential energy is mgh and the speed is v. Hence,

$$E_{mech\,f} = K_f + U_f = \tfrac{1}{2}mv^2 + mgh$$

Setting $E_{mech\,f} = E_{mech\,i}$ we find

$$\tfrac{1}{2}mv^2 + mgh = mgh_0$$

or

$$v = \sqrt{2g(h_0 - h)}$$

Your speed is the same as if you had undergone free-fall through a distance $h_0 - h$. However, by skiing down the hill, you travel farther and take more time than you would if you were in free-fall and falling straight down.

FIGURE 7-1

E X A M P L E   7 - 1

Standing near the edge of the roof of a 12-m-high building, you kick a ball with an initial speed of $v_i = 16$ m/s at an angle of $60°$ above the horizontal (Figure 7-1). Neglecting air resistance, find (a) how high above the height of the building the ball rises and (b) its speed just before it hits the ground.

PICTURE THE PROBLEM We choose the ball and the earth as the system. We consider this system during the interval from just after the kick to just before impact with the ground. There are no external forces doing work on the system, and there are no internal nonconservative forces doing work, so the mechanical energy of the system is conserved. At the top of its flight, the ball is moving horizontally with a speed v_{top}, equal to the horizontal component of its initial horizontal velocity v_{ix}. We choose $y = 0$ at the roof of the building.

(a) 1. Conservation of mechanical energy relates the height h above the top of the building to the initial speed v_i and the speed at the top of its flight v_{top}:

$$E_{\text{mech top}} = E_{\text{mech i}}$$

$$\tfrac{1}{2}mv_{top}^2 + mgy_{top} = \tfrac{1}{2}mv_i^2 + mgy_i$$

$$\tfrac{1}{2}mv_{top}^2 + mgh_{top} = \tfrac{1}{2}mv_i^2 + 0$$

2. Solve for h_{top}:

$$h_{top} = \frac{v_i^2 - v_{top}^2}{2g}$$

3. The velocity at the top of its flight equals its initial horizontal velocity:

$$v_{top} = v_{ix} = v_i \cos\theta$$

4. Substitute the step 3 result into the step 2 result and solve for h_{top}:

$$h_{top} = \frac{v_i^2 - v_{top}^2}{2g} = \frac{v_i^2 - v_i^2\cos^2\theta}{2g} = \frac{v_i^2(1 - \cos^2\theta)}{2g}$$

$$= \frac{(16 \text{ m/s})^2 (1 - \cos^2 60°)}{2(9.81 \text{ m/s}^2)} = \boxed{9.79 \text{ m}}$$

(b) 1. If v_f is the speed of the ball just before it hits the ground, where $y = y_f = -12$ m, its energy is:

$$E_{\text{mech f}} = \tfrac{1}{2}mv_f^2 + mgy_f$$

2. Set the final mechanical energy equal to the initial mechanical energy:

$$\tfrac{1}{2}mv_f^2 + mgy_f = \tfrac{1}{2}mv_i^2 + 0$$

3. Solve for v_f, and set $y = -12$ m to find the final velocity:

$$v_f = \sqrt{v_i^2 - 2gy_f}$$

$$= \sqrt{(16 \text{ m/s})^2 - 2(9.81 \text{ m/s}^2)(-12 \text{ m})} = \boxed{22.2 \text{ m/s}}$$

E X A M P L E   7 - 2

A pendulum consists of a bob of mass m attached to a string of length L. The bob is pulled aside so that the string makes an angle θ_0 with the vertical, and is released from rest. As it passes through the bottom of the arc, find expressions for (a) the speed and (b) the tension. Air resistance is negligible.

PICTURE THE PROBLEM The system consists of the pendulum, the earth, and the pendulum mount and its supports. There are no external forces on this system. The two internal forces acting on the bob (air resistance being negligible) are the force of gravity $m\vec{g}$, which is conservative, and the tension force $\vec{T}$. The rate at which $\vec{T}$ does work is $\vec{T} \cdot \vec{v}$. Because $\vec{T}$ is perpendicular to $\vec{v}$, we know that $\vec{T} \cdot \vec{v} = 0$. Since only $m\vec{g}$ does work, we know that the mechanical energy of the bob–earth system is conserved. To find the speed of the bob, we equate the initial and final mechanical energies. The tension in the string is obtained using Newton's second law.

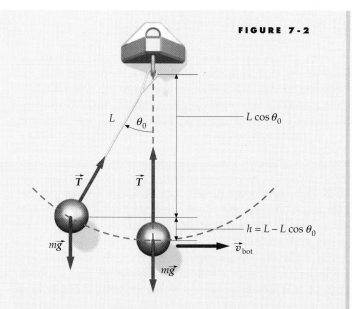

FIGURE 7-2

(a) 1. Make a sketch of the system in its initial and final configurations (Figure 7-2). We choose $y = 0$ at the bottom of the swing and $y = h$ at the initial position:

2. Apply conservation of mechanical energy. Initially the bob is at rest:

$$E_{\text{mech f}} = E_{\text{mech i}}$$

$$\tfrac{1}{2}mv_{\text{f}}^2 + mgy_{\text{f}} = \tfrac{1}{2}mv_{\text{i}}^2 + mgy_{\text{i}}$$

$$\tfrac{1}{2}mv_{\text{bot}}^2 + 0 = 0 + mgh$$

3. Conservation of mechanical energy thus relates the speed v_{bot} to the height h:

$$\tfrac{1}{2}mv_{\text{bot}}^2 = mgh$$

4. Solve for the speed v_{bot}:

$$v_{\text{bot}} = \sqrt{2gh}$$

5. To express speed in terms of the initial angle θ_0, we need to relate h to θ_0. This relation is illustrated in Figure 7-2:

$$L = L \cos \theta_0 + h$$

so

$$h = L - L \cos \theta_0 = L(1 - \cos \theta_0)$$

6. Substitute this value for h to express the speed at the bottom in terms of θ_0:

$$\boxed{v_{\text{bot}} = \sqrt{2gL(1 - \cos \theta_0)}}$$

(b) 1. When the bob is at the bottom of the circle, the forces on it are $m\vec{g}$ and $\vec{T}$. Apply $\Sigma F_y = ma_y$:

$$T - mg = ma_y$$

2. At the bottom, the bob has an acceleration v_{bot}^2/L toward the center of the circle, which is upward:

$$a_y = \frac{v_{\text{bot}}^2}{L} = \frac{2gL(1 - \cos \theta_0)}{L}$$

$$= 2g(1 - \cos \theta_0)$$

3. Substitute for a_y the Part (b) step 1 result and solve for T:

$$T = mg + ma_y = m(g + a_y)$$

$$= m[g + 2g(1 - \cos \theta_0)]$$

$$= \boxed{(3 - 2 \cos \theta_0)mg}$$

REMARKS 1. The tension at the bottom is greater than the weight of the bob because the bob is accelerating upward. 2. Step 3 in Part (b) shows that for $\theta_0 = 0$, $T = mg$, the expected result for a stationary bob hanging from a string. 3. Step 4 in Part (a) shows that the speed at the bottom is the same as if the bob had dropped in free-fall from a height h. The speed of the bob at the bottom of the arc can also be found using Newton's laws directly (see Problem 7-92), but such a solution is more challenging because the acceleration component tangential to the circle varies with position, and therefore with time, so the constant-acceleration formulas do not apply.

A 2-kg block on a frictionless horizontal surface is pushed against a spring that has a force constant of 500 N/m, compressing the spring by 20 cm. The block is then released, and the spring projects it along the surface and then up a frictionless incline of angle 45°. How far up the incline does the block travel before momentarily coming to rest?

PICTURE THE PROBLEM Let the system include the block, the spring, the earth, the horizontal surface, the ramp, and the wall to which the spring is attached. After the block is released there are no external forces on this system. The only forces that do work are the force exerted by the spring on the block and the force of gravity, both of which are conservative. Thus, the total mechanical energy of the system is conserved. Find the maximum height h from the conservation of mechanical energy, and then the distance up the incline s from $\sin 45° = h/s$.

Cover the column to the right and try these on your own before looking at the answers.

Answers

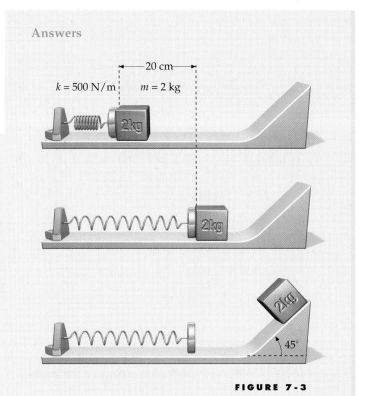

FIGURE 7-3

Steps

1. Sketch the system with both the initial and final configurations (Figure 7-3).

2. Write the initial mechanical energy in terms of the compression distance x.

$$E_{\text{mech i}} = U_{\text{s i}} + U_{\text{g i}} + K_{\text{i}} = \tfrac{1}{2}kx^2 + 0 + 0$$

3. Write the final mechanical energy in terms of the height h.

$$E_{\text{mech f}} = U_{\text{s f}} + U_{\text{g f}} + K_{\text{f}} = 0 + mgh + 0$$

4. Apply conservation of mechanical energy and solve for h.

$$mgh = \tfrac{1}{2}kx^2$$

$$h = \frac{kx^2}{2mg} = 0.51 \text{ m}$$

5. Find the distance s from h and the angle of inclination.

$$h = s \sin \theta$$

$$s = \boxed{0.72 \text{ m}}$$

REMARKS In this problem, the initial mechanical energy of the system is the potential energy of the spring. This energy is converted first into kinetic energy and then into gravitational potential energy.

EXERCISE Find the speed of the block just after it leaves the spring. (*Answer* 3.16 m/s)

A spring with a force constant of k hangs vertically. A block of mass m is attached to the unstretched spring and allowed to fall from rest. Find an expression for the maximum distance the block falls before it begins moving upward.

PICTURE THE PROBLEM As the block drops, its speed first increases, then reaches some maximum value, and then decreases until it is again zero when the block is at its lowest point. There are only conservative forces present, so we apply the conservation of mechanical energy to the earth–spring–block system.

1. Sketch the system showing the initial and final positions of the block and spring (Figure 7-4). Include a y axis with up as the positive y direction. Choose the gravitational potential energy of the system to be zero at the original position $y = 0$ when the spring is unstretched. The potential energy of the spring is zero when the spring is unstretched. Let d be the distance the block falls.

2. Apply conservation of mechanical energy to relate the initial ($y = 0$) and final ($y = -d$) positions:

$$mgy_f + \tfrac{1}{2}ky_f^2 + \tfrac{1}{2}mv_f^2 = mgy_i + \tfrac{1}{2}ky_i^2 + \tfrac{1}{2}mv_i^2$$

$$mg(-d) + \tfrac{1}{2}k(-d)^2 + 0 = 0 + 0 + 0$$

3. Solve for d. There are two solutions. One gives $d = 0$ and the other is the solution we want—the maximum distance the block falls before it begins moving upward:

$$\tfrac{1}{2}kd^2 - mgd = 0$$

$$(\tfrac{1}{2}kd - mg)d = 0$$

Since $d \neq 0$, $d = \boxed{\dfrac{2mg}{k}}$

FIGURE 7-4

REMARKS Gravitational potential energy is converted into the kinetic energy of the block plus the potential energy of the spring. At the lowest point, where the block is momentarily at rest, the elastic potential energy of the spring equals the decrease in the gravitational potential energy of the system.

BACK TO THE FUTURE   **EXAMPLE   7 - 5**   **Put It in Context**

Imagine that you have time-traveled back to the late 1800s and are watching your great-great-great-grandparents on their honeymoon taking a ride on the Flip Flap Railway, a Coney Island roller coaster with a circular loop-the-loop. The car they are in is about to enter the loop-the-loop when a 100-lb sack of sand falls from a construction-site platform and lands in the back seat of the car. No one is hurt, but the impact causes the car to lose 25 percent of its speed. The car started from rest at a high point 2 times as high as the top of the loop-the-loop. Neglect any losses due to either friction or air drag. Will their car make it over the top of the loop-the-loop without falling off?

PICTURE THE PROBLEM The car has to have enough speed at the top of the loop-the-loop to maintain contact with the track. We can use conservation of mechanical energy to determine the speed just before the sandbag hits the car, and we can use it again to determine the speed it has at the top of the track. Then we can use Newton's second law to determine the normal force exerted on the car by the track.

FIGURE 7-5

1. Draw a picture of the car and track, with the car at the starting point, at the bottom of the track, and at the top of the loop-the-loop (Figure 7-5):

2. Apply Newton's second law to relate the speed at the top of the loop-the-loop to the normal force:

$$F_n + mg = m\frac{v_{top}^2}{R}$$

3. Using conservation of mechanical energy, find the speed just prior to impact. The initial height is $4R$, where R is the radius of the loop-the-loop:

$$U_1 + K_1 = U_0 + K_0$$

$$0 + \tfrac{1}{2}mv_1^2 = mg4R + 0$$

so

$$v_1 = \sqrt{8Rg}$$

4. The impact with the sandbag results in a 25 percent decrease in speed. Find the speed after impact:

$$v_2 = 0.75v_1 = 0.75\sqrt{8Rg}$$

5. Using conservation of mechanical energy, find the speed just at the top of the loop-the-loop:

$$U_{top} + K_{top} = U_2 + K_2$$

$$mg\,2R + \tfrac{1}{2}mv_{top}^2 = 0 + \tfrac{1}{2}m(0.75^2\,8Rg)$$

so

$$v_{top}^2 = (0.75^2\,8 - 4)Rg = 0.5Rg$$

6. Substituting for v_{top}^2 in the step 2 result gives:

$$F_n + mg = 0.5mg$$

7. Solve for F_n:

$$F_n = -0.5mg$$

8. F_n is the magnitude of the normal force. It cannot be negative:

> Oops! The car has left the track.

REMARKS A loss of 25 percent of your speed means losing almost 44 percent of your kinetic energy. Fortunately, there were safety devices to prevent the cars from falling, so your ancestors likely would have survived. The biggest concern for riders on the Flip Flap Railway was a broken neck. In the loop-the-loop, the riders were subjected to up to $12g$s. This was the last of the circular loop-the-loop roller coasters. Loop-the-loops on more recent rides are oval shaped, higher than they are wide.

TWO BLOCKS ON A STRING   **EXAMPLE 7 - 6   Try It Yourself**

Two blocks are attached to a light string that passes over a massless, frictionless pulley. The two blocks have masses m_1 and m_2, where $m_2 > m_1$, and are initially at rest. Find the speed of either block when m_2 falls a distance h. Find the magnitude of the acceleration of the blocks as m_2 falls.

PICTURE THE PROBLEM Let the system be the two blocks, the string, the pulley, its supports, and the earth. There are no external forces on this system, hence no work is done on the system by external forces. Also, there is no kinetic friction. Thus the mechanical energy of the system is conserved.

Cover the column to the right and try these on your own before looking at the answers.

Steps

Answers

1. Make a sketch of the system showing it in both its initial and final configurations (Figure 7-6). Let h be the distance m_2 falls.

FIGURE 7-6

2. Initially, the total kinetic energy of the system is zero. Write the total kinetic energy K of the system when the blocks are moving with speed v.

$$K = \tfrac{1}{2}m_1v^2 + \tfrac{1}{2}m_2v^2 = \tfrac{1}{2}(m_1 + m_2)v^2$$

3. Initially, choose the total gravitational potential energy of the system U to be zero. Write an expression for U when m_1 has moved up a distance h (and m_2 has moved down the same distance).

$$U = m_1 gh - m_2 gh = (m_1 - m_2)gh$$

4. Add U and K to obtain the total mechanical energy E.

$$E = K + U$$
$$= \tfrac{1}{2}(m_1 + m_2)v^2 + (m_1 - m_2)gh$$

5. Apply conservation of mechanical energy.

$$E = E_i$$
$$\tfrac{1}{2}(m_1 + m_2)v^2 + (m_1 - m_2)gh = 0$$

6. Solve for v.

$$v = \boxed{\sqrt{\frac{2(m_2 - m_1)}{(m_1 + m_2)}gh}}$$

7. Find the acceleration a by dv/dt. Use the step 6 result and the chain rule for differentiation. Note that $v = dh/dt$.

$$a = \frac{dv}{dt} = \frac{dv}{dh}\frac{dh}{dt} = \boxed{\frac{(m_2 - m_1)}{(m_1 + m_2)}g}$$

PLAUSIBILITY CHECK This device, called an *Atwood's machine,* is analyzed in terms of forces in Problems 84–90 in Chapter 4. There we see that the acceleration is given by the step 7 result.

EXERCISE What is the magnitude of the acceleration of either block if the masses are $m_1 = 3$ kg and $m_2 = 5$ kg? (*Answer* $a = 0.25g = 2.45$ m/s²)

h

(a)

We've seen that mechanical energy conservation can be used as an alternative to Newton's laws for solving certain problems in mechanics. When we are not interested in the time t, the conservation of mechanical energy is often much easier to use than Newton's second law (Figure 7-7). Because mechanical energy conservation is derived from Newton's laws, any problem that can be solved using it can also be solved directly from Newton's laws, though often with much more difficulty.

7-2 The Conservation of Energy

In the macroscopic world, dissipative nonconservative forces, like kinetic friction, are always present to some extent. Such forces tend to decrease the mechanical energy of a system. However, any such decrease in mechanical energy is accompanied by a corresponding increase in thermal energy. (Consider how the tires of a car feel warm after a long trip.) Another type of nonconservative force is that involved in the deformations of objects. When you bend a metal coat hanger back and forth, you do work on the coat hanger but the work you do does not appear as mechanical energy. Instead the coat hanger becomes warm. The work done in deforming the hanger is dissipated as thermal energy. Similarly, when a falling ball of putty lands on the floor, it becomes warm as it deforms. The dissipated kinetic energy appears as thermal energy. For the ball–floor–earth system, the total energy is the sum of the thermal energy and the mechanical energy. The total energy of the system is conserved even though neither the total mechanical energy nor the total thermal energy are individually conserved.

A third type of nonconservative force is associated with chemical reactions. When we include systems in which chemical reactions take place, the sum of

h

(b)

FIGURE 7-7 (*a*) One can easily find the speed of a block sliding down a frictionless incline of constant slope by applying Newton's second law or by using conservation of mechanical energy. However, if the incline is frictionless but not of constant slope, as in (*b*), the problem can still be solved easily using conservation of mechanical energy, whereas it can be solved using Newton's second law only if the slope of the incline is known at each point, and then the calculation is quite tedious.

mechanical energy plus thermal energy is not conserved. For example, suppose that you begin running from rest. Initially you have no kinetic energy. When you begin to run, internal chemical energy in your muscles is converted to kinetic energy of your body and thermal energy is produced. It is possible to identify and measure the chemical energy that is used. In this case, the sum of mechanical, thermal, and chemical energy is conserved.

Even when thermal energy and chemical energy are included, the total energy of the system does not always remain constant, because energy can be converted to radiation energy, such as sound waves or electromagnetic waves. But *the increase or decrease in the total energy of a system can always be accounted for by the appearance or disappearance of energy somewhere else.* This experimental result, known as the **law of conservation of energy,** is one of the most important laws in all of science. Let E_{sys} be the total energy of a given system, E_{in} be the energy that enters the system, and E_{out} be the energy that leaves the system. The law of conservation of energy then states

$$E_{in} - E_{out} = \Delta E_{sys} \qquad\qquad 7\text{-}8$$

LAW OF CONSERVATION OF ENERGY

Alternatively,

> The total energy of the universe is constant. Energy can be converted from one form to another, or transmitted from one region to another, but energy can never be created or destroyed.

LAW OF CONSERVATION OF ENERGY

The total energy E of many systems familiar from everyday life can be accounted for completely by mechanical energy E_{mech}, thermal energy E_{therm}, and chemical energy E_{chem}. To be comprehensive and include other possible forms of energy, such as electromagnetic or nuclear energy, we include E_{other}, and write generally

$$E_{sys} = E_{mech} + E_{therm} + E_{chem} + E_{other} \qquad\qquad 7\text{-}9$$

The Work–Energy Theorem

A common way to transfer energy into or out of a system is to do work on the system from the outside. If this is the only method of energy transfer,[†] the law of conservation of energy becomes

$$W_{ext} = \Delta E_{sys} = \Delta E_{mech} + \Delta E_{therm} + \Delta E_{chem} + \Delta E_{other} \qquad\qquad 7\text{-}10$$

WORK–ENERGY THEOREM

where W_{ext} is the work done on the system by external forces and ΔE_{sys} is the change in the system's total energy. This work–energy theorem for systems, which we will call simply the work–energy theorem, is a powerful tool for studying a wide variety of systems. Note that if the system is just a single particle its energy can only be kinetic, so Equation 7-10 is equivalent to the work–kinetic energy theorem studied in Chapter 6.

EXPLORING

Devices that convert one form of energy to another are called transducers. To find out what your inner ear, a strain gauge, a microphone, and your sense of touch have in common, go to
www.whfreeman.com/tipler5e.

---

† Energy can also be transferred when heat is exchanged between a system and its surroundings. Exchanges of heat energy, which occur when there is a temperature difference between a system and its surroundings, are discussed in Chapter 18.

FALLING CLAY   **E X A M P L E   7 - 7**

A ball of modeling clay with mass m is released from rest from a height h and falls to the perfectly rigid floor. Discuss the application of the law of conservation of energy to (*a*) the system consisting of the ball alone, and (*b*) the system consisting of the earth, the floor, and the ball.

PICTURE THE PROBLEM There are two forces that act on the ball: gravity and the contact force of the floor. Since the floor does not move, the contact force it exerts on the ball does no work. There are no chemical or other energy changes so we can neglect ΔE_{chem} and ΔE_{other}. We also neglect the sound energy radiated when the ball hits the floor. Thus, the only energy transferred to or from the ball is the work done by the force of gravity. Apply the work–energy theorem.

(*a*) 1. Write the work–energy theorem for the ball:

$$W_{ext} = \Delta E_{sys} = \Delta E_{mech} + \Delta E_{therm}$$

2. The two external forces on the system are gravity and the force exerted by the floor. The floor does not move and therefore does no work. The only work done on the ball is by gravity:

$$W_{ext} = mgh$$

3. Since the ball alone is our system, its mechanical energy is entirely kinetic, which is zero both initially and finally. Thus, the change in mechanical energy is zero:

$$\Delta E_{mech} = 0$$

4. Substitute mgh for W_{ext} and 0 for ΔE_{mech} in step 1:

$$W_{ext} = \Delta E_{mech} + \Delta E_{therm}$$
$$mgh = 0 + \Delta E_{therm}$$

so

$$\boxed{\Delta E_{therm} = mgh}$$

(*b*) 1. There are no external forces acting on the ball–earth–floor system (the force of gravity and the force of the floor are now internal to the system), so there is no external work done:

$$W_{ext} = 0$$

2. Write the work–energy theorem with $W_{ext} = 0$:

$$W_{ext} = \Delta E_{sys} = \Delta E_{mech} + \Delta E_{therm}$$
$$0 = \Delta E_{mech} + \Delta E_{therm}$$

3. The original mechanical energy of the ball–earth system is the original gravitational potential energy, and the final mechanical energy is zero:

$$E_{mech\,i} = mgh$$
$$E_{mech\,f} = 0$$

4. The change in mechanical energy of the ball–earth system is thus:

$$\Delta E_{mech} = 0 - mgh = -mgh$$

5. The work–energy theorem thus gives the same result as in Part (*a*):

$$\Delta E_{therm} = \boxed{-\Delta E_{mech} = mgh}$$

REMARKS In (*a*), energy is transferred to the ball by the work done on it by gravity. This energy appears as the kinetic energy of the ball before it hits the floor and as thermal energy after. The ball warms slightly and the energy is eventually transferred to the surroundings via heat. In (*b*), no energy is transferred to the ball–earth–floor system. The original potential energy of the system is converted to kinetic energy of the ball just before it hits, and then into thermal energy.

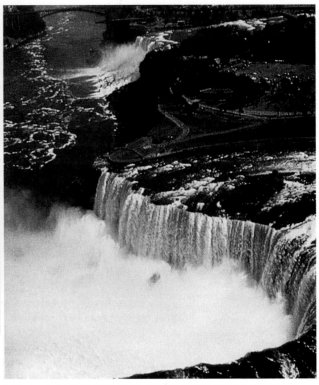

In this power plant in Kansas, energy stored in the fossil fuel coal (black area at the right) is released by burning the coal to produce steam; the steam is then used to drive turbines to produce electricity. The excess heat is dissipated by the cooling towers.

The potential energy of the water at the top of Niagara Falls is used to produce electrical energy.

Problems Involving Kinetic Friction

When surfaces slide across each other kinetic friction decreases the mechanical energy of the system and increases the thermal energy. Consider a block that begins with initial velocity v_i and slides along a table until it stops (Figure 7-8). We choose the block and table to be our system. Then $\Delta E_{chem} = \Delta E_{other} = 0$ and there is no external work done on this system. The work–energy theorem gives

$$0 = \Delta E_{mech} + \Delta E_{therm}$$

The mechanical energy lost is the initial kinetic energy of the block

$$\Delta E_{mech} = -\tfrac{1}{2}mv_i^2 \qquad\qquad 7\text{-}11$$

We can relate the loss in mechanical energy to frictional force. If f is the magnitude of the frictional force, Newton's second law gives

$$-f = ma$$

Multiplying both sides of this equation by Δs, we find

$$-f\Delta s = ma\,\Delta s = m(\tfrac{1}{2}v_f^2 - \tfrac{1}{2}v_i^2) = -\tfrac{1}{2}mv_i^2 \qquad\qquad 7\text{-}12$$

where Δs is the displacement of the block, and we have used the constant-acceleration formula $2a\,\Delta s = v_f^2 - v_i^2$ (with $v_f = 0$). Equating the left sides of Equations 7-11 and 7-12 we obtain

$$f\,\Delta s = -\Delta E_{mech} \qquad\qquad 7\text{-}13$$

Note that the quantity $-f\,\Delta s$ is *not* the work done by friction on the sliding block, because careful analysis shows the actual displacement of the kinetic frictional

FIGURE 7-8 A block sliding on a table. The force of friction reduces the mechanical energy of the block–table system.

force on the block is not in general equal to the displacement of the block.[†] However, it can be shown that $f\Delta s$ does equal the increase in thermal energy of the block–table system due to the dissipation of the system's mechanical energy. This thermal energy is produced both on the bottom surface of the block and on the upper surface of the tabletop as the block slides across the tabletop. Thus,

$$f\,\Delta s = \Delta E_{\text{therm}} \qquad\qquad 7\text{-}14$$

ENERGY DISSIPATED BY FRICTION

Substituting this result into the work–energy theorem (with $E_{\text{chem}} = E_{\text{other}} = 0$), we obtain

$$W_{\text{ext}} = \Delta E_{\text{mech}} + \Delta E_{\text{therm}} = \Delta E_{\text{mech}} + f\,\Delta s \qquad\qquad 7\text{-}15$$

WORK–ENERGY THEOREM FOR PROBLEMS WITH KINETIC FRICTION

where Δs is the distance one contacting surface slides relative to the other. When there is no external work done on the system, the energy dissipated by friction equals the decrease in mechanical energy:

$$\Delta E_{\text{therm}} = f\,\Delta s = -\Delta E_{\text{mech}} \qquad (W_{\text{ext}} = 0) \qquad\qquad 7\text{-}16$$

Irish wind farm now under construction. The Arklow Banks project in the Irish Sea will have 200 turbines and produce 520 MW of electrical power—three times the combined capacity of all offshore wind farms currently in production in the world.

PUSHING A BOX   **EXAMPLE 7 - 8**

You push a 4-kg box, which is initially at rest on a horizontal table, a distance of 3 m with a horizontal force of 25 N. The coefficient of kinetic friction between the box and table is 0.35. Find (*a*) the external work done on the block–table system, (*b*) the energy dissipated by friction, (*c*) the final kinetic energy of the box, and (*d*) the speed of the box.

PICTURE THE PROBLEM The box plus table is the system (Figure 7-9). You are external to this system, so the force you push with is an external force. The final speed of the box is found from its final kinetic energy, which we find using the work–energy theorem with $\Delta E_{\text{chem}} = 0$ and $\Delta E_{\text{therm}} = f\,\Delta s$. The energy of the system is increased by the external work. Some of the energy increase is kinetic energy and some is thermal energy.

FIGURE 7-9

(*a*) There are four external forces acting on the system. However, only one of them does work. The total external work done is the product of the push force and the distance traveled:

$$\Sigma W_{\text{ext}} = W_{\text{by push on block}} + W_{\text{by gravity on block}}$$
$$+ W_{\text{by gravity on table}} + W_{\text{by floor on table}}$$
$$= F_{\text{push}}\,\Delta x + 0 + 0 + 0 = (25\text{ N})(3\text{ m})$$
$$= \boxed{75\text{ J}}$$

(*b*) The energy dissipated by friction is $f\,\Delta x$ (the magnitude of the normal force equals mg):

$$\Delta E_{\text{therm}} = f\,\Delta x = \mu_k F_n\,\Delta x = \mu_k mg\,\Delta x$$
$$= (0.35)(4\text{ kg})(9.81\text{ N/kg})(3\text{ m}) = \boxed{41.2\text{ J}}$$

---

† The work done by kinetic friction is examined in detail in B. A. Sherwood and W. H. Bernard, "Work and Heat Transfer in the Presence of Sliding Friction," *American Journal of Physics,* **52**, 1001 (1984).

(c) 1. Apply the work–energy theorem to find the final kinetic energy:

$$W_{ext} = \Delta E_{mech} + \Delta E_{therm}$$

2. There are no internal conservative forces doing work, so the change in potential energy is zero. The change in kinetic energy equals the change in mechanical energy:

$$\Delta E_{mech} = \Delta U + \Delta K$$
$$= 0 + (K_f - 0) = K_f$$

3. Substitute this result into the work–energy theorem:

$$W_{ext} = K_f + \Delta E_{therm}$$

so

$$K_f = W_{ext} - \Delta E_{therm}$$
$$= 75.0\,J - 41.2\,J = \boxed{33.8\,J}$$

(d) The final speed of the box is related to its kinetic energy. Solve for the final speed of the box:

$$K_f = \tfrac{1}{2}mv_f^2$$

so

$$v_f = \sqrt{\frac{2K_f}{m}} = \sqrt{\frac{2(33.8\,J)}{4\,kg}} = \boxed{4.11\,m/s}$$

A MOVING SLED

EXAMPLE 7-9 **Try It Yourself**

A sled is coasting on a horizontal snow-covered surface with an initial speed of 4 m/s. If the coefficient of friction between the sled and the snow is 0.14, how far will the sled go before coming to rest?

PICTURE THE PROBLEM We choose the sled and snow as our system and then apply the work–energy theorem.

Cover the column to the right and try these on your own before looking at the answers.

Steps

Answers

1. Sketch the system in its initial and final configurations (Figure 7-10).

FIGURE 7-10

2. Apply the work–energy theorem.

$$W_{ext} = \Delta E_{mech} + \Delta E_{therm}$$
$$= (\Delta U + \Delta K) + f\,\Delta s$$

3. Solve for f. The normal force is equal to the weight.

$$f = \mu_k F_n = \mu_k\, mg$$

4. There are no conservative forces doing work and there are no external forces acting on the system. Use these observations to eliminate two terms from the step 2 result.

$$W_{ext} = \Delta U = 0$$

and

$$W_{ext} = \Delta U + \Delta K + f\,\Delta s$$
$$0 = 0 + \Delta K + \mu_k\, mg\,\Delta s$$

5. Express the change in kinetic energy in terms of the mass and the initial speed, and solve for Δs.

$$\Delta s = \frac{v_i^2}{2\mu_k g} = \boxed{5.82\,m}$$

A PLAYGROUND SLIDE   **EXAMPLE 7-10**

A child of mass 40 kg goes down an 8.0-m-long slide inclined at 30° with the horizontal. The coefficient of kinetic friction between the child and the slide is 0.35. If the child starts from rest at the top of the slide, how fast is she traveling when she reaches the bottom?

PICTURE THE PROBLEM As the child slides down, some of her potential energy is converted into kinetic energy and some into thermal energy because of friction. We choose the child–slide–earth as our system and apply the conservation of energy theorem.

1. Make a sketch of the system showing both its initial and final configurations (Figure 7-11).

$m = 40$ kg

$|\Delta h| = 4$ m

$\Delta s = 8$ m

30°

FIGURE 7-11

2. Write out the conservation of energy equation:

$$W_{ext} = \Delta E_{mech} + \Delta E_{therm}$$
$$= (\Delta U + \Delta K) + f\,\Delta s$$

3. The initial kinetic energy is zero. The speed at the bottom is related to the final kinetic energy:

$$\Delta K = K_f - 0 = \tfrac{1}{2}mv_f^2$$

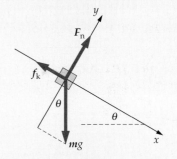

FIGURE 7-12

4. There are no external forces acting on the system:

$$W_{ext} = 0$$

5. The change in potential energy is related to the change in height Δh (which is negative).

$$\Delta U = mg\,\Delta h$$

6. To find f_k we apply Newton's second law to the child. First we draw a free-body diagram (Figure 7-12):

7. Next we apply Newton's second law. The normal component of the acceleration is zero. To find F_n we take components in the normal direction. Then we solve for f_k using $f_k = \mu_k F_n$:

$$F_n - mg\cos\theta = 0$$

so

$$f_k = \mu_k F_n = \mu_k mg\cos\theta$$

8. We use trigonometry to relate Δs to Δh:

$$|\Delta h| = \Delta s\sin\theta$$

9. Substituting into the step 2 result gives:

$$0 = mg\,\Delta h + \tfrac{1}{2}mv_f^2 + \mu_k mg\cos\theta\,\Delta s$$
$$= -mg\,\Delta s\sin\theta + \tfrac{1}{2}mv_f^2 + \mu_k mg\cos\theta\,\Delta s$$

10. Solving for v_f gives:

$$v_f^2 = 2g\,\Delta s(\sin\theta - \mu_k\cos\theta)$$
$$= 2(9.81\ \text{m/s}^2)(8\ \text{m})(\sin 30° - 0.35\cos 30°)$$
$$= 30.9\ \text{m}^2/\text{s}^2$$

so

$$v_f = \boxed{5.60\ \text{m/s}}$$

REMARKS Note that the result is independent of the mass of the child.

EXERCISE For the earth–child–slide system, calculate (*a*) the initial mechanical energy, (*b*) the final mechanical energy, and (*c*) the energy dissipated by friction. (*Answer*   (*a*) 1570 J, (*b*) 618 J, (*c*) 952 J)

A 4-kg block hangs by a light string that passes over a massless, frictionless pulley and is connected to a 6-kg block that rests on a shelf. The coefficient of kinetic friction is 0.2. The 6-kg block is pushed against a spring, compressing it 30 cm. The spring has a force constant of 180 N/m. Find the speed of the blocks after the 6-kg block is released and the 4-kg block has fallen a distance of 40 cm.

PICTURE THE PROBLEM The speed of the blocks is obtained from their final kinetic energy. Consider the system to be everything shown in Figure 7-13 plus the earth. This system has both gravitational and elastic potential energy. Apply the work–energy theorem. Knowing the kinetic energy of the blocks means you can solve for their speed.

FIGURE 7-13

Cover the column to the right and try these on your own before looking at the answers.

Steps

Answers

1. Write out the equation for the conservation of energy of the system.

$$W_{ext} = \Delta E_{mech} + \Delta E_{therm}$$
$$= (\Delta U_s + \Delta U_g + \Delta K) + f\,\Delta s$$

2. There are no external forces on the system.

$$W_{ext} = 0$$

3. Make a table of the mechanical energy terms both initially, when the spring is compressed 30 cm, and finally, when each block has moved a distance $\Delta s = 40$ cm and the spring is relaxed. Let the gravitational potential energy of the initial configuration equal zero. Also, write down the difference (final minus initial) between each initial and final expression.

	U_s	U_g	K
Final	0	$-m_2 g\,\Delta s$	$\frac{1}{2}(m_1 + m_2)v_f^2$
Initial	$\frac{1}{2}kx_i^2$	0	0
Difference	$-\frac{1}{2}kx_i^2$	$-m_2 g\,\Delta s$	$\frac{1}{2}(m_1 + m_2)v_f^2$

4. Find an expression for f_k that includes μ_k. Substitute it, the step 2 result, and the step 3 results into the step 1 result.

$$0 = -\tfrac{1}{2}kx_i^2 - m_2 g\,\Delta s + \tfrac{1}{2}(m_1 + m_2)v_f^2 + \mu_k m_1 g\,\Delta s$$

5. Solve the step 4 result for v_f^2. Then substitute numerical values and solve for v_f:

$$v_f^2 = \frac{kx_i^2 + 2m_2 g\,\Delta s - 2\mu_k m_1 g\,\Delta s}{m_1 + m_2}$$

so

$$v_f = \boxed{1.95 \text{ m/s}}$$

REMARKS This solution assumes that the string remains taut at all times. This will be true if the acceleration of block 1 remains less than g, that is, if the net force on block 1 is less than $m_1 g = (6 \text{ kg})(9.81 \text{ N/kg}) = 58.9$ N. Initially the force exerted by the spring on block 1 has the magnitude $kx_1 = (180 \text{ N/m})(0.3 \text{ m}) = 54.0$ N and the frictional force has magnitude $f_k = \mu_k m_1 g = 0.2(58.9 \text{ N}) = 11.8$ N. These forces combine to produce a net force of 42.2 N directed to the right. Since the spring force decreases as block 1 moves following release, the acceleration of the 6-kg block will never exceed g, so the string will remain taut.

This pizza contains about 16 MJ of energy, about half the energy in 1 L (0.26 U.S. gal) of gasoline.

Systems With Chemical Energy

Sometimes a system's internal chemical energy is converted into mechanical energy and thermal energy with no work being done on the system by external forces. For example, at the beginning of this section we described the energy conversions that take place when you start running. In order to move forward, you push back on the floor and the floor pushes forward on you with a static frictional force. This force causes you to accelerate, but it does not do work. It does no work because the displacement of the point of application of the force is zero (assuming your shoes do not slip on the floor). Because no work is done, no energy is transferred from the floor to your body. The kinetic-energy increase of your body comes from the conversion of internal chemical energy derived from the food you eat. Consider the following example.

CLIMBING STAIRS   **E X A M P L E   7 - 1 2**

Suppose that you have mass m and you walk up a flight of stairs of height h. Discuss the application of energy conservation to a system consisting of you alone.

PICTURE THE PROBLEM   There are two forces that act on you: gravity and the force of the stair treads on your feet. The force of gravity does negative work on you. To determine the work done by the force of the stair treads on your feet, consider the force of one stair tread on one of your feet. This force is a contact force. It does no work because the sole of your foot does not move while it is in contact with the tread. Thus, no work is done by the force of the stair treads on your feet.

1. Write the work–energy equation for the you-alone system:

$$W_{ext} = \Delta E_{sys} = \Delta E_{mech} + \Delta E_{therm} + \Delta E_{chem}$$

2. There are two external forces acting on this system, the force of gravity and the force of the stair treads on your feet. The force of gravity does negative work because this force is directed opposite to your displacement. The force of the stair treads does no work because the point of application, the soles of your feet, do not move while this force is applied:

$$W_{ext} = -mgh$$

3. As stated in the problem, the system consists of you alone. Because your configuration does not change, any change in your mechanical energy is entirely a change in your kinetic energy, which is the same initially and finally:

$$\Delta E_{mech} = 0$$

4. Substitute these results into the work–energy equation:

$$-mgh = \boxed{\Delta E_{therm} + \Delta E_{chem}}$$

REMARKS   If there were no change in thermal energy, then your chemical energy would decrease by mgh. Because the human body is relatively inefficient, the increase in the amount of thermal energy will be considerably greater than mgh. The decrease in stored chemical energy equals mgh plus any thermal energy, which eventually is transferred from your body to your surroundings via heat.

EXERCISE   Discuss the energy conservation for a system consisting of both you and the earth. (*Answer*   For the you–earth system no external work is done, so the total energy, which now includes gravitational potential energy, is conserved. The change in mechanical energy is mgh, so the work–energy theorem gives $0 = mgh + \Delta E_{therm} + \Delta E_{chem}$.)

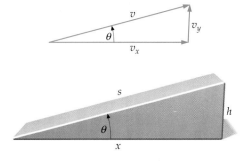

AN UPHILL DRIVE

EXAMPLE 7-13

You are driving a 1000-kg gasoline-powered car at a constant speed of 100 km/h (= 27.8 m/s = 62.2 mi/h) up a 10-percent grade (Figure 7-14). (A 10-percent grade means that the road rises 1 m for each 10 m of horizontal distance—that is, the angle of inclination θ is given by tan θ = 0.1.) (a) If the efficiency is 15 percent, what is the rate at which the chemical energy of the car–earth–atmosphere system changes? (The efficiency is the fraction of the chemical energy consumed that appears as mechanical energy.) (b) What is the rate at which thermal energy is created?

FIGURE 7-14 $\tan \theta = h/x \sim \sin \theta = h/s$

PICTURE THE PROBLEM Some of the chemical energy goes into increasing the potential energy of the car as it climbs the hill, and some goes into an increase in thermal energy, much of which is expelled by the car as exhaust. Apply the work–energy theorem to a system consisting of the car, the hill, the air, and the earth.

(a) 1. The rate of loss of chemical energy equals the negative of change in chemical energy per unit time:

$$\text{chemical energy loss rate} = \frac{-\Delta E_{\text{chem}}}{\Delta t}$$

2. The increase in mechanical energy equals 15% of the decrease in chemical energy:

$$\Delta E_{\text{mech}} = 0.15|\Delta E_{\text{chem}}| = -0.15\,\Delta E_{\text{chem}}$$

3. Solve for the loss rate of chemical energy:

$$\frac{-\Delta E_{\text{chem}}}{\Delta t} = \frac{1}{0.15}\frac{\Delta E_{\text{mech}}}{\Delta t}$$

4. The car moves at constant speed, so $\Delta K = 0$ and $\Delta E_{\text{mech}} = \Delta U$. Relate the change in mechanical energy to the change in height Δh and substitute it into the step 3 result:

$$\Delta E_{\text{mech}} = mg\,\Delta h$$

so

$$\frac{-\Delta E_{\text{chem}}}{\Delta t} = \frac{mg}{0.15}\frac{\Delta h}{\Delta t}$$

5. Convert the changes to time derivatives. That is, take the limit of both sides as Δt approaches zero:

$$-\frac{dE_{\text{chem}}}{dt} = \frac{mg}{0.15}\frac{dh}{dt}$$

6. The rate of change of h equals v_y, which is related to the speed v as shown in Figure 7-14:

$$\frac{dh}{dt} = v_y = v\sin\theta$$

7. We can approximate sin θ by tan θ because the angle is small:

$$\sin\theta \approx \tan\theta$$

8. Solve for the loss rate of chemical energy:

$$-\frac{dE_{\text{chem}}}{dt} = \frac{mg}{0.15}v\sin\theta \approx \frac{mg}{0.15}v\tan\theta$$

$$= \frac{(1000\text{ kg})(9.81\text{ N/kg})}{0.15}(27.8\text{ m/s})0.1$$

$$= \boxed{182\text{ kW}}$$

(b) 1. Write out the work–energy relation:

$$W_{\text{ext}} = \Delta E_{\text{mech}} + \Delta E_{\text{therm}} + \Delta E_{\text{chem}}$$

2. Convert to derivatives and solve for dE_{therm}/dt:

$$0 = \frac{dE_{\text{mech}}}{dt} + \frac{dE_{\text{therm}}}{dt} + \frac{dE_{\text{chem}}}{dt}$$

so

$$\frac{dE_{\text{therm}}}{dt} = -\frac{dE_{\text{mech}}}{dt} - \frac{dE_{\text{chem}}}{dt} = 0.15\frac{dE_{\text{chem}}}{dt} - \frac{dE_{\text{chem}}}{dt}$$

$$= -0.85\frac{dE_{\text{chem}}}{dt} = 0.85(182\text{ kW}) = \boxed{155\text{ kW}}$$

REMARKS Gasoline-powered cars are typically only about 15 percent efficient. About 85 percent of the chemical energy of the gasoline goes to thermal energy, most of which is expelled as heat exhaust. Additional thermal energy is created by rolling friction and wind resistance. The energy content of gasoline is about 31.8 MJ/L.

7-3   Mass and Energy

In 1905, Albert Einstein published his special theory of relativity, a result of which is the famous equation

$$E_0 = mc^2 \qquad\qquad 7\text{-}17$$

where $c = 3 \times 10^8$ m/s is the speed of light in vacuum. We will study this theory in some detail in later chapters.

According to Equation 7-17, a particle or system of mass m has "rest" energy mc^2. This energy is intrinsic to the particle. Consider the positron, a particle emitted in a nuclear process called beta decay. Positrons and electrons have identical masses, but equal and opposite electrical charge. When a positron encounters an electron in matter, electron–positron annihilation occurs, a process in which the electron and positron disappear and their energy appears as electromagnetic radiation. If the two particles are initially at rest, the energy of the electromagnetic radiation equals the rest energy of the electron plus that of the positron.

Energies in atomic and nuclear physics are usually expressed in units of electron volts (eV) or mega-electron-volts (1 MeV = 10^6 eV). A convenient unit for the masses of atomic particles is eV/c^2 or MeV/c^2. Table 7-1 lists rest energies (and therefore the masses) of some elementary particles and light nuclei. The total rest energy of a positron plus an electron is 2(0.511 MeV), which is the radiation energy emitted upon annihilation.

TABLE 7-1

Rest Energies of Some Elementary Particles and Light Nuclei

Particle	Symbol	Rest Energy (MeV)	
Electron	e^-	0.5110	
Positron	e^+	0.5110	
Proton	p	938.272	
Neutron	n	939.565	
Deuteron	d	1875.613	
Triton	t	2808.410	
Helium-3	^{3}He	2808.39	
Alpha particle	α	3727.379	

The rest energy of a *system* can consist of the potential energy of the system or other internal energies of the system, in addition to the intrinsic rest energies of the particles in the system. If the system at rest absorbs energy ΔE and remains at rest, its rest energy increases by ΔE and its mass increases by

$$\Delta M = \frac{\Delta E}{c^2} \qquad\qquad 7\text{-}18$$

Consider two 1-kg blocks connected by a spring of force constant k. If we stretch the spring a distance x, the potential energy of the system increases by $\Delta U = \frac{1}{2}kx^2$. According to Equation 7-18, the mass of the system has also increased by $\Delta M = \Delta U/c^2$. Because c is such a large number, this increase in mass cannot be observed in macroscopic systems. For example, suppose $k = 800$ N/m and $x = 10$ cm $= 0.1$ m. The potential energy of the spring system is then $\frac{1}{2}kx^2 = \frac{1}{2}(800$ N/m$)(0.1$ m$)^2 = 4$ J. The increase in mass of the system is

$$\Delta M = \frac{\Delta U}{c^2} = \frac{4 \text{ J}}{(3 \times 10^8 \text{ m/s})^2} = 4.44 \times 10^{-17} \text{ kg}$$

The relative mass increase $\Delta M/M \approx 2 \times 10^{-17}$ is much too small to be observed.

Nuclear Energy

In nuclear reactions, the energy changes are often an appreciable fraction of the rest energy of the system. Consider the deuteron, which is the nucleus of deuterium, an isotope of hydrogen also called heavy hydrogen. The deuteron consists of a proton and neutron bound together. From Table 7-1 we see that the mass of the proton is 938.28 MeV/c^2 and the mass of the neutron is 939.57 MeV/c^2. The sum of these two masses is 1877.85 MeV/c^2. But the mass of the deuteron is 1875.63 MeV/c^2, which is less than the sum of the masses of the proton and neutron by 2.22 MeV/c^2. Note that this mass difference $\Delta M/M \approx 1.2 \times 10^{-3}$, much greater than any uncertainties in the measurement of these masses, and almost 14 orders of magnitude greater than the 2×10^{-17} discussed above for the spring–blocks system.

Heavy water (deuterium oxide) molecules are produced in the primary cooling water of a nuclear reactor when neutrons collide with the hydrogen nuclei (protons) of the water molecules. If a slow moving neutron is captured by a proton, 2.22 MeV of energy are released in the form of electromagnetic radiation. Thus, the mass of a deuterium atom is 2.22 MeV/c^2 less than the sum of the masses of an isolated hydrogen atom and an isolated neutron.

This process can be reversed by breaking a deuteron into its constituent parts if at least 2.22 MeV of energy is transferred to the deuteron with electromagnetic radiation or by collisions with other energetic particles. Any transferred energy in excess of 2.22 MeV appears as kinetic energy of the resulting proton and neutron.

The energy needed to completely separate a nucleus into individual neutrons and protons is called the **binding energy.** The binding energy of a deuteron is 2.22 MeV.

The deuteron is an example of a bound system. Its rest energy is less than the rest energy of its parts, so energy must be put into the system to break it apart. If the rest energy of a system is greater than the rest energy of its parts, the system is unbound. An example is uranium-236, which breaks apart or **fissions** into two smaller nuclei.[†] The sum of the masses of the resultant parts is less than the mass of the original nucleus. Thus the mass of the system decreases, and energy is released.

In nuclear fusion, two very light nuclei such as a deuteron and a triton (the nucleus of the hydrogen isotope tritium) fuse together. The rest mass of the resultant nucleus is less than that of the original parts and again energy is released. In a chemical reaction that produces energy, such as burning coal, the mass decrease is of the order of 1 eV/c^2 per atom. This is more than a million times smaller than the mass changes per nucleus in many nuclear reactions, and is not readily observable.

---

† Uranium-236, written ^{236}U, is made in a nuclear reactor when the stable isotope uranium-235 absorbs a neutron. This reaction will be discussed in Chapter 34.

BINDING ENERGY   **E X A M P L E   7 - 1 4**

A hydrogen atom consisting of a proton and an electron has a binding energy of 13.6 eV. By what percentage is the mass of a proton plus the mass of an electron greater than that of the hydrogen atom?

1. The fractional difference between the mass of the hydrogen atom and the masses of its parts is the ratio of the binding energy E_b divided by c^2 to $m_e + m_p$:

$$\text{Fractional difference} = \frac{E_b/c^2}{m_e + m_p} = \frac{13.6\ \text{eV}/c^2}{m_e + m_p}$$

2. Obtain the rest masses of the proton and electron from Table 7-1:

$$m_p = 938.28\ \text{MeV}/c^2; \quad m_e = 0.511\ \text{MeV}/c^2$$

3. Add to find the sum of these masses:

$$m_p + m_e = 938.79\ \text{MeV}/c^2$$

4. The rest mass of the hydrogen atom is less than this by 13.6 eV/c^2. The percentage difference is:

$$\text{Fractional difference} = \frac{13.6\ \text{eV}/c^2}{938.79 \times 10^6\ \text{eV}/c^2}$$
$$= 1.45 \times 10^{-8}$$
$$= \boxed{1.45 \times 10^{-6}\ \%}$$

REMARKS This mass difference is too small to be measured directly. However, binding energies can be accurately measured, so the mass difference can be found from $E_b = (\Delta m)c^2$.

NUCLEAR FUSION   **E X A M P L E   7 - 1 5**   **Try It Yourself**

In a typical nuclear fusion reaction, a tritium nucleus (^{3}H) and a deuterium nucleus (^{2}H) fuse together to form a helium nucleus (^{4}He) plus a neutron. The reaction is written ^{2}H + ^{3}H → ^{4}He + n. If the initial kinetic energy of the particles is negligible, how much energy is released in this fusion reaction?

PICTURE THE PROBLEM Because energy is released, the total rest energy of the initial particles must be greater than that of the final particles. This difference equals the energy released.

Cover the column to the right and try these on your own before looking at the answers.

Steps

1. Write down the rest energies of ^{2}H and ^{3}H from Table 7-1 and add to find the total initial rest energy.

2. Do the same for ^{4}He and n to find the final rest energy.

3. Find the energy released from $E_{\text{released}} = E_0\ (\text{initial}) - E_0\ (\text{final})$.

Answers

$E_0\ (\text{initial}) = 1875.628\ \text{MeV} + 2808.944\ \text{MeV}$
$= 4684.572\ \text{MeV}$

$E_0\ (\text{final}) = 3727.409\ \text{MeV} + 939.573\ \text{MeV}$
$= 4666.982\ \text{MeV}$

$E_{\text{released}} = 4684.572\ \text{MeV} - 4666.982\ \text{MeV}$
$= \boxed{17.59\ \text{MeV} \approx 17.6\ \text{MeV}}$

REMARKS This and other fusion reactions occur in the sun. The energy that is released bathes the earth and is ultimately responsible for all life on the planet. The energy emitted by the sun is accompanied by a continuous decrease in the sun's rest mass.

Newtonian Mechanics and Relativity

When the speed of a particle approaches the speed of light, Newton's second law breaks down, and we must modify Newtonian mechanics according to Einstein's theory of relativity. The criterion for the validity of Newtonian mechanics can also be stated in terms of the energy of a particle. In Newtonian mechanics, the kinetic energy of a particle moving with speed v is

$$K = \tfrac{1}{2}mv^2 = \tfrac{1}{2}mc^2\frac{v^2}{c^2} = \tfrac{1}{2}E_0\frac{v^2}{c^2}$$

where $E_0 = mc^2$ is the rest energy of the particle. Then

$$\frac{v}{c} = \sqrt{\frac{2K}{E_0}}$$

Newtonian mechanics is valid if the speed of the particle is much less than the speed of light, or alternatively, the kinetic energy of a particle is much less than its rest energy.

7-4 Quantization of Energy

When energy is put into a system that remains at rest, the internal energy of the system increases. While it might seem that we could put any amount of energy into a system, this is found not to be true for microscopic systems such as atoms or molecules. The internal energy of a microscopic system can increase only by discrete increments.

If we have two blocks attached to a spring and we pull the blocks apart, we do work on the block–spring system, and its potential energy increases. If we then release the blocks, they oscillate back and forth. The energy of oscillation E—the kinetic energy of motion of the blocks plus the potential energy due to the stretching of the spring—equals the initial potential energy. In time, the energy of the system decreases because of various damping effects such as friction and air resistance. As closely as we can measure, the energy decreases continuously. Eventually all the energy is dissipated and the energy of oscillation is zero.

Now consider a diatomic molecule such as molecular oxygen, O_2. The force of attraction between the two oxygen atoms varies approximately linearly with the change in separation (for small changes), like that of two blocks connected by a spring. If a diatomic molecule is set oscillating with some energy E, the energy decreases with time as the molecule radiates or interacts with its surroundings, but the decrease is *not continuous*. The energy decreases in finite steps, and the lowest energy state, called the ground state, is not zero. The vibrational energy of a diatomic molecule is said to be **quantized;** that is, the molecule can absorb or release energies only in certain amounts, known as quanta.

When either blocks on a spring or diatomic molecules oscillate, the time for one oscillation is called the period T. The reciprocal of the period is the frequency of oscillation $f = 1/T$. We will see in Chapter 14 that the period and frequency of an oscillator do not depend on the energy of oscillation. As the energy decreases, the frequency remains the same. Figure 7-15 shows an **energy-level diagram** for an oscillator. The allowed energies are approximately equally spaced, and are given by[†]

$$E_n = (n + \tfrac{1}{2})hf, \qquad n = 0, 1, 2, 3, \ldots \qquad\qquad 7\text{-}19$$

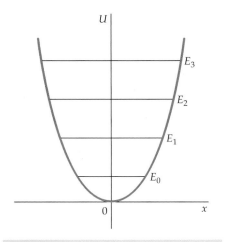

FIGURE 7-15 Energy-level diagram for an oscillator.

---

† A diatomic molecule can also have rotational energy. The rotational energy is also quantized, but the energy levels are not equally spaced, and the lowest possible energy is zero. We will study rotational energy in Chapters 9 and 10.

where f is the frequency of oscillation and h is a fundamental constant of nature called Planck's constant:[†]

$$h = 6.626 \times 10^{-34} \text{ J·s} \qquad 7\text{-}20$$

The integer n is called a **quantum number.** The lowest possible energy is the **ground state energy** $E_0 = \frac{1}{2}hf$.

Microscopic systems often gain or lose energy by absorbing or emitting electromagnetic radiation. By conservation of energy, if E_i and E_f are the initial and final energies of a system, the energy of the radiation emitted or absorbed is

$$E_{\text{rad}} = E_i - E_f$$

Since the system energies E_i and E_f are quantized, the radiated energy is also quantized.[‡] The quantum of radiation energy is called a **photon.** The energy of a photon is given by

$$E_{\text{photon}} = hf \qquad 7\text{-}21$$

The quasar 3C 273 is shown imaged via X-ray energy. The X-ray energy emitted by this quasar is more than a million times that emitted by the entire Milky Way galaxy.

where f is the frequency of the electromagnetic radiation.[§]

As far as we know, all bound systems exhibit energy quantization. For macroscopic bound systems, the steps between energy levels are so small that they are unobservable. For example, typical oscillation frequencies for two blocks on a spring are 1 to 10 times per second. If $f = 10$ oscillations per second, the spacing between allowed levels is $hf = (6.626 \times 10^{-34} \text{ J·s}) \times (10/\text{s}) \approx 6 \times 10^{-33}$ J. Since the energy of a macroscopic system is of the order of joules, a quantum step of 10^{-33} J is too small to be noticed. To put it another way, if the energy of a system is 1 J, the value of n is of the order of 10^{32} and changes of one or two quantum units will not be observable.

For a diatomic molecule, a typical frequency of vibration is 10^{14} vibrations per second, and a typical energy is 10^{-19} J. The spacing between allowed levels is then

$$E_{n+1} - E_n = hf \approx (6.63 \times 10^{-34} \text{ J·s})(10^{14} \text{ s}) \approx 6 \times 10^{-20} \text{ J}$$

Thus, changes in the energy of oscillation are of the same order of magnitude as the energy of the molecule, and quantization is definitely not negligible.

SUMMARY

1. The conservation of mechanical energy is an important relation derived from Newton's laws for conservative forces. It is useful in solving many problems.

2. The work–energy theorem and the conservation of energy are fundamental laws of nature that have applications in all areas of physics.

3. Einstein's equation $E_0 = mc^2$ is a fundamental relation between mass and energy.

4. Quantization is a fundamental property of the energy in bound systems.

† In 1900, the German physicist Max Planck introduced this constant as a calculational device to explain discrepancies between the theoretical curves and experimental data on the spectrum of blackbody radiation. The significance of Planck's constant was not appreciated by Planck or anyone else until Einstein postulated in 1905 that the energy of electromagnetic radiation is not continuous, but occurs in packets of size hf where f is the frequency of the radiation.

‡ Historically, the quantization of electromagnetic radiation, as proposed by Max Planck and Albert Einstein, was the first "discovery" of energy quantization.

§ Electromagnetic radiation includes light, microwaves, radio waves, television waves, X rays, and gamma rays. These differ from one another in their frequencies.

Topic	Relevant Equations and Remarks
1. Mechanical Energy	The sum of the kinetic and potential energy of a system is called the total mechanical energy $$E_{mech} = K_{sys} + U_{sys} \qquad \text{7-5}$$
Conservation of mechanical energy	If no external forces do work on a system, and if the internal forces that do work are all conservative, the total mechanical energy of the system remains constant: $$E_{mech} = K_{sys} + U_{sys} = \text{constant} \qquad \text{7-6}$$ $$K_f + U_f = K_i + U_i \qquad \text{7-7}$$
2. Total Energy of a System	The energy of a system consists of mechanical energy E_{mech}, thermal energy E_{therm}, chemical energy E_{chem}, and other types of energy E_{other}, such as sound radiation and electromagnetic radiation. $$E_{sys} = E_{mech} + E_{therm} + E_{chem} + E_{other} \qquad \text{7-9}$$
3. Conservation of Energy	
Universe	The total energy of the universe is constant. Energy can be converted from one form to another, or transmitted from one region to another, but energy can never be created or destroyed.
System	The energy of a system can be changed by work being done on the system and by energy transfer via heat. (These include the emission or absorption of radiation.) The increase or decrease in the energy of the system can always be accounted for by the appearance or disappearance of some kind of energy somewhere else: $$E_{in} - E_{out} = \Delta E_{sys} \qquad \text{7-8}$$
Work–energy theorem	$$W_{ext} = \Delta E_{sys} = \Delta E_{mech} + \Delta E_{therm} + \Delta E_{chem} + \Delta E_{other} \qquad \text{7-10}$$
4. Energy Dissipated by Friction	For a system that involves a pair of sliding surfaces, the total energy dissipated by friction on both surfaces equals the increase in thermal energy of the system and is given by $$f \, \Delta s = \Delta E_{therm} \qquad \text{7-14}$$ where Δs is the distance one surface slides relative to the other.
5. Problem Solving	The conservation of mechanical energy and the work–energy theorem can be used as an alternative to Newton's laws to solve mechanics problems that require the determination of the speed of a particle as a function of its position.
6. Mass and Energy	A particle with mass m has an intrinsic rest energy E_0 given by $$E_0 = mc^2 \qquad \text{7-17}$$ where $c = 3 \times 10^8$ m/s is the speed of light in vacuum. A system with mass M also has a rest energy $E_0 = Mc^2$. If a system gains or loses internal energy ΔE, it simultaneously gains or loses mass $\Delta M = \Delta E / c^2$.
Binding energy	The energy required to separate a bound system into its constituent parts is called its binding energy. The binding energy is $\Delta M c^2$, where ΔM is the sum of the masses of the constituent parts, less the mass of the bound system.
8. Newtonian Mechanics and Special Relativity	When the speed of a particle approaches the speed of light c (when the kinetic energy of the particle approaches its rest energy), Newtonian mechanics breaks down, and must be replaced by Einstein's special theory of relativity.

9. Energy Quantization — The internal energy of a system is found to have only a discrete set of possible values. For a system oscillating with frequency f, the allowed energy values are separated by an amount hf, where h is Planck's constant:

$$h = 6.626 \times 10^{-34}\,\text{J·s} \qquad \text{7-20}$$

Photons — Microscopic systems often exchange energy with their surroundings by emitting or absorbing electromagnetic radiation, which is also quantized. The quantum of energy of radiation is called the photon:

$$E_{\text{photon}} = hf \qquad \text{7-21}$$

where f is the frequency of the electromagnetic radiation.

PROBLEMS

- • Single-concept, single-step, relatively easy
- •• Intermediate-level, may require synthesis of concepts
- ••• Challenging
- SSM Solution is in the *Student Solutions Manual*
- iSOLVE Problems available on iSOLVE online homework service
- iSOLVE ✓ These "Checkpoint" online homework service problems ask students additional questions about their confidence level, and how they arrived at their answer

In a few problems, you are given more data than you actually need; in a few other problems, you are required to supply data from your general knowledge, outside sources, or informed estimates.

Take $g = 9.81$ N/kg = 9.81 m/s² and neglect friction in all problems unless otherwise stated.

Conceptual Problems

1 • SSM Two objects of unequal mass are connected by a massless cord passing over a frictionless peg. After the objects are released from rest, which of the following statements are true? (U = gravitational potential energy, K = kinetic energy of the system.) (a) $\Delta U < 0$ and $\Delta K > 0$. (b) $\Delta U = 0$ and $\Delta K > 0$. (c) $\Delta U < 0$ and $\Delta K = 0$. (d) $\Delta U = 0$ and $\Delta K = 0$. (e) $\Delta U > 0$ and $\Delta K < 0$.

2 • Two stones are thrown with the same initial speed at the same instant from the roof of a building. One stone is thrown at an angle of 30° above the horizontal, the other is thrown horizontally. (Neglect air resistance.) Which statement below is true?

(a) The stones strike the ground at the same time and with equal speeds.
(b) The stones strike the ground at the same time with different speeds.
(c) The stones strike the ground at different times with equal speeds.
(d) The stones strike the ground at different times with different speeds.

3 • True or false:

(a) The total energy of a system cannot change.
(b) When you jump into the air, the floor does work on you increasing your mechanical energy.

4 • You stand on roller skates next to a rigid wall. To get started, you push off against the wall. Discuss the energy changes pertinent to this situation.

5 • SSM In *Surely You're Joking, Mr. Feynman*,[†] Richard Feynman described his annoyance at how the concept of energy was portrayed in a children's textbook in the following way: "There was a book which started out with four pictures: first, there was a wind-up toy; then there was an automobile; then there was a boy riding a bicycle; then there was something else. And underneath each picture it said 'What makes it go?' ... I turned the page. The answer was ... for everything, 'Energy makes it go' ... It's also not even true that 'energy makes it go' because if it stops, you could say 'energy makes it stop' just as well. ... Energy is neither increased or decreased in these examples; it's just changed from one form to another." Describe how energy changes from one form to another when a little girl pedals her bike up a hill, then freewheels down the hill, and brakes to a stop.

6 • SSM A body falling through the atmosphere (air resistance is present) gains 20 J of kinetic energy. The amount of gravitational potential energy that it lost is (a) 20 J, (b) more than 20 J, (c) less than 20 J, (d) impossible to tell without knowing the mass of the body, (e) impossible to tell without knowing how far the body falls.

† Richard P. Feynman and Ralph Leighton, *Surely You're Joking, Mr. Feynman*, New York: Bantam Books (1985).

7 •• Assume that, on applying the brakes, a constant frictional force acts on the wheels of a car. If that is so, it follows that (a) the distance the car travels before coming to rest is proportional to the speed of the car before the brakes are applied, (b) the car's kinetic energy diminishes at a constant rate, (c) the kinetic energy of the car is inversely proportional to the time that has elapsed since the application of the brakes, (d) none of the above are true.

8 • You are given two frictionless ramps and a block to slide down them (Figure 7-16). One ramp is in the form of an inclined plane with height H and length L. The other has a ramp cut in the form of a partial arc of a circle, but also has height H and length L. You slide the block down each ramp, releasing it from rest, and measure the time it takes to reach the bottom and the speed of the block upon getting there. You find that (a) the block takes the same time to slide down each ramp, (b) the block has the same speed on reaching the bottom of each ramp, (c) both (a) and (b) are correct, (d) neither (a) nor (b) is correct.

FIGURE 7-16 Problem 8

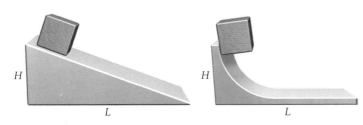

9 •• If a rock is attached to a massless, rigid rod and swung in a vertical circle at a constant speed, it will not have a constant total energy, as the kinetic energy of the rock will be constant, but the potential energy will be continually changing. Is any total work being done on the rock? Does the rod exert a tangential force on the rock?

Estimation and Approximation

10 •• **SSM** The *metabolic rate* is the rate at which the body uses chemical energy to sustain its life functions. Experimentally, the average metabolic rate is proportional to the total skin surface area of the body. The surface area for a 5-ft, 10-in. male weighing 175 lb is just about 2.0 m² and for a 5-ft, 4-in. female weighing 110 lb is approximately 1.5 m². There is about a 1 percent change in surface area for every 3 lb above or below the weights quoted here and a 1 percent change for every inch above or below the heights quoted. (a) Estimate your average metabolic rate over the course of a day using the following guide for physical activity: sleeping, metabolic rate = 40 W/m²; sitting, 60 W/m²; walking, 160 W/m²; moderate physical activity, 175 W/m²; and moderate aerobic exercise, 300 W/m². How does it compare to the power of a 100-W light bulb? (b) Express your average metabolic rate in terms of kcal/day (1 kcal = 4190 J). (A kcal is the "food calorie" used by nutritionists.) (c) An estimate used by nutritionists is that the "average person" must eat roughly 12–15 kcal/lb of body weight a day to maintain his or her weight. From the calculations in part (b), are these estimates reasonable?

11 • Assume that the maximum rate at which your body can expend energy is 250 W. Assuming a 20 percent efficiency for the conversion of chemical energy into mechanical energy, estimate how quickly you can run up four flights of stairs, with each flight 3.5 m high.

12 • How much rest mass is consumed in the core of a nuclear-fueled electric generating plant in producing (a) 1 J of thermal energy? (b) enough electrical energy to keep a 100-W light bulb burning for 10 y? (For each joule of electrical energy produced by the generator, the reactor core must produce 3 J of nuclear energy.)

13 • **SSM** The chemical energy released by burning a gallon of gasoline is approximately 2.6×10^5 kJ. Estimate the total energy used by all of the cars in the United States during the course of 1 y. What fraction does this represent of the total energy use by the United States in 1 y (about 5×10^{20} J)?

14 • The maximum efficiency of a solar energy panel in converting solar energy into useful electrical energy is about 12 percent. Using the known value of the solar intensity reaching the earth's surface (1.0 kW/m²), what area would have to be covered by solar panels in order to supply the energy requirements of the United States (approximately 5×10^{20} J/y)? Assume cloudless skies.

15 • Hydroelectric power plants convert gravitational potential energy into more useful forms by flowing water downhill through a turbine system to generate electrical energy. The Hoover Dam on the Colorado River is 211 m high and generates 4 billion kW·h/y (1 W · h = 3.6×10^3 J.) At what flow rate (in L/s) must water be flowing through the turbines to generate this power? The density of water is 1 kg/L. Assume a total efficiency of 20 percent in converting the water's potential energy into electrical energy.

The Conservation of Mechanical Energy

16 • A block of mass m is pushed against a spring, compressing it a distance x, and the block is then released. The spring projects the block along a frictionless horizontal surface, giving a speed v. The same spring projects a second block of mass $4m$, giving it a speed $3v$. What distance was the spring compressed in the second case?

17 • A bicyclist traveling at 10 m/s on a horizontal road stops pedaling as she starts up a hill inclined at 3.0° to the horizontal. Ignoring frictional forces, how far up the hill will she travel before stopping? (a) 5.1 m. (b) 30 m. (c) 97 m. (d) 10.2 m. (e) The answer depends on the mass of the person.

18 • **SSM** A pendulum of length L with a bob of mass m is pulled aside until the bob is at a height $L/4$ above its equilibrium position. The bob is then released. Find the speed of the bob as it passes the equilibrium position.

19 • **iSOLVE** A 3-kg block slides along a frictionless horizontal surface with a speed of 7 m/s (Figure 7-17). After sliding a distance of 2 m, the block makes a smooth transition to a frictionless ramp inclined at an angle of 40° to the horizontal. How far up the ramp does the block slide before coming momentarily to rest?

FIGURE 7-17 Problems 19, 46

20 • [ISOLVE ✓] The 3-kg object in Figure 7-18 is released from rest at a height of 5 m on a curved frictionless ramp. At the foot of the ramp is a spring of force constant $k = 400$ N/m. The object slides down the ramp and into the spring, compressing it a distance x before coming momentarily to rest. (a) Find x. (b) What happens to the object after it comes to rest?

FIGURE 7-18 Problem 20

21 • [ISOLVE ✓] A 15-g ball is shot from a spring gun whose spring has a force constant of 600 N/m. The spring can be compressed 5 cm. How high will the ball go if the gun is aimed vertically?

22 • [ISOLVE] At a dock, a crane lifts a 4000-kg container 30 m, swings it out over the deck of a freighter, and lowers the container into the hold of the freighter, which is 8 m below the level of the dock. How much work is done by the crane on the container? (Neglect friction losses.)

23 • [ISOLVE] A 16-kg child on a playground swing moves with a speed of 3.4 m/s when the 6-m long swing is at its lowest point. What is the angle that the swing makes with the vertical when the swing is at its highest point?

24 •• [SSM] The system shown in Figure 7-19 is initially at rest when the lower string is cut. Find the speed of the objects when they are at the same height. The frictionless pulley has negligible mass.

FIGURE 7-19 Problem 24

25 •• A block rests on an inclined plane as shown in Figure 7-20. A spring to which it is attached via a pulley is being pulled downward with gradually increasing force. The value of μ_s is known. Find the potential energy U of the spring at the moment when the block begins to move.

FIGURE 7-20 Problem 25

26 •• [ISOLVE] A 2.4-kg block is dropped onto a spring of spring constant 3955 N/m from a height of 5.0 m (Figure 7-21). When the block is momentarily at rest, the spring has been compressed by 25 cm. Find the speed of the block when the compression of the spring is 15 cm.

FIGURE 7-21 Problems 26, 91

27 •• [SSM] A ball at the end of a string moves in a vertical circle with constant mechanical energy E. What is the difference between the tension at the bottom of the circle and the tension at the top?

28 •• A girl of mass m is taking a picnic lunch to her grandmother. She ties a rope of length R to a tree branch over a creek and starts to swing from rest at a point that is a distance $R/2$ lower than the branch. What is the minimum breaking tension for the rope if it is not to break and drop the girl into the creek?

29 •• ISOLVE A roller coaster car of mass 1500 kg starts at a distance $H = 23$ m above the bottom of a loop 15 m in diameter (Figure 7-22). If friction is negligible, the downward force of the rails on the car when it is upside down at the top of the loop is (a) 4.6×10^4 N, (b) 3.1×10^4 N, (c) 1.7×10^4 N, (d) 980 N, (e) 1.6×10^3 N.

FIGURE 7-22 Problem 29

30 • A single-car roller coaster pushes off, and on the first section of track, descends a 5-m-deep valley, then climbs to the top of a hill that is 9.5 m above the valley floor. (a) What is the minimum initial speed required to carry the coaster beyond the first hill? Assume that the track is frictionless. (b) Can we affect this speed by changing the depth of the valley to make the coaster pick up more speed at the bottom?

31 •• The Gravitron single-car roller coaster has a loop-the-loop that is constructed so that riders on the coaster will feel perfectly weightless when they reach the top of the circular arc. How heavy will they feel when they reach the bottom of the arc (that is, what is the normal force pressing up into their seats at the bottom of the loop)? Express the answer as a multiple of their normal weight. Assume that the arc is perfectly circular and no frictional forces act on the car.

32 •• SSM ISOLVE A stone is thrown upward at an angle of 53° above the horizontal. Its maximum height during the trajectory is 24 m. What was the stone's initial speed?

33 •• ISOLVE ✔ A baseball of mass 0.17 kg is thrown from the roof of a building 12 m above the ground. Its initial velocity is 30 m/s at an angle of 40° above the horizontal. (a) What is the maximum height the ball reaches? (b) What is the work done by gravity as the ball moves from the roof to its maximum height? (c) What is the speed of the ball as it strikes the ground?

34 •• An 80-cm-long pendulum with a 0.6-kg bob is released from rest at initial angle of θ_0 with the vertical. At the bottom of the swing, the speed of the bob is 2.8 m/s. (a) What was the initial angle of the pendulum? (b) What angle does the pendulum make with the vertical when the speed of the bob is 1.4 m/s?

35 •• SSM ISOLVE ✔ The Royal Gorge bridge over the Arkansas River is $L = 310$ m above the river. A bungee jumper of mass 60 kg has an elastic cord of length $d = 50$ m attached to her feet. Assume the cord acts like a spring of force constant k. The jumper leaps, barely touches the water,

and after numerous ups and downs comes to rest at a height h above the water. (a) Find h. (b) Find the maximum speed of the jumper.

36 •• A pendulum consists of a 2-kg bob attached to a light string of length 3 m. The bob is struck horizontally so that it has an initial horizontal velocity of 4.5 m/s. For the point at which the string makes an angle of 30° with the vertical, what is (a) the speed, (b) the potential energy, and (c) the tension in the string? (d) What is the angle of the string with the vertical when the bob reaches its greatest height?

37 •• A pendulum consists of a string of length L and a bob of mass m. The string is brought to a horizontal position and the bob is given the minimum initial speed enabling the pendulum to make a full turn in the vertical plane. (a) What is the maximum kinetic energy K of the bob? (b) What is the tension in the string when the kinetic energy is maximum?

38 •• ISOLVE ✔ A child whose weight is 360 N swings out over a pool of water using a rope attached to the branch of a tree at the edge of the pool. The branch is 12 m above ground level and the surface of the water is 1.8 m below ground level. The child holds on to the rope at a point 10.6 m from the branch and moves back until the angle between the rope and the vertical is 23°. When the rope is in the vertical position, the child lets go and drops into the pool. Find the speed of the child at the surface of the water.

39 •• SSM ISOLVE Walking by a pond, you find a rope attached to a stout tree limb 5.2 m off the ground. You decide to use the rope to swing out over the pond. The rope is a bit frayed but supports your weight, which is about 650 N. You estimate that the rope might break if the tension is 80 N greater than your weight. You grab the rope at a point 4.6 m from the limb and move back to swing out over the pond. (a) What is the maximum safe initial angle between the rope and the vertical at which it will not break during the swing? (b) If you begin at this maximum angle, and the surface of the pond is 1.2 m below the level of the ground, with what speed will you enter the water if you let go of the rope when the rope is vertical?

40 •• A pendulum bob of mass m is attached to a light string of length L and is also attached to a spring of force constant k. With the pendulum in the position shown in Figure 7-23, the spring is at its unstretched length. If the bob is now pulled aside so that the string makes a *small* angle θ with the vertical and released, what is the speed of the bob as it passes through the equilibrium position? For θ in radians, if $|\theta| \ll 1$, then $\sin \theta \approx \tan \theta \approx \theta$ and $\cos \theta \approx 1$.

FIGURE 7-23 Problem 40

41 ••• A pendulum is suspended from the ceiling and attached to a spring fixed to the floor directly below the pendulum support (Figure 7-24). The mass of the pendulum bob is m, the length of the pendulum is L, and the spring constant is k. The unstretched length of the spring is $L/2$ and the distance between the floor and ceiling is $1.5L$. The pendulum is pulled aside so that it makes an angle θ with the vertical and is then released from rest. Obtain an expression for the speed of the pendulum bob when $\theta = 0$.

FIGURE 7-24 Problem 41

The Conservation of Energy

42 • [SOLVE] In a volcanic eruption, 4 km³ of mountain with a density of 1600 kg/m³ was lifted an average height of 500 m. (a) How much energy in joules was released in this eruption? (b) The energy released by thermonuclear bombs is measured in megatons of TNT, where 1 megaton of TNT = 4.2×10^{15} J. Convert your answer for (a) to megatons of TNT.

43 •• [SOLVE] An 80-kg physics student climbs a 120-m-high hill. (a) What is the increase in the gravitational potential energy of the student? (b) Where does this energy come from? (c) The student's body is 20 percent efficient; that is, for every 20 J that are converted to mechanical energy, 100 J of chemical energy are expended, with 80 J going into thermal energy. How much chemical energy is expended by the student during the climb?

Kinetic Friction

44 • [SOLVE] A 2000-kg car moving at an initial speed of 25 m/s along a horizontal road skids to a stop in 60 m. (a) Find the energy dissipated by friction. (b) Find the coefficient of kinetic friction between the tires and the road.

45 • An 8-kg sled is initially at rest on a horizontal road. The coefficient of kinetic friction between the sled and the road is 0.4. The sled is pulled a distance of 3 m by a force of 40 N applied to the sled at an angle of 30° above the horizontal. (a) Find the work done by the applied force. (b) Find the energy dissipated by friction. (c) Find the change in the kinetic energy of the sled. (d) Find the speed of the sled after it has traveled 3 m.

46 • [SSM] Returning to Problem 19 and Figure 7-17, suppose that the surfaces described are not frictionless and that the coefficient of kinetic friction between the block and the surfaces is 0.30. Find (a) the speed of the block when it reaches the ramp, and (b) the distance that the block slides up the ramp before coming momentarily to rest. (Neglect any energy dissipated along the transition curve.)

47 • The 2-kg block in Figure 7-25 slides down a frictionless curved ramp, starting from rest at a height of 3 m. The block then slides 9 m on a rough horizontal surface before coming to rest. (a) What is the speed of the block at the bottom of the ramp? (b) What is the energy dissipated by friction? (c) What is the coefficient of friction between the block and the horizontal surface?

FIGURE 7-25 Problem 47

48 •• [SOLVE] A 20-kg girl slides down a playground slide that is 3.2 m high. When she reaches the bottom of the slide, her speed is 1.3 m/s. (a) How much energy was dissipated by friction? (b) If the slide is inclined at 20°, what is the coefficient of friction between the girl and the slide?

49 •• In Figure 7-26, the coefficient of kinetic friction between the 4-kg block and the table is 0.35. (a) Find the energy dissipated by friction when the 2-kg block falls a distance y. (b) Find the total mechanical energy E of the two-block–earth system after the 2-kg block falls a distance y, assuming that $E = 0$ initially. (c) Use your result for (b) to find the speed of either block after the 2-kg block falls 2 m.

FIGURE 7-26 Problem 49

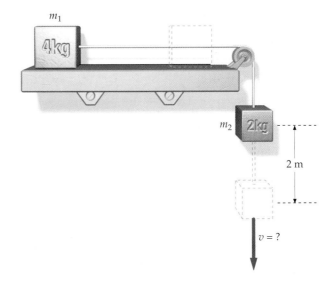

50 •• SSM A compact object of mass m moves in a horizontal circle of radius r on a rough table. It is attached to a horizontal string fixed at the center of the circle. The speed of the object is initially v_0. After completing one full trip around the circle, the speed of the object is $\frac{1}{2}v_0$. (a) Find the energy dissipated by friction during that one revolution in terms of m, v_0, and r. (b) What is the coefficient of kinetic friction? (c) How many more revolutions will the object make before coming to rest?

51 •• ISOLVE ✔ A 2.4-kg box has an initial velocity of 3.8 m/s upward along a plane inclined at 37° to the horizontal. The coefficient of kinetic friction between the box and the plane is 0.30. How far up the incline does the box travel? What is its speed when it passes its starting point on its way down the incline?

52 ••• A block of mass m rests on a plane inclined at θ with the horizontal. The block is attached to a spring of constant k as shown in Figure 7-27. The coefficients of static and kinetic friction between the block and plane are μ_s and μ_k, respectively. Very slowly, the spring is pulled upward along the plane until the block starts to move. (a) Obtain an expression for the extension d of the spring the instant the block moves. (b) Determine the value of μ_k such that the block comes to rest just as the spring is in its unstressed condition, that is, neither extended nor compressed.

FIGURE 7-27 Problem 52

Mass and Energy

53 • (a) Calculate the rest energy in 1 g of dirt. (b) If you could convert this energy into electrical energy and sell it for $0.10/kW·h, how much money would you get? (c) If you could power a 100-W light bulb with this energy, for how long could you keep the bulb lit?

54 • One kiloton of TNT can be detonated, yielding an explosive energy of roughly 5×10^{12} J. What is the mass equivalent of the energy of the explosion?

55 • A muon has a rest energy of 105.7 MeV. Calculate its rest mass in kilograms.

56 • SSM If a black hole and a "normal" star orbit each other, gases from the star falling into the black hole can be heated millions of degrees by frictional heating in the black hole's accretion disk. When the gases are heated that much, they begin to radiate light in the X-ray region of the spectrum. Cygnus X-1, the second brightest source in the X-ray sky, is thought to be one such binary system; it radiates an estimated

power of 4×10^{31} W. If we assume that 1 percent of the infalling mass-energy escapes as X rays, at what rate is the black hole gaining mass?

57 • For the fusion reaction in Example 7-15, calculate the number of reactions per second that are necessary to generate 1 kW of power.

58 • Use Table 7-1 to calculate how much energy is needed to remove one neutron from ^{4}He, leaving ^{3}He plus a neutron.

59 • A free neutron at rest decays into a proton plus an electron:

$$n \rightarrow p + e$$

Use Table 7-1 to calculate the energy released in this reaction.

60 •• In one nuclear fusion reaction, two ^{2}H nuclei combine to produce ^{4}He. (a) How much energy is released in this reaction? (b) How many such reactions must take place per second to produce 1 kW of power?

61 •• A large nuclear power plant produces 3000 MW of electrical power by nuclear fission, which converts matter into energy. (a) How many kilograms of matter does the plant consume in one year? (Assume an efficiency of 33 percent for a nuclear power plant.) (b) In a coal-burning power plant, each kilogram of coal releases 31 MJ of thermal energy when burned. How many kilograms of coal are needed each year for a 3000-MW plant? (Assume an efficiency of 38 percent for a coal-burning power plant.)

General Problems

62 •• SSM A block of mass m, starting from rest, is pulled up a frictionless inclined plane that makes an angle θ with the horizontal by a string parallel to the plane. The tension in the string is T. After traveling a distance L, the speed of the block is v. The work done by the tension T is (a) $mgL \sin \theta$, (b) $mgL \cos \theta + \frac{1}{2}mv^2$, (c) $mgL \sin \theta + \frac{1}{2}mv^2$, (d) $mgL \cos \theta$, (e) $TL \cos \theta$.

63 •• A block of mass m slides with constant velocity v down a plane inclined at θ with the horizontal. During the time interval Δt, the energy dissipated by friction is (a) $mgv \Delta t \tan \theta$, (b) $mgv \Delta t \sin \theta$, (c) $\frac{1}{2}mv^3 \Delta t$, (d) cannot be determined without knowing the coefficient of kinetic friction.

64 • ISOLVE A 3.5-kg box rests on a horizontal frictionless surface in contact with a spring of spring constant 6800 N/m. The spring is fixed at its other end and is initially at its uncompressed length. A constant horizontal force of 70 N is applied to the box so that the spring compresses. Determine the distance the spring is compressed when the box is momentarily at rest.

65 • SSM ISOLVE The average energy per unit time per unit area that reaches the upper atmosphere of the earth from the sun, called the solar constant, is 1.35 kW/m². Because of absorption and reflection by the atmosphere, about 1 kW/m² reaches the surface of the earth on a clear day. How much energy is collected during 8 h of daylight by a window 1 m by 2 m? The window is on a mount that rotates, keeping the window facing the sun so the sun's rays remain perpendicular to the window.

66 •• (a) Using the value of the solar constant given in Problem 65 and the known distance from the earth to the sun, find the total energy that the sun radiates every second. (b) This energy is due to fusion reactions such as the one detailed in Example 7-15. There are several such reactions taking place in the "Phoenix cycle" in the sun, but the overall result is that 4 hydrogen nuclei fuse together to form 1 helium nucleus, liberating 26.7 MeV. If we assume that the sun is primarily made up of hydrogen, and that it will continue burning until roughly 10% of its hydrogen fuel is used up, use the known mass of the sun to determine approximately how long it can keep burning (that is, determine the solar lifetime).

67 • ISOLVE ✓ After the 1250-kg jet-powered car *Spirit of America* went out of control during a test drive at Bonneville Salt Flats, Utah, it left skid marks about 9.5 km long. (a) If the car was moving initially at a speed of 708 km/h, estimate the coefficient of kinetic friction μ_k. (b) What was the kinetic energy K of the car 60 s after the brakes were applied?

68 •• ISOLVE A T-bar tow is required to pull 80 skiers up a 600-m slope inclined at 15° above the horizontal at a speed of 2.5 m/s. The coefficient of kinetic friction is 0.06. Find the motor power required if the mass of the average skier is 75 kg.

69 •• ISOLVE A 2-kg box is projected with an initial speed of 3 m/s up a rough plane inclined at 60° above the horizontal. The coefficient of kinetic friction is 0.3. (a) List all the forces acting on the box. (b) How far up the plane does the box slide before it stops momentarily? (c) What is the energy dissipated by friction as the box slides up the plane? (d) What is the speed of the box when it again reaches its initial position?

70 • SSM ISOLVE A 1200-kg elevator can safely carry a maximum load of 800 kg. What is the power provided by the electric motor powering the elevator when the elevator ascends with a full load at a speed of 2.3 m/s?

71 •• ISOLVE To reduce the power requirement of elevator motors, elevators are counterbalanced with weights connected to the elevator by a cable that runs over a pulley at the top of the elevator shaft. If the elevator in Problem 70 is counterbalanced with a mass of 1500 kg, what is the power provided by the motor when the elevator ascends fully loaded at a speed of 2.3 m/s? How much power is provided by the motor when the elevator ascends without a load at 2.3 m/s?

72 •• ISOLVE ✓ The spring constant of a toy dart gun is 5000 N/m. To cock the gun the spring is compressed 3 cm. The 7-g dart, fired straight upward, reaches a maximum height of 24 m. Determine the energy dissipated by air friction during the dart's ascent. Estimate the speed of the projectile when it returns to its starting point.

73 •• SSM ISOLVE ✓ In a volcanic eruption, a 2-kg piece of porous volcanic rock is thrown vertically upward with an initial speed of 40 m/s. It travels upward a distance of 50 m before it begins to fall back to the earth. (a) What is the initial kinetic energy of the rock? (b) What is the increase in thermal energy due to air friction during ascent? (c) If the increase in thermal energy due to air friction on the way down is 70% of that on the way up, what is the speed of the rock when it returns to its initial position?

74 •• A block of mass m starts from rest at a height h and slides down a frictionless plane inclined at θ with the horizontal as shown in Figure 7-28. The block strikes a spring of force constant k. Find the compression of the spring when the block is momentarily at rest.

FIGURE 7-28 Problem 74

75 • SSM A 1.5×10^4-kg stone slab rests on a steel girder. On a very hot day, you find that the girder has expanded, lifting the slab by 0.1 cm. (a) What work does the girder do on the slab? (b) Where does the energy come from to lift the slab? Give a microscopic picture of what is happening in the steel.

76 •• ISOLVE ✓ A 1500-kg car traveling at 24 m/s is at the foot of a hill that rises 120 m in 2.0 km. At the top of the hill, the speed of the car is 10 m/s. Assuming constant acceleration, find the average power delivered by the car's engine, neglecting any internal frictional losses.

77 •• SSM A block of mass m is suspended from the ceiling by a spring and is free to move vertically in the y direction as indicated (Figure 7-29). We are given that the potential energy as a function of position is $U = \frac{1}{2}ky^2 - mgy$. (a) Using a spreadsheet program or graphing calculator, make a graph of U as a function of y. For what value of y is the spring unstretched? (b) From the expression given for U, find the net force acting on m at any position y. (c) The block is released from rest at $y = 0$; if there is no friction, what is the maximum value of y, y_{max}, that will be reached by the mass? Indicate y_{max} on your graph. (d) Now consider the effect of friction. The mass ultimately settles into an equilibrium position y_{eq}. Find this point on your sketch. (e) Find the amount of thermal energy produced by friction from the start of the operation to the final equilibrium.

FIGURE 7-29 Problem 77

78 •• A spring-loaded gun is cocked by compressing a short, strong spring by a distance d. It fires a signal flare of mass m directly upward. The flare has speed v_0 as it leaves the spring and is observed to rise to a maximum height h above the point where it leaves the spring. After it leaves the spring, effects of drag force by the air on the flare are significant. (Express answers in terms of m, v_0, d, h, and g.) (*a*) How much work is done on the spring in the course of the compression? (*b*) What is the value of the spring constant k? (*c*) How much mechanical energy is converted to thermal energy because of the drag force of the air on the flare between the time of firing and the time at which maximum elevation is reached?

79 •• **ISOLVE** A roller-coaster car having a total mass (including passengers) of 500 kg travels freely along the winding frictionless track shown in Figure 7-30. Points A, E, and G are on horizontal straight sections, all at the same height of 10 m above the ground. Point C is at a height of 10 m above the ground on an inclined section of slope angle 30°. Point B is at the crest of a hill, while point D is at ground level at the bottom of a valley; the radius of curvature at both of these points is 20 m. Point F is at the middle of a banked horizontal curve with a radius of curvature of 30 m, and at the same height of 10 m above the ground as points A, E, and G. At point A the speed of the car is 12 m/s. (*a*) If the car is just barely able to make it over the hill at point B, what is the height of that point above the ground? (*b*) If the car is just barely able to make it over the hill at point B, what is the magnitude of the total force exerted on the car by the track at that point? (*c*) What is the acceleration of the car at point C? (*d*) What are the magnitude and direction of the total force exerted on the car by the track at point D? (*e*) What are the magnitude and direction of the total force exerted on the car by the track at point F? (*f*) At point G a constant braking force is applied to the car, bringing it to a halt in a distance of 25 m. What is the magnitude of the braking force?

FIGURE 7-30 Problem 79

80 • **SSM** **ISOLVE** An elevator (mass M = 2000 kg) is moving downward at v_0 = 1.5 m/s. A braking system prevents the downward speed from increasing. (*a*) At what rate (in J/s) is the braking system converting mechanical energy to thermal energy? (*b*) While the elevator is moving downward at v_0 = 1.5 m/s, the braking system fails and the elevator is in free-fall for a distance d = 5 m before hitting the top of a large safety spring with force constant k = 1.5 × 10⁴ N/m. After the elevator hits the top of the spring, we want to know the distance Δy that the spring is compressed before the elevator is brought to rest. Write an algebraic expression for the value of Δy in terms of the known quantities M, v_0, g, k, and d, and substitute the given values to find Δy.

81 • To measure the force of friction (rolling friction plus air drag) on a moving car, engineers turn off the engine and allow the car to coast down hills of known steepness. The engineers collect the following data:

1. On a 2.87° hill, the car can coast at a steady 20 m/s.
2. On a 5.74° hill, the steady coasting speed is 30 m/s.

The total mass of the car is 1000 kg. (*a*) What is the force of friction at 20 m/s (F_{20}) and at 30 m/s (F_{30})? (*b*) How much useful power must the engine deliver to drive the car on a level road at steady speeds of 20 m/s (P_{20}) and 30 m/s (P_{30})? (*c*) At full throttle, the engine delivers 40 kW. What is the angle of the steepest incline up which the car can maintain a steady 20 m/s? (*d*) Assume that the engine delivers the same total useful work from each liter of gas, no matter what the speed. At 20 m/s on a level road, the car gets 12.7 km/L. How many kilometers per liter does it get if it goes 30 m/s instead?

82 •• A 2-kg block is released 4 m from a massless spring with a force constant k = 100 N/m that is along a plane inclined at 30°, as shown in Figure 7-31. (*a*) If the plane is frictionless, find the maximum compression of the spring. (*b*) If the coefficient of kinetic friction between the plane and the block is 0.2, find the maximum compression. (*c*) For the plane in Part (*b*), how far up the incline will the block travel after leaving the spring?

FIGURE 7-31 Problem 82

83 •• **ISOLVE** A train with a total mass of 2 × 10⁶ kg rises 707 m while traveling a distance of 62 km at an average speed of 15.0 km/h. The frictional force is 0.8 percent of the weight. Find (*a*) the kinetic energy of the train, (*b*) the total change in its potential energy, (*c*) the energy dissipated by friction, and (*d*) the power output of the train's engines.

84 •• **SSM** While driving, one expects to expend more power when accelerating than when driving at a constant speed. (*a*) Neglecting friction, calculate the energy required to give a 1200-kg car a speed of 50 km/h. (*b*) If friction (rolling friction and air drag) results in a retarding force of 300 N at a speed of 50 km/h, what is the energy needed to move the car a distance of 300 m at a constant speed of 50 km/h? (*c*) Assuming the energy losses due to friction in Part (*a*) are 75 percent of those found in Part (*b*), estimate the ratio of the energy consumption for the two cases considered.

85 ••• [SSM] A pendulum consisting of a string of length L with a small bob on the end (mass M) is pulled horizontally, then released (Figure 7-32). At the lowest point of the swing, the string catches on a thin peg a distance R above the lowest point. Show that R must be smaller than $2L/5$ if the bob is to swing around the peg in a full circle.

FIGURE 7-32 Problem 85

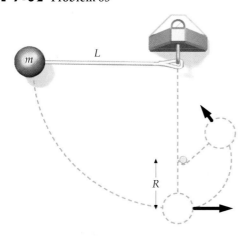

86 •• A rifle bullet is fired into a massive, stationary block of wood to determine its penetrating power. It is found that it penetrates the wood to a distance D. Another bullet that has the same mass, but double the velocity of the first bullet, is fired into an identical block. Assume that the average force acting on the bullet from the wood doesn't depend on the bullet's speed. The penetration depth is now found to be (a) D, (b) $2D$, (c) $4D$, (d) undeterminable from the information given.

87 •• A standard introductory physics lab to examine the conservation of energy and Newton's laws is shown in Figure 7-33. A glider is set up on a linear air track and is attached by a string over a massless pulley to a hanging weight. The mass of the glider is M, while the mass of the hanging weight is m. When the air supply to the air track is turned on, the track is essentially frictionless; you then release the hanging weight and measure the speed of the glider after the weight has fallen a given distance (Y). To show that the laws of physics are consistent, work out the speed of the glider in two different ways: (a) by conservation of mechanical energy; (b) by using Newton's second and third laws directly—make a free-body diagram for the two masses, find their acceleration, and use that to determine the speed of the glider.

FIGURE 7-33 Problem 87

88 •• [SSM] In one model of jogging, the energy expended is assumed to go into accelerating and decelerating the legs. If the mass of the leg is m and the running speed is v, the energy needed to accelerate the leg from rest to v is $\frac{1}{2}mv^2$, and the same energy is needed to decelerate the leg back to rest for the next stride. Thus the energy required for each stride is mv^2. Assume that the mass of a man's leg is 10 kg and that he jogs at a speed of 3 m/s with 1 m between one footfall and the next. Therefore, the energy he must provide to his legs in each second is $3 \times mv^2$. Calculate the rate of the man's energy expenditure using this model and assuming that his muscles have an efficiency of 20 percent.

89 •• A high school chemistry teacher once suggested measuring the magnitude of free-fall acceleration by the following method: Hang a weight on a very fine thread (length L) to make a pendulum, which is then put on the edge of a table so that the weight is a height H above the floor when at its lowest point. Pull the pendulum back so that the thread makes an angle θ_0 with the vertical. At the bottom point of the pendulum, place a razor blade that is positioned to cut through the thread at its lowest point. Once the thread is cut, the weight is projected horizontally, and is thrown a distance D from the edge of the table. The idea was that the measurement of D as a function of θ_0 should somehow determine g. Apart from some obvious experimental difficulties, the experimental had one fatal flaw: D doesn't depend on g! Show that this is true, and that D depends only on the angle θ_0.

90 •• A 5-kg block is held against a spring of force constant 20 N/cm, compressing it 3 cm. The block is released and the spring extends, pushing the block along a horizontal surface. The coefficient of friction between the surface and the block is 0.2. (a) Find the work done on the block by the spring as it extends from its compressed position to its equilibrium position. (b) Find the energy dissipated by friction while the block moves the 3 cm to the equilibrium position of the spring. (c) What is the speed of the block when the spring is at its equilibrium position? (d) If the block is not attached to the spring, how far will it slide along the surface before coming to rest?

91 •• A block of mass m is dropped onto the top of a vertical spring whose force constant is k (Figure 7-21, Problem 26). If the block is released from a height h above the top of the spring, (a) what is the maximum kinetic energy of the block? (b) What is the maximum compression of the spring? (c) At what compression is the block's kinetic energy half its maximum value?

92 ••• The bob of a pendulum of length L is pulled aside so that the string makes an angle θ_0 with the vertical, and the bob is then released. In Example 7-2 the conservation of energy was used to obtain the speed of the bob at the bottom of its swing. In this problem, you are to obtain the same result using Newton's second law. (a) Show that the tangential component of Newton's second law gives $dv/dt = -g\sin\theta$, where v is the speed and θ is the angle between the string and the vertical. (b) Show that v can be written $v = L\,d\theta/dt$ (c) Use this result and the chain rule for derivatives to obtain

$$\frac{dv}{dt} = \frac{dv}{d\theta}\frac{d\theta}{dt} = \frac{dv}{d\theta}\frac{v}{L}$$

(d) Combine the results of (a) and (c) to obtain $v\,dv = -gL\sin\theta\,d\theta$. (e) Integrate the left side of the equation in Part (d) from

$v = 0$ to the final speed v and the right side from $\theta = \theta_0$ to $\theta = 0$, and show that the result is equivalent to $v = \sqrt{2gh}$, where h is the original height of the bob above the bottom.

93 ••• A rock climber is rappelling down the face of a cliff when his hold slips and he slides down over the rock face, supported only by the bungee cord he attached to the top of the cliff. The cliff face is in the form of a smooth quarter-cylinder with height (and radius) $H = 300$ m (see Figure 7-34). Treat the bungee cord as a spring with spring constant $k = 5$ N/m and unstretched length $L = 60$ m. The climber's mass is 85 kg. (a) Using a spreadsheet program, make a graph of the rock climber's potential energy as a function of s, his distance from the top of the cliff *measured along the curved surface*. Use values of s between 60 m and 200 m. (b) If his fall began when he was a distance $s_i = 60$ m from the top of the cliff, and ends when he is a distance $s_f = 110$ m from the top, determine how much energy is dissipated by friction between the time he initially slipped and finally came to a stop.

FIGURE 7-34 Problem 93

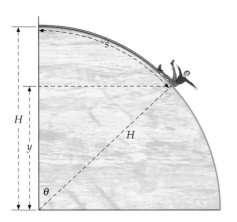

94 ••• [SSM] A block of wood (mass M) is connected to two massless springs as shown in Figure 7-35. Each spring has unstretched length L and spring constant k. (a) If the block is displaced a distance x, as shown, what is the change in the potential energy stored in the springs? (b) What is the magnitude of the force pulling the block back toward the equilibrium position? (c) Using a spreadsheet program or graphing calculator, make a graph of the potential energy U as a function of x for $0 \le x \le 0.2$ m. Assume $k = 1$ N/m, $L = 0.1$ m, and $M = 1$ kg. (d) If the block is displaced a distance $x = 0.1$ m and released, what is its speed as it passes the equilibrium point? Assume that the block is resting on a frictionless surface.

FIGURE 7-35 Problem 94

Systems of Particles and Conservation of Linear Momentum

THE MASS OF THIS GOLF CLUB TIMES ITS VELOCITY IS CALLED THE MOMENTUM OF THE CLUB. WHEN THE GOLFER STRIKES THE BALL WITH THE CLUB, SOME OF THE MOMENTUM OF THE CLUB IS TRANSFERRED TO THE BALL WITH DRAMATIC EFFECT.

? **If the golfer uses a more massive club, how will the distance traveled by the ball be affected? (See Example 8-12.)**

We saw in Chapters 6 and 7 that the work–kinetic energy theorem for particles and the work–energy theorem for systems of particles are useful tools for analyzing a variety of situations and solving a wide variety of problems. But in order to analyze the motion of extended objects, and to understand collisions between objects like the golf club and ball in the photo above, we need to add two more theorems to our problem-solving toolbox. These are **Newton's second law for systems** (Section 8-3) and **the impulse–momentum theorem** (Section 8-6). The impulse–momentum theorem for systems introduces the conservation of momentum, one of the universal laws of physics.

The mass of a particle times its velocity is called the **momentum** of the particle. The momentum of a system of particles is the vector sum of the momenta of the individual particles in the system. When the net external force acting on a system remains zero, the system's momentum remains constant. The momentum of an isolated system is a conserved quantity, just like the energy of an isolated system.

➤ In this chapter, we will use conservation of momentum to analyze collisions between billiard balls, cars, and subatomic particles, and the decay of radioactive nuclei. We will apply Newton's laws to the motion of extended objects, such as cars, rockets, and people, by realizing that there is one point of a system, the *center of mass*, that moves as if all the mass of the system were concentrated at that point and all the external forces acting on the system were acting exclusively on that point.

8-1 The Center of Mass

The motion of any object or system of particles can be described in terms of the motion of the center of mass (which may be thought of as the bulk motion of the system) plus the motion of individual particles in the system relative to the center of mass. Let's first consider a simple system of two particles in one dimension. If two point particles with masses m_1 and m_2 have coordinates x_1 and x_2 on the x axis, then the center-of-mass coordinate x_{cm} is defined by

$$Mx_{cm} = m_1x_1 + m_2x_2 \qquad \text{8-1}$$

where $M = m_1 + m_2$ is the total mass of the system. In the case of just two particles, the center of mass lies at some point on the line between the particles; if the particles have equal masses, then the center of mass is midway between them (Figure 8-1).

If the two particles are of unequal mass, then the center of mass is closer to the more massive particle (Figure 8-2).

If we choose the origin and direction of the x axis such that the position of m_1 is at the origin and that of m_2 is on the positive x axis, then $x_1 = 0$ and $x_2 = d$, where d is the distance between the particles (Figure 8-3) and the center of mass is given by

$$Mx_{cm} = m_1x_1 + m_2x_2 = m_1(0) + m_2d$$

$$x_{cm} = \frac{m_2}{M}d = \frac{m_2}{m_1 + m_2}d \qquad \text{8-2}$$

EXERCISE A 4-kg mass is at the origin and a 2-kg mass is at $x = 6$ cm. Find x_{cm}. (*Answer* $x_{cm} = 2$ cm)

We can generalize from two particles in one dimension to a system of many particles in three dimensions. For N particles,

$$Mx_{cm} = m_1x_1 + m_2x_2 + m_3x_3 + \cdots + m_Nx_N = \sum_i m_ix_i \qquad \text{8-3}$$

where again $M = \sum_i m_i$ is the total mass of the system. Similarly,

$$My_{cm} = \sum_i m_iy_i \quad \text{and} \quad Mz_{cm} = \sum_i m_iz_i \qquad \text{8-4}$$

In vector notation, $\vec{r}_i = x_i\hat{i} + y_i\hat{j} + z_i\hat{k}$ is the position vector of the ith particle. The position vector of the **center of mass**, $\vec{r}_{cm}$, is defined by

$$M\vec{r}_{cm} = m_1\vec{r}_1 + m_2\vec{r}_2 + \cdots = \sum_i m_i\vec{r}_i \qquad \text{8-5}$$

DEFINITION—CENTER OF MASS

where $\vec{r}_{cm} = x_{cm}\hat{i} + y_{cm}\hat{j} + z_{cm}\hat{k}$.

FIGURE 8-1

FIGURE 8-2

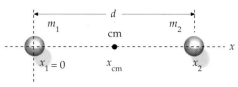

FIGURE 8-3

To find the center of mass of a continuous object we replace the sum in Equation (8-5) with an integral:

$$M\vec{r}_{cm} = \int \vec{r}\, dm \qquad\qquad 8\text{-}6$$

CENTER OF MASS, CONTINUOUS OBJECT

where dm is an element of mass located at position $\vec{r}$, as shown in Figure 8-4. Examples involving Equation 8-6 are presented in Section 8-2. For highly symmetric objects, the center of mass is at the center of symmetry. For example, the center of mass of a uniform cylinder is located at its geometric center. Consider the following examples.

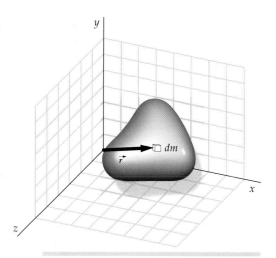

FIGURE 8-4 Mass element dm located at $\vec{r}$ for finding the center of mass by integration.

THE CENTER OF MASS OF A WATER MOLECULE **EXAMPLE 8-1**

FIGURE 8-5

A water molecule consists of an oxygen atom and two hydrogen atoms. Oxygen has a mass of 16 unified mass units (u) and each hydrogen has a mass of 1 u. The hydrogen atoms are each at an average distance of 96 pm (96×10^{-12} m) from the oxygen atom, and are separated from one another by an angle of 104.5°. Find the center of mass of a water molecule.

PICTURE THE PROBLEM The calculation is simplified if we place the origin at the location of the oxygen atom, with the x axis bisecting the angle between the hydrogen atoms (Figure 8-5). Then, given the symmetries of the molecule, the center of mass will be on the x axis, and the line from the oxygen atom to each hydrogen atom will make an angle of 52.2°.

1. The location of the center of mass is given by its coordinates, x_{cm} and y_{cm}:

$$x_{cm} = \frac{\Sigma\, m_i x_i}{M}, \qquad y_{cm} = \frac{\Sigma\, m_i y_i}{M}$$

2. Writing these out explicitly gives:

$$x_{cm} = \frac{m_{H1} x_{H1} + m_{H2} x_{H2} + m_O x_O}{m_{H1} + m_{H2} + m_O}$$

$$y_{cm} = \frac{m_{H1} y_{H1} + m_{H2} y_{H2} + m_O y_O}{m_{H1} + m_{H2} + m_O}$$

3. We have chosen the origin to be the location of the oxygen atom, so both the x and y coordinates of the oxygen atom are zero. The x and y coordinates of the hydrogen atoms are calculated from the 52.2° angle each hydrogen makes with the x axis:

$$x_O = y_O = 0$$

$$x_{H1} = x_{H2} = 96 \text{ pm} \cos 52.2° = 59 \text{ pm}$$

$$y_{H1} = 96 \text{ pm} \sin 52.2° = 76 \text{ pm}$$

$$y_{H2} = -96 \text{ pm} \sin 52.2° = -76 \text{ pm}$$

4. Substituting the coordinate and mass values into step 2 gives x_{cm}:

$$x_{cm} = \frac{(1\text{ u})(59\text{ pm}) + (1\text{ u})(59\text{ pm}) + (16\text{ u})0}{1\text{ u} + 1\text{ u} + 16\text{ u}}$$

$$= 6.6 \text{ pm}$$

$$y_{cm} = \frac{(1\text{ u})(76\text{ pm}) + (1\text{ u})(-76\text{ pm}) + (16\text{ u})0}{1\text{ u} + 1\text{ u} + 16\text{ u}}$$

$$= 0$$

5. The center of mass is on the x axis:

$$\vec{r}_{cm} = x_{cm}\hat{i} + y_{cm}\hat{j}$$
$$= 6.6 \text{ pm } \hat{i} + 0\hat{j} = \boxed{6.6 \text{ pm } \hat{i}}$$

REMARKS That $y_{cm} = 0$ can be seen from the symmetry of the mass distribution. Also, the center of mass is very close to the relatively massive oxygen atom, as expected.

We also can solve Example 8-1 by first finding the center of mass of just the two hydrogen atoms. For a system of three particles Equation 8-5 is

$$M\vec{r}_{cm} = m_1\vec{r}_1 + m_2\vec{r}_2 + m_3\vec{r}_3$$

The first two terms on the right side of this equation are related to the center of mass of the first two particles $\vec{r}'_{cm}$:

$$m_1\vec{r}_1 + m_2\vec{r}_2 = (m_1 + m_2)\vec{r}'_{cm}$$

The center of mass of the three-particle system can then be written

$$M\vec{r}_{cm} = (m_1 + m_2)\vec{r}'_{cm} + m_3\vec{r}_3$$

So we can first find the center of mass for two of the particles, the hydrogen atoms for example, and then replace them with a single particle of total mass $m_1 + m_2$ at that center of mass (Figure 8-6).

The same technique enables us to calculate centers of mass for more complex systems, for instance, two uniform rods (Figure 8-7). The center of mass of each rod separately is at the center of the rod. The center of mass of the two-rod system can be found by treating each rod as a point particle at its individual center of mass.

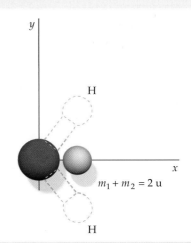

FIGURE 8-6 Example 8-1 with the two H atoms replaced by a single particle of mass $m_1 + m_2 = 2$ u on the x axis at the center of mass of the original atoms. The center of mass then falls between the oxygen atom at the origin and the calculated center of mass of the two hydrogen atoms.

FIGURE 8-7

THE CENTER OF MASS OF A PLYWOOD SHEET

EXAMPLE 8-2

Find the center of mass of the uniform sheet of plywood shown in Figure 8-8*a*.

FIGURE 8-8*a*

Try It Yourself

PICTURE THE PROBLEM The sheet can be divided into two symmetrical parts (Figure 8-8*b*). The center of mass of each part is at its geometric center. Let m_1 be the mass of part 1 and m_2 be the mass of part 2. The total mass is $M = m_1 + m_2$. The masses are proportional to the areas.

FIGURE 8-8*b*

Cover the column to the right and try these on your own before looking at the answers.

Steps

Answers

1. Write the x and y coordinates of the center of mass in terms of m_1 and m_2.

$$Mx_{cm} = m_1 x_{1,cm} + m_2 x_{2,cm}$$

$$My_{cm} = m_1 y_{1,cm} + m_2 y_{2,cm}$$

2. Divide through these equation by M and then substitute area ratios for the mass ratios.

$$x_{cm} = \frac{A_1}{A} x_{1,cm} + \frac{A_2}{A} x_{2,cm}$$

$$y_{cm} = \frac{A_1}{A} y_{1,cm} + \frac{A_2}{A} y_{2,cm}$$

3. Calculate the areas and the ratios of the areas.

$$A_1 = 0.32 \text{ m}^2, \qquad A_2 = 0.04 \text{ m}^2, \qquad A = 0.36 \text{ m}^2$$

$$\frac{A_1}{A} = \frac{8}{9} \qquad \frac{A_2}{A} = \frac{1}{9}$$

4. Write the x and y coordinates of the center-of-mass coordinates for each part by inspection of the figure.

$$x_{1,cm} = 0.4 \text{ m}, \qquad y_{1,cm} = 0.2 \text{ m}$$

$$x_{2,cm} = 0.7 \text{ m}, \qquad y_{2,cm} = 0.5 \text{ m}$$

5. Substitute these results to calculate x_{cm} and y_{cm}.

$$x_{cm} = \boxed{0.433 \text{ m}}, \qquad y_{cm} = \boxed{0.233 \text{ m}}$$

REMARKS 1. Note that the center of mass is very near the center of mass of part 1 because $m_1 = 8m_2$. 2. Placing the origin at the geometric center of part 1 of the sheet and drawing the x axis through the center of part 2 would make the calculation of the center of mass somewhat simpler.

Gravitational Potential Energy of a System

The gravitational potential energy of a system of particles in a uniform gravitational field is the same as if all the mass were concentrated at the center of mass. Let h_i be the height of the ith particle in a system above some reference level. The gravitational potential energy of the system is

$$U = \sum_i m_i g h_i = g \sum_i m_i h_i$$

But, by definition of the center of mass, the height of the center of mass is given by

$$Mh_{cm} = \sum_i m_i h_i$$

so

$$U = Mgh_{cm} \qquad\qquad 8\text{-}7$$

We can use this result to locate the center of mass of an object experimentally. For example, two objects connected by a light rod will balance if the pivot is at the center of mass (Figure 8-9). If we pivot the system at any other point, the system will rotate until the potential energy is at a minimum, which occurs when the center of mass is at its lowest possible point directly below the pivot (Figure 8-10).

FIGURE 8-9

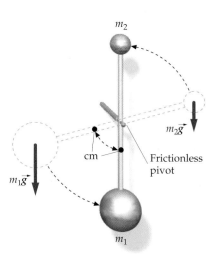

FIGURE 8-10

If we suspend any irregular object from a pivot, the object will hang so that its center of mass lies somewhere on the vertical line drawn directly downward from the pivot. Now suspend the object from another point and note where the vertical line now passes across the object. The center of mass will lie at the intersection of the two lines (Figure 8-11).

*8-2 Finding the Center of Mass by Integration

In this section we find the center of mass by integration (Equation 8-6):

$$M\vec{r}_{cm} = \int \vec{r}\, dm$$

We will use the simple problem of finding the center of mass of a uniform thin rod to illustrate the technique for setting up the integration. While we can find the answer to this problem by symmetry, here we will use integration.

Uniform Rod

We first choose a coordinate system. A good choice for a coordinate system is one with an x axis through the length of the rod, with the origin at one end of the rod (Figure 8-12). Shown on the figure is a mass element dm of length dx a distance x from the origin. Equation 8-6 thus gives

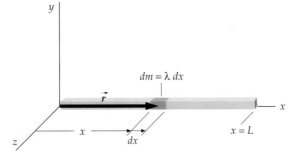

$dm = \lambda\, dx$

$x = L$

FIGURE 8-12

$$M\vec{r}_{cm} = \int \vec{r}\, dm = \int x\hat{i}\, dm$$

The mass is distributed on the x axis along the interval $0 \le x \le L$. To sweep dm along the mass distribution means the limits of the integral are 0 and L. (We sweep in the direction of increasing x.) The ratio dm/dx is the mass per unit length λ, so $dm = \lambda\, dx$:

$$M\vec{r}_{cm} = \hat{i} \int x\, dm = \hat{i} \int_0^L x\lambda\, dx$$

Because the rod is uniform, λ is constant and equal to M/L. Substituting for λ, we complete the calculation and obtain the expected result

$$\vec{r}_{cm} = \frac{1}{M}\hat{i}\,\lambda\,\frac{x^2}{2}\bigg|_0^L = \frac{1}{M}\hat{i}\,\frac{M}{L}\frac{L^2}{2} = \tfrac{1}{2}L\hat{i}$$

Semicircular Hoop

In calculating the center of mass of a uniform semicircular hoop, a good choice of coordinate axes is one with the origin at the center and with the x axis bisecting the semicircle (Figure 8-13). To find the center of mass we use Equation 8-6 $\left(M\vec{r}_{cm} = \int \vec{r}\, dm\right)$, where $\vec{r} = x\hat{i} + y\hat{j}$. The semicircular mass distribution suggests using polar coordinates, for which $x = r \cos\theta$ and $y = r \sin\theta$. With these substitutions we have

FIGURE 8-11 The center of mass of an irregular object can be found by suspending it from two points.

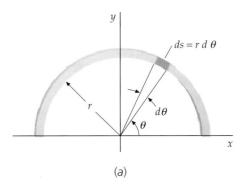

$ds = r\, d\theta$

(a)

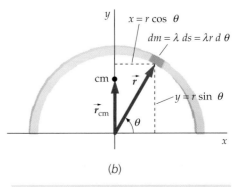

$x = r \cos\theta$

$dm = \lambda\, ds = \lambda r\, d\theta$

$y = r \sin\theta$

(b)

FIGURE 8-13 Geometry for calculating the center of mass of a semicircular hoop by integration.

$$M\vec{r}_{cm} = \int (x\hat{i} + y\hat{j})\,dm = \int r(\cos\theta\,\hat{i} + \sin\theta\,\hat{j})\,dm$$

Next we express dm in terms of $d\theta$. First, the mass element dm has length $ds = r\,d\theta$, so

$$dm = \lambda\,ds = \lambda r\,d\theta$$

where $\lambda = dm/ds$ is the mass per unit length. Thus, we have

$$M\vec{r}_{cm} = \int r(\cos\theta\,\hat{i} + \sin\theta\,\hat{j})\,\lambda r\,d\theta$$

Evaluating this integral involves sweeping dm along the semicircular mass distribution. This means $0 \le \theta \le \pi$. Sweeping in the direction of increasing θ, the integration limits go from 0 to π. That is,

$$\vec{r}_{cm} = \frac{r^2}{M}\int_0^\pi (\cos\theta\,\hat{i} + \sin\theta\,\hat{j})\lambda\,d\theta$$

Because the hoop is uniform, we know that $\lambda = M/\pi r$, where πr is the length of the semicircle. Substituting for λ and rearranging gives

$$\vec{r}_{cm} = \frac{r}{\pi}\hat{i}\int_0^\pi \cos\theta\,d\theta + \frac{r}{\pi}\hat{j}\int_0^\pi \sin\theta\,d\theta = \frac{r}{\pi}\hat{i}\sin\theta\,\Big|_0^\pi - \frac{r}{\pi}\hat{j}\cos\theta\,\Big|_0^\pi = \frac{2r}{\pi}\hat{j}$$

The center of mass is on the y axis a distance of $\dfrac{2}{\pi}r$ from the origin. Interestingly, it is outside of the material of the object.

8-3   Motion of the Center of Mass

Figure 8-14 is a multiflash photograph of a baton thrown into the air. Although the motion of the baton is complicated, the motion of the center of mass is simple. While the baton is in the air, the center of mass follows a parabolic path, the same path that would be followed by a point particle. We will show in general that the acceleration of the center of mass of a system of particles equals the net external force acting on the system divided by the total mass of the system. For the baton thrown into the air, the acceleration of the center of mass is g downward; to find it, we first find the velocity by differentiating Equation 8-5 with respect to time:

$$M\frac{d\vec{r}_{cm}}{dt} = m_1\frac{d\vec{r}_1}{dt} + m_2\frac{d\vec{r}_2}{dt} + \cdots = \sum_i m_i\frac{d\vec{r}_i}{dt}$$

FIGURE 8-14

Because the time derivative of the position is the velocity, this gives

$$M\vec{v}_{cm} = m_1\vec{v}_1 + m_2\vec{v}_2 + \cdots = \sum_i m_i\vec{v}_i \qquad\qquad 8\text{-}8$$

Differentiating again, we obtain the accelerations:

$$M\vec{a}_{cm} = m_1\vec{a}_1 + m_2\vec{a}_2 + \cdots = \sum_i m_i\vec{a}_i \qquad\qquad 8\text{-}9$$

However, in accord with Newton's second law, $m_i \vec{a}_i$ equals the sum of the forces acting on the *i*th particle, so

$$\sum_i m_i \vec{a}_i = \sum_i \vec{F}_i$$

where the term on the right is the sum of all the forces acting on each and every particle in the system. Some of these forces are *internal* forces (exerted on a particle in the system by some other particle in the system) and others are *external* forces (exerted on a particle in the system by a particle not in the system). Thus,

$$M\vec{a}_{cm} = \sum_i \vec{F}_{i,int} + \sum_i \vec{F}_{i,ext} \qquad \text{8-10}$$

According to Newton's third law, forces come in action–reaction pairs. Thus, for each internal force acting on a particle in the system there is an equal and opposite internal force acting on some other particle in the system. When we sum all the internal forces, each action–reaction force pair sums to zero, so $\sum \vec{F}_{i,int} = 0$. Equation 8-10 then becomes

$$\vec{F}_{net,ext} = \sum_i \vec{F}_{i,ext} = M\vec{a}_{cm} \qquad \text{8-11}$$

NEWTON'S SECOND LAW FOR A SYSTEM

That is, the net external force acting on the system equals the product of the total mass M of the system and the acceleration of the center of mass $\vec{a}_{cm}$ of the system. Thus,

> The center of mass of a system moves like a particle of mass $M = \sum m_i$ under the influence of the net external force acting on the system.

This theorem is important because it describes the motion of the center of mass for *any* system of particles: The center of mass moves exactly like a single point particle of mass M acted on by only the external forces. The individual motion of any particle in the system is typically much more complex and is not described by Equation 8-11. The baton thrown into the air in Figure 8-14 is an example. The only external force acting is gravity, so the center of mass of the baton moves in a simple parabolic path, as would a point particle. (Equation 8-11 does not describe the rotational motion of one end of the baton about the center of mass.)

AN EXPLODING PROJECTILE   **E X A M P L E   8 - 3**

A projectile is fired into the air over level ground on a trajectory that would result in it landing 55 m away. However, at its highest point it explodes into two fragments of equal mass. Immediately following the explosion one fragment has a momentary speed of zero and then falls straight down to the ground. Where does the other fragment land? Neglect air resistance.

FIGURE 8-15

PICTURE THE PROBLEM Let the projectile be the system. Then the forces of the explosion are all internal forces. Since the only *external* force acting on the system is that due to gravity, the center of mass, which is midway between the two fragments, continues on its parabolic path as if there had been no explosion (Figure 8-15).

1. Let $x = 0$ be the initial position of the projectile. The landing positions x_1 and x_2 of the fragments are related to the final position of the center of mass by:

$$(2m)x_{cm} = mx_1 + mx_2$$

or

$$2x_{cm} = x_1 + x_2$$

2. At impact, $x_{cm} = R$ and $x_1 = 0.5R$, where $R = 55$ m is the range for the unexploded projectile. Solve for x_2:

$$x_2 = 2x_{cm} - x_1 = 2R - 0.5R = 1.5R$$

$$= 1.5(55 \text{ m}) = \boxed{82.5 \text{ m}}$$

REMARKS In Figure 8-16 height versus distance is plotted for exploding projectiles when fragment 1 has a horizontal velocity of half of the initial horizontal velocity. As in the original example in which fragment 1 falls straight down, the center of mass follows a normal parabolic trajectory.

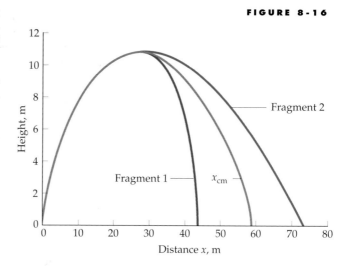

FIGURE 8-16

EXERCISE If the fragment that falls straight down has twice the mass of the other fragment, how far from the launch position does the lighter fragment land? (*Answer* 2R)

REMARKS If both fragments have the same vertical component of velocity after the explosion, they land at the same time. If just after the explosion the vertical component of the velocity of one fragment is less than that of the other, the fragment with the smaller vertical velocity component will hit the ground first. As soon as it does, the ground exerts a force on it and the net external force on the system is no longer just the gravitational force. From that moment on, our analysis is invalid.

EXERCISE A cylinder rests on a sheet of paper on a table (Figure 8-17). You pull the paper to the right, causing the cylinder to roll leftward *relative to the paper*. How does the cylinder's center of mass move relative to the table? (*Answer* It accelerates to the right, because the net external force acting on the cylinder is the frictional force to the right exerted on it by the paper. Try it. The cylinder may *appear* to move to the left, because relative to the paper it rolls leftward. However, relative to the table, which serves as an inertial reference frame, it moves to the right. If you mark the table with the original position of the cylinder, you will observe the center of mass move to the right *while the cylinder remains in contact with the paper*.)

FIGURE 8-17

A special case of a system's center of mass in motion is the system with zero net external force acting on it. Here, $\vec{a}_{cm} = 0$, so the center of mass either remains at rest or moves with constant velocity. The internal forces and motion may be complex, but the behavior of the center of mass is simple. Further, if the net external force is not zero, but if a component of it in a given direction, say the x direction, remains zero, then $a_{cmx} = 0$ and v_{cmx} remains constant. An example of this is a projectile in the absence of air drag. The net external force on the projectile is the gravitational force. This force acts straight downward, so its component in any horizontal direction remains zero. It follows that the horizontal component of the velocity of the center of mass remains constant.

Changing Places in a Rowboat   **E X A M P L E   8 - 4**

Pete (mass 80 kg) and Dave (mass 120 kg) are in a rowboat (mass 60 kg) on a calm lake. Dave is near the bow of the boat, rowing, and Pete is at the stern, 2 m from the center. Dave gets tired and stops rowing. Pete offers to row, and after the boat comes to rest, they change places. How far does the boat move as Pete and Dave change places? (Neglect any horizontal force exerted by the water.)

PICTURE THE PROBLEM Let the system be Dave, Pete, and the boat. There are no *external* forces in the horizontal direction, so the center of mass does not move horizontally relative to an inertial reference frame, such as the water. Relative to the boat, however, the center of mass does move. First find the displacement Δx_{cm} that the center of mass moves relative to the boat. Relative to the water, the boat must move the same distance Δx_{cm} in the opposite direction.

1. Make a sketch of the system in its initial and final configurations (Figure 8-18). Choose the origin at the back of the boat (Pete's initial position) and let $d = 2$ m so $x_{\text{Pete i}} = 0$ and $x_{\text{Dave i}} = d$. Let x_{boat} be the position of the center of mass of the boat. When Pete and Dave switch places they merely exchange positions.

FIGURE 8-18 (a) Pete and Dave changing places viewed from the reference frame of the water. (b) Pete and Dave changing places viewed from the reference frame of the boat.

2. Express the total mass M times the initial value of the center of mass. Repeat this for the final value of the center of mass.

$$Mx_{cm\,i} = m_{\text{Pete}}x_{\text{Pete i}} + m_{\text{Dave}}x_{\text{Dave i}} + m_{\text{boat}}x_{\text{boat i}}$$
$$= 0 + m_{\text{Dave}}d + m_{\text{boat}}x_{\text{boat}}$$

$$Mx_{cm\,f} = m_{\text{Pete}}x_{\text{Pete f}} + m_{\text{Dave}}x_{\text{Dave f}} + m_{\text{boat}}x_{\text{boat f}}$$
$$= m_{\text{Pete}}d + 0 + m_{\text{boat}}x_{\text{boat}}$$

3. Find the displacement of the center of mass relative to the boat by subtracting the first step 2 result from the second:

$$Mx_{cm\,f} - Mx_{cm\,i} = (m_{\text{Pete}} - m_{\text{Dave}})d + 0$$

so

$$x_{cm\,f} - x_{cm\,i} = \frac{(m_{\text{Pete}} - m_{\text{Dave}})d}{M}$$

4. Because the center of mass does not move relative to the water, the boat must move an equal distance in the opposite direction:

$$-(x_{cm\,f} - x_{cm\,i}) = \frac{(m_{\text{Dave}} - m_{\text{Pete}})d}{M}$$

$$= \frac{(120 \text{ kg} - 80 \text{ kg})}{120 \text{ kg} + 80 \text{ kg} + 60 \text{ kg}}(2 \text{ m})$$

$$= \boxed{0.308 \text{ m}}$$

REMARKS The location of the center of mass of the boat, relative to the boat, subtracted out in step 3. The displacement of the boat is independent of this location.

FIGURE 8-19

A SLIDING BLOCK   **E X A M P L E   8 - 5**

A wedge of mass m_2 sits at rest on a scale as shown in Figure 8-19. A small block of mass m_1 slides down the frictionless incline of the wedge. Find the scale reading while the block slides.

PICTURE THE PROBLEM We choose the wedge plus block to be the system. Because the block accelerates down the wedge, the center of mass has acceleration components to the right and downward. The external forces on the system are the weights of the block and wedge, the force F_x exerted by the scale on the wedge, and the normal force F_n exerted upward by the scale. The scale reading is equal to the magnitude of F_n.

1. Draw a free-body diagram for the wedge–block system (Figure 8-20):

2. Write the vertical component of Newton's second law for the system and solve for F_n:

$$F_n - m_1 g - m_2 g = Ma_{cm,y} = (m_1 + m_2)a_{cm,y}$$
$$F_n = (m_1 + m_2)g + (m_1 + m_2)a_{cm,y}$$

FIGURE 8-20

3. Using Equation 8-9, express $a_{cm,y}$ in terms of the acceleration of the block a_{1y}:

$$Ma_{cm,y} = m_1 a_{1y} + m_2 a_{2y} = m_1 a_{1y} + 0$$

$$a_{cm,y} = \frac{m_1}{m_1 + m_2} a_{1y}$$

4. From Example 4-7, a block sliding down a stationary incline has acceleration $g \sin \theta$ down the incline. Use trigonometry to find the y component of this acceleration and use it to find $a_{cm,y}$:

$$a_{1y} = -a_1 \sin \theta = -g \sin^2 \theta$$

so

$$a_{cm,y} = \frac{m_1}{m_1 + m_2} a_{1y}$$

$$= -\frac{m_1}{m_1 + m_2} g \sin^2 \theta$$

5. Substitute for $a_{cm,y}$ in the step 2 result and solve for F_n:

$$F_n = (m_1 + m_2)g + (m_1 + m_2)a_{cm,y}$$

$$= (m_1 + m_2)g - m_1 g \sin^2 \theta = [(1 - \sin^2 \theta)m_1 + m_2]g$$

$$= \boxed{(m_1 \cos^2 \theta + m_2)g}$$

EXERCISE What values of F_n do you intuitively expect for $\theta = 0$ and for $\theta = 90°$? Show that $(m_1 \cos^2 \theta + m_2)g$ gives these expected values. (*Answer* For $\theta = 0$ we expect $F_n = (m_1 + m_2)g$; for $\theta = 90°$ we expect $F_n = m_2 g$.)

EXERCISE Find the force component F_x exerted on the wedge by the scale. (*Answer* $F_x = m_1 g \sin \theta \cos \theta$.)

8-4 Conservation of Linear Momentum

A particle's **linear momentum** $\vec{p}$ is defined as the product of its mass and velocity:

$$\vec{p} = m\vec{v}$$

8-12

DEFINITION—MOMENTUM OF A PARTICLE

Linear momentum is a vector quantity that may be thought of as a measurement of the effort needed to bring a particle to rest. For example, a heavy truck has more momentum than a light car traveling at the same speed. It takes a greater force to stop the truck in a given time than it does to stop the car in the same time. (The quantity $m\vec{v}$ is sometimes referred to as the *linear momentum* of a particle to distinguish it from the *angular momentum*, which is discussed in Chapter 10. When there is no need to clarify that the momentum being referred to is the linear momentum, the adjective *linear* will be suppressed and we will just refer to the *momentum*. Throughout the remainder of this chapter momentum is used when referring to linear momentum.)

Newton's second law can be written in terms of the momentum of a particle. Differentiating Equation 8-12 we obtain

$$\frac{d\vec{p}}{dt} = \frac{d(m\vec{v})}{dt} = m\frac{d\vec{v}}{dt} = m\vec{a}$$

Then substituting the force $\vec{F}_{net}$ for $m\vec{a}$,

$$\vec{F}_{net} = \frac{d\vec{p}}{dt} \qquad\qquad 8\text{-}13$$

Thus, the net force acting on a particle equals the time rate of change of the particle's linear momentum. (Newton's original statement of his second law was in fact in this form.)

The total momentum $\vec{P}_{sys}$ of a system of particles is the sum of the momenta of the individual particles:

$$\vec{P}_{sys} = \sum_i m_i\vec{v}_i = \sum_i \vec{p}_i$$

According to Equation 8-8, $\Sigma m_i\vec{v}_i$ equals the total mass M times the velocity of the center of mass:

$$\vec{P}_{sys} = \sum_i m_i\vec{v}_i = M\vec{v}_{cm} \qquad\qquad 8\text{-}14$$

TOTAL MOMENTUM OF A SYSTEM

Differentiating this equation, we obtain

$$\frac{d\vec{P}_{sys}}{dt} = M\frac{d\vec{v}_{cm}}{dt} = M\vec{a}_{cm}$$

But according to Newton's second law (Equation 8-11), $M\vec{a}_{cm}$ equals the net external force acting on the system. Thus,

$$\sum_i \vec{F}_{ext} = \vec{F}_{net,ext} = \frac{d\vec{P}_{sys}}{dt} \qquad\qquad 8\text{-}15$$

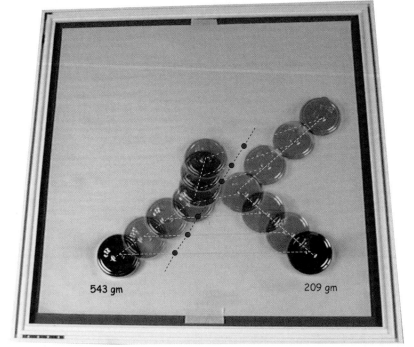

543 gm   209 gm

When the net external force acting on a system of particles remains zero, the rate of change of the total momentum remains zero and the total momentum of the system remains constant:

$$\vec{P}_{sys} = \sum_i m_i\vec{v}_i = M\vec{v}_{cm} = \text{constant} \qquad (\vec{F}_{net,ext} = 0) \qquad 8\text{-}16$$

CONSERVATION OF MOMENTUM

The two pucks are moving on an air cushion on a horizontal flat surface. (The hoses supplying the air are not shown.) The velocity of each puck changes in both magnitude and direction during the collision, but the velocity of the center of mass remains constant—unaffected by the internal forces of the collision.

This result is known as the **law of conservation of momentum:**

If the net external force on a system remains zero, the total momentum of the system remains constant.

This law is one of the most important in physics. It is more widely applicable than the law of conservation of mechanical energy because internal forces exerted by one particle in a system on another are often not conservative. Thus, these internal forces can change the total mechanical energy of the system, though they have no effect on the system's total momentum. If the total momentum of a system is constant, then the velocity of the center of mass of the system is constant. The law of conservation of momentum is a vector relation so it is valid component by component. For example, if the x component of the net external force on a system remains zero, then the x component of the total momentum of the system remains constant.

A SPACE REPAIR   **EXAMPLE 8-6**

During repair of the Hubble Space Telescope, an astronaut replaces a solar panel whose frame is bent. Pushing the detached panel away into space, she is propelled in the opposite direction. The astronaut's mass is 60 kg and the panel's mass is 80 kg. The astronaut is at rest relative to her spaceship when she shoves away the panel, and she shoves it at 0.3 m/s relative to the spaceship. What is her subsequent velocity relative to the space ship? (During this operation the astronaut is tethered to the ship; for our calculation, assume that the tether remains slack.)

PICTURE THE PROBLEM The velocity of the astronaut can be found from the velocity of the panel using conservation of momentum. Choose the direction of motion of the panel to be positive.

1. Apply conservation of momentum to find the velocity of the astronaut. Since the total momentum is initially zero, it remains zero:

$$p_p + p_a = m_p v_p + m_a v_a = 0$$

2. Solve for the astronaut's velocity:

$$v_a = -\frac{m_p}{m_a} v_p$$

$$= -\frac{80 \text{ kg}}{60 \text{ kg}}(0.3 \text{ m/s}) = \boxed{-0.4 \text{ m/s}}$$

REMARKS Although momentum is conserved, the mechanical energy of this system increased because chemical energy of the astronaut was converted to kinetic energy.

EXERCISE Find the final kinetic energy of the astronaut–panel system. (*Answer* 8.4 J)

A RUNAWAY RAILROAD CAR   **EXAMPLE 8-7**

A runaway 14,000-kg railroad car is rolling horizontally at 4 m/s toward a switchyard. As it passes by a grain elevator, 2000 kg of grain are suddenly dropped into the car. How long does it take the car to cover the 500-m distance from the elevator to the switchyard? Assume that the grain falls straight down and that slowing due to rolling friction or air drag is negligible.

FIGURE 8-21

PICTURE THE PROBLEM We find the travel time that we seek from the distance traveled and the speed of the car. Consider the car and the grain as our system (Figure 8-21). There are no horizontal external forces acting on this system, so the horizontal component of the momentum of the system is conserved. The final speed of the grain-filled car is found from its final momentum, which equals the car's initial momentum. (The grain initially has no horizontal momentum.) Let m_c and m_g be the masses of the car and grain, respectively.

$m_g = 2000$ kg

$m_c = 14{,}000$ kg

$v_i = 4$ m/s

1. The time for the car to travel from the elevator to the yard is the distance to the yard d divided by the car's speed v_f following the grain dump:

$$\Delta t = \frac{d}{v_f}$$

2. Apply conservation of momentum to relate the final velocity v_f to the initial velocity v_i. Be careful, only the horizontal component of the system's momentum is conserved:

$$(m_c + m_g)v_f = m_c v_i + m_g(0)$$

3. Solve for v_f:

$$v_f = \frac{m_c v_i}{m_c + m_g}$$

4. Substitute the result for v_f into step 1 and solve for the time:

$$\Delta t = \frac{d}{v_f} = \frac{(m_c + m_g)d}{m_c v_i}$$

$$= \frac{(14000 \text{ kg} + 2000 \text{ kg})(500 \text{ m})}{(14000 \text{ kg})(4 \text{ m/s})}$$

$$= \boxed{143 \text{ s}}$$

REMARKS Mechanical energy of the system is converted to thermal energy. Let K_{gi} be the kinetic energy of the grain just as it hits the car. The initial mechanical energy is $K_{gi} + \frac{1}{2}m_c v_i^2 = K_{gi} + \frac{1}{2}(14{,}000 \text{ kg})(4 \text{ m/s})^2 = K_{gi} + 112$ kJ. The final kinetic energy is $\frac{1}{2}(m_c + m_g)v_f^2$, where $v_f = m_c v_i/(m_c + m_g) = 14(4 \text{ m/s})/16 = 3.5$ m/s, so the final kinetic energy is $\frac{1}{2}(16{,}000 \text{ kg})(3.5 \text{ m/s})^2 = 98$ kJ. The final kinetic energy is less than the initial kinetic energy by 14 kJ.

EXERCISE Suppose that there is a small vertical chute in the bottom of the car so that the grain leaks out at 10 kg/s. Now how long does it take the car to cover the 500 m? (*Answer* 143 s. The grain leaking out does not impart any momentum to the rest of the system. If the ground were frictionless and flat, all of the grain initially in the car would arrive at the switchyard along with the car.)

A Skateboard Workout   **EXAMPLE 8 - 8**

A 40-kg skateboarder on a 3-kg board is training with two 5-kg weights. Beginning from rest, she throws the weights horizontally, one at a time from her board. The speed of each weight is 7 m/s relative to her and the board after it is thrown. Assume the board rolls without friction. (*a*) How fast is she propelled in the opposite direction after throwing the first weight? (*b*) the second weight?

PICTURE THE PROBLEM Because no external forces act on the skateboarder–weights–board system with horizontal components, the horizontal component of its momentum is conserved. We need to find the velocity of the skateboarder after throwing each weight (Figure 8-22).

(a) 1. The mass m of each weight is 5 kg and the mass M of the skateboard and skateboarder is 43 kg. Choose the direction of the skateboarder's motion to be the positive direction. Then $v_{ws} = -7$ m/s is the velocity of the thrown weight relative to the skateboarder. Let V_{sg1} and v_{wg1} be the respective velocities of the skateboarder and the thrown weight relative to the ground. Apply conservation of momentum for the first throw.

$$P_{sys1} = P_{sys0}$$

$$(M + m)V_{sg1} + mv_{wg1} = 0$$

(The numbers in the subscripts stand for the time. Time 0 is before the first throw, time 1 is between the two throws, and time 2 is following the second throw.)

FIGURE 8-22

Time 0

5 kg

v_{wg1}

40 kg

V_{sg1}

Time 1

3 kg

x

2. The velocity of the thrown weight relative to the ground equals the velocity of the weight relative to the skateboarder plus the velocity of the skateboarder relative to the ground:

$$v_{wg1} = v_{ws1} + V_{sg1}$$

Time 2

5 kg

v_{wg2}

V_{sg1}

V_{sg2}

40 kg

3 kg

3. Substitute for v_{wg1} in the step 1 result and solve for V_{sg1}:

$$(M + m)V_{sg1} + m(v_{ws1} + V_{sg1}) = 0$$

so

$$V_{sg1} = -\frac{m}{M + 2m}v_{ws1} = -\frac{5\text{ kg}}{43\text{ kg} + 10\text{ kg}}(-7\text{ m/s})$$

$$= \boxed{0.66\text{ m/s}}$$

(b) 1. Repeat step 1 of Part (a) for the second throw. Let V_{sg2} and $v_{w'g2}$ be the respective velocities of the skateboarder and the second thrown weight relative to the ground:

$$P_{sys2} = P_{sys1}$$

$$MV_{sg2} + mv_{w'g2} = (M + m)V_{sg1}$$

2. Repeat step 2 of Part (a) for the second throw.

$$v_{w'g2} = v_{w's2} + V_{sg2}$$

3. Substitute for $v_{w'g2}$ in the Part (b) step 1 result and solve for V_{sg2}:

$$MV_{sg2} + m(v_{w's2} + V_{sg2}) = (M + m)V_{sg1}$$

so

$$V_{sg2} = \frac{(M + m)V_{sg1} - mv_{w's2}}{M + m} = V_{sg1} - \frac{m}{M + m}v_{w's2}$$

$$= 0.66\text{ m/s} - \frac{5\text{ kg}}{48\text{ kg}}(-7\text{ m/s}) = \boxed{1.39\text{ m/s}}$$

REMARKS This example illustrates the principle of the rocket; a rocket moves forward by throwing its fuel out backward in the form of exhaust gases.

EXERCISE How fast is the skateboarder moving if, starting from rest, she throws both weights together, and the weights have speed 7 m/s relative to herself *after they are thrown*? (*Answer* 1.32 m/s)

RADIOACTIVE DECAY   **EXAMPLE 8 - 9** Try It Yourself

A thorium-227 nucleus at rest decays into a radium-223 nucleus (mass 223 u) by emitting an α particle (mass 4 u) (Figure 8-23). The kinetic energy of the α particle is measured to be 6.00 MeV. What is the kinetic energy of the recoiling radium nucleus?

Thorium-227   Radium-223

FIGURE 8 - 2 3

PICTURE THE PROBLEM Because the thorium nucleus before decay is at rest, its total momentum is zero. We can relate the velocity of the radium nucleus to that of the α particle using conservation of momentum.

Cover the column to the right and try these on your own before looking at the answers.

Steps

Answers

1. Write the kinetic energy of the radium nucleus K_{Ra} in terms of its mass m_{Ra} and speed v_{Ra}.

$K_{Ra} = \frac{1}{2}m_{Ra}v_{Ra}^2$

2. Write the kinetic energy of the alpha particle K_α in terms of its mass m_α and speed v_α.

$K_\alpha = \frac{1}{2}m_\alpha v_\alpha^2$

3. Use conservation of momentum to relate v_{Ra} to v_α.

$m_\alpha v_\alpha = m_{Ra}v_{Ra}$

4. Solve the step 1 and step 2 results for the speeds v_{Ra} and v_α, and substitute these expressions into the step 3 result.

$m_\alpha\left(\dfrac{2K_\alpha}{m_\alpha}\right)^{1/2} = m_{Ra}\left(\dfrac{2K_{Ra}}{m_{Ra}}\right)^{1/2}$

5. Solve the step 4 result for K_{Ra}.

$K_{Ra} = \dfrac{m_\alpha}{m_{Ra}}K_\alpha = \boxed{0.108\ \text{MeV}}$

REMARKS In this process, rest energy of the thorium nucleus is converted into kinetic energy of the alpha particle plus radium nucleus. The mass of the thorium nucleus is greater than that of the alpha particle plus radium nucleus by 6.1 MeV/c^2.

8-5 Kinetic Energy of a System

If the net external force on the system remains zero, then the total momentum of a system of particles must remain constant; however, the total mechanical energy of the system can change. As we saw in the examples of the previous section, internal forces that cannot change the total momentum may be nonconservative and thus change the total mechanical energy of the system. There is an important theorem concerning the kinetic energy of a system of particles that allows us to treat the energy of complex systems more easily and gives us insight into energy changes within a system:

The kinetic energy of a system of particles can be written as the sum of two terms: (1) the kinetic energy associated with the motion of the center of mass, $\frac{1}{2}Mv_{cm}^2$, where M is the total mass of the system; and (2) the kinetic energy associated with the motion of the particles of the system relative to the center of mass, $\sum \frac{1}{2}m_i u_i^2$, where $\vec{u}_i$ is the velocity of the ith particle relative to the center of mass.

THEOREM FOR THE KINETIC ENERGY OF A SYSTEM

The kinetic energy K of a system of particles is the sum of the kinetic energies of the individual particles:

$$K = \sum_i K_i = \sum_i \tfrac{1}{2}m_i v_i^2 = \sum_i \tfrac{1}{2}m_i (\vec{v}_i \cdot \vec{v}_i)$$

The velocity of each particle can be written as the sum of the velocity of the center of mass $\vec{v}_{cm}$ and the velocity of the particle relative to the center of mass $\vec{u}_i$:

$$\vec{v}_i = \vec{v}_{cm} + \vec{u}_i \qquad\qquad 8\text{-}17$$

Then

$$K = \sum_i \tfrac{1}{2}m_i(\vec{v}_i \cdot \vec{v}_i) = \sum_i \tfrac{1}{2}m_i(\vec{v}_{cm} + \vec{u}_i) \cdot (\vec{v}_{cm} + \vec{u}_i)$$

$$= \sum_i \tfrac{1}{2}m_i(v_{cm}^2 + 2\vec{v}_{cm} \cdot \vec{u}_i + u_i^2)$$

$$= \sum_i \tfrac{1}{2}m_i v_{cm}^2 + \vec{v}_{cm} \cdot \sum_i m_i \vec{u}_i + \sum_i \tfrac{1}{2}m_i u_i^2$$

where in the middle term we have factored $\vec{v}_{cm}$ from the sum because it is the same for each term in the sum. That is, $\vec{v}_{cm}$ does not change from particle to particle. The quantity $\Sigma m_i \vec{u}_i$ is the total momentum of the system *relative to the center of mass*. This quantity, which equals $M\vec{u}_{cm}$ is necessarily zero. (Relative to the center of mass, the velocity of the center of mass $\vec{u}_{cm}$ is zero, so the total momentum $M\vec{u}_{cm}$ is also zero.) Thus,

$$K = \sum_i \tfrac{1}{2}m_i v_{cm}^2 + \sum_i \tfrac{1}{2}m_i u_i^2 = \tfrac{1}{2}Mv_{cm}^2 + K_{rel} \qquad\qquad 8\text{-}18$$

KINETIC ENERGY OF A SYSTEM OF PARTICLES

where M is the total mass and K_{rel} is the kinetic energy of the particles *relative to the center of mass*. If the net external force is zero, $\vec{v}_{cm}$ remains constant and the kinetic energy associated with bulk motion ($\tfrac{1}{2}Mv_{cm}^2$) does not change. Only the relative kinetic energy can change in an isolated system.

8-6 Collisions

In a **collision,** two objects approach and interact strongly for a very short time. During the brief time of collision, any external forces on the objects are assumed to be much weaker than the forces of interaction between the objects. Thus, during the collision the only important forces acting on the two-object system are the interaction forces, which are equal and opposite, so the total momentum of the system remains unchanged. Also, the collision time is so short that during the collision any displacements of the colliding objects can be neglected. Both prior to and following the collision, the interaction of the two objects is weak compared with their interaction during the collision. Examples of collisions are a cue ball hitting a billiard ball, a bat hitting a baseball, or a dart colliding with a dart board. **A comet swinging around the sun can be considered a collision, even though the two do not physically touch, and in the lab you may see a collision between two carts with magnetic bumpers that repel without touching.**

When the total kinetic energy of the two-object system is the same after the collision as before, the collision is called an **elastic collision.** Otherwise it is called an **inelastic collision.** An extreme case is the **perfectly inelastic collision,** in which all of the kinetic energy relative to the center of mass is converted to thermal or internal energy of the system, and the two objects stick together after the collision.

Impulse and Average Force

Figure 8-24 shows the time variation of the magnitude of a typical force exerted by one object on another during a collision. During the collision time $\Delta t = t_f - t_i$ the force is large. For other times the force is negligibly small. The **impulse** $\vec{I}$ of the force is a vector defined as

$$\vec{I} = \int_{t_i}^{t_f} \vec{F} \, dt$$

8-19

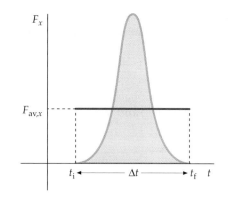

FIGURE 8-24 Typical time variation of force during a collision. The area under the F_x-versus-t curve is the x component of the impulse, I_x. $F_{av,x}$ is the average force for time interval Δt. The rectangular area $F_{av,x} \Delta t$ is the same as the area under the F_x-versus-t curve.

The magnitude of the impulse of the force is the area under its F-versus-t curve. Units of impulse are N·s. The net force $\vec{F}_{net}$ acting on a particle is related to the rate of change of momentum of the particle by Newton's second law: $\vec{F}_{net} = d\vec{p}/dt$, so the impulse of the net force equals the total change in momentum $\Delta \vec{p}$ during the time interval:

$$\vec{I}_{net} = \int_{t_i}^{t_f} \vec{F}_{net} \, dt = \int_{t_i}^{t_f} \frac{d\vec{p}}{dt} \, dt = \vec{p}_f - \vec{p}_i = \Delta \vec{p}$$

8-20

IMPULSE–MOMENTUM THEOREM FOR A PARTICLE

Also, the net impulse on a system due to external forces equals the change in the total momentum of the system:

$$\vec{I}_{net,ext} = \int_{t_i}^{t_f} \vec{F}_{net,ext} \, dt = \Delta \vec{P}_{sys}$$

8-21

IMPULSE–MOMENTUM THEOREM FOR A SYSTEM

The **average force** for the interval $\Delta t = t_f - t_i$ is defined as

$$\vec{F}_{av} = \frac{1}{\Delta t} \int_{t_i}^{t_f} \vec{F} \, dt = \frac{\vec{I}}{\Delta t}$$

8-22

The average force is the constant force that gives the same impulse as the actual force in the time interval Δt, as shown by the rectangle in Figure 8-24. The average net force can be calculated from the change in momentum if a collision time is known. This time can often be estimated using the displacement of one of the bodies during the collision.

A *KARATE COLLISION*

E X A M P L E   8 - 1 0

With an expert karate blow, you shatter a concrete block. Consider your fist to have a mass 0.70 kg, to be moving 5.0 m/s as it strikes the block, and to stop within 6 mm of the point of contact. (*a*) What impulse does the block exert on your fist? (*b*) What is the approximate collision time and the average force the block exerts on your fist?

PICTURE THE PROBLEM The net impulse equals the change in momentum $\Delta\vec{p}$. We find $\Delta\vec{p}$ from the mass and velocity of the fist. The time of collision for Part (b) comes from the given displacement $\Delta y = -6$ mm and the average velocity v_{av} during the collision, which we can estimate by assuming constant acceleration. Make a sketch of the fist and block. Include a vertical coordinate axis on the sketch (Figure 8-25).

FIGURE 8-25

(a) 1. Set the impulse equal to the change in momentum:

$$\vec{I} = \Delta\vec{p} = \vec{p}_f - \vec{p}_i$$

2. The initial momentum is that of the fist just before it hits the block with velocity $\vec{v}$, and the final momentum is zero:

$$\vec{p}_i = m\vec{v} = (0.7\ kg)(-5.0\ m/s)\hat{j}$$
$$= -3.5\ kg\cdot m/s\,\hat{j}$$
$$\vec{p}_f = 0$$

3. Find the impulse exerted by the block on the fist:

$$\vec{I} = \vec{p}_f - \vec{p}_i = 0 - (-3.5\ kg\cdot m/s\,\hat{j})$$
$$= \boxed{3.5\ N\cdot s\,\hat{j}}$$

(b) 1. The collision time is the displacement divided by the average velocity:

$$\Delta t = \frac{\Delta y}{v_{av}}$$

2. We estimate the average speed by assuming constant acceleration, $v_{av} = \frac{1}{2}v$. Since we have chosen up to be positive, both Δy and v_{av} are negative. Calculate Δt:

$$\Delta t = \frac{\Delta y}{\frac{1}{2}v} = \frac{-0.006\ m}{-2.5\ m/s} = 0.0024\ s$$
$$= 2.4\ ms$$

3. The average force is the impulse divided by the collision time. It is upward, as expected:

$$\vec{F}_{av} = \frac{\vec{I}}{\Delta t} = \frac{3.5\ N\cdot s\,\hat{j}}{0.0024\ s} = \boxed{1.46\ kN\,\hat{j}}$$

REMARKS Note that the average force is large—about 212 times the weight of the fist.

A Crash Test

EXAMPLE 8-11 Try It Yourself

A car equipped with an 80-kg crash-test dummy (Figure 8-26) drives into a wall at 25 m/s (about 56 mi/h). Estimate the force that the seatbelt exerts on the dummy upon impact.

PICTURE THE PROBLEM Assume that the car and dummy travel about 1 m as the front end of the car crumples, and that the acceleration is constant during the crash. To find the force, calculate the impulse I, then divide it by the collision time Δt. Choose forward as the positive x direction.

FIGURE 8-26

Cover the column on the right and try these on your own before looking at the answers.

Steps

Answers

1. Relate the average force to the impulse, and thus to the change in momentum.

$$\vec{F}_{av}\Delta t = \vec{I} = \Delta\vec{p},\quad \text{so}\quad \vec{F}_{av} = \frac{\Delta\vec{p}}{\Delta t}$$

2. Find the change in the dummy's momentum.

$$\Delta\vec{p} = m\vec{v}_f - m\vec{v}_i = -2000\ N\cdot s\,\hat{i}$$

3. Relate the time to the displacement, assuming constant acceleration.

$$\Delta t = \frac{\Delta x}{v_{av}}$$

4. Find the average velocity and use it and the step 3 result to find the time.

$$\vec{v}_{av} = \tfrac{1}{2}(\vec{v}_f + \vec{v}_i) = 12.5 \text{ m/s } \hat{i}, \quad \text{so}$$

$$\Delta t = 0.08 \text{ s}$$

5. Substitute the step 2 and step 4 results into the step 1 result and solve for the force.

$$\vec{F}_{av} = \boxed{-25 \text{ kN } \hat{i}}$$

REMARKS The magnitude of the average acceleration is $a_{av} = \Delta v/\Delta t = 313 \text{ m/s}^2$, or roughly $32g$. Such an acceleration means a net force about 32 times the weight of the dummy, clearly enough to cause serious injuries. An air bag increases the stopping distance somewhat, which helps to reduce the force and also allows the force to be distributed over a much larger area.

REMARKS In Figure 8-27, plot (a) shows the average force on the dummy as a function of the stopping distance. With no seat belt or air bag, you either fly though the windshield, or are stopped in a fraction of a meter by the dashboard or steering wheel. Plot (b) shows the force as a function of the initial velocity for three stopping distances: 2 m, 1.5 m, and 1 m.

FIGURE 8-27

(a)

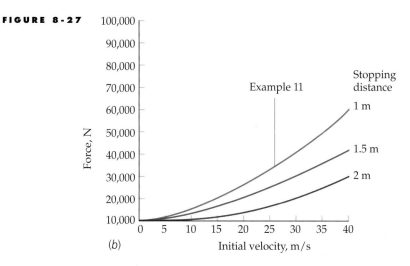

(b)

A GOLF BALL   **EXAMPLE 8-12**

You strike a golf ball with a driving iron. What are reasonable estimates for the (a) impulse I, (b) collision time Δt, and (c) average force F_{av}? A typical golf ball has a mass $m = 45$ g and a radius $r = 2$ cm. For a typical drive, the range R is roughly 192 m (210 yd) (Figure 8-28).

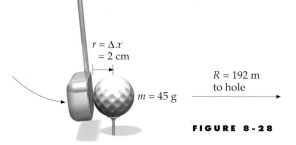

FIGURE 8-28

PICTURE THE PROBLEM Let v_0 denote the speed of the ball as it leaves the club face. The impulse on the ball equals its change in momentum, which is mv_0. We estimate v_0 from the range. We estimate the collision time from the distance traveled Δx and the average speed $\tfrac{1}{2}v_0$, assuming constant acceleration. Taking $\Delta x = 2$ cm, the average force is then obtained from the impulse I and collision time Δt.

(a) 1. Set the impulse equal to the change in momentum of the ball:

$$I = F_{av} \Delta t = \Delta p$$

2. The initial speed is related to the range R, which is given by Equation 2-23:

$$R = \frac{v_0^2}{g} \sin 2\theta_0$$

3. Take $\theta_0 = 13°$ and calculate the initial speed:

$$v_0 = \sqrt{\frac{Rg}{\sin 2\theta_0}}$$

$$= \sqrt{\frac{(192 \text{ m})(9.81 \text{ m/s}^2)}{\sin 26°}} = 65.5 \text{ m/s}$$

4. Use this value of v_0 to calculate the magnitude of the impulse:

$$I = \Delta p = m(v_0 - 0) = (0.045 \text{ kg})(65.5 \text{ m/s})$$

$$= 2.95 \text{ kg·m/s} = \boxed{2.95 \text{ N·s}}$$

(b) Calculate the collision time Δt using $\Delta x = 2$ cm and $v_{av} = \frac{1}{2}(v_f - v_i)$:

$$\Delta t = \frac{\Delta x}{v_{av}} = \frac{\Delta x}{\frac{1}{2}v_0}$$

$$= \frac{0.02 \text{ m}}{\frac{1}{2}(65.5 \text{ m/s})} = \boxed{0.610 \times 10^{-3} \text{ s}}$$

(c) Use the calculated values of I and Δt to find the magnitude of the average force:

$$F_{av} = \frac{I}{\Delta t} = \frac{2.95 \text{ N·s}}{6.10 \times 10^{-4} \text{ s}} = \boxed{4.83 \text{ kN}}$$

REMARKS Again we see that forces exerted during a collision are very large. Here the force exerted on the golf ball by the club is roughly 10,000 times the weight of the ball, giving it an average acceleration of $10,000g$ for 0.61 ms. Also, the force of the air on the ball has been left out of our analysis. For an actual golf shot the effects of the air are not negligible.

Collisions in One Dimension (Head-on Collisions)

Consider an object of mass m_1 with initial velocity v_{1i} approaching a second object of mass m_2 that is moving in the same direction with initial velocity v_{2i}. If $v_{2i} < v_{1i}$, the objects collide. Let v_{1f} and v_{2f} be their final velocities after the collision. (These velocities can be positive or negative, depending on whether the objects are moving to the right or to the left.) Conservation of momentum gives one relation between the two unknown velocities v_{1f} and v_{2f}:

$$m_1 v_{1f} + m_2 v_{2f} = m_1 v_{1i} + m_2 v_{2i} \qquad \text{8-23}$$

To determine v_{1f} and v_{2f}, we must have a second relation. That second relation, which we will develop here, depends on the type of collision.

Perfectly Inelastic Head-on Collisions In perfectly inelastic collisions, the particles stick together after the collision. For example, the collision between a catcher's mitt and a baseball is perfectly inelastic if the catcher does not drop the ball. The second kind of relation between the velocities is that the final velocities are equal to each other and to the velocity of the center of mass:

$$v_{1f} = v_{2f} = v_{cm}$$

Substituting this result combined with conservation of momentum gives

$$(m_1 + m_2)v_{cm} = m_1 v_{1i} + m_2 v_{2i} \qquad \text{8-24}$$

Perfectly inelastic collision of two cars.

A CATCH IN SPACE

EXAMPLE 8 - 1 3

FIGURE 8-29

An astronaut of mass 60 kg is on a space walk to repair a communications satellite. Suddenly she needs to consult her physics book. You happen to have it with you, so you throw it to her with speed 4 m/s relative to your spacecraft. She is at rest relative to the spacecraft before catching the 3.0-kg book (Figure 8-29). Find (*a*) her velocity just after she catches the book, (*b*) the initial and final kinetic energies of the book–astronaut system, and (*c*) the impulse exerted by the book on the astronaut.

4 m/s

3 kg

60 kg   63 kg

PICTURE THE PROBLEM (*a*) The final velocity of the book and astronaut is the velocity of their center of mass. We find this using conservation of momentum, as expressed in Equation 8-24. The initial and final kinetic energies are calculated from the initial and final velocities. Because the book and astronaut move with the same final velocity, the collision is perfectly inelastic. (*b*) The kinetic energies of the book and astronaut are calculated directly from their masses and speeds. (*c*) The impulse exerted by the book on the astronaut equals the change in momentum of the astronaut.

(*a*) 1. Use conservation of momentum to relate the final velocity of the system v_{cm} to the initial velocities:

$$m_b v_b + m_a v_a = (m_b + m_a)v_{cm}$$

2. Solve for v_{cm}:

$$v_{cm} = \frac{m_b v_b + m_a v_a}{m_b + m_a}$$

$$= \frac{(3.0 \text{ kg})(4 \text{ m/s}) + 0}{60 \text{ kg} + 3 \text{ kg}} = \boxed{0.19 \text{ m/s}}$$

(*b*) 1. The initial mechanical energy of the book–astronaut system is the initial kinetic energy of the book:

$$K_i = K_b = \tfrac{1}{2}m_b v_b^2$$

$$= \tfrac{1}{2}(3.0 \text{ kg})(4 \text{ m/s})^2 = \boxed{24 \text{ J}}$$

2. The final kinetic energy is the kinetic energy of the book and astronaut moving together at v_{cm}:

$$K_f = \tfrac{1}{2}(m_b + m_a)v_{cm}^2$$

$$= \tfrac{1}{2}(63 \text{ kg})(0.19 \text{ m/s})^2 = \boxed{1.14 \text{ J}}$$

(*c*) Set the impulse exerted on the astronaut equal to the change in momentum of the astronaut:

$$I = \Delta p_a = m_a \Delta v_a$$

$$= (60 \text{ kg})(0.19 \text{ m/s} - 0)$$

$$= 11.4 \text{ kg·m/s} = \boxed{11.4 \text{ N·s}}$$

REMARKS Most of the initial kinetic energy in this collision is lost by conversion to thermal energy. The impulse exerted by the book on the astronaut is equal and opposite to that exerted by the astronaut on the book, so the total change in momentum of the book–astronaut system is zero.

It is useful to express the kinetic energy K of a particle in terms of its momentum p. For a mass m moving with speed v, we have

$$K = \tfrac{1}{2}mv^2 = \frac{(mv)^2}{2m}$$

Since $p = mv$,

$$K = \frac{p^2}{2m}$$

8-25

We can apply this to a perfectly inelastic collision where one object is initially at rest. The momentum of the system is that of the incoming object:

$$P_{sys} = p_{1i} = m_1 v_{1i}$$

The initial kinetic energy is

$$K_i = \frac{P^2_{sys}}{2m_1} \qquad\qquad 8\text{-}26$$

After colliding, the objects move together as a single mass $m_1 + m_2$ with v_{cm}. Momentum is conserved, so the final momentum equals P_{sys}. The final kinetic energy is then

$$K_f = \frac{P^2_{sys}}{2(m_1 + m_2)} \qquad\qquad 8\text{-}27$$

Comparing Equations 8-26 and 8-27, we see that the final kinetic energy is less than the initial kinetic energy.

A BALLISTIC PENDULUM   **E X A M P L E   8 - 1 4**

In a feat of public marksmanship, you fire a bullet into a hanging wood block. The block, with bullet embedded, swings upward. Noting the height reached at the top of the swing, you immediately inform the crowd of the bullet's speed. How fast was the bullet traveling?

PICTURE THE PROBLEM The initial speed of the bullet v_{1i} is related to the postcollision speed of the bullet–block system v_f by conservation of momentum. The speed v_f is related to the height h by conservation of mechanical energy (Figure 8-30). Let m_1 be the mass of the bullet and m_2 be the mass of the target.

1. Using conservation of mechanical energy *after* the collision we relate the postcollision speed v_f to the maximum height h:

$$\tfrac{1}{2}(m_1 + m_2)v_f^2 = (m_1 + m_2)gh$$

FIGURE 8-30

2. Using conservation of momentum *during* the collision we relate v_{1i} to v_f:

$$m_1 v_{1i} = (m_1 + m_2)v_f$$

3. Solving the step 1 result for v_f and then substituting for v_f in the step 2 result, we can solve for v_{1i}:

$$v_f = \sqrt{2gh}$$

so

$$v_{1i} = \frac{m_1 + m_2}{m_1}v_f = \boxed{\frac{m_1 + m_2}{m_1}\sqrt{2gh}}$$

REMARKS We assumed that the time of the collision is so short that the displacement of the block during the collision is negligible. This meant the block had the postcollision speed while still at the lowest point in the arc.

Devices such as the one pictured are called *ballistic pendulums*.

EXERCISE Find the initial speed of the bullet if the mass of the bullet is 12 g, the mass of the block on the ballistic pendulum is 2 kg, and the final height is 10.4 cm. (*Answer* 240 m/s)

EXERCISE Can this example be solved by equating the initial kinetic energy of the bullet with the potential energy of the block–bullet composite at maximum height? That is, is mechanical energy conserved both during the inelastic collision and during the rise of the pendulum? (*Answer* No)

EXERCISE A 2000-kg car moving 25 m/s runs head-on into a 1500-kg car initially at rest. If the collision is perfectly inelastic, find (*a*) each car's speed after the collision, and (*b*) the ratio of the system's final kinetic energy to its initial kinetic energy. (*Answer* (*a*) 14.3 m/s, (*b*) 0.57)

COLLISION WITH AN EMPTY BOX **EXAMPLE 8-15** Try It Yourself

You repeat your feat of Example 8-14, this time with an empty box as the target. The bullet strikes the box and passes through it completely. A laser ranging device indicates that the bullet emerged with half its initial velocity. Hearing this, you correctly report how high the target must have swung. How high did it swing?

FIGURE 8-31

PICTURE THE PROBLEM The height h is related to the box's speed v_2 after colliding by conservation of mechanical energy (Figure 8-31). This speed can be determined using conservation of momentum.

Cover the column to the right and try these on your own before looking at the answers.

Steps

Answers

1. Use conservation of mechanical energy to relate the final height h to the speed v_2 of the box after the collision.

$$m_2gh = \tfrac{1}{2}m_2v_2^2$$

2. Use conservation of momentum to write an equation relating v_{1i} to the postcollision speed of the box v_2.

$$m_2v_2 + m_1\left(\tfrac{1}{2}v_{1i}\right) = m_1v_{1i}$$

3. Eliminate v_2 and solve for h.

$$h = \frac{m_1^2v_{1i}^2}{8m_2^2g}$$

REMARKS The collision of the bullet and the box is an inelastic collision, but not a perfectly inelastic collision because the two objects do not have the same velocity after the collision. Inelastic collisions also occur in microscopic systems. For example, when an electron collides with an atom, the atom is sometimes excited to a higher internal energy state. As a result, the total kinetic energy of the atom and the electron is less after the collision.

Elastic Head-on Collisions In some collisions, called **elastic collisions,** the initial and final kinetic energies are equal. Elastic collisions are an ideal that is sometimes approached but never realized in the macroscopic world. If a ball dropped onto a concrete platform bounces back to its original height, then the collision between the ball and the concrete would be elastic. That has never been observed. At the microscopic level, elastic collisions are common. For example, the collisions between air molecules are almost always elastic.

A bullet traveling 850 m/s collides inelastically with an apple, which moments later disintegrates completely. Exposure time is less than a millionth of a second.

For all elastic collisions we have:

$$\tfrac{1}{2}m_1 v_{1f}^2 + \tfrac{1}{2}m_2 v_{2f}^2 = \tfrac{1}{2}m_1 v_{1i}^2 + \tfrac{1}{2}m_2 v_{2i}^2 \qquad\qquad 8\text{-}28$$

This equation, together with the head-on conservation of momentum equation (Equation 8-23), is sufficient to determine the final velocities of the two objects. However, the quadratic nature of Equation 8-28 often complicates the solution of an elastic collision problem. Such problems can be treated more easily if we express the velocity of the two particles relative to each other after the collision in terms of their relative velocity before the collision. Rearranging Equation 8-28 gives

$$m_2(v_{2f}^2 - v_{2i}^2) = m_1(v_{1i}^2 - v_{1f}^2)$$

or

$$m_2(v_{2f} - v_{2i})(v_{2f} + v_{2i}) = m_1(v_{1i} - v_{1f})(v_{1i} + v_{1f}) \qquad\qquad 8\text{-}29$$

From conservation of momentum, we know that

$$m_1 v_{1f} + m_2 v_{2f} = m_1 v_{1i} + m_2 v_{2i}$$

or

$$m_2(v_{2f} - v_{2i}) = m_1(v_{1i} - v_{1f}) \qquad\qquad 8\text{-}30$$

Then dividing Equation 8-29 by Equation 8-30, we get

$$v_{2f} + v_{2i} = v_{1i} + v_{1f}$$

Rearranging, we obtain

$$v_{2f} - v_{1f} = -(v_{2i} - v_{1i}) \qquad\qquad 8\text{-}31$$

RELATIVE VELOCITIES IN AN ELASTIC COLLISION

If two objects are to collide, then $v_{2i} - v_{1i}$ must be negative (Figure 8-32), making their **speed of approach** $-(v_{2i} - v_{1i})$. After colliding, the objects' **speed of recession** is $v_{2f} - v_{1f}$. (Both of these terms give the speed of one object relative to the other.) Equation 8-31 states

In elastic collisions, the speed of recession equals the speed of approach.

Solving elastic-collision problems is usually easier using Equation 8-31 rather than Equation 8-28. But beware! Equation 8-31 is valid only if the initial and final kinetic energies are equal, so it applies *only* to elastic collisions.

FIGURE 8-32 Closing (approaching) and separating (receding) in a head-on elastic collision.

ELASTIC COLLISION OF TWO BLOCKS **EXAMPLE 8-16**

A 4-kg block moving right at 6 m/s collides elastically with a 2-kg block moving right at 3 m/s (Figure 8-33). Find their final velocities.

FIGURE 8-33

PICTURE THE PROBLEM Conservation of momentum and the equality of the initial and final kinetic energies (expressed as a reversal of relative velocities) give two equations for the two unknown final velocities. Let subscript 1 denote the 4-kg block, subscript 2 the 2-kg block.

1. Apply conservation of momentum and simplify to obtain an equation relating the two final velocities:

$$m_1 v_{1f} + m_2 v_{2f} = m_1 v_{1i} + m_2 v_{2i}$$

$$(4 \text{ kg})v_{1f} + (2 \text{ kg})v_{2f} = (4 \text{ kg})(6 \text{ m/s}) + (2 \text{ kg})(3 \text{ m/s})$$

so

$$2v_{1f} + v_{2f} = 15 \text{ m/s}$$

2. The equality of the initial and final kinetic energies provides a second equation relating the two final velocities. This is implemented by equating the speeds of recession and approach:

$$v_{2f} - v_{1f} = -(v_{2i} - v_{1i})$$

$$= -(3 \text{ m/s} - 6 \text{ m/s}) = 3 \text{ m/s}$$

3. Subtract the step 2 result from the step 1 result and solve for v_{1f}:

$$2v_{1f} + v_{1f} = 12 \text{ m/s}$$

so

$$v_{1f} = \boxed{4 \text{ m/s}}$$

4. Substitute into the step 2 result and solve for v_{2f}:

$$v_{2f} - 4 \text{ m/s} = 3 \text{ m/s}$$

so

$$v_{2f} = \boxed{7 \text{ m/s}}$$

PLAUSIBILITY CHECK As a check, we calculate the initial and final kinetic energies.

$$K_i = \tfrac{1}{2}(4 \text{ kg})(6 \text{ m/s})^2 + \tfrac{1}{2}(2 \text{ kg})(3 \text{ m/s})^2 = 72 \text{ J} + 9 \text{ J} = 81 \text{ J}.$$

$$K_f = \tfrac{1}{2}(4 \text{ kg})(4 \text{ m/s})^2 + \tfrac{1}{2}(2 \text{ kg})(7 \text{ m/s})^2 = 32 \text{ J} + 49 \text{ J} = 81 \text{ J} = K_i.$$

ELASTIC COLLISION OF A NEUTRON AND A NUCLEUS **EXAMPLE 8-17**

A neutron of mass m_n and speed v_{ni} collides elastically with a carbon nucleus of mass m_C initially at rest (Figure 8-34). (*a*) What are the final velocities of both particles? (*b*) What fraction f of its initial kinetic energy does the neutron lose?

FIGURE 8-34

PICTURE THE PROBLEM Conservation of momentum and conservation of energy allow us to find the final velocities. Since the initial kinetic energy of the carbon nucleus is zero, its final kinetic energy equals the energy lost by the neutron.

(*a*) 1. Use conservation of momentum to obtain one relation for the final velocities:

$$m_n v_{ni} = m_n v_{nf} + m_C v_{Cf}$$

2. The equality of the initial and final kinetic energies provides a second equation relating the two final velocities. This is implemented by equating the speeds of recession and approach:

$$v_{Cf} - v_{nf} = -(v_{Ci} - v_{ni}) = 0 + v_{ni}$$

so

$$v_{Cf} = v_{ni} + v_{nf}$$

3. To eliminate v_{Cf}, substitute the expression for v_{Cf} from step 2 into the Step 1 result:

$$m_n v_{ni} = m_n v_{nf} + m_C(v_{ni} + v_{nf})$$

4. Solve for v_{nf}: (Note that v_{nf} is negative. The neutron m_n bounces back from the more massive carbon nucleus m_C.)

$$\boxed{v_{nf} = -\frac{m_C - m_n}{m_n + m_C} v_{ni}}$$

5. Substitute the step 4 result into the step 2 result and solve for v_{Cf}:

$$v_{Cf} = v_{ni} - \frac{m_C - m_n}{m_n + m_C} v_{ni} = \boxed{\frac{2m_n}{m_n + m_C} v_{ni}}$$

(b) 1. The collision is elastic, so the kinetic energy lost by the neutron is the final kinetic energy of the carbon nucleus:

$$f = \frac{-\Delta K_n}{K_{ni}} = \frac{K_{Cf}}{K_{ni}}$$

$$= \frac{\frac{1}{2}m_C v_{Cf}^2}{\frac{1}{2}m_n v_{ni}^2} = \frac{m_C}{m_n}\left(\frac{v_{Cf}}{v_{ni}}\right)^2$$

2. Solve the Part (a) step 5 result for the ratio of the velocities, substitute into the Part (b) step 1 result, and solve for the fractional energy loss of the neutron:

$$f = \frac{m_C}{m_n}\left(\frac{2m_n}{m_n + m_C}\right)^2 = \boxed{\frac{4m_n m_C}{(m_n + m_C)^2}}$$

REMARKS An important application of energy transfer in elastic collisions is the slowing down of neutrons in a nuclear reactor. High-energy neutrons are emitted in the fission of a uranium nucleus. If these neutrons are to cause another uranium nucleus to fission, their energy must be reduced; that is, they must be slowed down or "moderated." One mechanism for slowing down neutrons is the elastic scattering of the neutrons with the nuclei in the reactor. The fractional energy loss $f = -\Delta K_n/K_n$ depends on the ratio of the mass of the moderator nucleus to that of the neutron, as shown in Figure 8-35. For uranium, $m_U \approx 235m_n$ and $f \approx 0.017 = 1.7$ percent. For carbon, $m_C \approx 12m_n$ and $f = 0.28 = 28$ percent; for deuterium, $m_D \approx 2m_n$ and $f \approx 0.89 = 89$ percent; and for Hydrogen, $m_H \approx 1m_n$ and $f \approx 1 = 100$ percent. A moderator such as graphite or water is added to a reactor to slow down the neutrons so that they can be captured by uranium nuclei.

EXERCISE Show that for an elastic head-on collision between a moving object and a stationary object of equal mass, the fractional loss in kinetic energy is one and the final speed of the initially stationary object equals the initial speed of the initially moving object.

EXERCISE A 2-kg box moving at 3 m/s makes an elastic collision with a stationary 4-kg box. (a) What is the original kinetic energy? (b) How much energy is transferred to the 4-kg box? (*Answer*   (a) 9 J, (b) 8 J)

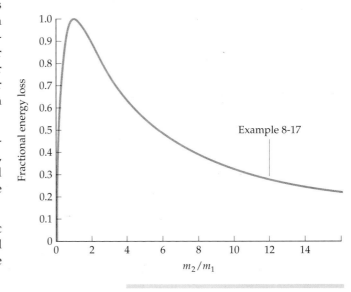

FIGURE 8-35 Fractional energy loss as a function of the ratio of the two masses. The maximum energy loss occurs when $m_1 = m_2$.

The results of Example 8-17 for the final velocities of an incoming particle colliding with a second particle initially at rest are worth noting. The final velocity of the incoming particle v_{1f} and that of the originally stationary particle v_{2f} are related to the initial velocity of the incoming particle by

$$v_{1f} = \frac{m_1 - m_2}{m_1 + m_2} v_{1i} \qquad\qquad 8\text{-}32a$$

and

$$v_{2f} = \frac{2m_1}{m_1 + m_2} v_{1i} \qquad\qquad 8\text{-}32b$$

When a very massive object (say a bowling ball) collides with a light stationery object (say a Ping-Pong ball), the massive object is essentially unaffected. Before the collision, the relative velocity of approach is v_{1i}. If the massive object continues with a velocity that is essentially v_{1i} after the collision, the velocity of the smaller object must be $2v_{1i}$ so that the speed of recession is equal to the speed of approach. This result also follows from Equations 8-32a and 8-32b if we take m_2 to be much smaller than m_1, in which case $v_{1f} \approx v_{1i}$ and $v_{2f} \approx 2v_{1i}$, as expected.

*The Coefficient of Restitution   Most collisions lie somewhere between the extreme cases of elastic, in which the relative velocities are reversed, and perfectly inelastic, in which there is no relative velocity after the collision. The **coefficient of restitution** e, is a measure of the elasticity of a collision. It is defined as the ratio of the speed of recession to the speed of approach.

$$ e = \frac{v_{rec}}{v_{app}} = -\frac{v_{2f} - v_{1f}}{v_{2i} - v_{1i}} \qquad\qquad 8\text{-}33 $$

DEFINITION—COEFFICIENT OF RESTITUTION

For an elastic collision, $e = 1$. For a perfectly inelastic collision, $e = 0$.

Collisions in Three Dimensions

Perfectly Inelastic Collisions in Three Dimensions   For collisions in three dimensions, the total initial momentum is the sum of the initial momentum vectors of each object involved in the collision. Because the objects stick together and because momentum is conserved we have $m_1\vec{v}_{1i} + m_2\vec{v}_{2i} = (m_1 + m_2)\vec{v}_f$. Because of this relation we know that the three velocity vectors, and thus the collision, are in the same plane. Also, from the definition of the center of mass we know that $\vec{v}_f = \vec{v}_{cm}$.

A CAR–TRUCK COLLISION   **EXAMPLE 8-18**   Put It in Context

You are at the wheel of a 1200-kg car traveling east through an intersection when a 3000-kg truck traveling north through the intersection crashes into your car, as shown in Figure 8-36. Your car and the truck stick together after impact. The driver of the truck claims you were at fault because you were speeding. You look for evidence to disprove this claim. First, there are no skid marks, indicating that neither you nor the truck driver saw the accident coming and braked hard; second, there is a sign reading "Speed Limit 80 km/h" on the road you were driving on; third, the speedometer of the truck was smashed with the needle stuck at 50 km/h; and fourth, the wreck initially skidded from the impact zone at an angle of no less than 59° north of east. Does this evidence support or undermine the claim that you were speeding?

FIGURE 8-36

PICTURE THE PROBLEM   Choose your coordinate system so that initially the car is traveling in the $+x$ direction and the truck is traveling in the $+y$ direction (Figure 8-37). Then write the momentum of each object in vector form, and use conservation of momentum.

1. Write out the conservation of momentum equation in vector form in terms of masses and velocities:

$$m_c\vec{v}_c + m_t\vec{v}_t = (m_c + m_t)\vec{v}_f$$

FIGURE 8-37

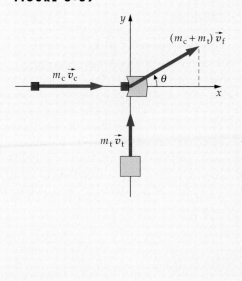

2. Equate the x component of the initial momentum to the x component of the final momentum:

$$m_c v_c + 0 = (m_c + m_t)v_f \cos\theta$$

3. Equate the y component of the initial momentum to the y component of the final momentum:

$$0 + m_t v_t = (m_c + m_t)v_f \sin\theta$$

4. Eliminate v_f by dividing the y component equation by the x component equation:

$$\frac{m_t v_t}{m_c v_c} = \frac{\sin\theta}{\cos\theta} = \tan\theta$$

so

$$v_c = \frac{m_t v_t}{m_c \tan\theta}$$

$$= \frac{(3000\ \text{kg})(50\ \text{km/h})}{(1200\ \text{kg})\tan 59°}$$

$$= \boxed{75.1\ \text{km/h}}$$

5. Does this undermine the truck driver's claim that you were speeding?

> Because 75.1 km/h is less than the 80 km/h speed limit, the truck driver's claim is undermined by the careful application of physics.

REMARKS A court would probably require that these arguments be presented by an expert witness.

*Elastic Collisions in Three Dimensions Elastic collisions in three dimensions are more complicated than those we have already covered. Figure 8-38 shows an off-center collision between an object of mass m_1 moving with velocity $\vec{v}_{1i}$ parallel to the x axis toward an object of mass m_2 that is initially at rest at the origin. The distance b between the centers measured perpendicular to the direction of $\vec{v}_{1i}$ is called the **impact parameter.** After the collision, object 1 moves off with velocity $\vec{v}_{1i}$, making an angle θ_1 with its initial velocity, and object 2 moves with velocity $\vec{v}_{2f}$, making an angle θ_2 with $\vec{v}_{1f}$. Conservation of momentum gives

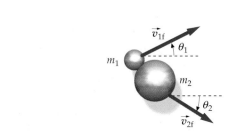

$$\vec{P}_{sys} = m_1\vec{v}_{1i} = m_1\vec{v}_{1f} + m_2\vec{v}_{2f}$$

We can see from this equation that the vector $\vec{v}_{2f}$ must lie in the plane formed by $\vec{v}_{1i}$ and $\vec{v}_{1f}$, which we will take to be the xy plane. Assuming that we know the initial velocity $\vec{v}_{1i}$, we have four unknowns: the x and y components of both final velocities; or alternatively, the two final speeds and the two angles of deflection. The x and y components of the conservation-of-momentum equation give us two of the needed relations among these quantities. Conservation of mechanical energy gives a third relation. To find the four unknowns, we need another relation. The fourth relation depends on the impact parameter b and on the type of interacting force exerted by the bodies on each other. In practice, the fourth relation is often found experimentally, by measuring the angle of deflection or the angle of recoil. Such a measurement can then give us information about the type of interacting force between the bodies.

FIGURE 8-38 Off-center collision. The final velocities depend on the impact parameter b and on the type of force exerted by one object on the other.

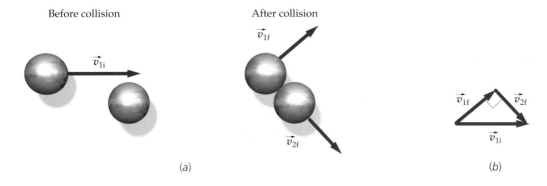

Before collision

After collision

$\vec{v}_{1f}$

$\vec{v}_{1i}$

$\vec{v}_{2f}$

$\vec{v}_{1f}$ $\vec{v}_{2f}$

$\vec{v}_{1i}$

(a)

(b)

FIGURE 8-39 (a) Off-center elastic collision of two spheres of equal mass when one sphere is initially at rest. After the collision, the spheres move off at right angles to each other. (b) The velocity vectors for this collision form a right triangle.

Consider the interesting special case of the off-center elastic collision of two objects *of equal mass* when one is initially at rest (Figure 8-39a). If $\vec{v}_{1i}$ and $\vec{v}_{1f}$ are the initial and final velocities of object 1, and $\vec{v}_{2f}$ is the final velocity of object 2, conservation of momentum gives

$$m\vec{v}_{1i} = m\vec{v}_{1f} + m\vec{v}_{2f}$$

or

$$\vec{v}_{1i} = \vec{v}_{1f} + \vec{v}_{2f}$$

These vectors form the triangle shown in Figure 8-39b. Because energy is conserved in the collision,

$$\tfrac{1}{2}mv_{1i}^2 = \tfrac{1}{2}mv_{1f}^2 + \tfrac{1}{2}mv_{2f}^2$$

or

$$v_{1i}^2 = v_{1f}^2 + v_{2f}^2 \qquad\qquad 8\text{-}34$$

Equation 8-34 is the Pythagorean theorem for a right triangle formed by the vectors $\vec{v}_{1f}$, $\vec{v}_{2f}$, and $\vec{v}_{1i}$, with the hypotenuse of the triangle being $\vec{v}_{1i}$. So for this special case, the final velocity vectors $\vec{v}_{1f}$ and $\vec{v}_{2f}$ are perpendicular to each other, as shown in Figure 8-39b.

Multiflash photograph of an off-center elastic collision of two balls of equal mass. The dotted ball, entering from the left, strikes the striped ball, which is initially at rest. The final velocities of the two balls are perpendicular to each other.

Proton–proton collision in a liquid-hydrogen bubble chamber. A proton entering from the left interacts with a stationary proton. The two then move off at right angles. The slight curvature of the tracks is due to a magnetic field.

*8-7   The Center-of-Mass Reference Frame

If the net external force on a system remains zero, the velocity of the center of mass remains constant. It is often convenient to choose a coordinate system with the origin at the center of mass. Then, relative to the original coordinate system, this coordinate system moves with a constant velocity $\vec{v}_{cm}$. The frame of reference that moves with the center of mass is called the **center-of-mass reference frame.** If a particle has velocity $\vec{v}$ in the original reference frame, then its velocity relative to the center of mass is $\vec{u} = \vec{v} - \vec{v}_{cm}$. In the center-of-mass frame, the velocity of the center of mass is zero. Since the total momentum of a system equals the total mass times the velocity of the center of mass, the total momentum is also zero in the center-of-mass frame. Thus, the center-of-mass reference frame is also a **zero-momentum reference frame.**

The mathematics of collisions is greatly simplified when considered within the center-of-mass reference frame. The momenta of the two incoming objects are equal and opposite. After a perfectly inelastic collision, the objects remain at rest. A perfectly elastic collision in one dimension between two objects reverses the direction of each velocity vector without changing the magnitudes of these vectors (you will derive this in *Problem 8-101*).

Consider a simple two-particle system in a reference frame in which one particle of mass m_1 is moving with a velocity $\vec{v}_1$ and a second particle of mass m_2 is moving with a velocity $\vec{v}_2$ (Figure 8-40). In this frame, the velocity of the center of mass is

$$\vec{v}_{cm} = \frac{m_1\vec{v}_1 + m_2\vec{v}_2}{m_1 + m_2}$$

We can transform the velocities of the two particles to their velocities in the center-of-mass reference frame by subtracting $\vec{v}_{cm}$. The velocities of the particles in the center-of-mass frame are $\vec{u}_1$ and $\vec{u}_2$, given by

$$\vec{u}_1 = \vec{v}_1 - \vec{v}_{cm} \qquad\qquad 8\text{-}35a$$

and

$$\vec{u}_2 = \vec{v}_2 - \vec{v}_{cm} \qquad\qquad 8\text{-}35b$$

Since the total momentum is zero in the center-of-mass frame, the particles have equal and opposite momenta in this frame.

Original reference frame

(a)

Center-of-mass reference frame

(b)

FIGURE 8-40 (*a*) Two particles viewed from a frame in which the center of mass has a velocity $\vec{v}_{cm}$. (*b*) The same two particles viewed from a reference frame for which the center of mass is at rest.

THE ELASTIC COLLISION OF TWO BLOCKS   **E X A M P L E   8 - 1 9**

Find the final velocities for the elastic head-on collision in Example 8-16 (in which a 4-kg block moving right at 6 m/s collides elastically with a 2-kg block moving right at 3 m/s) by transforming their velocities to the center-of-mass reference frame.

PICTURE THE PROBLEM We transform to the center-of-mass reference frame by first finding v_{cm} and subtracting it from each velocity. We then solve the collision by reversing the velocities and transforming back to the original frame.

1. Calculate the velocity of the center of mass v_{cm} (Figure 8-41):

$$v_{cm} = \frac{m_1 v_{1i} + m_2 v_{2i}}{m_1 + m_2}$$

$$= \frac{(4 \text{ kg})(6 \text{ m/s}) + (2 \text{ kg})(3 \text{ m/s})}{4 \text{ kg} + 2 \text{ kg}}$$

$$= 5 \text{ m/s}$$

FIGURE 8-41   Initial conditions

$v_{cm} = 5 \text{ m/s}$

2. Transform the initial velocities to the center-of-mass reference frame by subtracting v_{cm} from the initial velocities (Figure 8-42):

$$u_{1i} = v_{1i} - v_{cm}$$

$$= 6 \text{ m/s} - 5 \text{ m/s} = 1 \text{ m/s}$$

$$u_{2i} = v_{2i} - v_{cm}$$

$$= 3 \text{ m/s} - 5 \text{ m/s} = -2 \text{ m/s}$$

FIGURE 8-42   Transform to the center-of-mass frame by subtracting v_{cm}

$v_{cm} = 0$

3. Solve the collision in the center-of-mass reference frame by reversing the velocity of each object (Figure 8-43):

$$u_{1f} = -u_{1i} = -1 \text{ m/s}$$

$$u_{2f} = -u_{2i} = +2 \text{ m/s}$$

FIGURE 8-43   Solve collision

$v_{cm} = 0$

4. To find the final velocities in the original frame, add v_{cm} to each final velocity (Figure 8-44).

$$v_{1f} = u_{1f} + v_{cm}$$

$$= -1 \text{ m/s} + 5 \text{ m/s} = \boxed{4 \text{ m/s}}$$

$$v_{2f} = u_{2f} + v_{cm}$$

$$= 2 \text{ m/s} + 5 \text{ m/s} = \boxed{7 \text{ m/s}}$$

FIGURE 8-44   Transform back to the original frame by adding v_{cm}

$v_{cm} = 5 \text{ m/s}$

REMARKS This is the same result found in Example 8-16.

EXERCISE Show that the total momentum of the system both before the collision and after the collision is zero in the center-of-mass reference frame. (*Answer* before: $P_{\text{sys i}} = (4 \text{ kg})(1 \text{ m/s}) + (2 \text{ kg})(-2 \text{ m/s}) = 0$; after: $P_{\text{sys f}} = (4 \text{ kg})(-1 \text{ m/s}) + (2 \text{ kg})(2 \text{ m/s}) = 0$)

*8-8 Systems With Continuously Varying Mass: Rocket Propulsion

A creative and important step in solving physics problems is specifying the system. In this section we explore situations in which the system has a continuously changing mass. One example of such a system is a rocket. For a rocket, we specify the system to be the rocket plus any unspent fuel in it. As the spent fuel (the exhaust) spews out the back, the mass of the system decreases.

Another example is a stream of effluence from a volcano on the Jovian moon Io impacting on an asteroid passing through the stream. We specify the system at time t to be the asteroid, plus the volcanic effluence accumulated on the asteroid, plus the effluence that will impact the asteroid between time t and time $t + \Delta t$. As the effluence continues to accumulate, the mass of this system increases. To solve such problems we will apply the impulse–momentum theorem to the system during the short time interval Δt, and then take the limit as $\Delta t \to 0$.

Consider the following derivation of Newton's second law for objects with continuously varying mass. A continuous stream of matter moving at velocity $\vec{u}$ is impacting an object of mass M that is moving with velocity $\vec{v}$ (Figure 8-45). The mass of particles that impact the object during time Δt is ΔM, and these impacting particles stick to the object, increasing its mass by ΔM during time Δt. Also, during time Δt the velocity $\vec{v}$ changes by $\Delta \vec{v}$, as shown. At time t we define the system to be the object (including all matter that impacted the object prior to time t) plus all matter that will impact the object between time t and time $t + \Delta t$. Applying the impulse–momentum theorem to this system gives

$$\vec{F}_{net\ ext}\Delta t = \Delta \vec{P} = [(M + \Delta M)(\vec{v} + \Delta \vec{v})] - [M\vec{v} + \Delta M\vec{u}]$$

where the first term (in square brackets) is the momentum at time $t + \Delta t$ and the second term in square brackets is the momentum at time t. Reorganizing terms gives

$$\vec{F}_{net\ ext}\Delta t = M\,\Delta\vec{v} + \Delta M(\vec{v} - \vec{u}) + \Delta M\,\Delta\vec{v} \qquad \text{8-36}$$

(where the net external force $\vec{F}_{net\ ext}$ would likely include a gravitational force plus one or more contact forces). Dividing through Equation 8-36 by Δt gives

$$\vec{F}_{net\ ext} = M\frac{\Delta\vec{v}}{\Delta t} + \frac{\Delta M}{\Delta t}(\vec{v} - \vec{u}) + \frac{\Delta M}{\Delta t}\Delta\vec{v}$$

and taking the limit as $\Delta t \to 0$ (which also means as $\Delta M \to 0$ and as $\Delta \vec{v} \to 0$) gives

$$\vec{F}_{net\ ext} = M\frac{d\vec{v}}{dt} - \frac{dM}{dt}(\vec{u} - \vec{v}) + 0$$

Rearranging, we have

$$\vec{F}_{net\ ext} + \frac{dM}{dt}\vec{v}_{rel} = M\frac{d\vec{v}}{dt} \qquad \text{8-37}$$

NEWTON'S SECOND LAW—CONTINUOUSLY VARIABLE MASS

where $\vec{v}_{rel} = \vec{u} - \vec{v}$, the velocity of dM relative to M. Note that except for the term $(dM/dt)\vec{v}_{rel}$, Equation 8-37 is just Newton's second law for a system with constant mass.

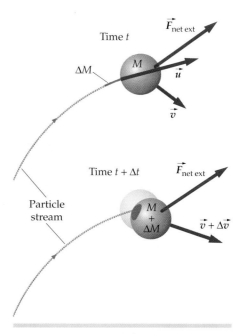

FIGURE 8-45 Particles in a continuous stream and moving at velocity $\vec{u}$ undergo perfectly inelastic collisions with an object of mass M moving at velocity $\vec{v}$.

A FALLING ROPE   **EXAMPLE 8 - 20**

A uniform rope of mass M and length L is held with its lower end just touching the surface of a scale. The rope is released and begins to fall. Find the force of the scale on the rope just as the midpoint of the rope first touches the scale.

PICTURE THE PROBLEM Apply Equation 8-37 to that portion of the rope on the scale. There are two external forces on that portion, the force of gravity and the normal force exerted by the scale. The impact velocities of the different points along the falling rope depend upon their initial heights above the scale.

1. Draw a sketch of the situation (Figure 8-46). Include the initial configuration and the configuration at an arbitrary time later. Include a coordinate axis.

2. Express Equation 8-37 in component form. Let m denote the mass of the system (that part of the rope on the scale). The velocity of the system remains zero, so the dv_y/dt is zero:

$$F_{\text{net ext } y} + \frac{dm}{dt} v_{\text{rel } y} = m\frac{dv_y}{dt}$$

$$F_n - mg + \frac{dm}{dt} v_{\text{rel } y} = 0$$

3. Let dm denote the mass of a short rope segment of length $d\ell$ that falls on the scale during time dt. Since the rope is uniform, the relation between dm and $d\ell$ is:

$$\frac{dm}{d\ell} = \frac{M}{L}$$

FIGURE 8-46 A very flexible rope of length L and mass M is released from rest and falls on the surface of a scale.

4. Solve for dm/dt by multiplying both sides of the step 3 result by $d\ell/dt$:

$$\frac{dm}{dt} = \frac{M}{L}\frac{d\ell}{dt}$$

5. $d\ell/dt$ is the impact speed of the segment, so $v_{\text{rel } y} = -d\ell/dt$. ($v_{\text{rel } y}$ is negative because up is the positive y direction and the rope is falling.) Substituting this into the step 4 result gives:

$$\frac{dm}{dt} = -\frac{M}{L}v_{\text{rel } y}$$

6. Substituting the step 5 result into the step 2 result and solving for F_n gives:

$$F_n = mg + \frac{M}{L}v_{\text{rel } y}^2$$

7. Until it touches the scale, each point along the rope falls with constant acceleration g. Using Equation 2-23 with $\Delta y = -L/2$ gives:

$$v_{\text{rel } y}^2 = v_{\text{rel } y 0}^2 + 2a_y\Delta y$$
$$= 0 + 2(-g)(-L/2) = gL$$

8. Substituting the step 7 result into the step 6 result, with $m = M/2$, gives:

$$F_n = \frac{M}{2}g + \frac{M}{L}gL = \boxed{\frac{3}{2}Mg}$$

REMARKS As the midpoint of the rope strikes the surface, the weight of the system (the rope already on the scale) is $\frac{1}{2}Mg$, which is one third of the normal force. This means $F_{\text{net ext } y} = Mg$ (directed upward). This is the force required to stop the momentum of the falling rope striking the scale at that instant.

EXERCISE Find the normal force exerted by the scale on the rope (a) just before the upper end of the rope reaches the scale and (b) just after the upper end of the rope reaches the scale. (*Answer* (a) $3Mg$, (b) Mg)

Rocket propulsion is a striking example of the conservation of momentum in action. The mass of the rocket changes continuously as it burns fuel and expels exhaust gas. Consider a rocket moving straight up with velocity $\vec{v}$ relative to the earth, as shown in Figure 8-47. Assuming that the fuel is burned at a constant rate R, the rocket's mass at time t is

$$M = M_0 - Rt \qquad\qquad\qquad 8\text{-}38$$

where M_0 is the initial mass of the rocket. The exhaust gas leaves the rocket with velocity $\vec{u}_{\text{ex}}$ relative to the rocket, and the rate at which the fuel is burned is the rate at which the mass M decreases. We choose the rocket and the unspent fuel within it as the system. Neglecting air drag, the only external force on the system is that of gravity. With $\vec{F}_{\text{net ext}} = M\vec{g}$ and $dM/dt = -R$, Equation 8-37 becomes the **rocket equation:**

$$M\vec{g} - R\vec{u}_{ex} = M\frac{d\vec{v}}{dt}$$

8-39

ROCKET EQUATION

The quantity $-R\vec{u}_{ex}$ is the force exerted on the rocket by the exhausting fuel. It is called the **thrust** $\vec{F}_{th}$:

$$\vec{F}_{th} = -R\vec{u}_{ex} = -\left|\frac{dM}{dt}\right|\vec{u}_{ex}$$

8-40

DEFINITION—ROCKET THRUST

The rocket is moving straight up, so we choose upward as the positive y direction and express Equation 8-39 as

$$-Mg + Ru_{ex} = M\frac{dv_y}{dt}$$

8-41

ROCKET EQUATION (VERTICAL COMPONENT)

Dividing through by M, rearranging terms, and substituting from Equation 8-38 gives

$$\frac{dv_y}{dt} = \frac{Ru_{ex}}{M} - g = \frac{Ru_{ex}}{M_0 - Rt} - g$$

8-42

ACCELERATION OF ROCKET (VERTICAL COMPONENT)

where M_0 is the initial value of M. Equation 8-42 is solved by integrating both sides with respect to time. For a rocket starting at rest at $t = 0$, the result is

$$v_y = u_{ex} \ln\left(\frac{M_0}{M_0 - Rt}\right) - gt$$

8-43

VELOCITY OF ROCKET (VERTICAL COMPONENT)

assuming the acceleration of gravity to be constant.

FIGURE 8-47

LIFTOFF **E X A M P L E   8 - 2 1** Try It Yourself

The Saturn V rocket used in the Apollo moon-landing program had an initial mass M_0 of 2.85×10^6 kg, a payload of 27%, a burn rate R of 13.84×10^3 kg/s, and a thrust F_{th} of 34×10^6 N. Find (*a*) the exhaust speed, (*b*) the burn time t_b, (*c*) the acceleration at liftoff, (*d*) the acceleration at burnout t_b, and (*e*) the final speed of the rocket.

PICTURE THE PROBLEM (*a*) The exhaust speed can be found from the thrust and burn rate. (*b*) To find the burn time, you need to find the total mass of fuel burned, which is the initial mass minus the payload. (*c*) The acceleration is found from Equation 8-42. (*d*) The final speed is given by Equation 8-43.

Cover the column to the right and try these on your own before looking at the answers.

Steps	Answers
(a) Calculate u_{ex} from the given thrust and burn rate.	$u_{ex} = \boxed{2.46 \text{ km/s}}$
(b) 1. Calculate the final mass M_f of the rocket.	$M_f = (0.27)M_0 = 7.70 \times 10^5 \text{ kg}$
2. Use your result to calculate the burn time t_b.	$t_b = \dfrac{M_0 - M_f}{R} = \boxed{150 \text{ s}}$
(c) Calculate dv_y/dt for $M = M_0$ and for $M = M_f$.	Initially, $\dfrac{dv_y}{dt} = 2.14 \text{ m/s}^2$;
	finally, $\dfrac{dv_y}{dt} = \boxed{34.3 \text{ m/s}^2}$
(d) Calculate the final speed from Equation 8-43.	$v_{yf} = \boxed{1.75 \text{ km/s}}$

REMARKS The initial acceleration is small—only $0.21g$. At burnout (also called flameout), the rocket's acceleration has increased to $3.5g$. Immediately following burnout the acceleration is $-g$. The speed of the rocket at burnout, after two and a half minutes of burning, is roughly 6,300 km/h (3900 mi/h).

SUMMARY

The conservation of momentum for an isolated system is a fundamental law of nature that has applications in all areas of physics.

Topic	Relevant Equations and Remarks	
1. Center of Mass		
System of particles	$M\vec{r}_{cm} = m_1\vec{r}_1 + m_2\vec{r}_2 + \cdots$	8-5
	$M\vec{v}_{cm} = m_1\vec{v}_1 + m_2\vec{v}_2 + \cdots$	8-8
	$M\vec{a}_{cm} = m_1\vec{a}_1 + m_2\vec{a}_2 + \cdots$	8-9
Continuous objects	$M\vec{r}_{cm} = \displaystyle\int \vec{r}\, dm$	8-6
Newton's second law for a system	$\vec{F}_{net\,ext} = \displaystyle\sum_i \vec{F}_{iext} = M\vec{a}_{cm}$	8-11
2. Momentum		
Definition for a particle	$\vec{p} = m\vec{v}$	8-12
Kinetic energy of a particle	$K = \dfrac{p^2}{2m}$	8-25
Momentum of a system	$\vec{P}_{sys} = \displaystyle\sum_i m_i\vec{v}_i = M\vec{v}_{cm}$	8-14

Newton's second law for a system	$$\vec{F}_{\text{net ext}} = \frac{d\vec{P}_{\text{sys}}}{dt}$$	8-15
Law of conservation of momentum	If the net external force acting on a system remains zero, the total momentum of the system is conserved.	

3. Energy of a System

Kinetic energy	The kinetic energy associated with the motion of the particles of a system relative to its center of mass is $K_{\text{rel}} = \Sigma \frac{1}{2} m_i u_i^2$, where u_i is the speed of the ith particle relative to the center of mass.	
	$$K = \tfrac{1}{2} M v_{\text{cm}}^2 + K_{\text{rel}}$$	8-18
Gravitational potential energy	$$U = M g h_{\text{cm}}$$	8-7

4. Collisions

Impulse	The impulse of a force is defined as the integral of the force over the time interval during which the force acts.	
	$$\vec{I} = \int_{t_i}^{t_f} \vec{F}\, dt$$	8-19
Impulse–momentum theorem	$$\vec{I}_{\text{net}} = \int_{t_i}^{t_f} \vec{F}_{\text{net}}\, dt = \Delta\vec{p}$$	8-20
Average force	$$\vec{F}_{\text{av}} = \frac{1}{\Delta t} \int_{t_i}^{t_f} \vec{F}\, dt = \frac{\vec{I}}{\Delta t}$$	8-22
Elastic collisions	An elastic collision is one in which the sum of the kinetic energies of the two objects is the same before and after the collision.	
Relative speeds of approach and recession	For an elastic collision, the speed of separation of the two objects following an elastic collision equals their speed of approach prior to the collision. For a head-on collision,	
	$$v_{2f} - v_{1f} = -(v_{2i} - v_{1i})$$	8-31
Perfectly inelastic collisions	Following a perfectly inelastic collision, the two objects stick together and move with the velocity of the center of mass.	
*Coefficient of restitution	The coefficient of restitution e is a measure of the elasticity. It is the ratio of the separation speed to the closing speed:	
	$$e = -\frac{v_{2f} - v_{1f}}{(v_{2i} - v_{1i})}$$	8-33
	For an elastic collision, $e = 1$; for a perfectly inelastic collision $e = 0$.	

*4. Continuously Variable Mass

Newton's second law	$$\vec{F}_{\text{net ext}} + \frac{dM}{dt}\vec{v}_{\text{rel}} = M\frac{d\vec{v}}{dt}$$	8-37
Rocket equation	$$M\vec{g} - R\vec{u}_{\text{ex}} = M\frac{d\vec{v}}{dt}$$	8-39
Thrust	$$\vec{F}_{\text{th}} = -R\vec{u}_{\text{ex}} = -\left|\frac{dM}{dt}\right|\vec{u}_{\text{ex}}$$	8-40

PROBLEMS

- Single-concept, single-step, relatively easy
- •• Intermediate-level, may require synthesis of concepts
- ••• Challenging
- **SSM** Solution is in the *Student Solutions Manual*
- **iSOLVE** Problems available on iSOLVE online homework service
- **iSOLVE** ✓ These "Checkpoint" online homework service problems ask students additional questions about their confidence level, and how they arrived at their answer

In a few problems, you are given more data than you actually need; in a few other problems, you are required to supply data from your general knowledge, outside sources, or informed estimates.

Take $g = 9.81$ N/kg $= 9.81$ m/s^2 and neglect friction in all problems unless otherwise stated.

Conceptual Problems

1 • Give an example of a three-dimensional object that has no matter at its center of mass.

2 • **SSM** A cannonball is dropped off a high tower while, simultaneously, an identical cannonball is launched directly upward into the air. The center of mass of the two cannonballs (a) stays in the same place, (b) initially rises, then falls, but begins to fall *before* the cannonball launched into the air starts falling, (c) initially rises, then falls, but begins to fall *at the same time* as the cannonball launched upward begins to fall, (d) initially rises, then falls, but begins to fall *after* the cannonball launched upward begins to fall.

3 • Two pucks of masses m_1 and m_2 lie unconnected on a frictionless table. A horizontal force F_1 is exerted on m_1 only. What is the magnitude of the acceleration of the center of mass of the two-puck system? (a) F_1/m_1. (b) $F_1/(m_1 + m_2)$. (c) F_1/m_2. (d) $(m_1 + m_2)F_1/m_1m_2$.

4 • The two pucks in Problem 3 are lying on a frictionless table and connected by a massless spring of force constant k. A horizontal force F_1 is again exerted only on m_1, along the spring and away from m_2. What is the magnitude of the acceleration of the center of mass? (a) F_1/m_1. (b) $F_1/(m_1 + m_2)$. (c) $(F_1 + kx)/(m_1 + m_2)$, where x is the amount the spring is stretched. (d) $(m_1 + m_2)F_1/m_1m_2$.

5 • **SSM** If two particles have equal kinetic energies, are the magnitudes of their momenta necessarily equal? Explain and give an example.

6 • True or false:

(a) The momentum of a heavy object is greater than that of a light object moving at the same speed.
(b) The momentum of a system may be conserved even when mechanical energy is not.
(c) The velocity of the center of mass of a system equals the total momentum of the system divided by its total mass.

7 • How is the recoil of a rifle related to momentum conservation?

8 • **SSM** A child jumps from a small boat to a dock. Why does she have to jump with more energy than she would need if she were jumping the same distance from one dock to another?

9 •• **SSM** Much early research in rocket motion was done by Robert Goddard, physics professor at Clark College in Worcester, Massachusetts. A quotation from a 1920 editorial in the *New York Times* illustrates the public opinion of his work: "That Professor Goddard with his 'chair' at Clark College and the countenance of the Smithsonian Institution does not know the relation between action and reaction, and the need to have something better than a vacuum against which to react—to say that would be absurd. Of course, he only seems to lack the knowledge ladled out daily in high schools."[†] The belief that a rocket needs something to push against was a prevalent misconception before rockets in space were commonplace. Explain why that belief is wrong.

10 • Two bowling balls are moving with the same velocity, but one just slides down the alley, whereas the other rolls down the alley. Which ball has more kinetic energy? Explain.

11 • A philospher tells you, "All motion is impossible. Forces always come in equal but opposite action–reaction pairs. Therefore, all forces must cancel out." Answer his argument.

12 • If two objects collide and one is initially at rest, is it still possible for both to be at rest after the collision? (Any external forces acting on the two-object system are negligibly small.) Is it possible for one to be at rest? Explain.

13 • From the standpoint of physical laws, what is the problem with superhero comics where the hero flies around with no jets or hovers in midair while tossing huge objects like cars at the villains?

14 •• **SSM** If only external forces can cause the center of mass of a system of particles to accelerate, how can a car move? We normally think of the car's engine as supplying the force needed to accelerate the car, but is this true? Where does the external force that accelerates the car come from?

[†] On page 43 of the July 17, 1969, edition of the *New York Times* "A Correction" to their editorial of 1920 was printed. This commentary, which was published three days before man's first walk on the moon, stated that "it is now definitely established that a rocket can function in a vacuum as well as in an atmosphere. *The Times* regrets the error."

15 •• When we push on the brake pedal to slow down a car, a brake pad is pressed against the rotor so that the friction of the pad slows the wheel's rotation. However, the friction of the pad against the rotor can't be the force that slows the car down, because it is an internal force—both the rotor and the wheel are parts of the car, so any forces between them are purely internal to the system. What is the external force that slows down the car? Give a detailed explanation of how this force operates.

16 • Explain why a safety net can save the life of a circus performer.

17 • How might you estimate the collision time of a baseball and bat?

18 • Why does a wine glass survive a fall onto a carpet but not onto a concrete floor?

19 • True or false:

(a) In any perfectly inelastic collision, all the kinetic energy of the bodies is lost.
(b) In a head-on elastic collision, the relative speed of recession after the collision equals the relative speed of approach before the collision.
(c) Mechanical energy is conserved in an elastic collision.

20 •• [SSM] Under what conditions can all the initial kinetic energy of an isolated system consisting of two colliding bodies be lost in a collision?

21 •• Consider a perfectly inelastic collision of two objects of equal mass. (a) Is the loss of kinetic energy greater if the two objects have oppositely directed velocities of equal magnitude $v/2$ or if one of the two objects is initially at rest and the other has an initial velocity of v? (b) In which situation is the percentage loss in kinetic energy the greatest?

22 •• [SSM] A double-barrelled pea shooter is shown in Figure 8-48. Air is blown from the left end of the straw, and identical peas A and B are positioned inside the straw as shown. If the shooter is held horizontally while the peas are shot off, which pea, A or B, will travel farther after leaving the straw? Why? *Hint: The answer has to do with the impulse–momentum theorem.*

FIGURE 8-48 Problem 22

23 •• A particle of mass m_1 traveling with a speed v makes a head-on elastic collision with a stationary mass m_2. In which scenario will the energy imparted to the particle of mass m_2 be greatest? (a) $m_2 \ll m_1$. (b) $m_2 = m_1$. (c) $m_2 \gg m_1$. (d) None of these.

Optional Section

24 • Describe a perfectly inelastic collision as viewed in the center-of-mass reference frame.

25 • Nozzles for a garden hose are often made with a right-angle shape as shown in Figure 8-49. If you open the nozzle and spray water out, you will find that the nozzle presses against your hand with a pretty strong force— much stronger than if you used a nozzle not bent into a right angle. Why is this?

FIGURE 8-49 Problem 25

26 • Why can friction and the force of gravity usually be neglected in collision problems?

27 • As a pendulum bob swings back and forth, is the momentum of the bob conserved? Explain.

28 •• [SSM] A railroad car is passing by a grain elevator, which is dumping grain into it at a constant rate. (a) Does momentum conservation imply that the railroad car should be slowing down as it passes the grain elevator? Assume that the track is frictionless and perfectly level. (b) If the car is slowing down, this implies that there is some external force acting on the car to slow it down. Where does this force come from? (c) After passing the elevator, the railroad car springs a leak, and grain starts leaking out of a vertical hole in its floor at a constant rate. Should the car speed up as it loses mass?

29 •• [SSM] To show that even really bright people can make mistakes, consider the following problem which was given to the freshman class at Caltech on an exam (paraphrased): *A sailboat is sitting in the water on a windless day. In order to make the boat move, a misguided sailor sets up a fan in the back of the boat to blow into the sails to make the boat move forward. Explain why the boat won't move.* The idea was that the net force of the wind pushing the sail forward would be counteracted by the force pushing the fan back (Newton's third law). However, as one of the students pointed out to his professor, the sailboat *could* in fact move forward. Why is that?

Estimation and Approximation

30 •• A 2000-kg car traveling at 90 km/h crashes into a concrete wall that does not give at all. (a) Estimate the time of collision, assuming that the center of the car travels halfway to the wall with constant deceleration. (Use any reasonable length for the car.) (b) Estimate the average force exerted by the wall on the car.

31 •• In hand-pumped railcar races, a speed of 32 km/h has been achieved by teams of four. A car of mass 350 kg is moving at that speed toward a river when Carlos, the chief pumper, notices that the bridge ahead is out. All four people (of mass 75 kg each) jump simultaneously backward off the car with a velocity that has a horizontal component of 4 m/s relative to the car after jumping. The car proceeds off the bank and falls into the water a horizontal distance of 25.0 m off the bank. (*a*) Estimate the time of the fall of the railcar. (*b*) What happens to the team of pumpers?

32 •• **SSM** A counterintuitive physics demonstration can be performed by firing a rifle bullet into a melon. (Don't try this at home!) When hit, nine times out of ten the melon will jump backward, toward the rifle, *opposite* to the direction in which the bullet was moving. (The tenth time, the melon simply explodes.) Doesn't this violate the laws of conservation of momentum? It doesn't, because we're not dealing simply with a two-body collision. Instead, a significant fraction of the energy of the bullet can be dumped into a jet of melon that is violently ejected out of the front of the melon. This jet can have a momentum greater than the momentum of the bullet, so that the rest of the melon must jump backward to conserve momentum. Let's make the following assumptions:

1. The mass of the melon is 2.50 kg.
2. The mass of the rifle bullet is 10.4 g and its velocity is 1800 ft/s.
3. 10 percent of the energy of the bullet is deposited as kinetic energy into a jet flying out of the front of the melon.
4. The mass of the matter in the jet is 0.14 kg.
5. All collisions occur in a straight line. What would be the speed of the melon's recoil? Compare this to a typical measured recoil speed of about 1.6 ft/s.

Finding the Center of Mass

33 • Three point masses of 2 kg each are located on the *x* axis at the origin, at $x = 0.20$ m, and at $x = 0.50$ m. Find the center of mass of the system.

34 • **SSM** Alley Oop's club-ax consists of a symmetrical 8-kg stone attached to the end of a uniform 2.5-kg stick that is 98 cm long. The dimensions of the club-ax are shown in Figure 8-50. How far is the center of mass from the handle end of the club-ax?

FIGURE 8-50 Problem 34

35 • **SOLVE** ✓ Three balls A, B, and C, with masses of 3 kg, 1 kg, and 1 kg, respectively, are connected by massless rods. The balls are located as in Figure 8-51. What are the coordinates of the center of mass?

FIGURE 8-51 Problems 35, 45

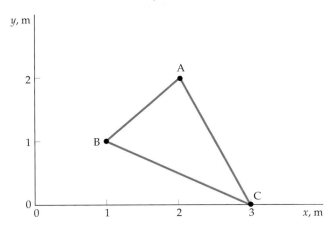

36 • **SOLVE** By symmetry, locate the center of mass of an equilateral triangle of side length *a* with one vertex on the *y* axis and the others at $(-a/2, 0)$ and $(+a/2, 0)$.

37 •• **SSM** Find the center of mass of the uniform sheet of plywood in Figure 8-52. We shall consider this as two sheets, a square sheet of 3-m edge length and mass m_1 and a rectangular sheet 1 m × 2 m with a mass of $-m_2$. Let the coordinate origin be at the lower left-hand corner of the sheet.

FIGURE 8-52 Problem 37

38 •• A can in the shape of a symmetrical cylinder with mass *M* and height *H* is filled with water. The initial mass of the water is *M*, the same mass as the can. A hole is punched in the bottom of the can, and the water drains out. (*a*) If the height of the water in the can is *x*, what is the height of the center of mass of the can + water? (*b*) What is the minimum height of the center of mass as the water drains out?

*Finding the Center of Mass by Integration

39 •• **SSM** Show that the center of mass of a uniform semicircular disk of radius *R* is at a point $4R/(3\pi)$ from the center of the circle.

40 ••• Find the center of mass of a homogeneous solid hemisphere of radius R and mass M.

41 ••• Find the center of mass of a thin homogeneous hemispherical shell.

42 ••• A sheet of metal is cut in the shape of a parabola (Figure 8-53). The curved edge of the sheet is specified by the equation $y = ax^2$, and y ranges from 0 to b. Find the center of mass in terms of a and b. (You will need to find the area first.)

FIGURE 8-53 Problem 42

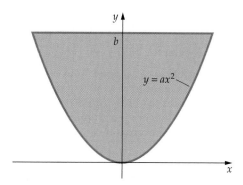

Motion of the Center of Mass

43 • ✓ Two 3-kg particles have velocities $\vec{v}_1 = 2 \text{ m/s } \hat{i} + 3 \text{ m/s } \hat{j}$ and $\vec{v}_2 = 4 \text{ m/s } \hat{i} - 6 \text{ m/s } \hat{j}$. Find the velocity of the center of mass for the system.

44 • SSM ✓ A 1500-kg car is moving westward with a speed of 20 m/s, and a 3000-kg truck is traveling east with a speed of 16 m/s. Find the velocity of the center of mass of the system.

45 • A force $\vec{F} = 12 \text{ N } \hat{i}$ is applied to the 3-kg ball in Figure 8-51 in Problem 35. What is the acceleration of the center of mass?

46 •• A block of mass m is attached to a string and suspended inside a hollow box of mass M. The box rests on a scale that measures the system's weight. (a) If the string breaks, does the reading on the scale change? Explain your reasoning. (b) Assume that the string breaks and the mass m falls with constant acceleration g. Find the acceleration of the center of mass of the box–block system, giving both direction and magnitude. (c) Using the result from (b), determine the reading on the scale while m is in free-fall.

47 •• SSM A massless, vertical spring of force constant k is attached at the bottom to a platform of mass m_p, and at the top to a massless cup, as in Figure 8-54. The platform rests on a scale. A ball of mass m_b is dropped into the cup from a negligible height. What is the reading on the scale (a) when the spring is compressed an amount $d = m_b g/k$; (b) when the ball comes to rest momentarily with the spring compressed; (c) when the ball again comes to rest in its original position?

FIGURE 8-54 Problems 47, 49, 136

48 •• SSM In the Atwood's machine in Figure 8-55, the string passes over a fixed cylinder of mass m_c. The cylinder does not rotate. Instead, the string slides on its frictionless surface. (a) Find the acceleration of the center of mass of the two-block-and-cylinder system. (b) Use Newton's second law for systems to find the force F exerted by the support. (c) Find the tension T in the string connecting the blocks and show that $F = m_c g + 2T$.

FIGURE 8-55 Problem 48

49 •• Repeat Problem 47(a) and 47(b) if the ball is dropped into the cup from a height h above the cup.

The Conservation of Momentum

50 • A woman of mass 55 kg jumps off the bow of a 75-kg canoe that is initially at rest. If her velocity is 2.5 m/s to the right, what is the velocity of the canoe after she jumps?

51 • A 5-kg object and a 10-kg object are connected by a massless compressed spring and rest on a frictionless table. After the spring is released, the object with the smaller mass has a velocity of 8 m/s to the left. What is the velocity of the object with the larger mass?

52 • SSM Figure 8-56 shows the behavior of a projectile just after it has broken up into three pieces. What was the speed of the projectile the instant before it broke up? (a) v_3. (b) $v_3/3$. (c) $v_3/4$. (d) $4v_3$. (e) $(v_1 + v_2 + v_3)/4$.

FIGURE 8-56
Problem 52

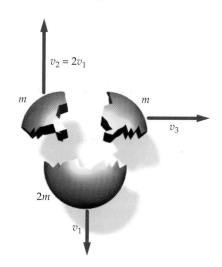

53 • A shell of mass m and speed v explodes into two identical fragments. If the shell was moving horizontally with respect to the earth, and one of the fragments is subsequently moving vertically with speed v, find the velocity $\vec{v}'$ of the other fragment.

54 •• **SSM** A block and a gun are firmly fixed to opposite ends of a long glider mounted on a frictionless air track (Figure 8-57). The block and gun are a distance L apart. The system is initially at rest. The gun is fired and the bullet leaves the muzzle with a velocity v_b and impacts the block, becoming imbedded in it. The mass of the bullet is m_b and the mass of the gun–glider–block system is m_p. (a) What is the velocity of the glider immediately after the bullet leaves the muzzle? (b) What is the velocity of the glider immediately after the bullet comes to rest in the block? (c) How far does the glider move while the bullet is in transit between the gun at rest and the block at rest?

FIGURE 8-57 Problem 54

55 •• A small object of mass m slides down a wedge of mass $2m$ and exits smoothly onto a frictionless table. The wedge is initially at rest on the table. If the object is initially at rest at a height h above the table, find the velocity of the wedge when the object leaves it.

56 •• **SSM** Each glider on a frictionless air track (Figure 8-58) supports a strong magnet. These magnets attract each other. The mass of glider 1 is 0.10 kg and the mass of glider 2 is 0.20 kg. (Each mass includes the mass of the glider plus the supported magnet.) The origin is at the left end of the track, the center of glider 1 is at $x_1 = 0.10$ m and the center of glider 2 is at $x_2 = 1.60$ m. Glider 1 is 10 cm long, while glider 2 is 20 cm long. Each glider has its center of mass at its center. (a) When the two gliders are released from rest, they move toward each other and stick together. What is the position of the center of each glider when they first touch? (b) Will the two gliders continue to move after they stick together? Explain.

FIGURE 8-58 Problem 56

Kinetic Energy of a System of Particles

57 • **SSM** **iSOLVE** A 3-kg block is traveling to the right at 5 m/s, and a second 3-kg block is traveling to the left at 2 m/s. (a) Find the total kinetic energy of the two blocks. (b) Find the velocity of the center of mass of the two-block system. (c) Find the velocities of each block relative to the center of mass. (d) Find the kinetic energy of the motion of the blocks relative to the center of mass. (e) Show that your answer for

part (a) is greater than your answer for part (d) by an amount equal to the kinetic energy associated with the motion of the center of mass.

58 • Repeat Problem 57 with the second 3-kg block replaced by a 5-kg block moving to the right at 3 m/s.

Impulse and Average Force

59 • You kick a soccer ball of mass 0.43 kg. The ball leaves your foot with an initial speed of 25 m/s. (a) What is the impulse imparted to the ball by you? (b) If your foot is in contact with the ball for 0.008 s, what is the average force exerted by your foot on the ball?

60 • A 0.3-kg brick is dropped from a height of 8 m. It hits the ground and comes to rest. (a) What is the impulse exerted by the ground on the brick? (b) If it takes 0.0013 s from the time the brick first touches the ground until it comes to rest, what is the average force exerted by the ground on the brick?

61 • **SSM** A meteorite of mass 30.8 tonne (1 tonne = 1000 kg) is exhibited in the American Museum of Natural History in New York City. Suppose that the kinetic energy of the meteorite as it hit the ground was 617 MJ. Find the impulse I experienced by the meteorite up to the time its kinetic energy was halved (which took about $t = 3.0$ s). Find also the average force F exerted on the meteorite during this time interval.

62 •• **iSOLVE** When a 0.15-kg baseball is hit, its velocity changes from +20 m/s to −20 m/s. (a) What is the magnitude of the impulse delivered by the bat to the ball? (b) If the baseball is in contact with the bat for 1.3 ms, what is the average force exerted by the bat on the ball?

63 •• **SSM** **iSOLVE** ✓ A 60-g handball moving with a speed of 5.0 m/s strikes the wall at an angle of 40° and then bounces off with the same speed at the same angle. It is in contact with the wall for 2 ms. What is the average force exerted by the ball on the wall?

64 •• **iSOLVE** You throw a 150-g ball to a height of 40 m. (a) Use a reasonable value for the distance the ball moves while it is in your hand to calculate the average force exerted by your hand and the time the ball is in your hand while you throw it. (b) Is it reasonable to neglect the weight of the ball while it is being thrown?

65 •• A handball of mass 60 g is thrown straight against a wall with a speed of 10 m/s. It rebounds with a speed of 8 m/s. (a) What impulse is delivered to the wall? (b) If the ball is in contact with the wall for 0.003 s, what average force is exerted on the wall by the ball? (c) The ball is caught by a player who brings it to rest. In the process, her hand moves back 0.5 m. What is the impulse received by the player? (d) What average force was exerted on the player by the ball?

66 ••• **iSOLVE** The great limestone caverns were formed by dripping water. (a) If water droplets of 0.03 mL fall from a height of 5 m at a rate of 10 per minute, what is the average force exerted on the limestone floor by the droplets of water during a 1-min period? (b) Compare this force to the weight of a water droplet.

Collisions in One Dimension

67 • **SSM** A 2000-kg car traveling to the right at 30 m/s is chasing a second car of the same mass that is traveling to the right at 10 m/s. (a) If the two cars collide and stick together, what is their speed just after the collision? (b) What fraction of the initial kinetic energy of the cars is lost during this collision? Where does it go?

68 • An 85-kg running back moving at 7 m/s makes a perfectly inelastic collision with a 105-kg linebacker who is initially at rest. What is the speed of the players just after their collision?

69 • A 5.0-kg object with a speed of 4.0 m/s collides head-on with a 10-kg object moving toward it with a speed of 3.0 m/s. The 10-kg object stops dead after the collision. (a) What is the final speed of the 5-kg object? (b) Is the collision elastic?

70 • A ball of mass m moves with speed v to the right toward a much heavier bat that is moving to the left with speed v. Find the speed of the ball after it makes an elastic collision with the bat.

71 •• **SSM** SOLVE A proton of mass m undergoes a head-on elastic collision with a stationary carbon nucleus of mass $12m$. The speed of the proton is 300 m/s. (a) Find the velocity of the center of mass of the system. (b) Find the velocity of the proton after the collision.

72 •• A 3-kg block moving at 4 m/s makes a head-on elastic collision with a stationary block of mass 2 kg. Use conservation of momentum and the fact that the relative speed of recession equals the relative speed of approach to find the velocity of each block after the collision. Check your answer by calculating the initial and final kinetic energies of each block.

73 •• SOLVE A block of mass $m_1 = 2$ kg slides along a frictionless table with a speed of 10 m/s. Directly in front of it, and moving in the same direction with a speed of 3 m/s, is a block of mass $m_2 = 5$ kg. A massless spring with spring constant $k = 1120$ N/m is attached to the second block as in Figure 8-59. (a) Before m_1 makes contact with the spring, what is the velocity of the center of mass of the system? (b) During the collision, the spring is compressed by a maximum amount Δx. What is the value of Δx? (c) The blocks will eventually separate again. What are the final velocities of the two blocks measured in the reference frame of the table?

FIGURE 8-59 Problem 73

74 •• **SSM** SOLVE A bullet of mass m is fired vertically from below into a thin sheet of plywood of mass M that is initially at rest, supported by a thin sheet of paper. The bullet blasts through the plywood, which rises to a height H above the paper before falling back down. The bullet continues rising to a height h above the paper. (a) Express the upward velocity of the bullet and the plywood immediately after the bullet exits the plywood in terms of h and H. (b) Use conservation of momentum to express the speed of the bullet before it enters the sheet of plywood in terms of m, h, M, and H. (c) Obtain expressions for the mechanical energy of the system before and after the inelastic collision. (d) Express the energy dissipated in terms of m, h, M, and H.

75 •• A proton of mass m is moving with initial speed v_0 directly toward the center of an α particle of mass $4m$, which is initially at rest. Because both particles carry positive electrical charge, they repel each other. Find the speed v' of the α particle (a) when the distance between the two particles is least and (b) later when the two particles are far apart.

76 • An electron collides elastically with a hydrogen atom initially at rest. All motion occurs along a straight line. What fraction of the electron's initial kinetic energy is transferred to the atom? (Take the mass of the hydrogen atom to be 1840 times the mass of an electron.)

77 •• SOLVE ✓ A 16-g bullet is fired into the bob of a ballistic pendulum of mass 1.5 kg. When the bob is at its maximum height, the strings make an angle of 60° with the vertical. The length of the pendulum is 2.3 m. Find the speed of the bullet.

78 •• **SSM** Show that in a one-dimensional elastic collision, if the mass and velocity of object 1 are m_1 and v_{1i}, and if the mass and velocity of object 2 are m_2 and v_{2i}, then the final velocities v_{1f} and v_{2f} are given by

$$v_{1f} = \frac{m_1 - m_2}{m_1 + m_2}v_{1i} + \frac{2m_2}{m_1 + m_2}v_{2i}$$

$$v_{2f} = \frac{2m_1}{m_1 + m_2}v_{1i} + \frac{m_2 - m_1}{m_1 + m_2}v_{2i}$$

79 •• Use the results of Problem 78 to investigate one-dimensional elastic collisions in the following limits: (a) When the two masses are equal, show that the particles "swap" velocities: $v_{1f} = v_{2i}$ and $v_{2f} = v_{1i}$. (b) If $m_2 \gg m_1$, show that $v_{1f} \approx -v_{1i} + 2v_{2i}$ and $v_{2f} \approx v_{2i}$.

Perfectly Inelastic Collisions and the Ballistic Pendulum

80 •• A bullet of mass m_1 is fired with a speed v into the bob of a ballistic pendulum of mass m_2. The bob is attached to a very light rod of length L that is pivoted at the other end. The bullet is stopped in the bob. Find the minimum v such that the bob will swing through a complete circle.

81 •• **SSM** A bullet of mass m_1 is fired with a speed v into the bob of a ballistic pendulum of mass m_2. Find the maximum height h attained by the bob if the bullet passes through the bob and emerges with a speed $v/2$.

82 • If we take a heavy wooden block normally used as a ballistic pendulum and put it on a flat table before firing a bullet into it, about how far will it slide before coming to a stop? Assume that the mass of the rifle bullet is 10.5 g, the mass of the wooden block is 10.5 kg, the bullet's velocity is 750 m/s, and the coefficient of kinetic friction of the block against the table is 0.22. Also assume that the bullet doesn't cause the block to rotate.

83 •• A ball with mass $M = 0.425$ kg and speed $V = 1.3$ m/s rolls across a level table into an open box with mass $m = 0.327$ kg. The box with the ball inside it then slides across the table for a distance $x = 0.520$ m. What is the coefficient of kinetic friction of the table?

84 •• [SSM] Tarzan is in the path of a pack of stampeding elephants when Jane swings in to the rescue on a rope vine, hauling him off to safety. The length of the vine is 25 m, and Jane starts her swing with the rope horizontal. If Jane's mass is 54 kg, and Tarzan's is 82 kg, to what height above the ground will the pair swing after she grabs him?

Exploding Objects and Radioactive Decay

85 •• [SOLVE] The beryllium isotope ^{8}Be is unstable and decays into two α particles (helium nuclei of mass $m = 6.68 \times 10^{-27}$ kg) with the release of 1.5×10^{-14} J of energy. Determine the velocities of the two α particles that arise from the decay of a ^{8}Be nucleus at rest, assuming that all the energy appears as kinetic energy of the particles.

86 •• The light isotope of lithium, ^{5}Li, is unstable and breaks up spontaneously into a proton (hydrogen nucleus) and an α particle (helium nucleus). In this process, a total energy of 3.15×10^{-13} J is released, appearing as the kinetic energy of the two decay products. Determine the velocities of the proton and α particle that arise from the decay of a ^{5}Li nucleus at rest. (*Note:* The masses of the proton and alpha particle are $m_p = 1.67 \times 10^{-27}$ kg and $m_\alpha = 4m_p = 6.68 \times 10^{-27}$ kg.)

87 ••• [SOLVE] A projectile of mass $m = 3$ kg is fired with an initial speed of 120 m/s at an angle of 30° with the horizontal. At the top of its trajectory, the projectile explodes into two fragments of masses 1 kg and 2 kg. The 2-kg fragment lands on the ground directly below the point of explosion 3.6 s after the explosion. (*a*) Determine the velocity of the 1-kg fragment immediately after the explosion. (*b*) Find the distance between the point of firing and the point at which the 1-kg fragment strikes the ground. (*c*) Determine the energy released in the explosion.

88 ••• [SSM] The boron isotope ^{9}B is unstable and disintegrates into a proton and two α particles. The total energy released as kinetic energy of the decay products is 4.4×10^{-14} J. In one such event, with the ^{9}B nucleus at rest prior to decay, the velocity of the proton is measured as 6.0×10^6 m/s. If the two α particles have equal energies, find the magnitude and the direction of their velocities with respect to that of the proton.

*Coefficient of Restitution

89 • The coefficient of restitution for steel on steel is measured by dropping a steel ball onto a steel plate that is rigidly attached to the earth. If the ball is dropped from a height of 3 m and rebounds to a height of 2.5 m, what is the coefficient of restitution?

90 • [SSM] According to the official rules of racquetball, a ball acceptable for tournament play must bounce to a height of between 173 and 183 cm when dropped from a height of 254 cm at room temperature. What is the acceptable range of values for the coefficient of restitution for the racquetball–floor system?

91 • [SOLVE ✓] A ball bounces to 80 percent of its original height. (*a*) What fraction of its mechanical energy is lost each time it bounces? (*b*) What is the coefficient of restitution of the ball–floor system?

92 •• [SOLVE] A 2-kg object moving at 6 m/s collides with a 4-kg object that is initially at rest. After the collision, the 2-kg object moves backward at 1 m/s. (*a*) Find the velocity of the 4-kg object after the collision. (*b*) Find the energy lost in the collision. (*c*) What is the coefficient of restitution for this collision?

93 •• A 2-kg block moving to the right with speed 5 m/s collides with a 3-kg block that is moving in the same direction at 2 m/s, as in Figure 8-60. After the collision, the 3-kg block moves at 4.2 m/s. Find (*a*) the velocity of the 2-kg block after the collision and (*b*) the coefficient of restitution for the collision.

FIGURE 8-60 Problem 93

*Collisions in Three Dimensions

94 •• [SSM] In Section 8-6 it was proven by geometrical means that when a particle elastically collides with another particle of equal mass that is initially at rest, the two separate at right angles. In this problem we will examine another way of proving this, one that illustrates the power of abstract vector notation. (*a*) Given $\vec{A} = \vec{B} + \vec{C}$, show that $A^2 = B^2 + C^2 + 2\vec{B} \cdot \vec{C}$. (*b*) Let the momentum of the initially moving particle be $\vec{P}$ and the momenta of the particles after the collision be $\vec{p}_1$ and $\vec{p}_2$. By writing the vector equation for the conservation of momentum and obtaining the dot product of each side with itself, and then comparing it to the equation for the conservation of energy, show that $\vec{p}_1 \cdot \vec{p}_2 = 0$.

95 • [SOLVE] In a pool game, the cue ball, which has an initial speed of 5 m/s, makes an elastic collision with the eight ball, which is initially at rest. After the collision, the eight ball moves at an angle of 30° to the original direction of the cue ball. (*a*) Find the direction of motion of the cue ball after the collision. (*b*) Find the speed of each ball. Assume that the balls have equal mass.

96 •• Object A with mass m and velocity $v_0\hat{i}$ collides head-on with object B of mass $2m$ and velocity $\frac{1}{2}v_0\hat{j}$. Following the collision, object B has a velocity of $\frac{1}{4}v_0\hat{i}$. (a) Determine the velocity of object A after the collision. (b) Is the collision elastic? If not, express the change in the kinetic energy in terms of m and v_0.

97 •• SSM ISOLVE A puck of mass 5 kg moving at 2 m/s approaches an identical puck that is stationary on frictionless ice. After the collision, the first puck leaves with a speed v_1 at 30° to the original line of motion; the second puck leaves with speed v_2 at 60°, as in Figure 8-61. (a) Calculate v_1 and v_2. (b) Was the collision elastic?

FIGURE 8-61 Problem 97

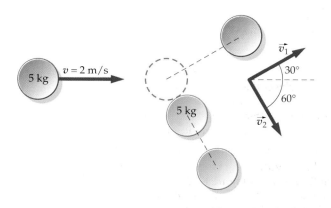

98 •• Figure 8-62 shows the result of a collision between two objects of unequal mass. (a) Find the speed v_2 of the larger mass after the collision and the angle θ_2. (b) Show that the collision is elastic.

FIGURE 8-62 Problem 98

99 •• SSM ISOLVE A ball moving at 10 m/s makes an off-center elastic collision with another ball of equal mass that is initially at rest. The incoming ball is deflected at an angle of 30° from its original direction of motion. Find the velocity of each ball after the collision.

100 •• A particle has initial speed v_0. It collides with a second particle *with the same mass* that is at rest, and is deflected through an angle ϕ. Its speed after the collision is v. The second particle recoils, and its velocity makes an angle θ with the initial direction of the first particle. (a) Show that

$$\tan\theta = \frac{v\sin\phi}{(v_0 - v\cos\phi)}$$

(b) Show that if the collision is elastic, $v = v_0\cos\phi$.

Center-of-Mass Frame

101 •• In the center-of-mass reference frame a particle with mass m_1 and momentum p_1 makes an elastic head-on collision with a second particle of mass m_2 and momentum $p_2 = -p_1$. After the collision its momentum is p_1'. Write the total initial energy in terms of m_1, m_2, and p_1 and the total final energy in terms of m_1, m_2, and p_1', and show that $p_1' = \pm p_1$. If $p_1' = -p_1$, the particle is merely turned around by the collision and leaves with the speed it had initially. What is the significance of the plus sign in your solution?

102 •• SSM ISOLVE A 3-kg block is traveling in the negative x direction at 5 m/s, and a 1-kg block is traveling in the positive x direction at 3 m/s. (a) Find the velocity v_{cm} of the center of mass. (b) Subtract v_{cm} from the velocity of each block to find the velocity of each block in the center-of-mass reference frame. (c) After they make a head-on elastic collision, the velocity of each block is reversed (in this frame). Find the velocity of each block in the center-of-mass frame after the collision. (d) Transform back into the original frame by adding v_{cm} to the velocity of each block. (e) Check your result by finding the initial and final kinetic energies of the blocks in the original frame and comparing them.

103 •• Repeat Problem 102 with the second block having a mass of 5 kg and moving to the right at 3 m/s.

*Systems With Continuously Varying Mass: Rocket Propulsion

104 •• ISOLVE A rocket burns fuel at a rate of 200 kg/s and exhausts the gas at a relative speed of 6 km/s. Find the thrust of the rocket.

105 •• A rocket has an initial mass of 30,000 kg, of which 80 percent is the fuel. It burns fuel at a rate of 200 kg/s and exhausts its gas at a relative speed of 1.8 km/s. Find (a) the thrust of the rocket, (b) the time until burnout, and (c) its speed at burnout assuming it moves straight upward near the surface of the earth where the gravitational field g is constant.

106 •• SSM The *specific impulse* of a rocket propellant is defined as $I_{sp} = F_{th}/(Rg)$, where F_{th} is the thrust of the propellant, g the magnitude of free-fall acceleration at the surface of the earth, and R the rate at which the propellant is burned. The rate depends predominantly on the type and exact mixture of the propellant. (a) Show that the specific impulse has the dimension of time. (b) Show that $u_{ex} = gI_{sp}$, where u_{ex} is the exhaust velocity of the propellant. (c) What is the specific impulse (in seconds) of the propellant used in the Saturn V rocket of Example 8-21?

107 ••• [SSM] The initial *thrust-to-weight ratio* τ_0 of a rocket is $\tau_0 = F_{th}/(m_0 g)$, where F_{th} is the rocket's thrust and m_0 the initial mass of the rocket, including the propellant. (*a*) For a rocket launched straight up from the earth's surface, show that $\tau_0 = 1 + (a_0/g)$, where a_0 is the initial acceleration of the rocket. For manned rocket flight, τ_0 cannot be made much larger than 4 for the comfort and safety of the astronauts. (The astronauts will feel that their weight as the rocket lifts off is equal to τ_0 times their normal weight.) (*b*) Show that the final velocity of a rocket launched from the earth's surface, in terms of τ_0 and I_{sp} (see Problem 106) can be written as

$$ v_f = gI_{sp}\left[\ln\left(\frac{m_0}{m_f}\right) - \frac{1}{\tau_0}\left(1 - \frac{m_f}{m_0}\right) \right] $$

where m_f is the mass of the rocket (not including the spent propellant). (*c*) Using a spreadsheet program or graphing calculator, graph v_f as a function of the mass ratio m_0/m_f for $I_{sp} = 250$ s and $\tau_0 = 2$ for values of the mass ratio from 2 to 10. (Note that the mass ratio cannot be less than 1.) (*d*) To lift a rocket into orbit, a final velocity after burnout of $v_f = 7$ km/s is needed. Calculate the mass ratio required of a single stage rocket to do this, using the values of specific impulse and thrust ratio given in Part (*b*). For engineering reasons, it is difficult to make a rocket with a mass ratio much greater than 10. Can you see why multistage rockets are usually used to put payloads into orbit around the earth?

108 •• The height that a model rocket launched from earth can reach can be estimated by assuming that the burn time is short compared to the total flight time, so for most of the flight the rocket is in free-fall. (This estimate neglects the burn time in calculations of both time and displacement.) For a model rocket with specific impulse $I_{sp} = 100$ s, mass ratio $m_0/m_f = 1.2$, and initial thrust-to-weight ratio $\tau_0 = 5$ (these parameters are defined in Problems 106 and 107), estimate (*a*) the height the rocket can reach, and (*b*) the total flight time. (*c*) Justify the assumption used in the estimates by comparing the flight time from Part (*b*) to the time it takes for the fuel to be spent.

General Problems

109 • A model-train car of mass 250 g traveling with a speed of 0.50 m/s links up with another car of mass 400 g that is initially at rest. What is the speed of the cars immediately after they have linked together? Find the initial and final kinetic energies.

110 • (*a*) Find the total kinetic energy of the two model-train cars of Problem 109 before they couple. (*b*) Find the initial velocities of the two cars relative to the center of mass of the system, and use them to calculate the initial center-of-mass kinetic energy of the system. (*c*) Find the kinetic energy of the center of mass. (*d*) Compare the sum of your answers for (*b*) and (*c*) with that for (*a*).

111 • [SSM] A 4-kg fish is swimming at 1.5 m/s to the right. He swallows a 1.2-kg fish swimming toward him at

3 m/s. Neglecting water resistance, what is the velocity of the larger fish immediately after his lunch?

112 • A 3-kg block moves at 6 m/s to the right while a 6-kg block moves at 3 m/s to the right. Find (*a*) the total kinetic energy of the two-block system, (*b*) the velocity of the center of mass, (*c*) the center-of-mass kinetic energy, and (*d*) the kinetic energy relative to the center of mass.

113 • A 1500-kg car traveling north at 70 km/h collides at an intersection with a 2000-kg car traveling west at 55 km/h. The two cars stick together. (*a*) What is the total momentum of the system before the collision? (*b*) Find the magnitude and direction of the velocity of the wreckage just after the collision.

114 •• [SSM] A 60-kg woman stands on the back of a 6-m-long, 120-kg raft that is floating at rest in still water with no friction. The raft is 0.5 m from a fixed pier, as shown in Figure 8-63. (*a*) The woman walks to the front of the raft and stops. How far is the raft from the pier now? (*b*) While the woman walks, she maintains a constant speed of 3 m/s relative to the raft. Find the total kinetic energy of the system (woman plus raft), and compare with the kinetic energy if the woman walked at 3 m/s on a raft tied to the pier. (*c*) Where does this energy come from, and where does it go when the woman stops at the front of the raft? (*d*) On land, the woman puts a lead shot 6 m. She stands at the back of the raft, aims forward, and puts the shot so that just after it leaves her hand, it has the same velocity relative to her as it did when she threw it from the ground. Approximately, where does her shot land?

FIGURE 8-63 Problem 114

115 •• [SOLVE] ✔ A 1-kg steel ball and a 2-m cord of negligible mass make up a simple pendulum that can pivot without friction about the point *O*, as in Figure 8-64. This pendulum is released from rest in a horizontal position and when the ball is at its lowest point it strikes a 1-kg block sitting at rest on a shelf. Assume that the collision is perfectly elastic and take the coefficient of friction between the block and shelf to be 0.1. (*a*) What is the velocity of the block just after impact? (*b*) How far does the block move before coming to rest?

FIGURE 8-64 Problem 115

1 kg

2 m

O

1kg

m = 0.8 kg

500 m/s   m = 15 g

0.8 m

D

FIGURE 8-66 Problem 117

118 •• Two particles of masses m and $4m$ are moving in a vacuum at right angles as in Figure 8-67. A force $\vec{F}$ acts on the particle of mass m for a time T. As a result, the velocity of this particle becomes $4v$ in its original direction. A force equal to force $\vec{F}$ acts on the other particle for the same time T. Find the new velocity $\vec{v}\,'$ of the particle of mass $4m$.

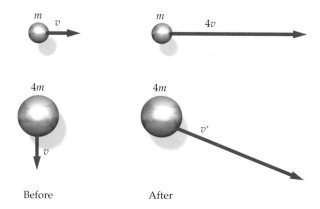

m   v

m   4v

4m

4m

v

v'

Before   After

FIGURE 8-67 Problem 118

116 •• **SSM** Figure 8-65 shows a World War I cannon mounted on a railcar so that it will project a shell at an angle of 30°. With the car initially at rest, the cannon fires a 200-kg projectile at 125 m/s. (All values are for the frame of reference of the track.) Now consider a system composed of a cannon, shell, and railcar, all on the frictionless track. (a) Will the total vector momentum of that system be the same (that is, "conserved") before and after the shell is fired? Explain your answer. (b) If the mass of the railcar plus cannon is 5000 kg, what will be the recoil velocity of the car along the track after the firing? (c) The shell is observed to rise to a maximum height of 180 m as it moves through its trajectory. At this point, its speed is 80 m/s. On the basis of this information, calculate the amount of thermal energy produced by air friction on the shell on its way from firing to this maximum height.

FIGURE 8-65 Problem 116

117 •• A 15-g bullet traveling at 500 m/s strikes an 0.8-kg block of wood that is balanced on a table edge 0.8 m above the ground (Figure 8-66). If the bullet buries itself in the block, find the distance D at which the block hits the floor.

119 •• One popular, if dangerous, classroom demonstration involves holding a baseball an inch or so directly above a basketball, holding the basketball a few feet above a hard floor, and dropping the two balls simultaneously. The two balls will collide just after the basketball bounces from the floor; the baseball will then rocket off into the ceiling tiles with a hard "thud" while the basketball will stop in midair. (The author of this problem once broke a light doing this.) (a) Assuming that the collision of the basketball with the floor is elastic, what is the relation between the velocities of the balls just before they collide? (b) Assuming the collision between the two balls is elastic, use the result of Part (a) and the conservation of momentum and energy to show that, if the basketball is three times as heavy as the baseball, the final velocity of the basketball will be zero. (This is approximately the true mass ratio, which is why the demonstration is so dramatic.) (c) If the speed of the top ball is v just before the collision, what is its speed just after the collision?

120 ••• (a) In Problem 119, if we held a third ball above the other two, and wanted both the lowest and middle balls to stop in mid-air, what should be the ratio of the mass of the top ball to the mass of the middle ball? (b) If the speed of the top ball is v just before the collision, what is its speed just after the collision?

121 •• [SSM] In the "slingshot effect," the transfer of energy in an elastic collision is used to boost the energy of a space probe so that it can escape from the solar system. All speeds are relative to an inertial frame in which the center of the sun remains at rest. Figure 8-68 shows a space probe moving at 10.4 km/s toward Saturn, which is moving at 9.6 km/s toward the probe. Because of the gravitational attraction between Saturn and the probe, the probe swings around Saturn and heads back in the opposite direction with speed v_f. (*a*) Assuming this collision to be a one-dimensional elastic collision with the mass of Saturn much greater than that of the probe, find v_f. (*b*) By what factor is the kinetic energy of the probe increased? Where does this energy come from?

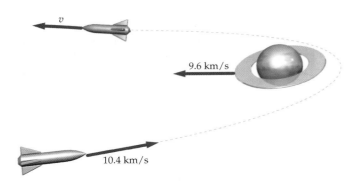

FIGURE 8-68 Problem 121

122 •• [SSM] Imagine that a flashlight is floating in intergalactic space, far from any planet or star. It has batteries in it that are charged with a total energy $E = 1.5$ kJ. When the flashlight is turned on, it loses energy in the form of light; because of this, it also loses mass slightly, from $E = mc^2$. Because the lost "mass" is traveling away at the speed of light, the flashlight should start moving in the opposite direction. (*a*) If the flashlight loses energy ΔE, argue that the change in momentum of the flashlight should be $P = \Delta E/c$. (*b*) If the mass of the flashlight is 1.5 kg, and it is turned on and left on until the batteries are discharged, what is its final velocity? Assume that the flashlight is 100 percent efficient in converting battery power into radiant power, and that it is initially at rest.

123 • A block has a mass $M = 1$ kg. If we could convert 1 g of the mass of the block into a well-collimated beam of light shining in one direction, calculate the speed attained by the rest of the block. (See Problem 122 for a discussion of the momentum carried by a light beam.)

124 •• You (mass 80 kg) and your friend (mass unknown) are in a rowboat (mass 60 kg) on a calm lake. You are at the center of the boat rowing and she is at the back, 2 m from the center. You get tired and stop rowing. She offers to row and after the boat comes to rest, you change places. You notice that after changing places the boat has moved 20 cm relative to a fixed log. What is your friend's mass?

125 •• [i-SOLVE ✓] A small car of mass 800 kg is parked behind a small truck of mass 1550 kg on a level road (Figure 8-69). The brakes of both the car and the truck are off so that they are free to roll with negligible friction. A 50-kg woman sitting on the tailgate of the truck shoves the car away

by exerting a constant force on the car with her feet. The car accelerates at 1.2 m/s². (*a*) What is the acceleration of the truck? (*b*) What is the magnitude of the force exerted on the car?

FIGURE 8-69 Problem 125

126 •• A 13-kg block is at rest on a level floor. A 400-g glob of putty is thrown at the block so that the putty travels horizontally, hits the block, and sticks to it. The block and putty slide 15 cm along the floor. If the coefficient of sliding friction is 0.4, what is the initial speed of the putty?

127 •• [SSM] A careless driver rear-ends a car that is halted at a stop sign. Just before impact, the driver slams on his brakes, locking the wheels. The driver of the struck car also has his foot solidly on the brake pedal, locking his brakes. The mass of the struck car is 900 kg, and that of the initially moving vehicle is 1200 kg. On collision, the bumpers of the two cars mesh. Police determine from the skid marks that after the collision the two cars moved 0.76 m together. Tests revealed that the coefficient of sliding friction between the tires and pavement was 0.92. The driver of the moving car claims that he was traveling at less than 15 km/h as he approached the intersection. Is he telling the truth?

128 •• A pendulum consists of a 0.4-kg bob attached to a string of length 1.6 m. A block of mass m rests on a horizontal frictionless surface (Figure 8-70). The pendulum is released from rest at an angle of 53° with the vertical and the bob collides elastically with the block. Following the collision, the maximum angle of the pendulum with the vertical is 5.73°. Determine the mass m.

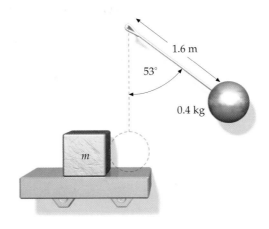

FIGURE 8-70 Problem 128

129 •• Initially, the block whose mass $m = 1.0$ kg and the block whose mass is M are both at rest on a frictionless inclined plane (Figure 8-71). The block of mass M rests against a spring that has a spring constant of 11,000 N/m. The distance along the plane between the two blocks is 4.0 m. The block of mass m is released, makes an elastic collision with the block of mass M, and rebounds a distance of 2.56 m back up the inclined plane. The block whose mass is M comes to rest momentarily 4.0 cm from its initial position. Find M.

FIGURE 8-71 Problem 129

130 •• [SSM] A circular plate of radius r has a circular hole of radius $r/2$ cut out of it (Figure 8-72). Find the center of mass of the plate. *Hint: The hole can be represented by two disks superimposed, one of mass* m *and the other of mass* $-$m.

FIGURE 8-72
Problem 130

131 •• Using the hint from Problem 130, find the center of mass of a solid sphere of radius r that has a spherical cavity of radius $r/2$, as in Figure 8-73.

FIGURE 8-73
Problem 131

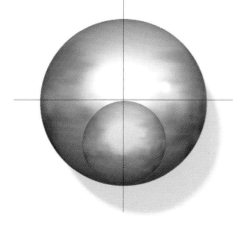

132 •• [SSM] A neutron of mass m makes an elastic head-on collision with a stationary nucleus of mass M. (a) Show that the energy of the nucleus after the collision is $K_{nucleus} = [4mM/(m + M)^2]K_n$, where K_n is the initial energy of the neutron. (b) Show that the fraction of energy lost by the neutron in this collision is

$$\frac{-\Delta K_n}{K_n} = \frac{4mM}{(m + M)^2} = \frac{4(m/M)}{(1 + m/M)^2}$$

133 •• The mass of a carbon nucleus is approximately 12 times the mass of a neutron. (a) Use the results of Problem 132 to show that after N head-on collisions of a neutron with carbon nuclei at rest, the energy of the neutron is approximately $0.716^N E_0$, where E_0 is its original energy. Neutrons emitted in the fission of a uranium nucleus have an energy of about 2 MeV. For such a neutron to cause the fission of another uranium nucleus in a reactor, its energy must be reduced to about 0.02 eV. (b) How many head-on collisions are needed to reduce the energy of a neutron from 2 MeV to 0.02 eV, assuming elastic head-on collisions with stationary carbon nuclei?

134 •• On average, a neutron loses 63 percent of its energy in an elastic collision with a hydrogen atom and 11 percent of its energy in an elastic collision with a carbon atom. (The numbers are lower than the ones we have been using in earlier problems because most collisions are not head-on.) Calculate the number of collisions, on average, needed to reduce the energy of a neutron from 2 MeV to 0.02 eV (a desirable outcome for reasons explained in Problem 133) if the neutron collides with (a) hydrogen atoms and (b) carbon atoms.

135 •• A rope of length L and mass M lies coiled on a table. Starting at $t = 0$, one end of the rope is lifted from the table with a force F such that it moves with a constant velocity v. (a) Find the height of the center of mass of the rope as a function of time. (b) Differentiate your result in (a) twice to find the acceleration of the center of mass. (c) Assuming that the force exerted by the table equals the weight of the rope still there, find the force F you exert on the top of the rope.

136 •• Repeat Problem 47 if the cup has a mass m_c and the ball collides with it inelastically.

137 •• Two astronauts at rest face each other in space. One, with mass m_1, throws a ball of mass m_b to the other, whose mass is m_2. She catches the ball and throws it back to the first astronaut. Following each throw the ball has a speed of v relative to the thrower. How fast are they moving after each has made one throw and one catch?

138 •• [SSM] The ratio of the mass of the earth to the mass of the moon is $M_e/m_m = 81.3$. The radius of the earth is about 6370 km, and the distance from the earth to the moon is about 384,000 km. (*a*) Locate the center of mass of the earth–moon system. Is it above or below the surface of the earth? (*b*) What external forces act on the earth–moon system? (*c*) In what direction is the acceleration of the center of mass of this system? (*d*) Assume that the center of mass of this system moves in a circular orbit around the sun. How far must the center of the earth move in the radial direction (toward or away from the sun) during the 14 days between the time the moon is farthest from the sun (full moon) and the time it is closest to the sun (new moon)?

139 •• [SOLVE✓] You want to enlarge a skating surface so you stand on the ice at one end and aim a hose horizontally to spray water on the schoolyard pavement. Water leaves the hose at 2.4 kg/s with a speed 30 m/s. If your mass is 75 kg, what is your recoil acceleration? (Neglect friction and the mass of the hose.)

140 ••• [SSM] A stream of glass beads, each with a mass of 0.5 g, comes out of a horizontal tube at a rate of 100 per second (Figure 8-74). The beads fall a distance of 0.5 m to a balance pan and bounce back to their original height. How much mass must be placed in the other pan of the balance to keep the pointer at zero?

FIGURE 8-74 Problem 140

141 ••• A dumbbell consisting of two balls of mass m connected by a massless rod of length L rests on a frictionless floor against a frictionless wall until it begins to slide down the wall as in Figure 8-75. Find the speed of the bottom ball at the moment when it equals the speed of the top ball.

FIGURE 8-75
Problem 141

Rotation

THESE BOWLING BALLS ROLL ON A
HORIZONTAL BALL RETURN WITHOUT
SLIPPING.

? **How might you design a
ball return to stop the balls'
rotation? (See Example 9-13.)**

In Chapters 4 and 5 we explored Newton's laws. In Chapters 6 and 7 we examined the conservation of energy, and in Chapter 8 we studied the conservation of momentum. In those chapters we discovered tools (laws and theorems) that are useful in analyzing new situations and in solving new problems. We will use those tools now as we explore rotational motion.

Rotational motion is all around us. The earth rotates about its axis. Wheels, gears, propellers, motors, the drive shaft in a car, a CD in its player, a pirouetting ice skater, all rotate.

➤ **In this chapter, we consider rotation about an axis that is fixed in space, as in a spinning top or a merry-go-round, or an axis that is moving parallel to itself, as in a rolling ball. (More general examples of rotational motion will be discussed in Chapter 10.)**

9-1 Rotational Kinematics: Angular Velocity and Angular Acceleration

Every point in a body rotating about a fixed axis moves in a circle whose center is on the axis and whose radius is the distance of that point from the axis of rotation. A line drawn from this axis to any point sweeps out the same angle in the same time. Imagine a disk spinning about a fixed axis perpendicular to the disk and through its center (Figure 9-1). Let r_i be the distance from the center of the disk to the ith particle (Figure 9-2), and let θ_i be the angle measured counterclockwise from a fixed reference line in space to a radial line from the center to the particle. As the disk rotates through an angle $d\theta$, the particle moves through a circular arc of length ds_i, such that

$$ds_i = r_i|d\theta| \qquad \qquad 9\text{-}1$$

where $d\theta$ is measured in radians. The distances ds_i and r_i vary from particle to particle, but their ratio, called the **angular displacement** $d\theta$, is the same for all particles of the disk. For one complete revolution, the arc length Δs_i is $2\pi r$ and the angular displacement $\Delta\theta$ is

$$\Delta\theta = \frac{2\pi r_i}{r_i} = 2\pi \text{ rad} = 360° = 1 \text{ rev}$$

The time rate of change of the angle is the same for all particles of the disk, and is called the **angular velocity** ω of the disk:

$$\omega = \frac{d\theta}{dt} \qquad \qquad 9\text{-}2$$

DEFINITION—ANGULAR VELOCITY

where ω is a lowercase Greek omega. For counterclockwise rotation, θ increases, so ω is positive. For clockwise rotation, θ decreases, and ω is negative. The units of ω are radians per second. (In Chapter 10 we will see that, for rotation in general, the angular velocity is a vector quantity that points along the axis of rotation.) Since radians are dimensionless, the dimension of angular velocity is that of reciprocal time, $[T]^{-1}$. The magnitude of the angular velocity is called the **angular speed.** We often use revolutions per minute (rev/min or RPM) to describe rotation. To convert between revolutions, radians, and degrees, we use

$$1 \text{ rev} = 2\pi \text{ rad} = 360°$$

EXERCISE A compact disk is rotating at 3000 rev/min. What is its angular speed in radians per second? (*Answer* 314 rad/s)

The time rate of change of angular velocity is called the **angular acceleration** α:

$$\alpha = \frac{d\omega}{dt} = \frac{d^2\theta}{dt^2} \qquad \qquad 9\text{-}3$$

DEFINITION—ANGULAR ACCELERATION

The units of α are radians per second per second (rad/s²). If ω is increasing, α is positive; if ω is decreasing, α is negative.

The angular displacement θ, angular velocity ω, and angular acceleration α are analogous to the linear displacement x, linear velocity v, and linear

FIGURE 9-1

FIGURE 9-2

acceleration a in one-dimensional motion. If the angular acceleration α is constant, we can integrate Equation 9-3 to find ω:

$$\omega = \omega_0 + \alpha t \qquad\qquad 9\text{-}4$$

where the constant of integration ω_0 is the initial angular velocity. This is the rotational analog of $v = v_0 + at$. Integrating again, we obtain

$$\theta = \theta_0 + \omega_0 t + \tfrac{1}{2}\alpha t^2 \qquad\qquad 9\text{-}5$$

which is the rotational analog of $x = x_0 + v_0 t + \tfrac{1}{2}at^2$ with θ replacing x, ω replacing v, and α replacing a. Similarly, by eliminating t from Equations 9-4 and 9-5, we get

$$\omega^2 = \omega_0^2 + 2\alpha(\theta - \theta_0) \qquad\qquad 9\text{-}6$$

which is the rotational analog of $v^2 = v_0^2 + 2a(x - x_0)$. The equations for constant angular acceleration have the same form as those for constant linear acceleration.

Star tracks in a time exposure of the night sky.

A CD PLAYER

EXAMPLE 9 - 1

A compact disk rotates from rest to 500 rev/min in 5.5 s. (a) What is its angular acceleration, assuming it is constant? (b) How many revolutions does the disk make in 5.5 s? (c) How far does a point on the rim 6 cm from the center of the disk travel during the 5.5 s it takes to get to 500 rev/min?

PICTURE THE PROBLEM Part (a) is analogous to the linear problem of finding the acceleration given the time and the final velocity. To find α in rad/s² we first convert ω to rad/s. Part (b) is analogous to finding the distance traveled given the time and the final velocity.

(a) 1. The angular acceleration is related to the initial and final angular velocities:

$$\omega = \omega_0 + \alpha t = 0 + \alpha t$$

2. Solve for α:

$$\alpha = \frac{\omega}{t}$$

$$= \frac{500 \text{ rev/min}}{5.5 \text{ s}} \times \frac{2\pi \text{ rad}}{1 \text{ rev}} \times \frac{1 \text{ min}}{60 \text{ s}}$$

$$= \boxed{9.52 \text{ rad/s}^2}$$

(b) 1. The angular displacement is related to the time by Equation 9-5:

$$\theta - \theta_0 = \omega_0 t + \tfrac{1}{2}\alpha t^2$$

$$= 0 + \tfrac{1}{2}(9.52 \text{ rad/s}^2)(5.5 \text{ s})^2$$

$$= 144 \text{ rad}$$

2. Convert radians to revolutions:

$$\theta - \theta_0 = 144 \text{ rad} \times \frac{1 \text{ rev}}{2\pi \text{ rad}} = \boxed{22.9 \text{ rev}}$$

(c) The distance traveled Δs is r times the angular displacement:

$$\Delta s = r\,\Delta\theta = (6 \text{ cm})(144 \text{ rad}) = \boxed{8.64 \text{ m}}$$

❶ PLAUSIBILITY CHECK The average angular velocity in revolutions per minute is 250 rev/min. In 5.5 s, the compact disk rotates (250 rev/60 s)(5.5 s) = 22.9 rev.

REMARKS A compact disk is scanned by a laser that begins at the inner radius of about 2.4 cm and moves out to the edge at 6.0 cm. As the laser moves outward, the angular velocity of the disk decreases from 500 rev/min to 200 rev/min so that the linear (tangential) velocity of the disk at the point where the laser beam strikes remains constant.

EXERCISE (a) Convert 500 rev/min to rad/s. (b) Check the result of Part (b) in the example using $\omega^2 = \omega_0^2 + 2\alpha(\theta - \theta_0)$. (*Answer* (a) 500 rev/min = 52.4 rad/s)

The linear velocity v_t of a particle on the disk is tangent to the circular path of the particle and has magnitude ds_i/dt. We can relate this "tangential" velocity to the angular velocity of the disk using Equations 9-1 and 9-2:

$$v_t = \frac{r_i\,d\theta}{dt}$$

so

$$v_t = r_i\omega \qquad\qquad\qquad\qquad 9\text{-}7$$

Similarly, the tangential acceleration of a particle on the disk is

$$a_t = \frac{dv_t}{dt} = r_i\frac{d\omega}{dt}$$

so

$$a_t = r\alpha \qquad\qquad\qquad\qquad 9\text{-}8$$

Each particle of the disk also has a centripetal acceleration, which points inward along the radial line and has the magnitude

$$a_c = \frac{v_t^2}{r_i} = \frac{(r_i\omega)^2}{r_i}$$

so

$$a_c = r_i\omega^2 \qquad\qquad\qquad\qquad 9\text{-}9$$

EXERCISE A point on the rim of a compact disk is 6.0 cm from the axis of rotation. Find the tangential speed v_t, tangential acceleration a_t, and centripetal acceleration a_c of the point when the disk is rotating at a constant angular speed of 300 rev/min. (*Answer* v_t = 188 cm/s, a_t = 0, a_c = 5.92 × 10³ cm/s²)

EXERCISE Find the linear speed of a point on the CD in Example 9-1 at (*a*) *r* = 2.4 cm when the disk rotates at 500 rev/min, and (*b*) *r* = 6.0 cm when the disk rotates at 200 rev/min. (*Answer* (*a*) 126 cm/s (*b*) 126 cm/s)

9-2 Rotational Kinetic Energy

The kinetic energy of a rigid object rotating about a fixed axis is the sum of the kinetic energies of the individual particles that collectively constitute the object. The kinetic energy of the *i*th particle, with mass m_i, is

$$K_i = \tfrac{1}{2}m_i v_i^2$$

Summing over all the particles and using $v_i = r_i \omega$ gives

$$K = \sum_i (\tfrac{1}{2}m_i v_i^2) = \tfrac{1}{2}\sum_i (m_i r_i^2 \omega^2) = \tfrac{1}{2}\left(\sum_i m_i r_i^2\right)\omega^2$$

The sum in the term on the right is the object's **moment of inertia** *I* for the axis of rotation.

$$I = \sum_i m_i r_i^2 \qquad\qquad 9\text{-}10$$

MOMENT OF INERTIA DEFINED

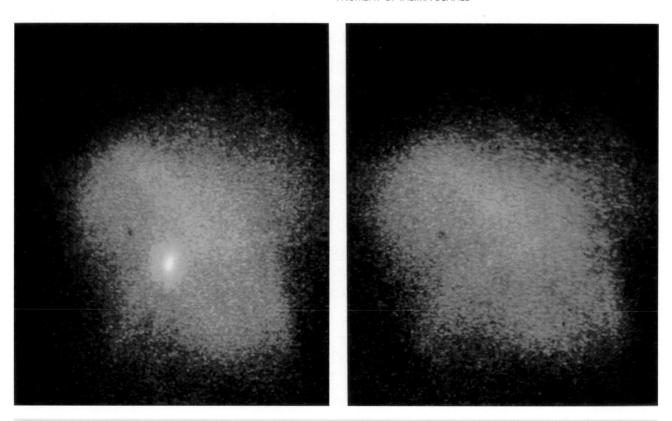

The Crab Pulsar is one of the fastest-rotating neutron stars known, but it is slowing down. It appears to blink on (left) and off (right) like the rotating lamp in a lighthouse, at the fast rate of about 30 times per second, but the period is increasing by about 10^{-5} s/y. The loss in rotational energy, which is equivalent to the power output of 100,000 suns, appears as light emitted by electrons accelerated in the magnetic field of the pulsar.

The kinetic energy is thus

$$K = \tfrac{1}{2}I\omega^2 \qquad 9\text{-}11$$

KINETIC ENERGY OF ROTATING OBJECT

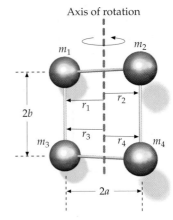

FIGURE 9-3

A ROTATING SYSTEM OF PARTICLES **EXAMPLE 9-2**

An object consists of four point particles, each of mass m, that are connected by rigid massless rods to form a rectangle of sides $2a$ and $2b$ as shown in Figure 9-3. The system rotates with angular speed ω about an axis in the plane of the figure through the center as shown. (*a*) Find the kinetic energy of this object using Equations 9-10 and 9-11. (*b*) Check your result by individually calculating the kinetic energy of each particle and then taking their sum.

PICTURE THE PROBLEM Because we are given that the objects are particles, we use Equation 9-10 to calculate I and then use Equation 9-11. In Equation 9-10, r_i is the radial distance from the rotation axis to the particle of mass m_i.

(*a*) 1. Apply the definition of moment of inertia (Equation 9-10):

$$I = \sum_i m_i r_i^2$$
$$= m_1 r_1^2 + m_2 r_2^2 + m_3 r_3^2 + m_4 r_4^2$$

2. The masses m_i and the distances r_i are given:

$$m_1 = m_2 = m_3 = m_4 = m$$
$$r_1 = r_2 = r_3 = r_4 = a$$

3. Substitution gives the moment of inertia:

$$I = ma^2 + ma^2 + ma^2 + ma^2 = 4ma^2$$

4. Using Equation 9-11, solve for the kinetic energy:

$$K = \tfrac{1}{2}I\omega^2 = \tfrac{1}{2}4ma^2\omega^2 = \boxed{2ma^2\omega^2}$$

(*b*) 1. To find the kinetic energy of the *i*th particle we must first find its speed:

$$K_i = \tfrac{1}{2}m_i v_i^2$$

2. The particles are all moving in circles of radius a. Find the speed of each particle:

$$v_i = r_i \omega = a\omega \qquad (i = 1, \ldots, 4)$$

3. Substitute into the Part (*b*) step 1 result:

$$K_i = \tfrac{1}{2}m_i v_i^2 = \tfrac{1}{2}ma^2\omega^2$$

4. Each particle has the same kinetic energy. Sum the kinetic energies to get the total:

$$K = \sum_{i=1}^{4} K_i = \tfrac{1}{2}m_1 v_1^2 + \tfrac{1}{2}m_2 v_2^2 + \tfrac{1}{2}m_3 v_3^2 + \tfrac{1}{2}m_4 v_4^2$$
$$= 4(\tfrac{1}{2}ma^2\omega^2) = 2ma^2\omega^2$$

5. Compare with the Part (*a*) result:

$$\boxed{\text{The two calculations give the same result.}}$$

REMARKS Notice that I is independent of the length b, which has no effect on how far the masses are from the axis of rotation.

EXERCISE Find the moment of inertia of this system for rotation about an axis parallel to the first axis but passing through two of the particles, as shown in Figure 9-4. (*Answer* $I = 8ma^2$)

9-3 Calculating the Moment of Inertia

The moment of inertia about an axis is a measure of the inertial resistance of the object to changes in its rotational motion about the axis. It is the rotational analog of mass. The moment of inertia about an axis depends on the distribution of mass

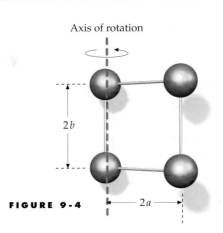

FIGURE 9-4

within the object relative to the axis. The farther an element of mass is from the axis, the greater its contribution to the moment of inertia about that axis. Thus, unlike the mass of an object, which is a property of the object itself, the moment of inertia depends on the location of the axis of rotation.

Systems of Discrete Particles

For systems consisting of discrete particles, we can compute the moment of inertia about a given axis directly from Equation 9-10.

Continuous Objects

To calculate the moment of inertia for continuous objects, we imagine the object to consist of a continuum of very small mass elements. Thus, the finite sum $\Sigma m_i r_i^2$ in Equation 9-10 becomes the integral

$$I = \int r^2 \, dm \qquad\qquad 9\text{-}12$$

where r is the radial distance from the axis to mass element dm.

MOMENT OF INERTIA OF A UNIFORM ROD   **EXAMPLE 9-3**

Find the moment of inertia of a uniform rod of length L and mass M about an axis perpendicular to the rod and through one end. Assume that the rod has negligible thickness.

FIGURE 9-5

$dm = \dfrac{M}{L} \, dx$

PICTURE THE PROBLEM Let the rod lie along the x axis with its end at the origin. To calculate I about the y axis, we choose a mass element dm at a distance x from the axis (Figure 9-5). Since the total mass M is uniformly distributed along the length L, the mass per unit length (linear mass density) is $\lambda = M/L$.

1. The moment of inertia is given by the integral:

$$I = \int_0^L x^2 \, dm$$

2. To compute the integral, we first relate dm to dx. Express dm in terms of the mass density λ and dx:

$$dm = \lambda \, dx = \frac{M}{L} \, dx$$

3. Substitute and perform the integration. We choose integration limits so that we sweep dm through the mass distribution in the direction of increasing x:

$$I_y = \int x^2 \, dm = \int_0^L x^2 \frac{M}{L} \, dx = \frac{M}{L} \int_0^L x^2 \, dx$$

$$= \frac{M}{L} \frac{1}{3} x^3 \Big|_0^L = \frac{M}{L} \frac{L^3}{3} = \boxed{\frac{1}{3} ML^2}$$

REMARKS The moment of inertia about the z axis is also $\frac{1}{3} ML^2$, and that about the x axis is zero (assuming that all of the mass is right on the x axis).

We can calculate I for continuous objects of various shapes, again using Equation 9-12 (see Table 9-1).

*Hoop About a Perpendicular Axis Through Its Center   Assume that a hoop has mass M and radius R (Figure 9-6). The axis of rotation is the axis of the hoop, which is perpendicular to the plane of the hoop. All the mass is at a distance $r = R$, and the moment of inertia is

$$I = \int r^2 \, dm = \int R^2 \, dm = R^2 \int dm = MR^2$$

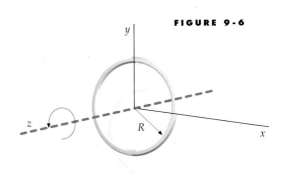

FIGURE 9-6

TABLE 9-1

Moments of Inertia of Uniform Bodies of Various Shapes

Thin cylindrical shell about axis

$$I = MR^2$$

Thin cylindrical shell about diameter through center

$$I = \tfrac{1}{2}MR^2 + \tfrac{1}{12}ML^2$$

Thin rod about perpendicular line through center

$$I = \tfrac{1}{12}ML^2$$

Thin spherical shell about diameter

$$I = \tfrac{2}{3}MR^2$$

Solid cylinder about axis

$$I = \tfrac{1}{2}MR^2$$

Solid cylinder about diameter through center

$$I = \tfrac{1}{4}MR^2 + \tfrac{1}{12}ML^2$$

Thin rod about perpendicular line through one end

$$I = \tfrac{1}{3}ML^2$$

Solid sphere about diameter

$$I = \tfrac{2}{5}MR^2$$

Hollow cylinder about axis

$$I = \tfrac{1}{2}M(R_1^2 + R_2^2)$$

Hollow cylinder about diameter through center

$$I = \tfrac{1}{4}M(R_1^2 + R_2^2) + \tfrac{1}{12}ML^2$$

Solid rectangular parallelepiped about axis through center perpendicular to face

$$I = \tfrac{1}{12}M(a^2 + b^2)$$

A disk is a cylinder whose length L is negligible. By setting $L = 0$, the above formulas for cylinders hold for disks.

*Uniform Disk About a Perpendicular Axis Through Its Center For the case of a uniform disk, we expect that I will be smaller than MR^2 since the mass is uniformly distributed from $r = 0$ to $r = R$ rather than being concentrated at $r = R$ as it is in a hoop. In Figure 9-7, each mass element is a hoop of radius r and thickness dr. The moment of inertia of any given mass element is $r^2 dm$. Since the disk is uniform, mass per unit area σ is constant. $\sigma = M/A$, where $A = \pi R^2$ is the area of the disk. Since the area of each mass element is $dA = 2\pi r\, dr$, the mass of each element is

$$dm = \sigma\, dA = \frac{M}{A} 2\pi r\, dr$$

We thus have

$$I = \int r^2\, dm = \int_0^R r^2 \sigma 2\pi r\, dr = 2\pi\sigma \int_0^R r^3\, dr$$

$$= \frac{2\pi M}{A} \frac{R^4}{4} = \frac{\pi M}{2\pi R^2} R^4 = \frac{1}{2} MR^2$$

FIGURE 9-7

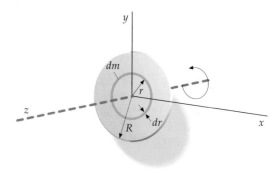

*Uniform Solid Cylinder About Its Axis We consider a cylinder to be a set of disks, each with mass dm and moment of inertia $\frac{1}{2}dm\,R^2$ (Figure 9-8). The moment of inertia of the complete cylinder is then

$$I = \int \tfrac{1}{2}dm\,R^2 = \tfrac{1}{2}R^2 \int dm = \tfrac{1}{2}MR^2$$

where M is the total mass of the cylinder.

The Parallel-Axis Theorem

We can often simplify the calculation of moments of inertia for various bodies by using the **parallel-axis theorem,** which relates the moment of inertia about an axis through the center of mass of an object to the moment of inertia about a second, parallel axis (Figure 9-9). Let I_{cm} be the moment of inertia about an axis through the center of mass of an object of total mass M, and let I be that about a parallel axis a distance h away. The parallel-axis theorem states that

$$I = I_{cm} + Mh^2 \qquad\qquad 9\text{-}13$$

PARALLEL-AXIS THEOREM

Example 9-2 and the exercise following it illustrate a special case of this theorem with $h = a$, $M = 4m$, and $I_{cm} = 4ma^2$. A proof of the parallel-axis theorem is given at the end of this section.

FIGURE 9-9 An object rotating about an axis parallel to an axis through the center of mass and a distance h from it.

FIGURE 9-8

MOMENT OF INERTIA ABOUT AN AXIS THROUGH THE CENTER OF **E X A M P L E   9 - 4**   **Try It Yourself**
MASS OF A UNIFORM ROD

Find the moment of inertia of a uniform rod about the y' axis through the center of mass (Figure 9-10).

PICTURE THE PROBLEM Here you know that $I = \frac{1}{3}ML^2$ about one end and want to find I_{cm}. Use the parallel-axis theorem with $h = \frac{1}{2}L$.

FIGURE 9-10

Cover the column to the right and try these on your own before looking at the answers.

Steps

1. Apply the parallel-axis theorem to write I about the end in terms of I_{cm}.

2. Substitute $I = \frac{1}{3}ML^2$ about the end and solve for I_{cm}.

Answers

$$I = I_{cm} + Mh^2$$

$$I_{cm} = I - Mh^2 = \frac{1}{3}ML^2 - M\left(\frac{L}{2}\right)^2$$

$$= \boxed{\frac{1}{12}ML^2}$$

REMARKS The moment of inertia is least when an object is rotated about its center of mass, as in this example. Compare this result to that of Example 9-3, where the uniform rod is rotated about an axis through one end.

EXERCISE Using the parallel-axis theorem, show that when comparing the moments of inertia of an object about two parallel axes, the moment of inertia is least about the axis that is nearest to the center of mass.

*Proof of the Parallel-Axis Theorem

To prove the parallel-axis theorem we start with an object and an axis A (Figure 9-11). We choose our origin to be at the center of mass with the z axis parallel with A. Then I is the moment of inertia about A and I_{cm} is the moment of inertia about the z axis. The mass m_i is located at (x_i, y_i, z_i), and the intersection of axis A with the xy plane is at coordinates x_A, y_A. The distances r_i and R_i are the perpendicular distances of m_i from A and the z axis, respectively, and h is the perpendicular distance between axes. It follows that r_i is the distance from $(x_A, y_A, 0)$ to $(x_i, y_i, 0)$, R_i is the distance from $(0, 0, 0)$ to $(x_i, y_i, 0)$, and h is the distance from $(0, 0, 0)$ to $(x_A, y_A, 0)$. Thus, $r_i^2 = (x_i - x_A)^2 + (y_i - y_A)^2$, $R_i^2 = x_i^2 + y_i^2$, and $h^2 = x_A^2 + y_A^2$. Consequently,

$$I_{cm} = \sum m_i R_i^2 = \sum m_i(x_i^2 + y_i^2)$$

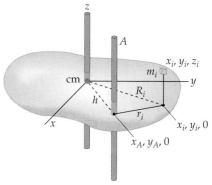

FIGURE 9-11

and

$$I = \sum m_i r_i^2 = \sum m_i[(x_i - x_A)^2 + (y_i - y_A)^2]$$
$$= \sum m_i(x_i^2 + y_i^2) - \sum m_i 2x_i x_A - \sum m_i 2y_i y_A + \sum m_i(x_A^2 + y_A^2)$$

By factoring the common factors from these sums we have

$$I = \sum m_i(x_i^2 + y_i^2) - 2x_A \sum m_i x_i - 2y_A \sum m_i y_i + \left(\sum m_i\right)(x_A^2 + y_A^2)$$

The first term is I_{cm}. The second and third terms can be simplified using $\sum m_i x_i = Mx_{cm}$ and $\sum m_i y_i = My_{cm}$, and because $x_{cm} = y_{cm} = 0$, both the second and the third terms are equal to zero. The fourth term is Mh^2. Thus,

$$I = I_{cm} + Mh^2$$

which is the parallel-axis theorem.

A FLYWHEEL–POWERED CAR   **EXAMPLE 9 - 5**   Put It in Context

You are driving an experimental hybrid vehicle that is designed for use in stop-and-go traffic. In a conventional car, each time you brake to a stop, the kinetic energy is dissipated as heat. In this hybrid vehicle, the braking mechanism transforms the translational kinetic energy of the vehicle's motion to the rotational kinetic energy of a massive flywheel. When the car returns to cruising speed this energy is transferred back into the translational energy of the car. The 100-kg flywheel is a hollow cylinder with an inner diameter R_1 of 25 cm, an outer diameter R_2 of 40 cm, and a maximum angular speed of 30,000 rev/min. On a dark and dreary night, the car runs out of gas 15 mi from home with the flywheel spinning at maximum speed. Is there sufficient energy stored in the flywheel for you and your nervous grandmother to make it home? (When driving at the minimum highway speed of 40 mi/h, air drag and rolling friction dissipate energy at 10 kW.)

PICTURE THE PROBLEM The kinetic energy is calculated directly from $K = \frac{1}{2}I\omega^2$.

1. The kinetic energy of rotation is:

$$K = \tfrac{1}{2}I\omega^2$$

2. Calculate the moment of inertia of the hollow cylinder:

$$I = \tfrac{1}{2}m(R_1^2 + R_2^2) = 11.1 \text{ kg·m}^2$$

3. Convert ω to rad/s:

$$\omega = 30{,}000 \text{ rev/min} = 3140 \text{ rad/s}$$

4. Substitute these values to find the kinetic energy:

$$K = \tfrac{1}{2}I\omega^2 = 54.8 \text{ MJ}$$

5. Energy is dissipated at 10 kW at a speed of 40 mi/h. To find the energy dissipated during the 15-mi trip, we first need to find the time required for the trip:

$$\Delta x = v\,\Delta t, \quad \text{so} \quad \Delta t = 1350 \text{ s}$$

6. The energy is dissipated at 10 kW for 1350 s. The total energy dissipated is:

13.5 MJ

7. Is there enough energy stored in the flywheel?

54.8 MJ are available and 13.5 MJ are dissipated.

$$\boxed{\text{Yes, there is more than enough energy stored.}}$$

REMARKS There are 130 MJ of energy in a gallon of gasoline. If the engine is 10% efficient, only 13 MJ/gal are available to move the car. We estimate the mileage for the car in this example to be about 15 mi/gal.

THE PIVOTED ROD   **EXAMPLE 9 - 6**

A uniform thin rod of length L and mass M, pivoted at one end as shown in Figure 9-12, is held horizontal and then released from rest. Assuming the pivot to be frictionless, find (a) the angular velocity of the rod when it reaches its vertical position, and (b) the force exerted by the pivot at this time. (c) What initial angular velocity is needed for the rod to reach a vertical position at the top of its swing?

PICTURE THE PROBLEM (a) As the rod swings down, its potential energy decreases and its kinetic energy increases. Since the pivot is frictionless, we use conservation of mechanical energy. The angular velocity of the rod is then found from its rotational kinetic energy. (b) To find the force of the pivot use Newton's second law for a system. (c) As in Part (a), use conservation of mechanical energy.

(a) 1. Make a diagram of the rod showing both its initial and its final configuration (Figure 9-12). Put a vertical coordinate axis on the figure with up as the positive direction and with its origin at the rotation axis.

FIGURE 9-12

2. Apply conservation of mechanical energy to relate the initial and final mechanical energies:

$$K_f + U_f = K_i + U_i$$

$$\tfrac{1}{2}I\omega_f^2 + Mgy_{cm\,f} = \tfrac{1}{2}I\omega_i^2 + Mgy_{cm\,i}$$

$$\frac{1}{2}I\omega_f^2 + Mg\left(-\frac{L}{2}\right) = 0 + 0$$

3. Solve for ω_f:

$$\omega_f = \sqrt{\frac{MgL}{I}}$$

4. Obtain I from Table 9-1 and substitute into the step 3 result:

$$\omega_f = \sqrt{\frac{MgL}{\tfrac{1}{3}ML^2}} = \boxed{\sqrt{\frac{3g}{L}}}$$

(b) 1. Make a free-body diagram of the rod as it passes through the vertical position at the bottom of its swing (Figure 9-13).

2. Apply Newton's second law for a system to the rod. At the bottom of the swing the acceleration of the center of mass is in the centripetal (upward) direction:

$$\Sigma F_{ext\,y} = Ma_{cm}$$
$$F_p - Mg = Ma_{cm}$$

3. Relate the acceleration of the center of mass to the angular speed using $a_c = r\omega^2$. Substitute the Part (a) step 4 result for ω and solve for a_{cm}:

$$a_c = r\omega^2$$
$$a_{cm} = \frac{L}{2}\frac{3g}{L} = \frac{3}{2}g$$

4. Substitute into the Part (b) step 2 result and calculate F_p:

$$F_p = Mg + Ma_{cm} = Mg + M\tfrac{3}{2}g$$
$$= \boxed{\tfrac{5}{2}Mg}$$

FIGURE 9-13

(c) 1. The initial angular velocity ω_i is related to the initial kinetic energy:

$$K_i = \tfrac{1}{2}I\omega_i^2$$

2. Make a diagram of the rod showing both its initial and its final configuration (Figure 9-14). Put a vertical coordinate axis on the figure with up as the positive direction and with its origin at the rotation axis.

3. Apply conservation of mechanical energy with $K_f = 0$ and $U_i = 0$ to relate the initial kinetic energy to the final position:

$$K_f + U_f = K_i + U_i$$
$$\tfrac{1}{2}I\omega_f^2 + Mgy_{cm\,f} = \tfrac{1}{2}I\omega_i^2 + Mgy_{cm\,i}$$
$$0 + Mg\frac{L}{2} = \frac{1}{2}I\omega_i^2 + 0$$

FIGURE 9-14

4. Solve for the initial angular velocity:

$$\omega_i = \sqrt{\frac{MgL}{I}} = \sqrt{\frac{MgL}{\tfrac{1}{3}ML^2}} = \boxed{\sqrt{\frac{3g}{L}}}$$

REMARKS It is no coincidence that the answers to Part (a) and Part (c) are identical. The decrease in potential energy in Part (a) is equal to the increase in potential energy in Part (c). Thus the increase in kinetic energy in Part (a) is equal to the decrease in kinetic energy in Part (c).

Try It Yourself

A WINCH AND A BUCKET   **EXAMPLE 9-7**

A winch is at the top of a deep well. The drum of the winch has mass m_w and radius R. Virtually all its mass is concentrated a distance R from the axis. A cable wound around the drum suspends a bucket of water of mass m_b. The entire cable has mass m_c and length L. Just when you have the bucket at the highest point your hand slips and the bucket falls back down the well, unwinding the winch cable as it does. How fast is the bucket moving after it has fallen a distance d, where d is less than L?

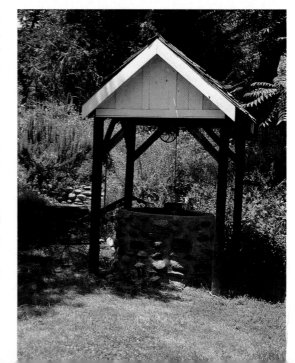

PICTURE THE PROBLEM As the load falls, mechanical energy is conserved. Choose the initial potential energy to be zero. When the load has fallen a distance d, the center of mass of the hanging cable has dropped a distance $d/2$. Since the hanging part of the cable moves with speed v and the cable does not stretch or become slack, the entire cable must move at speed v. We find v from the conservation of mechanical energy.

Cover the column to the right and try these on your own before looking at the answers.

Steps

Answers

1. Make a diagram of the winch–cable–bucket system in both its initial and its final configuration, as shown in Figure 9-15. Include a y axis with the origin at the height of the center of the winch.

FIGURE 9-15

2. Apply conservation of mechanical energy. Choose the potential energy to be zero when the bucket of water is at the highest point.

$U_f + K_f = U_i + K_i$

$= 0 + 0 = 0$

3. Write an expression for the total potential energy when the bucket has fallen a distance d. Let m'_c denote the mass of the hanging part of the cable.

$U_f = U_{bf} + U_{cf} + U_{wf} = m_b g(-d) + m'_c g\left(-\dfrac{d}{2}\right) + 0$

$= -(m_b + \frac{1}{2}m'_c)gd$

4. Express the total kinetic energy when the bucket is falling with speed v. All the cable and the entire mass of the drum move with the same speed v as the bucket.

$K_f = K_{fc} + K_{fb} + K_{fw} = \frac{1}{2}m_c v^2 + \frac{1}{2}m_b v^2 + \frac{1}{2}m_w v^2$

$= \frac{1}{2}(m_c + m_b + m_w)v^2$

5. Substitute into the conservation of mechanical energy equation (step 2) and solve for v.

$-(m_b + \frac{1}{2}m'_c)gd + \frac{1}{2}(m_c + m_b + m_w)v^2 = 0$

so

$v = \sqrt{\dfrac{(2m_b + m'_c)gd}{m_c + m_b + m_w}}$

6. Assume the cable is uniform and express m'_c in terms of m_c, d, and L.

$\dfrac{m'_c}{d} = \dfrac{m_c}{L} \Rightarrow m'_c = \dfrac{d}{L}m_c$

7. Substitute the step 6 result into the step 5 result.

$v = \sqrt{\dfrac{(2m_b L + m_c d)gd}{(m_c + m_b + m_w)L}}$

REMARKS Because the entire mass of the drum is moving at the same speed v we can express its kinetic energy as $\frac{1}{2}m_w v^2$. However, we can express it as $\frac{1}{2}I_w \omega^2$, where $I_w = m_w R^2$ and $\omega = v/R$. With these substitutions $K_w = \frac{1}{2}I_w \omega^2 = \frac{1}{2}m_w R^2(v^2/R^2) = \frac{1}{2}m_w v^2$.

9-4 Newton's Second Law for Rotation

To set a top spinning, you twist it. In Figure 9-16, a disk is set spinning by the forces $\vec{F}_1$ and $\vec{F}_2$ exerted at the edges of the disk in the tangential direction. The directions of these forces are important. If the same forces are applied in the radial direction (Figure 9-17), the disk will not start to spin.

Figure 9-18 shows a particle of mass m attached to one end of a massless rigid rod of length r. There is an axis perpendicular to the rod and passing through its other end, and the rod is free to rotate about this axis. Consequently, the particle is constrained to move in a circle of radius r. A single force $\vec{F}$ is applied to the particle as shown. Applying Newton's second law to the particle and taking components in the tangential direction gives

$$F_t = ma_t$$

We wish to obtain an equation involving angular quantities. Substituting $r\alpha$ for a_t (Equation 9-8) and multiplying both sides by r gives

$$rF_t = mr^2\alpha \qquad\qquad 9\text{-}14$$

The product rF_t is the **torque** τ associated with the force. That is,

$$\tau = F_t r \qquad\qquad 9\text{-}15$$

TORQUE

Substituting into Equation 9-14 gives

$$\tau = mr^2\alpha \qquad\qquad 9\text{-}16$$

A rigid object that rotates about a fixed axis is just a collection of individual particles, each of which is constrained to move in a circular path with the same angular velocity ω and acceleration α. Applying Equation 9-16 to the ith of these particles gives

$$\tau_{i\,\text{net}} = m_i r_i^2 \alpha$$

where $\tau_{i\,\text{net}}$ is the torque due to the net force on the ith particle. Summing both sides over all particles gives

$$\sum \tau_{i\,\text{net}} = \sum m_i r_i^2 \alpha = \left(\sum m_i r_i^2\right)\alpha = I\alpha \qquad\qquad 9\text{-}17$$

In Chapter 8 we saw that the net force acting on a system of particles is equal to the net *external* force acting on the system because the internal forces (those exerted by the particles within the system on one another) cancel in pairs. The treatment of internal torques exerted by the particles within a system on one another leads to a similar result, that is, the net torque acting on a system equals the net *external* torque acting on the system. (We discuss this further in Chapter 10.) We can thus write Equation 9-17 as

$$\tau_{\text{net ext}} = \sum \tau_{\text{ext}} = I\alpha \qquad\qquad 9\text{-}18$$

NEWTON'S SECOND LAW FOR ROTATION

This is the rotational analog of Newton's second law for linear motion, $\sum \vec{F} = m\vec{a}$.

FIGURE 9-16

FIGURE 9-17

FIGURE 9-18

Calculating Torques

Figure 9-19 shows a force $\vec{F}$ acting on an object constrained to rotate about a fixed axis A, not shown, which passes through O and is perpendicular to the page. The positive tangential direction is shown at the point of application of the force, and r is the radial distance of this point of application from A. The torque τ due to this force about axis A is $\tau = F_t r$ (Equation 9-15). In principle, the expression $F_t r$ is all that is needed to calculate torques. However, in practice, calculations are often simpler if alternative expressions for torque are used. From the figure we can see that

$$F_t = F \sin \phi$$

where ϕ is the angle between the radial direction and the direction of the force. Thus, we can express the torque as $\tau = F_t r = (F \sin \phi)r$. The *line of action* of a force is the line parallel to the force and passing through the point of application of the force. From Figure 9-20 we can see that $r \sin \phi = \ell$, where the **lever arm** ℓ is the perpendicular distance between A and the line of action. Consequently, the torque is also given by $\tau = F\ell$. Putting all three equivalent expressions for the torque in one place, we have

$$\tau = F_t r = Fr \sin \phi = F\ell \qquad\qquad 9\text{-}19$$

EQUIVALENT EXPRESSIONS FOR TORQUE

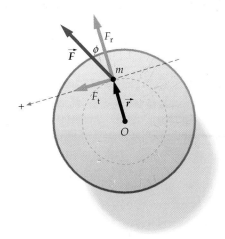

FIGURE 9-19 The force $\vec{F}$ produces a torque $F_t r$ about the center.

Torque Due to Gravity

We can model an extended object as an assembly of microscopic point particles, and there is a microscopic gravitational force on each particle. Each of these microscopic gravitational forces exerts a microscopic torque about a given axis, and the net gravitational torque on the object is the sum of these microscopic torques. The net gravitational torque can be calculated by considering the entire weight (the sum of the microscopic gravitational forces) to act at a single point—**the center of gravity.** Consider an object (Figure 9-21) constrained to rotate about a horizontal axis A coming out of the page. We choose the z axis of our coordinate system to coincide with axis A, and choose the x axis direction to be horizontal and the y axis vertical as shown. The torque on a particle of mass m_i due to gravity is $m_i g x_i$, where x_i is the lever arm of the force $m_i \vec{g}$. The net gravitational torque on the object is the sum of the gravitational torques on the particles that make it up. That is, $\tau_{\text{grav}} = \Sigma m_i g x_i$. If $\vec{g}$ is the same throughout the object, then g can be factored out of the sum. Factoring g out of the sum gives $\tau_{\text{grav}} = (\Sigma m_i x_i)g$. You should recognize the sum in the parentheses as $M x_{\text{cm}}$ (see Equation 8-1). Substituting this for the sum gives

$$\tau_{\text{grav}} = Mgx_{\text{cm}} \qquad\qquad 9\text{-}20$$

TORQUE DUE TO GRAVITY

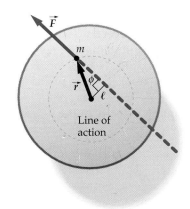

FIGURE 9-20 The force $\vec{F}$ produces a torque $F\ell$ about the center.

FIGURE 9-21

(a)

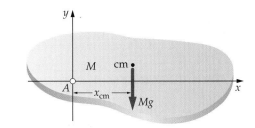

(b)

The torque due to a uniform gravitational field is calculated as if the entire gravitational force is applied at the center of mass. For an object in a uniform gravitational field, the center of gravity coincides with the center of mass.

9-5 Applications of Newton's Second Law for Rotation

In this section we give several applications of Newton's second law for rotation as expressed in Equation 9-18.

Problem-Solving Guidelines for Applying Newton's Second Law for Rotation

Free-Body Diagram Draw the free-body diagram with the object shown as a picture, not just a dot. Draw each force vector along the line of action of the force. On the diagram indicate the positive rotational direction (clockwise or counterclockwise).

A STATIONARY BIKE   **EXAMPLE 9-8**

To get some exercise without going anywhere, you set your bike on a stand so that the rear wheel is free to turn. As you pedal, the chain applies a force of 18 N to the sprocket at a distance of $r_s = 7$ cm from the axle of the wheel. Consider the wheel to be a hoop ($I = MR^2$) of radius $R = 35$ cm and mass 2.4 kg. What is the angular velocity of the wheel after 5 s?

PICTURE THE PROBLEM The angular velocity is found from the angular acceleration, which is found from Newton's second law for rotation. Since the forces are constant, the torques are also constant and the constant angular acceleration equations apply. Note that $\vec{F}$ acts in the direction of the chain, so the line of force is tangent to the sprocket and the lever arm is the radius r_s of the sprocket (see Figure 9-22).

FIGURE 9-22

$\vec{F} = 18$ N

35 cm

$M = 2.4$ kg

1. The angular velocity is related to the angular acceleration and the time:

$$\omega = \omega_0 + \alpha t = 0 + \alpha t$$

2. Apply Newton's second law for rotational motion to relate α to the net torque and the moment of inertia:

$$\Sigma \tau_{\text{ext}} = I\alpha$$

3. The only torque acting on the system is the applied force F with lever arm r_s:

$$\tau_{\text{ext}} = Fr_s$$

4. Substitute this value for the torque and $I = MR^2$ for the moment of inertia:

$$\alpha = \frac{\Sigma \tau_{\text{ext}}}{I} = \frac{Fr_s}{MR^2}$$

5. Substitute into the step 1 result and solve for the angular velocity after 5 s:

$$\omega = \alpha t = \frac{Fr_s}{MR^2}t = \frac{(18\ \text{N})(0.07\ \text{m})}{(2.4\ \text{kg})(0.35\ \text{m})^2}5\ \text{s}$$

$$= \boxed{21.4\ \text{rad/s}}$$

A Uniform Rod, Pivoted at One End   **EXAMPLE 9-9**

A uniform thin rod of length L and mass M is pivoted at one end. It is held horizontal and released. The pivot is frictionless. Find (a) the angular acceleration of the rod immediately following its release and (b) the force F_A exerted on the rod by the pivot at this instant.

PICTURE THE PROBLEM The angular acceleration is found from Newton's second law for rotation (Equation 9-18). The force F_A is found from Newton's second law for a system (Equation 8-10). The tangential acceleration of the center of mass is related to the angular acceleration (Equation 9-5) and the centripetal acceleration of the center of mass is related to the angular speed (Equation 9-6).

FIGURE 9-23

(a) 1. Sketch a free-body diagram of the rod (Figure 9-23).

2. Write Newton's second law for rotation:

$$\Sigma \tau_{ext} = I\alpha$$

3. Compute the torque due to gravity about the given axis. The rod is uniform so its center of mass is at its center, a distance $L/2$ from the axis:

$$\tau_{grav} = Mg\frac{L}{2}$$

4. Find the moment of inertia about the end of the rod from Table 9.1:

$$I = \tfrac{1}{3}ML^2$$

5. Substitute these values into the step 2 equation to compute α:

$$\alpha = \frac{\tau_{grav}}{I} = \frac{Mg(L/2)}{(1/3)ML^2} = \boxed{\frac{3g}{2L}}$$

(b) 1. Write Newton's second law for a system for the rod:

$$\Sigma F_{ext\,y} = Ma_{cm\,y}$$
$$Mg - F_A = Ma_{cm\,y}$$

2. Use the relation $a_c = r\omega^2$ to find $a_{cm\,c}$. Just following release, $\omega = 0$:

$$a_c = r\omega^2$$
$$a_{cm\,c} = r_{cm}\omega^2 = \frac{L}{2}\omega^2 = 0$$

3. We now have two equations and three unknowns, α, $a_{cm\,y}$, and F_A. Use the relation $a_t = r\alpha$ to obtain an equation relating $a_{cm\,y}$ to α:

$$a_t = r\alpha$$
$$a_{cm\,y} = r_{cm}\alpha = \frac{L}{2}\alpha$$

4. Substitute both the Part (a) step 5 result and the Part (b) step 1 result into the Part (b) step 3 result and solve for F_A:

$$\frac{Mg - F_A}{M} = \frac{L}{2}\frac{3g}{2L}$$

so

$$F_A = \boxed{\tfrac{1}{4}Mg}$$

REMARKS Just after the rod is released, the acceleration of the center of mass is directed straight down. Since the net external force and the acceleration must be in the same direction, it follows that $\vec{F}_A$ can not have a horizontal component at this moment.

EXERCISE A small pebble of mass $m \ll M$ is placed on top of the rod at its center. Find (a) the acceleration of the pebble and (b) the force it exerts on the rod just after the rod is released. (*Answer* (a) $a = 3g/4$ downward (b) $f = mg/4$ downward)

Nonslip Conditions

There are many situations in which a string is wrapped around a rotating wheel or cylinder. The string must move with a tangential velocity v_t that is equal to the tangential velocity of the rim of the wheel, provided the string remains taut and does not slip:

$$v_t = R\omega \qquad\qquad 9\text{-}21$$

NONSLIP CONDITION FOR v AND ω

For a point on a string, the velocity component tangent to the string is its *tangential* velocity, and the velocity perpendicular to the tangent is its *transverse* velocity. For example, if a bosun is being lifted straight upward by a rope and pulley system, both he and all points on the rope between him and the pulley have the same tangential velocity as the rim of the pulley wheel, but the transverse velocities of these points all equal zero. However, if, while being raised, the bosun is swinging back and forth, then he and the points along the swinging rope above him will have a transverse velocity in addition to their tangential velocity. In the discussions that follow it is assumed that there is no transverse motion (unless stated otherwise). Equation 9-21 is called a nonslip condition. Differentiating it with respect to time relates the tangential acceleration of the string to the angular acceleration of the rim.

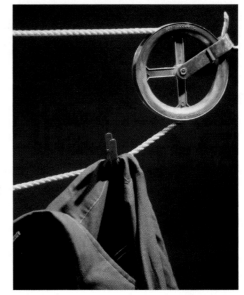

$$a_t = R\alpha \qquad\qquad 9\text{-}22$$

a AND α UNDER NONSLIP CONDITIONS

Problem-Solving Guidelines for Applying Newton's Second Law for Rotation

Tension   Because of friction and inertia, if a string goes around a pulley wheel, the tension in it is greater on one side of the wheel than on the other. (This is necessary for the string to exert a torque on the wheel.) Use two different labels, T_1 and T_2, for these two tensions.

TENSION IN A STRING   **EXAMPLE   9-10**

An object of mass m is tied to a light string wound around a pulley wheel that has a moment of inertia I and radius R. The wheel bearing is frictionless and the string does not slip on the rim. Find the tension in the string and the acceleration of the object.

FIGURE 9-24

PICTURE THE PROBLEM   In this system, the object descends with a downward acceleration a, while the wheel turns with an angular acceleration α (Figure 9-24). We apply Newton's second law for rotation to the wheel to determine α, and Newton's second law to the object to obtain a. Relate a_t and α, using the nonslip condition.

1. Draw a free-body diagram of the pulley wheel, drawing each force vector with its tail at the point of application of the force. Put labels on the diagram and indicate the positive rotational direction as shown in Figure 9-25:

FIGURE 9-25

2. The only force that exerts a torque on the wheel is the tension T, which has lever arm R. Apply Newton's second law for rotational motion to relate T and the angular acceleration α:

$$\Sigma \tau_{ext} = I\alpha$$

$$TR = I\alpha$$

3. Draw a free-body diagram for the suspended object, and apply Newton's second law to relate T to the tangential acceleration a_t (Figure 9-26):

$$\Sigma F_{exty} = ma_y$$

$$mg - T = ma_t$$

FIGURE 9-26

4. We have two equations for three unknowns, T, a_t, and α. A third equation is the relation between a_t and α for the nonslip condition:

$$a_t = R\alpha$$

5. We now have three equations enabling us to determine T, a_t, and α. To solve for T use the step 2 result to obtain an expression for α and use the step 3 result to obtain an expression for a_t. Substitute these expressions into the step 4 result and solve for T.

$$\frac{mg - T}{m} = R\frac{TR}{I}$$

so

$$\boxed{T = \frac{mg}{1 + (mR^2/I)}}$$

6. Substitute this result for T into the step 3 result and solve for a_t:

$$mg - \frac{mg}{1 + (mR^2/I)} = ma_t$$

so

$$\boxed{a_t = \frac{1}{1 + \dfrac{I}{mR^2}}\, g}$$

PLAUSIBILITY CHECK Let's check a couple of extreme limits. If $I = 0$, the object should fall freely, and the string should be slack; our results give $T = 0$, $a_t = g$. What happens if I approaches ∞? For $I \gg mR^2$, our equations give $T \approx mg$ and $a_t \approx 0$.

A PULLEY

EXAMPLE 9-11   Try It Yourself

Two blocks are connected by a string that passes over a pulley of radius R and moment of inertia I. The block of mass m_1 slides on a frictionless, horizontal surface; the block of mass m_2 is suspended from the string (Figure 9-27). Find the acceleration a of the blocks and the tensions T_1 and T_2 assuming the string does not slip on the pulley.

FIGURE 9-27

PICTURE THE PROBLEM In this problem, the tensions T_1 and T_2 are not equal because the pulley has mass and because there is static friction between the string and the pulley (Figure 9-27). (Otherwise the pulley would not turn.) Note that T_2 exerts a clockwise torque and T_1 exerts a counterclockwise torque on the pulley. Use Newton's second law for each block and Newton's second law for rotational motion for the pulley. Relate α and a using the nonslip condition.

Cover the column to the right and try these on your own before looking at the answers.

Steps   **Answers**

FIGURE 9-28

1. Draw a free-body diagram for each block and for the pulley, as shown in Figure 9-28. Note that the center of mass of the pulley does not accelerate, so the support must exert a force on the axle F_s that balances the resultant of the gravitational force and the forces exerted by the string.

2. Apply Newton's second law to each block.

$$T_1 = m_1 a; \qquad m_2 g - T_2 = m_2 a$$

3. Apply Newton's second law for rotation to the pulley wheel.

$$T_2 R - T_1 R = I\alpha$$

4. We have three equations and four unknowns. To get a fourth equation, use the nonslip condition to relate a and α.

$$a = R\alpha$$

5. Now we have four equations and four unknowns, so the rest is algebra. Do the algebra and obtain expressions for a, T_1, and T_2. (*Hint: To find a, obtain expressions for T_1 and T_2 from the step 2 results. Substitute these into the step 3 result to obtain an equation with unknowns a and α. Use the step 4 result to eliminate α and solve for a.*)

$$T_1 = m_1 a = \boxed{\frac{m_2}{m_1 + m_2 + I/R^2} m_1 g}$$

$$T_2 = m_2(g - a) = \boxed{\frac{(m_1 + I/R^2)}{m_1 + m_2 + I/R^2} m_2 g}$$

❶ **PLAUSIBILITY CHECK** If $I = 0$, $T_1 = T_2$, and the acceleration is $a = m_2 g/(m_1 + m_2)$, as expected. If I is very large ($I/R^2 \gg m_1 + m_2$), then $T_1 \approx 0$, $T_2 \approx m_2 g$, and $a \approx 0$.

Power

When you spin an object you do work on it, increasing its kinetic energy. Consider a force F_i acting on a rotating object. As the object rotates through an angle $d\theta$, the point of application of the force moves a distance $ds_i = r_i \, d\theta$, and the force does work

$$dW_i = F_{it} \, ds_i = F_{it} r_i \, d\theta = \tau_i \, d\theta$$

where τ_i is the torque exerted by the force F_i and F_{it} is the tangential component of F_i. In general, the work done by a torque τ when an object turns through a small angle $d\theta$ is

$$dW = \tau \, d\theta \qquad\qquad\qquad 9\text{-}23$$

The rate at which the torque does work is the power input of the torque:

$$P = \frac{dW}{dt} = \tau \frac{d\theta}{dt}$$

or

$$P = \tau\omega \qquad\qquad\qquad 9\text{-}24$$

POWER

Equations 9-23 and 9-24 are the rotational analogs of $dW = F_s \, ds$ and $P = F_s v_s$.

TORQUE EXERTED BY AN AUTOMOBILE ENGINE   E X A M P L E   9 - 1 2   Try It Yourself

The maximum torque produced by the 8.0-L V10 engine of a 2002 Dodge Viper is 675 N·m of torque at 3700 rev/min. Find the power output of the engine operating at these maximum torque conditions.

PICTURE THE PROBLEM The power equals the product of the torque and angular velocity, which are given. You must express ω in radians per second to obtain the power in watts.

Cover the column on the right and try these on your own before looking at the answers.

Steps	Answers
1. Write the power in terms of τ and ω.	$P = \tau\omega$
2. Convert rev/min to rad/s.	$\omega = 387 \text{ rad/s}$
3. Calculate the power.	$P = \boxed{262 \text{ kW}}$

REMARKS This power output is about 350 hp.

EXERCISE The maximum power produced by the Viper engine is 450 hp at 5200 rev/min. What is the torque when the engine is operating at maximum horsepower? (*Answer*   616 N·m)

There are many parallels between one-dimensional linear motion and rotational motion about a fixed axis. The similarities of the formulas can be seen in Table 9-2. The formulas are the same, but the symbols are different.

TABLE 9-2

Analogs in Rotational and Linear Motion

Rotational Motion		Linear Motion	
Angular displacement	$\Delta\theta$	Displacement	Δx
Angular velocity	$\omega = \dfrac{d\theta}{dt}$	Velocity	$v = \dfrac{dx}{dt}$
Angular acceleration	$\alpha = \dfrac{d\omega}{dt} = \dfrac{d^2\theta}{dt^2}$	Acceleration	$a = \dfrac{dv}{dt} = \dfrac{d^2x}{dt^2}$
Constant angular acceleration equations	$\omega = \omega_0 + \alpha t$ $\Delta\theta = \omega_{av}\,\Delta t$ $\omega_{av} = \frac{1}{2}(\omega_0 + \omega)$ $\theta = \theta_0 + \omega_0 t + \frac{1}{2}\alpha t^2$ $\omega^2 = \omega_0^2 + 2\alpha\,\Delta\theta$	Constant acceleration equations	$v = v_0 + at$ $\Delta x = v_{av}\,\Delta t$ $v_{av} = \frac{1}{2}(v_0 + v)$ $x = x_0 + v_0 t + \frac{1}{2}at^2$ $v^2 = v_0^2 + 2a\,\Delta x$
Torque	τ	Force	F
Moment of inertia	I	Mass	m
Work	$dW = \tau\,d\theta$	Work	$dW = F_s\,ds$
Kinetic energy	$K = \frac{1}{2}I\omega^2$	Kinetic energy	$K = \frac{1}{2}mv^2$
Power	$P = \tau\omega$	Power	$P = Fv$
Angular momentum[†]	$L = I\omega$	Momentum	$p = mv$
Newton's second law	$\tau_{net} = I\alpha = \dfrac{dL}{dt}$	Newton's second law	$F_{net} = ma = \dfrac{dp}{dt}$

† Angular momentum is introduced in Chapter 10.

9-6 Rolling Objects

Rolling Without Slipping

When a spool rolls without slipping down an incline (Figure 9-29), the points in contact with the incline are instantaneously at rest and the spool rotates about a rotation axis through the contact point. This can be observed because rapid motion causes blurring, so the part of the spool that is moving slowest is blurred the least. In Figure 9-30 a wheel of radius R is rolling without slipping along a flat surface. Point P on the wheel moves as shown with speed

$$v = r\omega \qquad\qquad 9\text{-}25$$

NONSLIP CONDITION FOR SPEED

FIGURE 9-29

where r is the perpendicular distance from P to the rotation axis. The center of mass of the wheel moves with speed

$$v_{cm} = R\omega \qquad\qquad 9\text{-}26$$

NONSLIP CONDITION FOR v_{CM}

Interestingly, for a point on the top of the wheel, $r = 2R$, so the top of the wheel is moving at twice the speed of the center of mass.

Differentiating each side of Equation (9-26) gives

$$a_{cm} = R\alpha \qquad\qquad 9\text{-}27$$

NONSLIP CONDITION FOR ACCELERATION

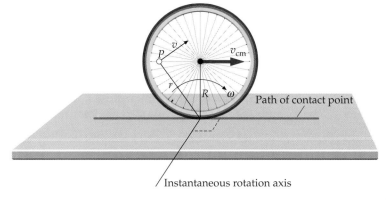

Path of contact point

Instantaneous rotation axis

FIGURE 9-30

A falling yo-yo that is unwinding from a string, the top end of which is held fixed, follows the same nonslip conditions as the wheel.

A wheel of radius R is rolling without slipping along a straight path. As the wheel rotates through angle ϕ (Figure 9-31), the point of contact between the wheel and the plane moves a distance s that is related to ϕ by

$$s = R\phi \qquad\qquad 9\text{-}28$$

NONSLIP CONDITION FOR DISTANCE

FIGURE 9-31

If the wheel is rolling on a flat surface, the wheel's center of mass remains directly over the point of contact, so it also moves through s.

We saw in Chapter 8 that the kinetic energy of a system can be written as the sum of the kinetic energy of motion of the center of mass plus the kinetic energy relative to the center of mass. For a rotation object, the relative kinetic energy is $\frac{1}{2}I_{cm}\,\omega^2$. Thus, the kinetic energy of a rotating object is

$$K = \tfrac{1}{2}Mv_{cm}^2 + \tfrac{1}{2}I_{cm}\omega^2 \qquad\qquad 9\text{-}29$$

KINETIC ENERGY OF A ROTATING OBJECT

A BOWLING BALL   **EXAMPLE 9 - 1 3**   Try It Yourself

A bowling ball of radius 11 cm and mass $M = 7.2$ kg is rolling without slipping on a horizontal ball return at 2 m/s. It then rolls without slipping up a hill to a height h before momentarily stopping before rolling back down the hill. Find h.

PICTURE THE PROBLEM Mechanical energy is conserved. The initial kinetic energy, which is the translational kinetic energy of the center of mass, $\frac{1}{2}mv_{cm}^2$, plus the kinetic energy of rotation about the center of mass, $\frac{1}{2}I_{cm}\omega^2$, is converted to potential energy mgh. Because the sphere rolls without slipping, the initial linear and angular speeds are related by $v_{cm} = R\omega$. Make a labeled sketch showing the ball in both its initial and final positions (Figure 9-32).

FIGURE 9-32

Cover the column to the right and try these on your own before looking at the answers.

Steps

1. Apply conservation of mechanical energy with $U_i = 0$ and $K_f = 0$. Write the total initial kinetic energy K_i in terms of the speed v_{cm} and the angular speed ω.

2. Substitute from $\omega = v_i/R$ and $I_{cm} = \frac{2}{5}MR^2$ and solve for h.

Answers

$U_f + K_f = U_i + K_i$

$Mgh + 0 = 0 + \frac{1}{2}Mv_i^2 + \frac{1}{2}I_{cm}\omega_i^2$

$Mgh = \frac{1}{2}Mv_i^2 + \frac{1}{2}\left(\frac{2}{5}MR^2\right)\frac{v_i^2}{R^2}$

$= \frac{7}{10}Mv_i^2$

so

$h = \dfrac{7v_i^2}{10g} = 0.285 \text{ m} = \boxed{28.5 \text{ cm}}$

REMARKS The height h is independent of the mass or radius of the ball.

EXERCISE Find the initial kinetic energy of the ball. (*Answer* 20.2 J)

PLAYING POOL   **EXAMPLE 9 - 1 4**

FIGURE 9-33

A cue stick hits a cue ball horizontally a distance x above the center of the ball (Figure 9-33). Find the value of x for which the cue ball will roll without slipping from the beginning. Express your answer in terms of the radius R of the ball.

PICTURE THE PROBLEM The lines of action of the weight and normal forces pass through the center of mass and thus exert no torque about it. The frictional force is much smaller than the collision force of the stick and can be neglected. If the stick hits the ball at the level of the ball's center, the ball initially translates with no rotation. If the stick hits below the center, the ball initially has backspin. At a certain value of x, the ball has just the right forward spin and forward acceleration to satisfy the nonslip condition. The value of x determines the torque to force ratio on the ball, and hence its angular acceleration α to linear acceleration ratio a. The linear acceleration a is F/m, independent of x. For the ball to roll without slipping from the start, we find α and a, then set $a = R\alpha$ (nonslip condition) to find x.

1. Sketch a free-body diagram of the ball (Figure 9-34). We are assuming friction between the ball and the table is negligible, so do not include a frictional force:

2. The torque about the axis through the center of the ball equals F times x:

$$\tau = Fx$$

3. Apply Newton's second law for a system and Newton's second law for rotational motion about the center of the ball:

$$F = ma_{cm} \quad \text{and} \quad \tau = I_{cm}\alpha$$

FIGURE 9-34

4. The nonslip condition relates a and α:

$$a_{cm} = R\alpha$$

5. Substitute from steps 2 and 3 into step 4:

$$\frac{F}{m} = R\frac{Fx}{I_{cm}}$$

6. Find the moment of inertia from Table 9.1 and solve for x:

$$x = \frac{I_{cm}}{mR} = \frac{(2/5)mR^2}{mR} = \boxed{(2/5)R}$$

REMARKS Striking the ball at a point either higher or lower than $2R/5$ from the center will result in the ball rolling *and* slipping (skidding). Skidding is often desirable in the game of pool. Rolling and slipping is discussed in the next subsection.

When an object rolls down an incline, its center of mass is accelerated. The analysis of such a problem is simplified by an important theorem concerning the center of mass:

If the torques are computed from a reference frame moving with the center of mass, then Newton's second law for rotation holds even when the center of mass is accelerating, and the torques are computed in the center-of-mass reference frame. That is,

$$\tau_{net,cm} = I_{cm}\alpha \qquad\qquad 9\text{-}30$$

This is the same as Equation 9-18 except that here the torques and the moment of inertia are computed from a reference frame moving with the center of mass. (A derivation may be found at www.whfreeman.com/tipler.) When the center of mass is accelerating (a ball rolling down an incline, for example), the center-of-mass reference frame is a noninertial one, where we would not necessarily expect our equations for Newton's second law for rotation to be valid. Nevertheless, they are.

ACCELERATION OF A BALL THAT IS ROLLING WITHOUT SLIPPING **EXAMPLE 9-15**

A uniform solid ball of mass m and radius R rolls without slipping down a plane inclined at an angle ϕ above the horizontal. Find the acceleration of the center of mass and the frictional force.

PICTURE THE PROBLEM From Newton's second law, the acceleration of the center of mass equals the net force divided by the mass. The forces acting are the weight $m\vec{g}$ downward, the normal force $\vec{F}_n$ that balances the normal component of the weight, and the force of friction $\vec{f}$ acting up the incline (Figure 9-35). As the object accelerates down the incline, its angular velocity must increase to maintain the nonslip condition. This angular acceleration requires a net external torque about the axis through the center of mass. We apply Newton's second law for rotation to find α. The nonslip condition relates α and a_{cm}.

FIGURE 9-35

1. Apply Newton's second law for a system along the x axis:

$$\Sigma F_x = ma_{cmx}$$

$$mg \sin \phi - f_s = ma_{cm}$$

2. Apply Newton's second law for rotational motion, for rotations about an axis parallel to the instantaneous rotation axis and passing through the center of mass:

$$\Sigma \tau_i = I_{cm}\alpha$$

$$f_s R + 0 + 0 = I_{cm}\alpha$$

3. Relate a_{cm} and α using the nonslip condition:

$$a_{cm} = R\alpha$$

4. We now have three equations and three unknowns. Solve the step 1 result for f_s and the step 3 result for α, substitute for these quantities in the step 2 result and solve for a_{cm}:

$$f_s R = I_{cm}\alpha$$

$$(mg \sin \phi - ma_{cm})R = I_{cm}\frac{a_{cm}}{R}$$

so

$$a_{cm} = \frac{g \sin \phi}{1 + \dfrac{I_{cm}}{mR^2}}$$

5. Substitute the step 4 result into the step 1 result and solve for f_s:

$$f_s = mg \sin \phi - ma_{cm}$$

$$= mg \sin \phi - \frac{mg \sin \phi}{1 + \dfrac{I_{cm}}{mR^2}} = \frac{mg \sin \phi}{1 + \dfrac{mR^2}{I_{cm}}}$$

6. For a solid sphere, $I_{cm} = \frac{2}{5}mR^2$ (see Table 9.1):

$$a_{cm} = \frac{g \sin \phi}{1 + \frac{2}{5}} = \boxed{\frac{5}{7}g \sin \phi}$$

$$f_s = \frac{mg \sin \phi}{\frac{7}{2}} = \boxed{\frac{2}{7}mg \sin \phi}$$

REMARKS Because the ball rolls without slipping, the friction is static friction. Note that the result seems independent of the coefficient of static friction. However, we have assumed that the coefficient of static friction was large enough to prevent slipping.

The results of steps 5 and 6 in Example 9-15 apply to any round object with the center of mass at the geometric center that is rolling without slipping. For such objects, $I_{cm} = \beta mR^2$, where $\beta = \frac{2}{5}$ for a sphere, $\frac{1}{2}$ for a rolling solid cylinder, 1 for a thin cylindrical shell, and so forth. For such objects the step 5 and step 6 results can be expressed

$$f_s = \frac{mg \sin \phi}{1 + \beta^{-1}} \qquad\qquad\qquad 9\text{-}31$$

$$a_{cm} = \frac{g \sin \phi}{1 + \beta} \qquad\qquad\qquad 9\text{-}32$$

The linear acceleration of any object rolling down an incline is less than $g \sin \phi$ because of the frictional force directed up the incline. Note that these accelerations are independent of both the mass and the radius of the objects. That is, all solid spheres rolling without slipping down the same incline will have identical accelerations. However, if we release a sphere, a cylinder, and a hoop at the top of an incline, and if they all roll without slipping, the sphere will reach the bottom first because it has the greatest acceleration. The cylinder will be second and the hoop last (Figure 9-36). A block that slides without friction down the incline will arrive at the bottom ahead of all three rolling objects.

Because the friction is static, it does no work, and there is no dissipation of mechanical energy. We can therefore use the conservation of mechanical energy to find the speed of an object rolling without slipping down an incline. At the top of the incline, the total energy is the potential energy mgh. At the bottom, the total energy is kinetic energy. Conservation of mechanical energy therefore gives

$$\tfrac{1}{2}mv_{cm}^2 + \tfrac{1}{2}I_{cm}\omega^2 = mgh$$

We can use the nonslip condition to eliminate either v_{cm} or ω. Substituting $I_{cm} = \beta mR^2$ and $\omega = v_{cm}/R$, we obtain $\tfrac{1}{2}mv_{cm}^2 + \tfrac{1}{2}\beta mR^2 (v_{cm}^2/R^2) = mgh$. Solving for v_{cm}^2 gives

$$v_{cm}^2 = \frac{2gh}{1 + \beta} \qquad\qquad 9\text{-}33$$

For a cylinder, with $\beta = \tfrac{1}{2}$, we obtain $v_{cm} = \sqrt{\tfrac{4}{3}gh}$. Note that the speed is independent of the mass and radius of the cylinder, and is less than $\sqrt{2gh}$, the speed of an object sliding with no friction down the incline.

For an object rolling without slipping down an incline, the frictional force f_s is less than or equal to its maximum value; that is, $f_s \le \mu_s F_n$, where $F_n = mg \cos \phi$. Substituting the expression from Equation 9-31 for the frictional force, we have

$$\frac{mg \sin \phi}{1 + \beta^{-1}} \le \mu_s mg \cos \phi$$

or

$$\tan \phi \le (1 + \beta^{-1})\mu_s \qquad\qquad 9\text{-}34$$

(For a uniform cylinder, $I_{cm} = \tfrac{1}{2}MR^2$, so $\beta = \tfrac{1}{2}$ and $\tan \phi \le 3\mu_s$.) If the tangent of the angle of incline is greater than $(1 + \beta^{-1})\mu_s$, the object will slip as it moves down the incline.

EXERCISE A uniform cylinder rolls down a plane inclined at $\theta = 50°$. What is the minimum value of the coefficient of static friction for which the cylinder will roll without slipping? (*Answer* 0.40)

EXERCISE For a hoop rolling down an incline, (*a*) what is the force of friction and (*b*) what is the maximum value of $\tan \theta$ for which the hoop will roll without slipping? (*Answer* (*a*) $f = \tfrac{1}{2}mg \sin \theta$ (*b*) $\tan \theta \le 2\mu_s$)

*Rolling With Slipping

When an object slides as it rolls, the nonslip condition $v_{cm} = R\omega$ does not hold. Suppose a bowling ball is thrown with no initial rotation. As the ball slides along the bowling lane, $v_{cm} > R\omega$. However, the frictional force both reduces its linear speed v_{cm} (Figure 9-37) and increases its angular speed ω until the nonslip condition $v_{cm} = R\omega$ is reached, after which the ball rolls without slipping.

Another example of rolling with slipping is a ball with topspin, such as a cue ball struck at a point higher than $(2/5)R$ above the center (see Example 9-14) so that $v_{cm} < R\omega$. Then the frictional force both increases v and decreases ω until the nonslip condition $v_{cm} = R\omega$ is reached (Figure 9-38).

FIGURE 9-36 A sphere, a cylinder, and a hoop are released together from rest at the top of an incline. The sphere reaches the bottom first, followed by the cylinder and then the hoop.

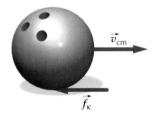

FIGURE 9-37 A bowling ball moving with no initial rotation. The frictional force $\vec{f}_k$ exerted by the floor reduces the speed v_{cm} and increases the angular speed ω until $v_{cm} = R\omega$.

FIGURE 9-38 Ball with excess topspin. The frictional force accelerates the ball in the direction of motion.

A SLIDING BOWLING BALL   **EXAMPLE   9 - 1 6**

A bowling ball of mass M and radius R is thrown so that the instant it touches the floor it is moving horizontally with speed $v_0 = 5$ m/s and is not rotating. The coefficient of kinetic friction between the ball and the floor is $\mu_k = 0.08$. Find (*a*) the time the ball slides and (*b*) the distance the ball slides before it rolls without slipping.

PICTURE THE PROBLEM   We calculate v_{cm} and ω as functions of time, set $v_{cm} = R\omega$, and solve for t. The linear and angular accelerations are found from $\Sigma F = ma$ and $\tau = I\alpha$. Let the direction of motion be positive. There is slipping and kinetic friction, so mechanical energy is dissipated. Therefore, conservation of mechanical energy cannot be used to solve this problem.

FIGURE 9-39

(*a*) 1. Sketch a free-body diagram of the ball (Figure 9-39).

2. The net force on the ball is the force of kinetic friction f_k, which acts in the negative x direction. Apply Newton's second law:

$$\Sigma F_x = Ma_{cmx}$$

$$-f_k = Ma_{cmx}$$

3. The acceleration is in the negative x direction and $a_{cmy} = 0$. Find f_k by first finding F_n:

$$\Sigma F_y = Ma_{cmy} = 0 \;\Rightarrow\; F_n = Mg$$

so

$$f_k = \mu_k F_n = \mu_k Mg$$

4. Find the acceleration using the step 2 and step 3 results:

$$-\mu_k Mg = Ma_{cmx} \;\Rightarrow\; a_{cmx} = -\mu_k g$$

5. Relate the linear velocity to the acceleration and the time using a kinematic equation:

$$v_{cmx} = v_0 + a_{cmx}t = v_0 - \mu_k gt$$

6. Find α by applying Newton's second law for rotational motion to the ball. Compute the torques about the axis through the center of mass. Note that the free-body diagram has clockwise as positive:

$$\Sigma \tau = I_{cm}\alpha$$

$$\mu_k MgR + 0 + 0 = \tfrac{2}{5}MR^2\alpha$$

so

$$\alpha = \frac{5}{2}\frac{\mu_k g}{R}$$

7. Relate the angular velocity to the angular acceleration and the time using a kinematic equation:

$$\omega = \omega_0 + \alpha t = 0 + \alpha t = \frac{5}{2}\frac{\mu_k g}{R}t$$

8. Solve for the time t at which $v_{cm} = R\omega$:

$$v_{cm} = R\omega$$

$$v_0 - \mu_k gt = \tfrac{5}{2}\mu_k gt$$

so

$$t = \frac{2v_0}{7\,\mu_k g} = \frac{2(5\text{ m/s})}{7(0.08)(9.81\text{ m/s}^2)} = \boxed{1.82\text{ s}}$$

(*b*) The distance traveled in 1.82 s is:

$$\Delta x = v_0 t + \tfrac{1}{2}a_{cm}t^2$$

$$= v_0 \frac{2v_0}{7\,\mu_k g} + \frac{1}{2}(-\mu_k g)\left(\frac{2v_0}{7\,\mu_k g}\right)^2 = \frac{12}{49}\frac{v_0^2}{\mu_k g}$$

$$= \frac{12}{49}\frac{(5\text{ m/s})^2}{(0.08)(9.81\text{ m/s}^2)} = \boxed{7.80\text{ m}}$$

EXERCISE Find the speed of the bowling ball when it begins to roll without slipping. (*Answer* $v_{cm} = \frac{5}{7}v_0$). This result is independent of the coefficient of kinetic friction. The rolling speed is $\frac{5}{7}v_0$ whether μ_k is large or small. The total mechanical energy lost is thus independent of μ_k. (The time and the distance are sensitive to the value of μ_k, however.)

EXERCISE Find the total kinetic energy of the ball after it begins to roll without slipping. (*Answer* $K = \frac{5}{14}mv_0^2$)

REMARKS In a well-maintained bowling alley, the lanes are lightly oiled and very slick so that the ball slides over a great distance, giving the bowler added control.

SUMMARY

1. Angular displacement, angular velocity, and angular acceleration are fundamental defined quantities in rotational kinematics.

2. Torque and moment of inertia are important derived dynamic concepts. Torque is a measure of the effect of a force in causing an object to start or stop rotating. Moment of inertia is the measure of an object's inertial resistance to angular accelerations. The moment of inertia depends on the distribution of the mass relative to the rotation axis.

3. The parallel-axis theorem, which follows from the definition of the moment of inertia, often simplifies the calculation of I.

4. Newton's second law for rotation, $\Sigma \tau_{ext} = I\alpha$, is derived from Newton's second law and the definitions of τ, I, and α. It is an important relation for problems involving the rotation of a rigid object about a rotation axis of fixed direction.

Topic	Relevant Equations and Remarks	
1. Angular Velocity and Angular Acceleration		
Angular velocity	$\omega = \dfrac{d\theta}{dt}$ (Definition)	9-2
Angular acceleration	$\alpha = \dfrac{d\omega}{dt} = \dfrac{d^2\theta}{dt^2}$ (Definition)	9-3
Tangential velocity	$v_t = r\omega$	9-7
Tangential acceleration	$a_t = r\alpha$	9-8
Centripetal acceleration	$a_c = \dfrac{v^2}{r} = r\omega^2$	9-9

2. Equations for Rotation With

Constant Angular Acceleration

$$\omega = \omega_0 + \alpha t \qquad \text{9-4}$$

$$\theta = \theta_0 + \omega_0 t + \tfrac{1}{2}\alpha t^2 \qquad \text{9-5}$$

$$\omega^2 = \omega_0^2 + 2\alpha(\theta - \theta_0) \qquad \text{9-6}$$

3. Moment of Inertia

System of particles

$$I = \Sigma\, m_i r_i^2 \qquad \text{(Definition)} \qquad \text{9-10}$$

Continuous object

$$I = \int r^2\, dm \qquad \text{9-12}$$

Parallel-axis theorem

The moment of inertia about an axis a distance h from a parallel axis through the center of mass is

$$I = I_{cm} + Mh^2 \qquad \text{9-13}$$

where I_{cm} is the moment of inertia about the axis through the center of mass and M is the total mass of the object.

4. Energy

Kinetic energy for rotation about a fixed axis

$$K = \tfrac{1}{2}I\omega^2 \qquad \text{9-11}$$

Kinetic energy for rotating object

$$K = \tfrac{1}{2}Mv_{cm}^2 + \tfrac{1}{2}I_{cm}\omega^2 \qquad \text{9-29}$$

Power

$$P = \tau\omega \qquad \text{9-24}$$

5. Torque

The torque due to a force equals the product of the tangential component of the force and the radial distance from the axis:

$$\tau = F_t r = Fr\sin\phi = F\ell \qquad \text{9-19}$$

6. Newton's Second Law for Rotation

$$\tau_{net,ext} = \sum_i \tau_{i,ext} = I\alpha \qquad \text{9-18}$$

If torques about an axis through the center of mass are computed from a reference frame moving with the center of mass, Newton's second law for rotation holds for rotation about an axis through the center of mass, even if the reference frame is noninertial.

7. Nonslip Conditions

When a string that is wrapped around a pulley or disk does not slip, the linear and angular quantities are related by

$$v_t = R\omega \qquad \text{9-21}$$

$$a_t = R\alpha \qquad \text{9-22}$$

8. Rolling Objects

Rolling without slipping

$$v_{cm} = R\omega \qquad \text{9-26}$$

*Rolling with slipping

When an object rolls and slips, $v_{cm} \neq R\omega$. Kinetic friction exerts a force that tends to change v_{cm}, and also exerts a torque that changes ω until $v_{cm} = R\omega$ and pure rolling sets in.

- Single-concept, single-step, relatively easy
- •• Intermediate-level, may require synthesis of concepts
- ••• Challenging
- **SSM** Solution is in the *Student Solutions Manual*
- **iSOLVE** Problems available on iSOLVE online homework service
- **iSOLVE** ✓ These "Checkpoint" online homework service problems ask students additional questions about their confidence level, and how they arrived at their answer

In a few problems, you are given more data than you actually need; in a few other problems, you are required to supply data from your general knowledge, outside sources, or informed estimates.

Take $g = 9.81$ N/kg $= 9.81$ m/s^2 and neglect friction in all problems unless otherwise stated.

Conceptual Problems

1 • **SSM** Two points are on a disk turning at constant angular velocity, one point on the rim and the other halfway between the rim and the axis. Which point moves the greater distance in a given time? Which turns through the greater angle? Which has the greater speed? The greater angular velocity? The greater tangential acceleration? The greater angular acceleration? The greater centripetal acceleration?

2 • True or false:
(a) Angular velocity and linear velocity have the same dimensions.
(b) All parts of a rotating wheel must have the same angular velocity.
(c) All parts of a rotating wheel must have the same angular acceleration.

3 •• Starting from rest, a disk takes 10 revolutions to reach an angular velocity ω at constant angular acceleration. How many additional revolutions are required to reach an angular velocity of 2ω? (a) 10 rev. (b) 20 rev. (c) 30 rev. (d) 40 rev. (e) 50 rev.

4 • **SSM** The dimension of torque is the same as that of (a) impulse, (b) energy, (c) momentum, (d) none of these.

5 • The moment of inertia of an object of mass M (a) is an intrinsic property of the object, (b) depends on the choice of axis of rotation, (c) is proportional to M regardless of the choice of axis, (d) both (b) and (c) are correct.

6 • **SSM** Can an object continue to rotate in the absence of torque?

7 • Does an applied net torque always increase the angular speed of an object?

8 • True or false:
(a) If the angular velocity of an object is zero at some instant, the net torque on the object must be zero at that instant.
(b) The moment of inertia of an object depends on the location of the axis of rotation.
(c) The moment of inertia of an object depends on the angular velocity of the object.

9 • A disk is free to rotate about an axis. A tangential force applied a distance d from the axis causes an angular acceleration α. What angular acceleration is produced if the same force is applied a distance $2d$ from the axis? (a) α. (b) 2α. (c) $\alpha/2$. (d) 4α. (e) $\alpha/4$.

10 • **SSM** The moment of inertia of an object about an axis that does not pass through its center of mass is _____ the moment of inertia about a parallel axis through its center of mass. (a) always less than, (b) sometimes less than, (c) sometimes equal to, (d) always greater than.

11 • A constant torque acts on a merry-go-round. The power input of the torque is (a) constant, (b) proportional to the angular speed of the merry-go-round, (c) zero, (d) none of these.

12 • True or false: When an object rolls without slipping, friction does no work on the object.

13 • Why do we put the knob on a door about as far away from the hinges as possible?

14 • **SSM** A wheel of radius R is rolling without slipping on a flat stationary surface. The velocity of the point on the rim that is in contact with the surface is (a) equal to $R\omega$ in the direction of motion of the center of mass, (b) equal to $R\omega$ opposite to the direction of motion of the center of mass, (c) zero, (d) equal to the velocity of the center of mass and in the same direction, (e) equal to the velocity of the center of mass but in the opposite direction.

15 •• A solid uniform cylinder and a solid uniform sphere have equal masses. Both roll without slipping on a horizontal surface. If their kinetic energies are the same, then (a) the translational speed of the cylinder is greater than that of the sphere, (b) the translational speed of the cylinder is less than that of the sphere, (c) the translational speeds of the two objects are the same, (d) (a), (b), or (c) could be correct depending on the radii of the objects.

16 • **SSM** Two identical-looking 1-m-long pipes enclose slugs of lead whose total mass is 10 kg (much larger than the mass of the pipe). In the first pipe the lead is concentrated at the center of the pipe, while in the second the lead is divided into two equal masses placed at opposite ends of the pipe. Without opening either pipe, how could you determine which is which?

17 •• Starting from rest at the same time, a coin and a ring roll down an incline without slipping. Which of the following is true? (*a*) The ring reaches the bottom first. (*b*) The coin reaches the bottom first. (*c*) The coin and ring arrive at the bottom simultaneously. (*d*) The race to the bottom depends on their relative masses. (*e*) The race to the bottom depends on their relative diameters.

18 •• For a hoop of mass M and radius R that is rolling without slipping, which is larger, its translational kinetic energy or its rotational kinetic energy? (*a*) Translational kinetic energy is larger. (*b*) Rotational kinetic energy is larger. (*c*) Both are the same size. (*d*) The answer depends on the radius. (*e*) The answer depends on the mass.

19 •• For a disk of mass M and radius R that is rolling without slipping, which is larger, its translational kinetic energy or its rotational kinetic energy? (*a*) Translational kinetic energy is larger. (*b*) Rotational kinetic energy is larger. (*c*) Both are the same size. (*d*) The answer depends on the radius. (*e*) The answer depends on the mass.

20 •• A perfectly rigid ball rolls without slipping along a perfectly rigid horizontal plane. Show that the frictional force acting on the ball must be zero. *Hint: Consider a possible direction for the action of the frictional force and what effects such a force would have on the velocity of the center of mass and on the angular velocity.*

21 • True or false: When a sphere rolls and slips on a rough surface, mechanical energy is dissipated.

22 • A cue ball is hit very near the top so that it starts to move with topspin. As it slides, the force of friction (*a*) increases v_{cm}, (*b*) decreases v_{cm}, (*c*) has no effect on v_{cm}.

Estimation and Approximation

23 •• A bicycle of mass 14 kg has 1.2-m-diameter wheels, each of mass 3 kg. The mass of the rider is 38 kg. Estimate the fraction of the total kinetic energy of bicycle and rider associated with rotation of the wheels.

24 •• Why does toast falling off a table always land jelly-side down? The question may sound silly, but it has been a subject of serious scientific enquiry. The analysis is too complicated to reproduce here, but R. D. Edge and Darryl Steinert showed that a piece of toast, pushed gently over the edge of a table until it tilts off, typically falls off the table when it makes an angle of about 30° with the horizontal (see Figure 9-40) and has an angular velocity of $\omega = 0.956\sqrt{g/\ell}$, where ℓ is the length of one side of the piece of toast (assumed square). Assuming that it starts jelly-side up, what side will it land on if it falls from a table of height 0.5 m? How about a height of 1 m? Assume that the toast has a side length $\ell = 0.1$ m, and that it will land jelly-side down if the angle is between 180° and 270°. (If the toast rotates through an angle greater than 360°, we will need to reduce it to the range 0°–360°.) Ignore any forces due to air resistance. For readers interested in this problem and a host of others, we highly recommend Robert Erlich's wonderful book, *Why Toast Lands Jelly-Side Down: Zen and the Art of Physics Demonstrations.*[†]

FIGURE 9-40 Problem 24

25 •• SSM Consider the moment of inertia of an average adult man about an axis running vertically through the center of his body when he is standing straight up with arms flat at his sides and again when he is standing straight up holding his arms straight out. Estimate the ratio of his moment of inertia with his arms straight out to the moment of inertia with his arms flat against his sides.

Angular Velocity and Angular Acceleration

26 • A particle moves in a circle of radius 90 m with a constant speed of 25 m/s. (*a*) What is its angular velocity in radians per second about the center of the circle? (*b*) How many revolutions does it make in 30 s?

27 • ✓ A wheel starts from rest with constant angular acceleration of 2.6 rad/s^2. After 6 s: (*a*) What is its angular velocity? (*b*) Through what angle has the wheel turned? (*c*) How many revolutions has it made? (*d*) What is the speed and acceleration of a point 0.3 m from the axis of rotation?

28 • SSM When a turntable rotating at $33\frac{1}{3}$ rev/min is shut off, it comes to rest in 26 s. Assuming constant angular acceleration, find (*a*) the angular acceleration, (*b*) the average angular velocity of the turntable, and (*c*) the number of revolutions it makes before stopping.

29 • A disk of radius 12 cm, initially at rest, begins rotating about its axis with a constant angular acceleration of 8 rad/s^2. At $t = 5$ s, what are (*a*) the angular velocity of the disk and (*b*) the tangential acceleration a_t and the centripetal acceleration a_c of a point on the edge of the disk?

30 • A Ferris wheel of radius 12 m rotates once in 27 s. (*a*) What is its angular velocity in radians per second? (*b*) What is the linear speed of a passenger? What is the centripetal acceleration of a passenger?

31 • A cyclist accelerates from rest. After 8 s, the wheels have made 3 revolutions. (*a*) What is the angular acceleration of the wheels? (*b*) What is the angular velocity of the wheels after 8 s?

32 • What is the angular velocity of the earth in radians per second as it rotates about its axis?

33 • A wheel rotates through 5.0 rad in 2.8 s as it is brought to rest with constant angular acceleration. The initial angular velocity of the wheel before braking began was (*a*) 0.6 rad/s, (*b*) 0.9 rad/s, (*c*) 1.8 rad/s, (*d*) 3.6 rad/s, (*e*) 7.2 rad/s.

34 • A bicycle has wheels of 1.2-m diameter. The bicyclist accelerates from rest with constant acceleration to 24 km/h in 14.0 s. What is the angular acceleration of the wheels?

35 •• **SSM** **SOLVE** ✓ The tape in a standard VHS video-tape cassette has a length $L = 246$ m; the tape plays for 2.0 h (Figure 9-41). As the tape starts, the full reel has an outer radius of about $R = 45$ mm and an inner radius of about $r = 12$ mm. At some point during the play, both reels have the same angular speed. Calculate this angular speed in radians per second and revolutions per minute.

FIGURE 9-41 Problem 35

45 mm
12 mm

Torque, Moment of Inertia, and Newton's Second Law for Rotation

36 • **SOLVE** A disk-shaped grindstone of mass 1.7 kg and radius 8 cm is spinning at 730 rev/min. After the power is shut off, a woman continues to sharpen her ax by holding it against the grindstone for 9 s until the grindstone stops rotating. (a) What is the angular acceleration of the grindstone? (b) What is the torque exerted by the ax on the grindstone? (Assume constant angular acceleration and a lack of other frictional torques.)

37 • **SSM** A 2.5-kg cylinder of radius 11 cm, initially at rest, is free to rotate about the axis of the cylinder. A rope of negligible mass is wrapped around it and pulled with a force of 17 N. Find (a) the torque exerted by the rope, (b) the angular acceleration of the cylinder, and (c) the angular velocity of the cylinder at $t = 5$ s.

38 •• **SOLVE** ✓ A wheel free to rotate about its axis that is not frictionless is initially at rest. A constant external torque of 50 N·m is applied to the wheel for 20 s, giving the wheel an angular velocity of 600 rev/min. The external torque is then removed, and the wheel comes to rest 120 s later. Find (a) the moment of inertia of the wheel and (b) the frictional torque, which is assumed to be constant.

39 •• A pendulum consisting of a string of length L attached to a bob of mass m swings in a vertical plane. When the string is at an angle θ to the vertical, (a) what is the tangential component of acceleration of the bob? (b) What is the torque exerted about the pivot point? (c) Show that $\tau = I\alpha$ with $a_t = L\alpha$ gives the same tangential acceleration as found in Part (a).

40 ••• **SSM** A uniform rod of mass M and length L is pivoted at one end and hangs as in Figure 9-42 so that it is free to rotate without friction about its pivot. It is struck by a horizontal force F_0 for a short time Δt at a distance x below the pivot as shown. (a) Show that the speed of the center of mass of the rod just after being struck is given by $v_0 = 3F_0 x \, \Delta t/2ML$. (b) Find the horizontal component of the force delivered by the pivot, and show that this force component is zero if $x = \frac{2}{3}L$. (Note: The point $x = \frac{2}{3}L$ is called the center of percussion of the rod.)

FIGURE 9-42
Problem 40

x

F_0

41 ••• A uniform horizontal disk of mass M and radius R is rotating about the vertical axis through its center with an angular velocity ω. When it is placed on a horizontal surface, the coefficient of kinetic friction between the disk and the surface is μ_k. (a) Find the torque $d\tau$ exerted by the force of friction on a circular element of radius r and width dr. (b) Find the total torque exerted by friction on the disk. (c) Find the time required for the disk to stop rotating.

Calculating the Moment of Inertia

42 • **SOLVE** A tennis ball has a mass of 57 g and a diameter of 7 cm. Find the moment of inertia about its diameter. Assume that the ball is a thin spherical shell.

43 • **SSM** **SOLVE** Four particles at the corners of a square with side length $L = 2$ m are connected by massless rods (Figure 9-43). The masses of the particles are $m_1 = m_3 = 3$ kg and $m_2 = m_4 = 4$ kg. Find the moment of inertia of the system about the z axis.

FIGURE 9-43 Problems 43–45

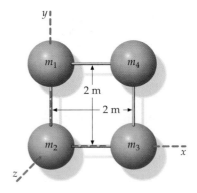

44 • Use the parallel-axis theorem and your results for Problem 43 to find the moment of inertia of the four-particle system in Figure 9-43 about an axis that is perpendicular to the plane of the configuration and passes through the center of mass of the system. Check your result by direct computation.

45 • For the four-particle system of Figure 9-43, (a) find the moment of inertia I_x about the x axis, which passes through m_3 and m_4, and (b) find I_y about the y axis, which passes through m_1 and m_4.

46 • [SOLVE] Use the parallel-axis theorem to find the moment of inertia of a solid sphere of mass M and radius R about an axis that is tangent to the sphere (Figure 9-44).

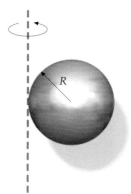

FIGURE 9-44 Problem 46

47 •• [SOLVE] ✔ A 1.0-m-diameter wagon wheel consists of a thin rim having a mass of 8 kg and six spokes, each having a mass of 1.2 kg. Determine the moment of inertia of the wagon wheel for rotation about its axis.

48 •• [SSM] [SOLVE] Two point masses m_1 and m_2 are separated by a massless rod of length L. (a) Write an expression for the moment of inertia about an axis perpendicular to the rod and passing through it at a distance x from mass m_1. (b) Calculate dI/dx and show that I is at a minimum when the axis passes through the center of mass of the system.

49 •• A uniform rectangular plate has mass m and sides of lengths a and b. (a) Show by integration that the moment of inertia of the plate about an axis that is perpendicular to the plate and passes through one corner is $m(a^2 + b^2)/3$. (b) What is the moment of inertia about an axis that is perpendicular to the plate and passes through its center of mass?

50 •• [SSM] Tracey and Corey are doing intensive research on baton-twirling. Each is using "The Beast" as a model baton: two uniform spheres, each of mass 500 g and radius 5 cm, mounted at the ends of a 30-cm uniform rod of mass 60 g (Figure 9-45). They want to calculate the moment of inertia of The Beast about an axis perpendicular to the rod and passing through its center. Corey uses the approximation that the two spheres can be treated as point particles that are 20 cm from the axis of rotation, and that the mass of the rod is negligible. Tracey, however, makes her calculations without approximations. (a) Compare the two results. (b) If the spheres retained the same mass but were hollow, would the rotational inertia increase or decrease? Justify your choice with a sentence or two. It is not necessary to calculate the new value of I.

FIGURE 9-45 Problem 50

10 cm $L = 30$ cm   10 cm

$m = 60$ g

500 g   500 g

Axis of rotation

51 •• The methane molecule (CH_4) has four hydrogen atoms located at the vertices of a regular tetrahedron of side length 0.18 nm, with the carbon atom at the center of the tetrahedron (Figure 9-46). Find the moment of inertia of this molecule for rotation about an axis that passes through the centers of the carbon atom and one of the hydrogen atoms.

FIGURE 9-46 Problem 51

52 •• A hollow cylinder has mass m, an outside radius R_2, and an inside radius R_1. Show that its moment of inertia about its symmetry axis is given by $I = \frac{1}{2}m\,(R_2^2 + R_1^2)$. In Section 9-3 the moment of inertia was calculated for a solid cylinder by direct integration, first finding the moment of inertia of a disk. This calculation is the same as that one, except for the value of one of the integration limits and the expression for the area of a cross section.

53 ••• Show that the moment of inertia of a spherical shell of radius R and mass m is $2mR^2/3$. This can be done by direct integration or, more easily, by finding the increase in the moment of inertia of a solid sphere when its radius changes. To do this, first show that the moment of inertia of a solid sphere of density ρ is $I = (8/15)\pi\rho R^5$. Then compute the change dI in I for a change dR, and use the fact that the mass of this shell is $dm = 4\pi R^2 \rho\, dR$.

54 ••• [SSM] The density of the earth is not quite uniform. It varies with the distance r from the center of the earth as $\rho = C\,[1.22 - (r/R)]$, where R is the radius of the earth and C is a constant. (a) Find C in terms of the total mass M and the radius R. (b) Find the moment of inertia of the earth. (See Problem 53.)

55 ••• Use integration to determine the moment of inertia of a right circular homogeneous solid cone of height H, base radius R, and mass density ρ about its symmetry axis.

56 ••• Use integration to determine the moment of inertia of a thin uniform disk of mass M and radius R for rotation about a diameter. Check your answer by referring to Table 9-1.

57 ••• A roadside ice-cream stand uses rotating solid cones to catch the eye of travelers. Each cone rotates about an axis perpendicular to its axis of symmetry and passing through its apex. The sizes of the cones vary, and the owner wonders if it would be more energy-efficient to use several smaller cones or a few big ones. To answer this, he must calculate the moment of inertia of a homogeneous right circular cone of height H, base radius R, and mass density ρ. What is the result?

Rotational Kinetic Energy

58 • The particles in Figure 9-47 are connected by a very light rod whose moment of inertia can be neglected. They rotate about the y axis with angular velocity $\omega = 2$ rad/s. (a) Find the speed of each particle, and use it to calculate the kinetic energy of this system directly from $\Sigma \frac{1}{2} m_i v_i^2$. (b) Find the moment of inertia about the y axis, and calculate the kinetic energy from $K = \frac{1}{2} I \omega^2$.

FIGURE 9-47 Problem 58

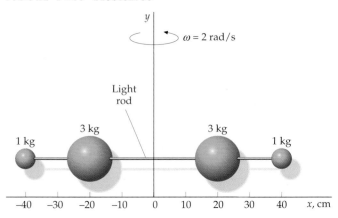

59 • **SSM** **iSOLVE** A solid ball of mass 1.4 kg and diameter 15 cm is rotating about its diameter at 70 rev/min. (a) What is its kinetic energy? (b) If an additional 2 J of energy are supplied to the rotational energy, what is the new angular speed of the ball?

60 • An engine develops 400 N·m of torque at 3700 rev/min. Find the power developed by the engine.

61 •• Two point particles with masses m_1 and m_2 are connected by a massless rod of length L to form a dumbbell that rotates about its center of mass with angular velocity ω. Show that the ratio of kinetic energies of the particles is $K_1/K_2 = m_2/m_1$.

62 •• **iSOLVE** Calculate the kinetic energy of rotation of the earth about its axis, and compare it with the kinetic energy of the orbital motion of the earth's center of mass about the sun. Assume the earth to be a homogeneous sphere of mass 6.0×10^{24} kg and radius 6.4×10^6 m. The radius of the earth's orbit is 1.5×10^{11} m.

63 •• **SSM** **iSOLVE** A 2000-kg block is lifted at a constant speed of 8 cm/s by a steel cable that passes over a massless pulley to a motor-driven winch (Figure 9-48). The radius of the winch drum is 30 cm. (a) What force must be exerted by the cable? (b) What torque does the cable exert on the winch drum? (c) What is the angular velocity of the winch drum? (d) What power must be developed by the motor to drive the winch drum?

FIGURE 9-48 Problem 63

64 •• A uniform disk of mass M and radius R can rotate freely about a horizontal axis through its center and perpendicular to the plane of the disk. A small particle of mass m is attached to the rim of the disk at the top, directly above the pivot. The system is given a gentle nudge, and the disk begins to rotate. (a) What is the angular velocity of the disk when the particle is at its lowest point? (b) At this point, what force must be exerted by the disk on the particle to keep it stuck to the disk?

65 •• A uniform ring 1.5 m in diameter is pivoted at one point on its perimeter so that it is free to rotate about a horizontal axis. Initially, the line joining the support point and the center is horizontal (Figure 9-49). (a) If the ring is released from rest, what is its maximum angular velocity? (b) What minimum initial angular velocity must it be given if it is to rotate a full 360°?

FIGURE 9-49
Problem 65

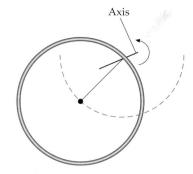

66 •• **iSOLVE** You set out to design a car that uses the energy stored in a flywheel consisting of a uniform 100-kg cylinder of radius R. The flywheel must deliver an average of 2 MJ/km of mechanical energy, with a maximum angular velocity of 400 rev/s. Find the smallest value of R for which the car can travel 300 km without the flywheel having to be recharged.

67 •• A ladder that is 8.6 m long and has a mass of 60 kg is placed in a nearly vertical position against the wall of a building. You stand on a rung with your center of mass at the top of the ladder. Assume that your mass is 80 kg. As you lean back slightly, the ladder begins to rotate about its base away from the wall. Is it better to quickly step off the ladder and drop to the ground or to hold onto the ladder and step off just before the top end hits the ground?

Pulleys, Yo-Yos, and Hanging Things

68 •• **SSM** **iSOLVE** ✔ A 4-kg block resting on a friction-less horizontal ledge is attached to a string that passes over a pulley and is attached to a hanging 2-kg block (Figure 9-50). The pulley is a uniform disk of radius 8 cm and mass 0.6 kg. (a) Find the speed of the 2-kg block after it falls from rest a distance of 2.5 m. (b) What is the angular velocity of the pulley at this time?

FIGURE 9-50
Problems 68–70

69 •• For the system in Problem 68, find the linear acceleration of each block and the tension in the string.

70 •• Solve Problem 68 for the case in which the coefficient of friction between the ledge and the 4-kg block is 0.25.

71 •• A 1200-kg car is being unloaded by a winch. At the moment shown in Figure 9-51, the gearbox shaft of the winch breaks and the car falls from rest. During the car's fall, there is no slipping between the (massless) rope, the pulley, and the winch drum. The moment of inertia of the winch drum is 320 kg·m² and that of the pulley is 4 kg·m². The radius of the winch drum is 0.80 m and that of the pulley is 0.30 m. Find the speed of the car as it hits the water.

FIGURE 9-51
Problem 71

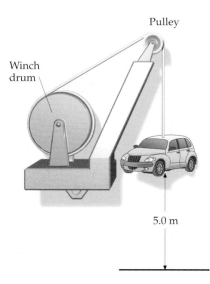

Pulley

Winch drum

5.0 m

72 •• **SSM** The system in Figure 9-52 is released from rest. The 30-kg block is 2 m above the ledge. The pulley is a uniform disk with a radius of 10 cm and a mass of 5 kg. Find (a) the speed of the 30-kg block just before it hits the ledge, (b) the angular speed of the pulley at that time, (c) the tensions in the strings, and (d) the time it takes for the 30-kg block to reach the ledge. Assume that the string does not slip on the pulley.

FIGURE 9-52
Problem 72

$m = 5$ kg
$r = 10$ cm

30kg

2 m

20kg

73 •• A uniform sphere of mass M and radius R is free to rotate about a horizontal axis through its center. A string is wrapped around the sphere and is attached to an object of mass m as shown in Figure 9-53. Find (a) the acceleration of the object and (b) the tension in the string.

FIGURE 9-53 Problem 73

$\vec{T}$

m

74 •• **iSOLVE** An Atwood's machine has two objects of masses $m_1 = 500$ g and $m_2 = 510$ g, connected by a string of negligible mass that passes over a pulley (Figure 9-54) with frictionless bearings. The pulley is a uniform disk with a mass of 50 g and a radius of 4 cm. The string does not slip on the pulley. (a) Find the acceleration of the objects. (b) What is the tension in the string supporting m_1? In the string supporting m_2? By how much do they differ? (c) What would your answers have been if you had neglected the mass of the pulley?

FIGURE 9-54 Problem 74

$m = 50$ g
$r = 4$ cm

500g

510g

$m_2\vec{g}$

$m_1\vec{g}$

75 •• **SSM** Two objects are attached to ropes that are attached to wheels on a common axle as shown in Figure 9-55. The two wheels are glued together so that they form a single object. The total moment of inertia of the object is 40 kg·m². The radii of the wheels are $R_1 = 1.2$ m and $R_2 = 0.4$ m. (*a*) If $m_1 = 24$ kg, find m_2 such that there is no angular acceleration of the wheels. (*b*) If 12 kg is gently added to the top of m_1, find the angular acceleration of the wheels and the tensions in the ropes.

R_1
R_2

m_1

m_2

FIGURE 9-55 Problem 75

76 •• The string wrapped around the cylinder in Figure 9-56 is held by a hand that is accelerated upward so that the center of mass of the cylinder does not move. Find (*a*) the tension in the string, (*b*) the angular acceleration of the cylinder, and (*c*) the acceleration of the hand.

FIGURE 9-56
Problems 76, 80

$\vec{T}$

77 •• A uniform cylinder of mass m_1 and radius R is pivoted on frictionless bearings. A massless string wrapped around the cylinder is connected to a block of mass m_2 that is on a frictionless incline of angle θ as shown in Figure 9-57.

The system is released from rest with m_2 at a height h above the bottom of the incline. (*a*) What is the acceleration of the block? (*b*) What is the tension in the string? (*c*) What is the total energy of the cylinder–block–earth system when the block is at height h? (*d*) What is the total energy of the system when the block is at the bottom of the incline and has a speed v? (*e*) What is the speed v? (*f*) Evaluate your answers for the extreme cases of $\theta = 0°$, $\theta = 90°$, and $m_1 = 0$.

FIGURE 9-57
Problem 77

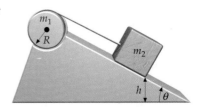

m_1
R
m_2
h
θ

78 •• **SSM** A device for measuring the moment of inertia of an object is shown in Figure 9-58. A circular platform has a concentric drum of radius 10 cm about which a string is wound. The string passes over a frictionless pulley to a weight of mass M. The weight is released from rest, and the time required for it to drop a distance D is measured. The system is then rewound, the object placed on the platform, and the system is again released from rest. The time required for the weight to drop the same distance D then provides the data needed to calculate I. With $M = 2.5$ kg and $D = 1.8$ m, the time is 4.2 s. (*a*) Find the combined moment of inertia of the platform, drum, shaft, and pulley. (*b*) With the object placed on the platform, the time is 6.8 s for $D = 1.8$ m. Find I of that object about the axis of the platform.

FIGURE 9-58
Problem 78

M

Objects Rolling Without Slipping

79 •• **ISOLVE** ✓ In 1993, a giant yo-yo of mass 400 kg measuring about 1.5 m in radius was dropped from a crane 57 m high. One end of the string was tied to the top of the crane, so the yo-yo unwound as it descended. Assuming that the axle of the yo-yo had a radius of $r = 0.1$ m, find the velocity of descent v at the end of the fall.

80 •• A uniform cylinder of mass M and radius R has a string wrapped around it. The string is held fixed, and the cylinder falls vertically as shown in Figure 9-56. (*a*) Show that the acceleration of the cylinder is downward with a magnitude $a = 2g/3$. (*b*) Find the tension in the string.

81 •• A 0.1-kg yo-yo consists of two solid disks of radius 10 cm joined together by a massless rod of radius 1 cm and a string wrapped around the rod. One end of the string is held fixed and is under constant tension T as the yo-yo is released. Find the acceleration of the yo-yo and the tension T.

82 • **SSM** A homogeneous solid cylinder rolls without slipping on a horizontal surface. The total kinetic energy is K. The kinetic energy due to rotation about its center of mass is (a) $\frac{1}{2}K$, (b) $\frac{1}{3}K$, (c) $\frac{4}{7}K$, (d) none of these.

83 • **iSOLVE** A homogeneous cylinder of radius 18 cm and mass 60 kg is rolling without slipping along a horizontal floor at 5 m/s. How much work was required to give it this motion?

84 • Find the percentages of the total kinetic energy associated with rotation and translation for an object that is rolling without slipping if the object is (a) a uniform sphere, (b) a uniform cylinder, or (c) a hoop.

85 • A hoop of radius 0.40 m and mass 0.6 kg is rolling without slipping at a speed of 15 m/s toward an incline of slope 30°. How far up the incline will the hoop roll, assuming that it rolls without slipping?

86 •• **SSM** A uniform sphere rolls without slipping down an incline. What must be the angle of the incline if the linear acceleration of the center of mass of the sphere is $0.2g$?

87 •• Repeat Problem 86 for a thin spherical shell.

88 •• A basketball rolls without slipping down an incline of angle θ. The coefficient of static friction is μ_s. Find (a) the acceleration of the center of mass of the ball, (b) the frictional force acting on the ball, and (c) the maximum angle of the incline for which the ball will roll without slipping.

89 •• Repeat Problem 88 for a solid cylinder of wood.

90 •• **SSM** A hollow sphere and uniform sphere of the same mass m and radius R roll down an inclined plane from the same height H without slipping (Figure 9-59). Each is moving horizontally as it leaves the ramp. When the spheres hit the ground, the range of the hollow sphere is L. Find the range L' of the uniform sphere.

FIGURE 9-59 Problem 90

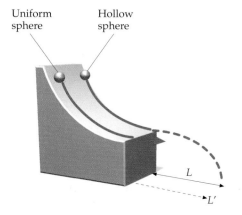

Uniform sphere   Hollow sphere

L

L'

91 •• A uniform thin-walled cylinder and a uniform solid cylinder are rolling horizontally without slipping. The speed of the thin-walled cylinder is v. The cylinders encounter an incline that they climb without slipping. If the maximum height they reach is the same, find the initial speed v' of the solid cylinder.

92 •• A hollow, thin-walled cylinder and a solid sphere start from rest and roll without slipping down an inclined plane of length 3 m. The cylinder arrives at the bottom of the plane 2.4 s after the sphere. Determine the angle between the inclined plane and the horizontal.

93 ••• A wheel has a thin 3.0-kg rim and four spokes, each of mass 1.2 kg. Find the kinetic energy of the wheel when it rolls at 6 m/s on a horizontal surface.

94 •• **iSOLVE** ✓ Two uniform 20-kg disks of radius 30 cm are connected by a short rod of radius 2 cm and mass 1 kg. When the rod is placed on a plane inclined at 30°, so that the disks hang over the sides, the assembly rolls without slipping. Find (a) the linear acceleration of the system and (b) the angular acceleration of the system. (c) Find the kinetic energy of translation of the system after it has rolled 2 m down the incline, starting from rest. (d) Find the kinetic energy of rotation of the system at the same point.

95 ••• A wheel of radius R rolls without slipping at a speed V. The coordinates of the center of the wheel are X, Y. (a) Show that the x and y coordinates of point P in Figure 9-60 are $X + r_0 \cos \theta$ and $R + r_0 \sin \theta$, respectively. (b) Show that the total velocity $\vec{v}$ of point P has the components $v_x = V + (r_0 V \sin \theta)/R$ and $v_y = -(r_0 V \cos \theta)/R$. (c) Show that at the instant that $X = 0$, $\vec{v}$ and $\vec{r}$ are perpendicular to each other by calculating $\vec{v} \cdot \vec{r}$. (d) Show that $v = r\omega$, where $\omega = V/R$ is the angular velocity of the wheel. These results demonstrate that, in the case of rolling without slipping, the motion is the same as if the rolling object were instantaneously rotating about the point of contact with an angular speed $\omega = V/R$.

FIGURE 9-60
Problem 95

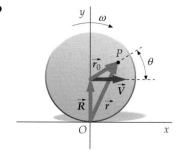

96 ••• **SSM** A uniform cylinder of mass M and radius R is at rest on a block of mass m, which in turn rests on a horizontal, frictionless table (Figure 9-61). If a horizontal force $\vec{F}$ is applied to the block, it accelerates and the cylinder rolls without slipping. Find the acceleration of the block.

FIGURE 9-61
Problems 96–98

M

m

$\vec{F}$

97 ••• (a) Find the angular acceleration of the cylinder in Problem 96. Is the cylinder rotating clockwise or counterclockwise? (b) What is the cylinder's linear acceleration relative to the table? Let the direction of $\vec{F}$ be the positive direction. (c) What is the linear acceleration of the cylinder relative to the block?

98 ••• If the force in Problem 96 acts over a distance d, find (a) the kinetic energy of the block and (b) the kinetic energy of the cylinder. (c) Show that the total kinetic energy is equal to the work done on the system.

99 •• Two large gears are shown in Figure 9-62; each is free to rotate about a fixed axis running through its center. The radius of the first gear is $R_1 = 0.5$ m and the radius of the second gear is $R_2 = 1$ m. The moment of inertia of gear 1 is $I_1 = 1$ kg·m² and the moment of inertia of gear 2 is $I_2 = 16$ kg·m². The lever attached to gear 1 is 1 m long. (a) If a force of 2 N is applied to the end of the lever, what will the angular accelerations of gears 1 and 2 be? (b) What force must be applied tangentially to the edge of gear 2 to keep the gear system from rotating?

FIGURE 9-62
Problem 99

100 •• SSM A marble of radius 1 cm rolls from rest from the top of a large sphere of radius 80 cm, which is held fixed. (a) Assuming that the marble rolls without slipping while it is in contact with the sphere (which is unrealistic), find the angle from the top of the sphere to the point where the marble breaks contact with the sphere. (b) Why is it unrealistic to assume that the marble rolls without slipping all the way down to the point where it breaks contact?

Rolling With Slipping

101 • A bowling ball of mass M and radius R is released so that at the instant it touches the floor it is moving horizontally with a speed v_0 and is not rotating. It slides for a time t_1 a distance s_1 before it begins to roll without slipping. (a) If μ_k is the coefficient of kinetic friction between the ball and the floor, find s_1, t_1, and the final speed v_1 of the ball. (b) Find the ratio of the final mechanical energy to the initial mechanical energy of the ball. (c) Evaluate these quantities for $v_0 = 8$ m/s and $\mu_k = 0.06$.

102 •• SSM A cue ball of radius r is initially at rest on a horizontal pool table (Figure 9-63). It is struck by a horizontal cue stick that delivers a force of magnitude F_0 for a very short time Δt. The stick strikes the ball at a point h above the ball's point of contact with the table. Show that the ball's initial angular velocity ω_0 is related to the initial linear velocity of its center of mass v_0 by $\omega_0 = (5/2)v_0(h - r)/r^2$.

FIGURE 9-63
Problem 102

103 •• A uniform solid sphere is set rotating about a horizontal axis with an angular speed ω_0 and is placed on the floor. If the coefficient of sliding friction between the sphere and the floor is μ_k, find the speed of the center of mass of the sphere when it begins to roll without slipping.

104 •• A uniform solid ball resting on a horizontal surface has a mass of 20 g and a radius of 5 cm. A sharp force is applied to the ball in a horizontal direction 9 cm above the horizontal surface. The force increases linearly from 0 to a peak value of 40,000 N in 10^{-4} s and then decreases linearly to 0 in 10^{-4} s. (a) What is the velocity of the ball after impact? (b) What is the angular velocity of the ball after impact? (c) What is the velocity of the ball when it begins to roll without sliding? (d) For how long does the ball slide on the surface? Assume that $\mu_k = 0.5$.

105 •• ISOLVE A 0.16-kg billiard ball of radius 3 cm is given a sharp blow by a cue stick. The applied force is horizontal and passes through the center of the ball. The initial velocity of the ball is 4 m/s. The coefficient of kinetic friction is 0.6. (a) For how many seconds does the ball slide before it begins to roll without slipping? (b) How far does it slide? (c) What is its velocity once it begins rolling without slipping?

106 •• A billiard ball initially at rest is given a sharp blow by a cue stick. The force is horizontal and is applied at a distance $2R/3$ below the centerline, as shown in Figure 9-64. The initial speed of the ball is v_0 and the coefficient of kinetic friction is μ_k. (a) What is the initial angular speed ω_0? (b) What is the speed of the ball once it begins to roll without slipping? (c) What is the initial kinetic energy of the ball? (d) What is the frictional work done as it slides on the table?

FIGURE 9-64
Problem 106

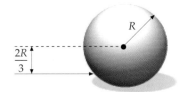

107 •• A bowling ball of radius R is given an initial velocity v_0 down the lane and a forward spin $\omega_0 = 3v_0/R$. The coefficient of kinetic friction is μ_k. (a) What is the speed of the ball when it begins to roll without slipping? (b) For how long does the ball slide before it begins to roll without slipping? (c) What distance does the ball slide down the lane before it begins rolling without slipping?

108 •• [SSM] A solid cylinder of mass M resting on its side on a horizontal surface is given a sharp blow by a cue stick. The applied force is horizontal and passes through the center of the cylinder so that the cylinder begins translating with initial velocity v_0. The coefficient of sliding friction between the cylinder and surface is μ_k. (a) What is the translational velocity of the cylinder when it is rolling without slipping? (b) How far does the cylinder travel before it rolls without slipping? (c) What fraction of its initial mechanical energy is dissipated in friction?

109 •• Consider a ball of radius r and total mass m, with a nonuniform but radially symmetric mass distribution inside it, so that it can have an almost arbitrary moment of inertia I.

(a) Show that if this ball is projected across a floor with an initial velocity v, and is at first purely sliding across the floor (that is, its initial rotational velocity $\omega = 0$), its final velocity when it is rolling without slipping will be

$$v_f = \frac{1}{1 + I/mr^2} v,$$

independent of the coefficient of kinetic friction of the floor.

(b) Show that the total kinetic energy of the ball will be

$$K = \frac{1}{2} \frac{m}{(1 + I/mr^2)} v^2$$

(Note that I/mr^2 is independent of the mass and radius of the ball, but depends only on the distribution of mass inside the ball.)

General Problems

110 • [SSM] The moon rotates as it revolves around the earth so that we always see the same side. Use this fact to find the angular velocity in radians per second of the moon about its axis. (The period of revolution of the moon about the earth is 27.3 d.)

111 • Find the moment of inertia of a hoop about an axis perpendicular to the plane of the hoop and through its edge.

112 •• The radius of a park merry-go-round is 2.2 m. To start it rotating, you wrap a rope around it and pull with a force of 260 N for 12 s. During this time, the merry-go-round makes one complete rotation. (a) Find the angular acceleration of the merry-go-round. (b) What torque is exerted by the rope on the merry-go-round? (c) What is the moment of inertia of the merry-go-round?

113 • A uniform stick of length 2 m is raised at an angle of 30° to the horizontal above a sheet of ice. The bottom end of the stick rests on the ice. When dropped from rest, the bottom of the stick remains in contact with the ice at all times. How far will the bottom of the stick have moved when the stick falls to the ice? Assume that the ice is frictionless.

114 •• [ISOLVE ✓] A uniform disk of radius 0.12 m and mass 5 kg is pivoted so that it rotates freely about its central axis (Figure 9-65). A string wrapped around the disk is pulled with a force of 20 N. (a) What is the torque exerted on the disk? (b) What is the angular acceleration of the disk? (c) If the

disk starts from rest, what is its angular velocity after 5 s? (d) What is its kinetic energy after 5 s? (e) What is the total angle θ that the disk turns through in 5 s? (f) Show that the work done by the torque, $\tau \Delta\theta$, equals the kinetic energy.

FIGURE 9-65 Problem 114

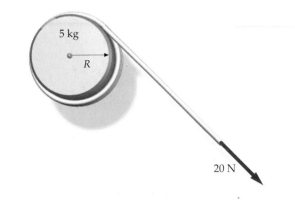

5 kg

R

20 N

115 •• A 0.25-kg rod of length 80 cm is suspended by a frictionless pivot at one end. It is held horizontal and released. Immediately after it is released, what is (a) the acceleration of the center of the rod and (b) the initial acceleration of a point on the end of the rod? (c) Find the linear velocity of the center of mass of the rod when it is vertical.

116 •• A marble of mass M and radius R rolls without slipping down the track on the left from a height h_1 as shown in Figure 9-66. The marble then goes up the *frictionless* track on the right to a height h_2. Find h_2.

FIGURE 9-66
Problem 116

h_1

h_2

117 •• [SSM] [ISOLVE] A uniform disk with a mass of 120 kg and a radius of 1.4 m rotates initially with an angular speed of 1100 rev/min. (a) A constant tangential force is applied at a radial distance of 0.6 m. How much work must this force do to stop the wheel? (b) If the wheel is brought to rest in 2.5 min, what torque does the force produce? What is the magnitude of the force? (c) How many revolutions does the wheel make in these 2.5 min?

118 •• [ISOLVE] A park merry-go-round consists of a 240-kg circular wooden platform 4.00 m in diameter. Four children running alongside push tangentially along the platform's circumference until, starting from rest, the merry-go-round reaches a steady speed of one complete revolution every 2.8 s. (a) If each child exerts a force of 26 N, how far does each child run? (b) What is the angular acceleration of the merry-go-round? (c) How much work does each child do? (d) What is the kinetic energy of the merry-go-round?

119 •• [SOLVE] A hoop of mass 1.5 kg and radius 65 cm has a string wrapped around its circumference and lies flat on a horizontal frictionless table. The string is pulled with a constant force of 5 N. (a) How far does the center of the hoop travel in 3 s? (b) What is the angular velocity of the hoop about its center of mass after 3 s?

120 •• A vertical grinding wheel is a uniform disk of mass 60 kg and radius 45 cm. It has a handle of radius 65 cm of negligible mass. A compact 25-kg load is attached to the handle when it is in the horizontal position. Neglecting friction, find (a) the initial angular acceleration of the wheel and (b) the maximum angular velocity of the wheel.

121 •• [SSM] Consider two uniform blocks of wood, identical in shape and composition, where one is larger than the other by a factor S in all dimensions. (a) What is the ratio of the surface areas of the two blocks? (b) What is the ratio of the masses of the two blocks? (c) What is the ratio of the moments of inertia about some axis running through the block (in the same relative position and orientation in each)? These are examples of *scaling laws:* How do surface area, mass, and moment of inertia vary with the size of an object?

122 •• In this problem, you are to derive the perpendicular-axis theorem for planar objects, which relates the moments of inertia about two perpendicular axes in the plane of Figure 9-67 to the moment of inertia about a third axis that is perpendicular to the plane of the figure. Consider the mass element dm for the figure shown in the xy plane. (a) Write an expression for the moment of inertia of the figure about the z axis in terms of dm and r. (b) Relate the distance r of dm to the distances x and y and show that $I_z = I_y + I_x$. (c) Apply your result to find the moment of inertia of a uniform disk of radius R about a diameter of the disk.

FIGURE 9-67
Problem 122

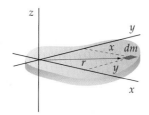

123 •• [SOLVE] A uniform disk of radius R and mass M is pivoted about a horizontal axis parallel to its symmetry axis and passing through its edge so that it can swing freely in a vertical plane (Figure 9-68). It is released from rest with its center of mass at the same height as the pivot. (a) What is the angular velocity of the disk when its center of mass is directly below the pivot? (b) What force is exerted by the pivot at this time?

FIGURE 9-68
Problem 123

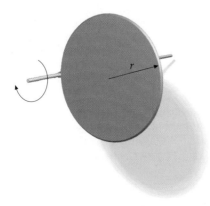

124 •• The roof of the student dining hall at St. Mary's College is supported by high cross-braced wooden beams attached together in the shape of an upside-down L (see Figure 9-69). Each vertical beam is 12 ft high and 2 ft wide and the horizontal cross-member is 6 ft long. The mass of the vertical beam is 350 kg and the mass of the horizontal beam is 175 kg. When workers were building the hall, one of the structures started to fall over before it was anchored into place. (Luckily, the workers stopped it before it fell.) (a) If it started falling from an upright position, what was the initial angular acceleration of the structure? Assume that the bottom didn't slide across the floor and that it didn't fall "out of plane," that is, that the fall occurred in the vertical plane defined by the vertical and horizontal beams. (b) If a sparrow were sitting on the right end of the horizontal beam, what would the magnitude of its initial linear acceleration be when the structure started falling? (c) What would the horizontal component of the sparrow's initial linear acceleration be?

FIGURE 9-69 Problem 124

125 •• A spool of mass M rests on an inclined plane at a distance D from the bottom. The ends of the spool have radius R, the center has radius r, and the moment of inertia of the spool about its axis is I. A long string of negligible mass is wound many times around the center of the spool. The other end of the string is fastened to a hook at the top of the inclined plane so that the string always pulls parallel to the slope as shown in Figure 9-70. (a) Suppose that initially the slope is so icy that there is *no* friction. How does the spool move as it slips down the slope? Use energy considerations to determine the speed of the center of mass of the spool when it reaches the bottom of the slope. Give your answer in terms of M, I, r, R, g, D, and θ. (b) Now suppose that the ice is gone and that when the spool is set up in the same way, there is enough friction to keep it from slipping on the slope. What is the direction and magnitude of the friction force in this case?

FIGURE 9-70 Problem 125

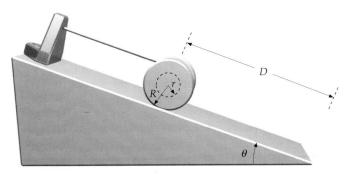

126 •• ![ISOLVE] A solid metal rod 1.5 m long is free to rotate without friction about a fixed horizontal axis perpendicular to the rod and passing through one end. The other end is held in a horizontal position. Small coins of mass m are placed on the rod 25 cm, 50 cm, 75 cm, 1 m, 1.25 m, and 1.5 m from the bearing. If the free end is now released, calculate the initial force exerted on each coin by the rod. Assume that the mass of the coins may be neglected in comparison to the mass of the rod.

127 •• **SSM** A popular classroom demonstration involves taking a meterstick and holding it horizontally at one end with a number of pennies spaced evenly along the meterstick. If the hand is relaxed so that the meterstick pivots about the hand under the influence of gravity, an interesting thing is seen: Pennies near the pivot point stay on the meterstick, while those farther away than a certain distance from the pivot are left behind by the falling meterstick. (This is often called the "faster than gravity" demonstration.) (a) What is the acceleration of the far end of the meterstick? (b) How far should a penny be from the end of the meterstick in order for it to be "left behind"?

128 •• ![ISOLVE] Figure 9-71 shows a hollow cylinder of length 1.8 m, mass 0.8 kg, and radius 0.2 m. The cylinder is free to rotate about a vertical axis that passes through its center and is perpendicular to the cylinder's axis. Inside the cylinder are two masses of 0.2 kg each, attached to springs of spring constant k and unstretched lengths 0.4 m. The inside walls of the cylinder are frictionless. (a) Determine the value of the spring constant if the masses are located 0.8 m from the center of the cylinder when the cylinder rotates at 24 rad/s. (b) How much work was needed to bring the system from $\omega = 0$ to $\omega = 24$ rad/s?

FIGURE 9-71 Problem 128

129 •• Suppose that for the system described in Problem 128, the spring constants are each $k = 60$ N/m. The system starts from rest and slowly accelerates until the masses are 0.8 m from the center of the cylinder. How much work was done in the process?

130 •• A string is wrapped around a uniform cylinder of radius R and mass M that rests on a horizontal frictionless surface. The string is pulled horizontally from the top with force F. (a) Show that the angular acceleration of the cylinder is twice that needed for rolling without slipping, so that the bottom point on the cylinder slides backward against the table. (b) Find the magnitude and direction of the frictional force between the table and cylinder needed for the cylinder to roll without slipping. What is the acceleration of the cylinder in this case?

131 •• **SSM** Let's calculate the position y of the falling load attached to the winch in Example 9-7 as a function of time by numerical integration. To do this, note that $v(y) = dy/dt$, or

$$t = \int_0^y \frac{1}{v(y')}\, dy' \approx \sum_{i=1}^{N} \frac{1}{v(y'_i)} \Delta y'$$

where $\Delta y'$ is some (small) increment, $y'_i = i(\Delta y)$, and $y = y'_N = N(\Delta y)$. Hence, we can calculate t as a function of y by numerical summation. Make a graph of y' versus t for t between 0 s and 2 s. Assume $M = 10$ kg, $R = 0.5$ m, $m = 5$ kg, $L = 10$ m, and $m_c = 3.5$ kg. Use $\Delta y' = 0.1$ m. Compare this to the position of the falling load if it were in free-fall.

132 •• Figure 9-72 shows a solid cylinder of mass M and radius R to which a hollow cylinder of radius r is attached. A string is wound about the hollow cylinder. The solid cylinder rests on a horizontal surface. The coefficient of static friction between the cylinder and surface is μ_s. If a light tension is applied to the string in the vertical direction, the cylinder will roll to the left; if the tension is applied with the string horizontally to the right, the cylinder rolls to the right. Find the angle of the string with the horizontal that will allow the cylinder to remain stationary when a small tension is applied to the string.

FIGURE 9-72
Problem 132

133 •• | SSM | In problems dealing with a pulley with a nonzero moment of inertia, the magnitude of the tensions in the ropes hanging on either side of the pulley are not equal. The difference in the tension is due to the static frictional force between the rope and the pulley; however, the static frictional force cannot be made arbitrarily large. If you consider a massless rope wrapped partly around a cylinder through an angle $\Delta\theta$ (measured in radians), then you can show that if the tension on one side of the pulley is T, while the tension on the other side is T' ($T' > T$), the maximum value of T' in relation to T that can be maintained without the rope slipping is $T'_{max} = Te^{\mu_s \Delta\theta}$, where μ_s is the coefficient of static friction. Consider the Atwood's machine in Figure 9-73: the pulley has a radius $r = 0.15$ m, the moment of inertia $I = 0.35$ kg·m², and the coefficient of static friction $\mu_s = 0.30$. (a) If the tension on one side of the pulley is 10 N, what is the maximum tension on the other side that will prevent the rope from slipping on the pulley? (b) If the mass of one of the hanging blocks is 1 kg, what is the maximum mass of the other block if, after the blocks are released, the pulley is to rotate without slipping? (c) What is the acceleration of the blocks in this case?

FIGURE 9-73 Problem 133

134 ••• A heavy, uniform cylinder has a mass m and a radius R (Figure 9-74). It is accelerated by a force $\vec{T}$ that is applied through a rope wound around a light drum of radius r that is attached to the cylinder. The coefficient of static friction is sufficient for the cylinder to roll without slipping. (a) Find the frictional force. (b) Find the acceleration a of the center of the cylinder. (c) Show that it is possible to choose r so that a is greater than T/m. (d) What is the direction of the frictional force in the circumstances of Part (c)?

FIGURE 9-74
Problem 134

135 ••• A uniform rod of length L and mass M is free to rotate about a horizontal axis through one end as shown in Figure 9-75. The rod is released from rest at $\theta = \theta_0$. Show that the force exerted by the axis on the rod is given by $F_\parallel = \frac{1}{2}Mg(5\cos\theta - 3\cos\theta_0)$ and $F_\perp = \frac{1}{4}Mg\sin\theta$, where $F_\parallel$ is the component of the force parallel with the rod and $F_\perp$ is the component of the force perpendicular to the rod.

FIGURE 9-75
Problem 135

Conservation of Angular Momentum

ANGULAR MOMENTUM IS THE
ROTATIONAL ANALOG OF LINEAR
MOMENTUM.

? When the riders move closer
to the center, the merry-go-round
gains speed. Why does this
happen? (See Example 10-4.)

➤ In this chapter, we extend our study of rotational motion to situations in which the direction of the axis of rotation may change.

We begin with an examination of the vector properties of angular velocity and torque and then introduce the concept of angular momentum, which is the rotational analog of linear momentum. We then show that the net torque acting on a system equals the rate of change of its angular momentum, a result that is equivalent to Newton's second law for rotational motion. Angular momentum is therefore conserved in systems with zero net external torque. Like conservation of linear momentum, conservation of angular momentum is a fundamental law of nature, applying even in the atomic domain where Newtonian mechanics fails.

10-1 The Vector Nature of Rotation

In Chapter 9, we indicated the direction of rotation about an axis with a fixed direction by assigning plus and minus signs to indicate the direction of the angular velocity, just as we used them to indicate the direction of the velocity in one-dimensional motion. But when the direction of the axis of rotation is *not* fixed in space, plus and minus signs are not adequate to describe the direction of the angular velocity. This inadequacy is overcome by treating the angular velocity as a vector $\vec{\omega}$ directed along the rotation axis. Consider, for example, the rotating disk in Figure 10-1. We determine the direction of $\vec{\omega}$ by a convention known as the

FIGURE 10-1

right-hand rule, which is illustrated in Figure 10-2. Thus, if the rotation is directed as shown in Figure 10-1, $\vec{\omega}$ is directed as shown; if the direction of the rotation is reversed, so is the direction of $\vec{\omega}$.

We apply similar considerations to the torque. Figure 10-3 shows a force $\vec{F}$ acting on a particle at some position $\vec{r}$ relative to the origin O. The torque $\vec{\tau}$ exerted by this force relative to the origin O is defined as a vector that is perpendicular to the plane formed by $\vec{F}$ and $\vec{r}$ and has magnitude $Fr \sin \phi$, where ϕ is the angle between $\vec{F}$ and $\vec{r}$. (The angle between two vectors refers to the angle *between the directions* of the two vectors.) If $\vec{F}$ and $\vec{r}$ are in the xy plane, as in Figure 10-3, the torque vector is along the z axis. If $\vec{F}$ is applied to the rim of a disk of radius r, as shown in Figure 10-4, the torque vector has the magnitude Fr, and is along the axis of rotation as shown.

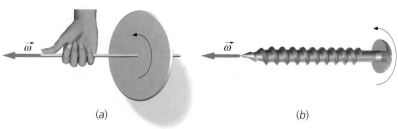

(a)   (b)

FIGURE 10-2 (*a*) When the fingers of the right hand curl in the direction of rotation, the thumb points in the direction of $\vec{\omega}$. (*b*) Looked at another way, the direction of $\vec{\omega}$ is that of the advance of a rotating right-hand screw.

The Cross Product

Torque is expressed mathematically as the **cross product** (or **vector product**) of $\vec{r}$ and $\vec{F}$:

$$\vec{\tau} = \vec{r} \times \vec{F} \qquad \text{10-1}$$

The cross product of two vectors $\vec{A}$ and $\vec{B}$ is defined to be a vector $\vec{C} = \vec{A} \times \vec{B}$ whose magnitude equals the area of the parallelogram formed by the two vectors (Figure 10-5). The vector $\vec{C}$ is perpendicular to the plane containing $\vec{A}$ and $\vec{B}$ in the direction given by the right-hand rule, that is, as the fingers curl from the direction of $\vec{A}$ toward the direction of $\vec{B}$ (Figure 10-6). If ϕ is the angle between the two vectors and $\hat{n}$ is a unit vector that is perpendicular to each vector in the direction of $\vec{C}$, the cross product of $\vec{A}$ and $\vec{B}$ is

$$\vec{A} \times \vec{B} = AB \sin \phi \, \hat{n} \qquad \text{10-2}$$

DEFINITION—CROSS PRODUCT

If $\vec{A}$ and $\vec{B}$ are parallel, $\vec{A} \times \vec{B}$ is zero. It follows from the definition of the cross product that

$$\vec{A} \times \vec{A} = 0 \qquad \text{10-3}$$

and

$$\vec{A} \times \vec{B} = -\vec{B} \times \vec{A} \qquad \text{10-4}$$

FIGURE 10-3

FIGURE 10-4

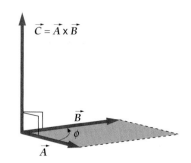

FIGURE 10-5 The cross product $\vec{A} \times \vec{B}$ is a vector $\vec{C}$ that is perpendicular to both $\vec{A}$ and $\vec{B}$ and has a magnitude $AB \sin \phi$, which equals the area of the parallelogram shown.

FIGURE 10-6 The direction of $\vec{A} \times \vec{B}$ is given by the right-hand rule when the fingers are rotated from the direction of $\vec{A}$ toward $\vec{B}$ through the angle ϕ.

Note that the order in which two vectors are multiplied in a cross product makes a difference. Below are some properties of the cross product of two vectors:

1. The cross product obeys a distributive law under addition:

$$\vec{A} \times (\vec{B} + \vec{C}) = (\vec{A} \times \vec{B}) + (\vec{A} \times \vec{C})$$ 10-5

2. If $\vec{A}$ and $\vec{B}$ are functions of some variable such as t, the derivative of $\vec{A} \times \vec{B}$ follows the usual product rule for derivatives:

$$\frac{d}{dt}(\vec{A} \times \vec{B}) = \left(\vec{A} \times \frac{d\vec{B}}{dt}\right) + \left(\frac{d\vec{A}}{dt} \times \vec{B}\right)$$ 10-6

3. The unit vectors $\hat{i}$, $\hat{j}$, and $\hat{k}$ (Figure 10-7), which are mutually perpendicular, have cross products given by

$$\hat{i} \times \hat{j} = \hat{k}, \qquad \hat{j} \times \hat{k} = \hat{i}, \qquad \text{and} \qquad \hat{k} \times \hat{i} = \hat{j}$$ 10-7a

Furthermore,

$$\hat{i} \times \hat{i} = \hat{j} \times \hat{j} = \hat{k} \times \hat{k} = 0$$ 10-7b

FIGURE 10-7

10-2   Torque and Angular Momentum

Figure 10-8 shows a particle of mass m moving with a velocity $\vec{v}$ at a position $\vec{r}$ relative to the origin O. The linear momentum of the particle is $\vec{p} = m\vec{v}$. The **angular momentum** $\vec{L}$ of the particle relative to the origin O is defined to be the cross product of $\vec{r}$ and $\vec{p}$:

$$\vec{L} = \vec{r} \times \vec{p}$$ 10-8

ANGULAR MOMENTUM OF A PARTICLE DEFINED

If $\vec{r}$ and $\vec{p}$ are in the xy plane, as in Figure 10-8, $\vec{L}$ is parallel with the z axis and is given by $\vec{L} = \vec{r} \times \vec{p} = mvr \sin \phi \, \hat{k}$. Like torque, angular momentum is defined *relative to a point in space*.

Figure 10-9 shows a particle of mass m attached to a circular disk of negligible mass in the xy plane with its center at the origin. The disk is spinning about its axis with angular speed ω. The speed v of the particle and its angular speed are related by $v = r\omega$. The angular momentum of the particle relative to the center of the disk is

$$\vec{L} = \vec{r} \times \vec{p} = \vec{r} \times m\vec{v} = rmv \sin 90° \, \hat{k} = rmv\hat{k} = mr^2\omega\hat{k} = mr^2\vec{\omega}$$

FIGURE 10-8

FIGURE 10-9

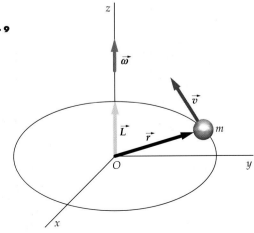

The angular momentum vector is in the same direction as the angular velocity vector.

Since mr^2 is the moment of inertia for a single particle about the z axis, we have

$$\vec{L} = mr^2\vec{\omega} = I\vec{\omega}$$

This result does not hold for the angular momentum about a general point on the z axis. Figure 10-10 shows the angular momentum vector $\vec{L}'$ for the same particle attached to the same disk but with $\vec{L}'$ computed about a point on the z axis that is not at the center of the circle. In this case, the angular momentum is not parallel to the angular velocity vector $\vec{\omega}$, which is parallel with the z axis.

In Figure 10-11, we attach a second particle of equal mass to the spinning disk. The angular momentum vectors $\vec{L}_1'$ and $\vec{L}_2'$ are

FIGURE 10-10

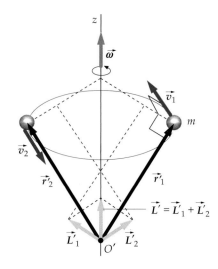

FIGURE 10-11

shown relative to the same point O'. The total angular momentum $\vec{L}_1' + \vec{L}_2'$ of the two-particle system is again parallel to the angular velocity vector $\vec{\omega}$. In this case, the axis of rotation, the z axis, passes through the center of mass of the two-particle system, and the mass distribution is symmetric about this axis. Such an axis is called a **symmetry axis.** For any system of particles that rotates about a symmetry axis, the total angular momentum (which is the sum of the angular momenta of the individual particles) is parallel to the angular velocity and is given by

$$\vec{L} = I\vec{\omega}$$

10-9

ANGULAR MOMENTUM OF A SYSTEM ROTATING ABOUT A SYMMETRY AXIS

ANGULAR MOMENTUM ABOUT THE ORIGIN   **EXAMPLE 10-1**

Find the angular momentum about the origin for the following situations. (*a*) A car of mass 1200 kg moves in a circle of radius 20 m with a speed of 15 m/s. The circle is in the xy plane, centered at the origin. When viewed from a point on the positive z axis, the car moves counterclockwise. (*b*) The same car moves in the xy plane with velocity $\vec{v} = -(15 \text{ m/s})\hat{i}$ along the line $y = y_0 = 20$ m parallel to the x axis. (*c*) A disk in the xy plane of radius 20 m and mass 1200 kg rotates at 0.75 rad/s about its axis, which is also the z axis. When viewed from a point on the positive z axis, the disk rotates counterclockwise.

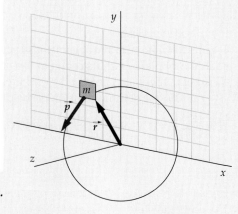

PICTURE THE PROBLEM For (*a*) and (*b*) use $\vec{L} = \vec{r} \times \vec{p}$. For (*c*) use $\vec{L} = I\vec{\omega}$. Draw a figure and apply the right-hand rule to find the direction of $\vec{L}$.

FIGURE 10-12

(*a*) $\vec{r}$ and $\vec{p}$ are perpendicular and $\vec{r} \times \vec{p}$ is along the z axis (Figure 10-12):

$\vec{L} = \vec{r} \times \vec{p} = rmv \sin 90° \, \hat{k}$

$= (20 \text{ m})(1200 \text{ kg})(15 \text{ m/s})\hat{k}$

$= \boxed{3.6 \times 10^5 \text{ kg·m}^2/\text{s} \, \hat{k}}$

(*b*) 1. For the same car moving in the direction of decreasing x along the line $y = 20$ m, we express $\vec{r}$ and $\vec{p}$ in terms of unit vectors:

$\vec{r} = x\hat{i} + y\hat{j} = x\hat{i} + y_0\hat{j}$

$\vec{p} = m\vec{v} = -mv\hat{i}$

2. Now compute $\vec{r} \times \vec{p}$ (Figure 10-13):

$$\vec{L} = \vec{r} \times \vec{p} = (x\hat{i} + y_0\hat{j}) \times (-mv\hat{i})$$
$$= -xmv(\hat{i} \times \hat{i}) - y_0\, mv(\hat{j} \times \hat{i})$$
$$= 0 - y_0mv(-\hat{k}) = y_0mv\hat{k}$$
$$= (20\text{ m})(1200\text{ kg})(15\text{ m/s})\hat{k}$$
$$= \boxed{3.6 \times 10^5\text{ kg·m}^2/\text{s}\,\hat{k}}$$

FIGURE 10-13

(c) Use $\vec{L} = I\vec{\omega}$
(Figure 10-14):

$$\vec{L} = I\vec{\omega} = I\omega\hat{k} = \tfrac{1}{2}mR^2\omega\,\hat{k}$$
$$= \tfrac{1}{2}(1200\text{ kg})(20\text{ m})^2\,(0.75\text{ rad/s})\hat{k}$$
$$= \boxed{1.8 \times 10^5\text{ kg·m}^2/\text{s}\,\hat{k}}$$

FIGURE 10-14

REMARKS The angular momentum of the car moving in a circle in (a) is the same as that of the car moving along a straight line in (b). In (c), the velocity of a point on the rim is $v = R\omega = (20\text{ m})(0.75\text{ rad/s}) = 15\text{ m/s}$, the same as the velocity of the car in Parts (a) and (b). The moment of inertia of a 1200-kg disk of radius 20 m is less than that of a 1200-kg car at 20 m from the axis because much of the mass of the disk is closer to the axis of rotation.

There are several additional results concerning torque and angular momentum for a system of particles. The first of these is

$$\vec{\tau}_{\text{net ext}} = \frac{d\vec{L}_{\text{sys}}}{dt}$$

10-10

The net external torque acting on a system equals the rate of change of the angular momentum of the system.

NEWTON'S SECOND LAW FOR ANGULAR MOTION

In Equation 10-10 the net external torque is the vector sum of the external torques acting on the system. Integrating both sides of this equation with respect to time gives

$$\Delta\vec{L}_{\text{sys}} = \int_{t_i}^{t_f} \vec{\tau}_{\text{net ext}}\, dt$$

10-11

ANGULAR IMPULSE–ANGULAR MOMENTUM EQUATION

It is often useful to split the total angular momentum of a system about an arbitrary point O into orbital angular momentum and spin angular momentum:

$$\vec{L}_{sys} = \vec{L}_{orbit} + \vec{L}_{spin} \qquad \text{10-12}$$

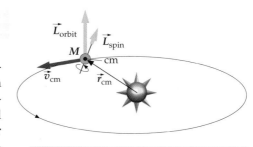

The earth has spin angular momentum due to its spinning motion about its rotational axis and it has orbital angular momentum about the center of the sun due to its orbital motion around the sun (Figure 10-15). The total angular momentum of the earth relative to the sun is the vector sum of the spin and orbital angular momenta. $\vec{L}_{spin}$ is the angular momentum of a system about its center of mass and $\vec{L}_{orbit}$ is the angular momentum that a point particle of mass M located at the center of mass and moving at the velocity of the center of mass would have. That is,

$$\vec{L}_{orbit} = \vec{r}_{cm} \times M\vec{v}_{cm} \qquad \text{10-13}$$

FIGURE 10-15 The angular momentum of the earth about the center of the sun is the sum of the orbital and the spin angular momenta.

In Chapter 9 torques are computed about axes instead of about points. The relation between the torque about an axis and the torque about a point is straightforward. If point O is the origin and if force $\vec{F}$ exerts torque $\vec{\tau}$ about O, then the corresponding torque exerted about the z axis is τ_z (the z component of $\vec{\tau}$.) Taking components of cross products requires some care. If $\vec{\tau} = \vec{r} \times \vec{F}$, then

$$\vec{\tau}_z = \vec{r}_{rad} \times \vec{F}_{xy} \qquad \text{10-14}$$

TORQUE ABOUT Z AXIS

FIGURE 10-16

where $\vec{\tau}_z$, $\vec{r}_{rad}$, and $\vec{F}_{xy}$ (see Figure 10-16) are vector components of $\vec{\tau}$, $\vec{r}$, and $\vec{F}$. (The vector component in a given direction is the scalar component in that direction times the unit vector in that direction. For example, $\vec{\tau}_z = \tau_z\hat{k}$.) $\vec{r}_{rad}$ is the component of $\vec{r}$ directed radially away from the z axis and $\vec{F}_{xy}$ is the component of $\vec{F}$ in the xy plane (the plane perpendicular to the z axis). The relation between angular momentum about an axis and angular momentum about a point is also straightforward. The angular momentum about an axis is

$$\vec{L}_z = \vec{r}_{rad} \times \vec{p}_{xy} \qquad \text{10-15}$$

ANGULAR MOMENTUM ABOUT Z AXIS

where $\vec{p}_{xy}$ is the projection of the linear momentum $\vec{p}$ in the xy plane. Taking the z vector components of both sides of Equation 10-10 gives

$$\vec{\tau}_{net\,ext,z} = \frac{d\vec{L}_{sys,z}}{dt} \qquad \text{10-16}$$

For a rigid object rotating about the z axis, $\vec{L}_{sys,z} = I_z\vec{\omega}$, where I_z is the moment of inertia about the z axis. Substituting this into Equation 10-16 gives

$$\vec{\tau}_{net\,ext,z} = \frac{d\vec{L}_{sys,z}}{dt} = \frac{d}{dt}(I_z\vec{\omega}) = I_z\vec{\alpha} \qquad \text{10-17}$$

where the angular acceleration vector $\vec{\alpha}$ is defined by $\vec{\alpha} = d\vec{\omega}/dt$. (Equation 10-17 is the vector form of Equation 9-18.)

For a system of particles, the total angular momentum about the z axis equals the sum of the angular momenta about the z axis of the individual particles. Also, the total torque about the z axis is the sum of the torques about the z axis acting on the system.

THE ATWOOD'S MACHINE REVISITED   **EXAMPLE 10-2**

An Atwood's machine has two blocks with masses m_1 and m_2 ($m_1 > m_2$) connected by a string of negligible mass that passes over a pulley with frictionless bearings. The pulley is a uniform disk of mass M and radius R. The string does not slip on the pulley. Apply Equation 10-16 to the system consisting of the two blocks, the string, and the pulley, to find the angular acceleration of the pulley and the linear acceleration of the blocks.

PICTURE THE PROBLEM Let the pulley and blocks be in the xy plane with the z axis out of the page through the center of the pulley at point O as shown in Figure 10-17. We compute the torques and angular momenta about the z axis and apply Newton's second law for angular motion (Equation 10-10). Since m_1 is greater than m_2, the disk will rotate counterclockwise corresponding to $\vec{\omega}$ out of the page in the positive z direction. All the forces are in the xy plane so all torques are along the z axis. Also, all the velocities are in the xy plane so all the angular momentum vectors are along the z axis. Since the torque, angular velocity, and angular momentum vectors are all along the z axis, we can treat this as a one-dimensional problem with positive assigned to counterclockwise motion and negative to clockwise motion. The acceleration a of the blocks is related to the angular acceleration α of the pulley by the nonslip condition $a = R\alpha$.

FIGURE 10-17

1. Draw a free-body diagram of the system (Figure 10-18). The only thing touching the system is the pulley bearings. The external forces on the system are the normal force of the pulley bearings on the pulley and the gravity forces on the two blocks and the pulley:

2. Express Newton's second law for rotation, z components only (Equation 10-16):

$$\sum \tau_{\text{ext},z} = \frac{dL_z}{dt}$$

3. The total external torque about the z axis is the sum of the torques exerted by the external forces. The lever arms F_1 and F_2 each equal R. (The lever arms of the other two forces are each zero.) $F_1 = m_1 g$ and $F_2 = m_2 g$:

$$\sum \tau_{\text{ext},z} = \tau_n + \tau_p + \tau_1 + \tau_2$$

$$= 0 + 0 + m_1 gR - m_2 gR$$

FIGURE 10-18

4. The total angular momentum about the z axis equals the angular momentum of the pulley $\vec{L}_p$ plus the angular momenta of block 1, $\vec{L}_1$, and block 2, $\vec{L}_2$, each in the positive z direction. The pulley has spin angular momentum but no orbital angular momentum, whereas each block has orbital angular momentum but no spin angular momentum.

$$L_z = L_p + L_1 + L_2$$

$$= I\omega + m_1 vR + m_2 vR$$

5. Substitute these results into Newton's second law for rotation in step 2:

$$\sum \tau_{\text{ext},z} = \frac{dL_z}{dt}$$

$$m_1 gR - m_2 gR = \frac{d}{dt}(I\omega + m_1 vR + m_2 vR)$$

$$m_1 gR - m_2 gR = I\alpha + (m_1 + m_2)Ra$$

6. Relate I to M and R, and use the nonslip condition to relate α to a and solve for both a and α:

$$m_1 gR - m_2 gR = \frac{1}{2}MR^2 \frac{a}{R} + (m_1 + m_2)Ra$$

so

$$a = \boxed{(m_1 - m_2)g/(\tfrac{1}{2}M + m_1 + m_2)}, \text{ and}$$

$$\alpha = a/R = \boxed{(m_1 - m_2)g/[(\tfrac{1}{2}M + m_1 + m_2)R]}$$

REMARKS 1. This problem could be solved by writing the tensions T_1 on the left and T_2 on the right and using $\tau = I\alpha$ (Equation 10-17) for the pulley and $\Sigma F_y = ma_y$ for each block. However, using angular momentum (Equation 10-16) is easier, and once you have solved for the acceleration, it is straightforward to solve for the two tensions. 2. Because $\vec{L}_2 = \vec{r}_2 \times m_2\vec{v}_2$ (Figure 10-19), the direction of $\vec{L}_2$ is gotten by applying the right-hand rule (Figure 10-6). And because $\vec{\tau}_2 = \vec{r}_2 \times \vec{F}_2$ (Figure 10-19), the direction of $\vec{\tau}_2$ also is gotten by applying the right-hand rule.

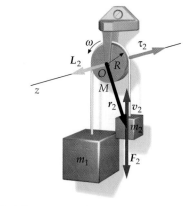

FIGURE 10-19

There are many problems in which the forces, position vectors, and velocities are all parallel to a plane, so that the torques, angular velocities, and angular momentum vectors are all along an axis of rotation that remains fixed in space. In such cases we can assign positive and negative values to counterclockwise or clockwise rotations, as we did in Example 10-2, and treat the case like a one-dimensional problem. However, there are other situations, such as the motion of a gyroscope, where torque, angular velocity, and angular momentum must be considered as multidimensional vectors.

The Gyroscope

A gyroscope is a common example of motion in which the axis of rotation changes direction. Figure 10-20 shows a gyroscope consisting of a bicycle wheel that is free to turn on its axle. The axle is pivoted at a point a distance D from the center of the wheel, and is free to rotate about the pivot in any direction. We can give a qualitative understanding of the complex motion of such a system by using Newton's second law for rotation,

$$\vec{\tau}_{net} = \frac{d\vec{L}}{dt} \qquad (\text{or} \quad d\vec{L} = \vec{\tau}_{net}dt)$$

along with the relations

$$\vec{\tau}_{net} = \vec{r}_{cm} \times M\vec{g}$$

and

$$\vec{L} = I_s\vec{\omega}_s$$

FIGURE 10-20

where I_s and ω_s are the moment of inertia and angular velocity of the wheel about its spin axis. All we need to remember in order to describe the motion of a gyroscope is that the *change* in angular momentum of the wheel must be in the direction of the net torque acting on it.

Suppose the axle is held horizontally and then released. If the wheel isn't spinning, it simply falls, rotating about a horizontal axis through O and perpendicular to $\vec{r}$. The torque vector is horizontal, into the page. For this case the initial angular momentum is zero, so the *change* in angular momentum equals the angular momentum itself, which, in this case, is also horizontal and into the page. However, if the wheel *is* spinning and has a large angular momentum along its axle, it does not fall when the axle is released. If it were to fall, the axle would point downward, resulting in a large component of angular momentum in the downward direction. But there is no torque in the downward direction. The torque is horizontal. What actually happens is that the axle moves horizontally (into the paper in the figure). The wheel must move this way so that the *change* in angular momentum is in the direction of the net torque. This is illustrated in Figure 10-21a, where we see a large angular momentum $\vec{L}$ along the axis of the wheel and a change in angular momentum $d\vec{L}$ in the direction of the torque $\vec{\tau}$. This motion, which is always surprising when first encountered, is called **precession.** We can calculate the angular velocity ω_p of precession. In a small time interval dt, the change in the angular momentum has a magnitude dL:

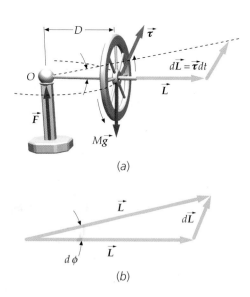

(a)

(b)

FIGURE 10-21

$$dL = \tau\, dt = MgD\, dt$$

where MgD is the magnitude of the torque about the pivot point. From Figure 10-21b, the angle $d\phi$ through which the axle moves is

$$d\phi = \frac{dL}{L} = \frac{\tau\, dt}{L} = \frac{MgD\, dt}{L}$$

The angular velocity of the precession is thus

$$\omega_{\mathrm{p}} = \frac{d\phi}{dt} = \frac{MgD}{L} = \frac{MgD}{I_{\mathrm{s}}\omega_{\mathrm{s}}} \qquad\qquad 10\text{-}18$$

If the angular momentum due to the spin of the wheel is large, the precession can be very slow.

In the preceding analysis it is assumed that the spin angular momentum of the wheel is very large compared to the orbital angular momentum associated with the precessional motion.

If you prevent a spinning gyroscope from precessing using your hand, upon release it will initiate precessional motion with an up and down bouncing motion called nutation. This initial bouncing motion can be avoided by giving the gyroscope axis an initial angular velocity equal to that associated with its precessional motion.

10-3 Conservation of Angular Momentum

When the net external torque acting on a system remains zero, we have

$$\frac{d\vec{L}_{\mathrm{sys}}}{dt} = 0$$

or

$$\vec{L}_{\mathrm{sys}} = \text{constant} \qquad\qquad 10\text{-}19$$

Equation 10-19 is a statement of the **law of conservation of angular momentum.**

If the net external torque acting on a system is zero, the total angular momentum of the system is constant.

CONSERVATION OF ANGULAR MOMENTUM

This is the rotational analog of the law of conservation of linear momentum. If a system is isolated from its surroundings, so that there are no external forces or torques acting on it, three quantities are conserved: energy, linear momentum, and angular momentum. The law of conservation of angular momentum is a fundamental law of nature. Even on the microscopic scale of atomic and nuclear physics, where Newtonian mechanics does not hold, the angular momentum of an isolated system is found to be constant over time.

Although conservation of angular momentum is an experimental law, independent of Newton's laws of motion, the fact that the internal torques of a system cancel is suggested by Newton's third law. Consider the two particles shown in Figure 10-22. Let $\vec{F}_{1,2}$ be the force exerted by particle 1 on particle 2 and $\vec{F}_{2,1}$ be

The Segway™ HT, a Dean Kamen invention, has been described as "the world's first self-balancing human transporter." When the person leans forward, the rotational motion of the device is detected by silicon "gyroscopes," which detect the rotational motion and then supply the internal control system with a signal. The control system interprets the signal and then activates motors that drive the wheels at just the right speed to maintain balance.

FIGURE 10-22

that exerted by particle 2 on particle 1. By Newton's third law, $\vec{F}_{2,1} = -\vec{F}_{1,2}$. The sum of the torques exerted by these forces about the origin O is

$$\vec{\tau}_1 + \vec{\tau}_2 = \vec{r}_1 \times \vec{F}_{2,1} + \vec{r}_2 \times \vec{F}_{1,2} = \vec{r}_1 \times \vec{F}_{2,1} + \vec{r}_2 \times (-\vec{F}_{2,1}) = (\vec{r}_1 - \vec{r}_2) \times \vec{F}_{2,1}$$

The vector $\vec{r}_1 - \vec{r}_2$ is along the line joining the two particles. If $\vec{F}_{2,1}$ acts parallel to the line joining m_1 and m_2, $\vec{F}_{2,1}$ and $\vec{r}_1 - \vec{r}_2$ are either parallel or antiparallel and

$$(\vec{r}_1 - \vec{r}_2) \times \vec{F}_{2,1} = 0$$

If this is true for all the internal forces, the internal torques cancel in pairs.

There are many examples of the conservation of angular momentum in everyday life. Figures 10-23 and 10-24 illustrate angular momentum conservation in diving and ice skating.

FIGURE 10-23 Multiflash photograph of a diver. The diver's center of mass moves along a parabolic path after he leaves the board. The angular momentum is provided by the initial external torque due to the force of the board, which does not pass through the diver's center of mass if he leans forward as he jumps. If the diver wanted to undergo one or more somersaults in the air, he would draw in his arms and legs, decreasing his moment of inertia to increase his angular velocity.

FIGURE 10-24 A spinning skater. Because the torque exerted by the ice is small, the angular momentum of the skater is approximately constant. When she reduces her moment of inertia by drawing in her arms, her angular velocity increases.

A ROTATING DISK   **EXAMPLE 10-3**

Frictionless shaft

FIGURE 10-25

A disk is rotating with an initial angular velocity ω_i about a frictionless shaft through its symmetry axis as shown in Figure 10-25. Its moment of inertia about this axis is I_1. It drops onto another disk of moment of inertia I_2 that is initially at rest on the same shaft. Because of surface friction, the two disks eventually attain a common angular velocity ω_f. Find ω_f.

PICTURE THE PROBLEM We find the final angular velocity from the final angular momentum, which is equal to the initial angular momentum because there are no external torques acting on the two-disk system. Note that we do *not* use conservation of mechanical energy. The angular speed of the upper disk is reduced while that of the lower disk is increased by the forces of kinetic friction. Because kinetic friction dissipates mechanical energy, we expect that the total mechanical energy is decreased.

1. The final angular velocity is related to the initial angular velocity by conservation of angular momentum:

$$L_f = L_i$$
$$(I_1 + I_2)\omega_f = I_1\omega_i$$

2. Solve for the final angular velocity:

$$\boxed{\omega_f = \frac{I_1}{I_1 + I_2}\omega_i}$$

PLAUSIBILITY CHECK If $I_2 \ll I_1$, the collision should have little effect on disk 1. Our results agree, and give $\omega_f \to \omega_i$. If $I_2 \gg I_1$, then disk 1 should slow to a stop without causing disk 2 to rotate appreciably. Our results give $\omega_f \to 0$, as expected.

In the collision of the two disks in Example 10-3, mechanical energy is not conserved. We can see this by writing the energy in terms of the angular momentum. An object rotating with an angular velocity ω has kinetic energy

$$K = \frac{1}{2} I\omega^2 = \frac{(I\omega)^2}{2I}$$

Using $L = I\omega$ gives

$$K = \frac{L^2}{2I} \qquad\qquad 10\text{-}20$$

(Compare this result with that for linear motion $K = p^2/2m$, Equation 8-25.) The initial kinetic energy in Example 10-3 is

$$K_i = \frac{L_i^2}{2I_1}$$

and the final kinetic energy is

$$K_f = \frac{L_f^2}{2(I_1 + I_2)}$$

Since $L_f = L_i$, the ratio of the final to the initial kinetic energy is

$$\frac{K_f}{K_i} = \frac{I_1}{I_1 + I_2}$$

which is less than one. This interaction of the disks is analogous to a one-dimensional perfectly inelastic collision of two objects.

The rotating plates in the transmission of a truck make inelastic collisions when engaged.

MUD IN YOUR EYE   **EXAMPLE 10-4**   **Put It in Context**

Benny, a high school physics student, has been the schoolyard bully for many years, so four of his fellow students decide to teach him a lesson using conservation of angular momentum. Here is their plan. The schoolyard has a small merry-go-round (Figure 10-26), a 3-m diameter turntable with a 130-kg·m² moment of inertia, and Benny loves to ride it. Initially, they need to get all five students to stand on the merry-go-round next to the rim while the merry-go-round is rotating at a modest 20 rev/min. When the signal is given, the four students will quickly move to the center of the merry-go-round leaving Benny near the rim. The merry-go-round will speed up, throwing Benny off and into the mud. (They plan to do this after a heavy rain.) Benny is very quick and very strong, so throwing him off will require that the centripetal acceleration of the rim be at least 4g. Will this plan work? (Assume that each student has a mass of 60 kg.)

FIGURE 10-26

PICTURE THE PROBLEM By moving to the center of the merry-go-round, the four students are decreasing the moment of inertia of the students–merry-go-round system. No external torques act on the system, so the angular momentum about the axis remains constant. The angular momentum is the moment of inertia times the angular velocity, so a decrease in the moment of inertia means an increase in the angular velocity. The angular velocity can be used to find the centripetal acceleration at the rim.

1. The centripetal acceleration depends on the angular speed ω and the radius R:

$$a_c = \omega^2 R$$

2. Angular momentum is conserved. For rotations about a fixed axis, $L = I\omega$:

$$L_f = L_i$$
$$I_f\omega_f = I_i\omega_i$$

3. The moment of inertia of the system is the sum of the moments of inertia of the students plus that of the merry-go-round. Each student has mass m:

$$I_i = 5 \times mR^2 + I_{mgr}$$
$$= 5(60 \text{ kg})(1.5 \text{ m})^2 + 130 \text{ kg} \cdot \text{m}^2$$
$$= 805 \text{ kg} \cdot \text{m}^2$$

4. To find the final moment of inertia, assume that the four students are 30 cm ($\approx$1 ft) from the center:

$$I_f = mR^2 + 4 \times mr^2 + I_{mgr}$$
$$= (60 \text{ kg})(1.5 \text{ m})^2 + 4(60 \text{ kg})(0.3 \text{ m})^2 + 130 \text{ kg} \cdot \text{m}^2$$
$$= 287 \text{ kg} \cdot \text{m}^2$$

5. Using conservation of angular momentum, solve for the final angular velocity:

$$\omega_f = \frac{I_i}{I_f}\omega_i = \frac{805 \text{ kg} \cdot \text{m}^2}{287 \text{ kg} \cdot \text{m}^2} 20 \text{ rev/min}$$
$$= 56.2 \text{ rev/min} = 5.88 \text{ rad/s}$$

6. Solve for the centripetal acceleration of the rim:

$$a_c = \omega^2 R = (5.88 \text{ rad/s})^2(1.5 \text{ m}) = 51.9 \text{ m/s}^2$$

7. Convert to gs:

$$a_c = 51.9 \text{ m/s}^2 \times \frac{1 g}{9.81 \text{ m/s}^2} = 5.29g$$

8. Does Benny end up in the mud?

> Success! The acceleration is much greater than $4g$, so Benny flies off and lands in the mud.

REMARKS The linear speed of the rotating merry-go-round is greatest at the rim and decreases to zero at the center. At the rim the students are moving in a circle. As they walk toward the center they are stepping onto a part of the merry-go-round that is moving more slowly than they are, so they drag the merry-go-round in the direction of its motion, thus speeding it up. Also, the merry-go-round drags the students back, slowing their motion. The static frictional forces exerted by their feet exerts a net torque on the merry-go-round, increasing its angular momentum about the rotation axis. The merry-go-round exerts equal and opposite static frictional forces on the students' feet, so the torques associated with these forces decreases their angular momentum. The two torques are equal and opposite, as are the associated angular momentum changes. Thus the angular momentum of the students–merry-go-round system remains constant.

The moment of inertia of the students–merry-go-round system decreases as the students walk toward the center. Thus, the system's moment of inertia decreases while its angular momentum remains constant. As a result, we can see from Equation 10-20 that the kinetic energy of the students–merry-go-round system increases. The energy for this kinetic energy increase comes from the internal energy of the students.

ANOTHER RIDE ON THE MERRY-GO-ROUND   **EXAMPLE 10-5** Try It Yourself

A 25-kg child in a playground runs with an initial speed of 2.5 m/s along a path *tangent* to the rim of a merry-go-round with a moment of inertia of 500 kg·m² that is initially at rest, and then jumps on (Figure 10-27). Find the final angular velocity of the child and the merry-go-round together.

FIGURE 10-27

PICTURE THE PROBLEM Once the child's feet leave the ground, no external torques act on the child–merry-go-round system, hence the total angular momentum of the system about the rotation axis of the merry-go-round is conserved. The mass of the child is $m = 25$ kg, her initial speed is $v = 2.5$ m/s, the moment of inertia of the merry-go-round is $I = 500$ kg $\cdot$ m^2, and the radius of the merry-go-round is $R = 2.0$ m. The initial angular speed of the merry-go-round is $\omega_i = 0$.

Cover the column at the right and solve it yourself before looking at the answers.

Steps	Answers
1. Write an expression for the initial angular momentum of the child running about the axis of the merry-go-round.	$L_i = mvR$
2. Write an expression for the total final angular momentum of the child–merry-go-round system in terms of the final angular velocity ω_f.	$L_f = (mR^2 + I_m)\omega_f$
3. Set your expressions in steps 1 and 2 equal and solve for ω_f.	$\omega_f = \dfrac{mvR}{mR^2 + I_m} = \boxed{0.208 \text{ rad/s}}$

EXERCISE Calculate the initial and final kinetic energies of the child–merry-go-round system. (*Answer* $K_i = 78.2$ J, $K_f = 13.0$ J)

The Hubble Space Telescope is aimed by regulating the spin rates of 45-kg flywheels arranged off-axis from each other and spinning at up to 3000 rpm. Software-controlled changes in the spin rates create angular momentum that causes the satellite to slew into new positions. This aiming mechanism can achieve and hold a target to within 0.005 arcsec—equivalent to holding a flashlight beam in Los Angeles on a dime in San Francisco.

PULLING THROUGH A HOLE   **EXAMPLE 10-6**

A particle of mass m moves with speed v_0 in a circle of radius r_0 on a frictionless tabletop. The particle is attached to a string that passes through a hole in the table, as shown in Figure 10-28. The string is slowly pulled downward so that the particle moves in a smaller circle of radius r_f. (*a*) Find the final velocity in terms of r_0, v_0, and r_f. (*b*) Find the tension when the particle is moving in a circle of radius r in terms of m, r, and the angular momentum $\vec{L}$. (*c*) Calculate the work done on the particle by the tension $\vec{T}$ by integrating $\vec{T} \cdot d\vec{\ell}$. Express your answer in terms of r and L_0.

FIGURE 10-28

PICTURE THE PROBLEM The speed of the particle is related to its angular momentum. The net torque is equal to the rate of change of the angular momentum. Since the net force acting on the particle is the tension force $\vec{T}$ exerted by the string, which is always directed toward the hole, the torque about the axis through the hole is zero. Thus, the angular momentum about this axis remains constant.

An astronaut examines the flywheel of the Hubble Space Telescope.

(a) Conservation of angular momentum relates the final speed to the initial speed and the initial and final radii:

$$L_f = L_0$$

$$mv_f r_f = mv_0 r_0$$

so

$$v_f = \boxed{\frac{r_0}{r_f} v_0}$$

(a)

(b) 1. Apply Newton's second law to relate T to v and r:

$$T = m\frac{v^2}{r}$$

2. Obtain a relation between L, r, and v using the definition of angular momentum:

$$\vec{L} = \vec{r} \times \vec{p}$$

$$L = rmv \sin 90° = mvr$$

3. Eliminate v by solving the Part (b) step 2 result for v and then substituting into the Part (a) result:

$$T = m\frac{v^2}{r} = \frac{m}{r}\left(\frac{L}{mr}\right)^2 = \boxed{\frac{L^2}{mr^3}}$$

(c) 1. Make a drawing of the particle as it moves closer to the hole (Figure 10-29). When the particle undergoes displacement $d\vec{\ell}$, its distance r from the axis changes by dr. Since r is decreasing, dr is negative. Thus:

$$dr = -|dr|$$

2. Using step 1, write $dW = \vec{T} \cdot d\vec{\ell}$ in terms of T and dr, with $T = L^2/mr^3$ from Part (b):

$$dW = \vec{T} \cdot d\vec{\ell} = T\, d\ell \cos\phi$$

Since

$$|dr| = d\ell \cos\phi$$

$$dW = T|dr| = -T\, dr$$

(b)

FIGURE 10-29

3. Integrate from r_0 to r_f after substituting for T from the Part (b) step 3 result:

$$W = -\int_{r_0}^{r_f} T\, dr = -\int_{r_0}^{r_f} \frac{L^2}{mr^3} dr$$

$$= -\frac{L^2}{m}\int_{r_0}^{r_f} r^{-3}\, dr = -\frac{L^2}{m}\frac{r^{-2}}{-2}\Big|_{r_0}^{r_f}$$

$$= \boxed{\frac{L^2}{2m}\left(\frac{1}{r_f^2} - \frac{1}{r_0^2}\right)}$$

PLAUSIBILITY CHECK Note that work must be done to pull the string downward. Since r_f is less than r_0, the work is positive. This work is converted into increased kinetic energy. We can calculate the change in kinetic energy of the particle directly. Using $K = L^2/2I$, with $L_0 = L_f = L$, and $I = mr^2$, the change in kinetic energy is $K_f - K_i = (L^2/2mr_f^2) - (L^2/2mr_0^2) = (L^2/2m)(r_f^{-2} - r_0^{-2})$, which is the same as the result found by direct integration.

REMARKS The increment of work dW can also be obtained by expressing the increment of displacement $d\vec{\ell}$ as $d\vec{r}$, the change in the position vector $\vec{r}$. The dot product $\vec{T} \cdot d\vec{r}$ is then expanded via components giving $dW = \vec{T} \cdot d\vec{r} = T_r\, dr = -T\, dr$. In this expansion $T_r = -T$ is the radial component of $\vec{T}$ and dr is the radial component of $d\vec{r}$.

EXERCISE At what radius r_N is the tension N times the tension at radius r_0? (*Answer* $r_N = r_0/\sqrt[3]{N}$)

In Figure 10-30, a puck on a frictionless plane is given an initial speed v_0. The puck is attached to a string that wraps around a vertical post. This situation looks similar to Example 10-6, but it is not the same. **There is no agent that can do work on the puck, nor is there any mechanism for energy dissipation.** Thus, mechanical energy must be conserved. Since $K = L^2/2I$ is constant and I decreases as r_0 decreases, L must also decrease. Note that the tension force does not act toward the axis of the post. The tension force on the puck produces a torque vector about the axis of the post in the downward direction, which reduces the angular momentum vector of the puck, which is in the upward direction.

FIGURE 10-30

THE BALLISTIC PENDULUM REVISITED   **EXAMPLE   10-7**   **Try It Yourself**

A thin rod of mass M and length d is attached to a pivot at the top. A piece of clay of mass m and speed v hits the rod a distance x from the pivot and sticks to it (Figure 10-31). Find the ratio of the system's kinetic energy just after the collision to the kinetic energy just before the collision.

PICTURE THE PROBLEM The collision is inelastic, so we do not expect mechanical energy to be conserved. During the collision, the pivot exerts a large force on the rod, so linear momentum is also not conserved. However, there are no external torques about the pivot point of the clay–rod system, so angular momentum about the pivot point is conserved. The kinetic energy after the inelastic collision can be written in terms of the angular momentum L_{sys} and the moment of inertia I' of the combined clay–rod system. Conservation of angular momentum allows you to relate L_{sys} to the mass m and velocity v of the clay.

FIGURE 10-31

Cover the column on the right and try these on your own before looking at the answers.

Steps   **Answers**

1. Before the collision the kinetic energy of the system is that of the moving clay ball.

$$K_i = \tfrac{1}{2}mv^2$$

2. After the collision it is that of the swinging clay–rod object. Write the kinetic energy after the collision in terms of the angular momentum L_{sys} and the moment of inertia I' of the clay–rod object.

$$K_f = \frac{L_{\text{sys}}^2}{2I'}$$

3. During the collision angular momentum is conserved. Write the angular momentum L_{sys} in terms of m, v, and x.

$$L_{\text{sys}} = mvx$$

4. Write I' in terms of m, x, M, and d.

$$I' = mx^2 + \tfrac{1}{3}Md^2$$

5. Substitute these expressions for L_f and I' into your equation for K_f.

$$K_f = \frac{L_{\text{sys}}^2}{2I'} = \frac{(mvx)^2}{2(mx^2 + \tfrac{1}{3}Md^2)}$$

$$= \frac{3}{2}\frac{m^2x^2v^2}{(3mx^2 + Md^2)}$$

6. Divide the kinetic energy after the collision by the initial kinetic energy of the clay.

$$\frac{K_f}{K_i} = \frac{\dfrac{3}{2}\dfrac{m^2x^2v^2}{(3mx^2 + Md^2)}}{\tfrac{1}{2}mv^2} = \boxed{\frac{1}{1 + \dfrac{Md^2}{3mx^2}}}$$

REMARKS This example is the rotational analog of the ballistic pendulum discussed in Example 8-14. In that example we used conservation of linear momentum to find the energy of the pendulum after the collision.

Figure 10-32 shows a student demonstrating conservation of angular momentum about an axis. She is sitting on a turntable, holding a bicycle wheel. We choose the z axis to coincide with the axis of the turntable. Because the turntable bearings are frictionless, the vertical component of the torque exerted on the turntable–student–wheel system remains zero. Consequently, the vertical component of the total angular momentum of the system remains constant. Initially, the turntable is stationary and the bicycle wheel is spinning rapidly about its axis, which is initially horizontal, as shown in Figure 10-32*a*. The system's initial angular momentum $\vec{L}_{\text{sys i}}$ is just the initial spin angular momentum of the wheel $\vec{L}_{\text{w,s i}}$, which is horizontal. This means that the vertical component of $\vec{L}_{\text{sys i}}$ is zero. If the student now (Figure 10-32*b*) tips the axis of the spinning wheel upward, the wheel's spin angular momentum vector $\vec{L}_{\text{w,s}}$ tips upward with it. As she tips the wheel upward the system starts rotating clockwise (viewed from above) about the turntable axis. The angular momentum vector associated with this clockwise rotation of the system about the turntable axis has a z component $L_{\text{t}z}$ directed downward. Since the vertical component of $\vec{L}_{\text{sys}}$ remains zero, we can conclude that $L_{\text{t}z}\hat{k}$ is equal and opposite to $L_{\text{w,s}z}\hat{k}$.

Unless you have seen this demonstration before, the rotation of the turntable is unexpected. Let's examine what causes it. The student must exert an upward torque on the spinning wheel in order to tip its angular momentum vector upward. (Due to the cross product, an upward torque requires horizontal forces.) The wheel exerts a downward torque (also horizontal forces) of equal magnitude back on the student. It is this downward torque on the student that causes the system to start rotating clockwise about the turntable axis.

EXERCISE The bicycle wheel is spinning with its axis vertical and with its angular momentum vector upward when it is handed to the student on the turntable. In what direction will the turntable rotate when the student tips and then rotates the axis of the wheel toward the horizontal? (*Answer* Counterclockwise, if viewed from above)

Proofs of Equations 10-10, 10-12, 10-13, 10-14, and 10-15

Proof of Equation 10-10 We will now show that Newton's second law implies that the rate of change of the angular momentum of a particle equals the net torque acting on the particle. If more than one force acts on a particle, then the net torque relative to the origin O is the sum of the torques due to each force:

$$\vec{\tau}_{\text{net}} = \vec{r} \times \vec{F}_1 + \vec{r} \times \vec{F}_2 + \cdots = \vec{r} \times \sum_i \vec{F}_i = \vec{r} \times \vec{F}_{\text{net}}$$

According to Newton's second law, the net force equals the rate of change of the linear momentum $d\vec{p}/dt$. Thus

$$\vec{\tau}_{\text{net}} = \vec{r} \times \text{F}_{\text{net}} = \vec{r} \times \frac{d\vec{p}}{dt} \qquad\qquad 10\text{-}21$$

We now compare this with the expression for the time rate of change of the angular momentum. The definition of the angular momentum of a particle (Equation 10-8) is

$$\vec{L} = \vec{r} \times \vec{p}$$

We can compute $d\vec{L}/dt$ using the product rule for derivatives:

$$\frac{d\vec{L}}{dt} = \frac{d}{dt}(\vec{r} \times \vec{p}) = \left(\frac{d\vec{r}}{dt} \times \vec{p}\right) + \left(\vec{r} \times \frac{d\vec{p}}{dt}\right)$$

The second term from the right of this equation is zero because $\vec{p} = m\vec{v}$ and $\vec{v} = d\vec{r}/dt$, so

$$\frac{d\vec{r}}{dt} \times \vec{p} = \vec{v} \times m\vec{v} = 0$$

$$\vec{L}_{\text{sys i}} = \vec{L}_{\text{w,s i}}$$

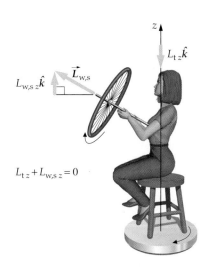

$$L_{\text{t}z} + L_{\text{w,s}z} = 0$$

FIGURE 10-32

Thus

$$\frac{d\vec{L}}{dt} = \vec{r} \times \frac{d\vec{p}}{dt}$$

Comparing with Equation 10-21 gives

$$\vec{\tau}_{net} = \frac{d\vec{L}}{dt} \qquad\qquad 10\text{-}22$$

The net torque acting on a system of particles is the sum of the individual torques. The generalization of Equation 10-22 to a system of particles is then

$$\sum_i \vec{\tau} = \sum_i \frac{d\vec{L}_i}{dt} = \frac{d}{dt} \sum_i \vec{L}_i = \frac{d\vec{L}_{sys}}{dt}$$

In this equation, the sum of the torques may include internal as well as external torques. The sum of the internal torques equals zero, so

$$\vec{\tau}_{net\ ext} = \frac{d\vec{L}_{sys}}{dt} \qquad\qquad 10\text{-}10$$

NEWTON'S SECOND LAW FOR ANGULAR MOTION

*Proofs of Equations 10-12 and 10-13 We will now show that the angular momentum of a system of particles can be written as the sum of the orbital angular momentum and the spin angular momentum.

Figure 10-33 shows a system of particles. The angular momentum $\vec{L}_i$ of the ith particle about arbitrary point O is given by

$$\vec{L}_i = \vec{r}_i \times \vec{p}_i = \vec{r}_i \times m_i \vec{v}_i \qquad\qquad 10\text{-}23$$

and the angular moment of the system about O is

$$\vec{L} = \Sigma \vec{L}_i = \Sigma (\vec{r}_i \times m_i \vec{v}_i)$$

The angular momentum about the center of mass is given by

$$\vec{L}_{cm} = \Sigma (\vec{r}_i' \times m_i \vec{u}_i)$$

where $\vec{r}_i'$ and $\vec{u}_i$ are the position and velocity of the ith particle relative to the center of mass. It can be seen from the figure that

$$\vec{r}_i = \vec{r}_{cm} + \vec{r}_i'$$

Differentiating both sides gives

$$\vec{v}_i = \vec{v}_{cm} + \vec{u}_i$$

Substituting these into Equation 10-23, we have

$$\vec{L}_i = \vec{r}_i \times m_i \vec{v}_i = (\vec{r}_{cm} + \vec{r}_i') \times m_i (\vec{v}_{cm} + \vec{u}_i)$$

Expanding the right side, we obtain

$$\vec{L}_i = (\vec{r}_{cm} \times m_i \vec{v}_{cm}) + (\vec{r}_{cm} \times m_i \vec{u}_i) + (m_i \vec{r}_i' \times \vec{v}_{cm}) + (\vec{r}_i' \times m_i \vec{u}_i)$$

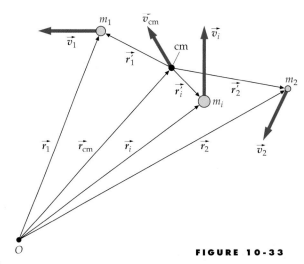

FIGURE 10-33

Summing both sides and factoring common terms out of the sums gives

$$\vec{L}_{sys} = \Sigma \vec{L}_i = \left(\vec{r}_{cm} \times \Sigma (m_i) \, \vec{v}_{cm} \right) + \left(\vec{r}_{cm} \times \Sigma (m_i \vec{u}_i) \right) + \left(\Sigma (m_i \vec{r}_i') \times \vec{v}_{cm} \right) + \left(\Sigma (\vec{r}_i' \times m_i \vec{u}_i) \right)$$

Because $\Sigma(m_i \vec{r}_i')$ and $\Sigma(m_i \vec{u}_i)$ are both zero, and because $\Sigma m_i = M$ and $\Sigma(\vec{r}_i' \times m_i \vec{u}_i) = \vec{L}_{cm}$, we have $\vec{L}_{sys} = \vec{r}_{cm} \times M \vec{v}_{cm} + \vec{L}_{cm}$, or

$$\vec{L}_{sys} = \vec{L}_{orbit} + \vec{L}_{spin} \qquad\qquad 10\text{-}12$$

where $\vec{L}_{spin} = \vec{L}_{cm}$ and $\vec{L}_{orbit} = \vec{v}_{cm} \times M \vec{v}_{cm}$. \qquad 10-13

*Proofs of Equations 10-14 and 10-15 We will now take the z components of the vectors for the torque and the angular momentum about a point to obtain the formulas for the torque and angular momentum about a fixed axis. The angular momentum of a particle about the origin is $\vec{L} = \vec{r} \times \vec{p}$, so finding the z component of the angular momentum means finding the z component of the product $\vec{r} \times \vec{p}$. To do this we express $\vec{r}$ and $\vec{p}$ as

$$\vec{r} = \vec{r}_{rad} + \vec{r}_z \qquad \text{and} \qquad \vec{p} = \vec{p}_{xy} + \vec{p}_z$$

where $\vec{r}_{rad}, \vec{r}_z, \vec{p}_{xy}$, and $\vec{p}_z$ are vector components (see Figure 10-34) of $\vec{r}$ and $\vec{p}$.
Substituting $\vec{r}$ and $\vec{p}$ gives

$$\vec{L} = \vec{r} \times \vec{p} = (\vec{r}_{rad} + \vec{r}_z) \times (\vec{p}_{xy} + \vec{p}_z)$$

and expanding the right side, we have

$$\vec{L} = (\vec{r}_{rad} \times \vec{p}_{xy}) + (\vec{r}_{rad} \times \vec{p}_z) + (\vec{r}_z \times \vec{p}_{xy}) + (\vec{r}_z \times \vec{p}_z)$$

The cross product of any two vectors is perpendicular to both vectors, so the product $\vec{r}_{rad} \times \vec{p}_{xy}$ is parallel to the z axis. In each of the other three products at least one of the two vectors is parallel to the z axis, so the z component of each of these cross products is zero. Therefore,

$$\vec{L}_z = \vec{r}_{rad} \times \vec{p}_{xy} \qquad\qquad 10\text{-}14$$

ANGULAR MOMENTUM ABOUT Z AXIS

The torque about the origin associated with a force acting on the particle is given by $\vec{\tau} = \vec{r} \times \vec{F}$ (Equation 10-1). Following the same procedure with the torque that we followed with the angular momentum gives

$$\vec{\tau}_z = \vec{r}_{rad} \times \vec{F}_{xy} \qquad\qquad 10\text{-}15$$

TORQUE ABOUT Z AXIS

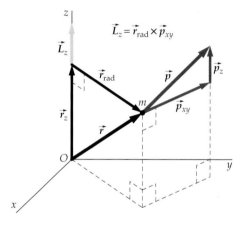

FIGURE 10-34 The vector components $\vec{r}_{rad}, \vec{r}_z, \vec{p}_{xy}$, and $\vec{p}_z$ of $\vec{r}$ and $\vec{p}$ that are used for calculating the angular momentum about the z axis $\vec{L}_z$.

*10-4 Quantization of Angular Momentum

Angular momentum plays an important role in the description of atoms, molecules, nuclei, and elementary particles. Like energy, angular momentum is **quantized,** that is, changes in angular momentum occur only in discrete amounts.

The angular momentum of a particle due to its motion is its orbital angular momentum. The magnitude of the orbital angular momentum L of a particle can have only the values

$$L = \sqrt{\ell(\ell + 1)} \, \hbar, \qquad \ell = 0, 1, 2, \ldots \qquad\qquad 10\text{-}24$$

where $\hbar$ (read "h-bar") is the **fundamental unit of angular momentum,** which is related to Planck's constant h:

$$\hbar = \frac{h}{2\pi} = 1.05 \times 10^{-34}\,\text{J}\cdot\text{s} \qquad\qquad 10\text{-}25$$

The component of orbital angular momentum along any line in space is also quantized and can have only the values $\pm m\hbar$, where m is a nonnegative integer that is less than or equal to ℓ. For example, if $\ell = 2$, m can equal 2, 1, or 0.

Because the quantum of angular momentum $\hbar$ is so small, the quantization of angular momentum is not noticed in the macroscopic world. Consider a particle of mass 1 g $= 10^{-3}$ kg moving in a circle of radius 1 cm with a period of 1 s. Its orbital angular momentum is

$$L = mvr = mr^2\omega = mr^2\frac{2\pi}{T} = (10^{-3}\,\text{kg})(10^{-2}\,\text{m})^2\frac{2\pi}{1\,\text{s}} = 6.28 \times 10^{-7}\,\text{J}\cdot\text{s}$$

If we divide by $\hbar$, we obtain

$$\frac{L}{\hbar} = \frac{6.28 \times 10^{-7}\,\text{J}\cdot\text{s}}{1.05 \times 10^{-34}\,\text{J}\cdot\text{s}} = 6 \times 10^{27}$$

Thus this typical macroscopic angular momentum contains 6×10^{27} units of the fundamental unit of angular momentum. Even if we could measure L to one part in a billion, we would never notice the quantization of this macroscopic angular momentum.

The quantization of orbital angular momentum leads to the quantization of rotational kinetic energy. Consider a molecule rotating about its center of mass with angular momentum L (Figure 10-35). Let I be its moment of inertia. Its kinetic energy is

$$K = \frac{L^2}{2I} \qquad\qquad 10\text{-}26$$

But L^2 is quantized to the values $L^2 = \ell(\ell + 1)\hbar^2$ with $\ell = 0, 1, 2,\ldots$. Thus, the kinetic energy is quantized to the values K_ℓ given by

$$K_\ell = \frac{L^2}{2I} = \frac{\ell(\ell + 1)\hbar^2}{2I} = \ell(\ell + 1)E_{0\text{r}} \qquad\qquad 10\text{-}27a$$

where

$$E_{0\text{r}} = \frac{\hbar^2}{2I} \qquad\qquad 10\text{-}27b$$

Figure 10-36 shows an energy-level diagram for a rotating molecule with constant moment of inertia I. Note that, unlike the energy levels for a vibrating system (Section 7-4), the rotational energy levels are not equally spaced, and the lowest level is zero.

Stable matter contains just three kinds of particles: electrons, protons, and neutrons. In addition to its orbital angular momentum each of these particles also has an intrinsic angular momentum called its **spin.** The spin angular momentum of a particle, like its mass and electric charge, is a fundamental property of the particle that cannot be changed. The magnitude of the spin angular momentum vector for electrons and other fermions is $s = \sqrt{\frac{1}{2}(\frac{1}{2} + 1)}\,\hbar$ and the component along any line in space can have just two values: $+\frac{1}{2}\hbar$ and $-\frac{1}{2}\hbar$. Such particles are called "spin one-half" particles. Electrons, protons, neutrons, and other spin-one-half particles are called **fermions.** Other particles such as photons and α particles, called

FIGURE 10-35 Model of a rigid diatomic molecule rotating about the z axis.

FIGURE 10-36 Energy-level diagram for a rotating molecule.

bosons, have zero spin or integral spin. Spin is a quantum property of the particle that has nothing to do with the motion of the particle.

The picture of an electron as a spinning ball that orbits the nucleus in an atom (like the spinning earth orbiting the sun) is often a useful visualization. However, the angular momentum of a spinning ball can be increased or decreased, whereas the spin of the electron is a fixed property like its charge and mass that does not change. Furthermore, as far as we know, electrons are point particles that have no size.

SUMMARY

1. Angular momentum is an important derived dynamic quantity in macroscopic physics. In microscopic physics, angular momentum is an intrinsic, fundamental property of elementary particles.
2. Conservation of angular momentum is a fundamental law of nature.
3. Quantization of angular momentum is a fundamental law of nature.

Topic	Relevant Equations and Remarks	
1. Vector Nature of Rotation	When the direction of the axis of rotation is not fixed in space, plus and minus signs are inadequate to describe the direction of the angular velocity direction.	
Angular velocity $\vec{\omega}$	The direction of the angular velocity $\vec{\omega}$ is along the axis of rotation in the sense given by the right-hand rule.	
Torque $\vec{\tau}$	$\vec{\tau} = \vec{r} \times \vec{F}$	10-1
2. Vector Product	$\vec{A} \times \vec{B} = AB \sin \phi \, \hat{n}$	10-2
	where ϕ is the angle between the vectors and $\hat{n}$ is a unit vector perpendicular to the plane of $\vec{A}$ and $\vec{B}$ in the sense given by the right-hand rule as $\vec{A}$ is rotated into $\vec{B}$.	
Properties	$\vec{A} \times \vec{A} = 0$	10-3
	$\vec{A} \times \vec{B} = -\vec{B} \times \vec{A}$	10-4
	$\frac{d}{dt}(\vec{A} \times \vec{B}) = \left(\vec{A} \times \frac{d\vec{B}}{dt}\right) + \left(\frac{d\vec{A}}{dt} \times \vec{B}\right)$	10-6
	$\hat{i} \times \hat{j} = \hat{k} \quad \hat{j} \times \hat{k} = \hat{i} \quad \hat{k} \times \hat{i} = \hat{j}$	10-7a
3. Angular Momentum		
For a particle	$\vec{L} = \vec{r} \times \vec{p}$	10-8
For a system rotating about a symmetry axis	$\vec{L} = I\vec{\omega}$	10-9
For any system	The angular momentum about any point O is the angular momentum about the center of mass (spin angular momentum) plus the angular momentum associated with center-of-mass motion about O (orbital angular momentum).	
	$\vec{L} = \vec{L}_{orbit} + \vec{L}_{spin} = \left(\vec{r}_{cm} \times M\vec{v}_{cm}\right) + \left(\sum_i \vec{r}_i \times m_i\vec{u}_i\right)$	10-12
Newton's second law for angular motion	$\vec{\tau}_{net\ ext} = \frac{d\vec{L}}{dt}$	10-10

| Conservation of angular momentum | If the net external torque is zero, the angular momentum of the system is conserved. (If the component of the net external torque in a given direction is zero, the component of the angular momentum of the system in that direction is conserved.) | |

| Kinetic energy of a rotating object | $$K = \frac{L^2}{2I}$$ | 10-20 |

| Quantization of angular momentum | The magnitude of the orbital angular momentum of a particle can have only the values $$L = \sqrt{\ell(\ell + 1)}\hbar, \qquad \ell = 0, 1, 2, \ldots$$ where | |

| Fundamental unit of angular momentum | $$\hbar = \frac{h}{2\pi} = 1.05 \times 10^{-34} \, \text{J·s}$$ is the fundamental unit of angular momentum and h is Planck's constant. | 10-25 |

| *Quantization of any component of orbital angular momentum | The component of orbital angular momentum along any line in space is also quantized and can have only the values $\pm m\,\hbar$, where m is a nonnegative integer that is less than or equal to ℓ. | |

| Spin | Electrons, protons, and neutrons have an intrinsic angular momentum called spin. The magnitude of the spin angular momentum vector for these particles is $$s = \sqrt{\tfrac{1}{2}(\tfrac{1}{2} + 1)}\,\hbar$$ and the component along any line in space can have just two values, $+\tfrac{1}{2}\hbar$ and $-\tfrac{1}{2}\hbar$. | |

PROBLEMS

- Single-concept, single-step, relatively easy
- • Intermediate-level, may require synthesis of concepts
- • • Challenging, for advanced students
- SSM Solution is in the *Student Solutions Manual*
- iSOLVE Problems available on iSOLVE online homework service
- iSOLVE ✓ These "Checkpoint" online homework service problems ask students additional questions about their confidence level, and how they arrived at their answer

In a few problems, you are given more data than you actually need; in a few other problems, you are required to supply data from your general knowledge, outside sources, or informed estimates.

Take $g = 9.81$ N/kg $= 9.81$ m/s^2 and neglect friction in all problems unless otherwise stated.

Conceptual Problems

1 • SSM True or false:

(a) If two vectors are parallel, their cross product must be zero.
(b) When a disk rotates about its symmetry axis, $\vec{\omega}$ is along the axis.
(c) The torque exerted by a force is always perpendicular to the force.

2 • Consider two nonzero vectors $\vec{A}$ and $\vec{B}$. Their cross product has the greatest magnitude if $\vec{A}$ and $\vec{B}$ are (a) parallel,
(b) equal in magnitude, (c) perpendicular, (d) antiparallel, (e) at an angle of 45° to each other.

3 • What is the angle between a particle's linear momentum $\vec{p}$ and its angular momentum $\vec{L}$?

4 • A particle of mass m is moving with speed v along a line that passes through point P. What is the angular momentum of the particle about point P? (a) mv. (b) Zero. (c) It changes sign as the particle passes through point P. (d) It depends on the distance of point P from the origin of the coordinates.

5 •• SSM A particle travels in a circular path and point P is at the center of the circle. (*a*) If its linear momentum $\vec{p}$ is doubled, how is its angular momentum about P affected? (*b*) If the radius of the circle is doubled but the speed is unchanged, how is the angular momentum of the particle about P affected?

6 •• A particle moves along a straight line at constant speed. How does its angular momentum about any fixed point vary over time?

7 • True or false: If the net torque on a rotating system is zero, the angular velocity of the system cannot change.

8 •• SSM Standing on a turntable that is initially not rotating, can you rotate yourself through 180°? Assume that no external torques act on the you–turntable system. *Hint: While you cannot change your mass easily, there are ways to change your moment of inertia.*

9 • If the angular momentum of a system is constant, which of the following statements must be true?

(*a*) No torque acts on any part of the system.
(*b*) A constant torque acts on each part of the system.
(*c*) Zero net torque acts on each part of the system.
(*d*) A constant external torque acts on the system.
(*e*) Zero net external torque acts on the system.

10 •• In Example 10-4, does the force exerted by the merry-go-round on Benny do work?

11 •• Is it easier to crawl radially outward or radially inward on a rotating merry-go-round? Why?

12 •• SSM A block sliding on a frictionless table is attached to a string that passes through a hole in the table. Initially, the block is sliding with speed v_0 in a circle of radius r_0. A student under the table pulls slowly on the string. What happens as the block spirals inward? Give supporting arguments for your choice. (*a*) Its energy and angular momentum are conserved. (*b*) Its angular momentum is conserved and its energy increases. (*c*) Its angular momentum is conserved and its energy decreases. (*d*) Its energy is conserved and its angular momentum increases. (*e*) Its energy is conserved and its angular momentum decreases.

13 •• SSM How can you tell a hardboiled egg from an uncooked one without breaking it? One way is to lay the egg flat on a hard surface and try to spin it. A hardboiled egg will spin easily, while it takes a lot of effort to make an uncooked egg spin. However, once spinning, the uncooked egg will do something unusual: If you stop it with your finger, it may start spinning again. Explain the difference in the behavior of the two types of eggs.

14 • True or false:

When a gyroscope is not spinning, $\vec{\tau} = \dfrac{d\vec{L}}{dt}$ does not hold.

15 • [SOLVE] An object of mass M is rotating about a fixed axis with angular momentum L. Its moment of inertia about this axis is I. What is its kinetic energy? (*a*) $IL^2/2$. (*b*) $L^2/2I$. (*c*) $ML^2/2$. (*d*) $IL^2/2M$.

16 • Explain why a helicopter with just one main rotor has a second smaller rotor mounted on a horizontal axis at the rear as in Figure 10-37. Describe the resultant motion of the helicopter if this rear rotor fails during flight.

FIGURE 10-37 Problem 16

17 •• The angular momentum vector for a spinning wheel lies along its axle and is pointed east. To make this vector point south, it is necessary to exert a force on the east end of the axle in which direction? (*a*) Up. (*b*) Down. (*c*) North. (*d*) South. (*e*) East.

18 •• SSM You are walking north and with your left hand you are carrying a suitcase that contains a spinning gyroscope mounted on an axle attached to the front and back of the case. The angular velocity of the gyroscope points north. You now begin to turn to walk east. As a result, the front end of the suitcase will (*a*) resist your attempt to turn and will try to remain pointed north, (*b*) fight your attempt to turn and will pull to the west, (*c*) rise upward, (*d*) dip downward, (*e*) show no effect whatsoever.

19 •• The angular momentum of the propeller of a small single-engine airplane points forward. The propeller rotates clockwise if viewed from behind. (*a*) As the plane takes off, the nose lifts and the airplane tends to veer to one side. To which side does it veer and why? (*b*) If the plane is flying horizontally and suddenly turns to the right, does the nose of the plane tend to move up or down? Why?

20 •• A car is powered by the energy stored in a single flywheel with an angular momentum $\vec{L}$. Discuss the problems that would arise for various orientations of $\vec{L}$ and various maneuvers of the car. For example, what would happen if $\vec{L}$ points vertically upward and the car travels over a hilltop or through a valley? What would happen if $\vec{L}$ points forward or to one side and the car attempts to turn to the left or right? In each case that you examine, consider the direction of the torque exerted on the car by the road.

21 •• [SOLVE] You sit on a spinning piano stool with your arms folded. When you extend your arms out to the side, your kinetic energy (*a*) increases, (*b*) decreases, (*c*) remains the same.

22 •• SSM In tetherball, a ball is attached to a string that is attached to a pole. When the ball is hit, the string wraps around the pole and the ball spirals inward. Neglecting air resistance, what happens as the ball swings around the pole? Give supporting arguments for your choice. (*a*) The mechanical energy and angular momentum of the ball are conserved. (*b*) The angular momentum of the ball is conserved, but the mechanical energy of the ball increases. (*c*) The angular momentum of the ball is conserved, and the mechanical energy of the ball decreases. (*d*) The mechanical energy of the ball is conserved and the angular momentum of the ball increases. (*e*) The mechanical energy of the ball is conserved and the angular momentum of the ball decreases.

23 •• A uniform rod of mass M and length L lies on a horizontal frictionless table. A piece of putty of mass $m = M/4$ moves along a line perpendicular to the rod, strikes the rod near its end, and sticks to the rod. Describe qualitatively the subsequent motion of the rod and putty.

24 • **SSM** True or false:

(a) The rate of change of a system's angular momentum is always parallel to the net external torque.

(b) If the net torque on a body is zero, the angular momentum must be zero.

Estimation and Approximation

25 •• **SSM** **iSOLVE** ✔ An ice skater starts her pirouette with arms outstretched, rotating at 1.5 rev/s. Estimate her rotational speed (in revolutions per second) when she brings her arms flat against her body.

26 •• **SOLVE** The polar ice caps contain about 2.3×10^{19} kg of ice. This mass contributes negligibly to the moment of inertia of the earth because it is located at the poles, close to the axis of rotation. Estimate the change in the length of the day that would be expected if the polar ice caps were to melt and the water were distributed uniformly over the surface of the earth. (The moment of inertia of a spherical shell of mass m and radius r is $2mr^2/3$.)

27 • A 2-g particle moves at a constant speed of 3 mm/s around a circle of radius 4 mm. (a) Find the magnitude of the angular momentum of the particle. (b) If $L = \sqrt{\ell(\ell + 1)}\hbar$, where ℓ is an integer, find the value of $\ell(\ell + 1)$ and the approximate value of ℓ. (c) Explain why the quantization of angular momentum is not noticed in macroscopic physics.

28 •• **SSM** One problem in astrophysics in the 1960s was explaining pulsars—extremely regular astronomical sources of radio pulses whose periods ranged from seconds to milliseconds. At one point, these radio sources were given the acronym LGM, standing for "Little Green Men," a reference to the idea that they might be signals of extraterrestrial civilizations. The explanation given today is no less interesting: The sun, which is a fairly typical star, has a mass of 1.99×10^{30} kg and a radius of 6.96×10^8 m. While it doesn't rotate uniformly, because it isn't a solid body, its average rate of rotation can be taken as about 1 rev/25 d. Stars somewhat larger than the sun can end their life in spectacular explosions—supernovae—leaving behind a collapsed remnant of the star called a neutron star. These neutron-star remnants have masses comparable to the original mass of the star, but radii of only a few kilometers! The high rotation rate is due to the conservation of angular momentum during the collapse. These stars emit beams of radio waves. Because of the rapid angular speed of the stars, the beam sweeps past the earth at regular intervals. To produce the observed radio-wave pulses, the star has to rotate at rates from about 1 rev/s to 1 rev/ms. (a) Using data from the textbook, estimate the rotation rate of the sun if it were to collapse into a neutron star of radius 10 km. Because the sun is not a uniform sphere of gas, its moment of inertia is given by the formula $I = 0.059MR^2$. Assume that the neutron star is spherical and has a uniform mass distribution. (b) Is the rotational kinetic energy of the sun greater or smaller after the collapse? By what factor does it change, and where does the energy go to or come from?

29 •• The moment of inertia of the earth about its spin axis is approximately 8.03×10^{37} kg·m². (a) Since the earth is nearly spherical, assume that the moment of inertia can be written as $I = CMR^2$, where C is a dimensionless constant, $M = 5.98 \times 10^{24}$ kg is the mass of the earth, and $R = 6370$ km is its radius. Determine C. (b) If the earth's mass were distributed uniformly, C would equal 2/5. From the value of C calculated in Part (a), is the earth's mass density greater near the core or near the crust?

30 •• **SSM** Estimate the angular velocity and angular momentum of the diver in Figure 10-23 (page 318) about his center of mass. Make any approximations that you think reasonable.

31 •• Estimate the angular velocity of the diver in Figure 10-23 (page 318) if he curled himself into a ball in middive.

32 •• **SSM** Estimate Timothy Goebel's initial takeoff speed, rotational velocity, and angular momentum when he performs a quadruple Lutz (Figure 10-38). Make any assumptions you think reasonable, but be prepared to justify them. Goebel's mass is about 60 kg and the height of the jump is about 0.6 m. Note that the angular velocity will change quite a bit during the jump, as he begins with arms outstretched and pulls them in. Your answer should be accurate to within a factor of 2 if you're careful.

FIGURE 10-38
Problem 32

Vector Nature of Rotation

33 • A force of magnitude F is applied horizontally in the negative x direction to the rim of a disk of radius R as shown in Figure 10-39. Write $\vec{F}$ and $\vec{r}$ in terms of the unit vectors $\hat{i}$, $\hat{j}$, and $\hat{k}$, and compute the torque produced by the force about the origin at the center of the disk.

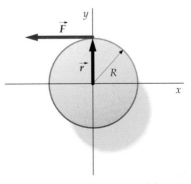

FIGURE 10-39 Problem 33

34 • Compute the torque about the origin for the force $\vec{F} = -mg\hat{j}$ acting on a particle at $\vec{r} = x\hat{i} + y\hat{j}$ and show that this torque is independent of the y coordinate.

35 • **ISOLVE** ✓ Find $\vec{A} \times \vec{B}$ for (a) $\vec{A} = 4\hat{i}$ and $\vec{B} = 6\hat{i} + 6\hat{j}$, (b) $\vec{A} = 4\hat{i}$ and $\vec{B} = 6\hat{i} + 6\hat{k}$, and (c) $\vec{A} = 2\hat{i} + 3\hat{j}$ and $\vec{B} = 3\hat{i} + 2\hat{j}$.

36 • **SSM** Under what conditions is the magnitude of $\vec{A} \times \vec{B}$ equal to $\vec{A} \cdot \vec{B}$?

37 •• A particle moves in a circle that is centered at the origin. The particle has position $\vec{r}$ and angular velocity $\vec{\omega}$. (a) Show that its velocity is $\vec{v} = \vec{\omega} \times \vec{r}$. (b) Show that its centripetal acceleration is $\vec{a}_c = \vec{\omega} \times \vec{v} = \vec{\omega} \times (\vec{\omega} \times \vec{r})$.

38 •• **ISOLVE** ✓ If $\vec{A} = 4\hat{i}$, $B_z = 0$, $|\vec{B}| = 5$, and $\vec{A} \times \vec{B} = 12\hat{k}$, determine $\vec{B}$.

39 • If $\vec{A} = 3\hat{j}$, $\vec{A} \times \vec{B} = 9\hat{i}$, and $\vec{A} \cdot \vec{B} = 12$, find $\vec{B}$.

40 •• Use the rules for evaluating a determinant to show that if

$$\vec{A} = a_x\hat{i} + a_y\hat{j} + a_z\hat{k}$$

$$\vec{B} = b_x\hat{i} + b_y\hat{j} + b_z\hat{k}$$

$$\vec{C} = c_x\hat{i} + c_y\hat{j} + c_z\hat{k}$$

then

$$\vec{A} \cdot (\vec{B} \times \vec{C}) = \begin{vmatrix} a_x & a_y & a_z \\ b_x & b_y & b_z \\ c_x & c_y & c_z \end{vmatrix} = \vec{C} \cdot (\vec{A} \times \vec{B}) = \vec{B} \cdot (\vec{C} \times \vec{A})$$

41 •• Given three noncoplanar vectors $\vec{A}$, $\vec{B}$, and $\vec{C}$, show that $\vec{A} \cdot (\vec{B} \times \vec{C})$ is the volume of the parallelopiped formed by the three vectors.

42 •• **SSM** Using the cross product, prove the *law of sines* for the triangle shown in Figure 10-40: if A, B, and C are the lengths of each side of the triangle, show that $A/\sin a = B/\sin b = C/\sin c$.

FIGURE 10-40
Problem 42

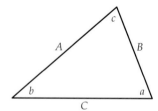

Angular Momentum

43 • A particle moving at constant velocity has zero angular momentum about a particular point. Show that the particle is moving either directly toward the point, directly away from the point, or through the point.

44 • A 2-kg particle moves at a constant speed of 3.5 m/s around a circle of radius 4 m. (a) What is its angular momentum about the center of the circle? (b) What is its moment of inertia about an axis through the center of the circle and per-

pendicular to the plane of the motion? (c) What is the angular speed of the particle?

45 • **ISOLVE** ✓ A 2-kg particle moves at constant speed of 4.5 m/s along a straight line. (a) What is the magnitude of its angular momentum about a point 6 m from the line? (b) Describe qualitatively how its angular speed about that point varies with time.

46 •• **SSM** A particle is traveling with a constant velocity $\vec{v}$ along a line that is a distance b from the origin O (Figure 10-41). Let dA be the area swept out by the position vector from O to the particle in time dt. Show that dA/dt is constant and is equal to $\frac{1}{2}L/m$, where L is the angular momentum of the particle about the origin.

FIGURE 10-41 Problem 46

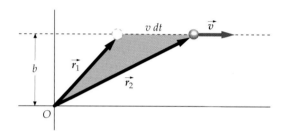

47 •• **ISOLVE** A 15-g coin of diameter 1.5 cm is spinning at 10 rev/s about a vertical diameter at a fixed point on a tabletop. (a) What is the angular momentum of the coin about its center of mass? (A coin is a cylinder of length L and radius R, where L is negligible compared to R. You can obtain its moment of inertia about an axis through a diameter in Table 9-1.) (b) What is its angular momentum about a point on the table 10 cm from the coin? If the coin spins about a vertical diameter at 10 rev/s while its center of mass travels in a straight line across the tabletop at 5 cm/s, (c) what is the angular momentum of the coin about a point on the line of motion of the center of mass? (d) What is the angular momentum of the coin about a point 10 cm to either side of the line of motion of the center of mass? (There are two answers to this question. Explain why and give both.)

48 •• Two particles of masses m_1 and m_2 are located at $\vec{r}_1$ and $\vec{r}_2$ relative to some origin O as in Figure 10-42. They exert equal and opposite forces on each other. Calculate the resultant torque exerted by these internal forces about the origin O and show that it is zero if the forces $\vec{F}_1$ and $\vec{F}_2$ lie along the line joining the particles.

FIGURE 10-42
Problem 48

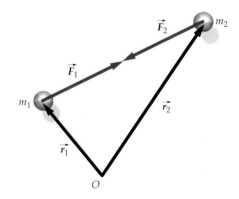

Torque and Angular Momentum

FIGURE 10-44 Problem 52

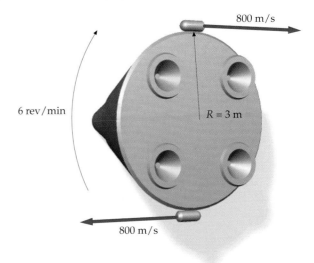

49 • [SOLVE] A 1.8-kg particle moves in a circle of radius 3.4 m. The magnitude of its angular momentum relative to the center of the circle depends on time according to $L = (4 \text{ N·m})t$. (a) Find the magnitude of the torque acting on the particle. (b) Find the angular speed of the particle as a function of time.

50 •• [SOLVE ✓] A uniform cylinder of mass 90 kg and radius 0.4 m is mounted so that it turns without friction on its fixed symmetry axis. It is rotated by a drive belt that wraps around its perimeter and exerts a constant torque. At time $t = 0$, its angular velocity is zero. At time $t = 25$ s, its angular velocity is 500 rev/min. (a) What is its angular momentum at $t = 25$ s? (b) At what rate is the angular momentum increasing? (c) What is the torque acting on the cylinder? (d) What is the magnitude of the frictional force acting on the rim of the cylinder?

51 •• [SSM] In Figure 10-43, the incline is frictionless and the string passes through the center of mass of each block. The pulley has a moment of inertia I and a radius R. (a) Find the net torque acting on the system (the two masses, string, and pulley) about the center of the pulley. (b) Write an expression for the total angular momentum of the system about the center of the pulley when the masses are moving with a speed v. (c) Find the acceleration of the masses from your results for Parts (a) and (b) by setting the net torque equal to the rate of change of the angular momentum of the system.

53 •• A projectile (mass M) is launched at an angle θ with an initial speed V. Considering the torque and angular momentum about the launch point, explicitly show that $d\mathbf{L}/dt = \tau$. Ignore the effects of air resistance. (The equations for projectile motion are found in Chapter 3.)

Conservation of Angular Momentum

54 • [SSM] [SOLVE] A planet moves in an elliptical orbit about the sun with the sun at one focus of the ellipse as in Figure 10-45. (a) What is the torque about the center of the sun due to the gravitational force of attraction of the sun for the planet? (b) At position A, the planet is a distance r_1 from the sun and is moving with a speed v_1 perpendicular to the line from the sun to the planet. At position B, it is at distance r_2 and is moving with speed v_2, again perpendicular to the line from the sun to the planet. What is the ratio of v_1 to v_2 in terms of r_1 and r_2?

FIGURE 10-43 Problem 51

FIGURE 10-45 Problem 54

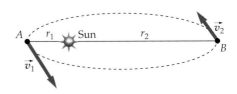

52 •• [SOLVE ✓] Figure 10-44 shows the rear view of a space capsule that is rotating about its longitudinal axis at 6 rev/min. The occupants want to stop this rotation. They have small jets mounted tangentially at a distance $R = 3$ m from the axis, as indicated, and can eject 10 g/s of gas from each jet with a nozzle velocity of 800 m/s. For how long must they turn on these jets to stop the rotation? The moment of inertia of the ship about its axis (assumed to be constant) is 4000 kg·m².

55 •• [SOLVE] You stand on a frictionless platform that is rotating with an angular speed of 1.5 rev/s. Your arms are outstretched, and you hold a heavy weight in each hand. The moment of inertia of you, the extended weights, and the platform is 6 kg·m². When you pull the weights in toward your body, the moment of inertia decreases to 1.8 kg·m². (a) What is the resulting angular speed of the platform? (b) What is the change in kinetic energy of the system? (c) Where did this increase in energy come from?

56 •• [SSM] [SOLVE] ✓ A small blob of putty of mass m falls from the ceiling and lands on the outer rim of a turntable of radius R and moment of inertia I_0 that is rotating freely with angular speed ω_i about its vertical fixed symmetry axis. (a) What is the postcollision angular speed of the turntable plus putty? (b) After several turns, the blob flies off the edge of the turntable. What is the angular speed of the turntable after the blob flies off?

57 •• A Lazy Susan consists of a heavy plastic cylinder mounted on a frictionless bearing resting on a vertical shaft. The cylinder has a radius $R = 15$ cm and mass $M = 0.25$ kg. A cockroach (mass $m = 0.015$ kg) is on the Lazy Susan, at a distance of 8 cm from the center. Both the cockroach and the Lazy Susan are initially at rest. The cockroach then walks along a circular path concentric with the center of the Lazy Susan at a distance $r = 8$ cm from the center of the shaft. If the speed of the cockroach with respect to the Lazy Susan is $v = 0.01$ m/s, what is the speed of the cockroach with respect to the room?

58 •• [SSM] [SOLVE] Two disks of identical mass but different radii (r and $2r$) are spinning on frictionless bearings at the same angular speed ω_0 but in opposite directions (Figure 10-46). The two disks are brought slowly together. The resulting frictional force between the surfaces eventually brings them to a common angular velocity. What is the magnitude of that final angular velocity in terms of ω_0?

FIGURE 10-46 Problem 58

59 •• [SOLVE] ✓ A block of mass m sliding on a frictionless table is attached to a string that passes through a hole in the table. Initially, the block is sliding with speed v_0 in a circle of radius r_0. Find (a) the angular momentum of the block, (b) the kinetic energy of the block, and (c) the tension in the string. A student under the table now slowly pulls the string downward. How much work is required to reduce the radius of the circle from r_0 to $r_0/2$?

60 •• [SSM] A 0.2-kg point mass moving on a frictionless horizontal surface is attached to a rubber band whose other end is fixed at point P. The rubber band exerts a force $F = bx$ toward P, where x is the length of the rubber band and b is an unknown constant. The mass moves along the dotted line in Figure 10-47. When it passes point A, its velocity is 4 m/s, directed as shown. The distance AP is 0.6 m and BP is 1.0 m. (a) Find the velocity of the mass at points B and C. (b) Find b.

FIGURE 10-47 Problem 60

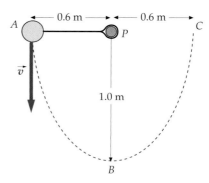

*Quantization of Angular Momentum

61 • [SSM] The z component of the spin of an electron is $\frac{1}{2}\hbar$, but the magnitude of the spin vector is $\sqrt{0.75}\hbar$. What is the angle between the electron's spin angular momentum vector and the z axis?

62 •• Show that the energy difference between one rotational state and the next higher state is proportional to $\ell + 1$ (see Equation 10-27a).

63 •• In the HBr molecule, the mass of the bromine nucleus is 80 times that of the hydrogen nucleus (a single proton); consequently, in calculating the rotational motion of the molecule, one may, to a good approximation, assume that the Br nucleus remains stationary as the H atom (mass 1.67×10^{-27} kg) revolves around it. The separation between the H atom and bromine nucleus is 0.144 nm. Calculate (a) the moment of inertia of the HBr molecule about the bromine nucleus and (b) the rotational energies for $\ell = 1$, $\ell = 2$, and $\ell = 3$.

64 •• [SOLVE] The equilibrium separation between the nuclei of the nitrogen molecule is 0.11 nm. The mass of each nitrogen nucleus is 14 u, where u $= 1.66 \times 10^{-27}$ kg. We wish to calculate the energies of the three lowest angular momentum states of the nitrogen molecule. (a) Approximate the nitrogen molecule as a rigid dumbbell of two equal point masses and calculate the moment of inertia about its center of mass. (b) Find the rotational energy levels using the relation $E\ell = \ell(\ell + 1)\hbar^2/(2I)$.

65 •• [SSM] How fast would a nitrogen molecule have to be moving for its translational kinetic energy to be equal to the rotational kinetic energy of its $l = 1$ quantum state?

Collision Problems

66 •• [SOLVE] A 16.0-kg, 2.4-m-long rod is supported on a knife edge at its midpoint. A 3.2-kg ball of clay is dropped from rest from a height of 1.2 m and makes a perfectly inelastic collision with the rod 0.9 m from the point of support (Figure 10-48). Find the angular momentum of the rod and clay system about the point of support immediately after the inelastic collision.

FIGURE 10-48 Problem 66

3.2 kg

16.0 kg ←0.9 m→ 1.2 m

2.4 m

67 •• **SSM** Figure 10-49 shows a thin uniform bar of length L and mass M and a small blob of putty of mass m. The system is supported by a frictionless horizontal surface. The putty moves to the right with velocity v, strikes the bar at a distance d from the center of the bar, and sticks to the bar at the point of contact. Obtain expressions for the velocity of the system's center of mass and for the angular velocity of the system about its center of mass.

FIGURE 10-49
Problems 67, 68

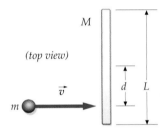

M

(top view)

$\vec{v}$

m

d L

68 •• In Problem 67, replace the blob of putty with a small hard sphere of negligible size that collides elastically with the bar. Find d so that the sphere is at rest after the collision.

69 •• **ISOLVE** Figure 10-50 shows a uniform rod of length d and mass M pivoted at the top. The rod, which is initially at rest, is struck by a particle of mass m at a point $x = 0.8d$ below the pivot. Assume that the particle sticks to the rod. What must be the speed v of the particle so that the maximum angle between the rod and the vertical is 90°?

FIGURE 10-50 Problems 69, 70

M x

d

m $\vec{v}$

70 •• If, for the system of Problem 69, $d = 1.2$ m, $M = 0.8$ kg, $m = 0.3$ kg, and the maximum angle between the rod and the vertical is 60°, find the speed of the particle before impact.

71 •• A uniform rod is resting on a frictionless table when it is suddenly struck near one end by a horizontal blow in a direction perpendicular to the rod. The mass of the rod is M and the impulse applied by the blow is K. Immediately after the rod is struck, (a) what is the velocity of the center of mass of the rod, (b) what is the velocity of the end that is struck, (c) what is the velocity of the other end of the rod, and (d) is there a point on the rod that is motionless?

72 •• A projectile of mass m_p is traveling at a constant velocity $\vec{v}_0$ toward a stationary disk of mass M and radius R that is free to rotate about a pivot through its axis O (Figure 10-51). Before impact, the projectile is traveling along a line displaced a distance b below the axis. The projectile strikes the disk and sticks to point B. Treat the projectile as a point mass. (a) Before impact, what is the total angular momentum L_0 of the projectile and disk about the O axis? (b) What is the angular speed ω of the disk and projectile system just after the impact? (c) What is the kinetic energy of the disk and projectile system after impact? (d) How much mechanical energy is lost in this collision?

FIGURE 10-51 Problem 72

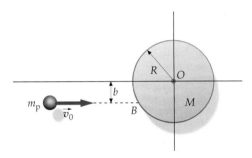

R O

b

m_p

$\vec{v}_0$ B M

73 •• **SSM** A uniform rod of length L_1 and mass $M = 0.75$ kg is supported by a hinge at one end and is free to rotate in the vertical plane (Figure 10-52). The rod is released from rest in the position shown. A particle of mass $m = 0.5$ kg is supported by a thin string of length L_2 from the hinge. The particle sticks to the rod on contact. What should be the ratio L_2/L_1 so that $\theta_{max} = 60°$ after the collision?

FIGURE 10-52
Problems 73–76

L_1

M

L_2

θ

m

74 •• **ISOLVE** A uniform rod of length $L_1 = 1.2$ m and mass $M = 2.0$ kg is supported by a hinge at one end and is free to rotate in the vertical plane (Figure 10-52). The rod is released from rest in the position shown. A particle of mass m is supported by a thin string of length $L_2 = 0.8$ m from the hinge. The particle sticks to the rod on contact, and after the collision the rod continues to rotate until $\theta = \theta_{max} = 37°$. (a) Find m. (b) How much energy is dissipated during the collision?

75 •• A uniform rod of length $L_1 = 1.2$ m and mass $M = 2.0$ kg is supported by a hinge at one end and is free to rotate in the vertical plane (Figure 10-52). The rod is given a sharp downward blow by a hammer, giving the rod an initial angular velocity. A particle of mass $m = 0.4$ kg is supported by a thin string of length $L_2 = 0.8$ m from the hinge. The particle sticks to the rod on contact, and after the collision the rod continues to rotate through a complete revolution. What is the minimum angular velocity given to the rod by the blow? How much energy is dissipated during the collision?

76 ••• Repeat Problem 74, assuming that the collision between the rod and the particle is elastic.

77 •• ▮SOLVE A bicycle wheel of radius 28 cm is mounted at the middle of an axle 50 cm long. The tire and rim weigh 30 N. The wheel is spun at 12 rev/s, and the axle is then placed in a horizontal position with one end resting on a pivot. (a) What is the angular momentum due to the spinning of the wheel? (Treat the wheel as a hoop.) (b) What is the angular velocity of precession? (c) How long does it take for the axle to swing through 360° around the pivot? (d) What is the angular momentum associated with the motion of the center of mass, that is, due to the precession? In what direction is this angular momentum?

78 •• SSM ▮SOLVE A uniform disk of mass 2.5 kg and radius 6.4 cm is mounted in the center of a 10-cm-long axle and spun at 700 rev/min. The axle is then placed in a horizontal position with one end resting on a pivot. The other end is given an initial horizontal velocity such that the precession is smooth with no nutation. (a) What is the angular velocity of precession? (b) What is the speed of the center of mass during the precession? (c) What are the magnitude and direction of the acceleration of the center of mass? (d) What are the vertical and horizontal components of the force exerted by the pivot?

General Problems

79 • ▮SOLVE✓ A particle of mass 3 kg moves with velocity $\vec{v} = 3$ m/s$\hat{i}$ along the line $z = 0$, $y = 5.3$ m. (a) Find the angular momentum $\vec{L}$ relative to the origin when the particle is at $x = 12$ m, $y = 5.3$ m. (b) A force $\vec{F} = -3$ N$\hat{i}$ is applied to the particle. Find the torque relative to the origin due to this force.

80 • ▮SOLVE The position vector of a particle of mass 3 kg is given by $\vec{r} = 4\hat{i} + 3t^2\hat{j}$, where $\vec{r}$ is in meters and t is in seconds. Determine the angular momentum and torque about the origin acting on the particle.

81 •• ▮SOLVE Two ice skaters hold hands and rotate, making one revolution in 2.5 s. Their masses are 55 kg and 85 kg, and they are separated by 1.7 m. Find (a) the angular momentum of the system about their center of mass and (b) the total kinetic energy of the system.

82 •• SSM A 2-kg ball attached to a string of length 1.5 m moves counterclockwise (as viewed from above) in a horizontal circle (Figure 10-53). The string makes an angle $\theta = 30°$ with the vertical. (a) Show that the angular momentum of the ball about the point of support P has a horizontal component

away from the center of the circle as well as a vertical component, and find these components. (b) Find the magnitude of $d\vec{L}/dt$ and show that it equals the magnitude of the torque exerted by gravity about the point of support.

FIGURE 10-53
Problem 82

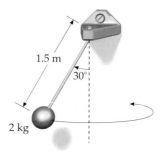

1.5 m

30°

2 kg

83 •• An object of mass m on a horizontal, frictionless surface is attached to a string that wraps around a vertical cylindrical post so that when the object is set into motion, it follows a path that spirals inward. (a) Is the angular momentum of the object about the point on the axis of the post in the plane of the spiral conserved? (b) Is the energy of the object conserved? (c) If the speed of the object is v_0 when the unwrapped length of the string is r, what is its speed when the unwrapped length has shortened to $r/2$?

84 •• Figure 10-54 shows a hollow cylindrical tube of mass M, length L, and moment of inertia $ML^2/10$. Inside the cylinder are two disks of mass m, separated by a distance ℓ and tied to a central post by a thin string. The system can rotate about a vertical axis through the center of the cylinder. With the system rotating at ω, the strings holding the disks suddenly break. When the disks reach the ends of the cylinder, they stick. Obtain expressions for the final angular velocity and the initial and final energies of the system. Assume that the inside walls of the cylinder are frictionless.

FIGURE 10-54 Problems 84–88

L

ℓ

m m

85 •• Repeat Problem 84, this time allowing for friction between the disks and the walls of the cylinder. However, the coefficient of friction is not great enough to prevent the disks from reaching the ends of the cylinder. Can the final energy of the system be determined without knowing the coefficient of kinetic friction?

86 •• [i·SOLVE ✔] Suppose that in Figure 10-54, $\ell = 0.6$ m, $L = 2.0$ m, $M = 0.8$ kg, and $m = 0.4$ kg. The system rotates at ω such that the tension in the string is 108 N just before it breaks. Determine the initial and final angular velocities and initial and final energies of the system. Assume that the inside walls of the cylinder are frictionless.

87 •• For Problem 84, determine the radial velocity of each disk just before it reaches the end of the cylinder.

88 •• Given the numerical values of Problem 86, suppose that the coefficient of friction between the disks and the walls of the cylinder is such that the disks cease sliding 0.2 m from the ends of the cylinder. Determine the initial and final angular velocities of the system and the energy dissipated in friction.

89 •• [SSM] Kepler's second law states: *The radius vector from the sun to a planet sweeps out equal areas in equal times.* Show that this law follows directly from the law of conservation of angular momentum and the fact that the force of gravitational attraction between a planet and the sun acts along the line joining the two celestial objects.

90 •• Figure 10-55 shows a hollow cylindrical shell of length 1.8 m, mass 0.8 kg, and radius 0.2 m that is free to rotate about a vertical axis through its center and perpendicular to the cylinder's axis. Inside the cylinder are two thin disks of 0.2 kg each, attached to springs of spring constant k and unstretched lengths 0.4 m. The system is brought to a rotational speed of 8 rad/s with the springs clamped so they do not stretch. The springs are then suddenly unclamped. When the disks have stopped their radial motion due to friction between the disks and the wall, they come to rest 0.6 m from the central axis. What is the angular velocity of the cylinder when the disks have stopped their radial motion? How much energy was dissipated in friction between the disks and cylinder wall?

FIGURE 10-55 Problem 90

91 •• You are given a heavy metal disk (like a coin, but larger; Figure 10-56). (Objects like this called *Euler disks* are sold through Tangent Toys.) Placing it on a turntable, you spin the disk about a vertical axis through a diameter of the disk and the center of the turntable. As you do this, you hold the turntable still with your other hand, letting it go immediately after you spin the disk. The mass of the disk is 0.500 kg and it has a radius of 0.125 m and negligible thickness. The turntable is a uniform solid cylinder with a radius of 0.250 m and a mass of 0.735 kg and rotates on a frictionless bearing.

The disk has an initial angular speed of 30 rev/min. (a) If the disk spins down to rest on the turntable with its symmetry axis coinciding with the turntable's, what is the final angular speed of the turntable? (b) What will be the final angular speed if the disk's axis is 0.10 m from the axis of the turntable when it comes to rest?

FIGURE 10-56 Problem 91

92 •• (a) Assuming the earth to be a homogeneous sphere of radius r and mass m, show that the period T of the earth's rotation about its axis is related to its radius by $T = br^2$, where $b = (4/5) \, \pi m/L$, and where L is the spin angular momentum of the earth. (b) Suppose that the radius r changes by a very small amount Δr due to some internal cause such as thermal expansion. Show that the fractional change in the period ΔT is given approximately by $\Delta T/T = 2\Delta r/r$. *Hint: Use the differentials* dr *and* dT *to approximate the changes in these quantities.* (c) By how many kilometers would the earth need to expand for the period to change by 0.25 d/y so that leap years would no longer be necessary?

93 •• [SSM] The Precession of the Equinoxes refers to the fact that the direction of the earth's spin axis does not stay fixed in the sky, but moves in a circle of radius 23° with a period of about 26,000 y. This is why our pole star, Polaris, will not remain the pole star forever. The reason for this is that the earth is a giant gyroscope, with the torque on the earth provided by the gravitational forces of the sun and moon. Calculate an approximate value for this torque, given that the period of rotation of the earth is 1 d and its moment of inertia is 8.03×10^{37} kg·m².

94 ••• [i·SOLVE] Figure 10-57 shows a cylindrical shell of mass $M = 1.2$ kg and length $L = 1.6$ m that is free to rotate about a vertical axis through its center. Inside the cylinder are two disks, each of mass 0.4 kg, that are tied to a central post by a thin string and separated by a distance $\ell = 0.8$ m. The string breaks if the tension exceeds 100 N. Starting from rest, a torque is applied to the system until the string breaks. Assuming that the disks are point masses and the radius of the cylindrical shell is negligible, find the amount of work done up to that instant. Suppose that, at that instant, the applied torque is removed and the walls of the cylindrical shell are frictionless. Obtain an expression for the angular velocity of the system as a function of x for $x < L/2$, where x is the distance between each mass and the central post.

FIGURE 10-57 Problems 94–96

95 ••• For the system of Problem 94, find the angular velocity of the system just before and just after the point masses pass the ends of the cylindrical shell.

96 ••• Repeat Problem 94 with the radius of the cylindrical shell equal to 0.4 m and the masses treated as thin disks rather than point masses.

97 ••• [SSM] Figure 10-58 shows a pulley in the form of a uniform disk with a heavy rope hanging over it. The circumference of the pulley is 1.2 m and its mass is 2.2 kg. The rope is 8.0 m long and its mass is 4.8 kg. At the instant shown in the figure, the system is at rest and the difference in height of the two ends of the rope is 0.6 m. (*a*) What is the angular velocity of the pulley when the difference in height between the two ends of the rope is 7.2 m? (*b*) Obtain an expression for the angular momentum of the system as a function of time while neither end of the rope is above the center of the pulley. There is no slippage between rope and pulley wheel.

FIGURE 10-58
Problem 97

0.6 m

Special Relativity

STS109-335-020 (1-12 MARCH 2002)—
ASTRONAUT SCOTT D. ALTMAN,
STS-109 MISSION COMMANDER,
SLEEPS ON THE FLIGHT DECK OF THE
SPACE SHUTTLE COLUMBIA. THE
ORBITAL SPEED OF THE SHUTTLE IS
ABOUT 7.7 KM/S (5 MI/S). THIS IS A
SMALL FRACTION OF THE SPEED OF
LIGHT, WHICH IS 3×10^5 KM/S.

? **If an astronaut on a spaceship that is traveling at 0.6c relative to the earth takes a one-hour-long nap, does that nap take one hour according to observers on earth? (See Example R-1.)**

The theory of relativity consists of two rather different theories, the special theory and the general theory. The special theory, developed by Albert Einstein and others in 1905, concerns the comparison of measurements made in different inertial reference frames moving with constant velocity relative to one another. Its consequences, which can be derived with a minimum of mathematics, are applicable in a wide variety of situations encountered in physics and engineering. On the other hand, the general theory, also developed by Einstein and others around 1916, is concerned with accelerated reference frames and gravity. A thorough understanding of the general theory requires sophisticated mathematics, and the applications of this theory are chiefly in the area of gravitation.

➤ In this chapter we concentrate on the special theory of relativity (often referred to as *special relativity*). In the early 1900s the special theory of relativity was accepted with enthusiasm by some, but by many it was either reluctantly accepted or dismissed as folly. Today it is not only widely accepted, but embraced as a window into the workings of nature. We will see how this theory challenges our everyday experience of time and distance, as we describe the slowing down of moving clocks, the shortening of moving sticks, the relativity of

simultaneity for events that occur in different locations, and the relativity of momentum and energy relation.

R-1 The Principle of Relativity and the Constancy of the Speed of Light

The principle of relativity can be stated as follows:

It is impossible to devise an experiment that determines whether you are at rest or moving uniformly.

POSTULATE I, THE PRINCIPLE OF RELATIVITY

Moving uniformly means moving at constant velocity relative to an inertial reference frame. For example, suppose that you are in your seat on board a high-speed airplane moving uniformly relative to the surface of the earth. If you drop your fork, it will fall to the floor in exactly the same way that it would if the plane were parked on a runway. When the airplane is in flight, you can consider yourself and the airplane to be at rest and the surface of the earth below you to be moving. There is nothing to distinguish whether you and the plane are moving and the surface of the earth is at rest, or vice versa.

Any reference frame in which a particle with no forces acting on it moves with constant velocity is, by definition, an inertial reference frame.[†] The surface of the earth is, to a good approximation, an inertial reference frame. The airplane is also an inertial reference frame as long as it moves with constant velocity relative to the surface of the earth. As long as you remain seated or standing still on the airplane you can consider yourself and the airplane to be at rest and the surface of the earth to be moving, or you can consider the surface of the earth to be at rest and yourself and the airplane to be moving.

In the nineteenth century the existence of a preferred frame of reference that could be considered to be at rest was widely accepted. This was thought to be the reference frame of the *ether,* the medium filling all of space through which light was thought to propagate. (It was then accepted that light waves needed a medium to propagate through, just as it is now accepted that sound waves need air or some other material medium through which to propagate.) The ether was considered to be the preferred "at rest" reference frame.

A carefully devised series of measurements to measure the orbital speed of the earth relative to the ether were carried out in 1887 by Albert Michelson and Edward Morley. These measurements were considered challenging because the orbital speed of the earth is less than 1/10,000 the speed of light in vacuum. Much to the surprise of nearly everyone, the observations always found the speed of the earth relative to the ether to be zero. It was Albert Einstein who came up with a theory that was consistent with these observations. His explanation was that light is capable of traveling through empty space and that the ether was an unnecessary construct that did not exist. Einstein also postulated:

The speed of light is independent of the speed of the light source.

POSTULATE II

The *speed of light* refers to the speed at which light travels through the vacuum of empty space.

A consequence of Postulate II and the principle of relativity is that all inertial observers measure the same value for the speed of light. (An inertial observer

---

† Further discussion on reference frames can be found in Chapters 2 and 4.

is one that remains at rest in an inertial reference frame.) To establish that all inertial observers measure the same value for the speed of light, we consider inertial observers A and B, where observer A is moving relative to observer B. The principle of relativity states that it is impossible to devise an experiment that determines whether an inertial observer is at rest or moving uniformly. If observer A measures a different value for the speed of light than observer B, then observers A and B could not both consider themselves to be at rest—a result in direct contradiction with the principle of relativity. Thus, a consequence of both the principle of relativity and Postulate II (that the speed of light is independent of the speed of the source) leads to the **constancy of the speed of light:**

The speed of light c is the same in any inertial reference frame.

THE CONSTANCY OF THE SPEED OF LIGHT

That is, anything (not just light) that travels at speed c relative to one inertial reference frame travels at the same speed c relative to any inertial reference frame.

Suppose you are in your backyard here on earth and Bob is on a spaceship moving away from you at half the speed of light ($\frac{1}{2}c$). You point a flashlight in Bob's direction and turn it on. The light leaves the flashlight, traveling at speed c (relative to the flashlight), and passes by your neighbor Keisha who is standing on the roof of her house next door. Keisha measures the speed of the light going by and finds it to be traveling at speed c. A few minutes later the light travels past Bob and his spaceship. Like Keisha, Bob measures the speed of the light going by him and also finds it to be traveling at speed c. This surprises Bob because he expected the light to be traveling past him at speed $\frac{1}{2}c$ rather than at speed c—after all, Bob is moving at speed $\frac{1}{2}c$ relative to the source of the light (the flashlight in your backyard). Like many people, Bob finds the constancy of the speed of light to be counterintuitive. This leaves him with a dilemma. Should he trust his measuring instruments or trust his intuition? It turns out that it is Bob's intuition that needs adjusting, not his instruments. Bob must change his concepts of both space and time.

Suppose that instead of pointing a flashlight you point a high-speed particle beam in his direction, where by "high speed" we mean a speed very close to the speed of light c. (A particle such as an electron or proton cannot travel at the speed of light, but it can travel at speeds extremely close to the speed of light.) If Keisha measures the particles going by her to be traveling at $0.9999c$ (relative to her), then how fast will Bob measure the particles going by him? Bob's intuition tells him that, because he is moving away from the source of the particles, they will be traveling past him at the slower speed of $0.4999c$, but that is not the case. When Bob measures the speed of the particles (relative to him) he finds it to be extremely close to $0.9999c$. (The actual value is $0.9997c$.)[†]

We tend to think of distances between cities as fixed. However, this too is not the case. According to a certain road map the distance between Baltimore and Philadelphia is 160 km. However, if you travel from Baltimore to Philadelphia at a significant fraction of the speed of light, the distance between the two cities will be much shorter than it is if you travel at 100 km/h (62 mi/h). For someone driving at 100 km/h, the distance between Baltimore and Philadelphia is very close to 160 km. However, for someone traveling at a speed of $0.866c$ (relative to the earth's surface), the distance is only 80 km, and for someone traveling at $0.9999c$, the distance is only 2.2 km.

The fastest speed that a human being has ever traveled relative to the earth (which occurred during the Apollo missions to the moon) is only about 10 km/s = $3.3 \times 10^{-5}c$. This speed is so slow compared to the speed of light, that for someone traveling from Baltimore to Philadelphia at that speed, the distance between those cities would be shorter by less than the diameter of a human hair. The logic explaining how this is determined is presented in the next three sections.

† Relative velocity in special relativity is covered in Chapter 37.

R-2 Moving Sticks

We wish to show that if a stick moves perpendicular to its length, its length does not change. We do this by showing that any increase or decrease in length contradicts the principle of relativity. Showing that a stick does not change its length may seem mundane. However, we show it because an immediate consequence is that moving clocks run slow.

Suppose that we have two identical metersticks, stick A and stick B. We verify that the sticks are identical by placing them side by side in a reference frame in which they are both at rest and examining them visually. We then give stick B to Bob before he takes off on another trip in his spaceship. On this trip Bob makes sure to hold the stick at right angles to the velocity of the spaceship relative to the earth. Is stick B now shorter than stick A, which remains back on the earth with us?

To answer this question we conduct a thought experiment. We attach felt-tipped marking pens to stick A, one at the 20-cm mark, the other at the 80-cm mark. Then Bob and his spaceship execute a flyby during which Bob holds stick B out a porthole, keeping it at right angles to the ship's velocity. During the flyby we hold up our stick (stick A), keeping it parallel with stick B. As the sticks pass by each other, two marks are drawn on stick B (Figure R-1). Bob then returns to earth with stick B and the two sticks are again placed side by side (Figure R-2), and the distance between the two marks on stick B and the distance between the two marking pens on stick A are compared. Let us assume that a stick moving perpendicular to its length is shorter than is an identical stationary stick. Then the distance between the two pens will be less than the distance between the two marks (Figure R-2)—clear evidence that during the flyby the moving stick (stick B) was shorter than the stationary stick. However, according to the principle of relativity it is equally valid to think of stick B as stationary and stick A as moving during the flyby. From this perspective, the same evidence (Figure R-2) demonstrates that the moving stick—now stick A—is longer than the stationary stick. Thus our assumption—that a stick moving perpendicular to its length is shorter than an identical stationary stick—leads to a contradiction and must be rejected. The assumption that a stick moving perpendicular to its length is longer than is an identical stationary stick also leads to a contradiction, as can be shown using an analogous argument. Thus we conclude:

> A stick moving perpendicular to its length has the same length as an identical stationary stick.

This rule is established without any consideration of the material from which the two sticks are made. Thus, the rule does not reflect a property of sticks. Instead it reflects a property of space.

The frame of reference in which the stick is at rest is called its **proper frame** or **rest frame,** and the length of a stick in its proper reference frame is called its **proper length** or **rest length.**

R-3 Moving Clocks

Clocks are used to measure time. In this section we will show that clocks moving at high speeds run slow, so if a high-speed spaceship travels by us, we would observe that all the clocks on the ship run slower than our clocks. However, the people on the ship are free to consider themselves to be at rest and us to be moving, and they would observe our clocks to run slow compared to their clocks. Let's examine how these observations are consistent with the constancy of the speed of light and the principle of relativity.

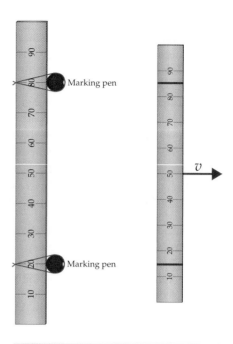

FIGURE R-1 During the flyby marks like this would be made on stick B by marking pens attached to stick A if the moving stick was shortened.

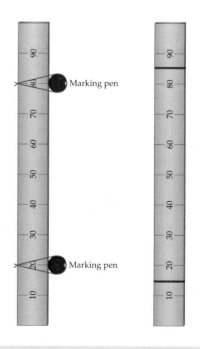

FIGURE R-2 If the distance between marks is greater than the distance between marker pens, this would demonstrate that stick B was shorter than stick A when the marks were made.

We construct a clock, called a *light clock*, using a stick of proper length L_0 and two mirrors (Figure R-3). The two mirrors face each other, and a light pulse bounces back and forth between them. Each time the light pulse strikes one of the mirrors, say the lower mirror, the clock is said to tick. Between successive ticks the light pulse travels a distance $2L_0$ in the proper reference frame of the clock. Thus the time between ticks T_0 is related to L_0 by

$$2L_0 = cT_0 \qquad \text{R-1}$$

Next we consider the time between ticks T of the same light clock, but this time we observe it from a reference frame in which the clock is moving with speed v perpendicular to the stick (Figure R-4). In this reference frame the clock moves a distance vT between ticks and the light pulse moves a distance cT between ticks. The distance the pulse moves in traveling from the bottom mirror to the top mirror is $\sqrt{L_0^2 + (\frac{1}{2}vT)^2}$. The light pulse travels the same distance in traveling from the top mirror to the bottom mirror. Thus,

$$2\sqrt{L_0^2 + (\tfrac{1}{2}vT)^2} = cT \qquad \text{R-2}$$

(Note that we have used the same symbol c for the speed of light in Equations R-1 and R-2.) Solving Equation R-1 for L_0 and substituting into Equation R-2 gives

$$\sqrt{(\tfrac{1}{2}cT_0)^2 + (\tfrac{1}{2}vT)^2} = \tfrac{1}{2}cT$$

Solving for T gives

$$T = \frac{T_0}{\sqrt{1 - (v^2/c^2)}} \qquad \text{R-3}$$

TIME DILATION

According to Equation R-3, the time between ticks in the reference frame in which the clock moves at speed v is greater than the time between ticks in the proper reference frame of the clock.

This raises the question, do other clocks run slow when they move with speed v according to Equation R-3 or is Equation R-3 valid only for light clocks? To answer this question we attach a conventional clock (with a conventional clock mechanism) to the lower mirror of the light clock (Figure R-5). The conventional clock has no minute or hour hands. Instead of a second hand it has an opaque disk with a narrow slot to indicate the time. The clock's face contains 60 equal-spaced marks (called tick marks) around its perimeter—one for each second. The clock ticks each time the slot passes over one of the tick marks. We adjust the length L_0 of the light-clock stick so that the time between ticks of both clocks is the same in the proper reference frame of the clocks. Next, we synchronize the clocks so each tick of the light clock occurs simultaneously with a tick of the conventional clock.

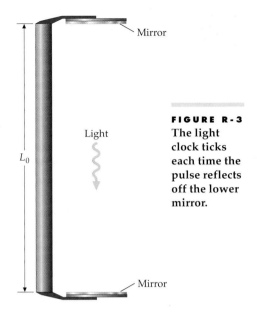

FIGURE R-3
The light clock ticks each time the pulse reflects off the lower mirror.

FIGURE R-4

FIGURE R-5

We then ask, "If the ticks of the two clocks occur simultaneously in the proper reference frame of the clocks, do they also occur simultaneously in a reference frame in which the clocks are moving at speed v?"

The answer is yes. To understand why this is so, consider the following experiment. In the proper reference frame of the clocks, the time between ticks of both clocks is exactly one second. A light-sensitive film is placed on the face of the conventional clock, behind the rotating disk. Each time the light pulse reflects off the lower mirror, the narrow region of the light-sensitive paper directly behind the slot gets exposed. These exposed regions will be aligned with the tick marks as shown in Figure R-6, and all observers must agree with this permanent record.

In reference frame A in which the clocks are both moving, the light pulse exposes the film behind the slot on the clock face each time the pulse reflects off the lower mirror. Because the light clock is moving, the time between these reflections is greater than 1 s, in accordance with Equation R-3. When an observer of reference frame A sees that the lines appearing where the film was exposed are aligned with the tick marks, she realizes that in her reference frame the conventional clock runs slow in exactly the same manner that the light clock runs slow—in accord with Equation R-3. Thus we conclude that all moving clocks run slow in exactly the same manner that a moving light clock runs slow. Because this is the case, we conclude that it is time itself that runs slow, a phenomenon known as **time dilation.**

Something that occurs at a specific instant in time and at a specific location in space is called a **spacetime event,** or just an **event.** Each reflection of the light pulse off the lower mirror of the light clock is a spacetime event. If we call one of these reflections event 1, and the next reflection event 2, then the time between events 1 and 2 in a frame of reference in which the two events occur at the same location is called the **proper time interval** T_0 between the two events. Let T be the time between the same two events in a reference frame in which they occur at different locations. Equation R-3 relates the time T between two events to the proper time T_0 between the same two events.

Each time the light pulse reflects off the lower mirror, the slot (second hand) of the conventional clock passes directly over a tick mark. In the proper frame of the two clocks, these two events—the arrival of the light pulse and the passing of the slot over a tick mark—occur at the same time *and* at the same place. Any two events that occur both at the same time and at the same place in one reference frame will occur both at the same time and at the same place in all reference frames. This is because such events can have lasting consequences—like producing lines on the light-sensitive film aligned with the tick marks on the clock face. We cannot have the marks aligned with the ticks marks in one reference frame and not aligned with the tick marks in another reference frame. After all, there is only one clock face and one set of marks. This conclusion can be generalized into a principle, called the **principle of invariance of coincidences:**

> If two events occur at the same time and at the same place in one reference frame, then they occur at the same time and at the same place in all reference frames.

PRINCIPLE OF INVARIANCE OF COINCIDENCES

We can better visualize this principle by considering two automobiles passing through an intersection at the same time. The two events are: (1) automobile A passes through the intersection and (2) automobile B passes through the intersection. If these two events occur at the same time in one reference frame, then they must occur at the same time in all reference frames. Either a fender becomes dented or it doesn't. That is, if the automobiles collide, then there is no question that the two cars were in the intersection at the same time. The lasting evidence dictates that observers in all reference frames must agree on this fact. Any pair of events that occur at the same time *and* at the same location are referred to as a **spacetime coincidence.**

Tick marks

Exposed lines

Conventional clock

FIGURE R-6

THE NAPPING ASTRONAUTS   **EXAMPLE R-1**   **Put It in Context**

Astronauts in a spaceship traveling at $v = 0.6c$ relative to the earth sign off from space control, saying that they are going to nap for 1 h and then will call back. How long does their nap last according to observers on the earth?

PICTURE THE PROBLEM Clock S on the ship reads t_0 when the nap begins (a spacetime coincidence) and reads $t_0 + 1$ h when the nap ends (also a spacetime co-incidence). Observers on the ship agree that, because clock S is stationary it does not run slow, so the nap lasted 1 h. In the reference frame of the ship the two events (the beginning of the nap and the end of the nap) occur at the same location, so the time interval between the events is the proper time interval between them. Observers on the earth agree that clock S reads t_0 when the nap begins and it reads $t_0 + 1$ h when the nap ends. However, the observers on the earth agree that because clock S is moving at speed v, it is running slow so the nap lasted more than 1 h. In the reference frame of the earth the ship is moving so the nap begins and ends at different locations. Therefore, in the reference frame of the earth the time interval between the events is not the proper time interval between the events.

1. Event 1 is the beginning of the nap and event 2 is the end of the nap. Clock S on the ship advances 1 h between these events. Determine the proper time interval T_0 between these events:

$$T_0 = 1 \text{ h}$$

2. Find the time interval T between events 1 and 2 for observers on earth:

$$T = \frac{T_0}{\sqrt{1 - (v^2/c^2)}} = 1 \text{ h}/\sqrt{1 - \frac{(0.6c)^2}{c^2}}$$

$$= \frac{1 \text{ h}}{\sqrt{1 - 0.36}} = \frac{1 \text{ h}}{\sqrt{0.64}} = \frac{1 \text{ h}}{0.8} = \boxed{1.25 \text{ h}}$$

REMARKS Clock S is an unnecessary construct as the astronauts themselves serve as clocks. What it is necessary to realize is that the proper time between the beginning and the end of the nap is 1 h, so the time T between the same events in a reference frame where the clocks (astronauts) are moving with speed v is given by Equation R-3.

EXERCISE A pion[†] has a mean proper lifetime of 26 ns (1 ns $= 1 \times 10^{-9}$ s) (measured when the pion is at rest). What is the mean lifetime if measured when the pion is moving at 0.995c? (*Answer*   260 ns)

EXERCISE A pion has a proper half life of 18 ns (measured when the pion is at rest). A beam of pions moving at 0.995c passes point P. How far from P do the pions travel before only half of the pions in the beam remain? (*Answer*   38 m)

R-4 Moving Sticks Again

In Section R-2 the length of a stick moving perpendicular to its length and the length of an identical stationary stick are compared and found to be equal. However, the technique used for this comparison works only if the velocity of the moving stick is perpendicular to its length. Here we apply a different technique to compare the length of a stick at rest to its length when it is moving parallel to its velocity.

    A light clock is shown in its proper frame in Figure R-7. This clock ticks each time the light pulse reflects off the mirror on the left. In its proper reference frame the length of the clock is L_0 and the time between ticks is $T_0 = 2L_0/c$ (Equation R-1). To find the

FIGURE R-7

Mirror   Mirror

Light

L_0

---

† A pion (short for pi meson) is a subatomic particle.

length of the clock in a reference frame in which it is moving to the right at speed v we consider three sequential events:

Event 0   Light pulse reflects off the mirror at the left end.

Event 1   Light pulse reflects off the mirror at the right end.

Event 2   Light pulse reflects off at the mirror at the left end.

In Figure R-8 the clock is shown at the time of each of these events in a reference frame in which the clocks move to the right with speed v. (The clocks are drawn lower down the page at later times to avoid visual overlap.) The times of occurrence for events 0, 1, and 2 in this reference frame are t_0', t_1', and t_2', respectively. In the time between events 0 and 1 the clock moves a distance $v(t_1' - t_0')$ and the light pulse travels a distance $c(t_1' - t_0')$. Thus,

$$c(t_1' - t_0') = L + v(t_1' - t_0') \qquad \text{R-4}$$

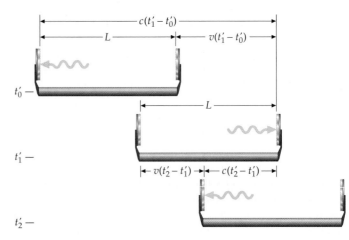

FIGURE R-8 A light clock moving to the right at speed v is shown at 3 instants of time.

In the time between events 1 and 2 the clock moves a distance $v(t_2' - t_1')$ and the light pulse travels a distance $c(t_2' - t_1')$, so

$$c(t_2' - t_1') = L - v(t_2' - t_1') \qquad \text{R-5}$$

Eliminating t_1' by solving Equation R-4 for t_1', substituting the result into Equation R-5, and then solving for $t_2' - t_0'$ gives

$$t_2' - t_0' = \frac{2L/c}{1 - (v^2/c^2)} \qquad \text{R-6}$$

The time interval $t_2' - t_0'$ is related to the proper time interval $t_2 - t_0$ between events 0 and 2 (Equation R-3) by

$$t_2' - t_0' = \frac{t_2 - t_0}{\sqrt{1 - (v^2/c^2)}} \qquad \text{R-7}$$

where $t_2 - t_0 = 2L_0/c$ (Equation R-1). Substituting $2L_0/c$ for $t_2 - t_0$ gives

$$t_2' - t_0' = \frac{2L_0/c}{\sqrt{1 - (v^2/c^2)}} \qquad \text{R-8}$$

Equating the right sides of Equations R-6 and R-8, and then solving for L gives

$$L = L_0\sqrt{1 - (v^2/c^2)} \qquad \text{R-9}$$

LENGTH CONTRACTION

Establishing this result did not involve any properties of the stick. Thus Equation R-9 reflects the nature of space and time, and not the nature of the sticks.

R-5 Distant Clocks and Simultaneity

We have established three useful relations: (1) that the length of a stick moving perpendicular to its length is the same as its rest length; (2) that the time T between two ticks of a moving clock is greater than the time between the two ticks of the same clock for an observer moving with the clock according to $T = T_0/\sqrt{1 - (v^2/c^2)}$; and (3) that the length L of a stick moving parallel to its length is less than its rest length L_0 according to $L = L_0\sqrt{1 - (v^2/c^2)}$. But in order to analyze events from the perspective of observers in reference frames moving at different velocities, we need one more relation, one that concerns the readings on clocks at different locations.

Clocks A and B (Figure R-9a) are at rest relative to each other, and in their rest frame the clocks are separated by a distance L_0. To synchronize these clocks there is a flash lamp on clock A and a light sensitive film on the face of clock B. The alarm on clock A is set to energize the flash lamp when the second hand on clock A passes zero. Like the conventional clock in Section R-3, the second hand on clock B is a rotating opaque disk with a slot to indicate the time. Behind the disk is a light-sensitive film. When the light from the flash reaches clock B, the film is illuminated on the narrow region behind the slot. This provides a lasting record of the reading on clock B when the light from the flash lamp reaches it. Let this reading be t_1. In the rest frame of the clocks, the time for the light to travel at speed c from clock A to clock B is L_0/c, so when the light arrives at clock B, clock A reads L_0/c and clock B reads t_1. To synchronize the two clocks we turn clock B back by $\Delta t = t_1 - L_0/c$.

With the two clocks synchronized in their rest frame (frame 1), we then determine whether they are also synchronized in a reference frame (frame 2) in which they are moving at speed v parallel to the line joining them, shown in Figure R-9b. We re-set the alarm to energize the flash lamp when clock A next reads zero. These two events—clock A reads zero and the lamp flashes—are a spacetime coincidence, so we know they occur simultaneously in all reference frames. Also, the light reaching clock B and clock B reading L_0/c, are a spacetime coincidence, so we know they occur simultaneously in all reference frames.

In frame 2, the distance L between the clocks is given by

$$L = L_0\sqrt{1 - (v^2/c^2)}$$

and clock B is moving toward the flash lamp. In this frame the light traveling from clock A to clock B travels a distance $L - vt$, where t is the time required for the light to travel the distance. The time t, the distance L, and the speed v are related by

$$ct = L - vt$$

Solving for the time gives $t = L/(c + v)$.

Moving clocks run slow, so during time t the readings on both clocks advance not by $L/(c + v)$ but by

$$\frac{L}{c + v}\sqrt{1 - (v^2/c^2)} = \frac{L_0}{c + v}\left(1 - \frac{v^2}{c^2}\right) = \frac{L_0}{(c + v)}\frac{(c + v)(c - v)}{c^2} = \frac{L_0}{c} - \frac{vL_0}{c^2}$$

Thus, when the light arrives at clock B, clock B reads L_0/c and clock A reads $L_0/c - vL_0/c^2$. Therefore, in frame 2 clock B is ahead of clock A by vL_0/c^2:

If two clocks are synchronized in their rest frame, then in a frame where they move with speed v parallel to the line joining them the clock in the rear is ahead of the clock in the front by vL_0/c^2.

THE RELATIVITY OF SIMULTANEITY

In this case L_0 is the distance between the clocks in their rest frame. It is also true that if two clocks are synchronized in their rest frame, they are also synchronized in a frame in which they move with speed v perpendicular to the line joining them. This follows from the symmetry of the situation. (For one thing, there is no way to state a rule specifying which of the two clocks is ahead.)

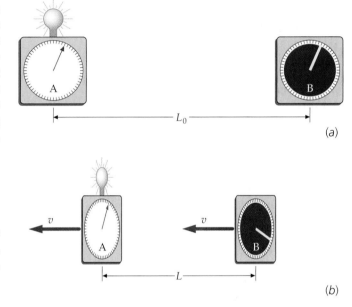

(a)

(b)

FIGURE R-9 (a) The clocks are synchronized in the reference frame in which they are at rest. (b) Are the clocks synchronized in the reference frame in which they are moving with speed v parallel to the line joining them?

R-6 Applying the Rules

EXAMPLE R-2

FIGURE R-10

A high-speed train is about to enter a tunnel through a mountain. The tunnel has a proper length of 1.2 km. The length of the train in the reference frame of the mountain is also 1.2 km, and the proper length of the train is 2.0 km. Clock A is attached to the mountain at the entrance to the tunnel, and clock B is attached to the mountain at the exit to the tunnel. In the reference frame of the mountain, when the front of the train enters the tunnel both clocks read zero. (a) In the reference frame of the mountain, what is the speed of the train and what is the reading of both clocks at the instant the front of the train exits the tunnel (Figure R-10a)? (b) In the reference frame of the train, what is the length of the tunnel, what is the reading of both clocks at the instant the front of the train enters the tunnel (Figure R-10b), and what is the reading of both clocks at the instant the front of the train exits the tunnel? (c) For a passenger on the train, how long does it take for the front of the train to pass through the tunnel?

PICTURE THE PROBLEM The speed of the train and the length of the train are related by the length-contraction formula. Some of the clock readings in the two reference frames can be equated because they are event pairs that form spacetime coincidences. Other clock readings can be related by the relativity–of–simultaneity relation.

(a) 1. Using the length-contraction formula, solve for the speed of the train:

$$L = L_0\sqrt{1 - (v^2/c^2)}$$

$$1.2 \text{ km} = 2.0 \text{ km}\sqrt{1 - (v^2/c^2)}$$

so

$$v = 0.8c = 0.8(3.00 \times 10^8 \text{ m/s}) = \boxed{2.4 \times 10^8 \text{ m/s}}$$

2. The length of the tunnel equals its proper length and because the clocks are not moving they do not run slow. The reading on both clocks is the time t that it takes for the front of the train to travel the length of the tunnel:

$$L_{\text{tunnel},0} = vt$$

so

$$t = \frac{L_{\text{tunnel},0}}{v} = \frac{1.2 \times 10^3 \text{ m}}{2.4 \times 10^8 \text{ m/s}} = 5 \times 10^{-6} \text{ s} = 5 \text{ } \mu\text{s}$$

3. The clocks are synchronized, so when the front of the train exits the tunnel both clocks read 5 μs:

Clock A reading = Clock B reading = $\boxed{5 \text{ } \mu\text{s}}$

(b) 1. In this frame the mountain is moving at 0.8c. Using the length-contraction formula, solve for the length of the tunnel:

$$L_{\text{tunnel}} = L_{\text{tunnel},0}\sqrt{1 - (v^2/c^2)} = 1.2 \text{ km}\sqrt{1 - \frac{(0.8c)^2}{c^2}}$$

$$= 1.2 \text{ km}\sqrt{1 - 0.8^2} = \boxed{0.72 \text{ km} = 720 \text{ m}}$$

2. The front of the train entering the tunnel and a zero reading on clock A are a spacetime coincidence:

$$\boxed{\text{Clock A reads zero}}$$

3. The two clocks are moving toward the train with clock B in the rear, so clock B is ahead of clock A by vL_0/c^2. When the train enters the tunnel clock A reads zero, so clock B reads vL_0/c^2:

$$\frac{vL_{\text{tunnel},0}}{c^2} = \frac{0.8cL_{\text{tunnel},0}}{c^2} = \frac{0.8L_{\text{tunnel},0}}{c}$$

$$= \frac{0.8(1.2 \times 10^3\ \text{m})}{3.0 \times 10^8\ \text{m/s}} = 3.2\ \mu s$$

$$\text{Clock B reading} = \boxed{3.2\ \mu s}$$

4. The front of the train exiting the tunnel and a 5-μs reading of clock B are a spacetime coincidence:

$$\text{Clock B reading} = \boxed{5\ \mu s}$$

5. Clock B is in the rear, so clock A lags behind clock B by vL_0/c^2.

$$\text{Clock A reading} = \text{Clock B reading} - \frac{vL_{\text{tunnel},0}}{c^2}$$

$$= 5\ \mu s - 3.2\ \mu s = \boxed{2.6\ \mu s}$$

(c) For an observer in the reference frame of the train, the mountain is traveling at 0.8c and the tunnel is 720-m long:

$$L_{\text{tunnel}} = vt$$

so

$$t = \frac{L_{\text{tunnel}}}{v} = \frac{L_{\text{tunnel}}}{0.8c} = \frac{720\ \text{m}}{2.4 \times 10^8\ \text{m/s}} = \boxed{3\ \mu s}$$

MASTER the CONCEPT   **REMARKS** In the frame of reference of the mountain, the length of the train and the length of the tunnel are both 1.2 km. In the frame of reference of the train, the length of the train is 2.0 km and the length of the tunnel is 720 m. In the frame of reference of the mountain, when both clocks read 5 μs the entire train is in the tunnel. In the reference frame of the train, the train is longer than the tunnel so at no time is the entire train within the tunnel.

EXERCISE Event 1 is the front of the train entering the tunnel, and event 2 is the front of the train exiting the tunnel. (a) In which reference frame do these two events occur at the same location? (b) What is the proper time interval between events 1 and 2? (*Answer* (a) The reference frame of the train, because both events occur at the front end of the train. (b) 3 μs)

It is often convenient to measure large distances in light-years, where a light-year is the distance traveled when traveling at the speed of light for a time of one year. That is,

1 light-year = 1 $c \cdot y$

where 1 $c \cdot y$ = c (1 y). This notation is particularly convenient when distances are divided by speeds. For example, the time T for a particle traveling at $v = 0.1c$ to travel a distance of $L = 25$ light-years is

$$T = \frac{L}{v} = \frac{25\ c \cdot y}{0.1c} = 250\ y$$

where the c's cancel.

EXPLORING

PDF files of two worked-out examples on the twin paradox are available at www.whfreeman.com/tipler5e.

EXERCISE In the reference frame of the earth, it takes 8 minutes for light to travel from the sun to the earth, so the distance between the sun and the earth is 8 $c \cdot$min. How many minutes does it take a particle from the sun to reach the earth if the particle travels at 0.1c? (*Answer* (8 $c \cdot$min)/0.1c = (8 min)/0.1 = 80 min)

R-7 Relativistic Momentum, Mass, and Energy

Momentum and Mass

In special relativity both momentum and energy are conserved, just as they are in classical physics. However, there are differences that have to be accounted for. The momentum of a particle moving with velocity v is given by

$$p = \frac{mv}{\sqrt{1 - (v^2/c^2)}} \qquad \text{R-10}$$

RELATIVISTIC MOMENTUM

where m is the mass of the particle. Equation R-10 is sometimes written $p = m_r v$, where m_r is called the relativistic mass: $m_r = m/\sqrt{1 - (v^2/c^2)}$. In the rest frame of the particle, $v = 0$ and the relativistic mass equals m. (The mass m is sometimes called the rest mass to differentiate it from the relativistic mass.) Relativistic momentum and mass are discussed further in Chapter 39.

Energy

In relativistic mechanics, as in classical mechanics, the net force on a particle is equal to the time rate of change of the momentum of the particle. Considering one-dimensional motion only, we have

$$F_{net} = \frac{dp}{dt} \qquad \text{R-11}$$

We wish to find an expression for the kinetic energy. To do this we will multiply both sides of Equation R-11 by the displacement ds. This gives

$$F_{net}\, ds = \frac{dp}{dt} ds \qquad \text{R-12}$$

where we identify the term on the left as the work and the term on the right as the change in kinetic energy dK. Substituting $v\, dt$ for ds in the term on the right we obtain

$$dK = \frac{dp}{dt} v\, dt = v\, dp$$

Integrating both sides gives

$$K = \int v\, dp \qquad \text{R-13}$$

This integral is evaluated in Chapter 37. The result is

$$K = \frac{mc^2}{\sqrt{1 - (v^2/c^2)}} - mc^2 \qquad \text{R-14}$$

Defining $\dfrac{mc^2}{\sqrt{1 - (v^2/c^2)}}$ as the **total relativistic energy** E gives

$$E = K + mc^2 = \frac{mc^2}{\sqrt{1 - (v^2/c^2)}} \qquad \text{R-15}$$

where $m_0 c^2$, called the **rest energy** E_0, is energy the particle has when it is at rest.

By multiplying both sides of Equation R-10 by c and then dividing the resulting equation by Equation R-15 we obtain

$$\frac{v}{c} = \frac{pc}{E}$$

R-16

which can be useful when trying to solve for the speed v. Eliminating v from Equations R-10 and R-16, and solving for E^2 (see Problem R-41) gives

$$E^2 = p^2c^2 + m^2c^4$$

R-17

The relation between mass and energy is discussed in Section 3 of Chapter 7.

SUMMARY

Topic	Relevant Equations and Remarks
1. Postulates of Special Relativity	
Postulate I: Principle of relativity	It is impossible to devise an experiment that determines whether you are at rest or moving uniformly, where moving uniformly means moving at constant velocity relative to an inertial reference frame.
Postulate II	The speed of light is independent of the speed of the source.
Constancy of the speed of light	It follows that the speed of light is the same in any inertial reference frame.
2. Moving Sticks	The length of a stick moving perpendicular to its length is equal to its proper length.
	The length of a stick moving with speed v parallel to its length is shorter than its proper length L_0 according to $$L = L_0\sqrt{1 - (v^2/c^2)} \qquad \text{R-9}$$
3. Moving Clocks	
Time dilation	The time between ticks of a clock moving with speed v is longer than the proper time between ticks of the same clock by $$T = \frac{T_0}{\sqrt{1 - (v^2/c^2)}} \qquad \text{R-3}$$
Relativity of simultaneity	If two clocks are synchronized in their rest frame, in a frame where they move with speed v parallel to the line joining them the clock in the rear is ahead of the clock in front by vL_0/c^2, where L_0 is the distance between them in their rest frame.
	If two clocks are synchronized in their rest frame they are also synchronized in a frame where they move with speed v perpendicular to the line joining them.
4. Spacetime Coincidence	If two events occur both at the same time and at the same place in one reference frame, they occur both at the same time and at the same place in any reference frame.
5. Momentum, Mass, and Energy	
Momentum	The momentum of a particle is given by $$p = \frac{mv}{\sqrt{1 - (v^2/c^2)}} \qquad \text{R-10}$$

Mass and energy	The total relativistic energy of a particle equals its rest energy plus its kinetic energy.	
	$$E = K + mc^2 = \frac{mc^2}{\sqrt{1 - (v^2/c^2)}}$$	R-15
	where mc^2 is the rest energy E_0.	

Momentum and Energy	$\dfrac{v}{c} = \dfrac{pc}{E}$ and $E^2 = p^2c^2 + m^2c^4$	R-16, R-17

PROBLEMS

- Single-concept, single-step, relatively easy
- •• Intermediate-level, may require synthesis of concepts
- ••• Challenging
- SSM Solution is in the *Student Solutions Manual*
- ISOLVE Problems available on iSOLVE online homework service
- ISOLVE✓ These "Checkpoint" online homework service problems ask students additional questions about their confidence level, and how they arrived at their answer

In a few problems, you are given more data than you actually need; in a few other problems, you are required to supply data from your general knowledge, outside sources, or informed estimates.

Conceptual Problems

1 • You are standing on a corner and a friend is driving past in an automobile. Each of you is wearing a wrist watch. Both of you note the times when the car passes two different intersections and determine from your watch readings the time that elapses between the two events. Which of you has determined the proper time interval?

2 • SSM If event A occurs before event B in some frame, might it be possible for there to be a reference frame in which event B occurs before event A?

3 • Two events are simultaneous in a frame in which they also occur at the same point in space. Are they simultaneous in all other reference frames?

4 •• Two inertial observers are in relative motion. In what circumstances can they agree on the simultaneity of two different events?

5 • The approximate total energy of a particle of mass $m = m_0$ moving at speed $v \ll c$ is (a) $mc^2 + \frac{1}{2}mv^2$, (b) mv^2, (c) cmv, (d) $\frac{1}{2}mc^2$, (e) cmv.

6 • SSM True or false:
(a) The speed of light is the same in all reference frames.
(b) Proper time is the shortest time interval between two events.
(c) Absolute motion can be determined by means of length contraction.
(d) The light-year is a unit of distance.
(e) For two events to form a spacetime coincidence they must occur at the same place.
(f) If two events are not simultaneous in one frame, they cannot be simultaneous in any other frame.

Estimation and Approximation

7 •• The satellites used for the Global Positioning System (GPS) have extremely accurate clocks on board. In fact, the positioning system is so accurate that the relativistic time dilation of the clocks with respect to earth-bound observers must be taken into account. If the GPS satellite orbits have an average radius of about 26,000 km, estimate the time-dilation factor of the satellite clocks with respect to an earth-bound observer.

8 •• It is said that exercising by riding a bike will extend your life. From the standpoint of relativity, this is certainly true. Estimate by what factor your life will be extended through relativistic time dilation with respect to a stationary observer if you ride a bike as part of your regular exercise program. Make any assumptions that you feel are reasonable, but be certain to state them in your answer.

9 •• SSM ISOLVE✓ In 1975, an airplane carrying an atomic clock flew back and forth for 15 h at an average speed of 140 m/s as part of a time-dilation experiment. The time on the clock was compared to the time on an atomic clock kept on the ground. How much time did the airborne clock "lose" with respect to the clock on the ground?

Length Contraction and Time Dilation

10 • ISOLVE✓ The proper mean lifetime of subnuclear particles called pions is 2.6×10^{-8} s. A beam of pions has a speed of $0.85c$ relative to a laboratory. (a) What would be their mean lifetime as measured in the laboratory? (b) How far would they travel, on average, before they decay? (c) What would be your answer to Part (b) if you neglect time dilation?

11 • ISOLVE (a) In the reference frame of the pion in Problem 10, how far does the laboratory travel in a typical lifetime of 2.6×10^{-8} s? (b) What is this distance in the laboratory's frame?

12 • SSM ISOLVE ✓ The proper mean lifetime of a subnuclear particle called a muon is $2 \ \mu s$. Muons in a beam are traveling at $0.999c$ relative to a laboratory. (a) What is their mean lifetime as measured in the laboratory? (b) How far do they travel, on average, before they decay?

13 • ISOLVE (a) In the reference frame of the muon in Problem 12, how far does the laboratory travel in a typical lifetime of $2 \ \mu s$? (b) What is this distance in the laboratory's frame?

14 • You have been posted to a remote region of space to monitor traffic. Toward the end of a quiet shift, a spacecraft goes by and you measure its length using a laser device, which reports a length of 85 m. You flip open your handy reference catalogue and identify the craft as a CCCNX-22, which has a proper length of 100 m. When you phone in your report, what speed should you give for this spacecraft?

15 • SSM A spaceship travels from earth to a star 95 light-years away at a speed of 2.2×10^8 m/s. How long does it take to get there (a) as measured on the earth and (b) as measured by a passenger on the spaceship?

16 • The average lifetime of a beam of subnuclear particles called pions traveling at high speed is measured to be 7.5×10^{-8} s. Their average lifetime when measured at rest is 2.6×10^{-8} s. How fast is the pion beam traveling?

17 • A meterstick moves with speed $0.8c$ relative to you in the direction parallel to the stick. (a) Find the length of the stick as measured by you. (b) How long does it take for the stick to pass you?

18 • The half life of charged subnuclear particles called pions in their rest frame is 1.8×10^{-8} s (that is, in the rest frame of the pions, if there are N pions at time $t = 0$, there will be only $N/2$ pions at time $t = 1.8 \times 10^{-8}$ s). Pions are produced in an accelerator and emerge with a speed of $0.998c$. How far do these particles travel in the accelerator laboratory before half of them have decayed?

19 •• ISOLVE Your friend, who is the same age as you, travels to the star Alpha Centauri, which is 4 light-years away, and returns immediately. He claims that the entire trip took just 6 y. How fast did he travel?

20 •• SSM Two spaceships pass each other traveling in opposite directions. A passenger in ship A, who happens to know that her ship is 100 m long, notes that ship B is moving with a speed of $0.92c$ relative to A and that the length of B is 36 m. What are the lengths of the two spaceships as measured by a passenger in ship B?

21 •• Supersonic jets achieve maximum speeds of about $3 \times 10^{-6}c$. (a) By what percentage would a jet traveling at this speed contract in length? (b) During a time of $1 \ y = 3.15 \times 10^7$ s on your clock, how much time would elapse on the pilot's clock? How many minutes are lost by the pilot's clock in 1 y of your time?

The Relativity of Simultaneity

Problems 22 through 26 refer to the following situation: Mary is a worker on a large space platform. She places a clock A at point A and clock B at point B, which is 100 light-minutes from point A (Figure R-12). She also places a flashbulb at a point midway between points A and B. Jamal, a worker on a different platform, is standing next to clock C. Each clock immediately starts at zero when the flash reaches it. Mary's platform moves at speed of $0.6c$ relative to Jamal's. As Mary's platform passes by, clock B, then the flashbulb, and then clock A pass directly over clock C—just missing it as they go by. As the flashbulb passes next to clock C, it flashes and clock C immediately starts at zero.

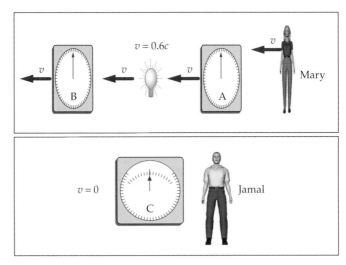

FIGURE R-12

22 • SSM According to Jamal, (a) what is the distance between the flashbulb and clock A, (b) how far does the flash travel to reach clock A, and (c) how far does clock A travel while the flash is traveling from the flashbulb to it?

23 •• According to Jamal, how long does it take the flash to travel to clock A, and what does clock C read as the flash reaches clock A.

24 •• Show that clock C reads 100 min as the light flash reaches clock B, which is traveling away from clock C with speed $0.6c$.

25 •• According to Jamal, the reading on clock C advances from 25 min to 100 min between the reception of the flashes by clocks A and B in Problems 23 and 24. Again according to Jamal, how much will the reading on clock A advance during this 75-min interval?

26 •• SSM The advance of clock A calculated in Problem 25 is the amount that clock A leads clock B according to Jamal. Compare this result with vL_0/c^2, where $v = 0.6c$.

27 •• In inertial reference frame S, event B occurs $2 \ \mu s$ after event A, which occurs 1.5 km from event A. How fast must an observer be moving along the line joining the two events so that the two events occur simultaneously? For an observer traveling fast enough is it possible for event B to precede event A?

28 •• A large flat space platform has an x axis painted on it. A firecracker explodes on the x axis at $x_1 = 480$ m, and a second firecracker explodes on the x axis 5 μs later at $x_2 = 1200$ m. In the reference frame of a train traveling alongside the x axis at speed v relative to the platform, these two explosions occur at the same place in space. What is the separation in time between the two explosions in the reference frame of the train?

29 • **ISOLVE** Herb and Randy are twin jazz musicians who perform as a trombone–saxophone duo. At the age of twenty, however, Randy got an irresistible offer to join a road trip to perform on a star 15 light-years away. To celebrate his good fortune, he bought a new vehicle for the trip—a deluxe space-coupe that travels at $0.999c$. Each of the twins promises to practice diligently, so they can reunite afterward. However, Randy's gig goes so well that he stays for a full 10 y before returning to Herb. After their reunion, (a) how many years of practice will Randy have had; (b) how many years of practice will Herb have had?

30 •• **SSM** Al and Bert are twins. Al travels at $0.6c$ to Alpha Centauri (which is 4 c·y from the earth as measured in the reference frame of the earth) and returns immediately. Each twin sends the other a light signal every 0.01 y as measured in his own reference frame. (a) At what rate does Bert receive signals as Al is moving away from him? (b) How many signals does Bert receive at this rate? (c) How many total signals are received by Bert before Al has returned? (d) At what rate does Al receive signals as Bert is receding from him? (e) How many signals does Al receive at this rate? (f) How many total signals are received by Al? (g) Which twin is younger at the end of the trip and by how many years?

Relativistic Energy and Momentum

31 • **ISOLVE** Find the ratio of the total energy to the rest energy of a particle of rest mass m_0 moving with speed (a) $0.1c$, (b) $0.5c$, (c) $0.8c$, and (d) $0.99c$.

32 • **ISOLVE**✓ A proton (rest energy 938 MeV) has a total energy of 1400 MeV. (a) What is its speed? (b) What is its momentum?

33 • **SSM** How much energy would be required to accelerate a particle of mass m from rest to (a) $0.5c$, (b) $0.9c$, and (c) $0.99c$? Express your answers as multiples of the rest energy.

34 • If the kinetic energy of a particle equals its rest energy, what error is made by using $p = mv$ for its momentum?

35 • What is the total energy of a proton whose momentum is $3mc$?

36 •• **SSM** Using a spreadsheet program or graphing calculator, make a graph of the kinetic energy of a particle with mass $m = 100$ MeV/c^2 for speeds between 0 and c. On the same graph, plot $\frac{1}{2}mv^2$ by way of comparison. Using the graph, estimate at about what velocity this formula is no longer a good approximation to the kinetic energy. As a suggestion, plot the energy in units of MeV and the velocity in the dimensionless form v/c.

37 •• Derive the equation $E^2 = p^2c^2 + m^2c^4$ (Equation R-17) by eliminating v from Equations R-10 and R-16.

38 •• Use the binomial expansion and Equation R-17 to show that when $pc \ll mc^2$, the total energy is given approximately by $E \approx mc^2 + p^2/(2m)$

39 •• (a) Show that the speed v of a particle of mass m and total energy E is given by

$$\frac{v}{c} = \left[1 - \frac{(mc^2)^2}{E^2}\right]^{1/2}$$

and that when E is much greater than mc^2, this can be approximated by

$$\frac{v}{c} \approx 1 - \frac{(mc^2)^2}{2E^2}.$$

Find the speed of an electron with kinetic energy of (b) 0.51 MeV and (c) 10 MeV.

40 •• **SSM** The rest energy of a proton is about 938 MeV. If its kinetic energy is also 938 MeV, find (a) its momentum and (b) its speed.

41 •• What percentage error is made in using $\frac{1}{2}m_0v^2$ for the kinetic energy of a particle if its speed is (a) $0.1c$ and (b) $0.9c$?

General Problems

42 • A spaceship departs from Earth for the star Alpha Centauri, which is 4 light-years away in the reference frame of Earth. The spaceship travels at $0.75c$. How long does it take to get there (a) as measured on Earth and (b) as measured by a passenger on the spaceship?

43 • The total energy of a particle is three times its rest energy. (a) Find v/c for the particle. (b) Show that its momentum is given by $p = \sqrt{8}mc$.

44 • **ISOLVE** A subnuclear particle called a muon has a mean lifetime of 2 μs when stationary. If you measure the mean lifetime of the muons coming out of a nuclear reactor port to be 46 μs, how fast are they moving?

45 •• **SSM** The rest mass of the neutrino (the "ghost particle" of physics) is known to have a small but as yet unmeasured value. Both high-energy neutrinos and light are produced by a supernova explosion, so it is possible to estimate the neutrino's mass by timing the relative arrival of light versus neutrinos from a supernova. If a supernova explodes 100,000 light-years from the earth, calculate the rest mass necessary for a neutrino with total energy of 100 MeV to trail the light by (a) 1 min, (b) 1 s, and (c) 0.01 s.

46 • **ISOLVE**✓ Relative to you, how fast must a meterstick travel in the direction parallel to itself so that its length as measured by you is 50 cm?

47 ••• **SSM** Keisha and Ernie are trying to fit a 15-ft-long ladder into a 10-ft-long shed with doors at each end. Recalling her physics lessons, Keisha suggests to Ernie that they open the front door to the shed and have Ernie run toward it with the ladder at a speed such that the length contraction of the ladder shortens it enough so that it fits in the shed. As soon as the back end of the ladder passes through the door, Keisha will slam it shut. (a) What is the minimum speed at which Ernie must run to fit the ladder into the shed? Express it as a fraction of the speed of light. (b) As Ernie runs toward the shed at a speed of $0.866c$, he realizes that in the reference frame of the ladder and himself, it is the shed which is shorter, not the ladder. How long is the shed in the rest frame of the ladder? (c) In the reference frame of the ladder is there any instant that both ends of the ladder are simultaneously inside the shed? Examine this from the point of view of relativistic simultaneity.

Gravity

ON A CLEAR AUTUMN NIGHT YOU ARE ON THE ROOF OF YOUR APARTMENT BUILDING IN NEW YORK CITY, TALKING ON YOUR CELL PHONE TO YOUR FRIEND IN KANSAS CITY, WHEN YOU SEE THE INTERNATIONAL SPACE STATION (ISS) PASS DIRECTLY OVERHEAD.

? **How might you use your understanding of gravity to determine when the ISS will next pass directly over Kansas City? (See Example 11-3.)**

ravity is the weakest of the four basic forces. It is negligible in the interactions of elementary particles and thus plays no role in the behavior of molecules, atoms, and nuclei. The gravitational attraction between objects of the size we ordinarily encounter, for example, the gravitational force of attraction exerted by a building on a car, is too small to be readily noticed. Yet when we consider objects of astronomical size, such as moons, planets, and stars, gravity is of primary importance. The gravitational force exerted by the earth on us and on the objects around us is a fundamental part of our experience. It is gravity that binds us to the earth and keeps the earth and the other planets on course within the solar system. The gravitational force plays an important role in the life history of stars and in the behavior of galaxies. On the largest of all scales, it is gravity that controls the evolution of the universe.

At the time of Newton many believed that nature followed different rules in the larger universe than here on earth. Newton's law of universal gravity, along with his three laws of motion, revealed that nature follows the same rules everywhere, and this revelation has had a profound effect on our view of the universe. ➤ **In this chapter we use the tools of conservation of angular momentum, conservation of energy, Newton's laws of motion, and Newton's law of gravity to predict the motion of the planets and other celestial bodies, including those we have put there.**

11-1 Kepler's Laws

The nighttime sky with its myriad stars and shining planets has always fascinated us on earth. Toward the end of the sixteenth century, the astronomer Tycho Brahe studied the motions of the planets and made observations that were considerably more accurate than those previously available. Using Brahe's data, Johannes Kepler discovered that the paths of the planets about the sun are ellipses (Figure 11-1). He also showed that each planet moves faster when its orbit brings it closer to the sun and slower when its orbit takes it farther away. Finally, Kepler developed a precise mathematical relation between the period of a planet and its average distance from the sun (see Table 11.1). He stated these results in three empirical laws of planetary motion. Ultimately, these laws provided the basis for Newton's discovery of the law of gravity. Kepler's three laws follow.

A mechanical model of the solar system, called an orrery, in the collection of Historical Scientific Instruments at Harvard University.

> Law 1. All planets move in elliptical orbits with the sun at one focus.

An ellipse is the locus of points for which the sum of the distances from two foci F is constant, as shown in Figure 11-2. Figure 11-3 shows a planet following an elliptical path with the sun at one focus. The earth's orbit is nearly circular, with the distance to the sun at perihelion (closest point) being 1.48×10^{11} m and at aphelion (farthest point) being 1.52×10^{11} m. The semimajor axis equals the average of these two distances, which is 1.50×10^{11} m (93 million miles) for the earth's orbit. The mean earth–sun distance defines the astronomical unit (AU):

$$1 \text{ AU} = 1.50 \times 10^{11} \text{ m} = 93.0 \times 10^6 \text{ mi} \qquad \text{11-1}$$

The AU is used frequently in problems dealing with the solar system.

TABLE 11-1

Mean Orbital Radii and Orbital Periods for the Planets

Planet	Mean Radius r ($\times 10^{10}$ m)	Period T (y)
Mercury	5.79	0.241
Venus	10.8	0.615
Earth	15.0	1.00
Mars	22.8	1.88
Jupiter	77.8	11.9
Saturn	143	29.5
Uranus	287	84
Neptune	450	165
Pluto	590	248

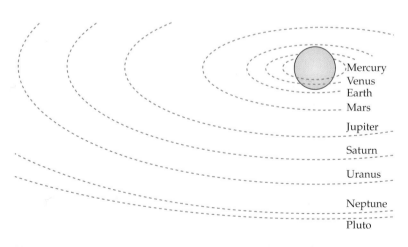

FIGURE 11-1 Orbits of the planets around the sun.

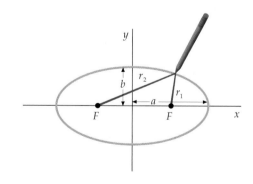

FIGURE 11-2 An ellipse is the locus of points for which $r_1 + r_2 = $ constant. The distance a is called the *semimajor* axis, and b is the *semiminor* axis. You can draw an ellipse with a piece of string by fixing each end at a focus F and using it to guide the pencil. Circles are special cases in which the two foci coincide.

Law 2. A line joining any planet to the sun sweeps out equal areas in equal times.

Figure 11-4 illustrates Kepler's second law, the law of equal areas. A planet moves faster when it is closer to the sun than when it is farther away, so that the area swept out by the radius vector in a given time interval is the same throughout the orbit. The law of equal areas is a consequence of the conservation of angular momentum, as we will see in the next section.

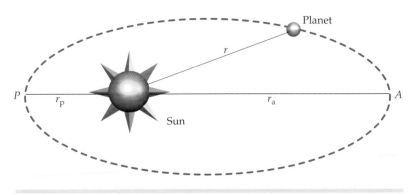

FIGURE 11-3 The elliptical path of a planet with the sun at one focus. Point P, where the planet is closest to the sun, is called the perihelion, and point A, where it is farthest, is called the aphelion. The average distance between the planet and the sun, defined as $(r_p + r_a)/2$, is equal to the semimajor axis.

Law 3. The square of the period of any planet is proportional to the cube of the semimajor axis of its orbit.

Kepler's third law relates the period of any planet to its mean distance from the sun, which equals the semimajor axis of its elliptical path. In algebraic form, if r is the mean distance between a planet and the sun and T is the planet's period of revolution, Kepler's third law states that

$$T^2 = Cr^3 \qquad\qquad 11\text{-}2$$

where the constant C has the same value for all the planets. This law is a consequence of the fact that the force exerted by the sun on a planet varies inversely with the square of the distance from the sun to the planet. We will demonstrate this in Section 11-2 for the special case of a circular orbit.

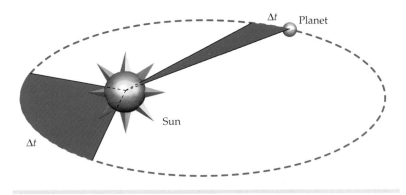

FIGURE 11-4 When a planet is close to the sun, it moves faster than when it is farther away. The areas swept out by the radius vector in a given time interval are equal.

JUPITER'S ORBIT   **EXAMPLE 1 1 - 1**

The mean distance from the sun to Jupiter is 5.20 AU. What is the period of Jupiter's orbit around the sun?

PICTURE THE PROBLEM We use Kepler's third law to relate the period of Jupiter to its mean orbital radius. The constant C can be obtained from the known mean distance and period of the earth. Let $T_E = 1$ y and $r_E = 1$ AU be the period and mean distance for the earth, and let T_J and $r_J = 5.20$ AU be the period and mean distance for Jupiter.

1. Kepler's third law relates Jupiter's period T_J and mean distance r_J: $\quad T_J^2 = Cr_J^3$

2. Apply Kepler's third law to the earth to obtain a second equation relating the same constant C to T_E and r_E: $\quad T_E^2 = Cr_E^3$

3. Divide the two equations, eliminating C, and solve for T_J: $\quad \dfrac{T_J^2}{T_E^2} = \dfrac{r_J^3}{r_E^3}$

so

$$T_J = T_E\left(\frac{r_J}{r_E}\right)^{3/2} = (1\text{ y})\left(\frac{5.20\text{ AU}}{1\text{ AU}}\right)^{3/2}$$

$$= \boxed{11.9\text{ y}}$$

REMARKS The periods of the planets Earth, Jupiter, Saturn, Uranus, and Neptune are plotted in Figure 11-5 as functions of their mean distances from the sun. In (a), periods are plotted versus mean distances from the sun. In (b), the squares of the periods are plotted versus the cubes of the mean distances from the sun. Here the points fall on a straight line.

FIGURE 11-5

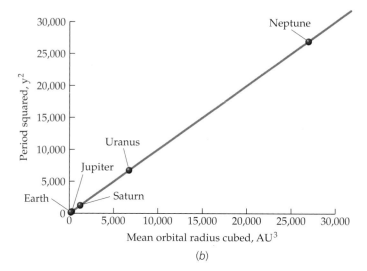

(a)

(b)

EXERCISE The period of Neptune is 164.8 y. What is its mean distance from the sun? (*Answer* 30.1 AU)

EXERCISE If the logs of the periods of the planets Earth, Jupiter, Saturn, Uranus, and Neptune are plotted versus the logs of their mean distances from the sun, the points fall on a curve. What is the shape of this curve? (*Answer* A straight line)

11-2 Newton's Law of Gravity

Although Kepler's laws were an important first step in understanding the motion of planets, they were nothing more than empirical rules obtained from the astronomical observations of Brahe. It remained for Newton to take the next giant step by attributing the acceleration of a planet in its orbit to a specific force exerted on it by the sun. Newton proved that a force that varies inversely with the square of the distance between the sun and a planet results in an elliptical orbit, as observed by Kepler. He then made the bold assumption that this force acts between any two objects in the universe. Before Newton, it was not even generally accepted that the laws of physics observed on earth were applicable to the heavenly bodies. **Newton's law of gravity** postulates that there is a force of attraction between each pair of point particles that is proportional to the product of the masses of the particles and inversely proportional to the square of the distance separating them. Let m_1 and m_2 be the masses of point particles 1 and 2 (at positions $\vec{r}_1$ and $\vec{r}_2$, respectively) and $\vec{r}_{1,2}$ be the vector pointing from particle 1 to particle 2 (Figure 11-6a).

The force $\vec{F}_{1,2}$ exerted by particle 1 on particle 2 is then

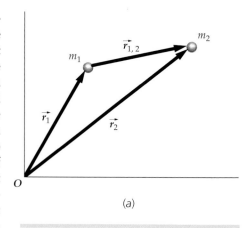

FIGURE 11-6 (*a*) Particles at $\vec{r}_1$ and $\vec{r}_2$. (*b*) The particles exert equal and opposite forces on each other.

$$\vec{F}_{1,2} = -\frac{Gm_1m_2}{r_{1,2}^2}\hat{r}_{1,2}$$

11-3

where $\hat{r}_{1,2} = \vec{r}_{1,2}/r_{1,2}$ is a unit vector pointing from 1 to 2 and G is the **universal gravitational constant,** which has the value

$$G = 6.67 \times 10^{-11} \text{ N·m}^2/\text{kg}^2 \qquad\qquad\text{11-4}$$

The force $\vec{F}_{2,1}$ exerted by 2 on 1 is the negative of $\vec{F}_{1,2}$, according to Newton's third law (Figure 11-6b). The magnitude of the gravitational force exerted by a point particle of mass m_1 on another point particle of mass m_2 a distance r away is thus given by

$$F = \frac{Gm_1m_2}{r^2} \qquad\qquad\text{11-5}$$

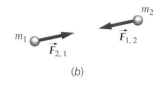

FIGURE 11-6 (b) The particles exert equal and opposite forces on each other.

Newton published his theory of gravitation in 1686, but it was not until a century later that an accurate experimental determination of G was made by Cavendish, as will be discussed in Section 11-4.

We can use the known value of G to compute the gravitational attraction between two ordinary objects.

EXERCISE Find the gravitational force that attracts a 65-kg man to a 50-kg woman when they are 0.5 m apart. Model them as point particles. (*Answer* 8.67×10^{-7} N)

This exercise demonstrates that the gravitational force exerted by an object of ordinary size on another such object is so small as to be unnoticeable. For comparison, a mosquito weighs about 1×10^{-7} N. The weight of a 50-kg person is 491 N, about half a billion times the force of attraction calculated in the exercise! Gravitational attraction is easily noticed only when at least one of the objects is astronomically massive. The gravitational attraction between the girl and the earth for example, is readily apparent.

To check the validity of the inverse-square nature of the gravitational force, Newton compared the acceleration of the moon in its orbit with the free-fall acceleration of objects near the surface of the earth (such as the legendary apple). He assumed that the gravitational attraction due to the earth causes both accelerations. He first assumed that the earth and moon could be treated as point particles with their total masses concentrated at their centers. The force on a particle of mass m a distance r from the center of the earth is

$$F = \frac{GM_{\text{E}}m}{r^2} \qquad\qquad\text{11-6}$$

If this is the only force acting on the particle, then its acceleration is

$$a = \frac{F}{m} = \frac{GM_{\text{E}}}{r^2} \qquad\qquad\text{11-7}$$

For objects near the surface of the earth, $r = R_{\text{E}}$ and the free-fall acceleration is g:

$$g = \frac{GM_{\text{E}}}{R_{\text{E}}^2} \qquad\qquad\text{11-8}$$

The distance to the moon is about 60 times the radius of the earth ($r = 60R_{\text{E}}$). Substituting this into Equation 11-7 gives $a = g/60^2$, so the acceleration of the moon in its near-circular orbit is the free-fall acceleration g near the surface of the earth divided by 60^2. That is, the acceleration of the moon a_{m} should be $(9.81 \text{ m/s}^2)/60^2$. The moon's acceleration can be calculated from its known distance from the center of the earth, $r = 3.84 \times 10^8$ m, and its period $T = 27.3$ d $= 2.36 \times 10^6$ s:

$$a_m = \frac{v^2}{r} = \frac{(2\pi r/T)^2}{r} = \frac{4\pi^2 r}{T^2} = \frac{4\pi^2(3.84 \times 10^8 \text{ m})}{(2.36 \times 10^6 \text{ s})^2} = 2.72 \times 10^{-3} \text{ m/s}^2$$

Then

$$\frac{g}{a_m} = \frac{9.81 \text{ m/s}^2}{2.72 \times 10^{-3} \text{ m/s}^2} = 3604 \approx 60^2$$

In Newton's words, "I thereby compared the force requisite to keep the Moon in her orb with the force of gravity at the surface of the Earth, and found them answer pretty nearly."

The assumption that the earth and moon can be treated as point particles in the calculation of the force on the moon is reasonable because the earth-to-moon distance is large compared with the radius of either the earth or the moon, but such an assumption is certainly questionable when applied to an object near the earth's surface. After considerable effort, Newton was able to prove that the force exerted by any object with a spherically symmetric mass distribution on a point mass either on or outside its surface is the same as if all the mass of the object were concentrated at its center. The proof involves integral calculus, which Newton developed to solve this problem.

Earth as seen from Apollo 11 orbiting the moon on July 16, 1969.

Because $g = 9.81 \text{ m/s}^2$ is readily measured and the radius of the earth is known, Equation 11-8 can be used to determine the value of the product GM_E. Newton estimated the value of G from an approximation of the mass of the earth. When Cavendish determined G some 100 years later by measuring the force between small spheres of known mass and separation, he called his experiment "weighing the earth." Knowing the value of G meant that the mass of the sun and the mass of any planet with a satellite could be determined. The method for doing this is described in Section 11-4.

FALLING TO EARTH   **EXAMPLE 1 1 - 2**

What is the free-fall acceleration of an object at the altitude of the space shuttle's orbit, about 400 km above the earth's surface?

PICTURE THE PROBLEM The force is given by Equation 11-6 with $r = R_E + 400$ km.

1. The acceleration is given by $a = F/m$, where F is given by Newton's law of gravity:

$$a = \frac{F}{m} = \frac{GmM_E/r^2}{m} = \frac{GM_E}{r^2}$$

2. The distance r is related to the radius of the earth R_E and the altitude h:

$$r = R_E + h = 6370 \text{ km} + 400 \text{ km}$$
$$= 6770 \text{ km}$$

3. The acceleration is then:

$$a = \frac{GM_E}{r^2}$$

$$= \frac{(6.67 \times 10^{-11} \text{ N·m}^2/\text{kg}^2)(5.98 \times 10^{24} \text{ kg})}{(6.77 \times 10^6 \text{ m})^2}$$

$$= \boxed{8.70 \text{ m/s}^2}$$

REMARKS This is also the acceleration of the "weightless" shuttle astronauts as they accelerate in their circular orbit.

The calculation in Example 11-2 can be simplified by using Equation 11-8 to eliminate GM_E from Equation 11-7. Then the acceleration at a distance r is

$$a = \frac{F}{m} = \frac{GM_E}{r^2} = g\frac{R_E^2}{r^2}$$ 11-9

EXERCISE At what distance h above the surface of the earth is the acceleration of gravity half its value at sea level? (*Answer* 2640 km)

Measurement of *G*

The universal gravitational constant G was first measured in 1798 by Henry Cavendish, who used the apparatus shown in Figure 11-7. Cavendish's measurement of G has been repeated by other experimenters with various improvements and refinements. All measurements of G are difficult because of the extreme weakness of the gravitational attraction. Consequently, the value of G is known today only to about 1 part in 10,000. Although G was one of the first physical constants ever measured, it remains one of the least accurately known.

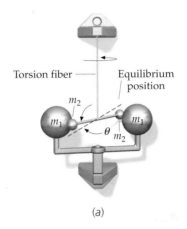

(a) (b)

Gravitational and Inertial Mass

The property of an object responsible for the gravitational force it exerts on another object is its *gravitational* mass. On the other hand, the property of an object that measures its resistance to acceleration is its *inertial* mass. We have used the same symbol m for these two properties because, experimentally, they are proportional. For convenience, units are judiciously defined to make the proportionality constant one. The fact that the gravitational force exerted on an object is proportional to its inertial mass is a characteristic unique to the force of gravity. One consequence is that all objects near the surface of the earth fall with the same acceleration if air resistance is neglected. The well-known story of Galileo dropping objects from the Leaning Tower of Pisa to demonstrate that the free-fall acceleration is the same for objects with different inertial masses is just one example of the excitement this discovery aroused in the sixteenth century.

We could easily imagine that the gravitational and inertial masses of an object were not the same. Suppose we write m_G for the gravitational mass and m for the inertial mass. The force exerted by the earth on an object near its surface would then be

$$F = \frac{GM_E m_G}{R_E^2}$$ 11-10

where M_E is the gravitational mass of the earth. The free-fall acceleration of the object near the earth's surface would then be

$$a = \frac{F}{m} = \left(\frac{GM_E}{R_E^2}\right)\frac{m_G}{m}$$ 11-11

If gravity were just another property of matter, like color or hardness, it might be

FIGURE 11-7 (*a*) Two small spheres, each of mass m_2, are at the ends of a light rod that is suspended by a fine fiber. Careful measurements determine the torque required to turn the fiber through a given angle. Two large spheres, each of mass m_1, are then placed near the small spheres. Because of the gravitational attraction of the large spheres of mass m_1 for the small spheres, the fiber is turned through a very small angle θ from its equilibrium position. (*b*) The apparatus as seen from above. After the apparatus comes to rest, the positions of the large spheres are reversed, as shown by the dashed lines, so that they are at the same distance from the equilibrium position of the balance but on the other side. If the apparatus is again allowed to come to rest, the fiber will turn through angle 2θ in response to the reversal of the torque. Once the torsion constant has been determined, the forces between the masses m_1 and m_2 can be determined from the measurement of this angle. Since the masses and their separations are known, G can be calculated. Cavendish obtained a value for G within about 1 percent of the presently accepted value given by Equation 11-4.

A gravitational torsion balance used in student labs for the measurement of G. A tiny angular deflection of the balance results in a large angular deflection of the laser beam that reflects from a mirror on the balance.

reasonable to expect that the ratio m_G/m would depend on such things as the chemical composition of the object or its temperature. The free-fall acceleration would then be different for different objects. **The experimental evidence, however, is that a is the same for all objects.** Thus we need not maintain the distinction between m_G and m and can set $m_G = m$. We must keep in mind, however, that the equivalence of gravitational and inertial mass is an empirical law that is limited by the accuracy of experiment. Experiments testing this equivalence were carried out by Simon Stevin in the 1580s. Galileo publicized this law widely, and his contemporaries made considerable improvements in the experimental accuracy with which the law was established.

The most precise early comparisons of gravitational and inertial mass were made by Newton. Through experiments using simple pendulums rather than falling bodies, Newton was able to establish the equivalence between gravitational and inertial mass to an accuracy of about 1 part in 1000. Experiments comparing gravitational and inertial mass have improved steadily over the years. Their equivalence is now established to about 1 part in 10^{12}. Thus, the equivalence of gravitational and inertial mass is one of the best established of all physical laws. It is the basis for the principle of equivalence, which is the foundation of Einstein's general theory of relativity.

Derivation of Kepler's Laws

Newton showed that when an object such as a planet or comet moves around a $1/r^2$ force center such as the sun, the object's path is a conic section (an ellipse, a parabola, or a hyperbola). The parabolic and hyperbolic paths apply to objects that make one pass by the sun and never return. Such orbits are not closed. The only closed orbits in an inverse-square force field are ellipses. Thus, Kepler's first law is a direct consequence of Newton's law of gravity. Kepler's second law, the law of equal areas, follows from the fact that the force exerted by the sun on a planet is directed toward the sun. Such a force is called a **central force.** Figure 11-8a shows a planet moving in an elliptical orbit around the sun. In time dt, the planet moves a distance $v\,dt$ and the radius vector $\vec{r}$ sweeps out the area shaded in the figure. This is half the area of the parallelogram formed by the vectors $\vec{r}$ and $\vec{v}\,dt$, which is $|\vec{r} \times \vec{v}\,dt|$. Thus the area dA swept out by the radius vector $\vec{r}$ in time dt is

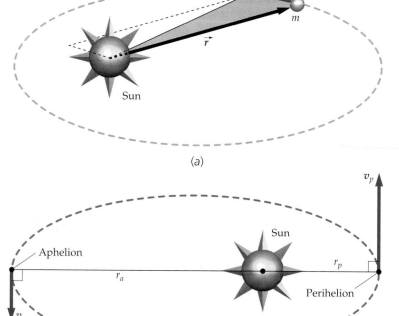

(a)

$$dA = \frac{1}{2}|\vec{r} \times \vec{v}\,dt| = \frac{1}{2m}|\vec{r} \times m\vec{v}|\,dt$$

or

$$\frac{dA}{dt} = \frac{1}{2m}L \qquad\qquad 11\text{-}12$$

(b)

FIGURE 11-8

where $L = |\vec{r} \times m\vec{v}|$ is the magnitude of the orbital angular momentum of the planet about the sun. The area dA swept out in a given time interval dt is therefore proportional to the magnitude of the angular momentum L. Since the force on a planet is along the line from the planet to the sun, it has no torque about the sun. Thus the magnitude of the angular momentum of the planet is conserved; that is, L is constant. Therefore, the rate at which area is swept out is the same for all parts of the orbit, which is Kepler's second law. Also, the fact that L is constant means that $rv \sin \phi$ is constant. At aphelion and perihelion $\phi = 90°$ (Figure 11-8b), so $r_a v_a = r_p v_p$.

We will now show that Newton's law of gravity implies Kepler's third law for the special case of a circular orbit. Consider a planet moving with speed v in a circular orbit of radius r about the sun. The gravitational force of attraction between the sun and the planet provides the centripetal acceleration v^2/r. Newton's second law gives

$$F = M_p a$$

$$\frac{GM_s M_p}{r^2} = M_p \frac{v^2}{r}$$ 11-13

where M_s is the mass of the sun and M_p is that of the planet. Solving for v^2,

$$v^2 = \frac{GM_s}{r}$$ 11-14

Because the planet moves a distance $2\pi r$ in time T, its speed is related to the period by

$$v = \frac{2\pi r}{T}$$ 11-15

Substituting this expression for v in Equation 11-14, we obtain

$$v^2 = \frac{4\pi^2 r^2}{T^2} = \frac{GM_s}{r}$$

or

$$T^2 = \frac{4\pi^2}{GM_s} r^3$$ 11-16

KEPLER'S THIRD LAW

Equation 11-16 is Kepler's third law, which is the same as Equation 11-2 with $C = 4\pi^2/GM_s$. Equation 11-16 also applies to the orbits of the satellites of any planet if we replace the mass of the sun M_s with the mass of the planet.

THE ORBITING SPACE STATION   **EXAMPLE   1 1 - 3**   **Put It in Context**

The International Space Station travels in a roughly circular orbit around the earth. If its altitude is 385 km above the earth's surface, how long do you have to wait between sightings? (Assume that air resistance can be neglected.)

PICTURE THE PROBLEM The sightings occur only at night and then only if the space station is above the horizon at your location. Thus, the minimum time between sightings is approximately equal to the orbital period. To find the orbital period we use Kepler's third law with M_s in Equation 11-16 replaced by the mass of the earth M_E. The numerical calculation is simplified somewhat by using $GM_E = R_E^2 g$ from Equation 11-8.

1. Apply Kepler's third law to the space station: $$T^2 = \frac{4\pi^2}{GM_E} r^3$$

2. At an altitude $h = 385$ km, $r = R_E + h = 6760$ km. Substitute $r = R_E + h$ and solve for the period:

$$T^2 = \frac{4\pi^2}{GM_E}(R_E + h)^3$$

3. Use $GM_E = R_E^2 g$ to write T in terms of g:

$$T^2 = \frac{4\pi^2}{R_E^2 g}(R_E + h)^3$$

so

$$T = \frac{2\pi}{R_E \sqrt{g}}(R_E + h)^{3/2}$$

$$= \frac{2\pi(6.755 \times 10^6 \text{ m})^{3/2}}{(6.37 \times 10^6 \text{ m})(9.81 \text{ m/s}^2)^{1/2}}$$

$$= 5529 \text{ s} = \boxed{92.1 \text{ min}}$$

EXERCISE How many degrees does the earth rotate in a single orbital period of the ISS? (*Answer* The earth rotates 360° in 24 h = 1440 min. In 92.1 min it rotates 23.0°.)

REMARKS The near-circular orbit of the ISS, which is inclined ≈52° to the equatorial plane, does not rotate with the earth. If the ISS is directly over your home at time t, 92.1 min later it will be directly over a location 23.0° due west of your home. For example, Kansas City is 23° due west of New York City. If the ISS passes over your home in New York City at midnight Eastern Time, you could tell your friend in Kansas City that it will pass over Kansas City at 12:32 A.M. Central Time (1:32 A.M. Eastern Time).

EXERCISE Find the radius of a circular orbit of a satellite that orbits the earth with a period of 1 d. (*Answer* $r = 6.63R_E = 4.22 \times 10^7$ m = 26,200 mi. If such a satellite is in orbit over the equator and moves in the same direction as the rotation of the earth, it appears stationary relative to the earth. Many satellites are "parked" in such an orbit, called a geosynchronous orbit.)

Because G is known, by using Equation 11-16 we can determine the mass of an astronomical object by measuring the period T and the mean orbital radius r of a satellite orbiting it. In establishing Equation 11-16 the mass of the satellite was assumed negligible compared to the mass of the central object. This means that the central object remains stationary as the satellite revolves around it. In fact, the central object and satellite both revolve around a common point, their center of mass. If the mass of the satellite is not assumed negligible, the result is

$$T^2 = \frac{4\pi^2}{G(M_1 + M_2)} r^3 \tag{11-17}$$

where r is the center-to-center separation of the objects. (For the more general elliptical orbits, the math is more challenging but the result is the same.) If the mass of the satellite is not negligible, as is the case with most binary star systems, then only the sum of the masses is determined, as revealed by Equation 11-17. The moon, along with planets Mercury and Venus, have no natural satellites, so their masses were not well known until the 1960s when artificial satellites were first placed in orbit around them.

EXERCISE The Martian moon Phobos has a period of 460 min and a mean orbital radius of 9400 km. What is the mass of Mars? (*Answer* 6.45×10^{23} kg = $0.108 M_E$)

EXPLORING

What is the difference between gravitational mass and inertial mass? Find out this, and more, at www.whfreeman.com/tipler5e.

11-3 Gravitational Potential Energy

Near the surface of the earth, the gravitational force exerted by the earth on an object is essentially uniform because the distance to the center of the earth, $r = R_E + h$, is always approximately R_E for $h \ll R_E$. The potential energy of an object near the earth's surface is $mgh = mg(r - R_E)$, where we have chosen $U = 0$ at the earth's surface, $r = R_E$. When we are far from the surface of the earth, we must take into account the fact that the gravitational force exerted by the earth is not uniform but decreases as $1/r^2$. The general definition of potential energy (Equation 6-19b) is

$$dU = -\vec{F} \cdot d\vec{s}$$

where $\vec{F}$ is the conservative force on a particle and $d\vec{s}$ is a general displacement of the particle. For the radial gravitational force $\vec{F}$ given by Equation 11-6 we have

$$dU = -\vec{F} \cdot d\vec{s} = -F\hat{r} \cdot d\vec{s} = -F_r\, dr = -\left(-\frac{GM_E m}{r^2}\right) dr = \frac{GM_E m}{r^2}\, dr \qquad 11\text{-}18$$

Integrating both sides of this equation we obtain

$$U = -\frac{GM_E m}{r} + U_0 \qquad\qquad 11\text{-}19$$

where U_0 is a constant of integration. Since only changes in potential energy are important, we can choose the potential energy to be zero at any position. The earth's surface is a good choice for many everyday problems, but it is not always a convenient choice. For example, when considering the potential energy associated with a planet and the sun, there is no reason to want the potential energy to be zero at the surface of the sun. In fact, it is nearly always more convenient to choose the gravitational potential energy of a two-object system to be zero when the separation of the objects is infinite. Thus, $U_0 = 0$ is often a convenient choice. Then

$$U(r) = -\frac{GMm}{r} \qquad\qquad 11\text{-}20$$

GRAVITATIONAL POTENTIAL ENERGY WITH $U = 0$ AT
INFINITE SEPARATION

where $U = 0$ if $r = \infty$. Figure 11-9 is a plot of $U(r)$ versus r for this choice of $U = 0$ at $r = \infty$ for an object of mass m and the earth of mass M_E. This function begins at the negative value $U = -GM_E m/R_E = -mgR_E$ at the earth's surface and increases as r increases, approaching zero at infinite r. The slope of this curve at $r = R_E$ is

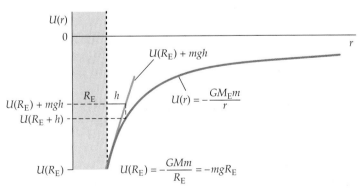

FIGURE 11-9

$GM_Em/R_E^2 = mg$, so the equation of the tangent line, drawn in blue, is $f(h) = U(R_E) + mgh$, where $h = r - R_E$ is the distance above the earth's surface. From the figure you can see that for small h, $U(R_E) + mgh \approx U(r)$.

Escape Speed

During the last half-century, the idea of escaping from the earth's gravity has changed from fantasy to reality. Space probes have been sent out to the far reaches of the solar system. Some of these probes orbit the sun, while others leave the solar system and drift on into outer space. We will see that there is a minimum initial speed, called the **escape speed,** that is required for an object in free-fall to escape from the earth.

If we project an object upward from the earth with some initial kinetic energy, the kinetic energy decreases and the potential energy increases as the object rises. But the maximum increase in potential energy is GM_Em/R_E. Therefore, this amount is the most that the kinetic energy can decrease. If the initial kinetic energy is greater than GM_Em/R_E, then the total energy E will be greater than zero (E_2 in Figure 11-10), and the object will still have some kinetic energy when r is very large (or even when r approaches infinity). Thus, if the initial kinetic energy is greater than GM_Em/R_E, the object will escape from the earth's gravity. Since the potential energy at the earth's surface is $-GM_Em/R_E$, the total energy $E = K + U$ must be greater than or equal to zero in order for the object to escape. The speed near the earth's surface corresponding to zero total energy is called the escape speed v_e. It is found from

$$K_f + U_f = K_i + U_i$$

$$0 + 0 = \frac{1}{2}mv_e^2 - \frac{GM_Em}{R_E}$$

so

$$v_e = \sqrt{\frac{2GM_E}{R_E}} = \sqrt{2gR_E} \qquad \text{11-21}$$

ESCAPE SPEED

Using $g = 9.81 \text{ m/s}^2$ and $R_E = 6.37 \times 10^6$ m, we obtain

$$v_e = \sqrt{2(9.81 \text{ m/s}^2)(6.37 \times 10^6 \text{ m})} = 11.2 \text{ km/s}$$

This is about 7 mi/s or 25,000 mi/h. An object with this speed will escape the earth's gravity but it will not escape the solar system because we have neglected the gravitational attraction of the sun and other planets (see Problems 46 and 48).

The escape speed for a planet or moon relative to the thermal speeds of gas molecules determines the kind of atmosphere a planet or moon can have. The average kinetic energy of gas molecules, $(\frac{1}{2}mv^2)_{av}$, is proportional to the absolute temperature T (Chapter 18). Near the surface of the earth, the speeds of nearly all of the oxygen and nitrogen molecules are much lower than the escape speed, so these gases are retained in our atmosphere. For the lighter molecules hydrogen and

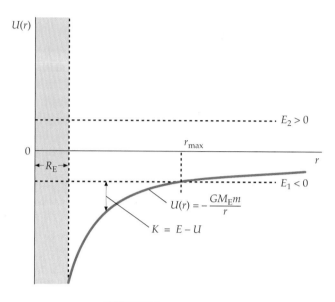

FIGURE 11-10 The kinetic energy of an object at a distance r from the center of the earth is $E - U(r)$. When the total energy is less than zero (E_1 in the figure), the kinetic energy is zero at $r = r_{max}$ and the object is bound to the earth. When the total energy is greater than zero (E_2 in the figure), the object can escape the earth.

helium, however, a significant fraction have speeds greater than the escape speed. Hydrogen and helium gases are therefore not found in our atmosphere. The escape speed at the surface of the moon is 2.3 km/s, which can be calculated from Equation 11-21 with the mass and radius of the moon replacing M_E and R_E. This is considerably smaller than the escape speed for earth and, in fact, is too small for any atmosphere to exist.

EXERCISE   Find the escape speed at the surface of Mercury, which has a mass $M = 3.31 \times 10^{23}$ kg and a radius $R = 2440$ km. (*Answer* $v_e = \sqrt{2GM/R} = 4.25$ km/s)

Classification of Orbits by Energy

In Figure 11-10, two possible values for the total energy E are indicated on a graph of U versus r: E_1, which is negative, and E_2, which is positive. A negative total energy simply means that the kinetic energy at the earth's surface is less than $+GM_Em/R_E$, so that $K + U$ is never greater than zero. From this figure we see that, if the total energy is negative, the total-energy line intersects the potential-energy curve at some maximum separation r_{max} and the system is bound. On the other hand, if the total energy is zero or positive, there is no such intersection and the system is unbound. The criteria for a bound or unbound system are simply stated:

>   If $E < 0$, the system is bound.
>
>   If $E \geq 0$, the system is unbound.

When E is negative, its absolute value $|E|$ is called the binding energy. The binding energy is the energy that must be added to the system to bring the total energy up to zero.

The potential energy of an object such as a planet or comet of mass m at a distance r from the sun is

$$U(r) = -\frac{GM_s m}{r} \qquad\qquad 11\text{-}22$$

where M_s is the mass of the sun. The kinetic energy of the object is $\frac{1}{2}mv^2$. If the total energy, kinetic plus potential, is less than zero, then the orbit will be an ellipse (possibly a circle), and the object will be bound to the sun. If, instead, the total energy is positive, then the orbit will be a hyperbola, and the object will make one swing around the sun and leave the solar system, never to return. If the total energy is exactly zero, the orbit will be a parabola, and again the object will make one pass and then escape. To summarize, when the total energy is zero or positive the object is not bound to the sun, but will escape. Curiously, there have not been any measurements of the energy E of a comet or an asteroid that are definitely nonnegative. Thus, all observed objects of this type are bound to the solar system.

HEIGHT OF A PROJECTILE   **EXAMPLE   1 1 - 4**

A projectile is fired straight up from the surface of the earth with an initial speed $v_i = 8$ km/s. Find the maximum height it reaches, neglecting air resistance.

PICTURE THE PROBLEM   The maximum height is found using energy conservation. We take the surface of the earth as the initial point, with $U_i = -GM_Em/R_E$ and $K_i = \frac{1}{2}mv_i^2$. At the greatest height, $K_f = 0$.

1. Apply conservation of mechanical energy:

$$K_f + U_f = K_i + U_i$$

$$0 - \frac{GM_E m}{r_f} = \frac{1}{2}mv_i^2 - \frac{GM_E m}{R_E}$$

2. Multiply through by $-1/(GM_E m)$, use $GM_E = gR_E^2$, and solve for r_f:

$$\frac{1}{r_f} = -\frac{v_i^2}{2GM_E} + \frac{1}{R_E} = -\frac{v_i^2}{2gR_E^2} + \frac{1}{R_E}$$

so

$$r_f = 1 \Big/ \left(\frac{1}{R_E} - \frac{v_i^2}{2gR_E^2}\right) = R_E \Big/ \left(1 - \frac{v_i^2}{2gR_E}\right)$$

3. Substitute numerical values to find r and $h = r - R_E$:

$$r_f = R_E \Big/ \left(1 - \frac{(8000 \text{ m/s})^2}{2(9.81 \text{ m/s}^2)(6.37 \times 10^6 \text{ m})}\right)$$

$$= 2.05R_E$$

$$h = r - R_E = \boxed{1.05R_E = 6.69 \times 10^6 \text{ m}}$$

SPEED OF A PROJECTILE

E X A M P L E   1 1 - 5   Try It Yourself

A projectile is fired straight up from the surface of the earth with an initial speed $v_i = 15$ km/s. Find the speed of the projectile when it is very far from the earth, neglecting air resistance.

PICTURE THE PROBLEM Very far from earth means $r \gg R_E$. The initial speed is greater than the escape speed of 11 km/s, so the total energy of the projectile is positive and the projectile will escape the earth with some final kinetic energy. Use conservation of mechanical energy to find this kinetic energy and then solve for the final speed.

v, km/s

h/R_E

FIGURE 11-11

Cover the column to the right and try these on your own before looking at the answers.

Steps

Answers

1. Apply conservation of mechanical energy, noting that $r_f = \infty$, so $U_f = 0$.

$$K_f + U_f = K_i + U_i$$

$$\frac{1}{2}mv_f^2 + 0 = \frac{1}{2}mv_i^2 - \frac{GM_E m}{R_E}$$

2. Solve for v_f^2 using $GM_E/R_E^2 = g$ to simplify.

$$v_f^2 = v_i^2 - \frac{2GM_E}{R_E} = v_i^2 - \frac{2gR_E^2}{R_E}$$

$$= v_i^2 - 2gR_E$$

3. Substitute known values for g and R_E to calculate v_f.

$$v_f = \sqrt{(15 \times 10^3 \text{ m/s})^2 - 2(9.81 \text{ m/s}^2)(6.37 \times 10^6 \text{ m})}$$

$$= 10^4 \text{ m/s} = \boxed{10 \text{ km/s}}$$

REMARKS In Figure 11-11, the speed of the projectile in kilometers per second is plotted versus h/R_E, where h is the height above the earth's surface. At very large values of h/R_E, the speed approaches the horizontal line $v = 10$ km/s.

TOTAL ENERGY OF A SATELLITE **E X A M P L E   1 1 - 6** Try It Yourself

Show that the total energy of a satellite in a circular orbit is half its potential energy.

PICTURE THE PROBLEM The total energy of a satellite is the sum of its potential and kinetic energies, $E = U + K$. Newton's second law will allow us to find the speed v of the satellite in terms of its orbital radius r. The kinetic energy depends on the speed, so we can find the kinetic energy in terms of r. Assume the mass of the earth is much greater than that of the satellite.

Cover the column to the right and try these on your own before looking at the answers.

Steps	Answers
1. Write the total energy equal to the sum of the potential energy and the kinetic energy.	$E = K + U$ $$= \frac{1}{2}mv^2 - \frac{GM_E m}{r}$$
2. Apply Newton's second law to the satellite and solve for the square of the speed.	$F = ma$ $$\frac{GM_E m}{r^2} = m\frac{v^2}{r}$$ so $$v^2 = \frac{GM_E}{r}$$
3. Substitute into the step 1 result and simplify.	$$E = \frac{1}{2}m\frac{GM_E}{r} - \frac{GM_E m}{r}$$ $$= \frac{GM_E m - 2GM_E m}{2r} = -\frac{GM_E m}{2r}$$
4. Compare the step 3 result with U in step 1.	$$E = -\frac{GM_E m}{2r} = \frac{1}{2}\left(-\frac{GM_E m}{r}\right) = \boxed{\frac{1}{2}U}$$

EXERCISE A satellite of mass 450 kg orbits the earth in a circular orbit at 6830 km above the earth's surface. Find (*a*) the potential energy, (*b*) the kinetic energy, and (*c*) the total energy of the satellite. (*Answer* Note that $r = R_E + h = 13{,}200$ km (*a*) $U = -13.6 \times 10^9$ J (*b*) $K = 6.80 \times 10^9$ J (*c*) $E = -6.80 \times 10^9$ J)

11-4 The Gravitational Field $\vec{g}$

The gravitational force exerted by a point particle of mass m_1 on a second point particle of mass m_2 a distance $r_{1,2}$ away is given by

$$\vec{F}_{1,2} = -\frac{Gm_1 m_2}{r_{1,2}^2}\hat{r}_{1,2}$$

where $\hat{r}_{1,2} = \vec{r}_{1,2}/r_{1,2}$ is a unit vector pointing from particle 1 to particle 2. The gravitational field at point P is determined by placing a point particle of mass m at P and calculating the gravitational force $\vec{F}$ on it due to all other particles. The gravitational force $\vec{F}$ divided by the mass m is the **gravitational field** $\vec{g}$ at P:

$$\vec{g} = \frac{\vec{F}}{m} \qquad\qquad 11\text{-}23$$

DEFINITION—GRAVITATIONAL FIELD

The point P is called a **field point.** The gravitational field at a field point due to the masses of a collection of point particles is the vector sum of the fields due to the individual masses at that point:

$$\vec{g} = \sum_i \vec{g}_i \qquad\qquad 11\text{-}24a$$

The points where these point particles are located are called **source points.** To find the gravitational field at a field point due to a continuous object, we find the field $d\vec{g}$ due to a small element of volume with mass dm and integrate over the entire mass distribution of the object (the entire set of source points).

$$\vec{g} = \int d\vec{g} \qquad\qquad 11\text{-}24b$$

The gravitational field of the earth at a distance $r \ge R_E$ points toward the earth and has the magnitude $g(r)$ given by

$$g(r) = \frac{F}{m} = \frac{GM_E}{r^2} \qquad\qquad 11\text{-}25$$

GRAVITATIONAL FIELD OF THE EARTH

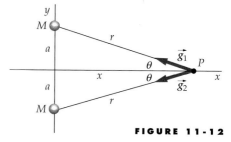

FIGURE 11-12

GRAVITATIONAL FIELD OF TWO POINT PARTICLES **EXAMPLE 1 1 - 7**

Two point particles, each of mass M, are fixed on the y axis at $y = +a$ and $y = -a$ (see Figure 11-12). Find the gravitational field at all points on the x axis as a function of x.

PICTURE THE PROBLEM Two particles of mass M each produce a gravitational field at point P located at $x\hat{i}$. The distance between P and either particle is $r = \sqrt{x^2 + a^2}$. The resultant field $\vec{g}$ is the vector sum of the fields $\vec{g}_1$ and $\vec{g}_2$ due to each particle.

1. Calculate the magnitude of either $\vec{g}_1$ or $\vec{g}_2$:

$$g_1 = \frac{GM}{r^2}$$

2. The y component of the resultant field is zero. The x component is the sum of g_{1x} and g_{2x}:

$$g_x = g_{1x} + g_{2x} = 2g_{1x} = 2g_1 \cos\theta$$

3. Express $\cos\theta$ in terms of x and r from the figure:

$$\cos\theta = \frac{x}{r}$$

4. Combining the last two results yields $\vec{g}$. To express $\vec{g}$ as a function of x, substitute $(x^2 + a^2)^{1/2}$ for r:

$$\vec{g} = g_x\hat{i} = -2\frac{GM}{r^2}\frac{x}{r}\hat{i} = -\frac{2GMx}{r^3}\hat{i}$$

$$= -\frac{2GMx}{(x^2 + a^2)^{3/2}}\hat{i}$$

PLAUSIBILITY CHECK For $x < 0$, $\vec{g}$ is in the positive x direction and for $x > 0$, $\vec{g}$ is in the negative direction, as expected. If $x = 0$, we find that $\vec{g} = 0$; the fields due to m_1 and m_2 are equal and opposite at $x = 0$, and hence they cancel. For $x \gg a$, $\vec{g} \approx -(2GM/x^2)\hat{i}$. The field is the same as if a single mass of $2M$ were at the origin.

GRAVITATIONAL FIELD OF A UNIFORM ROD   **EXAMPLE 11-8**

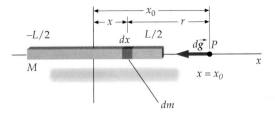

A uniform rod of mass M and length L is centered on the origin and lies along the x axis (Figure 11-13). Find the gravitational field due to the rod at all points on the x axis for $x > L/2$.

FIGURE 11-13

PICTURE THE PROBLEM Choose a mass element dm of length dx at $x\hat{i}$, and choose a field point P on the x axis at $x = x_0$ in the region $x > L/2$. Each mass element produces a gravitational field at P that points in the negative x direction. We can calculate the total field by integrating the x component of the field produced by dm from $x = -L/2$ to $x = +L/2$.

1. Find the x component of the field at P due to the element dm:
$$dg_x = -\frac{G\,dm}{r^2}$$

2. The mass dm is proportional to the size of the element dx:
$$dm = \frac{M}{L}\,dx$$

3. Write the distance r between dm and point P in terms of x and x_0:
$$r = x_0 - x$$

4. Substitute these results to express dg in terms of x:
$$dg_x = -\frac{G\,dm}{r^2} = -\frac{G(M/L)dx}{(x_0 - x)^2}$$

5. Integrate to find the x component of the total field:
$$g_x = \int dg_x = -\frac{GM}{L}\int_{-L/2}^{L/2}\frac{dx}{(x_0 - x)^2} = -\frac{GM}{L}\left[\frac{1}{x_0 - x}\right]_{-L/2}^{L/2}$$
$$= -\frac{GM}{L}\left(\frac{1}{x_0 - L/2} - \frac{1}{x_0 + L/2}\right) = -\frac{GM}{x_0^2 - (L/2)^2}$$

6. Express the resultant field as a vector:
$$\vec{g} = g_x\hat{i} = -\frac{GM}{x_0^2 - (L/2)^2}\hat{i}$$

7. x_0 is an arbitrary point on the x axis in the region $x > L/2$, so we can replace it with x:
$$\boxed{\vec{g} = -\frac{GM}{x^2 - (L/2)^2}\hat{i}}$$

PLAUSIBILITY CHECK For $x \gg L/2$, the field approaches that of a point particle of mass M, $\vec{g} = -(GM/x^2)\hat{i}$.

$\vec{g}$ of a Spherical Shell and of a Solid Sphere

One of Newton's motivations for developing calculus was to prove that the gravitational field outside a solid sphere is the same as if all the mass of the sphere were concentrated at its center. We will prove this in the next section. Here we merely discuss the results of this proof. We first consider a uniform spherical shell of mass M and radius R (Figure 11-14). We will show that the gravitational field due to the shell at a distance r from the center of the shell is given by

FIGURE 11-14 A uniform spherical shell of mass M and radius R.

$$\vec{g} = -\frac{GM}{r^2}\hat{r} \quad \text{for} \quad r > R \qquad\qquad 11\text{-}26a$$

$$\vec{g} = 0 \quad \text{for} \quad r < R \qquad\qquad 11\text{-}26b$$

GRAVITATIONAL FIELD OF A SPHERICAL SHELL

We can understand the result that $\vec{g} = 0$ inside the shell from Figure 11-15, which shows a point mass m_0 inside a spherical shell. In this figure, the masses of the shell segments with masses m_1 and m_2 are related by

$$\frac{m_1}{m_2} = \frac{r_1^2}{r_2^2} \quad \text{or} \quad \frac{m_1}{r_1^2} = \frac{m_2}{r_2^2}$$

Since the gravitational force falls off inversely as the square of the distance, the force due to the smaller mass on the left is exactly balanced by that due to the more distant, larger mass on the right.

The gravitational field outside a solid sphere is a simple extension of Equation 11-26a. We merely consider the solid sphere to consist of a continuous set of spherical shells. Since the field due to each shell is the same as if its mass were concentrated at the center of the shell, the field due to the entire sphere is the same as if the entire mass of the sphere were concentrated at its center:

$$g_r = -\frac{GM}{r^2} \quad \text{for} \quad r > R \qquad 11\text{-}27$$

This result holds whether or not the sphere has a constant density, as long as the density depends only on r so that spherical symmetry is maintained.

$\vec{g}$ Inside a Solid Sphere

We now use Equations 11-26a and 11-26b to find the gravitational field inside a solid sphere of constant density at a point a distance r from the center, where r is less than the radius R of the sphere. This would apply, for example, to finding the weight of an object at the bottom of a deep mine shaft. As we have seen, the field inside a spherical shell is zero. Thus, in Figure 11-16 the mass of that part of the sphere outside r exerts no force at or inside r. Therefore, only the mass M' within the radius r contributes to the gravitational field at r. This mass produces a field equal to that of a point mass M' at the center of the sphere. The fraction of the total mass of the sphere within r is equal to the ratio of the volume of a sphere of radius r to that of a sphere of radius R. Thus, for a uniform mass distribution, if M is the total mass of the sphere, M' is given by

$$M' = M\frac{\frac{4}{3}\pi r^3}{\frac{4}{3}\pi R^3} = M\frac{r^3}{R^3} \qquad 11\text{-}28$$

The gravitational field at the distance r is thus

$$g_r = -\frac{GM'}{r^2} = -\frac{GM}{r^2}\frac{r^3}{R^3}$$

or

$$g_r = -\frac{GM}{R^3}r \quad \text{for} \quad r < R \qquad 11\text{-}29$$

The magnitude of the field is zero at the center and increases with distance r inside the sphere. Figure 11-17 shows a plot of the field g_r as a function of r for a solid sphere of uniform mass density.

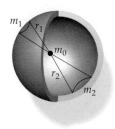

FIGURE 11-15 A point mass m_0 inside a uniform spherical shell feels no net force.

FIGURE 11-16 A uniform solid sphere of radius R and mass M. Only the mass M', which is inside the sphere of radius r, contributes to the gravitational field at the distance r.

M = total mass

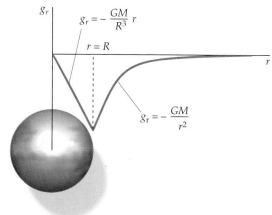

FIGURE 11-17 A plot of g_r versus r for a uniform solid sphere of mass M. The magnitude of the field increases linearly with r inside the sphere and decreases as $1/r^2$ outside the sphere.

A HOLLOW PLANET   **EXAMPLE 11-9**

A planet with a hollow core consists of a uniform, thick spherical shell with mass *M*, outer radius *R*, and inner radius *R/2*. (*a*) What amount of mass is closer than $\frac{3}{4}R$ to the center of the planet? (*b*) What is the gravitational field a distance $\frac{3}{4}R$ from the center?

PICTURE THE PROBLEM The mass of that part of the thick shell that is closer to the center than $\frac{3}{4}R$ is the density times the volume of the thick shell with outer radius $\frac{3}{4}R$ and inner radius $\frac{1}{2}R$. First find the density and the volume, then find the mass. The gravitational field at $r = \frac{3}{4}R$ is due only to the mass closer than this to the center.

(*a*) 1. The mass M' (the mass of the thick shell with outer radius $\frac{3}{4}R$ and inner radius $\frac{1}{2}R$) is the density ρ times the volume V':

$$M' = \rho V'$$

2. The density is the mass *M* divided by the volume *V*:

$$\rho = \frac{M}{V} = M / \left[\frac{4}{3}\pi R^3 - \frac{4}{3}\pi \left(\frac{R}{2}\right)^3 \right] = \frac{M}{7\pi R^3/6}$$

3. Find the volume V' of the thick shell with outer radius $\frac{3}{4}R$ and inner radius $\frac{1}{2}R$:

$$V' = \frac{4}{3}\pi \left(\frac{3R}{4}\right)^3 - \frac{4}{3}\pi \left(\frac{R}{2}\right)^3 = \frac{19}{48}\pi R^3$$

4. Find the mass M':

$$M' = \rho V' = \frac{6}{7}\frac{M}{\pi R^3}\frac{19}{48}\pi R^3 = \boxed{\frac{19}{56}M}$$

(*b*) The gravitational field at $r = \frac{3}{4}R$ is due only to the mass M':

$$\vec{g} = -\frac{GM'}{r^2}\hat{r} = -\frac{G\frac{19}{56}M}{(\frac{3}{4}R)^2}\hat{r}$$

$$= \boxed{-\frac{38}{63}\frac{GM}{R^2}\hat{r}}$$

⏻ PLAUSIBILITY CHECK We expect less than half the mass *M* to be in the region $\frac{1}{2}R < r < \frac{3}{4}R$, and that is the case here, 19/56 is less than half.

REMARKS Note that the volume *V* in the denominator of Part (*a*) step 2 is $7\pi R^3 16$. That is more than half the volume of a sphere of radius *R*. We expect the volume of the region $\frac{1}{2}R < r < R$ to be more than half the volume of a sphere of radius *R*. Also, note that the volume V' (Part (*a*) step 3) and mass M' (Part (*a*) step 4) are the same fraction of *V* and *M*. This is as expected because the shell is uniform.

RADIALLY DEPENDENT DENSITY   **EXAMPLE 11-10**   **Try It Yourself**

A solid sphere of radius *R* and mass *M* is spherically symmetric but not uniform. Its density ρ, defined as its mass per unit volume, is proportional to the distance from the center *r* for $r \le R$. That is, $\rho = Cr$ for $r \le R$, where *C* is a constant. (*a*) Find *C*. (*b*) Find g_r for $r \ge R$. (*c*) Find g_r at $r = R/2$.

PICTURE THE PROBLEM (*a*) You can find *C* by integrating the density over the volume of the sphere and setting the result equal to *M*. For a volume element, take a spherical shell of radius *r* and thickness *dr* (Figure 11-18). Its volume is $4\pi r^2\,dr$ and its mass is $dm = \rho\,dV = Cr(4\pi r^2\,dr)$. (*b*) The field at $r \ge R$ is the same as if the total mass *M* were at the center of the sphere. (*c*) The field at $r = R/2$ is

Area = $4\pi r^2$

dr

Area

r

R

dm

FIGURE 11-18

the same as if mass M' were at the center of the sphere, where M' is the amount of mass within the sphere of radius $R/2$. The mass between $r = R/2$ and $r = R$ produces zero field at $r = R/2$.

Cover the column to the right and try these on your own before looking at the answers.

Steps

Answers

(a) 1. Integrate dm from $r = 0$ to $r = R$.

$$M = \int dM = \int \rho\, dV = \int_0^R Cr(4\pi r^2\, dr) = C\pi R^4$$

2. Solve for C in terms of the given quantities M and R.

$$C = \boxed{\frac{M}{\pi R^4}}$$

(b) Write an expression for the field outside the sphere in terms of the mass M, the distance r from the center, and the unit vector in the $\hat{r}$ direction of increasing r.

$$\vec{g} = \boxed{-\frac{GM}{r^2}\hat{r}} \quad (r > R)$$

(c) 1. Compute the mass M' that is within the radius $R/2$ by integrating $dm = \rho\, dV$ from $r = 0$ to $r = R/2$ and use the value of C found in Part (a) step 2.

$$M' = \frac{M}{16}$$

2. Write an expression for the field at $r = R/2$ in terms of M and R.

$$\vec{g} = -\frac{GM'}{r^2}\hat{r} = \boxed{-\frac{GM}{4R^2}\hat{r}} \quad \text{at } r = \frac{R}{2}$$

PLAUSIBILITY CHECK For a uniform sphere, Equation 11-29 gives the field at $r = R/2$ as $g_r = -GM/(2R^2)$, twice as large as our Part (c) result. This is as expected since a uniform sphere has a larger fraction of its total mass in the region $0 < r < R/2$ than does the sphere in Equation 11-29.

REMARKS Note that the units for C are kg/m⁴, so the units for ρ are kg/m³, which is mass per volume.

*11-5 Finding the Gravitational Field of a Spherical Shell by Integration

We will derive the equation for the gravitational field of a spherical shell in two steps. First, we find the gravitational field on the axis of a ring of uniform mass. We then apply our result to a spherical shell, which we can consider to be a set of coaxial rings.

Figure 11-19 shows a ring of total mass m and radius a and a field point P on the axis of the ring a distance x from its center. We choose a mass element dm on the ring that is small enough to be considered a point particle. The distance from the element to P is s, and the line joining the element and P makes an angle α with the axis of the ring.

The field at P due to the element dm is toward the element and has magnitude dg given by

$$dg = \frac{G\, dm}{s^2}$$

From the symmetry of the figure, we can see that when we sum over all the elements of the ring, the net field will be along the axis of the ring; that is, the perpendicular components will sum to zero. For example, the perpendicular component of the field shown in the figure will be canceled by the perpendicular component due to another element of the ring directly opposite the one shown.

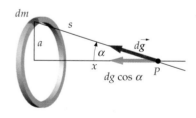

FIGURE 11-19 The gravitational field at a point P a distance x from a uniform ring. The field due to the element dm points toward the element. The total field due to the ring is along the axis of the ring.

The net field will therefore be in the negative x direction. The x component of the field due to the element dm is

$$dg_x = -dg \cos \alpha = -\frac{G\,dm}{s^2} \cos \alpha$$

We obtain the total field by summing over all the elements of the ring:

$$g_x = -\int \frac{G \cos \alpha}{s^2}\,dm$$

Since s and α are the same for all points on the ring, they are constants as far as the integration is concerned. Thus,

$$g_x = -\frac{G \cos \alpha}{s^2} \int dm = -\frac{Gm}{s^2} \cos \alpha \qquad\qquad 11\text{-}30$$

where $m = \int dm$ is the total mass of the ring.

We now use this result to calculate the gravitational field of a spherical shell of mass M and radius R at a point a distance r from the center of the shell. We first consider the case in which the field point P is outside the shell, as in Figure 11-20. By symmetry, the field must be radial. We choose for our element of mass the strip shown, which can be considered to be a ring of mass dM. The field due to this strip is given by Equation 11-30 with m replaced by dM.

$$dg_r = -\frac{G\,dM}{s^2} \cos \alpha \qquad\qquad 11\text{-}31$$

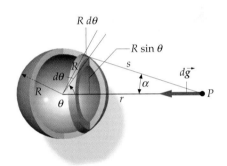

FIGURE 11-20 A uniform thin spherical shell of radius R and total mass M. The strip shown can be considered to be a ring of width $R\,d\theta$ and circumference $2\pi R \sin \theta$.

The mass dM is proportional to the area of the strip dA, which equals the circumference times the width. The radius of the strip is $R \sin \theta$, so the circumference is $2\pi R \sin \theta$. The width is $R\,d\theta$. If M is the total mass of the shell and $A = 4\pi R^2$ is its total area, the mass of the strip of area dA is

$$dM = \frac{M}{A}\,dA = \frac{M}{4\pi R^2}\,2\pi R^2 \sin \theta\,d\theta = \frac{1}{2}M \sin \theta\,d\theta \qquad\qquad 11\text{-}32$$

Substituting this result into Equation 11-31 gives

$$dg_r = -\frac{G\,dM}{s^2} \cos \alpha = -\frac{GM \sin \theta\,d\theta}{2s^2} \cos \alpha \qquad\qquad 11\text{-}33$$

Before integrating over the entire shell, we must eliminate two of the three related variables s, θ, and α. It turns out to be easiest to write everything in terms of s. By the law of cosines, we have

$$s^2 = r^2 + R^2 - 2rR \cos \theta$$

Differentiating gives

$$2s\,ds = +2rR \sin \theta\,d\theta$$

or

$$\sin \theta\,d\theta = \frac{s\,ds}{rR}$$

An expression for cos α can be obtained by again applying the law of cosines to the same triangle. We have

$$R^2 = s^2 + r^2 - 2sr \cos \alpha$$

or

$$\cos \alpha = \frac{s^2 + r^2 - R^2}{2sr}$$

Substituting these results into Equation 11-33 gives

$$dg_r = -\frac{GM \sin \theta \, d\theta}{2s^2} \cos \alpha = -\frac{GM}{2s^2}\left(\frac{s \, ds}{rR}\right)\frac{s^2 + r^2 - R^2}{2sr}$$

$$= -\frac{GM \, ds}{4s^2 r^2 R}(s^2 + r^2 - R^2) = -\frac{GM}{4r^2 R}\left(1 + \frac{r^2 - R^2}{s^2}\right)ds \qquad 11\text{-}34$$

To find the field at P we integrate this over the entire shell. The integration limits for this step depend on whether the field point P lies outside the shell or inside it. For P outside the shell s varies from $r - R$ ($\theta = 0$) to $s = r + R$ ($\theta = 180°$), so the field due to the entire shell is found by integrating from $s = r - R$ to $s = r + R$.

$$g_r = -\frac{GM}{4r^2 R}\int_{r-R}^{r+R}\left(1 + \frac{(r-R)(r+R)}{s^2}\right)ds = -\frac{GM}{4r^2 R}\left[s - \frac{(r-R)(r+R)}{s}\right]_{r-R}^{r+R}$$

Substitution of the upper and lower limits yields $4R$ for the quantity in brackets. Thus,

$$g_r = -\frac{GM}{r^2} \qquad \text{for} \quad r > R$$

which is the same result as Equation 11-26a.

If the field point P is inside the shell (Figure 11-21), the calculation is identical except that s now varies from $R - r$ to $R + r$. Thus,

$$g_r = -\frac{GM}{4r^2 R}\left[s - \frac{(r-R)(r+R)}{s}\right]_{R-r}^{R+r}$$

Substitution of these upper and lower limits yields 0. Therefore,

$$g_r = 0 \qquad \text{for} \quad r < R$$

which is the same as Equation 11-26b.

FIGURE 11-21

EXPLORING

Are tidal forces just as strong as gravitational forces? Find out this, and more, at www.whfreeman.com/tipler5e.

SUMMARY

1. Kepler's laws are *empirical* observations. They can also be derived from Newton's laws.

2. Newton's law of gravity is a *fundamental law* of physics.

3. The gravitational potential energy of a two-particle system relative to $U = 0$ at infinite separation is given by $U = -Gm_1m_2/r$. If the system is bound, its total energy is negative.

4. The gravitational field is a *fundamental physical concept* that describes the condition in space set up by a mass distribution.

Topic	Relevant Equations and Remarks
1. Kepler's Three Laws	Law 1. All planets move in elliptical orbits with the sun at one focus. Law 2. A line joining any planet to the sun sweeps out equal areas in equal times. Law 3. The square of the period of any planet is proportional to the cube of the planet's mean distance from the sun:

$$T^2 = Cr^3 \qquad \textbf{11-2}$$

where C has almost the same value for all planets; from Newton's law of gravity, C can be shown to be $4\pi^2/[G(M_s + M_p)]$. If $M_s \gg M_p$, this can be expressed as

$$T^2 = \frac{4\pi^2}{GM_s} r^3 \qquad \textbf{11-16}$$

Kepler's laws can be derived from Newton's law of gravity. The first and third laws follow from the fact that the force exerted by the sun on the planets varies inversely as the square of the separation distance. The second law follows from the fact that the force exerted by the sun on a planet is along the line joining them, so the orbital angular momentum of the planet is conserved. Kepler's laws also hold for any object orbiting another in an inverse-square field, such as a satellite orbiting a planet.

2. Newton's Law of Gravity	Every point particle exerts on every other point particle an attractive force that is proportional to the masses of the two particles and inversely proportional to the square of the distance separating them.

$$\vec{F}_{1,2} = -\frac{Gm_1m_2}{r_{1,2}^2}\, \hat{r}_{1,2} \qquad \textbf{11-3}$$

Universal gravitational constant	$G = 6.67 \times 10^{-11} \ \text{N·m}^2/\text{kg}^2 \qquad \textbf{11-4}$

3. Gravitational Potential Energy	The gravitational potential energy U for a system consisting of a particle of mass m outside a spherically symmetric object of mass M and at a distance r from its center is

$$U(r) = -\frac{GMm}{r} \qquad \textbf{11-20}$$

This potential-energy function approaches zero as r approaches infinity.

4. Mechanical Energy	The mechanical energy E for a system consisting of a particle of mass m outside a spherically symmetric object of mass M and at a distance r from its center is

$$E = \frac{1}{2}mv^2 - \frac{GMm}{r}$$

| Escape speed | | For a given value of r, the speed of the particle for which $E = 0$ is called the escape speed v_e. That is, if $v = v_e$, then $E = 0$. | |

| 5. **Classification of Orbits** | | If $E < 0$, the system is bound and the orbit is an ellipse (or circle, which is a type of ellipse). If $E \geq 0$, the system is unbound and the orbit is a hyperbola (or a parabola for $E = 0$). | |

6. **Gravitational Field**

| Definition | $\vec{g} = \dfrac{\vec{F}}{m}$ | 11-23 |

| Due to the earth | $\vec{g}(r) = \dfrac{\vec{F}}{m} = -\dfrac{GM_E}{r^2}\hat{r} \quad (r \geq R_E)$ | 11-29 |

| Due to thin spherical shell | Outside the shell the gravitational field is the same as if all the mass of the shell were concentrated at the center. The field inside the shell is zero. | |

$$\vec{g} = -\frac{GM}{r^2}\hat{r} \qquad \text{for} \quad r > R \qquad \text{11-26}a$$

$$\vec{g} = 0 \qquad \text{for} \quad r < R \qquad \text{11-26}b$$

PROBLEMS

- Single-concept, single-step, relatively easy
- •• Intermediate-level, may require synthesis of concepts
- ••• Challenging
- **SSM** Solution is in the *Student Solutions Manual*
- **iSOLVE** Problems available on iSOLVE online homework service
- **iSOLVE** ✔ These "Checkpoint" online homework service problems ask students additional questions about their confidence level, and how they arrived at their answer

In a few problems, you are given more data than you actually need; in a few other problems, you are required to supply data from your general knowledge, outside sources, or informed estimates.

Take $g = 9.81$ N/kg $= 9.81$ m/s^2 **and neglect friction in all problems unless otherwise stated.**

Conceptual Problems

1 • **SSM** True or false:

(a) Kepler's law of equal areas implies that gravity varies inversely with the square of the distance.

(b) The planet closest to the sun, on the average, has the shortest orbital period.

2 • If the mass of a satellite is doubled, the radius of its orbit can remain constant if the speed of the satellite (a) increases by a factor of 8, (b) increases by a factor of 2, (c) does not change, (d) is reduced by a factor of 8, (e) is reduced by a factor of 2.

3 •• **SSM** At the surface of the moon, the acceleration due to the gravity of the moon is a. At a distance from the center of the moon equal to four times the radius of the moon, the acceleration due to the gravity of the moon is (a) $16a$, (b) $a/4$, (c) $a/3$, (d) $a/16$, (e) none of the answers given.

4 • Why is G so difficult to measure?

5 • If the mass of a planet is doubled with no increase in its radius, the escape speed for that planet will be (a) increased by a factor of 1.4, (b) increased by a factor of 2, (c) unchanged, (d) reduced by a factor of 1.4, (e) reduced by a factor of 2.

6 •• A newly discovered comet-like object is observed to make a pass around the sun. How can we tell if the object will return many years later, or if it will never return?

7 •• Explain why the gravitational field is directly proportional to r rather than inversely proportional to r inside a solid sphere of uniform mass.

8 • **SSM** If K is the kinetic energy of Mercury in its orbit around the sun, and U is the potential energy of the Mercury–sun system, what is the relationship between K and U?

9 •• A student whose weight on the surface of the earth is w is lifted to a height that is two earth radii above the surface. Her weight there is (a) $w/2$, (b) $w/4$, (c) $w/3$, (d) $w/9$.

10 •• In the novel "First Men in the Moon" by H.G. Wells, the trip was accomplished in a slightly different manner than by Jules Verne's heroes. The novel's hero, Professor Cavour, invented a material, Cavourite, that shielded objects from the gravitational force. Simply by surrounding an object with Cavourite, the gravitational pull on it by its surroundings was nullified. The adventurers in this book built a ship whose bottom was covered by Cavourite, but whose top was unshielded. Then they pointed the top of the ship at the moon to let its attraction pull them up. As far as real physics goes, Cavourite is an impossibility, because it would allow the violation of the law of conservation of energy. To show this, invent a perpetual motion machine using Cavourite.

Estimation and Approximation

11 • Estimate the mass of the galaxy if the sun orbits the center of the galaxy with a period of 250 million years at a mean distance of 30,000 light-years. Express the mass in terms of multiples of the solar mass M_s. (Neglect the mass farther from the center than the sun, and assume that the mass closer to the center than the sun exerts the same force on the sun as would a point particle of the same mass.)

12 ••• [SSM] One of the great discoveries in astronomy in recent years is the detection of planets outside the solar system. Since 1996, 100 planets have been detected orbiting stars other than the sun. While the planets themselves cannot be seen directly, telescopes can detect the small periodic motion of the star as the star and planet orbit around their common center of mass. (This is measured using the *Doppler effect*, which will be discussed in Chapter 15.) Both the period of this motion and the variation in the velocity of the star over the course of time can be determined observationally. The mass of the star is found from its observed luminance and from the theory of stellar structure. Iota Draconis is the 8th brightest star in the constellation Draco. Observations show that there is a planet orbiting this star with an orbital period of 1.50 y. The mass of Iota Draconis is $1.05M_{sun}$. (a) What is the size (in AU) of the semimajor axis of this planet's orbit? (b) The radial velocity of the star is observed to vary by 592 m/s. Use conservation of momentum to find the mass of the planet. Assume the orbit is circular, we are observing the orbit edge-on, and no other planets orbit Iota Draconis. Express the mass as a multiple of the mass of Jupiter.

13 ••• One of the biggest unresolved problems in the theory of the formation of the solar system is that, while the mass of the sun is 99.9 percent of the total mass of the solar system, it carries only about 2 percent of the total angular momentum. The most widely accepted theory of solar system formation has as its central hypothesis the collapse of a cloud of dust and gas under the force of gravity, with most of the mass forming the sun. However, because the net angular momentum of this cloud is conserved, a simple theory would indicate that the sun should be rotating much more rapidly than it currently is. In this problem, you will show why it is important that most of the angular momentum was somehow transferred to the planets. (a) The sun is a cloud of gas held together by the force of gravity. If the sun were rotating too rapidly, gravity couldn't hold it together. Using the known mass of the sun (1.99×10^{30} kg) and its radius (6.96×10^8 m),

estimate the maximum angular velocity that the sun can have if it is to stay intact. What is the period of rotation corresponding to this rotation rate? (b) Calculate the orbital angular momentum of Jupiter and of Saturn from their masses (318 and 95.1 Earth masses, respectively), mean distances from the sun (778 and 1430 million km, respectively), and orbital periods (11.9 and 29.5 y, respectively). Compare them to the experimentally measured value of the sun's angular momentum of 1.91×10^{41} kg m^2/s. (c) If we were to somehow transfer all of Jupiter's and Saturn's angular momentum to the sun, what would be the sun's new rotational period? Because the sun is not a uniform sphere of gas, its moment of inertia is given by the formula $I = 0.059MR^2$. Compare this to the maximum rotation rate of Part(a).

Kepler's Laws

14 • [SSM] The new comet Alex-Casey is discovered on a very elliptical orbit with a period of 127.4 y. If the closest approach of Alex-Casey to the sun is 0.1 AU, what is its greatest distance from the sun?

15 • The radius of the earth's orbit is 1.496×10^{11} m and that of Uranus is 2.87×10^{12} m. What is the period of Uranus?

16 • [iSOLVE ✔] The asteroid Hektor, discovered in 1907, is in a nearly circular orbit of radius 5.16 AU about the sun. Determine the period of this asteroid.

17 •• The asteroid Icarus, discovered in 1949, was so named because its highly eccentric elliptical orbit brings it close to the sun at perihelion. The eccentricity e of an ellipse is defined by the relation $r_p = a(1 - e)$, where r_p is the perihelion distance and a is the semimajor axis. Icarus has an eccentricity of 0.83 and a period of 1.1 y. (a) Determine the semimajor axis of the orbit of Icarus. (b) Find the perihelion and aphelion distances of the orbit of Icarus.

18 •• There has been much discussion of a manned mission to Mars and the problems that it would entail due to the extremely long time spent in space by the astronauts. To examine this issue in a simple way, consider one possible trajectory for the spacecraft: the "Hohmann transfer orbit." This orbit consists of an elliptical orbit tangent to the orbit of Earth at its point nearest the sun and tangent to the orbit of Mars at the point farthest from the sun. Given that Mars has a mean distance from the sun of 1.52 times the mean sun–Earth distance, calculate the total time spent by the astronauts in transit.

19 •• [SSM] Kepler determined distances in the solar system from his data. For example, he found the relative distance from the sun to Venus (as compared to the distance from the sun to Earth) as follows. Because Venus's orbit is closer to the sun than is Earth's, Venus is a morning or evening star— its position in the sky is never very far from the sun (see Figure 11-22). If we consider the orbit of Venus as a perfect circle, then consider the relative orientation of Venus, Earth, and the sun at maximum extension—when Venus is farthest from the sun in the sky. (a) Under this condition, show that angle b in Figure 11-22 is 90°. (b) If the maximum elongation angle between Venus and the sun is 47°, what is the distance between Venus and the sun in AU?

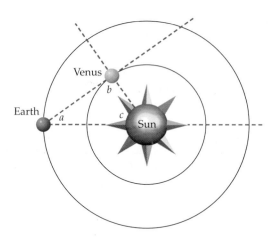

FIGURE 11-22 Problem 19

20 •• At apogee the moon is 406,395 km from the earth and at perigee it is 357,643 km. What is the velocity of the moon at perigee and apogee? Its orbital period is 27.3 d.

Newton's Law of Gravity

21 •• **SSM** Jupiter's satellite Europa orbits Jupiter with a period of 3.55 d at an average distance of 6.71×10^8 m. (a) Assuming that the orbit is circular, determine the mass of Jupiter from the data given. (b) Another satellite of Jupiter, Callisto, orbits at an average distance of 18.8×10^8 m with an orbital period of 16.7 d. Determine the acceleration of Callisto and Europa from the data given, and show that this data is consistent with an inverse square law for gravity [Note: Do NOT use the value of G anywhere in Part (b)].

22 • **SSM** Some people think that shuttle astronauts are "weightless" because they are "beyond the pull of Earth's gravity." In fact, this is completely untrue. (a) What is the magnitude of the acceleration of gravity for shuttle astronauts? A shuttle orbit is about 400 km above the ground. (b) Given the answer in Part (a), why are shuttle astronauts "weightless"?

23 • **SOLVE** The mass of Saturn is 5.69×10^{26} kg. (a) Find the period of its moon Mimas, whose mean orbital radius is 1.86×10^8 m. (b) Find the mean orbital radius of its moon Titan, whose period is 1.38×10^6 s.

24 • **SOLVE**✓ Calculate the mass of the earth from the period of the moon, T = 27.3 d; its mean orbital radius, $r_m = 3.84 \times 10^8$ m; and the known value of G.

25 • **SOLVE**✓ Use the period of the earth (1 y), its mean orbital radius (1.496×10^{11} m), and the value of G to calculate the mass of the sun.

26 • **SSM** **SOLVE**✓ An object is dropped from a height of 6.37×10^6 m above the surface of the earth. What is its initial acceleration?

27 • Suppose you leave the solar system and arrive at a planet that has the same mass-to-volume ratio as the earth but has 10 times the earth's radius. What would you weigh on this planet compared with what you weigh on the earth?

28 • Suppose that the earth retained its present mass but was somehow compressed to half its present radius. What

would be the value of g at the surface of this new, compact planet?

29 • A planet orbits a massive sun. When the planet is at perihelion, it has a speed of 5×10^4 m/s and is 1.0×10^{15} m from the sun. The orbital radius increases to 2.2×10^{15} m at aphelion. What is the planet's speed at aphelion?

30 • What is the acceleration of gravity at the surface of a neutron star whose mass is 1.60 times the mass of the sun and whose radius is 10.5 km?

31 •• **SSM** The speed of an asteroid is 20 km/s at perihelion and 14 km/s at aphelion. Determine the ratio of the aphelion to perihelion distances.

32 •• A satellite with a mass of 300 kg moves in a circular orbit 5×10^7 m above the earth's surface. (a) What is the gravitational force on the satellite? (b) What is the speed of the satellite? (c) What is the period of the satellite?

33 •• **SSM** **SOLVE** A superconducting gravity meter can measure changes in gravity of the order $\Delta g/g = 10^{-11}$. (a) You are hiding behind a tree holding the meter, and your 80-kg friend approaches the tree from the other side. How close to you can your friend get before the meter detects a change in g due to his presence? (b) You are in a hot air balloon and are using the meter to determine the rate of ascent (presumed constant). What is the smallest change in altitude that results in a detectable change in the gravitational field of the earth?

34 •• Suppose that the attractive interaction between a star of mass M and a planet of mass $m \ll M$ were of the form $F = KMm/r$, where K is the gravitational constant. What would be the relation between the radius of the planet's circular orbit and its period?

35 •• **SSM** The mass of the earth is 5.97×10^{24} kg and its radius is 6370 km. The radius of the moon is 1738 km. The acceleration of gravity at the surface of the moon is 1.62 m/s². What is the ratio of the average density of the moon to that of the earth?

Measurement of G

36 • **SOLVE** The large and small spheres in a Cavendish balance (Figure 11-7) have masses $m_1 = 10$ kg and $m_2 = 0.010$ kg, respectively, and the center-to-center separation of the two small spheres is 20 cm. The center-to-center separation of each large sphere with the small sphere adjacent to it is 6 cm. (a) What is the force by which each large sphere attracts the small sphere adjacent to it? (b) What torque must be exerted by the suspension to balance the torque due to these forces?

Gravitational and Inertial Mass

37 • **SOLVE** A standard object defined as having a mass of exactly 1 kg is given an acceleration of 2.6587 m/s² when a certain force is applied to it. A second object of unknown mass acquires an acceleration of 1.1705 m/s² when the same force is applied to it. (a) What is the mass of the second object? (b) Is the mass that you determined in Part (a) gravitational or inertial mass?

38 • ☒☑ The weight of a standard object defined as having a mass of exactly 1 kg is measured to be 9.81 N. In the same laboratory, a second object weighs 56.6 N. (a) What is the mass of the second object? (b) Is the mass you determined in Part (a) gravitational or inertial mass?

39 • SSM The principle of equivalence states that the free-fall acceleration of any object in a gravitational field is independent of the mass of the object. This can be seen from the form of the law of universal gravitation—but how well does it hold experimentally? The Roll-Krotkov-Dicke experiment performed in the 1960s indicates that the free-fall acceleration is independent of mass to at least 1 part in 10^{12}. Suppose two objects are simultaneously released from rest in a uniform gravitational field. Also, suppose one of the objects falls with a constant acceleration of exactly 9.8m/s^2 while the other falls with a constant acceleration that is greater than 9.8 m/s^2 by one part in 10^{12}. How far will the first object have fallen when the second object has fallen 1 mm further than it has? Note that this estimate is an upper bound on the difference in the accelerations; most physicists believe that there is no difference whatsoever.

Gravitational Potential Energy

40 • (a) Taking the potential energy to be zero at infinite separation, find the potential energy of a 100-kg object at the surface of the earth. (Use 6.37×10^6 m for the earth's radius.) (b) Find the potential energy of the same object at a height above the earth's surface equal to the earth's radius. (c) Find the escape speed for a body projected from this height.

41 • A point particle of mass m_0 is initially at the surface of a large sphere of mass M and radius R. How much work is needed to remove it to a very large distance away from the large sphere?

42 • Suppose that in space there is a duplicate earth, except that it has no atmosphere, is not rotating, and is not in motion around any sun. What initial velocity must a spacecraft on this planet's surface have to travel vertically upward a distance equal to one earth radius above the surface of the planet?

43 •• SSM ☒☑ An object is dropped from rest from a height of 4×10^6 m above the surface of the earth. If there is no air resistance, what is its speed when it strikes the earth?

44 •• ☒☑ An object is projected upward from the surface of the earth with an initial speed of 4 km/s. Find the maximum height it reaches.

45 •• ☒☑ A uniform thin spherical shell has a radius R and a mass M. (a) Write expressions for the force exerted by the shell on a point mass m_0 when m_0 is outside the shell and when it is inside the shell. (b) What is the potential-energy function $U(r)$ for this system when the mass m_0 is at a distance r ($r \geq R$) if $U = 0$ at $r = \infty$? Evaluate this function at $r = R$. (c) Using the general relation for $dU = -\vec{F} \cdot d\vec{r} = -F_r dr$, show that U is constant everywhere inside the shell. (d) Using the fact that U is continuous everywhere, including at $r = R$, find the value of the constant U inside the shell. (e) Sketch $U(r)$ versus r for all possible values of r.

46 • ☒☑ The planet Saturn has a mass 95.2 times that of the earth and a radius 9.47 times that of the earth. Find the escape speed for objects on the surface of Saturn.

47 • ☒☑ Find the escape speed for a rocket leaving the moon. The acceleration of gravity on the moon is 0.166 times that on earth and the moon's radius is $0.273R_E$.

48 • SSM The science fiction writer Robert Heinlein once said, "If you can get into orbit, then you're halfway to anywhere." Justify this statement by comparing the kinetic energy needed to place a satellite into low earth orbit ($h = 400$ km) to that needed to set it completely free from the bonds of earth's gravity.

49 •• A particle is projected from the surface of the earth with a speed twice the escape speed. When it is very far from the earth, what is its speed?

50 •• What initial speed should a particle be given if it is to have a final speed when it is very far from the earth that is equal to its escape speed?

51 •• ☒☑ (a) Calculate the energy in joules necessary to launch an object whose mass is 1 kg from the earth at escape speed. (b) Convert this energy to kilowatt-hours. (c) If energy can be obtained at $0.10/kW·h, what is the minimum cost of giving an 80-kg astronaut enough energy to escape the earth's gravitational field?

52 •• An object is projected vertically from the surface of the earth at less than the escape speed. Show that the maximum height reached by the object is $H = R_E H'/(R_E - H')$, where H' is the height that it would reach if the gravitational field were constant.

Orbits

53 •• ☒☑ A spacecraft of 100-kg mass is in a circular orbit about the earth at a height $h = 2R_E$. (a) What is the period of the spacecraft's orbit about the earth? (b) What is the spacecraft's kinetic energy? (c) Express the angular momentum L of the spacecraft about the center of the earth in terms of its kinetic energy K and find its numerical value.

54 • SSM We normally say that the moon orbits the earth, but this is not quite true: instead, both the moon and the earth orbit their common center of mass, which is not at the center of the earth. (a) The mass of the earth is 5.98×10^{24} kg and the mass of the moon is 7.36×10^{22} kg. The mean distance from the center of the earth to the center of the moon is 3.82×10^8 m. How far above the surface of the earth is the center of mass of the earth–moon system? (b) Estimate the mean "orbital speed" of the earth as it orbits the center of mass of the earth–moon system, using the moon's mass, period, and average distance from the earth. Ignore any external forces acting on the system. The orbital period of the moon is 27.3 d.

55 •• ☒☑ Many satellites orbit the earth about 1000 km above the earth's surface. Geosynchronous satellites orbit at a distance of 4.22×10^7 m from the center of the earth. How much more energy is required to launch a 500-kg satellite into a geosynchronous orbit than into an orbit 1000 km above the surface of the earth?

56 •• Using the period of the moon's orbit around the earth (27.3 d), its average distance from the center of the earth (3.82×10^8 m), the length of a year (365.25 d), and the average distance from the earth to the sun (1.50×10^{11} m), establish the ratio of the mass of the sun to the mass of the earth. Compare this to the measured ratio of 3.33×10^5.

The Gravitational Field

57 • **SOLVE** A 3-kg mass experiences a gravitational force of 12 N $\hat{i}$ at some point P. What is the gravitational field at that point?

58 • **SSM** **SOLVE** The gravitational field at some point is given by $\vec{g} = 2.5 \times 10^{-6}$ N/kg $\hat{j}$. What is the gravitational force on a mass of 4 g at that point?

59 •• A point mass m is on the x axis at $x = L$ and an identical point mass is on the y axis at $y = L$. (a) Find the gravitational field at the origin. (b) What is the magnitude of this field?

60 •• Five objects of mass M are equally spaced on the arc of a semicircle of radius R as in Figure 11-23. An object of mass m is located at the center of curvature of the arc. (a) If M is 3 kg, m is 2 kg, and R is 10 cm, what is the force on m due to the five objects? (b) If the object whose mass is m is removed, what is the gravitational field at the center of curvature of the arc?

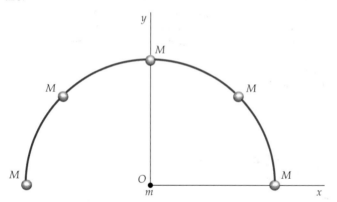

FIGURE 11-23 Problem 60

61 •• A point particle of mass $m_1 = 2$ kg is at the origin and a second point particle of mass $m_2 = 4$ kg is on the x axis at $x = 6$ m. Find the gravitational field at (a) $x = 2$ m and (b) $x = 12$ m. (c) Find the point on the x axis for which $g = 0$.

62 •• Show that the maximum value of $|g_x|$ for the field of Example 11-7 occurs at the points $x = \pm a/\sqrt{2}$.

63 •• A nonuniform rod of length L lies on the x axis with one end at the origin and the other at $x = L$. Its mass per unit length λ varies as $\lambda = Cx$, where C is a constant. (Thus an element of the rod has mass $dm = \lambda\,dx$.) (a) What is the total mass of the rod? (b) Find the gravitational field due to the rod on the x axis at a point x_0, where $x_0 > L$.

64 ••• A uniform rod of mass M and length L lies along the x axis with its center at the origin. Consider an element of length dx at point x, where $-\frac{1}{2}L < x < \frac{1}{2}L$. (a) Show that this

element produces a gravitational field at a point x_0 on the x axis ($x_0 > \frac{1}{2}L$) given by

$$dg_x = -\frac{GM}{L(x_0 - x)^2}\,dx$$

(b) Integrate this result over the length of the rod to find the total gravitational field at the point x_0 due to the rod. (c) What is the force on an object of mass m_0 at x_0? (d) Show that for $x_0 \gg L$, the field is approximately equal to that of a point particle of mass M at $x = 0$.

$\vec{g}$ due to Spherical Objects

65 • **SOLVE** A thin spherical shell has a radius of 2 m and a mass of 300 kg. What is the gravitational field at the following distances from the center of the shell: (a) 0.5 m, (b) 1.9 m, (c) 2.5 m?

66 • **SOLVE** A thin spherical shell has a radius of 2 m and a mass of 300 kg, and its center is located at the origin of a coordinate system. Another thin spherical shell with a radius of 1 m and mass 150 kg is inside the larger shell with its center at 0.6 m on the x axis. What is the gravitational force of attraction between the two shells?

67 • **SSM** Two solid spheres, S_1 and S_2, have equal radii R and equal masses M. The density of sphere S_1 is constant, whereas that of sphere S_2 depends on the radial distance according to $\rho(r) = C/r$. If the acceleration of gravity at the surface of sphere S_1 is g_1, what is the acceleration of gravity at the surface of sphere S_2?

68 •• Two homogeneous solid spheres, S_1 and S_2, have equal masses but different radii, R_1 and R_2. If the acceleration of gravity on the surface of sphere S_1 is g_1, what is the acceleration of gravity on the surface of sphere S_2?

69 •• Two concentric uniform thin spherical shells have masses M_1 and M_2 and radii a and $2a$ as in Figure 11-24. What is the magnitude of the gravitational force on a point particle of mass m located (a) a distance $3a$ from the center of the shells? (b) a distance $1.9a$ from the center of the shells? (c) a distance $0.9a$ from the center of the shells?

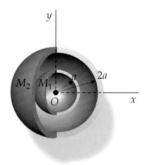

FIGURE 11-24 Problems 69, 70

70 •• The inner spherical shell in Problem 69 is shifted so that its center is now on the x axis at $x = 0.8a$. What is the magnitude of the gravitational force on a particle of point mass m located on the x axis at (a) $x = 3a$, (b) $x = 1.9a$, (c) $x = 0.9a$?

$\vec{g}$ Inside Solid Spheres

71 •• [SSM] Is your "weight" (as measured on a spring scale) at the bottom of a deep mine shaft greater than or less than your weight at the surface of the earth? Model the earth as a homogeneous sphere. Consider the effects in (a) and (b).
(a) Show that the force of gravity on you from a uniform-density, perfectly spherical planet is proportional to your distance from the center of the planet.
(b) Show that your effective "weight" increases linearly due to the effects of rotation as you approach the center. (Consider a mine shaft located on the equator.)
(c) Which effect is more important for the earth? Use $M = 5.98 \times 10^{24}$ kg, $R = 6370$ km, and $T = 24$ h.

72 •• Suppose the earth were a nonrotating uniform sphere. If there were an elevator shaft going 15,000 m into the earth, what would be the decrease in the weight of a student who weighs 800 N at the surface of the earth as he is lowered from the surface to the bottom of the shaft?

73 •• A solid sphere of radius R has its center at the origin. It has a uniform mass density ρ_0, except that there is a spherical cavity in it of radius $r = \frac{1}{2}R$ centered at $x = \frac{1}{2}R$ as in Figure 11-25. Find the gravitational field at points on the x axis for $|x| > R$. (*Hint: The cavity may be thought of as a sphere of mass $m = (4/3)\pi r^3 \rho_0$ plus a sphere of mass $-m$.*)

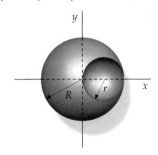

FIGURE 11-25 Problems 73, 74

74 ••• For the sphere with the cavity in Problem 73, show that the gravitational field inside the cavity is uniform and find its magnitude and direction.

75 ••• A straight, smooth tunnel is dug through a spherical planet whose mass density ρ_0 is constant. The tunnel passes through the center of the planet and is perpendicular to the planet's axis of rotation, which is fixed in space. The planet rotates with a constant angular velocity ω so that objects in the tunnel have no apparent weight. Find ω.

76 ••• [SOLVE] The density of a sphere is given by $\rho(r) = C/r$. The sphere has a radius of 5 m and a mass of 1011 kg. (a) Determine the constant C. (b) Obtain expressions for the gravitational field for (1) $r > 5$ m and (2) $r < 5$ m.

77 ••• [SSM] [SOLVE] A small diameter hole is drilled into the sphere of Problem 76 toward the center of the sphere to a depth of 2 m below the sphere's surface. A small mass is dropped from the surface into the hole. Determine the speed of the small mass as it strikes the bottom of the hole.

78 ••• [SOLVE]✓ The crust of the earth is about 40 km thick and has a density of about 3000 kg/m³. A spherical de-

posit of heavy metals with a density of 8000 kg/m³ and radius of 1000 m is centered 2000 m below the surface. Find $\Delta g/g$ at the surface directly above this deposit, where Δg is the increase in the gravitational field due to the deposit.

79 ••• [SSM] Two identical spherical cavities are made in a lead sphere of radius R. The cavities have a radius $R/2$. They touch the outside surface of the sphere and its center as in Figure 11-26. The mass of a solid uniform lead sphere is M. (a) Find the force of attraction of a point particle of mass m at the position shown in the figure to the cavitated lead sphere. (b) What is the attractive force on the point particle if it is located on the x axis at $x = R$?

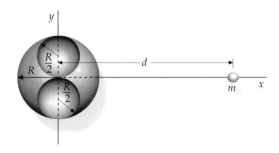

FIGURE 11-26 Problem 79

80 •• A globular cluster is a roughly spherical collection of up to millions of stars bound together by the force of gravity. Astronomers can measure the velocities of stars in the cluster to study its composition and to get an idea of the mass distribution within the cluster. Assuming that all of the stars have about the same mass and are distributed uniformly within the cluster, show that the mean speed of a star in a circular orbit around the center of the cluster should increase linearly with its distance from the center.

General Problems

81 • [SSM] The mean distance of Pluto from the sun is 39.5 AU. Find the period of Pluto.

82 • Calculate the mass of the earth using the known values of G, g, and R_E.

83 •• The force exerted by the earth on a particle of mass m at a distance r ($r > R_E$) from the center of the earth has the magnitude mgR_E^2/r^2, where $g = GM_E/R_E^2$. (a) Calculate the work you must do to move the particle from distance r_1 to distance r_2. (b) Show that when $r_1 = R_E$ and $r_2 = R_E + h$, the result can be written $W = mgR_E^2[(1/R_E) - 1/(R_E + h)]$. (c) Show that when $h \ll R_E$, the work is given approximately by $W = mgh$.

84 •• A uniform sphere of radius 100 m and density 2000 kg/m³ is in free space far from other massive objects. (a) Find the gravitational field outside of the sphere as a function of r. (b) Find the gravitational field inside the sphere as a function of r.

85 •• Jupiter has a mass 320 times that of Earth and a volume 1320 times that of Earth. A "day" on Jupiter is 9 h 50 min long. Find the height h above Jupiter at which a satellite must be orbiting to have a period equal to one Jovian day.

86 •• The average density of the moon is $\rho = 3340\ \text{kg/m}^3$. Find the minimum possible period T of a spacecraft orbiting the moon.

87 •• A satellite is circling around the moon (radius 1700 km) close to the surface at a speed v. A projectile is launched from the moon vertically up at the same initial speed v. How high will it rise?

88 •• [SSM] In the novel "A Voyage to the Moon" by Jules Verne, astronauts were launched by a giant cannon from the earth to the moon. (a) If we estimate the velocity needed to reach the moon as the earth's escape velocity, and the length of the cannon as 900 ft (as stated in the book), what is the probability that the astronauts would have survived "liftoff"? (b) In the book, the astronauts in their ship feel the effects of the earth's gravity until they reach the balance point where the moon's gravitational pull on the ship equals that of the earth. At this point, everything flips around and what was the ceiling in the ship becomes the floor. How far away from the center of the earth is this balance point? (c) Is this description of what the astronauts experienced reasonable? Is this what actually happened to the Apollo astronauts on their visit to the moon?

89 •• In a binary star system, two stars follow circular orbits about their common center of mass. If the stars have masses m_1 and m_2 and are separated by a distance r, show that the period of rotation is related to r by $T^2 = 4\pi r^3/[G(m_1 + m_2)]$.

90 •• Two particles of masses m_1 and m_2 are released from rest at a large separation distance. Find their speeds v_1 and v_2 when their separation distance is r. The initial separation distance is given as large, but large is a relative term. Relative to what distance is it large?

91 •• [SSM] Four identical planets are arranged in a square as shown in Figure 11-27. If the mass of each planet is M and the edge length of the square is a, what must be their speed if they are to orbit their common center under the influence of their mutual attraction?

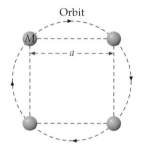

FIGURE 11-27 Problem 91

92 •• A hole is drilled from the surface of the earth to its center as in Figure 11-28. Ignore the earth's rotation and air resistance, and model the earth as a uniform sphere. (a) How much work is required to lift a particle of mass m from the center of the earth to the earth's surface? (b) If the particle is dropped from rest at the surface of the earth, what is its speed when it reaches the center of the earth? (c) What is the escape speed for a particle projected from the center of the earth? Express your answers in terms of m, g, and R_E.

FIGURE 11-28 Problem 92

93 •• A thick spherical shell of mass M and uniform density has an inner radius R_1 and an outer radius R_2. Find the gravitational field g_r as a function of r for $0 < r < \infty$. Sketch a graph of g_r versus r.

94 •• (a) Sketch a plot of the gravitational field g_x versus x due to a uniform ring of mass M and radius R whose axis is the x axis. (b) At what points is the magnitude of g_x a maximum?

95 ••• Find the magnitude of the gravitational field a distance r from an infinitely long wire whose mass per unit length is λ.

96 ••• One big question in planetary science is whether each of the rings of Saturn is solid or composed of many smaller satellites. There is a simple observation that can now be made to resolve this issue. Measure the velocity of the inner and outer portion of the ring: if the inner portion of the ring moves more slowly than the outer portion, then the ring is solid; if the opposite is true, then it is composed of many separate chunks. (a) If the thickness of the ring is r, the average distance of the ring from the center of Saturn is R, and the average velocity of the ring is v, show that $v_{\text{out}} - v_{\text{in}} \approx rv/R$ if the ring is solid. Here, v_{out} is the speed of the outermost portion of the ring, v_{in} is the speed of the innermost portion, and v is the average velocity of the ring. (b) If, however, the ring is composed of many small chunks, show that $v_{\text{out}} - v_{\text{in}} \approx -\frac{1}{2}rv/R$. (Assume that $r \ll R$.)

97 ••• In this problem you are to find the gravitational potential energy of the rod in Example 11-8 and a point mass m_0 that is on the x axis at x_0. (a) Show that the potential energy shared by an element of the rod of mass dm (shown in Figure 11-13) and a point particle of mass m_0 located on the x axis at $x_0 \geq \frac{1}{2}L$ is given by

$$dU = -\frac{Gm_0\,dm}{x_0 - x} = \frac{GMm_0}{L(x_0 - x)}\,dx$$

where $U = 0$ at $x_0 = \infty$. (b) Integrate your result for Part (a) over the length of the rod to find the total potential energy for the system. Write your result as a general function $U(x)$ by setting x_0 equal to a general point x. (c) Compute the force on m_0 at a general point x from $F_x = -dU/dx$ and compare your result with m_0g, where g is the field at x_0 calculated in Example 11-8.

98 ••• [SSM] A uniform sphere of mass M is located near a thin, uniform rod of mass m and length L as in Figure 11-29. Find the gravitational force of attraction exerted by the sphere on the rod.

FIGURE 11-29 Problem 98

99 ••• A uniform rod of mass $M = 20$ kg and length $L = 5$ m is bent into a semicircle. What is the gravitational force exerted by the rod on a point mass $m = 0.1$ kg located at the center of curvature of the circular arc?

100 ••• SSM Both the sun and the moon exert gravitational forces on the oceans of the earth, causing tides. (*a*) Show that the ratio of the force exerted on a point particle on earth by the sun to that exerted by the moon is $M_s r_m^2/M_m r_s^2$, where M_s and M_m are the masses of the sun and moon and r_s and r_m are the distances from the earth to the sun and to the moon. Evaluate this ratio. (*b*) Even though the sun exerts a much greater force on the oceans than does the moon, the moon has a greater ef-

fect on the tides because it is the difference in the force from one side of the earth to the other that is important. Differentiate the expression $F = Gm_1m_2/r^2$ to calculate the change in F due to a small change in r. Show that $dF/F = (-2\ dr)/r$. (*c*) During one full day, the rotation of the earth can cause the distance from the sun or moon to an ocean to change by, at most, the diameter of the earth. Show that for a small change in distance, the change in the force exerted by the sun is related to the change in the force exerted by the moon by $\Delta F_s/\Delta F_m \approx (M_s r_m^3)/(M_m r_s^3)$. Calculate this ratio.

101 •• United Federation Spaceship *Excelsior* is dropping two robot probes to the surface of a neutron star for exploration. The mass of the star is the same as that of the sun, but the star's diameter is only 10 km. The robot probes are linked together by a 1-m-long steel cord, and are dropped vertically (that is, one always above the other). (*a*) Explain why there seems to be a "force" trying to pull the robots apart. (See Problem 99 for a hint.) (*b*) How close will the robots be to the surface of the star before the cord breaks? Assume that the cord has a breaking tension of 25,000 N and that the robots each have a mass of 1 kg.

Static Equilibrium and Elasticity

THE PLANK IN THIS PHOTO IS IN STATIC EQUILIBRIUM. THE OVERHANG IS ONE-SIXTH THE LENGTH OF THE PLANK, AND THE PLANK REMAINS BALANCED WHEN A STUDENT STANDS WITH HIS CENTER OF MASS OVER THE END OF THE PLANK.

? **What is the maximum ratio of the mass of the student to the mass of the plank if the plank remains balanced when the student stands with his center of mass over the very end of the plank? (See Example 12-1.)**

If an object is stationary and remains stationary, it is said to be in static equilibrium. Being able to determine the forces acting on an object in static equilibrium has many important applications. For example, the forces exerted by the cables of a suspension bridge must be known so that the cables can be designed to be strong enough to support the bridge. Similarly, cranes must be designed so that they do not topple over when lifting a weight.

The forces exerted by the cables and beams in a structure are called elastic forces. They are the result of slight deformations—the stretching or compression of solid objects under stress from bearing loads.

➤ **In this chapter, we study the equilibrium of rigid bodies and then briefly consider the deformations and elastic forces that arise when real solids are under stress.**

12-1 Conditions for Equilibrium

A necessary condition for a particle at rest to remain at rest is that the net force acting on it remain zero. Similarly, a necessary condition for the center of mass of a rigid object to remain at rest is that the net force acting on the object remain zero. A rigid object can rotate, even when its center of mass is at rest, but then the object is not in static equilibrium. Therefore, a second necessary condition for a rigid object to remain in static equilibrium is that the net torque acting on it about *any* point must remain zero. This gives us the option to choose the point P when calculating torques, which greatly simplifies the solution of most static problems.

The two necessary conditions for a rigid body to be in static equilibrium are as follows:

1. The net external force acting on the body must remain zero:

$$\sum \vec{F} = 0 \qquad\qquad 12\text{-}1$$

2. The net external torque about *any* point must remain zero:

$$\sum \vec{\tau} = 0 \qquad\qquad 12\text{-}2$$

CONDITIONS FOR EQUILIBRIUM

12-2 The Center of Gravity

Figure 12-1a shows a rigid object in static equilibrium and a point O. We consider the object to be composed of many small particles. The weight of the ith small particle is $\vec{w}_i$, and the total weight of the object is $\vec{W} = \Sigma \vec{w}_i$. If $\vec{r}_i$ is the position vector of the ith particle relative to O, then $\vec{\tau}_i = \vec{r}_i \times \vec{w}_i$, where $\vec{\tau}_i$ the torque due to $\vec{w}_i$ about O. The net gravitational torque about O is then $\vec{\tau}_{net} = \Sigma(\vec{r}_i \times \vec{w}_i)$. Conveniently, the net torque due to gravity about any point can be calculated as if the entire weight $\vec{W}$ were applied at a single point, the **center of gravity** (see Figure 12-1b). That is,

$$\vec{\tau}_{net} = \vec{r}_{cg} \times \vec{W} \qquad\qquad 12\text{-}3$$

CENTER OF GRAVITY DEFINED

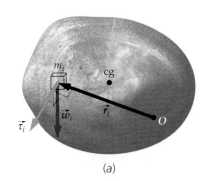

(a)

where $\vec{r}_{cg}$ is the position vector of the center of gravity relative to O.

If the gravitational field $\vec{g}$ is uniform over the object (as is nearly always the case for objects of less than astronomical size), we can write $\vec{w}_i = m_i \vec{g}$. Summing both sides of this gives $\vec{W} = M\vec{g}$, where $M = \Sigma m_i$ is the mass of the object. The net torque is the sum of the individual torques. That is,

$$\vec{\tau}_{net} = \sum_i (\vec{r}_i \times \vec{w}_i) = \sum_i (\vec{r}_i \times m_i \vec{g}) = \sum_i (m_i \vec{r}_i \times \vec{g})$$

Factoring $\vec{g}$ from the term on the right gives

$$\vec{\tau}_{net} = \left(\sum_i m_i \vec{r}_i \right) \times \vec{g}$$

and substituting $M\vec{r}_{cm}$ for $\Sigma m_i \vec{r}_i$ using the definition of center of mass ($M\vec{r}_{cm} = \Sigma m_i \vec{r}_i$), we obtain

$$\vec{\tau}_{net} = M\vec{r}_{cm} \times \vec{g} = \vec{r}_{cm} \times M\vec{g} = \vec{r}_{cm} \times \vec{W} \qquad\qquad 12\text{-}4$$

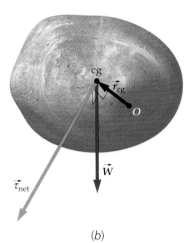

(b)

FIGURE 12-1

Equations 12-3 and 12-4 are valid for any choice of the point O only if $\vec{r}_{cg} = \vec{r}_{cm}$. That is, the center of gravity and the center of mass coincide if the object is in a uniform gravitational field.

If O is directly above the center of gravity, then $\vec{r}_{cg}$ and $\overrightarrow{W}$ are both in the same direction (downward), so $\vec{\tau}_{net} = \vec{r}_{cg} \times \overrightarrow{W} = 0$. For example, when a mobile is suspended with its center of gravity directly below its suspension point, the net torque on the mobile about the suspension point is zero, so it is in static equilibrium.

12-3 Some Examples of Static Equilibrium

For most examples and problems in this chapter, all the forces are perpendicular to the z axis. For such problems it is best to calculate torques about an axis parallel to the z axis. Out of the page is frequently chosen as the positive z direction. This is equivalent to choosing counterclockwise as positive and clockwise as negative.

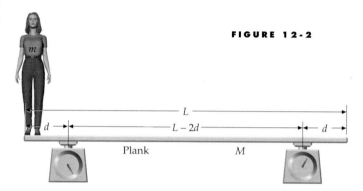

FIGURE 12-2

EXAMPLE 12-1

A uniform plank of length $L = 3$ m and mass $M = 35$ kg is supported by scales a distance $d = 0.5$ m from the ends of the board, as shown in Figure 12-2. (*a*) Find the reading on the scales when Mary, whose mass $m = 45$ kg, stands on the left end of the plank. (*b*) Sergio climbs onto the plank and walks toward Mary, who jumps to the floor when the plank starts to tip. Sergio keeps walking all the way to the left end of the plank, and when he gets there the scale supporting the right end of the plank reads zero. What is Sergio's mass?

PICTURE THE PROBLEM The readings on the scales are the magnitudes of the forces they exert on the boards. To find these magnitudes we apply the two conditions for equilibrium.

FIGURE 12-3

(*a*) 1. Draw a free-body diagram of the system consisting of Mary and the plank (Figure 12-3). Forces $\vec{F}_L$ and $\vec{F}_R$ are the forces exerted by the left and right scales:

2. Set the net force equal to zero, taking upward as positive:

$$\sum F_y = 0$$

$$F_L + F_R - Mg - mg = 0$$

3. Calculate the net torque about the axis directed out of the page (making counterclockwise positive) and through the point of application of $\vec{F}_L$:

$$\sum \tau = F_R(L - 2d) - Mg\frac{L - 2d}{2} + mgd$$

4. Set the net torque equal to zero and solve for F_R:

$$0 = F_R(L - 2d) - Mg\frac{L - 2d}{2} + mgd$$

so

$$F_R = \left(\frac{1}{2}M - \frac{d}{L - 2d}m\right)g$$

5. Substitute this result for F_R into step 2 and solve for F_L:

$$F_L = Mg + mg - F_R = \left(\frac{1}{2}M + \frac{L-d}{L-2d}m\right)g$$

6. Substitute numerical values to obtain numerical values for the forces:

$$F_R = \left(\frac{1}{2}(35\text{ kg}) - \frac{0.5\text{ m}}{2\text{ m}}45\text{ kg}\right)(9.81\text{ N/kg})$$

$$= \boxed{61.3\text{ N}}$$

$$F_L = \left(\frac{1}{2}(35\text{ kg}) + \frac{2.5\text{ m}}{2\text{ m}}45\text{ kg}\right)(9.81\text{ N/kg})$$

$$= \boxed{723.5\text{ N}}$$

(b) Using the Part (a) step 4 result, set $F_R = 0$ and solve for m:

$$0 = \left(\frac{1}{2}M - \frac{d}{L-2d}m\right)g$$

so

$$m = \frac{L-2d}{2d}M = \frac{2\text{ m}}{1\text{ m}}(35\text{ kg}) = \boxed{70\text{ kg}}$$

FIGURE 12-4

PLAUSIBILITY CHECK The sum of the two forces in the Part (a) step 6 results should equal Mary's weight plus the weight of the plank. The total weight is $(M + m)g = (35\text{ kg} + 45\text{ kg})(9.81\text{ N/kg}) = 785\text{ N}$. Also, $F_L + F_R = 723.5\text{ N} + 61.3\text{ N} = 785\text{ N}$. Also, $F_R < F_L$ as one would expect.

REMARKS Sergio is 0.5 m from the axis and the center of mass of the plank is 1 m from the axis when the system is balanced with $F_R = 0$. Thus, Sergio's mass is twice the mass of the plank.

Example 12-1 can be solved using an axis through the center of the plank, but in this case both F_L and F_R appear in the torque equation, hence the algebra is a bit more complex. In general, a statics problem can be simplified by computing the torques about an axis through the line of action of one of the unknown forces, as when we chose the axis through the point of application of force F_L in the example.

Obtain a simple solution by choosing an axis through the point of application of the force you have the least information about to calculate the torques.

PROBLEM SOLVING GUIDELINE

y

FIGURE 12-5

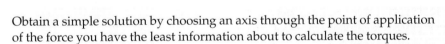

FORCE ON AN ELBOW   **EXAMPLE   12-2**

You hold a 6-kg weight in your hand with your forearm making a 90° angle with your upper arm, as shown in Figure 12-4. Your biceps muscle exerts an upward force $\vec{F}_m$ that acts 3.4 cm from the pivot point O at the elbow joint. Model the forearm and hand as a 30-cm-long uniform rod with a mass of 1 kg. (a) Find the magnitude of $\vec{F}_m$ if the distance from the weight to the pivot point (elbow joint) is 30 cm and (b) find the magnitude and direction of the force exerted on the elbow joint by the upper arm.

PICTURE THE PROBLEM To find the two forces, apply the two conditions for static equilibrium ($\Sigma F = 0$ and $\Sigma\tau = 0$) to the forearm.

(a) 1. Draw a free-body diagram of the forearm (Figure 12-5).
      Model the forearm as a horizontal rod.

2. The force we know least about is the force of the upper arm on the elbow joint $\vec{F}_{ua}$ (we know neither its magnitude nor its direction). Apply $\Sigma\tau = 0$ about an axis directed out of the page and through the point of application of $\vec{F}_{ua}$:

$$F_{ua}(0) - m_h g \frac{L}{2} + F_m d - mgL = 0$$

so

$$F_m = \left(\frac{1}{2}m_h + m\right)g\frac{L}{d}$$

$$= \left(\frac{1}{2}(1\text{ kg}) + 6\text{ kg}\right)(9.81\text{ N/kg})\frac{30\text{ cm}}{3.4\text{ cm}}$$

$$= \boxed{563\text{ N}}$$

(b) Apply $\Sigma F_x = 0$ and $\Sigma F_y = 0$ to obtain $\vec{F}_{ua}$:

$$F_{ua,x} + 0 + 0 + 0 = 0$$

and

$$F_{ua,y} + F_m - m_h g - mg = 0$$

so

$$F_{ua,x} = 0$$

and

$$F_{ua,y} = (m + m_h)g - F_m$$

$$= (7\text{ kg})(9.81\text{ N/kg}) - 563\text{ N}$$

$$= -494\text{ N}$$

Therefore

$$\vec{F}_{ua} = \boxed{494\text{ N, down}}$$

REMARKS The force that must be exerted by the muscle is 9.6 times the weight of the object! In addition, as the muscle pulls upward, the upper arm must push downward to keep the forearm in equilibrium. The force exerted by the upper arm is 8.4 times greater than the object's weight.

EXERCISE Show that F_{ua} can be found in one step by choosing the pivot point to be where the biceps attaches to the forearm. (*Answer* Setting net torque equal to zero gives $F_{ua}(3.4\text{ cm}) + F_m(0) - (6\text{ kg})(9.81\text{ N/kg})(30\text{ cm} - 3.4\text{ cm}) - (1\text{ kg})(9.81\text{ N/kg})(15\text{ cm} - 3.4\text{ cm}) = 0$. This yields $F_{ua} = 494\text{ N}$.)

REMARKS This example and exercise show that we can choose the pivot point wherever it is convenient for our calculation.

HANGING A SIGN **EXAMPLE 12-3** Try It Yourself

The manager of the campus bookstore has ordered a new sign to hang in front of the store. The sign has a mass of 20 kg, and the manager plans to hang the sign from the end of a rod that will be attached to the wall by a wire (Figure 12-6). The manager needs to know how strong a wire is needed, so she asks you, a physics student, to calculate the tension in the wire. She is also concerned about how much force the rod puts on the wall, so she asks you to calculate that as well. The rod has a length of 2 m and a mass of 4 kg, and the wire is attached to a point 1 m above the rod on the wall.

FIGURE 12-6

PICTURE THE PROBLEM There are three conditions for the rod to be in equilibrium: $\Sigma F_x = 0$, $\Sigma F_y = 0$, and $\Sigma \tau = 0$, and we have three unknowns: T and the components F_x and F_y of the force exerted by the wall on the rod. The force exerted by the rod on the wall is equal but opposite to the force exerted by the wall on the rod.

Cover the column to the right and try these on your own before looking at the answers.

Steps	Answers
1. Draw a free-body diagram for the rod (Figure 12-7).	
2. Set $\Sigma \tau = 0$ about point O.	$TL \sin \theta - MgL - mg\dfrac{L}{2} = 0$
3. Use trigonometry to solve for θ.	$\theta = \tan^{-1}\tfrac{1}{2} = 26.6°$
4. Solve the step 2 result for T.	$T = \boxed{483 \text{ N}}$
5. Set $\Sigma F_x = 0$ and $\Sigma F_y = 0$ and, using your values for T and θ, solve for F_x and F_y.	$F_x + T_x = 0$
	$F_y + T_y - Mg - mg = 0$
	so
	$F_x = 432 \text{ N}, \qquad F_y = 19.2 \text{ N}$
6. Solve for the force $\vec{F}'$ exerted by the rod on the wall. The force exerted by the rod on the wall and that by the wall on the rod constitute an action–reaction pair.	$\vec{F}' = -\vec{F} = \boxed{-432 \text{ N}\hat{i} - 19.2 \text{ N}\hat{j}}$

FIGURE 12-7

RAISING A WHEEL   **EXAMPLE 1 2 - 4   Try It Yourself**

A wheel of mass M and radius R (Figure 12-8) rests on a horizontal surface against a step of height h ($h < R$). The wheel is to be raised over the step by a horizontal force $\vec{F}$ applied to the axle of the wheel as shown. Find the minimum force $F_{\min}$ necessary to raise the wheel over the step.

PICTURE THE PROBLEM If the magnitude of F is less than $F_{\min}$, the surface at the bottom of the wheel exerts an upward normal force on the wheel. If F is increased, this normal force decreases. Apply the conditions for static equilibrium to find the value of F that will hold the wheel in place when the normal force is zero.

FIGURE 12-8

Cover the column to the right and try these on your own before looking at the answers.

Steps	Answers
1. Draw a free-body diagram of the wheel (Figure 12-9).	
2. Apply $\Sigma \tau = 0$ to the wheel. Both the direction and the magnitude of $\vec{F}'$ are unknown, so follow the guidelines and calculate torques about an axis through its point of application. Obtain expressions for the lever arms from the free-body diagram and solve for $F_{\min}$.	$F_{\min}(R - h) - Mgx = 0$
	so
	$F_{\min} = \dfrac{Mgx}{R - h}$

FIGURE 12-9

3. Use the Pythagorean theorem to express x in terms of h and R.

$$x = \sqrt{h(2R - h)}$$

4. Set the magnitudes of the torques equal to each other and solve for F.

$$F_{min} = \left| \frac{Mg\sqrt{h(2R - h)}}{R - h} \right|$$

REMARKS Applying $\Sigma\tau = 0$ about the axis through the center of the wheel shows that $\vec{F}'$ is directed toward the wheel's center; otherwise there would be a nonzero net torque.

Frictionless wall

Rough floor

5 m

4 m

$w = 60$ N

3 m

A LEANING LADDER   **EXAMPLE   12-5**

A uniform 5-m ladder weighing 60 N leans against a frictionless vertical wall, as shown in Figure 12-10. The foot of the ladder is 3 m from the wall. What is the minimum coefficient of static friction necessary between the ladder and the floor if the ladder is not to slip?

FIGURE 12-10

PICTURE THE PROBLEM There are three conditions for the ladder to be in equilibrium: $\Sigma F_x = 0$, $\Sigma F_y = 0$, and $\Sigma\tau = 0$. Apply these along with $f_s \leq \mu_s F_n$ to solve for μ_s.

1. Draw a free-body diagram of the ladder as shown in Figure 12-11. The forces acting on the ladder are the force due to gravity $\vec{w}$, the force $\vec{F}_1$ exerted by the wall (since the wall is frictionless, it exerts only a normal force), and the force exerted by the floor, which consists of a normal component F_n and a static frictional component f_s. The three conditions for static equilibrium determine F_1, f_s, and F_n.

$\vec{F}_1$

4 m

$\vec{w}$

$\vec{F}_n$

f_s

1.5 m   1.5 m

FIGURE 12-11

2. The minimum coefficient of static friction relates the frictional force f_s and normal force F_n:

$$\mu_s \geq \frac{f_s}{F_n} \quad \text{so} \quad \mu_{s,min} = \frac{f_s}{F_n}$$

3. Set $\Sigma F_x = 0$ and $\Sigma F_y = 0$:

$$f_s - F_1 = 0 \quad \text{and} \quad F_n - w = 0$$

4. Solve for f_s and F_n:

$$f_s = F_1 \quad \text{and} \quad F_n = w = 60 \text{ N}$$

5. Following the guideline, we set $\Sigma\tau = 0$ about an axis directed out of the page and through the foot of the ladder, the point of application of the force we know the least about:

$$F_1(4 \text{ m}) - w(1.5 \text{ m}) = 0$$

6. Solve for the force F_1:

$$F_1 = \frac{w(15 \text{ m})}{4 \text{ m}} = \frac{(60 \text{ N})(1.5 \text{ m})}{4 \text{ m}} = 22.5 \text{ N}$$

7. Use this result for F_1, and $f_s = F_1$ from step 4, to find f_s:

$$f_s = F_1 = 22.5 \text{ N}$$

8. Use the results for f_s and F_n to obtain the minimum value of μ_s from step 2:

$$\mu_{s,min} = \frac{f_s}{F_n} = \frac{22.5 \text{ N}}{60 \text{ N}} = \boxed{0.375}$$

REMARKS There is another way to look at this problem. In the free-body diagram for the ladder shown in Figure 12-12, the lines of action of the weight $\vec{w}$ and the force $\vec{F}_1$ exerted by the wall intersect at point P. The line of action of the resultant force exerted by the ground, $\vec{f}_s + \vec{F}_n$, must also go through point P or there would be an unbalanced torque about this point. The tangent of θ' equals $4 \text{ m}/1.5 \text{ m} = 2.67 = F_n/f_s$. Whenever an object is in static equilibrium under the influence of three nonparallel forces, the lines of action of the forces must intersect at one point.

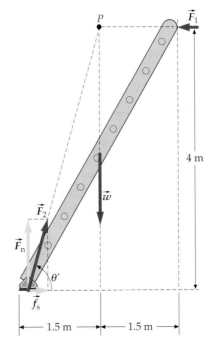

12-4 Couples

The forces $\vec{F}_n$ and $\vec{w}$ in Figure 12-11 of Example 12-5 are equal and opposite. Such a pair of forces, called a couple, tends to produce an angular acceleration, but its net force is zero. The forces $\vec{f}_s$ and $\vec{F}_1$ in those figures also constitute a couple. Figure 12-13 shows a couple consisting of forces $\vec{F}_1$ and $\vec{F}_2$ a distance D apart. The torque produced by this couple about an arbitrary point O is

$$\vec{\tau} = \vec{r}_1 \times \vec{F}_1 + \vec{r}_2 \times \vec{F}_2 = \vec{r}_1 \times \vec{F}_1 + \vec{r}_2 \times (-\vec{F}_1) = (\vec{r}_1 - \vec{r}_2) \times \vec{F}_1 \qquad 12\text{-}5$$

This result does not depend on the choice of the point O. The magnitude of this torque is

FIGURE 12-12

$$\tau = FD \qquad\qquad 12\text{-}6$$

where F is the magnitude of either force and D is the distance between their lines of action.

The torque produced by a couple is the same about all points in space.

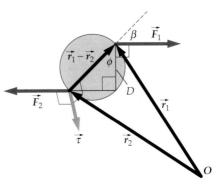

EXERCISE Show that the magnitude of the torque exerted by the couple in Figure 12-13 is given by FD, where D is the distance between the lines of action of the two forces and F is the magnitude of either force. (*Answer*   The angle β between $\vec{r}_1 - \vec{r}_2$ and $\vec{F}_1$ is $90° - \phi$, so $|\vec{\tau}| = |(\vec{r}_1 - \vec{r}_2) \times \vec{F}_1| = |\vec{r}_1 - \vec{r}_2| F \sin(90° - \phi) = F|\vec{r}_1 - \vec{r}_2| \cos \phi = FD$.)

FIGURE 12-13

12-5 Static Equilibrium in an Accelerated Frame

By an accelerated frame we mean a reference frame that is accelerating relative to an inertial frame of reference. The net force on an object that remains at rest relative to an accelerated reference frame is not equal to zero. An object at rest relative to the accelerated frame has the same acceleration as the frame. The two conditions for an object to be in static equilibrium in an accelerated reference frame are

1. $\Sigma \vec{F} = m\vec{a}_{cm}$

   where $\vec{a}_{cm}$ is the acceleration of the center of mass, which is the acceleration of the reference frame.

2. $\Sigma \vec{\tau}_{cm} = 0$

   The sum of the torques about the center of mass must be zero.

The second condition follows from the fact that Newton's second law for rotation, $\Sigma \vec{\tau}_{cm} = I_{cm}\vec{\alpha}$, holds for torques about the center of mass whether or not the center of mass is accelerating.[†]

[†] See the discussion surrounding Equation 9-30.

Moving a Box **EXAMPLE 12-6**

A truck (Figure 12-14*a*) carries a uniform box of mass *m*, height *h*, and square cross section of edge-length *L*. What is the greatest acceleration the truck can have without the box tipping over? Assume that the box tips before it slides.

(a)

PICTURE THE PROBLEM The acceleration of the box is due to the frictional force, as shown in Figure 12-14*b*. This force exerts a counterclockwise torque about the center of mass of the box. The only other force that exerts a torque about the center of mass of the box is the normal force. If the box is not accelerating, the normal force is distributed uniformly across the bottom of the box. If the acceleration is small, this distribution shifts and the effective normal force moves to the left to provide a balancing torque about the center of mass. The greatest balancing torque this force can exert is when the effective normal force is at the edge of the box, as shown.

(b)

FIGURE 12-14

1. Draw a free-body diagram of the box (Figure 12-15).

2. Apply $\Sigma F_y = ma_{cmy}$ to the box and then solve for the normal force:
$$F_n - mg = 0 \quad \text{so} \quad F_n = mg$$

3. Apply $\Sigma F_x = ma_{cmx}$ to the box:
$$f_s = ma$$

4. Apply $\Sigma \tau_{cm} = 0$:
$$f_s \frac{h}{2} - F_n d = 0$$

5. Substitute for d, f_s, and F_n and solve for *a*:
$$ma\frac{h}{2} - mg\frac{L}{2} = 0 \quad \text{so} \quad \boxed{a = \frac{L}{h}g}$$

FIGURE 12-15

REMARKS The maximum acceleration is proportional to L/h. This maximum acceleration is small for a tall, narrow box (L/h small) and large for a short, wide box (L/h large). Thus, a short, wide box is more stable.

12-6 Stability of Rotational Equilibrium

There are three categories of rotational equilibrium for an object: stable, unstable, or neutral. **Stable rotational equilibrium** occurs when the torques that arise from a small angular displacement of the object urge the object back toward its equilibrium position. Stable equilibrium is illustrated in Figure 12-16*a*. When the

FIGURE 12-16 If a slight rotation raises the center of gravity, as in (a), the equilibrium is stable. If a slight rotation lowers the center of gravity, as in (b), the equilibrium is unstable. If a slight rotation neither raises nor lowers the center of gravity, as in (c), the equilibrium is neutral.

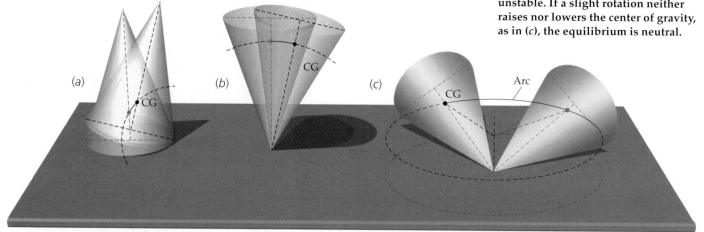

cone is tipped slightly as shown, the resulting torque about the pivot point tends to restore the cone to its original position. Note that this slight tipping lifts the center of gravity, increasing the potential energy of the cone.

Unstable rotational equilibrium, illustrated in Figure 12-16b, occurs when the torques that arise from a small angular displacement of the object urge the object away from its equilibrium position. A slight tipping of the cone causes it to fall over because the torque due to its weight tends to rotate it away from its original position. Here the rotation lowers the center of gravity and decreases the potential energy of the cone.

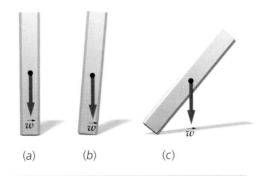

(a) (b) (c)

FIGURE 12-17 Stability of equilibrium is relative. If the rod in (a) is rotated slightly, as in (b), it returns to its original equilibrium position as long as the center of gravity lies over the base of support. (c) If the rotation is too great, the center of gravity is no longer over the base of support, and the rod falls over.

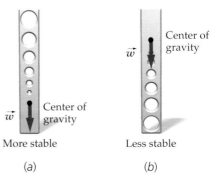

More stable   Less stable

(a)   (b)

FIGURE 12-18 When a nonuniform rod rests on its heavy end with its center of gravity low, as in (a), the equilibrium is more stable than when its center of gravity is high, as in (b).

The cone resting on a horizontal surface in Figure 12-16c illustrates **neutral rotational equilibrium.** If the cone is rolled slightly, there is no torque or force that urges it either back toward or away from its original position. As the cone rotates, the height of the center of gravity remains unchanged, so the potential energy does not change.

In summary, if a system is disturbed slightly from its equilibrium position, the equilibrium is stable if the system returns to its original position, unstable if it moves farther away, and neutral if there are no torques or forces tending to rotate it in either direction.

Because "disturbed slightly" is a relative term, stability is also relative. One example of equilibrium may be more or less stable than another. A rod is balanced on one end, as in Figure 12-17a. Here, if the disturbance is very small (Figure 12-17b), the rod will move back toward its original position, but if the disturbance is great enough so that the center of gravity no longer lies over the base of support (Figure 12-17c), the rod will fall.

We can improve the stability of a system by either lowering the center of gravity or widening the base of support. Figure 12-18 shows a nonuniform rod that is loaded so that its center of gravity is near one end. If it stands on its heavy end so that the center of gravity is low (Figure 12-18a), it is much more stable than if it stands on the other end so that the center of gravity is high (Figure 12-18b).

In Figure 12-19 the system is stable for any angular displacement because the resulting torque always rotates the system back toward its equilibrium position.

Standing or walking upright is difficult for a human because the center of gravity is high, and must be kept over a relatively small base of support, the feet. Human infants take about a year to learn to walk. A four-footed creature has a much easier time because its base of support is larger and its center of gravity is lower. Newborn kittens can walk almost immediately.

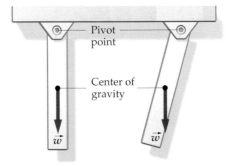

Pivot point

Center of gravity

FIGURE 12-19

12-7 Indeterminate Problems

When objects are not rigid, but deformable, we need more information to determine the forces required for equilibrium. Consider a car resting on a horizontal surface. Suppose there is a very heavy object on one side of the trunk.

We wish to find the vertical support force exerted by the road on each tire. Let the road be in the xy plane. If we choose one of the tires as our origin, the torque exerted by all the forces about that point has x and y components, but no z component because there are no horizontal forces. We thus obtain two equations by setting the net torque equal to zero, and a third equation by setting the net vertical force equal to zero. We need another equation to find the force exerted by the road on each of the four tires. If we let air out of one of the tires and pump up another tire to a greater pressure, the car remains in equilibrium, but the force exerted on each tire changes. Clearly, the forces on the tires in this problem are not determined by the information given. The tires are not rigid bodies. To some extent, every object is deformable.

(a)

12-8 Stress and Strain

If a solid object is subjected to forces that tend to stretch, shear, or compress the object, its shape changes. If the object returns to its original shape when the forces are removed, it is said to be **elastic**. Most objects are elastic for forces up to a certain maximum, called the **elastic limit**. If the forces exceed the elastic limit, the object does not return to its original shape but is permanently deformed.

Figure 12-20 shows a solid bar subjected to a stretching or **tensile force F** acting equally to the right and to the left. The bar is in equilibrium, but the forces acting on it tend to increase its length. The fractional change in the length $\Delta L/L$ of a segment of the bar is called the **strain**:

$$\text{Strain} = \frac{\Delta L}{L} \qquad 12\text{-}7$$

The ratio of the force F to the cross-sectional area A is called the **tensile stress**:

$$\text{Stress} = \frac{F}{A} \qquad 12\text{-}8$$

(b)

FIGURE 12-20 (a) A solid bar subjected to stretching forces of magnitude F acting on each end. (b) A small section of the bar of length L. The elements of the bar to the left and right of this section exert forces on the section. If the section is not too close to the end, these forces are distributed equally over the cross-sectional area. The force per unit area is the stress.

Figure 12-21 shows a graph of stress versus strain for a typical solid bar. The graph is linear until point A. Up to this point, known as the proportional limit, the strain is proportional to the stress. The result that strain varies linearly with stress is known as Hooke's law. Point B in Figure 12-21 is the elastic limit of the material. If the bar is stretched beyond this point, it is permanently deformed. If an even greater stress is applied, the material eventually breaks, shown happening at point C. The ratio of stress to strain in the linear region of the graph is a constant called **Young's modulus Y**:

$$Y = \frac{\text{stress}}{\text{strain}} = \frac{F/A}{\Delta L/L} \qquad 12\text{-}9$$

YOUNG'S MODULUS DEFINED

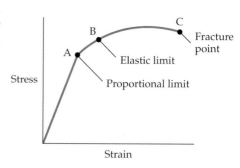

FIGURE 12-21 A graph of stress versus strain. Up to point A, the strain is proportional to the stress. Beyond the elastic limit at point B, the bar will not return to its original length when the stress is removed. At point C, the bar fractures.

The units of Young's modulus are newtons per square meter (or pounds per square inch). Approximate values of Young's modulus for various materials are listed in Table 12-1.

EXERCISE Suppose that the biceps muscle of your right arm has a maximum cross-sectional area of $12 \text{ cm}^2 = 1.2 \times 10^{-3} \text{ m}^2$. What is the stress in the muscle if it exerts a force of 300 N? (*Answer* Stress $= F/A = 2.5 \times 10^5 \text{ N/m}^2$. The maximum stress that can be exerted is approximately the same for all human muscles. Greater forces can be exerted by muscles with greater cross-sectional areas.)

TABLE 12-1

Young's Modulus Y and Strengths of Various Materials†

Material	Y, GN/m²‡	Tensile strength, MN/m²	Compressive strength, MN/m²
Aluminum	70	90	
Bone			
Tensile	16	200	
Compressive	9		270
Brass	90	370	
Concrete	23	2	17
Copper	110	230	
Iron (wrought)	190	390	
Lead	16	12	
Steel	200	520	520

† These values are representative. Actual values for particular samples may differ.
‡ 1 GN = 10^3 MN = 1×10^9 N.

If a bar is subjected to forces that tend to compress it rather than stretch it, the stress is called **compressive stress.** For many materials, Young's modulus for compressive stress is the same as that for tensile stress. Note that ΔL in Equation 12-7 is then taken to be the *decrease* in the length of the bar. If the tensile or compressive stress is too great, the bar breaks. The stress at which breakage occurs is called the **tensile strength,** or in the case of compression, the **compressive strength.** Approximate values of the tensile and compressive strengths of various materials are listed in Table 12-1. Note from the table that the compressive strength of bone is greater than the tensile strength. Also note that, for bone, Young's modulus is significantly larger for tensile stress than for compressive stress. These differences have biological significance, because the major job of bone is to resist the compressive load exerted by contracting muscles.

ELEVATOR SAFETY   **EXAMPLE 12-7**   **Put It in Context**

While working with an engineering company during the summer, you are assigned to check the safety of an elevator system in a new office building. The elevator has a maximum load of 1000 kg including its own mass, and is supported by a steel cable 3.0 cm in diameter and 300 m long at full extension. There will be safety concerns if the steel stretches more than 3.0 cm. Your job is to determine whether or not the elevator is safe as planned, given a maximum acceleration of the system of 1.5 m/s².

PICTURE THE PROBLEM L is the unstretched length of cable, F is the force acting on it, and A is its cross-sectional area. The stretch in the cable ΔL is related to Young's modulus by $Y = (F/A)/(\Delta L/L)$. From Table 12-1 we find the numerical value of Young's modulus for steel, $Y = 2.0 \times 10^{11}$ N/m².

1. The amount the cable is stretched, ΔL, is found from Young's modulus:

$$Y = \frac{F/A}{\Delta L/L} \quad \text{so} \quad \Delta L = \frac{FL}{AY}$$

2. To find the force acting on the cable we apply Newton's second law to the elevator. There are two forces on the elevator, the force F of the cable and the weight:

$$F - mg = ma_y$$

so

$$F_{max} = m(g + a_{y,max})$$
$$= (1000 \text{ kg})(9.81 \text{ N/kg} + 1.5 \text{ N/kg})$$
$$= 1.13 \times 10^4 \text{ N}$$

3. Substitute into the step 1 result and obtain the maximum amount of stretch:

$$\Delta L = \frac{F_{max}L}{AY} = \frac{F_{max}L}{\pi r^2 Y}$$
$$= \frac{(1.13 \times 10^4 \text{ N})(300 \text{ m})}{\pi(0.015 \text{ m})^2(2.0 \times 10^{11} \text{ N/m}^2)}$$
$$= 2.40 \text{ cm}$$

4. Report your results to your boss:

> According to my calculations, the most the cable will stretch is 2.4 cm, only 20% less than the 3.0-cm limit. However, in reading the footnote to the table, I note that the values given for Young's modulus are representative values, and that actual values vary from sample to sample. I recommend that you consult a engineer and get a professional evaluation.

EXERCISE A wire 1.5 m long has a cross-sectional area of 2.4 mm². It is hung vertically and stretches 0.32 mm when a 10-kg block is attached to it. Find (a) the stress, (b) the strain, and (c) Young's modulus for the wire. (*Answer* (a) $4.09 \times 10^7 \text{ N/m}^2$ (b) 2.13×10^{-4} (c) 192 GN/m^2)

In Figure 12-22, a force F_s is applied tangentially to the top of a book. Such a force is called a **shear force.** The ratio of the shear force F_s to the area A is called the **shear stress:**

$$\text{Shear stress} = \frac{F_s}{A} \qquad \text{12-10}$$

A shear stress tends to deform an object, as shown in Figure 12-22. The ratio $\Delta X/L$ is called the **shear strain:**

$$\text{Shear strain} = \frac{\Delta X}{L} = \tan \theta \qquad \text{12-11}$$

where θ is the shear angle shown in the figure. The ratio of the shear stress to the shear strain is called the **shear modulus M_s:**

$$M_s = \frac{\text{shear stress}}{\text{shear strain}} = \frac{F_s/A}{\Delta X/L} = \frac{F_s/A}{\tan \theta} \qquad \text{12-12}$$

DEFINITION — SHEAR MODULUS

The shear modulus is also known as the **torsion modulus.** The torsion modulus is approximately constant for small stresses, which implies that the shear strain varies linearly with the shear stress. This observation is known as Hooke's law for torsional stress. In a torsion balance, such as that used in Cavendish's apparatus for measuring the universal gravitational constant G, the torque (which is related to the stress) is proportional to the angle of twist (which equals the strain for small angles). Approximate values of the shear modulus for various materials are listed in Table 12-2.

FIGURE 12-22 The application of the horizontal force $\vec{F}_s$ to the book causes a shear stress defined as the force per unit area. The ratio $\Delta X/L = \tan \theta$ is the shear strain.

TABLE 12-2

Approximate Values of the Shear Modulus M_s of Various Materials

Material	M_s, GN/m²
Aluminum	30
Brass	36
Copper	42
Iron	70
Lead	5.6
Steel	84
Tungsten	150

Topic	Relevant Equations and Remarks
1. Equilibrium of a Rigid Object	

<table>
<tr><td>Conditions</td><td>

1. The net external force acting on the object must be zero:

$$\Sigma \vec{F} = 0 \qquad \text{12-1}$$

2. The net external torque about any point must be zero:

$$\Sigma \vec{\tau} = 0 \qquad \text{12-2}$$

If all the forces acting on a rigid object are perpendicular to an axis, then the sum of the torques about that axis equals zero.

</td></tr>
<tr><td>Stability</td><td>

The equilibrium of an object can be classified as stable, unstable, or neutral. An object resting on some surface will be in equilibrium if its center of gravity lies over its base of support. Stability can be improved by lowering the center of gravity or by increasing the size of the base.

</td></tr>
</table>

| **2. Center of Gravity** | The force of gravity exerted on the various parts of an object can be replaced by a single force, the total weight of the object $\vec{W}$, acting at the center of gravity. |

$$\vec{\tau}_{\text{net}} = \sum_i (\vec{r}_i \times \vec{w}_i) = \vec{r}_{\text{cg}} \times \vec{W} \qquad \text{12-3}$$

For an object in a uniform gravitational field, the center of gravity coincides with the center of mass.

| **3. Couples** | A pair of equal and opposite forces constitutes a couple. The torque produced by a couple is the same about any point in space. |

$$\vec{\tau} = (\vec{r}_1 - \vec{r}_2) \times \vec{F}_1, \quad \text{so} \quad \tau = FD \qquad \text{12-5, 12-6}$$

where D is the distance between the lines of action of the forces.

| **5. Accelerated Reference Frame** | The conditions for static equilibrium in an accelerated reference frame are |

1. $\Sigma \vec{F} = m\vec{a}_{\text{cm}}$

   where $\vec{a}_{\text{cm}}$ is the acceleration of the center of mass, which is the acceleration of the reference frame.

2. $\Sigma \vec{\tau}_{\text{cm}} = 0$

   The sum of the torques about the center of mass must be zero.

| **6. Stress and Strain** | |

| Young's modulus | $$Y = \frac{\text{stress}}{\text{strain}} = \frac{F/A}{\Delta L/L} \qquad \text{12-9}$$ |

| Shear modulus | $$M_s = \frac{\text{shear stress}}{\text{shear strain}} = \frac{F_s/A}{\Delta X/L} = \frac{F_s/A}{\tan \theta} \qquad \text{12-12}$$ |

PROBLEMS

- Single-concept, single-step, relatively easy
- •• Intermediate-level, may require synthesis of concepts
- ••• Challenging
- **SSM** Solution is in the *Student Solutions Manual*
- **iSOLVE** Problems available on iSOLVE online homework service
- **iSOLVE✓** These "Checkpoint" online homework service problems ask students additional questions about their confidence level, and how they arrived at their answer

Take $g = 9.81$ N/kg $= 9.81$ m/s^2 and neglect friction in all problems unless otherwise stated.

Conceptual Problems

1 • **SSM** True or false:

(a) $\sum_i \vec{F}_i = 0$ is sufficient for static equilibrium to exist.

(b) $\sum_i \vec{F}_i = 0$ is necessary for static equilibrium to exist.

(c) In static equilibrium, the net torque about any point is zero.

(d) An object is in equilibrium only when there are no forces acting on it.

2 • True or false: The center of gravity is always at the geometric center of a body.

3 • Must there be any material at the center of gravity of an object?

4 • **SSM** If the gravitational field $\vec{g}$ is not constant over an object, is it the center of mass or the center of gravity that is the pivot point when the object is balanced?

5 •• To find the center of mass of an irregular plane figure, the following method can be used: hang the figure from one corner, by a string and extend the line that the string makes downward across the figure. Repeat, hanging it from another corner. The intersection of the two lines will be at the center of mass. Explain why this technique works.

6 • **SSM** Is it possible to climb a ladder placed against a wall if the ground is frictionless but the wall is not? Explain.

7 • An aluminum wire and a steel wire of the same length L and diameter D are joined to form a wire of length $2L$. The wire is fastened to the ceiling and a weight W is attached to the other end. Neglecting the mass of the wires, which of the following statements is true? (a) The aluminum portion will stretch by the same amount as the steel portion. (b) The tensions in the aluminum portion and the steel portion are equal. (c) The tension in the aluminum portion is greater than that in the steel portion. (d) None of these statements is true.

8 • If the net torque about some point is zero, must it be zero about any other point? Explain.

9 • **SSM** The horizontal bar in Figure 12-23 will remain horizontal if (a) $L_1 = L_2$ and $R_1 = R_2$, (b) $L_1 = L_2$ and $M_1 = M_2$, (c) $R_1 = R_2$ and $M_1 = M_2$, (d) $L_1 M_1 = L_2 M_2$, (e) $R_1 L_1 = R_2 L_2$.

FIGURE 12-23 Problem 9

10 •• Sit in a chair with your back straight. Now try to stand up without leaning forward. Explain why you cannot do it.

11 •• **SSM** The great engineering feats of ancient times (Roman arch bridges, the great cathedrals, and the pyramids, to name a few) all have two things in common: they are made of stone, and they are compressive structures—that is, they are built so that all strains in the structure are compressive rather than tensile in nature. Look up the tensile and compressive strengths of stone and cement to explain why this is true.

Estimation and Approximation

12 •• A large crate weighing 4500 N rests on four 12-cm-high blocks on a horizontal surface (Figure 12-24). The crate is 2 m long, 1.2 m high, and 1.2 m deep. You are asked to lift one end of the crate using a long steel pry bar. The fulcrum on the pry bar is 10 cm from the end that lifts the crate. Estimate the length of the bar you will need to lift the end of the crate.

FIGURE 12-24 Problem 12

13 •• [SSM] Consider an atomic model for Young's modulus: assume that we have a large number of atoms arranged in a cubic array separated by distance a. Imagine that each atom is attached to its 6 nearest neighbors by little springs with spring constant k. (Atoms are not really attached by springs, but the forces between them act enough like springs to make this a good model.) (a) Show that this material, if stretched, will have a Young's modulus $Y = k/a$. (b) From Table 12-1, and assuming that $a \approx 1$ nm, estimate a typical value for the "atomic spring constant" k in a metal.

Conditions for Equilibrium

14 • A seesaw consists of a 4-m board pivoted at the center. A 28-kg child sits on one end of the board. Where should a 40-kg child sit to balance the seesaw?

15 • [SOLVE]✔ In Figure 12-25, Misako is about to do a push-up. Her center of gravity lies directly above point P on the floor, which is 0.9 m from her feet and 0.6 m from her hands. If her mass is 54 kg, what is the force exerted by the floor on her hands?

Center of gravity

0.9 m — 0.6 m

P

FIGURE 12-25 Problem 15

16 • [SSM] [SOLVE]✔ Misako wants to measure the strength of her biceps muscle by exerting a force on a test strap as shown in Figure 12-26. The strap is 28 cm from the pivot point at the elbow, and her biceps muscle is attached at a point 5 cm from the pivot point. If the scale reads 18 N when she exerts her maximum force, what force is exerted by the biceps muscle?

Strap

18 N

28 cm

5 cm

FIGURE 12-26 Problem 16

17 • A crutch is pressed against the sidewalk with a force $\vec{F}_c$ along its own direction as in Figure 12-27. This force is balanced by the normal force $\vec{F}_n$ and a frictional force $\vec{f}_s$. (a) Show that when the force of friction is at its maximum value, the coefficient of friction is related to the angle θ by $\mu_s = \tan\theta$. (b) Explain how this result applies to the forces on your foot when you are not using a crutch. (c) Why is it advantageous to take short steps when walking on ice?

FIGURE 12-27 Problem 17

The Center of Gravity

18 • [SOLVE] An automobile has 58 percent of its weight on the front wheels. The front and back wheels are separated by 2 m. Where is the center of gravity located with respect to the front wheels?

19 • [SSM] Each of the objects shown in Figure 12-28 is suspended from the ceiling by a thread attached to the point marked × on the object. Describe the orientation of each suspended object with a diagram.

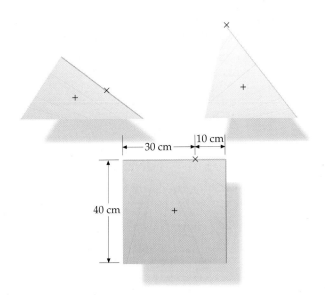

30 cm — 10 cm

40 cm

FIGURE 12-28 Problem 19

20 •• ![SOLVE]✓ A square plate is produced by welding together four smaller square plates, each of side a as shown in Figure 12-29. Plate 1 weighs 40 N; plate 2, 60 N; plate 3, 30 N; and plate 4, 50 N. Find the center of gravity (x_{cg}, y_{cg}).

FIGURE 12-29 Problem 20

21 •• ![SOLVE] A uniform rectangular plate has a circular section of radius R cut out as shown in Figure 12-30. Find the center of gravity of the system. *Hint: Do not integrate. Use superposition of a rectangular plate minus a circular plate.*

FIGURE 12-30 Problem 21

Some Examples of Static Equilibrium

22 • Figure 12-31 shows a lever with a force f being applied to lift a load F. (*a*) If we define the mechanical advantage of the lever as $M = F/f_{min}$, where f_{min} is the smallest force necessary to lift the load F, show that $M = x/X$, where x is the moment arm (distance to the pivot) for the applied force and X is the moment arm for the load. (*b*) To lift a heavy load, the moment arm for the applied force is usually larger than the load, so that the lifting force is smaller than the applied force. However, sometimes the moment arm for the applied force is much smaller than that for the load, so that the applied force must be much larger than the load (the arrangements of muscles in the forearm, as shown in Figure 12-4, is a good example). Why is this arrangement useful under some circumstances?

FIGURE 12-31 Problem 22

23 • ![SOLVE] Figure 12-32 shows a 25-foot sailboat. The mast is a uniform pole of 120 kg and is supported on the deck and held fore and aft by wires as shown. The tension in the forestay (wire leading to the bow) is 1000 N. Determine the tension in the backstay and the force that the deck exerts on the mast. Is there a tendency for the mast to slide forward or aft? If so, where should a block be placed to prevent the mast from moving?

FIGURE 12-32 Problem 23

24 •• ![SOLVE] A 10-m beam of mass 300 kg extends over a ledge as in Figure 12-33. The beam is not attached, but simply rests on the surface. A 60-kg student intends to position the beam so that he can walk to the end of it. How far from the edge of the ledge can the beam extend?

FIGURE 12-33
Problem 24

25 •• [SSM] A gravity board for locating the center of gravity of a person consists of a horizontal board supported by a fulcrum at one end and by a scale at the other end. A physics student lies horizontally on the board with the top of his head above the fulcrum point as shown in Figure 12-34. The scale is 2 m from the fulcrum. The student has a mass of 70 kg, and when he is on the gravity board, the scale advances 250 N. Where is the center of gravity of the student?

FIGURE 12-34 Problem 25

26 •• **iSOLVE** ✓ A 3-m board of mass 5 kg is hinged at one end. A force $\vec{F}$ is applied vertically at the other end to lift a 60-kg block, which rests on the board 80 cm from the hinge, as shown in Figure 12-35. (*a*) Find the magnitude of the force needed to hold the board stationary at $\theta = 30°$. (*b*) Find the force exerted by the hinge at this angle. (*c*) Find the magnitude of the force $\vec{F}$ and the force exerted by the hinge if $\vec{F}$ is exerted perpendicular to the board when $\theta = 30°$.

FIGURE 12-35
Problem 26

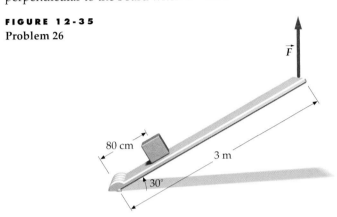

27 •• **SSM** **iSOLVE** A cylinder of weight W is supported by a frictionless trough formed by a plane inclined at 30° to the horizontal on the left and one inclined at 60° on the right as shown in Figure 12-36. Find the force exerted by each plane on the cylinder.

FIGURE 12-36
Problem 27

28 •• **iSOLVE** An 80-N weight is supported by a cable attached to a strut hinged at point A as in Figure 12-37. The strut is supported by a second cable under tension T_2. The mass of the strut is negligible. (*a*) What are the three nonnegligible forces acting on the strut? (*b*) Show that the vertical component of the tension T_2 must equal 80 N. (*c*) Find the force exerted on the strut by the hinge.

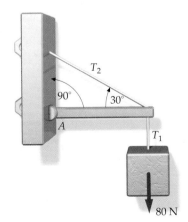

FIGURE 12-37 Problem 28

29 •• A horizontal board 8.0 m long is used by pirates to make their victims walk the plank. A pirate of mass 105 kg stands on the shipboard end of the plank to prevent it from tipping. Find the maximum distance the plank can overhang for a 63-kg victim to be able to walk to the end if (*a*) the mass of the plank is negligible and (*b*) the mass of the plank is 25 kg.

30 •• A uniform 18-kg door that is 2.0 m high by 0.8 m wide is hung from two hinges that are 20 cm from the top and 20 cm from the bottom. If each hinge supports half the weight of the door, find the magnitude and direction of the horizontal components of the forces exerted by the two hinges on the door.

31 •• Find the force exerted by the edge of the block on the wheel in Example 12-4, just as the wheel lifts off the surface.

32 •• **SSM** **iSOLVE** The diving board shown in Figure 12-38 has a mass of 30 kg. Find the force on the supports when a 70-kg diver stands at the end of the diving board. Give the direction of each support force as a tension or a compression.

FIGURE 12-38 Problem 32

33 •• Find the force exerted on the strut by the hinge at A for the arrangement in Figure 12-39 if (*a*) the strut is weightless and (*b*) the strut weighs 20 N.

FIGURE 12-39
Problem 33

34 •• Julie has been hired to help paint the trim of a building, but she is not convinced of the safety of the apparatus. A 5.0-m plank is suspended horizontally from the top of the building by ropes attached at each end. Julie knows from previous experience that the ropes being used will break if the tension exceeds 1 kN. Her 80-kg boss dismisses Julie's worries and begins painting while standing 1 m from the end of the plank. If Julie's mass is 60 kg and the plank has a mass of 20 kg, then over what range of positions can Julie stand to join her boss without causing the ropes to break?

35 •• A cylinder of mass M and radius R rolls against a step of height h as shown in Figure 12-40. When a horizontal force $\vec{F}$ is applied to the top of the cylinder, the cylinder remains at rest. (*a*) What is the normal force exerted by the floor on the cylinder? (*b*) What is the horizontal force exerted by the edge of the step on the cylinder? (*c*) What is the vertical component of the force exerted by the edge of the step on the cylinder?

FIGURE 12-40
Problems 35, 36

36 •• For the cylinder in Problem 35, find the minimum horizontal force $\vec{F}$ that will roll the cylinder over the step if the cylinder does not slide on the edge.

37 •• **SSM** Figure 12-41 shows a hand holding an epee, a weapon used in the sport of fencing. The center of mass of the epee, indicated in Figure 12-41, is 24 cm from the pommel; its total mass is 0.700 kg and its length is 110 cm. (*a*) Apply one of the conditions for static equilibrium to find the (total) force exerted by the hand on the epee. (*b*) Apply the other condition for static equilibrium to find the torque exerted by the hand on the epee. (*c*) Model the forces exerted by the hand as two oppositely directed forces whose lines of action are separated by the width of the fencer's hand ($\approx$10 cm). What are the magnitudes and directions of these two forces?

FIGURE 12-41 Problem 37

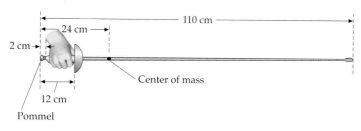

38 •• A large gate weighing 200 N is supported by hinges at the top and bottom and is further supported by a wire as shown in Figure 12-42. (*a*) What must be the tension in the wire for the force on the upper hinge to have no horizontal component? (*b*) What is the horizontal force on the lower hinge? (*c*) What are the vertical forces on the hinges?

FIGURE 12-42 Problem 38

39 ••• **SSM** A uniform log with a mass of 100 kg, a length of 4 m, and a radius of 12 cm is held in an inclined position, as shown in Figure 12-43. The coefficient of static friction between the log and the horizontal surface is 0.6. The log is on the verge of slipping to the right. Find the tension in the support wire and the angle the wire makes with the vertical wall.

FIGURE 12-43 Problem 39

40 ••• A tall, uniform, rectangular block sits on an inclined plane as shown in Figure 12-44. A cord is attached to the top of the block to prevent it from falling down the incline. What is the maximum angle θ for which the block will not slide on the incline? Let b/a be 4 and $\mu_s = 0.8$.

FIGURE 12-44 Problem 40

41 •• **SSM** A boat is moored at the end of a dock in a rapidly flowing river by a chain 5 m long, as shown in Figure 12-45. To give the chain some flexibility, a 100-N weight is attached in the center in the chain, to allow for variations in the force pulling the boat away from the dock. (*a*) If the drag force on the boat is 50 N, what is the tension in the chain? (*b*) How far will the chain sag? Ignore the weight of the chain itself. (*c*) How far is the boat from the dock? (*d*) If the maximum tension that the chain can support is 500 N, what is the maximum value of the force that the river can exert on the boat?

FIGURE 12-45 Problem 41

42 •• **iSOLVE** A thin rod of length 10 m and mass 20 kg is supported at a 30° incline. One support is 2 m and the other is 6 m from the lower end of the rod. Friction prevents the rod from sliding off the supports. Find the normal force exerted on the rod by each support.

43 • Two 80-N forces are applied to opposite corners of a rectangular plate as shown in Figure 12-46. Find the torque produced by this couple.

FIGURE 12-46 Problems 43, 45

44 •• **SSM** A uniform cube of side a and mass M rests on a horizontal surface. A horizontal force $\vec{F}$ is applied to the top of the cube as in Figure 12-47. This force is not sufficient to move or tip the cube. (a) Show that the force of static friction exerted by the surface and the applied force constitute a couple, and find the torque exerted by the couple. (b) This couple is balanced by the couple consisting of the normal force exerted by the surface and the weight of the cube. Use this fact to find the effective point of application of the normal force when $F = Mg/3$. (c) What is the greatest magnitude of $\vec{F}$ for which the cube will not tip?

FIGURE 12-47
Problem 44

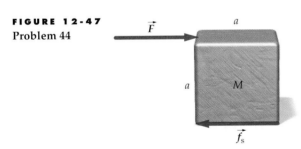

45 •• Resolve each force in Problem 43 into its horizontal and vertical components, producing two couples. The algebraic sum of the two component couples equals the resultant couple. Use this result to find the perpendicular distance between the lines of action of the two forces.

46 •• **SSM** A section of a cathedral wall is shown in Figure 12-48. The arch attached to the wall exerts a force of 2×10^5 N directed at an angle of 30° below the horizontal at a point 10 m above the ground and the mass of the wall itself is 30,000 kg. The coefficient of static friction between the wall and the ground is $\mu_k = 0.8$ and the base of the wall is 1.25 m long. (a) Calculate the effective normal force, the frictional force, and the point at which the effective normal force acts on the wall. (This is called the *thrust point* by architects and civil

engineers.) (b) If the thrust point ever moves outside the base of the wall, the wall will overturn. Apart from aesthetics, placing a heavy statue on top of the wall has good engineering practicality: it will move the thrust point toward the center of the wall. Explain why.

FIGURE 12-48
Problems 46, 47

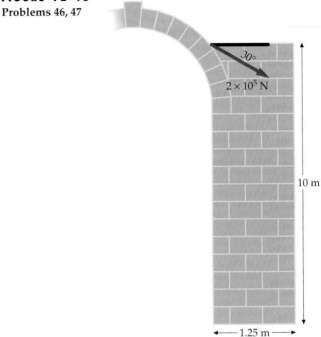

47 •• In civil engineering, the *thrust line* can be found by determining the thrust point at *any* height inside a wall or similar structure. Determine the thrust line for the wall in Problem 46 and graph it using a spreadsheet program or graphing calculator. (Despite its name, the thrust line is really a curve.)

Ladder Problems

48 •• **SSM** **iSOLVE** ✔ Romeo takes a uniform 10-m ladder and leans it against the smooth (frictionless) wall of the Capulet residence. The ladder's mass is 22.0 kg and the bottom rests on the ground 2.8 m from the wall. When Romeo, whose mass is 70 kg, gets 90 percent of the way to the top, the ladder begins to slip. What is the coefficient of static friction between the ground and the ladder?

49 •• **SSM** A massless ladder of length L leans against a smooth wall making an angle of θ with the horizontal floor. The coefficient of friction between the ladder and the floor is μ_s. A man of mass M climbs the ladder. What height h can he reach before the ladder slips?

50 •• A uniform ladder of length L and mass m leans against a frictionless vertical wall with its lower end on the ground. It makes an angle of 60° with the horizontal ground. The coefficient of static friction between the ladder and the ground is 0.45. If your mass is four times that of the ladder, how far up the ladder can you climb before it begins to slip?

51 •• A ladder making an angle θ with the horizontal, of mass m and length L, leans against a frictionless vertical wall. The center of mass is at a height h from the floor. A force F pulls horizontally against the ladder at the midpoint. Find the minimum coefficient of static friction μ_s for which the top end of the ladder will separate from the wall while the lower end does not slip.

52 •• [SOLVE]✓ A 900-N man sits on top of a stepladder of negligible weight that rests on a frictionless floor as in Figure 12-49. There is a cross brace halfway up the ladder. The angle at the apex is $\theta = 30°$. (a) What is the force exerted by the floor on each leg of the ladder? (b) Find the tension in the cross brace. (c) If the cross brace is moved down toward the bottom of the ladder (maintaining the same angle θ), will its tension be greater or less?

FIGURE 12-49
Problem 52

$w_m = 900$ N

$\theta = 30°$

2 m

2 m

53 •• [SSM] A uniform ladder rests against a frictionless vertical wall. The coefficient of static friction between the ladder and the floor is 0.3. What is the smallest angle at which the ladder will remain stationary?

Stress and Strain

54 • [SSM] A 50-kg ball is suspended from a steel wire of length 5 m and radius 2 mm. By how much does the wire stretch?

55 • [SOLVE]✓ Copper has a breaking stress of about 3×10^8 N/m². (a) What is the maximum load that can be hung from a copper wire of diameter 0.42 mm? (b) If half this maximum load is hung from the copper wire, by what percentage of its length will it stretch?

56 • [SOLVE] A 4-kg mass is supported by a steel wire of diameter 0.6 mm and length 1.2 m. How much will the wire stretch under this load?

57 • [SSM] [SOLVE] As a runner's foot pushes off on the ground, the shearing force acting on an 8-mm-thick sole is shown in Figure 12-50. If the force of 25 N is distributed over an area of 15 cm², find the angle of shear θ, given that the shear modulus of the sole is 1.9×10^5 N/m².

25 N

25 N

θ

FIGURE 12-50 Problem 57

58 •• [SOLVE] A steel wire of length 1.5 m and diameter 1 mm is joined to an aluminum wire of identical dimensions to make a composite wire of length 3.0 m. What is the length of the composite wire if it is used to support a mass of 5 kg?

59 •• A force F is applied to a long wire of length L and cross-sectional area A. Show that if the wire is considered to be a spring, the force constant k is given by $k = AY/L$ and the energy stored in the wire is $U = \frac{1}{2}F\Delta L$, where Y is Young's modulus and ΔL is the amount the wire has stretched.

60 •• [SOLVE] The steel E string of a violin is under a tension of 53 N. The diameter of the string is 0.20 mm and its length under tension is 35.0 cm. Find (a) the unstretched length of this string and (b) the work needed to stretch the string.

61 •• [SSM] When a rubber strip with a cross section of 3 mm $\times$ 1.5 mm is suspended vertically and various masses are attached to it, a student obtains the following data for length versus load:

Load, kg	0	0.1	0.2	0.3	0.4	0.5
Length, cm	5.0	5.6	6.2	6.9	7.8	10.0

(a) Find Young's modulus for the rubber strip for small loads.
(b) Find the energy stored in the strip when the load is 0.15 kg. (See Problem 59.)

62 •• A large mirror is hung from a nail as shown in Figure 12-51. The supporting steel wire has a diameter of 0.2 mm and an unstretched length of 1.7 m. The distance between the points of support at the top of the mirror's frame is 1.5 m. The mass of the mirror is 2.4 kg. What is the distance between the nail and the top of the frame when the mirror is hung?

FIGURE 12-51
Problems 62, 86

63 •• [SOLVE]✓ Two masses, M_1 and M_2, are supported by wires that have equal lengths when unstretched. The wire supporting M_1 is an aluminum wire 0.7 mm in diameter, and the one supporting M_2 is a steel wire 0.5 mm in diameter. What is the ratio M_1/M_2 if the two wires stretch by the same amount?

64 •• A 0.5-kg ball is attached to an aluminum wire having a diameter of 1.6 mm and an unstretched length of 0.7 m. The other end of the wire is fixed to a post. The ball rotates about the post in a horizontal plane at a rotational speed such that the angle between the wire and the horizontal is 5.0°. Find the tension in the wire and its length.

65 •• [SSM] An elevator cable is to be made of a new type of composite developed by Acme Laboratories. In the lab, a sample of the cable that is 2 m long and has a cross-sectional area of 0.2 mm² fails under a load of 1000 N. The cable in the elevator will be 20 m long and have a cross-sectional area of 1.2 mm². It will need to support a load of 20,000 N safely. Will it?

66 ••• [SSM] When a material is stretched in one direction, if its density remains constant, then (because its total volume remains constant), its length must decrease in one or both of the other directions. Take a rectangular block of length x, width y, and depth z, and pull on it so that its new length $x'' = x + \Delta x$. If $\Delta x \ll x$ and $\Delta y/y = \Delta z/z$, show that $\Delta y/y = -\frac{1}{2}\Delta x/x$.

67 •• If you are given a wire with a circular cross-section of radius r and a length L, show that $\Delta r/r = -\frac{1}{2}\Delta L/L$, assuming that $\Delta L \ll L$. (See Problem 66.)

68 ••• [SSM] For most materials listed in Table 12-1, the tensile strength is two to three orders of magnitude lower than Young's modulus. Consequently, most of these materials will break before their strain exceeds 1 percent. Of man-made materials, nylon has about the greatest extensibility—it can take strains of about 0.2 before breaking. But spider silk beats anything man-made. Certain forms of spider silk can take strains on the order of 10 before breaking! (a) If such a thread has a circular cross-section of radius r_0 and un-stretched length L_0, find its new radius r when stretched to a length $L = 10L_0$. (b) If the Young's modulus of the spider thread is Y, calculate the tension needed to break the thread in terms of Y and r_0.

General Problems

69 • A 90-N board 12 m long rests on two supports, each 1 m from the end of the board. A 360-N block is placed on the board 3 m from one end as shown in Figure 12-52. Find the force exerted by each support on the board.

FIGURE 12-52
Problem 69

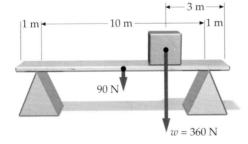

70 • The height of the center of gravity of a man standing erect is determined by weighing the man as he lies on a board of negligible weight supported by two scales as shown in Figure 12-53. If the man's height is 188 cm and the left scale reads 445 N while the right scale reads 400 N, where is his center of gravity relative to his feet? Would the reading on the scales be different if he were holding his head up—keeping it slightly above the board?

FIGURE 12-53 Problem 70

71 • [SSM] [SOLVE] ✔ Figure 12-54 shows a mobile consisting of four weights hanging on three rods of negligible mass. Find the value of each of the unknown weights if the mobile is to balance. *Hint: Find the weight w_1 first.*

FIGURE 12-54 Problem 71

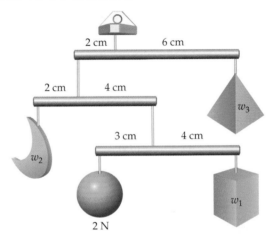

72 • A rope and pulley system, called a block and tackle, is used to support an object of mass m, as shown in Figure 12-55. The object is being raised at constant speed. When a segment of rope of length L passes over the uppermost pulley wheel, the height of the lower pulley is increased by h. (a) What is the ratio L/h? (b) Assume that the block and tackle are massless and frictionless. Show that $FL = mgh$ by applying the work–energy principle to the block–tackle object.

FIGURE 12-55 Problem 72

73 •• A plate of mass M in the shape of an equilateral triangle is suspended from one corner and a mass m is suspended from another of its corners. If the base of the triangle makes an angle of 6.0° with the horizontal, what is the ratio m/M?

74 •• A standard six-sided pencil is placed on a pad of paper (Figure 12-56). Find the minimum coefficient of static friction μ_s such that, if the pad is inclined, the pencil rolls down rather than slides.

FIGURE 12-56 Problem 74

75 •• [SSM] A uniform box of mass 8 kg that is twice as tall as it is wide rests on the floor of a truck. What is the maximum coefficient of static friction between the box and floor so that the box will slide toward the rear of the truck rather than tip when the truck accelerates on a level road?

76 •• A balance scale has unequal arms. A 1.5-kg block appears to have a mass of 1.95 kg on the left pan of the scale (Figure 12-57). Find its apparent mass if the 1.5-kg block is placed on the right pan.

FIGURE 12-57 Problem 76

77 •• [SSM] A cube of mass M leans against a frictionless wall making an angle of θ with the floor as shown in Figure 12-58. Find the minimum coefficient of static friction μ_s between the cube and the floor that allows the cube to stay at rest.

FIGURE 12-58 Problem 77

78 •• Figure 12-59 shows a steel meter stick hinged to a vertical wall and supported by a thin wire. The wire and meter stick make angles of 45° with the vertical. The mass of the meter stick is 5.0 kg. When a 10.0-kg block is suspended from the midpoint of the meter stick, the tension T in the supporting wire is 52 N. If the wire will break should the tension exceed 75 N, what is the maximum distance along the meter stick at which the block can be suspended?

FIGURE 12-59 Problem 78

79 •• [SOLVE ✓] Figure 12-60 shows a 20-kg ladder leaning against a frictionless wall and resting on a frictionless horizontal surface. To keep the ladder from slipping, the bottom of the ladder is tied to the wall with a thin wire; the tension in the wire is 29.4 N. The wire will break if the tension exceeds 200 N. (*a*) If an 80-kg person climbs halfway up the ladder, what force will be exerted by the ladder against the wall? (*b*) How far up can an 80-kg person climb this ladder?

FIGURE 12-60 Problem 79

80 •• [SSM] A uniform cube can be moved along a horizontal plane either by pushing the cube so that it slips or by turning it over ("rolling"). What coefficient of kinetic friction μ_k between the cube and the floor makes both ways equal in terms of the work needed?

81 •• A tall, uniform, rectangular block sits on an inclined plane as shown in Figure 12-61. If $\mu_s = 0.4$, does the block slide or fall over as the angle θ is slowly increased?

FIGURE 12-61 Problem 81

82 •• A 360-kg object is supported on a wire attached to a 15-m-long steel bar that is pivoted at a vertical wall and supported by a cable as shown in Figure 12-62. The mass of the bar is 85 kg. (a) With the cable attached to the bar 5.0 m from the lower end as shown, find the tension in the cable and the force exerted by the wall on the steel bar. (b) Repeat if a somewhat longer cable is attached to the steel bar 5.0 m from its upper end, maintaining the same angle between the bar and the wall.

FIGURE 12-62 Problem 82

83 •• Repeat Problem 75 if the truck accelerates up a hill that makes an angle of 9.0° with the horizontal.

84 •• **SSM** A thin rod 60 cm long is balanced 20 cm from one end when an object whose mass is (2m + 2 grams) is at the end nearest the pivot and an object of mass m is at the opposite end (Figure 12-63a). Balance is again achieved if the object whose mass is (2m + 2 grams) is replaced by the object of mass m and no object is placed at the other end (Figure 12-63b). Determine the mass of the rod.

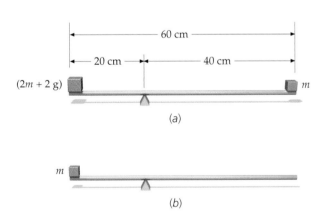

FIGURE 12-63 Problem 84

85 •• **SSM** If you balance a meter stick across two fingers (one from each hand) and slowly bring your hands together, the fingers will always meet in the middle of the stick, no matter where they are initially placed. (a) Explain why this is true. (b) As you move your fingers together, first one will move, and then the other; both will not move at the same time. Explain quantitatively why this is true; assume that the coefficient of static friction between the meter stick and your fingers is μ_s and the coefficient of kinetic friction is μ_k.

86 •• When a picture is hung on a smooth vertical wall using a wire and a nail, as in Figure 12-51, the picture almost always tips slightly forward, that is, the plane of the picture makes a small angle with the vertical. (a) Explain why pictures supported in this manner generally do not hang flush against the wall. (b) A framed picture 1.5 m wide and 1.2 m high with a mass of 8.0 kg is hung as in Figure 12-51 using a wire 1.7 m long. The ends of the wire are fastened to the sides of the frame at the rear, 0.4 m below the top. When the picture is hung, the angle between the plane of the frame and the wall is 5.0°. Determine the force that the wall exerts on the bottom of the frame.

87 •• **SSM** If a train travels around an unbanked bend in the rail bed too fast, the freight cars will tip over. Assume that the cargo portion of each freight car is a regular parallelepiped of uniform density and 1.5×10^4 kg mass, 10 m long, 3.0 m high, and 2.20 m wide, and that its base is 0.65 m above the rails. The car axles are 7.6 m apart, 1.2 m from each end of the boxcar. The separation between the rails is 1.55 m. Find the maximum safe speed of the train if the radius of curvature of the bend is (a) 150 m and (b) 240 m. (Neglect the mass of the flatcar.)

88 •• For balance, a tightrope walker uses a thin rod 8 m long and bowed in a circular arc shape. At each end of the rod is a lead mass of 20 kg. The tightrope walker, whose mass is 58 kg and whose center of gravity is 0.90 m above the rope, holds the rod tightly at its center 0.65 m above the rope. What should be the radius of curvature of the arc of the rod so that the tightrope walker will be in neutral equilibrium as he slowly makes his way across the rope? Neglect the mass of the rod.

89 ••• **SSM** We have a large number of identical uniform bricks, each of length L. If we stack one on top of another lengthwise (see Figure 12-64), the maximum offset that will allow the top brick to rest on the bottom brick is $L/2$. (a) Show that if we place this two-brick stack on top of a third brick, the maximum offset of the second brick on the third brick is $L/3$. (b) Show that, in general, if we have a stack of N bricks, the maximum offset of the $(n-1)$th brick (counting down from the top) on the nth brick is L/n, where $n \leq N$. (c) Write a spreadsheet program to calculate total offset (the sum of the individual offsets) for a stack of N bricks, and calculate this for $L = 1$ m and $N = 5$, 10, and 100. (d) Does the sum of the individual offsets approach a finite limit as $N \to \infty$? If so, what is that limit?

FIGURE 12-64 Problem 89

90 ••• A uniform sphere of radius R and mass M is held at rest on an inclined plane of angle θ by a horizontal string, as shown in Figure 12-65. Let $R = 20$ cm, $M = 3$ kg, and $\theta = 30°$. (*a*) Find the tension in the string. (*b*) What is the normal force exerted on the sphere by the inclined plane? (*c*) What is the frictional force acting on the sphere?

FIGURE 12-65
Problem 90

91 ••• The legs of a tripod make equal angles of 90° with each other at the apex, where they join together. A 100-kg block hangs from the apex. What are the compressional forces in the three legs?

92 •• Figure 12-66 shows a 20-cm-long uniform beam resting on a cylinder of 4-cm radius. The mass of the beam is 5.0 kg and that of the cylinder is 8.0 kg. The coefficient of static friction between beam and cylinder is zero, whereas the coefficient of static friction between the cylinder and the floor, and between the beam and the floor, are each greater than zero. Are there any values for these coefficients of static friction such that the system is in static equilibrium? If so, what are these values? If not, explain.

FIGURE 12-66
Problem 92

93 ••• Two solid smooth (frictionless) spheres of radius r are placed inside a cylinder of radius R as in Figure 12-67. The mass of each sphere is m. Find the force exerted by the bottom of the cylinder on the bottom sphere, the force exerted by the wall of the cylinder on each sphere, and the force exerted by one sphere on the other.

FIGURE 12-67
Problem 93

94 ••• [SSM] A solid cube of side length a balanced atop a cylinder of diameter d is in unstable equilibrium if $d \ll a$ and is in stable equilibrium if $d \gg a$ (Figure 12-68). Determine the minimum value of d/a for which the cube is in stable equilibrium.

FIGURE 12-68 Problem 94

Fluids

THE HOOVER DAM IS 660 FEET THICK AT THE BOTTOM, BUT ONLY 45 FEET THICK AT THE TOP OF THE SPAN.

? **Dams are designed to be thicker at the bottom than at the top. Why is this so? (See Example 13.2.)**

Fluids include both liquids and gases. Liquids flow under gravity until they occupy the lowest possible regions of their containers. Gases expand to fill their containers. Fluids permeate our environment as well as our bodies. To understand fluid behavior is to understand much about ourselves and our interactions with the world around us.

➤ **We begin this chapter by studying fluids at rest, followed by the study of steady-state flow with an emphasis on laminar flow.**

In a gas, the average distance between two molecules is large compared with the size of a molecule. The molecules have little influence on one another except during their brief collisions. In a liquid or solid, the molecules are close together and exert forces on one another that are comparable to the forces that bind atoms into molecules. Molecules in a liquid form temporary short-range bonds that are continually broken and reformed due to the thermal kinetic energy of the molecules. These bonds hold the liquid together; if the bonds were not present, the liquid would immediately evaporate and the molecules would escape as a vapor. The strength of the bonds in a liquid depends on the type of molecule that makes up the liquid. For example, the bonds between helium molecules are very weak and, for this reason, helium does not liquefy at atmospheric pressure unless the temperature is 4.2 K or lower.

13-1 Density

An important property of a substance is the ratio of its mass to its volume, which is called its **density:**

$$\text{Density} = \frac{\text{mass}}{\text{volume}}$$

The Greek letter ρ (rho) is usually used to denote density:

$$\rho = \frac{m}{V} \qquad\qquad 13\text{-}1$$

DEFINITION—DENSITY

Because the gram was originally defined as the mass of one cubic centimeter of water, the density of water in cgs units is 1 g/cm^3. Converting to SI units, we obtain for the density of water

$$\rho_w = \frac{1\,\text{g}}{\text{cm}^3} \times \frac{1\,\text{kg}}{10^3\,\text{g}} \times \left(\frac{100\,\text{cm}}{1\,\text{m}}\right)^3 = 10^3\,\text{kg/m}^3 \qquad 13\text{-}2$$

Precise measurements of density must take temperature into account, since the densities of most materials, including water, vary with temperature. Equation 13-2 gives the maximum value for the density of water, which occurs at 4°C. Table 13-1 lists the densities of some common materials.

A convenient unit of volume for fluids is the **liter** (L):

$$1\,\text{L} = 10^3\,\text{cm}^3 = 10^{-3}\,\text{m}^3$$

In terms of this unit, the density of water at 4°C is $1.00\text{ kg/L} = 1.00\text{ g/mL}$. When an object's density is greater than that of water, it sinks in water. When its density is less, it floats. The ratio of the density of a substance to that of water is called its **specific gravity.** For example, the specific gravity of aluminum is 2.7, meaning that a volume of aluminum has 2.7 times the mass of an equal volume of water. The specific gravities of objects that sink in water range from 1 to about 22.5 (for the densest element, osmium).

Most solids and liquids expand only slightly when heated, and contract slightly when subjected to an increase in external pressure. Since these changes in volume are relatively small, we often treat the densities of solids and liquids as approximately independent of temperature and pressure. The density of a gas, on the other hand, depends strongly on the pressure and temperature, so these variables must be specified when reporting the densities of gases. By convention, the standard conditions for the measurement of physical properties are atmospheric pressure at sea level and a temperature of 0°C. The densities for the substances listed in Table 13-1 are for these conditions. Note that the densities of liquids and solids are considerably greater than those of gases. For example, the density of water is about 800 times that of air under standard conditions.

TABLE 13-1

Densities of Selected Substances

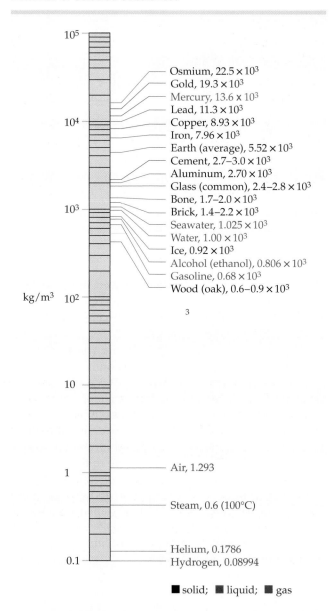

Osmium, 22.5×10^3
Gold, 19.3×10^3
Mercury, 13.6×10^3
Lead, 11.3×10^3
Copper, 8.93×10^3
Iron, 7.96×10^3
Earth (average), 5.52×10^3
Cement, $2.7–3.0 \times 10^3$
Aluminum, 2.70×10^3
Glass (common), $2.4–2.8 \times 10^3$
Bone, $1.7–2.0 \times 10^3$
Brick, $1.4–2.2 \times 10^3$
Seawater, 1.025×10^3
Water, 1.00×10^3
Ice, 0.92×10^3
Alcohol (ethanol), 0.806×10^3
Gasoline, 0.68×10^3
Wood (oak), $0.6–0.9 \times 10^3$

Air, 1.293

Steam, 0.6 (100°C)

Helium, 0.1786
Hydrogen, 0.08994

■ solid; ■ liquid; ■ gas

CALCULATING DENSITY

EXAMPLE 13-1

A 200-mL flask is filled with water at 4°C. When the flask is heated to 80°C, 6 g of water spill out. What is the density of water at 80°C? (Assume that the expansion of the flask is negligible.)

PICTURE THE PROBLEM The density of water at 80°C is $\rho' = m'/V$, where $V = 200$ mL $= 200$ cm^3 is the volume of the flask, and m' is the mass remaining in the flask after 6 g spill out. We find m' by first finding the mass of water originally in the flask.

1. Calculate the original mass m of water in the flask at 4°C using $\rho = 1.00$ g/cm: $\qquad m = \rho V = (1.00 \text{ g/cm}^3)(200 \text{ cm}^3) = 200 \text{ g}$

2. Calculate the mass of water remaining m' after 6 g spill out: $\qquad m' = m - 6 \text{ g} = 200 \text{ g} - 6 \text{ g} = 194 \text{ g}$

3. Use this value of m' to find the density of water at 80°: $\qquad \rho' = \dfrac{m'}{V} = \dfrac{194 \text{ g}}{200 \text{ cm}^3} = \boxed{0.97 \text{ g/cm}^3}$

EXERCISE A solid metal cube 8 cm on a side has a mass of 4.08 kg. (*a*) What is the density of the cube? (*b*) If the cube is made from a single element listed in Table 13-1, what is the element? (*Answer* (*a*) 7.97 kg/L (*b*) iron)

EXERCISE A gold brick is 5 cm × 10 cm × 20 cm. What is its mass? (*Answer* 19.3 kg)

13-2 Pressure in a Fluid

When a body is submerged in a fluid such as water, the fluid exerts a force perpendicular to the surface of the body at each point on the surface. This force per unit area is called the **pressure** P of the fluid:

$$P = \frac{F}{A} \qquad\qquad 13\text{-}3$$

DEFINITION — PRESSURE

The SI unit of pressure is the newton per square meter (N/m^2), which is called the **pascal** (Pa):

$$1 \text{ Pa} = 1 \text{ N/m}^2 \qquad\qquad 13\text{-}4$$

In the U.S. customary system, pressure is usually given in pounds per square inch (lb/in.^{2}). Another common unit of pressure is the atmosphere (atm), which equals approximately the air pressure at sea level. One atmosphere is now defined to be 101.325 kilopascals, which is approximately 14.70 lb/in.^{2}:

$$1 \text{ atm} = 101.325 \text{ kPa} \approx 14.70 \text{ lb/in.}^2 \qquad\qquad 13\text{-}5$$

Other units of pressure in common use will be discussed later in this chapter.

The pressure due to a fluid pressing in on an object tends to compress the object. The ratio of the increase in pressure ΔP to the fractional decrease in volume $(-\Delta V/V)$ is called the **bulk modulus**:

$$B = -\frac{\Delta P}{\Delta V/V} \qquad\qquad 13\text{-}6$$

DEFINITION — BULK MODULUS

(The minus sign in Equation 13-6 is introduced to make B positive, since all materials decrease in volume when subjected to an increase in external pressure.)

The more difficult it is to compress a material, the smaller is the fractional volume decrease $-\Delta V/V$ for a given pressure increase ΔP, and hence the greater the bulk modulus. The **compressibility** is the reciprocal of the bulk modulus. (The easier it is to compress a material, the larger the compressibility.) Liquids, gases, and solids all have a bulk modulus. Since liquids and solids are relatively incompressible, they have large values of B, and these values are relatively independent of temperature and pressure. Gases, on the other hand, are easily compressed, and their values for B depend strongly on pressure and temperature. Table 13-2 charts values for the bulk modulus of various materials.

As any scuba diver knows, the pressure in a lake or ocean increases with depth. Similarly, the pressure of the atmosphere decreases with altitude. For a liquid such as water, whose density is approximately constant throughout, the pressure increases linearly with depth. We can see this by considering a column of liquid of cross-sectional area A, as shown in Figure 13-1. To support the weight of the liquid in the column of height Δh, the pressure at the bottom of the column must be greater than the pressure at the top by the weight of the column. The weight of the liquid in the column is

$$w = mg = (\rho V)g = \rho A\, \Delta hg$$

If P_0 is the pressure at the top and P is the pressure at the bottom, the net upward force exerted by this pressure difference is $PA - P_0A$. Setting this net upward force equal to the weight of the column, we obtain

$$PA - P_0 A = \rho A \Delta hg$$

or

$$P = P_0 + \rho g \Delta h \qquad (\rho \text{ constant}) \qquad\qquad 13\text{-}7$$

TABLE 13-2

Approximate Values for the Bulk Modulus B of Various Materials

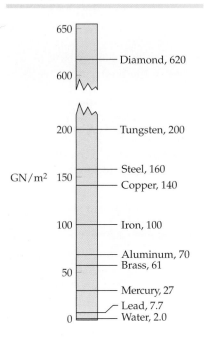

GN/m²

- Diamond, 620
- Tungsten, 200
- Steel, 160
- Copper, 140
- Iron, 100
- Aluminum, 70
- Brass, 61
- Mercury, 27
- Lead, 7.7
- Water, 2.0

FIGURE 13-1 Column of water of height Δh and cross-sectional area A. The pressure P at the bottom must be greater than the pressure P_0 at the top to balance the weight of the water.

EXERCISE How far below the surface of a lake is the pressure equal to 2 atm? (The pressure at the surface is 1 atm.) (*Answer* With $P_0 = 1$ atm $= 101$ kPa, $P = 2$ atm, $\rho = 1000$ kg/m³, and $g = 9.81$ N/kg, we have $\Delta h = \Delta P/\rho g = 10.3$ m. The pressure at a depth of 10.3 m is twice that at the surface.)

FORCE ON A DAM   **EXAMPLE 13-2**

A rectangular dam 30 m wide supports a body of water to a depth of 25 m. Find the total horizontal force on the dam.

PICTURE THE PROBLEM Because the pressure varies with depth, we cannot merely multiply the pressure times the area of the dam to find the force exerted by the water. Instead we can consider the force exerted on a strip of width $L = 30$ m, height dh, and area $dA = Ldh$ at a depth h (Figure 13-2), and then integrate from $h = 0$ to $h = H = 25$ m. The water pressure at depth h is $P_{atm} + \rho gh$. We can omit the atmospheric pressure because it is exerted on each side of the wall.

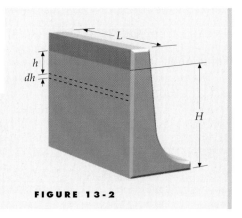

FIGURE 13-2

1. Express the force dF on the element of width L and height dh in terms of the net pressure ρhg

$$dF = P \, dA = \rho ghL \, dh$$

2. Integrate from $h = 0$ to $h = H$:

$$F = \int_{h=0}^{h=H} dF = \int_0^H \rho ghL \, dh = \rho gL \left.\frac{h^2}{2}\right|_0^H = \frac{1}{2}\rho gLH^2$$

3. Substitute the given values to find the numerical result:

$$F = \tfrac{1}{2}\rho gLH^2 = \tfrac{1}{2}(1000 \text{ kg/m}^3)(9.81 \text{ N/kg})(30 \text{ m})(25 \text{ m})^2$$

$$= \boxed{9.20 \times 10^7 \text{ N}}$$

REMARKS Dams typically are thicker at the bottom than at the top because the pressure on the dam increases with the depth of the water. Because the pressure on the dam is *proportional* to the depth of the water, the average pressure is $P_{\text{atm}} + \rho gH/2$ (the pressure at depth $h = H/2$).

The result that the pressure increases linearly with depth holds for a liquid in any container, independent of the shape of the container. Furthermore, the pressure is the same at all points at the same depth. We can see this by comparing the pressure at point 1 in Figure 13-3a with the pressure at point 2, which is inside an underwater cave. First we compare the pressure at points 1 and 3, a point directly below 1 at the same depth as 2, as shown in Figure 13-3b. Consider the vertical forces on the vertical column of water of height Δh and cross-sectional area A joining points 1 and 3. The upward force on the column, $P_3 A$, balances the two downward forces $P_1 A$ and mg, where $m = \rho A \Delta h$ is the mass of the water in the column ($A \Delta h$ is the volume of the column). That is, $P_3 A = P_1 A + \rho A \Delta hg$. Dividing through by A gives

$$P_3 = P_1 + \rho g \Delta h$$

Next consider the forces on the horizontal cylinder of water, also of cross-sectional area A, connecting points 2 and 3 (Figure 13-3c). There are two forces with components along the

FIGURE 13-3

(a)

(b)

(c)

cylinder's axis, P_3A and P_2A. The fact that these forces balance each other means that $P_3 = P_2$. It follows that

$$P_2 = P_1 + \rho g \Delta h$$

If we increase the pressure in a container of water by pressing down on the top surface with a piston, the increase in pressure is the same throughout the liquid. This is known as **Pascal's principle,** named after Blaise Pascal (1623–1662):

> A pressure change applied to an enclosed liquid is transmitted undiminished to every point in the liquid and to the walls of the container.

PASCAL'S PRINCIPLE

A common application of Pascal's principle is the hydraulic lift shown in Figure 13-4.

FIGURE 13-4 Hydraulic lift. A small force F_1 on the small piston produces a change in pressure F_1/A_1 that is transmitted by the liquid to the large piston. Since the pressures at the small and large pistons are the same, the forces are related by $F_2/A_2 = F_1/A_1$. Since the area of the large piston is much greater than that of the small piston, the force on the large piston $F_2 = (A_2/A_1)F_1$ is much greater than F_1.

A HYDRAULIC LIFT   **EXAMPLE 13-3**

The large piston in a hydraulic lift has a radius of 20 cm. What force must be applied to the small piston of radius 2 cm to raise a car of mass 1500 kg?

PICTURE THE PROBLEM The pressure P times the area A_2 of the large piston must equal the weight mg of the car. The force F_1 that must be exerted on the small piston is this pressure times the area A_1. (See Figure 13-4.)

1. The force F_1 is the pressure P times the area A_1:

$$F_1 = PA_1$$

2. The pressure P times the area A_2 equals the weight of the car:

$$PA_2 = mg \quad \text{so} \quad P = \frac{mg}{A_2}$$

3. Substitute this result for P into the step 1 result and calculate F_1:

$$F_1 = PA_1$$
$$= \frac{mg}{A_2}A_1 = mg\frac{A_1}{A_2} = mg\frac{\pi r_1^2}{\pi r_2^2}$$
$$= (1500 \text{ kg})(9.81 \text{ N/kg})\left(\frac{2 \text{ cm}}{20 \text{ cm}}\right)^2$$
$$= \boxed{147 \text{ N}}$$

Figure 13-5 shows water in a container with sections of different shapes. At first glance, it might seem that the pressure at the bottom of section 3 of the container would be greatest and that water would therefore be forced to a greater height in section 2, which is smaller. But that does not happen, a result known as the **hydrostatic paradox.** The pressure depends only on the depth of the water, not on the shape of the container, so at the same depth the pressure is the same in all parts of the container, a finding that can be shown experimentally. Although the water in section 4 of the container weighs more than that in section 2, the portion of the water in section 4 that is not above the opening at the bottom is supported by the horizontal shelf of the section. In fact, the water above the opening at the bottom of section 5 weighs less than the water above an opening of the same size at the bottom of section 1. However, the

FIGURE 13-5 The hydrostatic paradox. The water level is the same regardless of the shape of the vessel. The shaded portion of the water is supported by the sides of the container.

horizontal shelf of section 5 exerts a downward force on the water—exactly compensating for the shortfall of weight.

We can use the fact that the pressure difference is proportional to the depth of a liquid to measure unknown pressures. Figure 13-6 shows a simple pressure gauge, the open-tube manometer. The top of the tube is open to the atmosphere at pressure P_{at}. The other end of the tube is at pressure P, which is to be measured. The difference $P - P_{at}$, called the **gauge pressure** P_{gauge}, is equal to $\rho g h$, where ρ is the density of the liquid in the tube. The pressure you measure in your automobile tire is gauge pressure. When the tire is entirely flat, the gauge pressure is zero, and the absolute pressure in the tire is atmospheric pressure. The absolute pressure P is obtained from the gauge pressure by adding atmospheric pressure to it:

$$P = P_{gauge} + P_{at} \qquad \text{13-8}$$

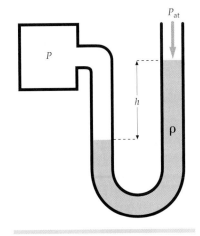

FIGURE 13-6 Open-tube manometer for measuring an unknown pressure P. The difference $P - P_{at}$ equals $\rho g h$.

Figure 13-7 shows a mercury barometer, which is used to measure atmospheric pressure. The top end of the tube has been closed off and evacuated so that the pressure there is zero. The other end is submerged in a pool of mercury that is open to the atmosphere at pressure P_{at}. The pressure P_{at} is $\rho g h$, where ρ is the density of mercury.

FIGURE 13-7

EXERCISE At 0°C, the density of mercury is 13.595×10^3 kg/m³. What is the height of the mercury column in a barometer if the pressure is 1 atm = 101.325 kPa? (*Answer* $h = P/\rho g = 0.760$ m = 760 mm)

In practice, pressure is often measured in millimeters of mercury, a unit called the **torr,** after the Italian physicist Evangelista Torricelli, or in inches of mercury (written inHg). The various units of pressure are related as follows:

$$1 \text{ atm} = 760 \text{ mmHg} = 760 \text{ torr} = 29.9 \text{ in.Hg}$$
$$= 101.325 \text{ kPa} = 14.7 \text{ lb/in.}^2 \qquad \text{13-9}$$

Other units commonly used on weather maps are the **bar** and the **millibar,** which are defined as

$$1 \text{ bar} = 10^3 \text{ millibars} = 100 \text{ kPa} \qquad \text{13-10}$$

A pressure of 1 atm is about 1.3% greater than a pressure of 1 bar.

Tire-pressure gauge. The piston pushes the rod to the right until the force of the spring plus the force due to atmospheric pressure balances the force due to the air pressure in the tire.

BLOOD PRESSURE IN THE AORTA   **EXAMPLE 13-4**

The average gauge pressure in the human aorta is about 100 mmHg. Convert this average blood pressure to pascals.

We use the conversion factors that can be obtained from Equation 13-9:

$$P = 100 \text{ mmHg} \times \frac{101.325 \text{ kPa}}{760 \text{ mmHg}}$$

$$= \boxed{13.3 \text{ kPa}}$$

EXERCISE Convert a pressure of 45 kPa to (*a*) millimeters of mercury and (*b*) atmospheres. (*Answer* (*a*) 338 mmHg (*b*) 0.444 atm)

The relation between pressure and altitude (or depth) is more complicated for a gas than for a liquid because the density of a gas is not constant like that of a liquid, but is approximately proportional to the pressure. As you go up from the surface of the earth, pressure in a column of air decreases, just as the pressure would decrease as you go up from the bottom in a water column. But the decrease in air pressure is not linear with distance. Instead, the air pressure decreases by a constant fraction for a given increase in height, as shown in Figure 13-8. At a height of about 5.5 km (18,000 ft), the air pressure is half its value at sea level. If we go up another 5.5 km to an altitude of 11 km (a typical altitude for airliners), the pressure is again halved so that it is one-fourth its value at sea level, and so on. This example of an *exponential decrease* is called the law of atmospheres. At the high altitudes at which commercial jets fly, the cabins must be pressurized. The density of air is approximately proportional to the pressure, so the density of air decreases with altitude. There is less oxygen available on a mountain top than at normal elevations. This makes exercising in the Rockies difficult, and climbing in the Himalayas dangerous.

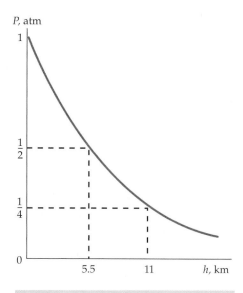

FIGURE 13-8 Variation in pressure with height above the earth's surface. For each 5.5-km increase in height, the pressure decreases by half.

13-3 Buoyancy and Archimedes' Principle

If a dense object submerged in water is weighed by suspending it from a spring scale (Figure 13-9a), the apparent weight of the object when submerged (the reading on the scale) is less than the weight of the object. This is so because the water exerts an upward force that partially balances the force of gravity. This upward force is even more evident when we submerge a piece of cork. When completely submerged, the cork experiences an upward force from the water pressure that is greater than the force of gravity, so when released it accelerates up toward the surface. The force exerted by a fluid on a body wholly or partially submerged in it is called the **buoyant force.**[†] It is equal to the weight of the fluid displaced by the body.

> A body wholly or partially submerged in a fluid is buoyed up by a force equal to the weight of the displaced fluid.
>
> ARCHIMEDES' PRINCIPLE

This result is known as **Archimedes' principle.**

We can derive Archimedes' principle from Newton's laws by considering the forces acting on a portion of a fluid and noting that in static equilibrium the net force must be zero. Figure 13-9b shows the vertical forces acting on an object being weighed while submerged. These are the force of gravity $\vec{w}$ acting down, the force of the spring scale $\vec{F}_s$ acting up, a force $\vec{F}_1$ acting down because of the fluid pressure on the top surface of the object, and a force $\vec{F}_2$ acting up because of the fluid pressing on the bottom surface of the object. Since the spring scale reads a force less than the weight, the force $\vec{F}_2$ must be greater in magnitude than the force $\vec{F}_1$. The difference in magnitude of these two forces is the buoyant force $\vec{B} = \vec{F}_2 - \vec{F}_1$ (Figure 13-9c). The buoyant force occurs because the pressure of the fluid at the bottom of the object is greater than that at the top.

(a)

(b)

$B = F_2 - F_1$

(c)

FIGURE 13-9 (a) Weighing an object submerged in a fluid. (b) Free-body diagram showing the weight $\vec{w}$, the force $\vec{F}_s$ of the spring, and the forces $\vec{F}_1$ and $\vec{F}_2$ exerted by the surrounding field. (c) The buoyant force $B = F_2 - F_1$ is the net force exerted on the object by the fluid.

---

† The definition of buoyant force is further refined later in this section.

In Figure 13-10, the spring scale has been eliminated and the submerged object has been replaced by an equal volume of fluid (outlined by the dashed lines). The buoyant force $\vec{B} = \vec{F}_2 - \vec{F}_1$ acting on this volume of fluid is the same as the buoyant force that acted on our original object since the fluid surrounding the space is the same. Because this volume of fluid is in equilibrium, the net force acting on it must be zero. The upward buoyant force thus equals the downward weight of this volume of the fluid:

$$B = w_f \tag{13-11}$$

FIGURE 13-10 Figure 13-9 with the submerged body replaced by an equal volume of fluid. The forces $\vec{F}_1$ and $\vec{F}_2$, due to the pressure of the fluid are the same as in Figure 13-9. The magnitude of the buoyant force is thus equal to the weight w_f of the displaced fluid.

Note that this result does not depend on the shape of the submerged object. If we consider any irregularly shaped portion of a static fluid, there must be a buoyant force acting on it by the surrounding fluid that exactly supports its weight. Thus, we have derived Archimedes' principle.

Archimedes (287–212 B.C.) had been given the task of determining whether a crown made for King Hieron II was of pure gold or had been adulterated with some cheaper metal such as silver, and to do this without destroying the crown. For Archimedes, the problem was to determine if the density of the irregularly shaped crown was the same as the density of gold. As the story goes, he came upon the solution while sinking himself into a bathtub and immediately rushed naked through the streets of Syracuse shouting "Eureka!" ("I have found it!"). This flash of insight preceded Newton's laws, which we used to derive Archimedes' principle, by some 1900 years. What Archimedes found was a simple and accurate way to compare the density of the crown with the density of gold, using a pan balance. He placed the balance in a large basin, and placed the crown on one pan and an equal mass of pure gold on the other pan. He then added water to the basin, submerging the crown and the pure gold. The balance tilted, with the crown rising—indicating that the buoyant force on the crown was greater than that on the pure gold because the volume of water displaced by the crown was greater than that displaced by the pure gold. The crown was less dense than the pure gold.

The specific gravity of an object equals the weight of the object divided by the weight of an equal volume of water. According to Archimedes' principle, the weight of an equal volume of water equals the buoyant force on the object when it is submerged in water. Therefore, the specific gravity is equal to the weight of the object divided by the buoyant force on it when it is submerged in water:

$$\text{Specific gravity} = \frac{\text{weight}}{\text{buoyant force when submerged in water}} = \frac{w}{B_{\text{water}}} \tag{13-12}$$

The apparent weight w_{app} of an object submerged in a fluid is the difference between its weight w and the buoyant force B:

$$w_{\text{app}} = w - B \tag{13-13}$$

(a)

(b)

(a) The crown and the gold nugget have equal weight. (b) The wreath displaces more water than the gold nugget.

Hot-air balloons rising in the night sky over Albuquerque during a balloon festival.

IS IT REALLY GOLD?   **EXAMPLE 1 3 - 5** **Put It in Context**

Your friend is concerned about a gold ring she bought on a recent trip. The ring was expensive, and she would like to know whether it is really made of gold or of something else. You decide to help her, using your knowledge of physics. You weigh the ring and find that it has a weight of 0.158 N. Using a string, you suspend the ring from the scale and, with the ring submerged in water, weigh it again to find a new reading of 0.150 N. Is the ring pure gold?

PICTURE THE PROBLEM If the ring is pure gold, its specific gravity (its density relative to that of water) is 19.3 (see Table 13-1). Using Equation 13-12 as a guide, determine the specific gravity of the ring.

1. Equation 13-12 relates the specific gravity of the ring to the ratio of its weight w to the buoyant force B when submerged in water:

$$\text{Specific gravity} = \frac{w}{B_{\text{water}}} = \frac{w}{B}$$

2. B equals the weight minus the apparent weight w_{app} when submerged:

$$B = w - w_{\text{app}}$$

3. Combine steps 1 and 2 and solve for the specific gravity:

$$\text{Specific gravity} = \frac{w}{B} = \frac{w}{w - w_{\text{app}}}$$

$$= \frac{0.158 \text{ N}}{0.158 \text{ N} - 0.150 \text{ N}} = 19.3$$

4. Compare the specific gravity of the ring with the specific gravity of gold, which is 19.3:

$19.3 \sim 19.3$

The ring is pure gold.

EXERCISE A block of an unknown material weighs 3 N and has an apparent weight of 1.89 N when submerged in water. What is the material? (*Answer* The specific gravity of the material is 2.70, which is the specific gravity of aluminum. The material is aluminum.)

EXERCISE An aluminum block has an apparent weight of 3 N when surrounded by air. What is the weight of the block? (*Answer* $w_{\text{app}} = w - B$ where $B = \rho_{\text{air}} Vg$ and $w = \rho_{\text{alum}} Vg$. From Table 13-1 we obtain $\rho_{\text{air}} = 1.293 \text{ kg/m}^3$ and $\rho_{\text{alum}} = 2.70 \times 10^3 \text{ kg/m}^3$. We have three equations and three unknowns: w, B, and V. Thus, $w = 3.0014$ N, which is only 0.048 percent greater than the apparent weight in air. Clearly, buoyant forces on solids and liquids due to air can usually be ignored.)

EXERCISE A piece of lead (specific gravity = 11.3) weighs 80 N in air. What does it weigh when submerged in water? (*Answer* 72.9 N)

The density of the block shown in Figure 13-11 is greater than the density of the surrounding fluid, and both the block and the scale pan are completely submerged in the fluid. The gravitation force on the block is its weight, $\vec{w}$, and the scale is adjusted, so it reads zero when the block is not being supported by the pan (Figure 11-13b). If the block is on the pan (Figure 13-11a), the scale reading is equal to the magnitude of the apparent weight w_{app} of the block. When the block is on the pan, the fluid is in direct contact with the entire surface of the block—except for those regions of the bottom surface of the block that are in direct contact with the pan. We assume the pan is not perfectly flat, but instead has some high and some low regions, and that the pan is in direct contact with the bottom surface of

FIGURE 13-11

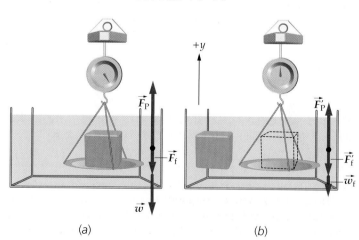

(a)   (b)

the block only at the high regions of the pan. (At the low regions of the pan, there is fluid in direct contact with bottom surface of the block.) We now analyze this situation to determine if the scale reading depends on the area of contact of the bottom surface of the block with the pan.

While resting on the pan, the net force exerted by the fluid on the block $\vec{F}_f$ is a combination of the downward force of the fluid on the top surface of the block and the upward force of the fluid on those regions of the bottom surface of the block that are in direct contact with the fluid. (We have drawn $\vec{F}_f$ downward. However, if the fluid were in direct contact with a large enough area of the bottom surface of the block, this force would be upward.) The two other vertical forces acting on the block are the gravitational force $\vec{w}$ and the upward force $\vec{F}_P$ exerted on the block by the pan at the regions of direct contact. In Figure 13-11b, the block has been moved off the pan, and in its place is a sample of fluid of identical size and shape (outlined by the dashed lines). The same regions of the surface of the pan are in direct contact with the bottom surface of this sample of fluid as if it were in direct contact with the block before it was moved. The forces acting on the fluid sample are the forces acting on it by the surrounding fluid $\vec{F}_f'$, the upward force on it by the pan $\vec{F}_P'$, and the gravitational force $\vec{w}_f$. The forces $\vec{F}_f$ and $\vec{F}_f'$ are equal because, at every point where the sample and the surrounding fluid are in direct contact, the pressure of the surrounding fluid is the same as it was at the same point before the block was moved off the pan.

When the submerged block rests on the pan, the block is in equilibrium, so $F_{f,y} + F_{P,y} + w_y = 0$, or

$$F_{f,y} + F_P - w = 0$$

and when the block is moved off the pan, the fluid sample in its place is in equilibrium, so

$$F_{f,y}' + F_P' - w_f = 0$$

Subtracting these equations, exploiting that $F_{f,y}' = F_{f,y}$, and rearranging the result gives

$$F_P - F_P' = w - w_f$$

where $F_P - F_P'$ is the change in the scale reading when the block is moved off the pan, so $F_P - F_P'$ is the magnitude of the apparent weight of the submerged block. Thus,

$$w_{app} = w - w_f$$

It is common parlance to refer to $w - w_{app}$ as the buoyant force B. Rearranging gives

$$B = w - w_{app} = w_f$$

This is the same expression for the buoyant force as is in Equation 13-13, which was established with the fluid in direct contact with 100 percent of surface of the submerged object.

MEASURING THE FAT   **E X A M P L E   1 3 - 6**

The percentage of body fat can be estimated by measuring the density of the body. Fat is less dense than muscle or bone. The density of fat $\approx 0.9 \times 10^3 \ kg/m^3$ and the density of lean tissue (everything except fat) $\approx 1.1 \times 10^3 \ kg/m^3$. Measuring the density of the body involves measuring the apparent weight while the body is submerged in water with the air completely exhaled from the lungs. (In practice, the amount of air remaining in the lungs is estimated and corrected for.) Suppose that your apparent weight when submerged in water is 5 percent of your weight. What percentage of your body mass is fat?

To determine the percentage of fat in this man's body, his density is measured by weighing him while he is submerged under water.

PICTURE THE PROBLEM For the person, the total volume equals the volume of the fat plus the volume of the lean, and the total mass equals the mass of the fat plus the mass of the lean. The volume and density are related to the mass by $m = \rho V$. The fraction of fat equals the mass of the fat divided by the total mass and the fraction of lean equals the mass of the lean divided by the total mass. Also, the fraction of fat plus the fraction of lean equals 1.

1. Using Equations 13-2 and 13-3, find the ratio of your body's density to the density of water:

$$\frac{\rho}{\rho_{water}} = \frac{w}{w - w_{app}} = \frac{w}{w - 0.05w} = 1.05$$

2. Your total body volume equals the volume of fat plus the volume of lean tissue:

$$V_{tot} = V_{fat} + V_{lean}$$

3. Because density equals mass times volume, volume equals mass divided by density. Substitute the corresponding mass-to-density ratio for each volume in the step 2 result:

$$\frac{m_{tot}}{\rho} = \frac{m_{fat}}{\rho_{fat}} + \frac{m_{lean}}{\rho_{lean}}$$

4. The ratio m_{fat}/m_{tot} is the fraction of fat f_{fat}, and m_{lean}/m_{tot} is the fraction of lean f_{lean}. Substitute for m_{fat} and m_{lean} in the step 3 result:

$$\frac{m_{tot}}{\rho} = \frac{f_{fat}m_{tot}}{\rho_{fat}} + \frac{f_{lean}m_{tot}}{\rho_{lean}}$$

5. The fraction of fat plus the fraction of lean tissue equals 1:

$$f_{fat} + f_{lean} = 1$$

6. Divide both sides of the step 4 result by m_{tot} and substitute $1 - f_{fat}$ for f_{lean}:

$$\frac{1}{\rho} = \frac{f_{fat}}{\rho_{fat}} + \frac{(1 - f_{fat})}{\rho_{lean}}$$

7. Solve the step 6 result for f_{fat}:

$$f_{fat} = \frac{1 - (\rho_{lean}/\rho)}{1 - (\rho_{lean}/\rho_{fat})}$$

8. Using the step 1 result, substitute for ρ in the step 7 result and solve for f_{fat}:

$$f_{fat} = \frac{1 - (\rho_{lean}/1.05\rho_{water})}{1 - (\rho_{lean}/\rho_{fat})} = \frac{1 - (1.1/1.05)}{1 - (1.1/0.9)} = 0.21$$

9. Convert to a percentage:

$$100\% \times f_{fat} = \boxed{21\%}$$

EXERCISE If Ed's apparent weight when submerged is zero, what is his body-fat percentage? (*Answer* 45 percent)

FIGURE 13-12

7.0 cm

FLOATING ON A RAFT   **EXAMPLE 13-7**

A raft of area A, thickness h, and mass $M = 600$ kg floats in calm water with 7 cm submerged (Figure 13-12). When Bob stands on the raft, 8.4 cm are submerged. What is Bob's mass m?

PICTURE THE PROBLEM Let A be the area of the raft. The weight of the displaced fluid is then $\rho_w A d_1 g$ with just the raft and $\rho_w A d_2 g$ with Bob on the raft, where ρ_w is the density of the water, $d_1 = 7$ cm, and $d_2 = 8.4$ cm. If we set the weight of displaced fluid equal to the weight of the floating objects in each case, we can eliminate A and ρ_w and solve for m.

8.4 cm

1. Set the buoyant force with $d_1 = 7$ cm submerged equal to the weight of the raft, and with $d_2 = 8.4$ cm submerged equal to the weight of the raft plus Bob:

$$\rho_w A d_1 g = Mg$$
$$\rho_w A d_2 g = (M + m)g$$

2. Divide these two equations to eliminate the unknowns, A and ρ_w:

$$\frac{d_2}{d_1} = \frac{M + m}{M}$$

3. Solve for m:

$$m = \left(\frac{d_2}{d_1} - 1\right)M = \left(\frac{8.4 \text{ cm}}{7.0 \text{ cm}} - 1\right)(600 \text{ kg}) = \boxed{120 \text{ kg}}$$

A BOBBING CORK   **EXAMPLE 13-8**

A cork has a density of 200 kg/m³. Find the fraction of the volume of the cork that is submerged when the cork floats in water.

PICTURE THE PROBLEM Let V be the volume of the cork and V_{sub} be the volume that is submerged when the cork floats. The weight of the cork is $\rho_c V g$ and the buoyant force due to the water is $\rho_w V_{sub} g$.

1. Since the cork is in equilibrium, the buoyant force equals its weight:

$$w = B$$
$$\rho_c V g = \rho_w V_{sub} g$$

2. Solve for V_{sub}/V:

$$\frac{V_{sub}}{V} = \frac{\rho_c}{\rho_w} = \frac{200 \text{ kg/m}^3}{1000 \text{ kg/m}^3} = \boxed{\frac{1}{5}}$$

REMARKS We see that only one-fifth of the cork is submerged. This result is independent of the shape of the cork.

If we replace ρ_w in the calculation above with ρ_f, the density of the fluid, we can determine the submerged fraction of an object floating in any fluid. From Example 13-8, the fraction of a floating object that is submerged equals the ratio of its density to the density of the fluid.

$$\frac{V_{sub}}{V} = \frac{\rho}{\rho_f}$$ 13-14

Because the density of ice is 920 kg/m³ and that of sea water is 1025 kg/m³, the fraction of an iceberg that is submerged in sea water is

$$\frac{V_{sub}}{V} = \frac{\rho}{\rho_f} = \frac{920 \text{kg/m}^3}{1025 \text{kg/m}^3} = 0.898$$

The great danger that icebergs pose to ships springs directly from the fact that only about 10 percent of an iceberg is visible above the water, and the visible portion gives little hint of where the submerged portion may extend, posing extreme danger to passing vessels.

Smoke from a burning cigarette. At first the smoke rises in a regular stream, but the simple streamlined flow quickly becomes turbulent and the smoke begins to swirl irregularly.

13-4 Fluids in Motion

The behavior of a fluid in motion can be complex. Consider, for example, the rise of smoke from a burning cigarette. At first the smoke rises in a regular stream of warm gas, but the simple streamlined flow quickly becomes turbulent and the smoke begins to swirl irregularly. Turbulent flow is very difficult to describe, even qualitatively. We will therefore consider only nonturbulent, steady-state flow of an "ideal" fluid, one that is nonviscous; that is, one that flows with no dissipation of mechanical energy. We also assume that the fluid is incompressible, which is an excellent approximation for liquids in most situations. In an incompressible fluid, the density is constant throughout the fluid.

Figure 13-13 shows a tube full of liquid. The tube contains a tapered section with decreasing cross-sectional area. The fluid is flowing from left to right, and the shaded portion on the left depicts the volume ΔV of fluid that flows past point 1 during time Δt. If the speed of the fluid at this point is v_1 and the cross-sectional area of the tube is A_1, then the volume flowing past point 1 in time Δt is

FIGURE 13-13

$$\Delta V = A_1 v_1 \, \Delta t$$

Since we assume that the fluid is incompressible, an equal volume of fluid must flow past any point along the tube during the same time Δt. The volume that flows past point 2 is depicted by the shaded portion on the right. If the speed of the fluid at this point is v_2 and the cross-sectional area is A_2, then the volume is $\Delta V = A_2 v_2 \, \Delta t$. Since these volumes must be equal, we have

$$A_1 v_1 \, \Delta t = A_2 v_2 \, \Delta t \quad \text{so} \quad A_1 v_1 = A_2 v_2 \qquad \text{13-15}$$

The quantity Av is called the **volume flow rate** I_v. The dimensions of I_v are volume divided by time. In the flow of an incompressible fluid, the volume flow rate is the same at any point in the fluid:

$$I_v = Av = \text{constant} \qquad \text{13-16}$$

CONTINUITY EQUATION FOR INCOMPRESSIBLE FLUID

Equation 13-16 is called the **continuity equation** for an incompressible fluid.

EXERCISE Blood flows in an aorta of radius 1.0 cm at 30 cm/s. What is the volume flow rate? (*Answer* $I_v = vA = 9.42 \times 10^{-5} \text{ m}^3/\text{s}$. It is customary to give the pumping rate of the heart in liters per minute. Using $1 \text{ m}^3 = 1000 \text{ L}$, we have $I_v = 5.65 \text{ L/min}$.)

EXERCISE Blood flows from a large artery of radius 0.3 cm, where its speed is 10 cm/s, into a region where the radius has been reduced to 0.2 cm because of thickening of the arterial walls (arteriosclerosis). What is the speed of the blood in the narrower region? (*Answer* If v_1 and v_2 are the initial and final speeds and A_1 and A_2 are the initial and final areas, Equation 13-16 gives

$$v_2 = \frac{A_1}{A_2} v_1 = \frac{\pi(0.3 \text{ cm})^2}{\pi(0.2 \text{ cm})^2}(10 \text{ cm/s}) = 22.5 \text{ cm/s} \Bigg)$$

Bernoulli's Equation

When a small sample of a fluid flows into a narrow region of a tube, it gains speed because the pressure behind it that pushes it forward is greater than the pressure in front of it opposing its motion. Bernoulli's equation relates the pressure, elevation, and speed of an *incompressible* fluid in *steady* flow. It follows from Newton's laws and is most easily derived by applying the work–energy theorem to a segment of the fluid.

Consider a fluid flowing in a tube that varies in elevation as well as in cross-sectional area, as shown in Figure 13-14. We apply the work–energy theorem to a sample of fluid that initially is contained between points 1 and 2 in Figure 13-14*a*.

During time Δt this sample moves along the tube to the region between points 1' and 2' as shown in Figure 13-14b. Let ΔV be the volume of fluid passing point 1' during time Δt. The same volume passes point 2 during the same time. Also, let $\Delta m = \rho \Delta V$ be the mass of the fluid with volume ΔV. The net effect on the sample during time Δt is that a mass Δm initially at height h_1 moving with speed v_1 is "transferred" to height h_2 with speed v_2. The change in the potential energy of the sample is thus

$$\Delta U = (\Delta m)gh_2 - (\Delta m)gh_1 = \rho \, \Delta V \, g(h_2 - h_1)$$

and the change in its kinetic energy is

$$\Delta K = \tfrac{1}{2}(\Delta m)v_2^2 - \tfrac{1}{2}(\Delta m)v_1^2 = \tfrac{1}{2}\rho \, \Delta V(v_2^2 - v_1^2)$$

The fluid behind the sample (to the sample's left in Figure 13-15) pushes on the sample with a force of magnitude $F_1 = P_1A_1$, where P_1 is the pressure at point 1. This force does work:

$$W_1 = F_1 \, \Delta x_1 = P_1A_1 \, \Delta x_1 = P_1 \, \Delta V$$

At the same time, the fluid in front of the sample (to the sample's right) exerts a force $F_2 = P_2A_2$ opposing the sample's motion (pushing it to the left). This force does negative work:

$$W_2 = -F_2 \, \Delta x_2 = -P_2A_2 \, \Delta x_2 = -P_2 \, \Delta V$$

The total work done by these forces is

$$W_{\text{total}} = P_1 \, \Delta V - P_2 \, \Delta V = (P_1 - P_2)\Delta V$$

The work–energy theorem gives

$$W_{\text{total}} = \Delta U + \Delta K$$

so

$$(P_1 - P_2)\Delta V = \rho \, \Delta V \, g(h_2 - h_1) + \tfrac{1}{2}\rho \, \Delta V(v_2^2 - v_1^2)$$

If we divide both sides by ΔV, we obtain

$$P_1 - P_2 = \rho gh_2 - \rho gh_1 + \tfrac{1}{2}\rho v_2^2 - \tfrac{1}{2}\rho v_1^2$$

When we collect all the quantities having a subscript 1 on one side and those having a subscript 2 on the other, this equation becomes

$$P_1 + \rho gh_1 + \tfrac{1}{2}\rho v_1^2 = P_2 + \rho gh_2 + \tfrac{1}{2}\rho v_2^2 \qquad \text{13-17}a$$

This result can be restated as

$$P + \rho gh + \tfrac{1}{2}\rho v^2 = \text{constant} \qquad \text{13-17}b$$

BERNOULLI'S EQUATION

(a)

(b)

FIGURE 13-14

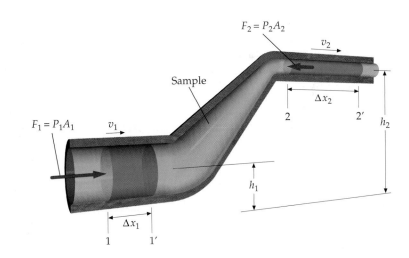

FIGURE 13-15 Fluid moving in a pipe that varies in both height and cross-sectional area. The total work done by the forces $F_1 = P_1A_1$ and $F_2 = P_2A_2$ has the effect of raising the shaded portion of the fluid from height h_1 to h_2 and changing its speed from v_1 to v_2.

which means that this combination of quantities has the same value at any point along the tube. Equations 13-17a and 13-17b are known as **Bernoulli's equation** for the steady, nonviscous flow of an incompressible fluid.

A special application of Bernoulli's equation is for a fluid at rest. Then $v_1 = v_2 = 0$, and we obtain

$$P_1 - P_2 = \rho g h_2 - \rho g h_1 = \rho g \Delta h$$

This is the same as Equation 13-7.

FIGURE 13-16

A LEAKING TANK   **EXAMPLE 13-9**

A large tank of water, open at the top, has a small hole through its side a distance h below the surface of the water. Find the speed of the water as it flows out the hole.

PICTURE THE PROBLEM We apply Bernoulli's equation to points a and b in Figure 13-16. Because the diameter of the hole is much smaller than the diameter of the tank, we can neglect the velocity of the water at the top (point a).

1. Bernoulli's equation with $v_a = 0$ gives:

$$P_a + \rho g h_a + 0 = P_b + \rho g h_b + \tfrac{1}{2}\rho v_b^2$$

2. The pressure at point a and at point b is the same, P_{at}, since both points are open to the atmosphere:

$$P_a = P_{at} \quad \text{and} \quad P_b = P_{at}$$

so

$$P_{at} + \rho g h_a + 0 = P_{at} + \rho g h_b + \tfrac{1}{2}\rho v_b^2$$

3. Solve the step 2 result for the speed v_b of the water flowing from the hole:

$$v_b^2 = 2g(h_a - h_b) = 2g\Delta h$$

so

$$v_b = \boxed{\sqrt{2g\Delta h}}$$

EXERCISE If the water flowing out of the hole is directed vertically upward, how high does it rise? (*Answer* The water shoots upward a distance h; that is, to the same level as the surface of the water in the tank.)

In Example 13-9, the water emerges from the hole with a speed equal to the speed it would have if it dropped in free-fall a distance h. This finding is known as *Torricelli's law*.

In Figure 13-17, water is shown flowing through a horizontal pipe that has a constricted section. Because both sections of the pipe are at the same elevation, $h_1 = h_2$ in Equation 13-17a. Then Bernoulli's equation becomes

$$P + \tfrac{1}{2}\rho v^2 = \text{constant} \qquad\qquad 13\text{-}18$$

When the fluid moves into the constriction, the area A gets smaller, so the speed v must get larger since Av remains constant. But because $P + \tfrac{1}{2}\rho v^2$ is constant, when the speed gets larger the pressure must get smaller. Thus, the pressure in the constriction is reduced.

FIGURE 13-17 Constriction in a pipe carrying a moving fluid. The pressure is lower in the narrow section of the pipe where the fluid is moving faster.

When the speed of a fluid increases, the pressure drops.

VENTURI EFFECT

This result is often referred to as the **Venturi effect.** Equation 13-18 is an important result that applies to many situations in which we can ignore changes in height.

In Figure 13-18, lines, called **streamlines,** are drawn to pictorially represent the flow of the fluid. The direction of the lines denotes the direction of flow and the spacing between lines represents the speed of the flow. The smaller the spacing, the greater the speed. For horizontal flow, where the speed increases, the pressure decreases, so a decrease in streamline spacing is accompanied by a decrease in pressure.

FIGURE 13-18

A VENTURI METER   **EXAMPLE 13-10** Try It Yourself

A *Venturi meter,* which is used to measure the flow rate of a fluid, is shown in Figure 13-19. The fluid of density ρ_F passes through a pipe of cross-sectional area A_1 that has a constriction of cross-sectional area A_2. The two parts of the pipe are connected with a U-tube manometer partially filled with a liquid of density ρ_L. Since the velocity of flow is greater in the constricted region, the pressure in that section is less than in the other portion of the pipe. The pressure difference is measured by the difference in the levels of the liquid in the U-tube, Δh. Express the velocity v_1 in terms of the measured height Δh and the known quantities ρ_F, ρ_L, and $r = A_1/A_2$.

PICTURE THE PROBLEM The pressures P_1 and P_2 in the two regions are related to the speeds v_1 and v_2 by Bernoulli's equation. The pressure difference is related to the height h by $P_1 - P_2 = \rho_L gh$. You can express v_2 in terms of v_1 and the areas A_1 and A_2 by the continuity equation.

FIGURE 13-19 A Venturi meter.

Cover the column to the right and try these on your own before looking at the answers.

Steps

1. Write Bernoulli's equation for constant elevation for the two regions.

2. Write the continuity equation for the two regions, and solve for v_2 in terms of v_1 and the areas A_1 and A_2.

3. Substitute your result for v_2 into the equation in step 1 and obtain an equation for $P_1 - P_2$.

4. Write $P_1 - P_2$ in terms of the difference in height Δh of the liquid in the arms of the U-tube. This pressure difference equals the pressure drop in a column of the liquid of height Δh and that in a column of the fluid of the same height.

5. Equate the two expressions for $P_1 - P_2$ and solve for v_1 in terms of Δh.

Answers

$P_1 + \tfrac{1}{2}\rho_F v_1^2 = P_2 + \tfrac{1}{2}\rho_F v_2^2$

$v_2 A_2 = v_1 A_1$

so

$v_2 = \dfrac{A_1}{A_2} v_1 = r v_1$

$P_1 - P_2 = \tfrac{1}{2}\rho_F(v_2^2 - v_1^2)$

$\quad\quad = \tfrac{1}{2}\rho_F(r^2 - 1)v_1^2$

$P_1 - P_2 = \rho_L g\Delta h - \rho_F g\Delta h$

$\quad\quad = (\rho_L - \rho_F)g\Delta h$

$\tfrac{1}{2}\rho_F(r^2 - 1)v_1^2 = (\rho_L - \rho_F)g\Delta h$

so

$$v_1 = \sqrt{\dfrac{2(\rho_L - \rho_F)g\Delta h}{\rho_F(r^2 - 1)}}$$

EXERCISE Find v_1 if $\Delta h = 3$ cm, $r = 4$, the fluid is air ($\rho_F = 1.29$ kg/m³), and the liquid in the U-tube portion of the Venturi meter is water ($\rho_w = 10^3$ kg/m³). (*Answer* $v_1 = 5.51$ m/s)

REMARKS Air is not an incompressible fluid, so the calculation in the Exercise is not as accurate as the calculation in the Example 13-9. Strictly speaking, Bernoulli's equation and the continuity equation hold only for incompressible fluids.

The Venturi effect can be used to give a qualitative understanding of the lift of an airplane wing and the path of a pitcher's curveball. An airplane wing is designed so that air moves faster over the top of the wing than it does under the wing, thus making the air pressure less on top than underneath. This difference in pressure results in a net force upward on the wing. Figure 13-20a shows a top view of the motion of a curveball. As the ball spins, it tends to drag air around with it. Figure 13-20b is drawn from the point of view of a stationary (but spinning) ball with the air rushing past it. The air movement caused by the spinning ball adds to the velocity of the air rushing by on the left side of the ball (the top in the figure) and subtracts from it on the right (the bottom in the figure). Thus, the air speed is greater on the left side of the ball than on the right, causing the pressure on the left to be less than that on the right. The ball therefore curves to the left. The atomizer shown in Figure 13-21 also works on the principle of the Venturi effect.

Airflow above and below the wing of this Indy race car creates greater pressure above the wing, increasing the effective weight of the car for better control at high speeds. An airplane wing is designed so that the flow creates greater pressure below the wing to lift the plane.

Although Bernoulli's equation is very useful for qualitative descriptions of many features of fluid flow, such descriptions are often grossly inaccurate when compared with the quantitative results of experiments. Prominent reasons for the discrepancies are that gases like air are hardly incompressible, and liquids like water have viscosity, which invalidates the assumption of the conservation of mechanical energy. In addition, it is often difficult to maintain steady-state, streamlined flow without turbulence, and the introduction of turbulence can greatly affect the results.

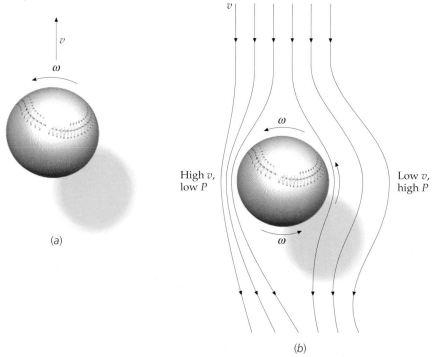

(a)

High v, low P Low v, high P

(b)

FIGURE 13-21 When the bulb of an atomizer is squeezed, the air is forced through the constriction in the horizontal tube, which reduces the pressure there below atmospheric pressure. Because of the resulting pressure difference, the liquid in the jar is pumped up through the vertical tube, enters the air stream, and emerges from the nozzle. A similar effect occurs in the carburetor of a gasoline lawnmower engine.

FIGURE 13-20 (a) Top view of a baseball thrown with a counterclockwise spin ω, like an overhead view of a curveball thrown by a right-handed pitcher. (b) In the frame moving with the ball, the ball is stationary (but spinning) and the air rushes past it. Because of its rough cover, the spinning ball drags the air around with it, making the air speed higher on the left side and lower on the right. The pressure is therefore lower on the left side, so the ball curves to the left.

*Viscous Flow

According to Bernoulli's equation, when a fluid flows steadily through a long, narrow, horizontal pipe of constant cross section, the pressure along the pipe will be constant. In practice, however, we observe a pressure drop as we move along the direction of the flow. Put another way, a pressure difference is required to push a fluid through a horizontal pipe. This pressure difference is needed because of the drag force that is exerted by the pipe on the layer of fluid in contact with the pipe, and because of the drag force exerted by each layer of the fluid on an adjacent layer that is moving with a slightly different velocity. These drag forces are called **viscous forces.** As a result of viscous forces, the velocity of the fluid is not constant across the diameter of the pipe. Instead, it is greatest near the center of the pipe and approaches zero where the fluid is in contact with the walls of the pipe (Figure 13-22). Let P_1 be the pressure at point 1 and P_2 be that at point 2, a distance L downstream from point 1. The pressure drop $\Delta P = P_1 - P_2$ is proportional to the volume flow rate:

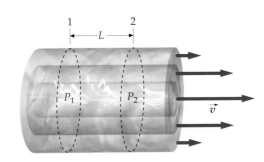

FIGURE 13-22 When a viscous fluid flows through a pipe, the speed is greatest at the center of the pipe. At the walls of the pipe, the fluid flow rate approaches zero.

$$\Delta P = P_1 - P_2 = I_V R \qquad \text{13-19}$$

where $I_V = vA$ is the volume flow rate and the proportionality constant R is the resistance to flow, which depends on the length of the pipe L, the radius r, and the viscosity of the fluid.

RESISTANCE TO BLOOD FLOW   **EXAMPLE 13-11**

Blood flows from the aorta through the major arteries, the small arteries, the capillaries, and the veins until it reaches the right atrium. In the course of that flow, the (gauge) pressure drops from about 100 torr to zero. If the volume flow rate is 0.8 L/s, find the total resistance of the circulatory system.

PICTURE THE PROBLEM The resistance is related to the pressure drop and volume flow rate by Equation 13-19. We can use Equation 13-9 to convert from torr to kPa.

Write the resistance in terms of the pressure drop and volume flow rate, and convert all terms to SI units:

$$R = \frac{\Delta P}{I_v}$$

$$= \frac{100 \text{ torr}}{0.8 \text{ L/s}} \times \frac{101 \text{ kPa}}{760 \text{ torr}} \times \frac{1 \text{ L}}{10^3 \text{ cm}^3} \times \frac{1 \text{ cm}^3}{10^{-6} \text{ m}^3}$$

$$= \boxed{16.6 \text{ kPa·s/m}^3}$$

REMARKS We could have used $1 \text{ Pa} = 1 \text{ N/m}^2$ to write the result as 16.6 kN·s/m^5.

To define the coefficient of viscosity of a fluid, we consider a fluid that is confined between two parallel plates, each of area A, separated by a distance z as shown in Figure 13-23. The upper plate is pulled at a constant speed v by a force $\vec{F}$ while the bottom plate is held at rest. A force is needed to pull the upper plate because the fluid next to the plate exerts a viscous drag force opposing its motion. The speed of the fluid between the plates approaches v near the upper plate and zero near the lower plate, and it varies linearly with separation between the plates. The force $\vec{F}$ is found to be directly

FIGURE 13-23 Two plates of equal area with a viscous fluid between them. When the upper plate is moved relative to the lower one, each layer of fluid exerts a drag force on adjacent layers. The force needed to pull the upper plate is directly proportional to v and the area A and inversely proportional to z, the separation between the plates.

proportional to v and A and inversely proportional to the plate separation z. The proportionality constant is the **coefficient of viscosity** η:

$$F = \eta \frac{vA}{z} \qquad\qquad 13\text{-}20$$

The SI unit of viscosity is the N·s/m² = Pa·s. An older cgs unit still in common use is the **poise,** named after the French physicist Jean Poiseuille. These units are related by

$$1\,\text{Pa·s} = 10\,\text{poise} \qquad\qquad 13\text{-}21$$

Table 13-3 lists the coefficients of viscosity for several fluids at various temperatures. Typically, the viscosity of a liquid increases as the temperature decreases. Thus, in cold climates, a less viscous grade of oil is used to lubricate automobile engines in the winter than in summer.

Poiseuille's Law   The resistance to flow R in Equation 13-19 for steady flow through a circular tube of radius r can be shown to be

$$R = \frac{8\eta L}{\pi r^4} \qquad\qquad 13\text{-}22$$

Equations 13-19 and 13-22 can be combined to give the pressure drop over a length L of a circular tube of radius r:

$$\Delta P = \frac{8\eta L}{\pi r^4} I_V \qquad\qquad 13\text{-}23$$

POISEUILLE'S LAW

Equation 13-23 is known as **Poiseuille's law.** Note the inverse r^4 dependence of the pressure drop. If the radius of the tube is halved, the pressure drop for a given volume flow rate is increased by a factor of 16; or a pressure 16 times as

TABLE 13-3

Coefficients of Viscosity for Various Fluids

Fluid	t, °C	η, mPa·s
Water	0	1.8
	20	1.00
	60	0.65
Blood (whole)	37	4.0
Engine oil (SAE 10W)	30	200
Glycerin	0	10,000
	20	1,410
	60	81
Air	20	0.018

great is needed to pump the fluid through the tube at the original volume flow rate. Thus, for example, if the diameter of a person's blood vessels or arteries is reduced for some reason, either the volume flow rate of the blood is greatly reduced, or the blood pressure must escalate to maintain the volume flow rate. For water flowing through a long garden hose, the pressure drop is pretty much fixed. It equals the difference in pressure between that at the water source and atmospheric pressure at the open end. The volume flow rate is then proportional to the fourth power of the radius. Thus, if the radius is halved, the volume flow rate drops by a factor of 16.

Poiseuille's law applies only to the laminar (nonturbulent) flow of a fluid of constant viscosity. In some fluids, viscosity changes with velocity, violating Poiseuille's law. Blood, for example, is a complex fluid consisting of solid particles of various shapes suspended in a liquid. Red blood cells are disk-shaped objects that are randomly oriented at low velocities but at high velocities tend to become oriented to facilitate the flow. Thus, the viscosity of blood decreases as the flow velocity increases, so Poiseuille's law cannot be strictly applied. Nevertheless, Poiseuille's law is a good approximation that is very useful for obtaining a qualitative understanding of blood flow.

In Chapter 25 the flow of electrical current I through metal wires is studied. One of the basic relations in that chapter is Ohm's law, $\Delta V = IR$, where ΔV is the potential difference and R is the electrical resistance of the wire. As we shall see, Ohm's law is analogous to Poiseuille's law $\Delta P = I_V R$.

Turbulence—Reynolds Number When the flow velocity of a fluid becomes sufficiently great, laminar flow breaks down and turbulence sets in. The critical velocity above which the flow through a tube is turbulent depends on the density and viscosity of the fluid and on the radius of the tube. The flow of a fluid can be characterized by a dimensionless number called the **Reynolds number** N_R, which is defined by

$$N_R = \frac{2r\rho v}{\eta} \qquad \text{13-24}$$

where v is the average velocity of the fluid. Experiments have shown that the flow will be laminar if the Reynolds number is less than about 2000 and turbulent if it is greater than 3000. Between these values, the flow is unstable and may change from one type to the other.

False-color view of turbulence of blood flowing into and out of the heart as seen by magnetic resonance imaging (MRI). Systolic ejection from the left ventricle into the aorta is seen in red, and diastolic filling of the ventricles in blue.

BLOOD FLOW IN THE AORTA **EXAMPLE 13-12**

Calculate the Reynolds number for blood flowing at 30 cm/s through an aorta of radius 1.0 cm. Assume that blood has a viscosity of 4 mPa·s and a density of 1060 kg/m³.

PICTURE THE PROBLEM Because N_R is dimensionless, we can use any set of units as long as we are consistent.

Write Equation 13-24 for the Reynolds number, expressing each quantity in SI units:

$$N_R = \frac{2r\rho v}{\eta}$$

$$= \frac{2(0.01\ \text{m})(1060\ \text{kg/m}^3)(0.3\ \text{m/s})}{4 \times 10^{-3}\ \text{Pa·s}} = \boxed{1590}$$

REMARKS Since the Reynolds number is less than 2000, this flow will be laminar rather than turbulent.

1. Density, specific gravity, and pressure are defined quantities that are important in fluid statics and dynamics.
2. Pascal's principle and Archimedes' principle are derived from Newton's laws.
3. Bernoulli's equation is derived from the conservation of mechanical energy.
*4. The Venturi effect is a special case of Bernoulli's equation.
*5. Poiseuille's law accounts for pressure drops due to viscosity, Reynolds number is used to predict whether flow is laminar or turbulent.

Topic	Relevant Equations and Remarks
1. Density	The density of a substance is the ratio of its mass to its volume:
	$$\rho = \frac{m}{V} \qquad \text{13-1}$$
	The densities of most solids and liquids are approximately independent of temperature and pressure, whereas those of gases depend strongly on these quantities.
2. Specific Gravity	The specific gravity of a substance is the ratio of its density to that of water. An object sinks or floats in a given fluid depending on whether its density is greater than or less than that of the fluid.
3. Pressure	$$P = \frac{F}{A} \qquad \text{13-3}$$
Units	$1\,\text{Pa} = 1\,\text{N/m}^2 \qquad \text{13-4}$
	$1\,\text{atm} = 760\,\text{mmHg} = 760\,\text{torr} = 29.9\,\text{in.Hg}$
	$\qquad = 101.325\,\text{kPa} = 14.7\,\text{lb/in.}^2 \qquad \text{13-9}$
	$1\,\text{bar} = 10^3\,\text{millibars} = 100\,\text{kPa} \qquad \text{13-10}$
Gauge pressure	Gauge pressure is the difference between the absolute pressure and atmospheric pressure:
	$$P = P_{\text{gauge}} + P_{\text{at}} \qquad \text{13-8}$$
In a liquid	$$P = P_0 + \rho g \Delta h \quad (\rho\ \text{constant}) \qquad \text{13-7}$$
In a gas	In a gas such as air, pressure decreases exponentially with altitude.
Bulk modulus	$$B = -\frac{\Delta P}{\Delta V / V} \qquad \text{13-6}$$
4. Pascal's Principle	Pressure changes applied to an enclosed liquid are transmitted undiminished to every point in the fluid and to the walls of the container.
5. Archimedes' Principle	A body wholly or partially submerged in a fluid is buoyed up by a force equal to the weight of the displaced fluid.
*6. Fluid Flow	
Volume flow rate	$I_v = Av$

Continuity equation for an incompressible fluid	$I_v = Av = $ constant	**13-16**
Bernoulli's equation	$P + \rho gh + \frac{1}{2}\rho v^2 = $ constant	**13-17b**
Venturi effect	When the speed of a fluid increases, the pressure drops.	
Resistance to fluid flow	$\Delta P_2 = I_V R$	**13-19**
Coefficient of viscosity	$F = \eta \dfrac{vA}{z}$	**13-20**
Poiseuille's law for viscous flow	$\Delta P = RI_V = \dfrac{8\eta L}{\pi r^4} I_V$	**13-23**
Laminar flow, turbulent flow, and the Reynolds number	The flow will be laminar if the Reynolds number N_R is less than about 2000 and turbulent if it is greater than 3000, where N_R is given by $$N_R = \frac{2r\rho v}{\eta}$$	**13-24**

PROBLEMS

- • Single-concept, single-step, relatively easy
- •• Intermediate-level, may require synthesis of concepts
- ••• Challenging
- **SSM** Solution is in the *Student Solutions Manual*
- Problems available on iSOLVE online homework service
- ✔ These "Checkpoint" online homework service problems ask students additional questions about their confidence level, and how they arrived at their answer

In a few problems, you are given more data than you actually need; in a few other problems, you are required to supply data from your general knowledge, outside sources, or informed estimates.

Conceptual Problems

1 • If the gauge pressure is doubled, the absolute pressure will be (*a*) halved, (*b*) doubled, (*c*) unchanged, (*d*) squared, (*e*) not enough information is given to determine the effect.

2 • **SSM** Does Archimedes' principle hold in a satellite orbiting the earth in a circular orbit? Explain.

3 •• A rock of mass M with a density three times that of water is suspended by a thread. Holding the free end of the thread in your hand, you lower the rock into an aquarium tank with water almost up to the rim. The tank is resting on the platform of a scale. When the rock is just above the bottom of the tank the thread breaks. From the moment the thread breaks to the time the rock is resting on the bottom of the tank the reading on the scale increases by (*a*) $2Mg$, (*b*) Mg, (*c*) $\frac{2}{3}Mg$, (*d*) $\frac{1}{3}Mg$, (*e*) zero.

4 •• A rock is thrown into a swimming pool filled with water of uniform temperature. Which of the following statements is true?

(*a*) The buoyant force on the rock is zero as it sinks.
(*b*) The buoyant force on the rock increases as it sinks.
(*c*) The buoyant force on the rock decreases as it sinks.
(*d*) The buoyant force on the rock is constant as it sinks.
(*e*) The buoyant force on the rock as it sinks is nonzero at first but becomes zero once the terminal velocity is reached.

5 •• A fishbowl rests on a scale. The fish suddenly swims upward to get food. What happens to the scale reading?

6 •• **SSM** Two objects are balanced as in Figure 13-24. The objects have identical volumes but different masses. Will the equilibrium be disturbed if the entire system is completely immersed in water? Explain.

FIGURE 13-24 Problem 6

7 •• A 200-g block of lead and a 200-g block of copper are completely under water. Each is suspended by a thread just above the bottom of an aquarium filled with water. Which of the following is true?

(a) The buoyant force is greater on the lead than on the copper.
(b) The buoyant force is greater on the copper than on the lead.
(c) The buoyant force is the same on both blocks.
(d) More information is needed to choose the correct answer.

8 •• A 20-cm³ block of lead and a 20-cm³ block of copper are completely under water. Each is suspended by a thread just above the bottom of an aquarium filled with water. Which of the following is true?

(a) The buoyant force is greater on the lead than on the copper.
(b) The buoyant force is greater on the copper than on the lead.
(c) The buoyant force is the same on both blocks.
(d) More information is needed to choose the correct answer.

9 • In a department store, a beach ball is supported by the airstream from a hose connected to the exhaust of a vacuum cleaner. Does the air blow under or over the ball to support it? Why?

10 • A horizontal pipe narrows from a diameter of 10 cm to 5 cm. For a nonviscous liquid flowing without turbulance from the larger diameter to the smaller, (a) the velocity and pressure both increase, (b) the velocity increases and the pressure decreases, (c) the velocity decreases and the pressure increases, (d) the velocity and pressure both decrease, (e) either the velocity or pressure changes but not both.

11 • **SSM** True or false: The buoyant force on a submerged object depends on the shape of the object.

12 • Figure 13-25 shows a Cartesian diver. The diver consists of a small tube, open at the bottom, with an air bubble at the top, inside a closed plastic soda bottle that is partly filled with water. Normally, the diver floats, but sinks when the bottle is squeezed hard. Explain why this happens.

FIGURE 13-25
Problem 12

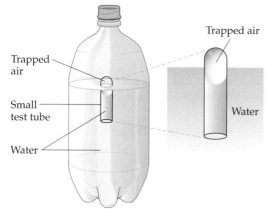

13 • A glass of water has ice cubes floating in it. What happens to the water level when the ice melts?

14 • Why do you float higher out of the water in salt water than in fresh water?

15 •• A certain object has a density just slightly less than that of water so that it floats almost completely submerged. However, the object is more compressible than water. What happens if the floating object is given a slight push to submerge it?

16 •• In Example 13-10, the fluid is accelerated to a greater speed as it enters the narrow part of the pipe. Identify the forces that act on the fluid to produce this acceleration.

17 •• A glass of water is accelerating to the right along a horizontal surface. What is the origin of the force that produces the acceleration on a small element of water in the middle of the glass? Explain with a picture.

18 •• **SSM** You are sitting in a boat floating on a very small pond. You take the anchor out of the boat and drop it into the water. What happens to the water level in the pond?

19 •• Figure 13-26 is a diagram of a prairie dog tunnel. The geometry of the two holes and their positioning ensures that the wind blowing above hole number 2 will always have a lower speed than that above hole number 1. Explain how Bernoulli's principle keeps the tunnel ventilated, and indicate in which direction air will flow through the tunnel.

FIGURE 13-26
Problem 19

20 • **SSM** Three bottles of different shapes are filled to the same level with water, as shown in Figure 13-27. The area of the bottom of each bottle is the same. The hydrostatic pressure is the same on the bottom of each of the bottles, but the total force must be different, as each bottle has a different amount of liquid in it. Explain this apparent paradox.

FIGURE 13-27
Problem 20

Density

21 • Find the mass of a copper cylinder that is 6 cm long with a radius of 2 cm.

22 • Find the mass of a lead sphere of radius 2 cm.

23 • **iSOLVE** Find the mass of air in a room 4 m × 5 m × 4 m.

24 •• [SSM] A 60-mL flask is filled with mercury at 0°C (Figure 13-28). When the temperature rises to 80°C, 1.47 g of mercury spill out of the flask. Assuming that the volume of the flask is constant, find the density of mercury at 80°C if its density at 0°C is 13,645 kg/m³.

0°C   80°C

FIGURE 13-28 Problem 24

Pressure

25 • [ISOLVE] Barometer readings are commonly given in inches of mercury. Find the pressure in inches of mercury equal to 101 kPa.

26 • The pressure on the surface of a lake is atmospheric pressure P_{at} = 101 kPa. (a) At what depth is the pressure twice atmospheric pressure? (b) If the pressure at the top of a deep pool of mercury is P_{at}, at what depth is the pressure $2P_{at}$?

27 • [SSM] [ISOLVE] A hydraulic lift is used to raise an automobile of mass 1500 kg. The radius of the shaft of the lift is 8 cm and that of the piston is 1 cm. How much force must be applied to the piston to raise the automobile?

28 •• When a woman in high-heeled shoes takes a step, she momentarily places her entire weight on one heel of her shoe. If her mass is 56 kg and if the area of the heel is 1 cm², what is the pressure exerted on the floor by her heel?

29 • [SSM] What pressure increase is required to compress the volume of 1 kg of water from 1.00 L to 0.99 L?

30 • [ISOLVE] A 1500-kg car rests on four tires, each of which is inflated to a gauge pressure of 200 kPa. What is the area of contact of each tire with the road, if the four tires support the weight equally?

31 •• In the seventeenth century, Blaise Pascal performed the experiment shown in Figure 13-29. A wine barrel filled with water was coupled to a long tube. Water was added to the tube until the barrel burst. (a) If the radius of the lid was 20 cm and the height of the water in the tube was 12 m, calculate the force exerted on the lid. (b) If the tube had an inner radius of 3 mm, what mass of water in the tube caused the pressure that burst the barrel?

FIGURE 13-29 Problem 31

32 •• [ISOLVE] Blood plasma flows from a bag through a tube into a patient's vein, where the blood pressure is 12 mmHg. The specific gravity of blood plasma at 37°C is 1.03. What is the minimum elevation the bag must have so the plasma flows into the vein?

33 •• Many people have imagined that if they were to float the top of a flexible snorkel tube out of the water, they would be able to breathe through it while walking underwater (Figure 13-30). However, they generally do not take into account just how much water pressure opposes the expansion of the chest and the inflation of the lungs. Suppose you can just breathe while lying on the floor with a 400-N (90-lb) weight on your chest. How far below the surface of the water could your chest be for you still to be able to breathe, assuming your chest has a frontal area of 0.09 m²?

FIGURE 13-30
Problem 33

34 •• In Example 13-3, a force of 147 N is applied to a small piston to lift a car that weighs 14,700 N. Demonstrate that this does not violate the law of conservation of mechanical energy by showing that, when the car is lifted some distance h, the work done by the 147-N force acting on the small piston equals the work done by the large piston on the car.

35 • [ISOLVE] A hollow cube with edge length a is half-filled with water of density ρ. Find the force exerted on a side of the cube by the water. (The edges of the cube are either horizontal or vertical.)

36 ••• [SSM] The volume of a cone of height h and base radius r is $V = \pi r^2 h/3$. A conical vessel of height 25 cm resting on its base of radius 15 cm is filled with water. (a) Find the volume and weight of the water in the vessel. (b) Find the force exerted by the water on the base of the vessel. Explain how this force can be greater than the weight of the water.

Buoyancy

37 • [SSM] [ISOLVE] A 500-g piece of copper (specific gravity 9.0) is suspended from a spring scale and is submerged in water (Figure 13-31). What force does the spring scale read?

FIGURE 13-31 Problem 37

38 • [SOLVE] When a 60-N stone is attached to a spring scale and is submerged in water, the spring scale reads 40 N. What is the density of the stone?

39 • [SOLVE] A block of an unknown material weighs 5 N in air and 4.55 N when submerged in water. (*a*) What is the density of the material? (*b*) Of what material is the block made?

40 • A solid piece of metal weighs 90 N in air and 56.6 N when submerged in water. Determine the density of this metal.

41 •• [SOLVE]✓ A homogeneous solid object floats on water with 80 percent of its volume below the surface. The same object when placed in a second liquid floats on that liquid with 72 percent of its volume below the surface. Determine the density of the object and the specific gravity of the liquid.

42 •• [SSM] A 5-kg iron block is suspended from a spring scale and is submerged in a fluid of unknown density. The spring scale reads 6.16 N. What is the density of the fluid?

43 •• A large piece of cork weighs 0.285 N in air. When held submerged underwater by a spring scale as shown in Figure 13-32, the spring scale reads 0.855 N. Find the density of the cork.

FIGURE 13-32 Problem 43

44 •• [SOLVE]✓ A helium balloon lifts a basket and cargo of total weight 2000 N under standard conditions, at which the density of air is 1.29 kg/m³ and the density of helium is 0.178 kg/m³. What is the minimum volume of the balloon?

45 •• [SSM] An object has neutral buoyancy when its density equals that of the liquid in which it is submerged, which means that it neither floats nor sinks. If the average density of an 85-kg diver is 0.96 kg/L, what mass of lead should be added to give him neutral buoyancy?

46 •• [SOLVE]✓ A beaker of mass 1 kg containing 2 kg of water rests on a scale. A 2-kg block of aluminum (density 2.70 × 10³ kg/m³) suspended from a spring scale is submerged in the water as in Figure 13-33. Find the readings of both scales.

FIGURE 13-33 Problem 46

Aluminum

47 ••• A ship sails from seawater (specific gravity 1.025) into freshwater and therefore sinks slightly. When its load of 600,000 kg is removed, it returns to its original level. Assuming that the sides of the ship are vertical at the water line, find the mass of the ship before it was unloaded.

48 ••• [SSM] The hydrometer shown in Figure 13-34 is a device for measuring the specific gravity of liquids. The bulb contains lead shot, and the specific gravity can be read directly from the liquid level on the stem after the hydrometer has been calibrated. The volume of the bulb is 20 mL, the stem is 15 cm long and has a diameter of 5.00 mm, and the mass of the glass is 6.0 g. (*a*) What mass of lead shot must be added so that the lowest specific gravity of liquid that can be measured is 0.9? (*b*) What is the maximum specific gravity of liquid that can be measured?

FIGURE 13-34 Problems 48, 98

49 • When cracks form at the base of a dam, the water seeping in the cracks exerts a buoyant force that tends to lift the dam. This can topple the dam. Estimate the buoyant force exerted on a 2-m thick by 5-m long dam wall by water seeping into cracks at its base. The water level is 5 m above the cracks.

50 •• A large helium weather balloon is spherical in shape, with a radius of 2.5 m and a total mass of 15 kg (balloon plus helium plus equipment). (*a*) What is the initial upward acceleration of the balloon when it is released from sea level? (*b*) If the drag force on the balloon is given by $F_D = \frac{1}{2}\pi r^2 \rho v^2$, where r is the balloon radius, ρ is the density of air, and v the balloon's ascension speed, calculate the terminal velocity of the ascending balloon. (*c*) Roughly how long will it take for the balloon to ascend to a height of 10 km?

Continuity and Bernoulli's Equation

51 •• [SSM] Water exits a circular tap moving straight down with a flow rate of 10.5 cm³/s. (*a*) If the diameter of the tap is 1.2 cm, what is the speed of the water? (*b*) As the fluid falls from the tap, the stream of water narrows. Find the new diameter of the stream at a point 7.5 cm below the tap. Assume that the stream still has a circular cross section and neglect any drag forces acting on the water. (*c*) If turbulent flows are characterized by Reynolds numbers above 2300 or so, how far does the water have to fall before it becomes turbulent? Does this match everyday experience?

52 • [iSOLVE✔] Water flows at 0.65 m/s through a 3-cm-diameter hose. At the end of the hose is a 0.30-cm-diameter nozzle. (*a*) At what speed does the water pass through the nozzle? (*b*) If the pump at one end of the hose and the nozzle at the other end are at the same height, and if the pressure at the nozzle is 1 atm, what is the pressure at the pump? Assume laminar nonviscous flow.

53 • [iSOLVE✔] Water is flowing at 3 m/s in a horizontal pipe under a pressure of 200 kPa. The pipe narrows to half its original diameter. (*a*) What is the speed of flow in the narrow section? (*b*) What is the pressure in the narrow section? (*c*) How do the volume flow rates in the two sections compare? Assume laminar nonviscous flow.

54 • [iSOLVE] The pressure in a section of horizontal pipe with a diameter of 2 cm is 142 kPa. Water flows through the pipe at 2.80 L/s. If the pressure at a certain point is to be reduced to 101 kPa by constricting a section of the pipe, what should the diameter of the constricted section be? Assume laminar nonviscous flow.

55 •• [SSM] Blood flows in an aorta of radius 9 mm at 30 cm/s. (*a*) Calculate the volume flow rate in liters per minute. (*b*) Although the cross-sectional area of a capillary is much smaller than that of the aorta, there are many capillaries, so their total cross-sectional area is much larger. If all the blood from the aorta flows into the capillaries and the speed of flow through the capillaries is 1.0 mm/s, calculate the total cross-sectional area of the capillaries.

56 •• [iSOLVE✔] A large tank of water is tapped a distance h below the water surface by a small pipe as in Figure 13-35. Find the distance x reached by the water flowing out of the pipe. Assume laminar nonviscous flow.

FIGURE 13-35 Problems 56, 61

57 •• [iSOLVE] The \$8-billion, 800-mile long Alaskan Pipeline has a capacity of 240,000 m³ of oil per day. Along most of the pipeline the radius is 60 cm. Find the pressure P' at a point where the pipe has a 30-cm radius. Take the pressure in the 60-cm-diameter sections to be $P = 180$ kPa and the density of oil to be 800 kg/m³. Assume laminar nonviscous flow.

58 •• [SSM] Water flows through a Venturi meter like that in Example 13-10 with a pipe diameter of 9.5 cm and a constriction diameter of 5.6 cm. The U-tube manometer is partially filled with mercury. Find the volume flow rate of the water if the difference in the mercury level in the U-tube is 2.40 cm.

59 •• A firefighter holds a hose with a bend in it as in Figure 13-36. Water is expelled from the hose in a stream of radius 1.5 cm at a speed of 30 m/s. (*a*) What mass of water emerges from the hose in 1 s? (*b*) What is the horizontal momentum of this water? (*c*) Before reaching the bend, the water has momentum upward, whereas afterward, its momentum is horizontal. Draw a vector diagram of the initial and final momentum vectors, and find the change in the momentum of the water at the bend during 1 s. From this, find the force exerted on the water by the hose.

FIGURE 13-36 Problem 59

60 •• [iSOLVE✔] A fountain designed to spray a column of water 12 m into the air has a 1-cm-diameter nozzle at ground level. The water pump is 3 m below the ground. The pipe to the nozzle has a diameter of 2 cm. Find the necessary pump pressure. Assume laminar nonviscous flow.

61 ••• In Figure 13-35, H is the depth of the liquid and h is the distance of the opening below the surface of the liquid. (*a*) Find the distance x at which the water strikes the ground as a function of h and H. (*b*) Show that, for a given value of H, there are two values of h (whose average value is $\frac{1}{2}H$) that give the same distance x. (*c*) Show that x is a maximum when $h = \frac{1}{2}H$. What is the value of this maximum distance x? Assume laminar nonviscous flow.

62 •• [SSM] Figure 13-37 shows a Pitot-static tube, a device used for measuring the velocity of a gas. The inner pipe faces the incoming fluid, while the ring of holes in the outer tube are parallel to the gas flow. Show that the speed of the gas is given by $v^2 = 2gh(\rho - \rho_g)/\rho_g$, where ρ is the density of the liquid used in the manometer and ρ_g is the density of the gas.

FIGURE 13-37 Problem 62

63 •• A siphon is a device for transferring a liquid from container to container. The tube shown in Figure 13-38 must be filled to start the siphon, but once this has been done, fluid will flow through the tube until the liquid surfaces in the containers are at the same level. (a) Using Bernoulli's equation, show that the velocity of water in the tube is $v = \sqrt{2gd}$ (b) What is the pressure at the highest part of the tube? Assume laminar nonviscous flow.

FIGURE 13-38 Problem 63

Flow

*Viscous Flow

64 • [SOLVE] A horizontal tube with an inside diameter of 1.2 mm and a length of 25 cm has water flowing through it at 0.30 mL/s. Find the pressure difference required to drive this flow if the viscosity of water is 1.00 mPa·s. Assume laminar flow.

65 • Find the diameter of a tube that would give double the flow rate for the pressure difference in Problem 64.

66 • [SSM] Blood takes about 1.0 s to pass through a 1-mm-long capillary of the human circulatory system. If the diameter of the capillary is 7 μm and the pressure drop is 2.60 kPa, find the viscosity of blood. Assume laminar flow.

67 • [SSM] At Reynolds numbers of about 3×10^5 there is an abrupt transition, where the drag on a sphere abruptly decreases. Estimate the velocity at which this drag crisis occurs for a baseball, and comment on whether or not it should play a role in the physics of the game.

68 ••• Stokes' law states that the drag force on a sphere at very low Reynolds number is given by $F_D = 6\pi\eta av$, where η is the viscosity of the surrounding fluid and a is the radius of the sphere. Using this, find the terminal velocity of ascent for a spherical carbon dioxide bubble of 1-mm diameter rising in a glass of soda (density 1.1×10^3 kg/m^3). How long should it take for the bubble to rise the height of a "typical" soda glass? Is this consistent with your experience?

General Problems

69 •• [SSM] Very roughly speaking, the mass of a person should increase as the cube of his or her height—that is, $M = C\rho h^3$, where M is the mass, h is the height, ρ is body density, and C is a person's "coefficient of roundness." Estimate C for an adult male and female, using "typical" values for height and weight. Assume $\rho = 1000$ kg/m^3.

70 • Using that weight is proportional to height cubed (see problem 69) what should be the difference in weight for two men, one of whom is 5' 9" tall and the other 6' 0"?

71 • The top of a card table is 80 cm × 80 cm. What is the force exerted on it by the atmosphere? Why doesn't the table collapse?

72 • [SOLVE]✓ A 4.0-g Ping-Pong ball is attached by a thread to the bottom of a beaker. When the beaker is filled with water so that the ball is totally submerged, the tension in the thread is 2.8×10^{-2} N. Determine the diameter of the ball.

73 • [SOLVE] Seawater has a bulk modulus of 2.3×10^9 N/m^2. Find the density of seawater at a depth where the pressure is 800 atm if the density at the surface is 1025 kg/m^3.

74 • A solid cubical block of edge length 0.6 m is suspended from a spring balance. When the block is submerged in water, the spring balance reads 80 percent of the reading when the block is in air. Determine the density of the block.

75 • [SSM] When submerged in water, a block of copper has an apparent weight of 56 N. What fraction of this copper block will be submerged if it is floated on a pool of mercury?

76 • [SOLVE] A 4.5-kg block of material floats on ethanol with 10 percent of its volume above the liquid surface. What fraction of the volume of this block will be submerged if it is floated on water?

77 • What is the buoyant force on your body when you are floating (a) in a freshwater lake (specific gravity = 1.00) and (b) in the ocean (specific gravity = 1.03)?

78 • Suppose that when you are floating in fresh water, 96 percent of your body volume is submerged. What is the volume of water your body displaces when it is fully submerged?

79 •• A block of wood of 1.5-kg mass floats on water with 68 percent of its volume submerged. A lead block is placed on the wood and the wood is then fully submerged with the lead entirely out of the water. Find the mass of the lead block.

80 •• [SSM] [SOLVE]✓ A Styrofoam cube, 25 cm on an edge, is placed on one pan of a balance. The balance is in equilibrium when a 20-g mass of brass is placed on the opposite pan of the balance. Find the mass of the Styrofoam cube.

81 •• [SOLVE] A spherical shell of copper with an outer diameter of 12 cm floats on water with half its volume above the water's surface. Determine the inner diameter of the shell.

82 •• A beaker filled with water is balanced on the left pan of a balance. A cube 4 cm on an edge is attached to a string and lowered into the water so that it is completely submerged. The cube is not touching the bottom of the beaker. A weight of mass m is added to the right pan to restore equilibrium. What is m?

83 •• [SSM] Crude oil has a viscosity of about 0.8 Pa·s at normal temperature. A 50-km pipeline is to be constructed from an oil field to a tanker terminal. The pipeline is to deliver oil at the terminal at a rate of 500 L/s and the flow through the pipeline is to be laminar to minimize the pressure needed to push the fluid through the pipeline. Assuming that the density of crude oil is 700 kg/m^3, estimate the diameter of the pipeline that should be used.

84 •• [SOLVE] Water flows through the pipe in Figure 13-39 and exits to the atmosphere at C. The diameter of the pipe is 2.0 cm at A, 1.0 cm at B, and 0.8 cm at C. The gauge pressure in the pipe at A is 1.22 atm and the flow rate is 0.8 L/s. The vertical pipes are open to the air. Find the level of the liquid–air interfaces in the two vertical pipes. Assume laminar nonviscous flow.

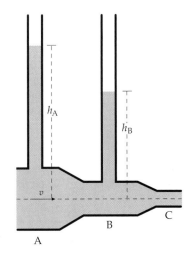

FIGURE 13-39 Problems 84, 85

85 •• Repeat Problem 84 with the flow rate reduced to 0.6 L/s and the size of the opening at C reduced so that the pressure in the pipe at A remains unchanged.

86 •• [SSM] [SOLVE] Figure 13-40 is a sketch of an *aspirator*, a simple device that can be used to achieve a partial vacuum in a reservoir connected to the vertical tube at B. An aspirator attached to the end of a garden hose may be used to deliver soap or fertilizer from the reservoir. Suppose that the diameter at A is 2.0 cm and at C, where the water exits to the atmosphere, it is 1.0 cm. If the flow rate is 0.5 L/s and

the gauge pressure at A is 0.187 atm, what diameter of the constriction at B will achieve a pressure of 0.1 atm in the container? Assume laminar nonviscous flow.

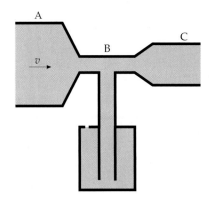

FIGURE 13-40 Problem 86

87 •• A cylindrical buoy at the entrance of a harbor has a diameter of 0.9 m and a height of 2.6 m. The mass of the buoy is 600 kg. It is attached to the bottom of the sea with a nylon cable of negligible mass. The specific gravity of the seawater is 1.025. (*a*) How much of the buoy is visible when the cable is slack? (*b*) If a tsunami completely submerges the buoy, what is the tension in the taut cable? (*c*) If the cable breaks, what is the initial upward acceleration of the buoy?

88 •• [SOLVE] Two connected vessels contain a liquid of density ρ_0 (Figure 13-41). The cross-sectional areas of the vessels are A and $3A$. Find the change in elevation of the liquid level if an object of mass m and density $\rho' = 0.8\rho_0$ is placed into one of the vessels.

FIGURE 13-41 Problem 88

89 •• If an oil-filled manometer ($\rho = 900$ kg/m^3) can be read to ± 0.05 mm, what is the smallest pressure change that can be detected?

90 •• [SOLVE]✓ A U-tube is filled with water until the liquid level is 28 cm above the bottom of the tube. An oil of specific gravity 0.78 is now poured into one arm of the U-tube until the level of the water in the other arm of the tube is 34 cm above the bottom of the tube. Find the levels of the oil–water and oil–air interfaces in the other arm of the tube.

91 •• A U-tube contains liquid of unknown specific gravity. An oil of density 800 kg/m^3 is poured into one arm of the tube until the oil column is 12 cm high. The oil–air interface is then 5.0 cm above the liquid level in the other arm of the U-tube. Find the specific gravity of the liquid.

92 •• A lead block is suspended from the underside of a 0.5-kg block of wood of specific gravity 0.7. If the upper surface of the wood is just level with the surface of the water, what is the mass of the lead block?

93 •• SSM A helium balloon can just lift a load of 750 N. The skin of the balloon has a mass of 1.5 kg. (a) What is the volume of the balloon? (b) If the volume of the balloon were twice that found in Part (a), what would be the initial acceleration of the balloon when it carried a load of 900 N?

94 •• A hollow sphere with an inner radius R and outer radius $2R$ is made of material of density ρ_0 and is floating in a liquid of density $2\rho_0$. The interior is now filled with material of density ρ' so that the sphere just floats completely submerged. Find ρ'.

95 •• SSM As mentioned in the discussion of *the law of atmospheres*, the fractional decrease in atmospheric pressure is proportional to the change in altitude. Expressed as a differential equation we have $dP/P = -C/dh$, where C is a constant. (a) Show that $P(h) = P_0 e^{-Ch}$ is a solution of the differential equation. (b) Show that if $\Delta h \ll h_0$ where $h_0 = 1/C$, then $P(h + \Delta h) \approx P(h)(1 - \Delta h/h_0)$. (c) Given that the pressure at $h = 5.5$ km is half that at sea level, find the constant C.

96 •• A submarine has a total mass of 2.4×10^6 kg, including crew and equipment. The vessel consists of two parts, the pressure hull, which has a volume of 2×10^3 m³, and the ballast tanks, which have a volume of 4×10^2 m³. When the sub cruises on the surface, the ballast tanks are filled with air; when it is cruising below the surface, seawater is admitted into the tanks. (a) What fraction of the submarine's volume is above the water surface when the tanks are filled with air? (b) How much water must be admitted into the tanks to give the submarine neutral buoyancy? Neglect the mass of air in the tanks and use 1.025 as the specific gravity of seawater.

97 •• A marine salvage crew raises a crate that measures 1.4 m × 0.75 m × 0.5 m. The average density of the empty crate is the same as seawater, 1.025×10^3 kg/m³, and its mass

when empty is 32 kg. The crate contains gold bullion that fills 36 percent of its volume; the remaining volume is filled with seawater. (a) What is the tension in the cable that raises the crate and bullion while the crate is below the surface of the sea? (b) What is the tension in the cable while the crate is lifted to the deck of the ship if (1) none of the seawater leaks out of the crate, and (2) the crate is lifted so slowly that all of the seawater leaks out of the crate?

98 ••• When the hydrometer of Problem 48 (Figure 13-34) is placed in a liquid whose specific gravity is greater than some minimum value, the device floats with part of the glass tube above the liquid level. Consider a hydrometer that has a spherical bulb 2.4 cm in diameter. The glass tube attached to the bulb is 20 cm long and has a diameter of 7.5 mm. The mass of the hydrometer before lead pellets are dropped into the bulb and the tube is sealed is 7.28 g. (a) What mass of lead should be placed in the bulb so that the hydrometer just floats in a liquid of specific gravity 0.78? (b) If the hydrometer is now placed in water, what is the length of the tube that shows above the surface of the water? (c) The hydrometer is placed in a liquid of unknown specific gravity; the length of the tube above the surface of the liquid is 12.2 cm. Determine the specific gravity of the liquid.

99 ••• A large root beer keg of height H and cross-sectional area A_1 is filled with root beer. The top is open to atmospheric pressure. At the bottom is a spigot opening of area A_2, which is much smaller than A_1. (a) Show that when the height of the root beer is h, the speed of the root beer leaving the spigot is approximately $\sqrt{2gh}$. (b) Show that for the approximation $A_2 \ll A_1$, the rate of change of the height h of the root beer is given by

$$\frac{dh}{dt} = -\frac{A_2}{A_1}(2gh)^{1/2}$$

(c) Find h as a function of time if $h = H$ at $t = 0$. (d) Find the total time needed to drain the keg if $H = 2$ m, $A_1 = 0.8$ m², and $A_2 = 10^{-4}A_1$. Assume laminar nonviscous flow.

APPENDIX A

SI Units and Conversion Factors

Basic Units

Length	The *meter* (m) is the distance traveled by light in a vacuum in 1/299,792,458 s.
Time	The *second* (s) is the duration of 9,192,631,770 periods of the radiation corresponding to the transition between the two hyperfine levels of the ground state of the ^{133}Cs atom.
Mass	The *kilogram* (kg) is the mass of the international standard body preserved at Sèvres, France.
Current	The *ampere* (A) is that current in two very long parallel wires 1 m apart that gives rise to a magnetic force per unit length of 2×10^{-7} N/m.
Temperature	The *kelvin* (K) is 1/273.16 of the thermodynamic temperature of the triple point of water.
Luminous intensity	The *candela* (cd) is the luminous intensity, in the perpendicular direction, of a surface of area 1/600,000 m^2 of a blackbody at the temperature of freezing platinum at a pressure of 1 atm.

Derived Units

Force	newton (N)	$1\,\text{N} = 1\,\text{kg·m/s}^2$
Work, energy	joule (J)	$1\,\text{J} = 1\,\text{N·m}$
Power	watt (W)	$1\,\text{W} = 1\,\text{J/s}$
Frequency	hertz (Hz)	$1\,\text{Hz} = \text{cy/s}$
Charge	coulomb (C)	$1\,\text{C} = 1\,\text{A·s}$
Potential	volt (V)	$1\,\text{V} = 1\,\text{J/C}$
Resistance	ohm (Ω)	$1\,\Omega = 1\,\text{V/A}$
Capacitance	farad (F)	$1\,\text{F} = 1\,\text{C/V}$
Magnetic field	tesla (T)	$1\,\text{T} = 1\,\text{N/(A·m)}$
Magnetic flux	weber (Wb)	$1\,\text{Wb} = 1\,\text{T·m}^2$
Inductance	henry (H)	$1\,\text{H} = 1\,\text{J/A}^2$

Conversion Factors

Conversion factors are written as equations for simplicity;
relations marked with an asterisk are exact.

Length

1 km = 0.6215 mi

1 mi = 1.609 km

1 m = 1.0936 yd = 3.281 ft = 39.37 in.

*1 in. = 2.54 cm

*1 ft = 12 in. = 30.48 cm

*1 yd = 3 ft = 91.44 cm

1 lightyear = 1 $c \cdot y$ = 9.461 × 10^{15} m

*1 Å = 0.1 nm

Area

*1 m^2 = 10^4 cm^2

1 km^2 = 0.3861 mi^2 = 247.1 acres

*1 $in.^2$ = 6.4516 cm^2

1 ft^2 = 9.29 × 10^{-2} m^2

1 m^2 = 10.76 ft^2

*1 acre = 43,560 ft^2

1 mi^2 = 640 acres = 2.590 km^2

Volume

*1 m^3 = 10^6 cm^3

*1 L = 1000 cm^3 = 10^{-3} m^3

1 gal = 3.786 L

1 gal = 4 qt = 8 pt = 128 oz = 231 in^3

1 in^3 = 16.39 cm^3

1 ft^3 = 1728 $in.^3$ = 28.32 L
      = 2.832 × 10^4 cm^3

Time

*1 h = 60 min = 3.6 ks

*1 d = 24 h = 1440 min = 86.4 ks

1 y = 365.24 d = 3.156 × 10^7 s

Speed

*1 m/s = 3.6 km/h

1 km/h = 0.2778 m/s = 0.6215 mi/h

1 mi/h = 0.4470 m/s = 1.609 km/h

1 mi/h = 1.467 ft/s

Angle and Angular Speed

*π rad = 180°

1 rad = 57.30°

1° = 1.745 × 10^{-2} rad

1 rev/min = 0.1047 rad/s

1 rad/s = 9.549 rev/min

Mass

*1 kg = 1000 g

*1 tonne = 1000 kg = 1 Mg

1 u = 1.6606 × 10^{-27} kg

1 kg = 6.022 × 10^{26} u

1 slug = 14.59 kg

1 kg = 6.852 × 10^{-2} slug

1 u = 931.50 MeV/c^2

Density

*1 g/cm^3 = 1000 kg/m^3 = 1 kg/L

(1 g/cm^3)g = 62.4 lb/ft^3

Force

1 N = 0.2248 lb = 10^5 dyn

*1 lb = 4.448222 N

(1 kg)g = 2.2046 lb

Pressure

*1 Pa = 1 N/m^2

*1 atm = 101.325 kPa = 1.01325 bars

1 atm = 14.7 $lb/in.^2$ = 760 mmHg
       = 29.9 in.Hg = 33.8 ftH_2O

1 $lb/in.^2$ = 6.895 kPa

1 torr = 1 mmHg = 133.32 Pa

1 bar = 100 kPa

Energy

*1 kW·h = 3.6 MJ

*1 cal = 4.1840 J

1 ft·lb = 1.356 J = 1.286 × 10^{-3} Btu

*1 L·atm = 101.325 J

1 L·atm = 24.217 cal

1 Btu = 778 ft·lb = 252 cal = 1054.35 J

1 eV = 1.602 × 10^{-19} J

1 u·c^2 = 931.50 MeV

*1 erg = 10^{-7} J

Power

1 horsepower = 550 ft·lb/s = 745.7 W

1 Btu/h = 1.055 kW

1 W = 1.341 × 10^{-3} horsepower
    = 0.7376 ft·lb/s

Magnetic Field

*1 T = 10^4 G

Thermal Conductivity

1 W/(m·K) = 6.938 Btu·in./(h·ft^2·F°)

1 Btu·in./(h·ft^2·F°) = 0.1441 W/(m·K)

APPENDIX B

Numerical Data

Terrestrial Data

Free-fall acceleration g	9.80665 m/s^2; 32.1740 ft/s^2
(Standard value at sea level at 45° latitude)[†]	
Standard value	
At sea level, at equator[†]	9.7804 m/s^2
At sea level, at poles[†]	9.8322 m/s^2
Mass of earth M_E	5.98×10^{24} kg
Radius of earth R_E, mean	6.37×10^6 m; 3960 mi
Escape speed $\sqrt{2R_E g}$	1.12×10^4 m/s; 6.95 mi/s
Solar constant[‡]	1.35 kW/m^2
Standard temperature and pressure (STP):	
Temperature	273.15 K
Pressure	101.325 kPa (1.00 atm)
Molar mass of air	28.97 g/mol
Density of air (STP), ρ_{air}	1.293 kg/m^3
Speed of sound (STP)	331 m/s
Heat of fusion of H_2O (0°C, 1 atm)	333.5 kJ/kg
Heat of vaporization of H_2O (100°C, 1 atm)	2.257 MJ/kg.

† Measured relative to the earth's surface.
‡ Average power incident normally on 1 m^2 outside the earth's atmosphere at the mean distance from the earth to the sun.

Astronomical Data[†]

Earth	
Distance to moon[‡]	3.844×10^8 m; 2.389×10^5 mi
Distance to sun, mean[‡]	1.496×10^{11} m; 9.30×10^7 mi; 1.00 AU
Orbital speed, mean	2.98×10^4 m/s
Moon	
Mass	7.35×10^{22} kg
Radius	1.738×10^6 m
Period	27.32 d
Acceleration of gravity at surface	1.62 m/s^2
Sun	
Mass	1.99×10^{30} kg
Radius	6.96×10^8 m

† Additional solar-system data is available from NASA at <http://nssdc.gsfc.nasa.gov/planetary/planetfact.html>.
‡ Center to center.

Physical Constants[†]

Gravitational constant	G	$6.673(10) \times 10^{-11}$ N·m^2/kg^2
Speed of light	c	$2.997\ 924\ 58 \times 10^8$ m/s
Fundamental charge	e	$1.602\ 1764\ 62(63) \times 10^{-19}$ C
Avogadro's number	N_A	$6.022\ 141\ 99(47) \times 10^{23}$ particles/mol
Gas constant	R	$8.314\ 472(15)$ J/(mol·K)
		$1.987\ 2065(36)$ cal/(mol·K)
		$8.205\ 746(15) \times 10^{-2}$ L·atm/(mol·K)
Boltzmann constant	$k = R/N_A$	$1.380\ 6503(24) \times 10^{-23}$ J/K
		$8.617\ 342(15) \times 10^{-5}$ eV/K
Stefan-Boltzmann constant	$\sigma = (\pi^2/60)k^4/(\hbar^3 c^2)$	$5.670\ 400(40) \times 10^{-8}$ W/(m^{2}k^4)
Atomic mass constant	$m_u = \frac{1}{12}m(^{12}C)$	$1.660\ 538\ 73(13) \times 10^{-27}$ kg = 1u
Coulomb constant	$k = 1/(4\pi\epsilon_0)$	$8.987\ 551\ 788\ ... \times 10^9$ N·m^2/C^2
Permittivity of free space	ϵ_0	$8.854\ 187\ 817\ ... \times 10^{-12}$ C^2/(N·m^2)
Permeability of free space	μ_0	$4\ \pi \times 10^{-7}$ N/A^2
		$1.256\ 637 \times 10^{-6}$ N/A^2
Planck's constant	h	$6.626\ 068\ 76(52) \times 10^{-34}$ J·s
		$4.135\ 667\ 27(16) \times 10^{-15}$ eV·s
	$\hbar = h/2\pi$	$1.054\ 571\ 596(82) \times 10^{-34}$ J·s
		$6.582\ 118\ 89(26) \times 10^{-16}$ eV·s
Mass of electron	m_e	$9.109\ 381\ 88(72) \times 10^{-31}$ kg
		$0.510\ 998\ 902(21)$ MeV/c^2
Mass of proton	m_p	$1.672\ 621\ 58(13) \times 10^{-27}$ kg
		$938.271\ 998(38) \times$ MeV/c^2
Mass of neutron	m_n	$1.674\ 927\ 16(13) \times 10^{-27}$ kg
		$939.565\ 330(38)$ MeV/c^2
Bohr magneton	$m_B = eh/2m_e$	$9.274\ 0008\ 99(37) \times 10^{-24}$ J/T
		$5.788\ 381\ 749(43) \times 10^{-5}$ eV/T
Nuclear magneton	$m_n = eh/2m_p$	$5.050\ 783\ 17(20) \times 10^{-27}$ J/T
		$3.152\ 451\ 238(24) \times 10^{-8}$ eV/T
Magnetic flux quantum	$\phi_0 = h/2e$	$2.067\ 833\ 636(81) \times 10^{-15}$ T·m^2
Quantized Hall resistance	$R_K = h/e^2$	$2.581\ 280\ 7572(95) \times 10^4\ \Omega$
Rydberg constant	R_H	$1.097\ 373\ 156\ 8549(83) \times 10^7$ m^{-1}
Josephson frequency-voltage quotient	$K_J = 2e/h$	$4.835\ 978\ 98(19) \times 10^{14}$ Hz/V
Compton wavelength	$\lambda_C = h/m_e c$	$2.426\ 310\ 215(18) \times 10^{-12}$ m

† The values for these and other constants may be found on the Internet at http://physics.nist.gov/cuu/Constants/index.html. The numbers in parentheses represent the uncertainties in the last two digits. (For example, 2.044 43(13) stands for 2.044 43 ± 0.000 13.) Values with without uncertainties are exact, including those values with ellipses (like the value of pi is exactly 3.1415...).

For additional data, see the following tables in the text.

APPENDIX C

Periodic Table of Elements

1																	18
1 **H** 1.00797	**2**												**13**	**14**	**15**	**16**	**17**	**2** **He** 4.003
3 **Li** 6.941	**4** **Be** 9.012												**5** **B** 10.81	**6** **C** 12.011	**7** **N** 14.007	**8** **O** 15.9994	**9** **F** 19.00	**10** **Ne** 20.179
11 **Na** 22.990	**12** **Mg** 24.31	**3**	**4**	**5**	**6**	**7**	**8**	**9**	**10**	**11**	**12**		**13** **Al** 26.98	**14** **Si** 28.09	**15** **P** 30.974	**16** **S** 32.064	**17** **Cl** 35.453	**18** **Ar** 39.948
19 **K** 39.102	**20** **Ca** 40.08	**21** **Sc** 44.96	**22** **Ti** 47.88	**23** **V** 50.94	**24** **Cr** 52.00	**25** **Mn** 54.94	**26** **Fe** 55.85	**27** **Co** 58.93	**28** **Ni** 58.69	**29** **Cu** 63.55	**30** **Zn** 65.38		**31** **Ga** 69.72	**32** **Ge** 72.59	**33** **As** 74.92	**34** **Se** 78.96	**35** **Br** 79.90	**36** **Kr** 83.80
37 **Rb** 85.47	**38** **Sr** 87.62	**39** **Y** 88.906	**40** **Zr** 91.22	**41** **Nb** 92.91	**42** **Mo** 95.94	**43** **Tc** (98)	**44** **Ru** 101.1	**45** **Rh** 102.905	**46** **Pd** 106.4	**47** **Ag** 107.870	**48** **Cd** 112.41		**49** **In** 114.82	**50** **Sn** 118.69	**51** **Sb** 121.75	**52** **Te** 127.60	**53** **I** 126.90	**54** **Xe** 131.29
55 **Cs** 132.905	**56** **Ba** 137.33	**57–71** **Rare Earths**	**72** **Hf** 178.49	**73** **Ta** 180.95	**74** **W** 183.85	**75** **Re** 186.2	**76** **Os** 190.2	**77** **Ir** 192.2	**78** **Pt** 195.09	**79** **Au** 196.97	**80** **Hg** 200.59		**81** **Tl** 204.37	**82** **Pb** 207.19	**83** **Bi** 208.98	**84** **Po** (210)	**85** **At** (210)	**86** **Rn** (222)
87 **Fr** (223)	**88** **Ra** (226)	**89–103** **Actinides**	**104** **Rf** (261)	**105** **Ha** (260)	**106** (263)	**107** (262)	**108** (265)	**109** (266)										

	57	58	59	60	61	62	63	64	65	66	67	68	69	70	71
Rare Earths (Lanthanides)	**La** 138.91	**Ce** 140.12	**Pr** 140.91	**Nd** 144.24	**Pm** (147)	**Sm** 150.36	**Eu** 152.0	**Gd** 157.25	**Tb** 158.92	**Dy** 162.50	**Ho** 164.93	**Er** 167.26	**Tm** 168.93	**Yb** 173.04	**Lu** 174.97
	89	90	91	92	93	94	95	96	97	98	99	100	101	102	103
Actinides	**Ac** 227.03	**Th** 232.04	**Pa** 231.04	**U** 238.03	**Np** 237.05	**Pu** (244)	**Am** (243)	**Cm** (247)	**Bk** (247)	**Cf** (251)	**Es** (252)	**Fm** (257)	**Md** (258)	**No** (259)	**Lr** (260)

The 1–18 group designation has been recommended by the International Union of Pure and Applied Chemistry (IUPAC).

Atomic Numbers and Atomic Masses[†]

Name	Symbol	Atomic Number	Mass	Name	Symbol	Atomic Number	Mass
Actinium	Ac	89	227.03	Mercury	Hg	80	200.59
Aluminum	Al	13	26.98	Molybdenum	Mo	42	95.94
Americium	Am	95	(243)	Neodymium	Nd	60	144.24
Antimony	Sb	51	121.75	Neon	Ne	10	20.179
Argon	Ar	18	39.948	Neptunium	Np	93	237.05
Arsenic	As	33	74.92	Nickel	Ni	28	58.69
Astatine	At	85	(210)	Niobium	Nb	41	92.91
Barium	Ba	56	137.3	Nitrogen	N	7	14.007
Berkelium	Bk	97	(247)	Nobelium	No	102	(259)
Beryllium	Be	4	9.012	Osmium	Os	76	190.2
Bismuth	Bi	83	208.98	Oxygen	O	8	15.9994
Boron	B	5	10.81	Palladium	Pd	46	106.4
Bromine	Br	35	79.90	Phosphorus	P	15	30.974
Cadmium	Cd	48	112.41	Platinum	Pt	78	195.09
Calcium	Ca	20	40.08	Plutonium	Pu	94	(244)
Californium	Cf	98	(251)	Polonium	Po	84	(210)
Carbon	C	6	12.011	Potassium	K	19	39.098
Cerium	Ce	58	140.12	Praseodymium	Pr	59	140.91
Cesium	Cs	55	132.905	Promethium	Pm	61	(147)
Chlorine	Cl	17	35.453	Protactinium	Pa	91	231.04
Chromium	Cr	24	52.00	Radium	Ra	88	(226)
Cobalt	Co	27	58.93	Radon	Rn	86	(222)
Copper	Cu	29	63.55	Rhenium	Re	75	186.2
Curium	Cm	96	(247)	Rhodium	Rh	45	102.905
Dysprosium	Dy	66	162.50	Rubidium	Rb	37	85.47
Einsteinium	Es	99	(252)	Ruthenium	Ru	44	101.1
Erbium	Er	68	167.26	Rutherfordium	Rf	104	(261)
Europium	Eu	63	152.0	Samarium	Sm	62	150.36
Fermium	Fm	100	(257)	Scandium	Sc	21	44.96
Fluorine	F	9	19.00	Selenium	Se	34	78.96
Francium	Fr	87	(223)	Silicon	Si	14	28.09
Gadolinium	Gd	64	157.25	Silver	Ag	47	107.870
Gallium	Ga	31	69.72	Sodium	Na	11	22.990
Germanium	Ge	32	72.59	Strontium	Sr	38	87.62
Gold	Au	79	196.97	Sulfur	S	16	32.064
Hafnium	Hf	72	178.49	Tantalum	Ta	73	180.95
Hahnium	Ha	105	(260)	Technetium	Tc	43	(98)
Helium	He	2	4.003	Tellurium	Te	52	127.60
Holmium	Ho	67	164.93	Terbium	Tb	65	158.92
Hydrogen	H	1	1.0079	Thallium	Tl	81	204.37
Indium	In	49	114.82	Thorium	Th	90	232.04
Iodine	I	53	126.90	Thulium	Tm	69	168.93
Iridium	Ir	77	192.2	Tin	Sn	50	118.69
Iron	Fe	26	55.85	Titanium	Ti	22	47.88
Krypton	Kr	36	83.80	Tungsten	W	74	183.85
Lanthanum	La	57	138.91	Uranium	U	92	238.03
Lawrencium	Lr	103	(260)	Vanadium	V	23	50.94
Lead	Pb	82	207.2	Xenon	Xe	54	131.29
Lithium	Li	3	6.941	Ytterbium	Yb	70	173.04
Lutetium	Lu	71	174.97	Yttrium	Y	39	88.906
Magnesium	Mg	12	24.31	Zinc	Zn	30	65.38
Manganese	Mn	25	54.94	Zirconium	Zr	40	91.22
Mendelevium	Md	101	(258)				

[†] More precise values for the atomic masses, along with the uncertainties in the masses, can be found at http://physics.nist.gov/PhysRefData/.

APPENDIX D

Review of Mathematics

In this appendix, we will review some of the basic results of algebra, geometry, trigonometry, and calculus. In many cases, we will merely state results without proof. Table D-1 lists some mathematical symbols.

Equations

The following operations can be performed on mathematical equations to facilitate their solution:

1. The same quantity can be added to or subtracted from each side of the equation.

2. Each side of the equation can be multiplied or divided by the same quantity.

3. Each side of the equation can be raised to the same power.

It is important to understand that the preceding rules apply to each *side* of the equation and not to each term in the equation.

EXAMPLE D-1

Solve the following equation for x: $(x - 3)^2 + 7 = 23$.

1. Subtract 7 from each side: $(x - 3)^2 = 16$

2. Take the square root of each side: $x - 3 = \pm 4$

3. Add 3 to each side: $x = 4 + 3 = 7$

      or

      $x = -4 + 3 = -1$

REMARKS Note that in step 2 we do not need to write $-(x - 3) = \pm 4$ because all possibilities are included in

   $x - 3 = \pm 4$.

CHECK THE RESULT We check our result by substituting each value into the original equation: $(7 - 3)^2 + 7 = 16 + 7 = 23$ and $(-1 - 3)^2 + 7 = 16 + 7 = 23$.

TABLE D-1

Mathematical Symbols

$=$	is equal to
$\neq$	is not equal to
$\approx$	is approximately equal to
$\sim$	is of the order of
$\propto$	is proportional to
$>$	is greater than
$\geq$	is greater than or equal to
$>>$	is much greater than
$<$	is less than
$\leq$	is less than or equal to
$<<$	is much less than
Δx	change in x
$\lvert x \rvert$	absolute value of x
$n!$	$n(n - 1)(n - 2) \ldots 1$
Σ	sum
lim	limit
$\Delta t \rightarrow 0$	Δt approaches zero
$\dfrac{dx}{dt}$	derivative of x with respect to t
$\dfrac{\partial x}{\partial t}$	partial derivative of x with respect to t
$\int$	integral

EXAMPLE D - 2

Solve the following equation for x:

$$\frac{1}{x} + \frac{1}{4} = \frac{1}{3}$$

1. Subtract from each side:

$$\frac{1}{x} = \frac{1}{3} - \frac{1}{4} = \frac{4}{12} - \frac{3}{12} = \frac{1}{12}$$

2. Multiply each side by $12x$:

$$x = 12$$

REMARKS This type of equation occurs both in geometric optics and in analyses of electric circuits. Although it is easy to solve, errors are often made. A typical mistake is to take the reciprocal of each *term*, obtaining $x + 4 = 3$. Taking the reciprocal of each term is not allowed; taking the reciprocal of each *side* of an equation is allowed. Note that multiplying each side by $12x$ in step 2 is equivalent to taking the reciprocal of each side of the equation.

Direct and Inverse Proportion

The relationships of direct proportion and inverse proportion are so important in physics that they deserve special consideration. Often much algebraic manipulation can be avoided through a simple knowledge of these relationships. Suppose, for example, that you work for 5 days at a certain pay rate and earn $400. How much would you earn at the same pay rate if you worked 8 days? In this problem, the money earned is *directly proportional* to the time worked. We can write an equation relating the money earned M to the time worked t using a constant of proportionality R:

$$M = Rt$$

The constant of proportionality in this case is the pay rate. We can express R in dollars per day. Since $400 was earned in 5 d, the value of R is $400/(5\ d) = \$80/d$. In 8 d, the amount earned is therefore

$$M = (\$80/d)(8\ d) = \$640$$

However, we do not have to find the pay rate explicitly to work the problem. Since the amount earned in 8 d is $\frac{8}{5}$ times that earned in 5 d, this amount is

$$M = \tfrac{8}{5}(\$400) = \$640$$

We can use a similar example to illustrate inverse proportion. If you get a 25% raise, how long would you need to work to earn $400? Here we consider R to be a variable and we wish to solve for t:

$$t = \frac{M}{R}$$

In this equation, the time t is *inversely proportional* to the pay rate R. Thus, if the new rate is $\frac{5}{4}$ times the old rate, the new time will be $\frac{4}{5}$ times the old time or 4 d.

There are some situations in which one quantity varies as the square or some other power of another quantity where the ideas of proportionality are also very useful. Suppose, for example, that a 10-in. diameter pizza costs $8.50. How much would you expect a 12-in. diameter pizza to cost? We expect the cost of a pizza to be approximately proportional to the amount of its contents, which is proportional to the area of the pizza. Since the area is in turn proportional to the square of the diameter, the cost should be proportional to the square of the diameter. If we increase the diameter by a factor of 12/10, the area increases by a factor of $(12/10)^2 = 1.44$, so we should expect the cost to be $(1.44)(\$8.50) = \$12.24$.

EXAMPLE D-3

The intensity of light from a point source varies inversely with the square of the distance from the source. If the intensity is 3.20 W/m² at 5 m from a source, what is it at 6 m from the source?

1. Write an equation expressing the fact that the intensity varies inversely with the square of the distance:

$$I = \frac{C}{r^2}$$

where C is some constant.

2. Let I_1 be the intensity at $r_1 = 5$ m and I_2 be the intensity at $r_2 = 6$ m, and express the ratio I_2/I_1 in terms of r_1 and r_2:

$$\frac{I_2}{I_1} = \frac{C/r_2^2}{C/r_1^2} = \frac{r_1^2}{r_2^2} = \left(\frac{r_1}{r_2}\right)^2 = \left(\frac{5}{6}\right)^2 = 0.694$$

3. Solve for I_2:

$$I_2 = 0.694 I_1 = (0.694)(3.20\ \text{W/m}^2) = 2.22\ \text{W/m}^2$$

Linear Equations

An equation in which the variables occur only to the first power is said to be linear. A linear equation relating y and x can always be put into the standard form

$$y = mx + b \qquad\qquad \text{D-1}$$

where m and b are constants that may be either positive or negative. Figure D-1 shows a graph of the values of x and y that satisfy (Equation D-1). The constant b, called the **intercept,** is the value of y at $x = 0$. The constant m is the **slope** of the line, which equals the ratio of the change in y to the corresponding change in x. In the figure, we have indicated two points on the line, x_1, y_1 and x_2, y_2, and the changes $\Delta x = x_2 - x_1$ and $\Delta y = y_2 - y_1$. The slope m is then

$$m = \frac{y_2 - y_1}{x_2 - x_1} = \frac{\Delta y}{\Delta x}$$

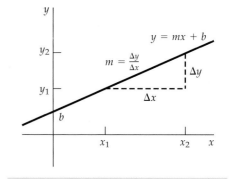

FIGURE D-1 Graph of the linear equation $y = mx + b$, where b is the intercept and $m = \Delta y/\Delta x$ is the slope.

If x and y are both unknown, there is no unique solution for their values. Any pair of values x_1, y_1 on the line in Figure D-1 will satisfy the equation. If we have two equations, each with the same two unknowns x and y, the equations can be solved simultaneously for the unknowns.

EXAMPLE D-4

Find the values of x and y that satisfy

$$3x - 2y = 8 \qquad\qquad \text{D-2}$$

and

$$y - x = 2 \qquad\qquad \text{D-3}$$

PICTURE THE PROBLEM Figure D-2 shows a graph of each of these equations. At the point where the lines intersect, the values of x and y satisfy both equations. We can solve two simultaneous equations by first solving either equation for one variable in terms of the other variable and then substituting the result into the other equation. An alternative method is to multiply one equation by a constant such that one of the unknown terms is eliminated when the equations are added or subtracted.

1. Solve (Equation D-3) for y: $y = x + 2$

2. Substitute this value for y into (Equation D-2): $3x - 2(x + 2) = 8$

3. Simplify and solve for x: $3x - 2x - 4 = 8$
   $x = 12$

4. To solve these equations using the alternative method, we first multiply (Equation D-3) by 2: $2y - 2x = 4$

5. Add this equation to (Equation D-2): $2y - 2x = 4$
   $\underline{3x - 2y = 8}$
   $3x - 2x = 12$
   $x = 12$

FIGURE D-2 Graph of Equations D-2 and D-3. At the point where the lines intersect, the values of x and y satisfy both equations.

Factoring

Equations can often be simplified by factoring. Three important examples are

1. Common factor: $2ax + 3ay = a(2x + 3y)$
2. Perfect square: $x^2 - 2xy + y^2 = (x - y)^2$
3. Difference of squares: $x^2 - y^2 = (x + y)(x - y)$

The Quadratic Formula

An equation that contains a variable to the second power is called a *quadratic equation*. The standard form for a quadratic equation is

$$ax^2 + bx + c = 0 \qquad \text{D-4}$$

where a, b, and c are constants. The general solution of this equation is

$$x = -\frac{b}{2a} \pm \frac{1}{2a}\sqrt{b^2 - 4ac} \qquad \text{D-5}$$

When b^2 is greater than $4ac$, there are two solutions corresponding to the + and − signs. Figure D-3 shows a graph of y versus x where $y = ax^2 + bx + c$. The curve, called a **parabola,** crosses the x axis twice. The values of x for which $y = 0$ are the solutions to (Equation D-4). When $b^2 < 4ac$, the graph of y versus x does not intersect the x axis, as is shown in Figure D-4, and there are no real solutions to (Equation D-4). When $b^2 = 4ac$, the graph of y versus x is tangent to the x axis at the point $x = -b/2a$.

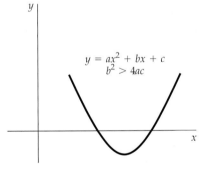

FIGURE D-3 Graph of y versus x when $y = ax^2 + bx + c$ for the case $b^2 > 4ac$. The two values of x for which $y = 0$ satisfy the quadratic equation (Equation D-4).

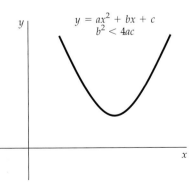

FIGURE D-4 Graph of y versus x when $y = ax^2 + bx + c$ for the case $b^2 < 4ac$. In this case, there are no (real) values of x for which $y = 0$.

Exponents

The notation x^n stands for the quantity obtained by multiplying x times itself n times. For example, $x^2 = x{\cdot}x$ and $x^3 = x{\cdot}x{\cdot}x$. The quantity n is called the **power,** or the **exponent,** of x. When two powers of x are multiplied, the exponents are added:

$$(x^m)(x^n) = x^{m+n} \qquad\qquad \text{D-6}$$

This can be readily seen from an example:

$$x^2 x^3 = (x \cdot x)(x \cdot x \cdot x) = x^5$$

Any number raised to the 0 power is defined to be 1:

$$x^0 = 1 \qquad\qquad \text{D-7}$$

Then

$$x^n x^{-n} = x^0 = 1$$
$$x^{-n} = \frac{1}{x^n} \qquad\qquad \text{D-8}$$

When two powers are divided, the exponents are subtracted:

$$\frac{x^n}{x^m} = x^n x^{-m} = n^{n-m} \qquad\qquad \text{D-9}$$

Using these rules, we have

$$x^{1/2} \cdot x^{1/2} = x$$

so

$$x^{1/2} = \sqrt{x}$$

When a power is raised to another power, the exponents are multiplied:

$$(x^n)^m = x^{nm} \qquad\qquad \text{D-10}$$

Logarithms

If y is related to x by $y = a^x$, the number x is said to be the logarithm of y to the base a and is written

$$x = \log_a y$$

then if $x = 1$ then $y = a^1 = a$ and

$$\log_a a = 1 \qquad\qquad \text{D-11}$$

and, if $x = 0$ then $y = a^0 = 1$ and

$$\log_a 1 = 0 \qquad\qquad \text{D-12}$$

Also, if $y_1 = a^n$ and $y_2 = a^m$, then

$$y_1 y_2 = a^n a^m = a^{n+m}$$

and

$$\log_a y_1 = n, \log_a y_2 = m, \text{ and } \log_a y_1 y_2 = n + m$$

so

$$\log_a y_1 y_2 = \log_a y_1 + \log_a y_2 \qquad \text{D-13}$$

It immediately follows that

$$\log_a y^n = n \log_a y \qquad \text{D-14}$$

There are two bases in common use: base 10, called **common logarithms,** and base $e (e = 2.728 \ldots)$, called **natural logarithms.** When no base is specified, the base is usually understood to be 10. Thus, $\log 100 = \log_{10} 100 = 2$ since $100 = 10^2$.

The symbol ln is used for natural logarithms. Thus,

$$\log_e x = \ln x \qquad \text{D-15}$$

and $y = \ln x$ implies

$$x = e^y \qquad \text{D-16}$$

Logarithms can be changed from one base to another. Suppose that

$$z = \log x \qquad \text{D-17}$$

Then

$$10^z = 10^{\log x} = x \qquad \text{D-18}$$

Taking the natural logarithm of both sides of (Equation D-18), we obtain

$$z \ln 10 = \ln x$$

Substituting $\log x$ for z (see Equation D-17) gives

$$\ln x = (\ln 10)\log x \qquad \text{D-19}$$

The Exponential Function

When the rate of change of a quantity is proportional to the quantity itself, the quantity increases or decreases exponentially. An example of *exponential decrease* is nuclear decay. If N is the number of radioactive nuclei at some time, then the change dN in some very small time interval dt will be proportional to N and to dt:

$$dN = -\lambda N \, dt$$

where the constant of proportionality λ is the decay rate. The function N satisfying this equation is

$$N = N_0 e^{-\lambda t} \qquad \text{D-20}$$

where N_0 is the number at time $t = 0$. Figure D-5 shows N versus t. A characteristic of exponential decay is that N decreases by a constant factor in a given time interval. The time interval for N to decrease to half its original value is its half-life $t_{1/2}$. The half life is obtained from (Equation D-20) by setting $N = \frac{1}{2}N_0$ and solving for the time. This gives

$$t_{1/2} = \frac{\ln 2}{\lambda} = \frac{0.693}{\lambda} \qquad \text{D-21}$$

An example of *exponential increase* is population growth. If the number of organisms is N, the change in N after a small time interval dt is given by

$$dN = +\lambda N \, dt$$

where λ is a constant that characterizes the rate of increase. The function N satisfying this equation is

$$N = N_0 e^{\lambda t} \qquad \text{D-22}$$

A graph of this function is shown in Figure D-6. An exponential increase is characterized by a doubling time T_2, which is related to λ by

$$T_2 = \frac{\ln 2}{\lambda} = \frac{0.693}{\lambda} \qquad \text{D-23}$$

If the rate of increase λ is expressed as a percentage, $r = \lambda/100\%$, the doubling time is

$$T_2 = \frac{69.3}{r} \qquad \text{D-24}$$

For example, if the population increases by 2 percent per year, the population will double every $69.3/2 \approx 35$ years. Table D-2 lists some useful relations for exponential and logarithmic functions.

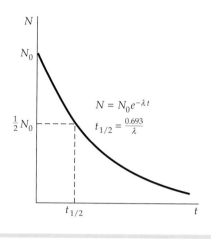

FIGURE D-5 Graph of N versus t when N decreases exponentially. The time $t_{1/2}$ is the time it takes for N to decrease by one-half.

TABLE D-2

Exponential and Logarithmic Functions

$e = 2.71828$;

$e^0 = 1$

If $y = e^x$, then $x = \ln y$.

$e^{\ln x} = x$

$e^x e^y = e^{(x+y)}$

$(e^x)^y = e^{xy} = (e^y)^x$

$\ln e = 1$; $\ln 1 = 0$

$\ln xy = \ln x + \ln y$

$\ln \dfrac{x}{y} = \ln x - \ln y$

$\ln e^x = x$; $\ln a^x = x \ln a$

$\ln x = (\ln 10) \log x$

$\quad = 2.3026 \log x$

$\log x = (\log e) \ln x = 0.43429 \ln x$

$e^x = 1 + x + \dfrac{x^2}{2!} + \dfrac{x^3}{3!} + \dots$

$\ln(1 + x) = x \pm \dfrac{x^2}{2} + \dfrac{x^3}{3} \pm \dfrac{x^4}{4} + \dots$

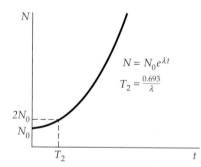

FIGURE D-6 Graph of N versus t when N increases exponentially. The time T_2 is the time it takes for N to double.

Geometry

The ratio of the circumference of a circle to its diameter is a natural number π, which has the approximate value

$$\pi = 3.141592$$

The circumference C of a circle is thus related to its diameter d and its radius r by

$$C = \pi d = 2\pi r \quad \text{circumference of circle} \qquad \text{D-25}$$

The area of a circle is

$$A = \pi r^2 \quad \text{area of circle} \qquad \text{D-26}$$

Area of parallelogram
$A = bh$

FIGURE D-7 Area of a parallelogram.

The area of a parallelogram is the base b times the height h (Figure D-7) and that of a triangle is one-half the base times the height (Figure D-8). A sphere of radius r (Figure D-9) has a surface area given by

$$A = 4\pi r^2 \quad \text{spherical surface area} \qquad \text{D-27}$$

and a volume given by

$$V = \tfrac{4}{3}\pi r^3 \quad \text{spherical volume} \qquad \text{D-28}$$

A cylinder of radius r and length L (Figure D-10) has surface area (not including the end faces) of

$$A = 2\pi rL \quad \text{cylindrical surface} \qquad \text{D-29}$$

and volume of

$$V = \pi r^2 L \quad \text{cylindrical volume} \qquad \text{D-30}$$

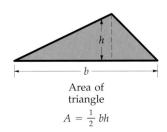

Area of
triangle
$A = \tfrac{1}{2}bh$

FIGURE D-8 Area of a triangle.

Trigonometry

The angle between two intersecting straight lines is measured as follows. A circle is drawn with its center at the intersection of the lines, and the circular arc is divided into 360 parts called **degrees**. The number of degrees in the arc between the lines is the measure of the angle between the lines. For very small angles, the degree is divided into minutes (′) and seconds (″) with $1' = 1°/60$ and $1'' = 1'/60 = 1°/3600$. For scientific work, a more useful measure of an angle is the radian (rad). In radian measure, the angle between two intersecting straight lines is found by again drawing a circle with its center at the intersection of the lines. The measure of the angle in radians is then defined as the length of the circular arc between the lines divided by the radius of the circle (Figure D-11). If s is the arc length and r is the radius of the circle, the angle θ measured in radians is

$$\theta = \frac{s}{r} \qquad \text{D-31}$$

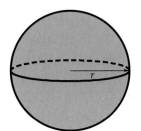

Spherical surface area
$A = 4\pi r^2$
Spherical volume
$V = \tfrac{4}{3}\pi r^3$

FIGURE D-9 Surface area and volume of a sphere.

Since the angle measured in radians is the ratio of two lengths, it is dimensionless. The relation between radians and degrees is

$$360° = 2\pi \text{ rad}$$

or

$$1 \text{ rad} = \frac{360°}{2\pi} = 57.3°$$

Cylindrical surface area
$A = 2\pi rL$
Cylindrical volume
$V = \pi r^2 L$

FIGURE D-10 Surface area (not including the end faces) and volume of a cylinder.

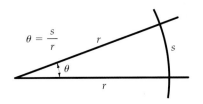

$\theta = \dfrac{s}{r}$

FIGURE D-11 The angle θ in radians is defined to be the ratio s/r, where s is the arc length intercepted on a circle of radius r.

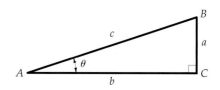

FIGURE D-12 Some useful relations for angles.

Figure D-12 shows some useful relations for angles.

Figure D-13 shows a right triangle formed by drawing the line BC perpendicular to AC. The lengths of the sides are labeled a, b, and c. The trigonometric functions $\sin\theta$, $\cos\theta$, and $\tan\theta$ for an acute angle θ are defined as

$$\sin\theta = \frac{a}{c} = \frac{\text{opposite side}}{\text{hypotenuse}} \qquad \text{D-33}$$

$$\cos\theta = \frac{b}{c} = \frac{\text{adjacent side}}{\text{hypotenuse}} \qquad \text{D-34}$$

$$\tan\theta = \frac{a}{b} = \frac{\text{opposite side}}{\text{adjacent side}} = \frac{\sin\theta}{\cos\theta} \qquad \text{D-35}$$

Three other trigonometric functions, defined as the reciprocals of these functions, are

$$\sec\theta = \frac{c}{b} = \frac{1}{\cos\theta} \qquad \text{D-36}$$

$$\csc\theta = \frac{c}{a} = \frac{1}{\sin\theta} \qquad \text{D-37}$$

$$\cot\theta = \frac{b}{a} = \frac{1}{\tan\theta} = \frac{\cos\theta}{\sin\theta} \qquad \text{D-38}$$

The angle θ whose sine is x is called the arcsine of x, and is written $\sin^{-1} x$. That is, if

$$\sin\theta = x$$

then

$$\theta = \arcsin x = \sin^{-1} x \qquad \text{D-39}$$

FIGURE D-13 A right triangle with sides of length a and b and a hypotenuse of length c.

The arcsine is the inverse of the sine. The inverse of the cosine and tangent are defined similarly. The angle whose cosine is y is the arccosine of y. That is, if

$$\cos \theta = y$$

then

$$\theta = \arccos y = \cos^{-1} y \qquad \text{D-40}$$

The angle whose tangent is z is the arctangent of z. That is, if

$$\tan \theta = z$$

$$\theta = \arctan z = \tan^{-1} z \qquad \text{D-41}$$

The Pythagorean theorem

$$a^2 + b^2 = c^2 \qquad \text{D-42}$$

gives some useful identities. If we divide each term in this equation by c^2, we obtain

$$\frac{a^2}{c^2} + \frac{b^2}{c^2} = 1$$

or, from the definitions of $\sin \theta$ and $\cos \theta$,

$$\sin^2 \theta + \cos^2 \theta = 1 \qquad \text{D-43}$$

Similarly, we can divide each term in (Equation D-42) by a^2 or b^2 and obtain

$$1 + \cot^2 \theta = \csc^2 \theta \qquad \text{D-44}$$

and

$$1 + \tan^2 \theta = \sec^2 \theta \qquad \text{D-45}$$

These and other useful trigonometric formulas are listed in Table D-3.

The following equation is actually two equations. For one of the equations the two upper signs are taken, and for the other equation the two lower signs are taken. The same sign rule applies to others of the following "equations":

$$\sin(A \pm B) = \sin A \cos B \pm \cos A \sin B$$

$$\cos(A \pm B) = \cos A \cos B \mp \sin A \sin B$$

$$\tan(A \pm B) = \frac{\tan A \pm \tan B}{1 \mp \tan A \tan B}$$

$$\sin A \pm \sin B = 2 \sin\left[\tfrac{1}{2}(A \pm B)\right]\cos\left[\tfrac{1}{2}(A \pm B)\right]$$

$$\cos A + \cos B = 2 \cos\left[\tfrac{1}{2}(A + B)\right]\cos\left[\tfrac{1}{2}(A - B)\right]$$

$$\cos A - \cos B = 2 \sin\left[\tfrac{1}{2}(A + B)\right]\sin\left[\tfrac{1}{2}(B - A)\right]$$

$$\tan A \pm \tan B = \frac{\sin(A \pm B)}{\cos A \cos B}$$

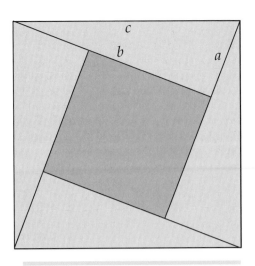

When this figure was first published the letters were absent and it was accompanied by the single word "Behold!" Using the drawing, establish the Pythagorean theorem ($a^2 + b^2 = c^2$).

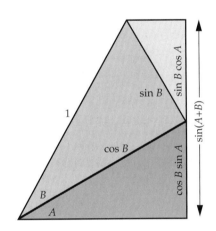

Using this drawing, establish the identity $\sin(A + B) = \sin A \cos B + \cos A \sin B$. You can also use it to establish the identity $\cos(A + B) = \cos A \cos B - \sin A \sin B$. Try it.

TABLE D-3

Trigonometric Formulas

$$\sin^2 \theta + \cos^2 \theta = 1; \quad \sec^2 \theta - \tan^2 \theta = 1; \quad \csc^2 \theta - \cot^2 \theta = 1$$

$$\sin 2\theta = 2 \sin \theta \cos \theta$$

$$\cos 2\theta = \cos^2 \theta - \sin^2 \theta = 2 \cos^2 \theta - 1 = 1 - 2 \sin^2 \theta$$

$$\tan 2\theta = \frac{2 \tan \theta}{1 - \tan^2 \theta}$$

$$\sin \frac{1}{2} \theta = \sqrt{\frac{1 - \cos \theta}{2}}; \quad \cos \frac{1}{2} \theta = \sqrt{\frac{1 + \cos \theta}{2}}; \quad \tan \frac{1}{2} \theta = \sqrt{\frac{1 - \cos \theta}{1 + \cos \theta}}$$

EXAMPLE D - 5

Use the isosceles right triangle shown in Figure D-14 to find the sine, cosine, and tangent of 45°.

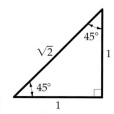

FIGURE D-14
An isosceles right triangle.

PICTURE THE PROBLEM It is clear from the figure that the two acute angles of this triangle are equal. Since the sum of the three angles in a triangle must equal 180°, and the right angle is 90°, each acute angle must be 45°. If we multiply each side of any triangle by a common factor, we obtain a similar triangle with the same angles as the first. We can therefore choose any convenient length for one side. We choose the equal sides to have a length of 1 unit.

1. Find the length of the hypotenuse from the Pythagorean theorem:

$$c = \sqrt{a^2 + b^2} = \sqrt{1^2 + 1^2} = \sqrt{2}\ \text{units}$$

2. Calculate sin 45° from its definition:

$$\sin 45° = \frac{a}{c} = \frac{1}{\sqrt{2}} = 0.707$$

3. Calculate cos 45° from its definition:

$$\cos 45° = \frac{b}{c} = \frac{1}{\sqrt{2}} = 0.707$$

4. Calculate tan 45° from its definition:

$$\tan 45° = \frac{a}{b} = \frac{1}{1} = 1$$

EXAMPLE D - 6

The sine of 30° is exactly 0.5. Find the ratios of the sides of a 30–60° right triangle.

FIGURE D-15 A
30–60° right triangle.

PICTURE THE PROBLEM This common triangle is shown in Figure D-15. We choose a length of 1 unit for the side opposite the 30° angle.

1. Calculate the hypotenuse from the definition of the sine and the choice of 1 unit for the opposite side:

$$\sin 30° = \frac{a}{c} = \frac{1}{c} = 0.5$$

2. Use the Pythagorean theorem to find the length b of the side opposite the 60° angle:

$$c = \frac{1}{0.5} = 2$$

$$b = \sqrt{c^2 - a^2} = \sqrt{2^2 - 1^2} = \sqrt{3}$$

3. Use these results to calculate cos 30°, tan 30°, sin 60°, cos 60°, and tan 60°:

$$\cos 30° = \frac{b}{c} = \frac{\sqrt{3}}{2} = 0.866$$

$$\tan 30° = \frac{a}{b} = \frac{1}{\sqrt{3}} = 0.577$$

$$\sin 60° = \frac{b}{c} = \cos 30° = 0.866$$

$$\cos 60° = \frac{a}{c} = \sin 30° = 0.500$$

$$\tan 60° = \frac{b}{a} = \frac{\sqrt{3}}{1} = 1.732$$

For small angles, the length a is nearly equal to the arc length s, as can be seen in Figure D-16. The angle $\theta = s/c$ is therefore nearly equal to $\sin \theta = a/c$:

$$\sin \theta \approx \theta \qquad \text{for small values of } \theta \qquad\qquad \text{D-46}$$

Similarly, the lengths c and b are nearly equal, so $\tan \theta = a/b$ is nearly equal to both θ and $\sin \theta$ for small values of θ:

$$\tan \theta \approx \sin \theta \approx \theta \qquad \text{for small values of } \theta \qquad\qquad \text{D-47}$$

FIGURE D-16 For small angles, $\sin \theta = a/c$, $\tan \theta = a/b$, and the angle $\theta = s/c$ are all approximately equal.

Equations D-46 and D-47 hold only if θ is measured in radians. Since $\cos \theta = b/c$ and these lengths are nearly equal for small values of θ, we have

$$\cos \theta \approx 1 \qquad \text{for small values of } \theta \qquad\qquad \text{D-48}$$

EXAMPLE D-7

By how much do $\sin \theta$, $\tan \theta$, and θ differ when $\theta = 15°$?

1. Convert 15° to radians:

$$\theta = 15° \frac{2\pi \text{ rad}}{360°} = 0.262 \text{ rad}$$

2. Find $\sin 15°$ and $\tan 15°$ using a calculator:

$$\sin 15° = 0.259$$

$$\tan 15° = 0.268$$

3. Compute the percentage difference between θ and $\sin \theta$:

$$\frac{|\sin \theta - \theta|}{\theta} = \frac{|0.259 - 0.262|}{0.262} = \frac{0.003}{0.262} = 0.011 \approx 1\%$$

4. Compute the percentage difference between θ and $\tan \theta$:

$$\frac{|\tan \theta - \theta|}{\theta} = \frac{|0.268 - 0.262|}{0.262} = \frac{0.006}{0.262} = 0.023 \approx 2\%$$

REMARKS For smaller angles, the approximation $\theta < \sin \theta < \tan \theta$ is even more accurate.

Example D-7 shows that if accuracy of a few percent is needed, small angle approximations can be used only for angles of about 15° or less. Figure D-17 shows graphs of θ, $\sin \theta$, and $\tan \theta$ versus θ, for small values of θ.

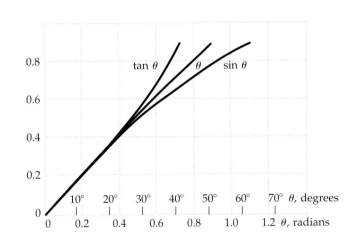

FIGURE D-17 Graphs of $\tan \theta$, θ, and $\sin \theta$ versus θ for small values of θ.

Figure D-18 shows an obtuse angle with its vertex at the origin and one side along the x axis. The trigonometric functions for a general angle such as this are defined by

$$\sin \theta = \frac{y}{c} \qquad \text{D-49}$$

$$\cos \theta = \frac{x}{c} \qquad \text{D-50}$$

$$\tan \theta = \frac{y}{x} \qquad \text{D-51}$$

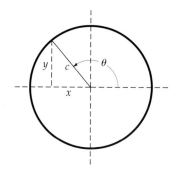

FIGURE D-18 Diagram for defining the trigonometric functions for an obtuse angle.

Figure D-19 shows plots of these functions versus θ. The sine function has a period of 2π rad. Thus, for any value of θ $\sin(\theta + 2\pi) = \sin \theta$ and so forth. That is, when an angle changes by 2π rad, the function returns to its original value. The tangent function has a period of π rad. Thus, $\sin(\theta + 2\pi) = \sin \theta$ and so forth. Some other useful relations are

$$\sin(\pi - \theta) = \sin \theta \qquad \text{D-52}$$

$$\cos(\pi - \theta) = -\cos \theta \qquad \text{D-53}$$

$$\sin(\pi/2 - \theta) = \cos \theta \qquad \text{D-54}$$

$$\cos(\pi/2 - \theta) = \sin \theta \qquad \text{D-55}$$

The trigonometric functions can be expressed as power series in θ. The series for $\sin \theta$ and $\cos \theta$ are

$$\sin \theta = \theta - \frac{\theta^3}{3!} + \frac{\theta^5}{5!} - \frac{\theta^7}{7!} + \dots \qquad \text{D-56}$$

$$\cos \theta = 1 - \frac{\theta^2}{2!} + \frac{\theta^4}{4!} - \frac{\theta^6}{6!} + \dots \qquad \text{D-57}$$

When θ is small, good approximations are obtained using only the first few terms in the series.

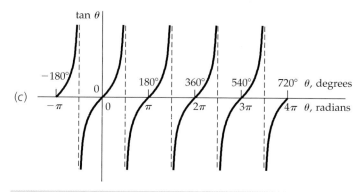

FIGURE D-19 The trigonometric functions sin θ, cos θ, and tan θ versus θ.

The Binomial Expansion

The binomial theorem is very useful for making approximations. One form of this theorem is

$$(1 + x)^n = 1 + nx + \frac{n(n-1)}{2!}x^2 + \frac{n(n-1)(n-2)}{3!}x^3 + \dots \qquad \text{D-58}$$

If n is a positive integer, there are $n + 1$ terms in this series. If n is a real number other than a positive integer, there are an infinite number of terms. The series is valid for any value of n if x^2 is less than 1. It is also valid for $x^2 = 1$ if n is positive. The series is particularly useful if x^2 is small compared to 1. Then each term is much smaller than the previous term and we can drop all but the first two or three terms in the equation. If x^2 is much less compared to 1, we have

$$(1 + x)^n \approx 1 + nx, \qquad |x| \ll 1 \qquad \text{D-59}$$

EXAMPLE D-8

Use (Equation D-59) to find an approximate value for the square root of 101.

1. Write $(101)^{1/2}$ so it is in the form $(1 + x)^n$ with x much less than 1: $(101)^{1/2} = (100 + 1)^{1/2} = (100)^{1/2}(1 + 0.01)^{1/2} = 10(1 + 0.01)^{1/2}$

2. Use Equation D-59 with $n = \frac{1}{2}$ and $x = 0.1$ to expand $(1 + 0.01)^{1/2}$. $(1 + 0.01)^{1/2} \approx 1 + \frac{1}{2}(0.01) = 1.005$

3. Substitute this result into the equation in step 1: $(101)^{1/2} = 10(1 + 0.01)^{1/2} \approx \boxed{10.05}$

REMARKS We can assess the accuracy of this result by computing the first term in Equation D-58 that was neglected. This term is

$$\frac{n(n - 1)}{2}x^2 = \frac{\frac{1}{2}(-\frac{1}{2})}{2}(0.01)^2 = -\frac{0.0001}{8} \approx -0.00001 = -0.001$$

We therefore expect our answer to be correct to within about 0.001%. The value of $(101)^{1/2}$ to eight significant figures is 10.049875, which differs from 10.05 by 0.000124 or about 0.001% of 10.05.

Complex Numbers

A general complex number z can be written

$$z = a + bi \qquad\qquad \text{D-60}$$

where a and b are real numbers and $i = \sqrt{-1}$. The quantity a is called the real part and the quantity ib is called the imaginary part of z. We can represent a complex number in a plane as shown in Figure D-20, where the x axis is the real axis and the y axis is the imaginary axis. We can use the relations $a = r \cos\theta$ and $b = r \sin\theta$ from Figure D-20 to write the complex number z in polar coordinates:

$$z = r \cos\theta + ir \sin\theta \qquad\qquad \text{D-61}$$

where $r = \sqrt{a^2 + b^2}$ is called the magnitude of z.

When complex numbers are added or subtracted, the real and imaginary parts are added or subtracted separately:

$$z_1 + z_2 = (a_1 + ib_1) + (a_2 + ib_2) = (a_1 + a_2) + i(b_1 + b_2) \qquad\qquad \text{D-62}$$

However, when two complex numbers are multiplied, each part of one number is multiplied by each part of the other number:

$$z_1 z_2 = (a_1 + ib_1)(a_2 + ib_2) = a_1 a_2 + i^2 b_1 b_2 + i(a_1 b_2 + a_2 b_1) \qquad\qquad \text{D-63}$$

$$= a_1 a_2 - b_1 b_2 + i(a_1 b_2 + a_2 b_1)$$

where we have used $i^2 = -1$.

The complex conjugate z^* of the complex number z is that number obtained by replacing i with $-i$. If $z = a + ib$ then

$$z^* = (a + ib)^* = a - ib \qquad\qquad \text{D-64}$$

The product of a complex number and its complex conjugate equals the square of the magnitude of the number:

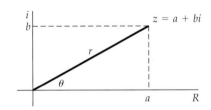

$$z = a + bi$$
$$= r \cos\theta + (r \sin\theta)i$$
$$= r(\cos\theta + i \sin\theta)$$

FIGURE D-20 Representation of a complex number in a plane. The real part of the complex number is plotted along the horizontal axis, and the imaginary part is plotted along the vertical axis.

$$zz^* = (a + ib)(a - ib) = a^2 + b^2 \qquad \text{D-65}$$

A particularly useful function of a complex number is the exponential $e^{i\theta}$. Using the expansion for e^x given in Table D-2, we have

$$e^{i\theta} = 1 + i\theta + \frac{(i\theta)^2}{2!} + \frac{(i\theta)^3}{3!} + \frac{(i\theta)^4}{4!} + \cdots$$

Using $i^2 = -1$, $i^3 = -i$, $i^4 = +1$, and so forth and separating the real parts from the imaginary parts, this expansion can be written

$$e^{i\theta} = \left(1 - \frac{\theta^2}{2!} + \frac{\theta^4}{4!} - \cdots\right) + i\left(\theta - \frac{\theta^3}{3!} + \cdots\right)$$

Comparing this result with Equations D-56 and D-57, we can see that

$$e^{i\theta} = \cos\theta + i\sin\theta \qquad \text{D-66}$$

Using this result, we can express a general complex number as an exponential:

$$z = a + ib = r\cos\theta + ir\sin\theta = re^{i\theta} \qquad \text{D-67}$$

where $r = \sqrt{a^2 + b^2}$.

Solving differential equations using complex numbers   Consider an equation of the form

$$a\frac{d^2x}{dt^2} + b\frac{dx}{dt} + cx = A\cos\omega t \qquad \text{D-68}$$

that represents a physical process, such as a damped harmonic oscillator driven by a sinusoidal force, or a series RLC combination being driven by a sunusoidal potential drop. Each of the parameters in (Equation D-68) is a real number. We wish to obtain the steady-state solution to this equation using complex numbers. To do so, we first construct the equation:

$$a\frac{d^2y}{dt^2} + b\frac{dy}{dt} + cy = A\sin\omega t \qquad \text{D-69}$$

Equation D-69 has no physical meaning and we have no interest in solving it. However, it is of use in solving Equation D-68. After multiplying both sides of (Equation D-69) by i we add it and (Equation D-68) to obtain

$$\left(a\frac{d^2x}{dt^2} + ai\frac{d^2y}{dt^2}\right) + \left(b\frac{dx}{dt} + bi\frac{dy}{dt}\right) + (cx + ciy) = A\cos\omega t + Ai\sin\omega t$$

We next combine terms to get

$$a\frac{d^2(x + iy)}{dt^2} + b\frac{d(x + iy)}{dt} + c(x + iy) = A(\cos\omega t + i\sin\omega t) \qquad \text{D-70}$$

whose validity depends on the derivative of a sum being equal to the sum of the derivatives. We simplify our result by defining $z = x + iy$ and by using the identity $e^{i\omega t} = \cos\omega t + i\sin\omega t$. Substituting these into (Equation D-70) we obtain

$$a\frac{d^2z}{dt^2} + b\frac{dz}{dt} + cz = Ae^{i\omega t}$$ D-71

which we now solve for z. Once z is obtained we can solve for x using $x = \text{Re}(z)$.

Since we are looking only for the steady state solution for (Equation D-68) we can assume its solution is of the form $x = x_0 \cos(\omega t - \phi)$, where ϕ is a constant. This is equivalent to assuming that the solution to (Equation D-71) is of the form $z = z_0 e^{i\omega t}$, where z_0 is a complex number. Then $dz/dt = i\omega z$ $d^2z/dt^2 = -\omega^2 z$, and $e^{i\omega t} = z/z_0$.

Substituting these into (Equation D-70) gives $-a\omega^2 z + i\omega bz + cz = A\dfrac{z}{z_0}$

Dividing both sides of this equation by z and solving for z_0 gives

$$z_0 = \frac{A}{-a\omega^2 + i\omega b + c}$$

Expressing the denominator in polar form gives $(-a\omega^2 + c) + i\omega b = \sqrt{(-a\omega^2 + c)^2 + \omega^2 b^2}\, e^{i\phi}$, where $\tan\phi = \omega^2 b^2 / (-a\omega^2 + c)$. Thus,

$$z_0 = \frac{A}{\sqrt{(-a\omega^2 + c)^2 + \omega^2 b^2}} e^{-i\phi}$$

so

$$z = z_0 e^{i\omega t} = \frac{A}{\sqrt{(-a\omega^2 + c)^2 + \omega^2 b^2}} e^{i(\omega t - \phi)}$$

$$= \frac{A}{\sqrt{(-a\omega^2 + c)^2 + \omega^2 b^2}}(\cos(\omega t - \delta) + i\sin(\omega t - \delta))$$ D-72

It follows that

$$x = \text{Re}(z) = \frac{A}{\sqrt{(-a\omega^2 + c)^2 + \omega^2 b^2}}\cos(\omega t - \phi)$$ D-73

Differential Calculus

When we say that x is a function of t, we mean that for each value of t there is a single corresponding value of x. An example is $x = At^2$, where A is a constant. To indicate that x is a function of t, we sometimes write $x(t)$ for x. Figure D-21 is a graph of x versus t for a typical function $x(t)$. At a particular value $t = t_1$, x has the value of x_1 as indicated. At another value t_2, x has the value x_2. The change in t, $t_2 - t_1$, is written $\Delta t = t_2 - t_1$ and the corresponding change in x is written $\Delta x = x_2 - x_1$. The ratio $\Delta x/\Delta t$ is the slope of the straight line connecting (x_1, t_1) and (x_2, t_2). If we make Δt smaller and smaller, the line connecting (x_1, t_1) and (x_2, t_2) approaches the line that is tangent to the curve at the point (x_1, t_1). The slope of this tangent line is called the derivative of x with respect to t and is written dx/dt:

$$\frac{dx}{dt} = \lim_{\Delta t \to 0} \frac{\Delta x}{\Delta t}$$ D-74

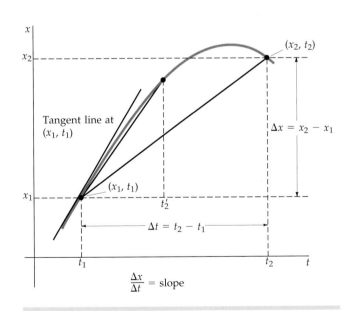

FIGURE D-21 Graph of a typical function $x(t)$. The points (x_1, t_1) and (x_2, t_2) are connected by a straight line. The slope of this line is $\Delta x/\Delta t$. As the time interval beginning at t_1 is decreased, the slope for that interval approaches the slope of the line tangent to the curve at time t_1, which is the derivative of x with respect to t.

The derivative of a function of t is generally another function of t. If x is a constant, the graph of x versus t is a horizontal line with zero slope. The derivative of a constant is thus zero. In Figure D-22, x is proportional to t:

$$x = Ct$$

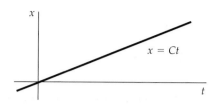

This function has a constant slope equal to C. Thus the derivative of Ct is C. Table D-4 lists some properties of derivatives and the derivatives of some particular functions that occur often in physics. It is followed by comments aimed at making these properties and rules clearer. More detailed discussion can be found in most calculus books.

FIGURE D-22 Graph of the linear function $x = Ct$. This function has a constant slope C.

TABLE D-4

Properties of Derivatives and Derivatives of Particular Functions

Linearity

1. The derivative of a constant times a function equals the constant times the derivative of the function:

$$\frac{d}{dt}[Cf(t)] = C\frac{df(t)}{dt}$$

2. The derivative of a sum of functions equals the sum of the derivatives of the functions:

$$\frac{d}{dt}[f(t) + g(t)] = \frac{df(t)}{dt} + \frac{dg(t)}{dt}$$

Chain rule

3. If f is a function of x and x is in turn a function of t, the derivative of f with respect to t equals the product of the derivative of f with respect to x and the derivative of x with respect to t:

$$\frac{d}{dt}f(t) = \frac{df}{dx}\frac{dx}{dt}$$

Derivative of a product

4. The derivative of a product of functions $f(t)g(t)$ equals the first function times the derivative of the second plus the second function times the derivative of the first:

$$\frac{d}{dt}[f(t)g(t)] = f(t)\frac{dg(t)}{dt} + \frac{df(t)}{dt}g(t)$$

Reciprocal derivative

5. The derivative of t with respect to x is the reciprocal of the derivative of x with respect to t, assuming that neither derivative is zero:

$$\frac{dx}{dt} = \left(\frac{dt}{dx}\right)^{-1} \quad \text{if} \quad \frac{dt}{dx} \neq 0$$

Derivatives of particular functions

6. $\dfrac{dC}{dt} = 0$ where C is a constant

7. $\dfrac{d(t^n)}{dt} = nt^{n-1}$

8. $\dfrac{d}{dt}\sin \omega t = \omega \cos \omega t$

9. $\dfrac{d}{dt}\cos \omega t = -\omega \sin \omega t$

10. $\dfrac{d}{dt}\tan \omega t = \omega \sec^2 \omega t$

11. $\dfrac{d}{dt}e^{bt} = be^{bt}$

12. $\dfrac{d}{dt}\ln bt = \dfrac{1}{t}$

EXAMPLE D-9

Find the derivative of $x = at^2 + bt + c$, where a, b, and c are constants.

PICTURE THE PROBLEM From rule 2, we can differentiate each term separately and add the results.

1. Use rules 1 and 7 to find the derivative of the first term:

$$\frac{d(at^2)}{dt} = 2at^1 = 2at$$

2. Compute the derivatives of the second and third terms:

$$\frac{d(bt)}{dt} = b, \quad \frac{d(c)}{dt} = 0$$

3. Add these results:

$$\frac{dx}{dt} = 2at + b$$

Comments on Rules 1 Through 5

Rules 1 and 2 follow from the fact that the limiting process is linear. We can understand rule 3, the chain rule, by multiplying $\Delta f / \Delta t$ by $\Delta x / \Delta x$ and noting that, since x is a function of t, both Δx and Δf approach zero as Δt approaches zero. Since the limit of a product of two functions equals the product of their limits, we have

$$\lim_{\Delta t \to 0} \frac{\Delta f}{\Delta t} = \lim_{\Delta t \to 0} \frac{\Delta f \, \Delta x}{\Delta x \, \Delta t} = \left(\lim_{\Delta t \to 0} \frac{\Delta f}{\Delta x} \right)\left(\lim_{\Delta t \to 0} \frac{\Delta x}{\Delta t} \right) = \frac{df}{dx} \frac{dx}{dt}$$

Rule 4 is not immediately apparent. The derivative of a product of functions is the limit of the ratio

$$\frac{f(t + \Delta t)g(t + \Delta t) - f(t)g(t)}{\Delta t}$$

If we add and subtract the quantity $f(t + \Delta t)g(t)$ in the numerator, we can write this ratio as

$$\frac{f(t + \Delta t)g(t + \Delta t) - f(t + \Delta t)g(t) + f(t + \Delta t)g(t) - f(t)g(t)}{\Delta t}$$

$$= f(t + \Delta t)\left[\frac{g(t + \Delta t) - g(t)}{\Delta t} \right] + g(t)\left[\frac{f(t + \Delta t) - f(t)}{\Delta t} \right]$$

As Δt approaches zero, the terms in square brackets become $dg(t)/dt$ and $df(t)/dt$, respectively, and the limit of the expression is

$$f(t)\frac{dg(t)}{dt} + g(t)\frac{df(t)}{dt}$$

Rule 5 follows directly from the definition:

$$\frac{dx}{dt} = \lim_{\Delta t \to 0} \frac{\Delta x}{\Delta t} = \lim_{\Delta x \to 0} \left(\frac{\Delta t}{\Delta x} \right)^{-1} = \left(\frac{dt}{dx} \right)^{-1}$$

Comments on Rule 7

We can obtain this important result using the binomial expansion. We have

$$f(t) = t^n$$

$$f(t + \Delta t) = (t + \Delta t)^n = t^n \left(1 + \frac{\Delta t}{t}\right)^n$$

$$= t^n \left[1 + n\frac{\Delta t}{t} + \frac{n(n-1)}{2!}\left(\frac{\Delta t}{t}\right)^2 + \frac{n(n-1)(n-2)}{3!}\left(\frac{\Delta t}{t}\right)^3 + \cdots\right]$$

Then

$$f(t - \Delta t) - f(t) = t^n \left[n\frac{\Delta t}{t} + \frac{n(n-1)}{2!}\left(\frac{\Delta t}{t}\right)^2 + \cdots\right]$$

and

$$\frac{f(t - \Delta t) - f(t)}{\Delta t} = nt^{n-1} + \frac{n(n-1)}{2!}t^{n-2}\,\Delta t + \cdots$$

The next term omitted from the last sum is proportional to $(\Delta t)^2$, the following to $(\Delta t)^3$, and so on. Each term except the first approaches zero as Δt approaches zero. Thus

$$\frac{df}{dt} = \lim_{\Delta x \to 0} \frac{f(t + \Delta t) - f(t)}{\Delta t} = nt^{n\pm 1}$$

Comments on Rules 8 to 10

We first write $\sin \omega t = \sin \theta$ with $\theta = \omega t$ and use the chain rule,

$$\frac{d \sin \theta}{dt} = \frac{d \sin \theta}{d\theta}\frac{d\theta}{dt} = \omega \frac{d \sin \theta}{d\theta}$$

We then use the trigonometric formula for the sine of the sum of two angles θ and $\Delta \theta$:

$$\sin(\theta + \Delta \theta) = \sin \Delta \theta \cos \theta + \cos \Delta \theta \sin \theta$$

Since $\Delta \theta$ is to approach zero, we can use the small-angle approximations

$$\sin \Delta \theta \approx \Delta \theta \qquad \text{and} \qquad \cos \Delta \theta \approx 1$$

Then

$$\sin(\theta + \Delta \theta) \approx \Delta \theta \cos \theta + \sin \theta$$

and

$$\frac{\sin(\theta + \Delta \theta) - \sin \theta}{\Delta \theta} \approx \cos \theta$$

Similar reasoning can be applied to the cosine function to obtain rule 9.

Rule 10 is obtained by writing $\tan \theta = \sin \theta / \cos \theta$ and applying rule 4 along with rules 8 and 9.

$$\frac{d}{dt}(\tan \theta) = \frac{d}{dt}(\sin \theta)(\cos \theta)^{-1} = \sin \theta \frac{d}{dt}(\cos \theta)^{-1} + \frac{d(\sin \theta)}{dt}(\cos \theta)^{-1}$$

$$= \sin \theta(-1)(\cos \theta)^{-2}(-\sin \theta) + (\cos \theta)(\cos \theta)^{-1}$$

$$= \frac{\sin^2 \theta}{\cos^2 \theta} + 1 = \tan^2 \theta + 1 = \sec^2 \theta$$

Comments on Rule 11

Again we use the chain rule

$$\frac{de^\theta}{dt} = \frac{bde^\theta}{bdt} = b\frac{de^\theta}{d(bt)} = b\frac{de^\theta}{d\theta} \qquad \text{with} \qquad \theta = bt$$

and the series expansion for the exponential function:

$$e^{\theta + \Delta\theta} = e^\theta e^{\Delta\theta} = e^\theta\left[1 + \Delta\theta + \frac{(\Delta\theta)^2}{2!} + \frac{(\Delta\theta)^3}{3!} + \cdots\right]$$

Then

$$\frac{e^{\theta + \Delta\theta} - e^\theta}{\Delta\theta} = e^\theta + e^\theta\frac{\Delta\theta}{2!} + e^\theta\frac{(\Delta\theta)^2}{3!} + \cdots$$

As $\Delta\theta$ approaches zero, the right side of the equation above approaches e^θ.

Comments on Rule 12

Let

$$y = \ln bt$$

Then

$$e^y = bt \qquad \text{and} \qquad \frac{dt}{dy} = \frac{1}{b}e^y = t$$

Then using rule 5, we obtain

$$\frac{dy}{dt} = \left(\frac{dt}{dy}\right)^{-1} = \frac{1}{t}$$

Integral Calculus

Integration is related to the problem of finding the area under a curve. It is also the inverse of differentiation. Figure D-23 shows a function $f(t)$. The area of the shaded element is approximately $f_i\Delta t_i$, where f_i is evaluated anywhere in the interval Δt_i. This approximation improves if Δt_i is very small. The total area from t_1 to t_2 is found by summing all the area elements from t_1 to t_2 and taking the limit as each Δt_i approaches zero. This limit is called the integral of f over t and is written

$$\int_{t_1}^{t_2} f\,dt = \text{Area} = \lim_{\Delta t_i \to 0} \sum_i f_i\,\Delta t_i$$

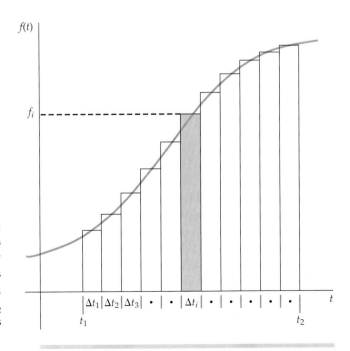

FIGURE D-23 A general function $f(t)$. The area of the shaded element is approximately $f_i\Delta t_i$, where f_i is evaluated anywhere in the interval.

If we integrate some function $f(t)$ from t_1 to some general value of t, we obtain another function of t. Let us call this function y:

$$y = \int_{t_1}^{t} f\, dt$$

The function y is the area under the f-versus-t curve from t_1 to a general value t. For a small interval Δt, the change in the area Δy is approximately $f\Delta t$.

$$\Delta y \approx f\, \Delta t$$

$$f \approx \frac{\Delta y}{\Delta t}$$

If we take the limit as Δt approaches 0, we can see that f is the derivative of y:

$$f = \frac{dy}{dt}$$

The relation between y and f is often written

$$y = \int f\, dt$$

where $\int f\, dt$ is called an **indefinite integral.** To evaluate an indefinite integral, we find the function y whose derivative is f. The definite integral of f from t_1 to t_2 is $y(t_1) - y(t_2)$, where $df/dt = y$:

$$\int_{t_1}^{t_2} f\, dt = y(t_2) - y(t_1)$$

EXAMPLE D-10

Find the indefinite integral of $f(t) = t$.

The function whose derivative is t is $\frac{1}{2}t^2$ plus any constant:

$$\int t\, dt = \tfrac{1}{2}t^2 + C$$

where C is any constant

Table D-5 lists some important integration formulas. More extensive lists of differentiation and integration formulas can be found on the Internet. They can be found by using a search engine and searching for *table of integrals,* and also in handbooks such as Herbert Dwight's *Tables of Integrals and Other Mathematical Data,* fourth edition, Macmillan Publishing Company, Inc., New York, 1961.

TABLE D-5

Integration Formulas†

1. $\displaystyle\int A\, dt = At$

2. $\displaystyle\int At\, dt = \tfrac{1}{2}At^2$

3. $\displaystyle\int At^n\, dt = A\frac{t^{n+1}}{n+1} \quad n \neq -1$

4. $\displaystyle\int At^{-1}\, dt = A \ln t$

5. $\displaystyle\int e^{bt}\, dt = \frac{1}{b}e^{bt}$

6. $\displaystyle\int \cos \omega t\, dt = \frac{1}{\omega}\sin \omega t$

7. $\displaystyle\int \sin \omega t\, dt = -\frac{1}{\omega}\cos \omega t$

8. $\displaystyle\int_0^\infty e^{-ax}\, dx = \frac{1}{a}$

9. $\displaystyle\int_0^\infty e^{-ax^2}\, dx = \frac{1}{2}\sqrt{\frac{\pi}{a}}$

10. $\displaystyle\int_0^\infty xe^{-ax^2}\, dx = \frac{2}{a}$

11. $\displaystyle\int_0^\infty x^2 e^{-ax^2}\, dx = \frac{1}{4}\sqrt{\frac{\pi}{a^3}}$

12. $\displaystyle\int_0^\infty x^3 e^{-ax^2}\, dx = \frac{4}{a^2}$

13. $\displaystyle\int_0^\infty x^4 e^{-ax^2}\, dx = \frac{3}{8}\sqrt{\frac{\pi}{a^5}}$

---

† In these formulas, A, b, and ω are constants. In formulas 1 through 7 an arbitrary constant C can be added to the right side of each equation. The constant a is greater than zero.

ILLUSTRATION CREDITS

Chapter 11

Opener p. 339 Stocktrek/Corbis; **p. 340** Collection of Historical Scientific Instruments, Harvard University; **p. 344 (top)** NASA; **(bottom)** NASA; **p. 345** Courtesy Central Scientific Company; **p. 353** NASA.

Chapter 12

Opener p. 370 Courtesy of Department of Physics, Purdue University; **p. 372** © 2002 Estate of Alexander Calder/Artists Rights Society (ARS), New York; **p. 379** Photodisk.

Chapter 13

Opener p. 395 Andy Pernick/Bureau of Reclamation; **p. 401** Vanessa Vick/Photo Researchers, Inc.; **p. 403** Chuck O'Rear/Woodfin Camp and Assoc.; **p. 405** David Burnett/Woodfin Camp and Assoc.; **p. 407 (top)** Estate of Harold E. Edgerton; **(bottom)** Takeski Takahara/Photo Researchers, Inc.; **p. 408** P. Motta/Photo Researchers, Inc.; **p. 412** Michael Dunn/The Stock Market; **p. 415** Picker International.

Chapter 14

Opener p. 425 Barry Slaven/Visuals Unlimited; **p. 427** Citibank; **p. 429** NASA; **p. 433** Institute for Marine Dynamics; **p. 442** Richard Menga/Fundamental Photographers; **p. 447 (top)** Monroe Auto Equipment; **(bottom)** David Wrobel/Visuals Unlimited; **p. 449** Eye Wire/Getty; **p. 451** Royal Swedish Academy of Music.

Chapter 15

Opener p. 465 John Cetrino/Check Six/Picture Quest; **p. 466 Figure 15-1** Richard Menga/Fundamental Photographs; **Figure 15-2** Richard Menga/Fundamental Photographs; **p. 478 (top right)** David Sacks/The Image Bank/Getty; **(left)** Maynard and Boucher/Visuals Unlimited; **p. 480** From Winston E. Cock, *Lasers and Holography,* Dover Publications, New York, 1981; **p. 484** Courtesy of Davies Symphony Hall; **p. 485 (top and bottom)** Fundamental Photographs; **p. 486 (top)** Bernard Benoit/Photo Researchers, Inc.; **(a)** Education Development Center; **p. 491 (top, a)** Sandia National Laboratory; **(top, b)** Robert de Gast/Photo Researchers; **(c)** Estate of Harold E. Edgerton/

Palm Press Inc.; **(bottom, b)** Education Development Center; **p. 496 Figure 15-30** Estate of Harold E. Edgerton/Palm Press Inc.

Chapter 16

Opener p. 503 David Yost/Steinway & Sons; **p. 507** Rubberball Productions; **p. 510 (a)** Berenice Abbott (8J 1328)/Photo Researchers; **p. 513 (left)** University of Washington; **(center)** University of Washington; **(right)** University of Washington; **p. 519** Professor Thomas D. Rossing, Northern Illinois University, DeKalb; **p. 525** Courtesy of Chuck Adler.

Chapter 17

Opener p. 532 Hoby Finn/PhotoDisk/Getty; **p. 535 (a)** Courtesy of Taylor Precision Products; **(b)** Courtesy Honeywell, Inc.; **p. 536** Richard Menga/Fundamental Photographs; **p. 538** NASA; **p. 554** Jet Propulsion Laboratory/NASA.

Chapter 18

Opener p. 558 Donna Day/PhotoDisk/Getty; **p. 559** Phoenix Pipe & Tube/Lana Berkovich; **p. 561** From Frank Press and Raymond Sievert, *Understanding Earth,* 3rd ed., W.H. Freeman and Co., 2001; **p. 562** From Donald Wink, Sharon Gislason, and Sheila McNicholas, *The Practice of Chemistry,* W.H. Freeman and Co., 2002; **p. 582** Will and Deni McIntyre/Photo Researchers.

Chapter 19

Opener p. 595 (top) Paul Chesley/National Geographic/Getty; **p. 595 (bottom)** Sandia National Laboratory; **p. 598** © 2002 Robert Briggs; **p. 601** Anderson Ross/PhotoDisk/Getty; **p. 605 (right)** Michael Collier/Stock, Boston; **(left)** Jean-Pierre Horlin/The Image Bank; **p. 606 (top left)** Sandia National Laboratory; **(top right)** Peter Miller/The Image Bank; **(bottom)** Sandia National Laboratory.

Chapter 20

Opener p. 628 Frank Siteman/Stock Boston, Inc./PictureQuest; **p. 637** Alfred Pasieka/Photo Researchers, Inc.; **p. 639** Courtesy of Eugene Mosca; **p. 643** Science Photo Library/Photo Researchers, Inc.

ANSWERS

Chapter 1

1. (c)
3. (c)
5. (a)
7. (e)
9. (a) True
   (b) False
   (c) True
11. 1.19×10^{57}
13. (a) 10^{-8} m
    (b) 20 atoms
15. (a) 3×10^{10} diapers
    (b) 1.5×10^7 m^3
    (c) 0.6 mi^2
17. \$177 M
19. (a) 0.000040 W
    (b) 0.000000004 s
    (c) 3,000,000 W
    (d) 25,000 m
21. (a) C_1 is in m; C_2 is in m/s
    (b) C_1 is in m/s^2
    (c) C_1 is in m/s^2
    (d) C_1 is in m; C_2 is in s^{-1}
    (e) C_1 is in m/s; C_2 is in s^{-1}
23. (a) 4×10^7 m
    (b) 6.37×10^6 m
    (c) 2.48×10^4 mi; 3.96×10^3 mi
25. 210 cm
27. 1.28 km
29. (a) 36.0 km/h·s
    (b) 10.0 m/s^2
    (c) 88.0 ft/s
    (d) 26.8 m/s
31. 4050 m^2
33. (a) m/s^2
    (b) s
    (c) m
35. T^{-1}
37. mv^2/r
41. M/L^3
43. (a) 30,000
    (b) 0.0062
    (c) 0.000004
    (d) 217,000
45. (a) 1.44×10^5
    (b) 2.55×10^{-8}
    (c) 8.27×10^3
    (d) 6.27×10^2
47. 4×10^6
49. (a) 1.69×10^3
    (b) 4.8
    (c) 5.6
    (d) 10
51. 31.7 y
53. 2.0×10^{23}
55. (a) 1.41×10^{17} kg/m^3
    (b) 216 m
57. (a) 4.85×10^{-6} parsec
    (b) 3.08×10^6 m
    (c) 9.47×10^{15} m
    (d) 6.33×10^4 AU
    (e) $3.25\ c \cdot$y
59. The claim is conservative as the actual weight of water used is closer to 55,000 tons.
61. (a) $n = 3/2; C = 17.0$ y/(Gm)$^{3/2}$
    (b) 0.510 Gm
63. 1.16×10^{19} lb

Chapter 2

1. 0
3. It is safer to land against the wind.
5. (a) Negative
   (b) During the last five steps, gradually slow the speed of walking, until the wall is reached.
   (c)

7. *(a)* True

    (b) True in one dimension

9. False

11. *(a)*

13. *(b)*

15. Yes. In any round-trip, A to B, and back to A, the average velocity is zero.

17. No. If the velocity is constant, a graph of position as a function of time is linear with a constant slope equal to the velocity.

19. *(b)*

21. *(a)* False

    (b) True

23. $v_{\text{top of flight}} = 0; a_{\text{top of flight}} = -g$

25. *(b)*

27. *(c)*

29. *(c)*

31. *(c)*

33. *(a)*

35. *(d)*

37. *(d)*

39. Velocity: *(a)* negative at t_0 and t_1; *(b)* positive at $t_3, t_4, t_6,$ and t_7; *(c)* zero at t_2 and t_5

    Acceleration: *(a)* negative at t_4; *(b)* positive at t_2 and t_6; *(c)* zero at $t_0, t_1, t_3, t_5,$ and t_7

41. *(a)* Graphs *(a)*, *(f)*, and *(i)*

    (b) Graphs *(c)* and *(d)*

    (c) Graphs *(a)*, *(d)*, *(e)*, *(f)*, *(h)*, and *(i)*

    (d) Graphs *(b)*, *(c)*, and *(g)*

    (e) Graphs *(a)* and *(i)* are mutually consistent. Graphs *(d)* and *(h)* are mutually consistent. Graphs *(f)* and *(i)* are also mutually consistent.

43. *(a)* 54.2 m/s; *(b)* $-123g$

45. 4.02 m/s^2

47. 14.2 ms

49. *(a)* 0.278 km/min

    (b) -0.0833 km/min

    (c) 0

    (d) 0.128 km/min

51. *(a)* 2.25 h

    (b) 4.99 h

    (c) 880 km/h

    (d) 611 km/h

53. *(a)* 4.33 y

    (b) 4.33×10^6 y; No

55. 35.8 m

57. *(a)* 0

    (b) 0.333 m/s

    (c) -2.00 m/s

    (d) 1.00 m/s

59. 122 km/h; $1.04v_{\text{av}}$

61. *(a)*

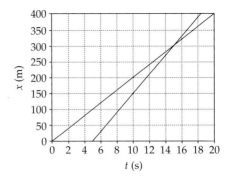

    (b) 15 s

    (c) 300 m

    (d) 100 m

63. 6 h

65. -2.00 m/s^2

67. *(a)* 2 m

    (b) $\Delta x = (2t - 5)\Delta t + (\Delta t)^2$, where x is in meters if t is in seconds.

    (c) $v = 2t - 5$, where v is in meters per second if t is in seconds.

69. *(a)* $a_{\text{av,AB}} = 3.33$ m/s^2; $a_{\text{av,BC}} = 0$; $a_{\text{av,CE}} = -7.50$ m/s^2

    (b) 75.0 m

    (c)

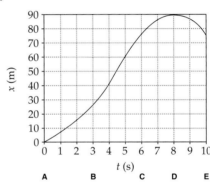

    (d) At point D, $t = 8$ s, the graph crosses the time axis and so $v = 0$.

71. *(a)* 80.0 m/s

    (b) 400 m

    (c) 40.0 m/s

73. 15.6 m/s^2

75. *(a)* 4.68 s

    (b) 20.4 m

    (c) 0.991 s and 3.09 s

77. *(a)*

    (b) 7.27 m

    (c) 1.73 s

    (d) 11.9 m/s

79. 43.6 m

81. 68.0 m/s

83. (a) 666 m

    (b) 13.6 m/s

85. (a) 10.4 s

    (b) 27.4 s

    (c) 12.8 s

87. (a) 19.0 km

    (b) 2 min 18 s

    (c) 610 m/s

89. 40.0 cm/s; -6.88 cm/s^2

91. (a) 4.76 m/s or 10.7 mi/h

    (b) 0.595

93. 10.9 m

95. 27.6 m

97. 4.59 km

99. 2.40 m; 1.40 s

103. (a) -25.7 m/s^2

     (b) 2.33 s

105. (a) 1.03 c·y/y^2 = 1.03 c/y

     (b) ≈ 2 d

107. 4.80 m/s

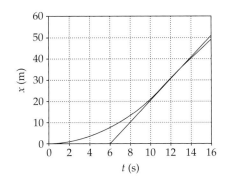

109. $\frac{1}{3} h$

111. (a) 34.7 s

     (b) 1.20 km

     (c)

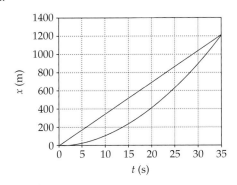

113. (a) $L_1 = \frac{2}{3} L$

     (b) $t = \frac{2}{3} t_{\text{fin}}$

115. (a) $A = 90$ m

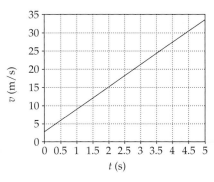

     (b) $x(t) = (3 \text{ m/s}^2)t^2 + (3 \text{ m/s})t; \Delta x = 90.0$ m

117. $x(t) = (\frac{7}{3} \text{ m/s}^3)t^3 - (5 \text{ m/s})t$

119. (a) 0.250 m/s per box

     (b) $v(1 \text{ s}) = 0.930$ m/s; $v(2 \text{ s}) = 3.20$ m/s; $v(3 \text{ s}) = 6.20$ m/s

     (c) $x(3 \text{ s}) = 7.00$ m

v vs. t

121.

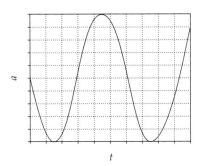

123. (a) $v(t) = (0.1 \text{ m/s}^3)t^2$

     (b) 2.23 m/s

125. 12.8 m/s^2; 30.5%

127. (a) $a = \omega v_{\text{max}} \cos(\omega t)$; Because a varies sinusoidally with time, it is *not* constant.

     (b) $x = x_0 + \dfrac{v_{\text{max}}}{\omega} [1 - \cos(\omega t)]$

129. (a) $x(t) = x_0 e^{(t-t_0)/b}$ (b = 1 s)

   (b) Because the numerical value of b, expressed in SI units, is one, the numerical values of a, v, and x are the same at each instant in time.

Chapter 3

1. The magnitude of the displacement of a particle is less than or equal to the distance it travels along its path.

3. The displacement for any trip around the track is ZERO. Thus we see that no matter how fast the race car travels, the average velocity is always ZERO at the end of each complete circuit.

5. No. The magnitude of a component of a vector must be less than or equal to the magnitude of the vector. If the angle θ shown in the figure is equal to 0° or multiples of 90°, then the magnitude of the vector and its component are equal.

7. No

9. (e)

11. (c)

13. (a) The velocity vector, as a consequence of always being in the direction of motion, is tangent to the path.

   (b)

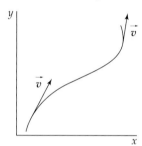

15. (a) A car moving along a straight road while braking

   (b) A car moving along a straight road while speeding up

   (c) A particle moving around a circular track at constant speed

17. (a)

   (b)

   (c)

19.

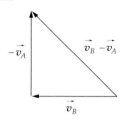

21. True

23. (d)

25. (a) False

   (b) True

27.

29. (a)

Path	Direction of velocity vector
AB	north
BC	northeast
CD	east
DE	southeast
EF	south

   (b)

Path	Direction of acceleration vector
AB	north
BC	southeast
CD	0
DE	southwest
EF	north

   (c) The magnitudes are approximately equal.

31. The droplet leaving the bottle has the same horizontal velocity as the ship. During the time the droplet is in the air, it is also moving horizontally with the same velocity as the rest of the ship. Because of this, it falls into the vessel, which has the same horizontal velocity. Because you have the same horizontal velocity as the ship, you see the same thing as if the ship were standing still.

33. True

35. The principal reason is aerodynamic drag; when moving through a fluid such as the atmosphere, the ball's acceleration will depend strongly on its velocity.

37. 14.8 m/s

39. $R = 22.2$ m; $\alpha = 22.5°$

41. (a)

(b)

(c)

(d)

(e)

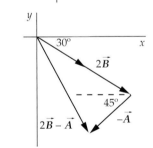

43. (b)

45.

	A	θ	A_x	A_y
(a)	10 m	30°	8.66 m	5 m
(b)	5 m	45°	3.54 m	3.54 m
(c)	7 km	60°	3.50 km	6.06 km
(d)	5 km	90°	0	5 km
(e)	15 km/s	150°	−13.0 km/s	7.50 km/s
(f)	10 m/s	240°	−5.00 m/s	−8.66 m/s
(g)	8 m/s²	270°	0	−8.00 m/s²

47. (a) 5.83; 31.0°

   (b) 122; −35.0°

   (c) 5.39; $\theta = 42.1°$; $\phi = 236°$

49. (a) $\vec{v} = (5 \text{ m/s})\hat{i} + (8.66 \text{ m/s})\hat{j}$

   (b) $\vec{A} = (-3.54 \text{ m})\hat{i} + (-3.54 \text{ m})\hat{j}$

   (c) $\vec{r} = (14 \text{ m})\hat{i} - (6 \text{ m})\hat{j}$

51. $\vec{D} = (3 \text{ m})\hat{i} + (3 \text{ m})\hat{j} + (3 \text{ m})\hat{k}$; $D = 5.20$ m

53. $\vec{v}_{av} = (14.1 \text{ km/h})\hat{i} + (-4.1 \text{ km/h})\hat{j}$

55. (b)

57. (a) $\vec{v}_{av} = (33.3 \text{ m/s})\hat{i} + (26.7 \text{ m/s})\hat{j}$

   (b) $\vec{a}_{av} = (-3.00 \text{ m/s}^2)\hat{i} + (-1.77 \text{ m/s}^2)\hat{j}$

59. $\vec{v} = (30 \text{ m/s})\hat{i} + [40 \text{ m/s} - (10 \text{ m/s}^2)t]\hat{j}$; $\vec{a} = (-10 \text{ m/s}^2)\hat{j}$

61. (a) $\vec{v}_{av} = (20 \text{ m/s})(-\hat{i} + \hat{j})$

   (b) $\vec{a}_{av} = (-2 \text{ m/s}^2)\hat{i}$

   (c) $\Delta\vec{r} = (600 \text{ m})(-\hat{i} + \hat{j})$

63. (a) 13.1° west of north

   (b) 300 km/h

65. 8.47°; 2.57 h

67. You should fly your plane across the wind.

69. (a) $\vec{r}_{AB}(6 \text{ s}) = (120 \text{ m})\hat{i} + (4 \text{ m})\hat{j}$

   (b) $\vec{v}_{AB}(6 \text{ s}) = (-20 \text{ m/s})\hat{i} - (12 \text{ m/s})\hat{j}$

   (c) $\vec{a}_{AB} = (-2 \text{ m/s}^2)\hat{j}$

71. 1.52×10^{-6} m/s²; $1.55 \times 10^{-7}g$

73. $3.44 \times 10^{-3}g$; $6.07 \times 10^{-4}g$

75. 33.4 min⁻¹

77. $h = \dfrac{(v_0 \sin \theta_0)^2}{2g}$

79. 33.8 m/s

81. 20.3 m/s; 36.2°

83. 69.3°

85. (a) 18.0 m/s

   (b) 14.0°

87. (a) 8.14 m/s

   (b) 23.2 m/s

89. −63.4°

91. 209 m

93. (a) 0.452 s

   (b) 22.6 m

95. (a) 485 km

   (b) 1.70 km/s

101. $L = \dfrac{2v_0^2 \tan \theta}{g \cos \theta}$

103. 10.8 m/s; $v = (6.50 \text{ m/s})\hat{i} + (-21.6 \text{ m/s})\hat{j}$

105. 40.5 m/s; 0.994 s

107. 7.41 m/s; 0.756 s; 15.9 m/s; 17.5 m/s; 25.0°

109. 0.785 m

111. 4.91 m/s²; 8.50 m/s²

113. (a)

y (m)

25

20

15

10

5

0   5   10   15

x (m)

   (b) $\vec{v} = (5 \text{ m/s})\hat{i} + (10 \text{ m/s})\hat{j}$; 11.2 m/s

115. $31.3°; 8.06$ m

117. Fourth step

119. (a) $v_{min} = \dfrac{x}{\cos\theta}\sqrt{\dfrac{g}{2(x\tan\theta - h)}}$

　(b) $v_{min} > 26.0$ m/s $= 58.0$ mi/h

　(c) $h_{max} < x\tan\theta$

Chapter 4

1. If an object with no net force acting on it is at rest or is moving with a constant speed in a straight line (i.e., with constant velocity) relative to the reference frame, then the reference frame is an inertial reference frame.

3. No. If the net force acting on an object is zero, its acceleration is zero. The only conclusion one can draw is that the *net* force acting on the object is zero.

5. No. Correctly predicting the direction of the subsequent motion requires knowledge of the initial velocity as well as the acceleration.

7. The mass of an object is an intrinsic property of the object whereas the weight of an object depends directly on the local gravitational field. Therefore, the mass of the object would not change and $w_{grav} = mg_{local}$. Note that if the gravitational field is zero then the gravitational force is also zero.

9. Your apparent weight would be greater than your true weight when observed from a reference frame that is accelerating upward. That is, when the surface on which you are standing has an acceleration a such that a_y is positive.

11. (a) $F_{n21} = m_1 g$
　(b) $F_{n12} = m_1 g$
　(c) $F_{nT2} = (m_1 + m_2) g$
　(d) $F_{n2T} = (m_1 + m_2) g$

13. (b)

15. (c)

17. (a)　　　　　(b)

19. (a) True
　(b) False
　(c) False
　(d) False

21. (d)

23. The velocity of the elevator has no effect on the person's apparent weight.

25. (a) 782 N; 62.6 N
　(b) Because there is no acceleration, the forces are the same going up and going down the incline.

27. (a) 6.00 m/s^2
　(b) $1/3$
　(c) 2.25 m/s^2

29. -3.75 kN

31. (a) 4.24 m/s^2 @ $45.0°$ from each force
　(b) 8.40 m/s^2 @ $14.6°$ from $2\vec{F}_0$

33. 12.0 kg

35. (a) 4.00 m/s^2
　(b) 2.40 m/s^2

37. (a) $\vec{a} = (1.50$ m/s$^2)\hat{i} + (-3.50$ m/s$^2)\hat{j}$
　(b) $\vec{v} = (4.50$ m/s$)\hat{i} + (-10.5$ m/s$)\hat{j}$
　(c) $\vec{r} = (6.75$ m$)\hat{i} + (-15.8$ m$)\hat{j}$

39. (a) 530 N
　(b) 119 lb

41. (a) 60.0 N
　(b) 57.7 N

43. $T_2 > T_1$

45. (a) $36.9°$
　(b) 4.08 N
　(c) 3.43 N; 2.40 N; 3.43 N

47. (a) $\vec{a} = (0.500$ m/s$^2)\hat{i} + (2.60$ m/s$^2)\hat{j}$
　(b) $\vec{F}_3 = (-5.00$ N$)\hat{i} + (-26.0$ N$)\hat{j}$

49. (a) $T = \dfrac{w}{2\sin\theta}; \theta = 90°; T \to T_{max}$ as $\theta \to 0°$
　(b) 19.6 N

51. (a) 11.8 kN
　(b) 9.81 kN
　(c) 7.81 kN

53. (a) 3.82 kN
　(b) 4.30 kN

55. 56.0 N

57. (d)

59. (a) 508 N; 508 N
　(b) $mg; 0$

61. 552 N

63. (a)

65. (e)

67. (a) 19.6 N
　(b) 19.6 N
　(c) 25.6 N
　(d) 14.6 N

69. (a) 1.31 m/s^2
　(b) 16.7 N; 21.3 N

71. (a) $a = \dfrac{F}{m_1 + m_2}; F_{2,1} = \dfrac{Fm_1}{m_1 + m_2}$
　(b) 0.400 m/s^2; 0.800 N

75. (a)

77. (a) $a = \dfrac{g(m_2 - m_1\sin\theta)}{m_1 + m_2}; T = \dfrac{gm_1m_2(1 + \sin\theta)}{m_1 + m_2}$
　(b) 2.45 m/s^2; 36.8 N

79. (a) 1.37 m/s^2; 61.4 N
　(b) 1.19; The answer is the ratio of two quantities with the same units and so has no units.

81. (a) 398 N
　(b) 368 N

83. (a) 5.00 cm
　(b) $a_{5\,kg} = 4.91$ m/s^2; $a_{20\,kg} = 2.45$ m/s^2; $T = 24.5$ N

85. 1.36 kg or 1.06 kg

87. $F = 2T = \dfrac{4m_1 m_2 g}{m_1 + m_2}$

91. (a) -100 m/s^2

    (b) 6.13 cm

    (c) 35.0 ms

93. 305 N; 1.55 kN

95. (a) 1.50 m/s

    (b) 1.50 m

    (c) 0.500 m/s

    (d) 12.0 N

97. (a) $a = \dfrac{F}{m_1 + m_2}$

    (b) $F_{net} = \dfrac{Fm_2}{m_1 + m_2}$

    (c) $T = \dfrac{Fm_1}{m_1 + m_2}$

    (d)

Yes . . . correct answers appear above.

99. (a) 55.0 g

    (b) 2.45 m/s^2; 2.03 N

101. (a) $\frac{1}{3}(F_2 + 2F_1)$

     (b) $\dfrac{3T_0}{4C}$

Chapter 5

1. The force of friction between the object and the floor of the truck must be the force that causes the object to accelerate.

3. (d)

5. (b)

7. As the spring is extended, the force exerted by the spring on the block increases. Once that force is greater than the maximum value of the force of static friction on the block, the block will begin to move. However, as it accelerates, it will shorten the length of the spring, decreasing the force that the spring exerts on the block. As this happens, the force of kinetic friction can then slow the block to a stop, which starts the cycle over again. One interesting application of this to the real world is the bowing of a violin string: the string under tension acts like the spring, while the bow acts as the block, so as the bow is dragged across the string, the string periodically sticks and frees itself from the bow.

9. (e)

11. Block 1 will hit the pulley before block 2 hits the wall.

13. (d)

15. (d)

17. For a rock, which has a relatively small surface area compared to its mass, the terminal speed will be relatively high; for a lightweight, spread-out object like a feather, the opposite is true.

    Another issue is that the higher the terminal velocity is, the longer it takes for a falling object to reach terminal velocity: from this, the feather will reach its terminal velocity quickly,

and fall at an almost constant speed very soon after being dropped; a rock, if not dropped from a great height, will have almost the same acceleration as if it were in free-fall for the duration of its fall, and thus be continually speeding up as it falls.

19. (a) M/T; kg/s

    (b) M/L; kg/m

    (c) ML/T^2

    (d) 56.9 m/s

    (e) 86.9 m/s

21. (b)

23. (a) 15.0 N

    (b) 12.0 N

25. 500 N

27. (a) 5.89 m/s^2

    (b) 76.4 m

29. (a) 49.1 N

    (b) 123 N

31. 4.57°

33. (a) 0.667

    (b) 2.16 m/s^2; 1.36 s

35. (a)

37. 2.36 m/s^2; 37.2 N

39. (a) 0.599

    (b) 9.25 m

    (c) 4.73 m/s

41. (a) 2.75 m/s^2

    (b) 10.1 s

43. (a) 0.965 m/s^2 directed up the incline

    (b) 0.184 N

45. (a) 25.0°

    (b) 0.118 N

47. (a) The static-frictional force opposes the motion of the object, and the maximum value of the static-frictional force is proportional to the normal force F_N. The normal force is equal to the weight minus the vertical component F_V of the force F. Keeping the magnitude F constant while increasing θ from 0 results in a decrease in F_v and thus a corresponding decrease in the maximum static-frictional force f_{max}. The object will begin to move if the horizontal component F_H of the force F exceeds f_{max}. An increase in θ results in a decrease in F_H. As θ increases from 0, the decrease in F_N is larger than the decrease in F_H, so the object is more and more likely to slip. However, as θ approaches 90°, F_H approaches 0, and no movement will be initiated. If F is large enough and if θ increases from 0, then at some value of θ, the block will start to move.

    (b)

49. (b)

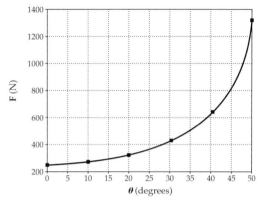

51. (a) 0.238
    (b) 1.40 m/s²
53. (a) 17.7 N
    (b) 1.47 m/s²; 5.88 N
    (c) 1.96 m/s²; 7.87 m/s²
55. (a) 0.163 m/s²
    (b) 0.0381 m
    (c) −0.254 m/s²
57. −8.41 glapp/plipp²; 0.191
59. (a) −1.57 N; 83.8 N
    (b) 6.49 N; 37.5 N
61. (a) −2.60 m/s²; 19.2 m
    (b) −2.11 m/s²; 23.7 m
63. (a) 0.297
    (b) 2.82 m/s
65. (c)
67. (a) 1.41 m/s
    (b) 8.50 N
69. (a) 8.33 m/s²; upward
    (b) 542 N; upward
    (c) 1.18 kN; upward
71. $T_2 = [m_2(L_1 + L_2)]\left(\dfrac{2\pi}{T}\right)^2$; $T_1 = [m_2(L_1 + L_2) + m_1L_1]\left(\dfrac{2\pi}{T}\right)^2$
73. 53.3°; 410 N
75. (a) 0.395 N
    (b) 0.644
77. $3.44 \times 10^{-3}g$; $6.07 \times 10^{-4}g$
79. 51.6°
81. (a) $a_c = \dfrac{v_0^2}{r}\left(\dfrac{1}{1 + \left(\dfrac{\mu_k v_0}{r}\right)t}\right)^2$
    (b) $a_t = -\mu_k a_c$
    (c) $a = a_c\sqrt{1 + \mu_k^2}$
83. 12.8 m/s
85. (a) 7.25 m/s
    (b) 0.536
87. 21.7°
89. (a) 7.832 kN
    (b) −766 kN
91. $v_{min} = 20.1$ km/h; $v_{max} = 56.1$ km/h
93. 2.79×10^{-4} kg/m
95. 88.2 km/h

97. 3.31 s; 100
99. $y(3.5$ s$) \approx 60.4$ m; $y_{max} \approx 60.6$ m @ $t = 3.3$ s; $t_{flight} \approx 7$ s; The ball spends a little longer coming down than it does going up.
101. 0.511
103. (a) 0.289
     (b) 600 N
105. 1.49 kN
107. $a = g(\sin\theta_1 - \tan\theta_0 \cos\theta_1)$
109. (a) 49.4 m/s²
     (b) 4.49 s
111. (a) 193 N
     (b) 51.8 N
     (c) The sled does not move.
     (d) μ_k is undetermined.
     (e) 536 N
113. 0.433
115. 23.6 rev/min
117. (a) Toward the earth's axis.
     (b) A stone dropped from a hand at a location on the earth. The effective weight of the stone is equal to $m\vec{a}_{st,\,surf}$, where $\vec{a}_{st,\,surf}$ is the acceleration of the falling stone (neglecting air resistance) relative to the local surface of the earth. The gravitational force on the stone is equal to $m\vec{a}_{st,\,iner}$, where $\vec{a}_{st,\,iner}$ is the acceleration of the stone relative to an inertial reference frame. These accelerations are related by $\vec{a}_{st,\,surf} + \vec{a}_{surf,\,iner} = \vec{a}_{st,\,iner}$, where $\vec{a}_{surf,\,iner}$ is the acceleration of the local surface of the earth relative to the inertial frame (the acceleration of the surface due to the rotation of the earth). Multiplying through this equation by m and rearranging gives $m\vec{a}_{st,\,surf} = m\vec{a}_{st,\,iner} - m\vec{a}_{surf,\,iner}$, which relates the apparent weight to the acceleration due to gravity and the acceleration due to the earth's rotation. A vector addition diagram can be used to show that the magnitude of $m\vec{a}_{st,\,surf}$ is slightly less than that of $m\vec{a}_{st,\,iner}$.
     (c) 983 cm/s²

Chapter 6

1. (a) False
   (b) True
   (c) True
3. False
5. No. The work done on any object by any force $\vec{F}$ is defined as $dW \equiv \vec{F} \cdot d\vec{r}$. The direction of $\vec{F}_{net}$ is toward the center of the circle in which the object is traveling, and $d\vec{r}$ is tangent to the circle. No work is done by the net force because $\vec{F}_{net}$ and $d\vec{r}$ are perpendicular, so the dot product is zero.
7. Because $W \propto x^2$, doubling the distance the spring is stretched will require four times as much work.
9. (d)
11. (a) False
    (b) False
    (c) True
13. (a) False
    (b) False
15. (a) 0.245 m
    (b) 120 J

17. $\approx 1\%$

19. 20.8 kJ

21. (a) 147 J

    (b) 266 J

23. 10.6 kJ

25. (a) 6.00 J

    (b) 12.0 J

    (c) 3.46 m/s

27. $W = -\frac{1}{2}kx_1^2 - \frac{1}{3}ax_1^3$

29. (a) $m(y) = 40 \text{ kg} - (1 \text{ kg/m})y$

    (b) 5.89 kJ

31. (a) 4.17 N

    (b) $\vec{T}, \vec{F}_g,$ and $\vec{F}_n$; Because all of these forces act perpendicularly to the direction of motion of the object, none of them do any work.

35. 180°

37. (a) -24

    (b) -10

    (c) 0

39. (a) 1.00 J

    (b) 0.213 N

43. No. Let $\vec{A} = \hat{i}, \vec{B} = 3\hat{i} + 4\hat{j}$ and $\vec{C} = 3\hat{i} - 4\hat{j}$ and form $\vec{A} \cdot \vec{B}$ and $\vec{A} \cdot \vec{C}$.

45. (b) The results of (a) and (b) tell us that $\vec{a}$ is perpendicular to $\vec{v}$ and parallel (or antiparallel) to $\vec{r}$.

47. (a) 98.1 W

    (b) 392 J

49. (a) $v = (\frac{5}{8} \text{ m/s}^2)t$

    (b) $P = 3.13t \text{ W/s}$

    (c) 9.38 W

51. 445 W

53. $v = \sqrt{v_0^2 + 2gH}$

55. 4.71 kJ

57. (a) 392 J

    (b) 2.45 m; 4.91 m/s

    (c) 24.1 J; 368 J

    (d) 392 J; 19.8 m/s

59. (a) 0.100 m

    (b) 0.141 m

61. (a) $U(\theta) = (m_2\ell_2 - m_1\ell_1)g \sin\theta$

    (b) U is a minimum at $\theta = -\pi/2$ and a maximum at $\theta = \pi/2$

    (c) $U = 0$ independently of θ

63. (a) $F_x = \dfrac{C}{x^2}$

    (b) F_x is positive for $x \neq 0$ and therefore $\vec{F}$ is directed away from the origin.

    (c) $U(x)$ decreases with increasing x.

    (d) F_x is negative for $x \neq 0$ and therefore $\vec{F}$ is directed toward the origin. $U(x)$ increases with increasing x.

65. $U(x) = \dfrac{a}{x} + U_0$

67. (a) $F_x = 4x(x + 2)(x - 2)$

    (b) $-2 \text{ m}, 0, 2 \text{ m}$

    (c) Unstable equilibrium at $x = -2$ m; stable equilibrium at $x = 0$; unstable equilibrium at $x = 2$ m

69. (a) 0 and 2 m; neutral equilibrium for $x > 3$ m

    (b)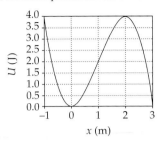

    (c) Stable equilibrium at $x = 0$; unstable equilibrium at $x = 2$ m

    (d) 2.00 m/s

71. (a) $U(y) = -mgy - 2Mg\left(L - \sqrt{y^2 + d^2}\right)$

    (b) $y = d\sqrt{\dfrac{m^2}{4M^2 - m^2}}$

    (c) Stable equilibrium

73. (a) 706 MJ

    (b) 11.8 MW

75. 0.500 m

77. (a) 34.4 N

    (b)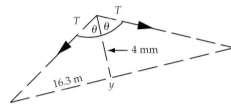

    1.68 N

    (c) 3.38 mJ

79. (a) $F(x) = mC^2x$

    (b) $W = \frac{1}{2}mC^2x_1^2$

81. In the following, if t is in seconds and m is in kilograms, then v is in m/s, a is in m/s², P is in W, and W is in J.

    (a) $v = (6t^2 - 8t); a = (12t - 8)$

    (b) $P = 8mt(9t^2 - 18t + 8)$

    (c) $W = 2mt_1^2(3t_1 - 4)^2$

83. 5.74 km

85. (a)

x	W
(m)	(J)
-4	6
-3	4
-2	2
-1	0.5
0	0
1	0.5
2	1.5
3	2.5
4	3

(b)

87. (b) $W = (10\pi\text{ m})F_0$ if the rotation is clockwise; $-(10\pi\text{ m})F_0$ if the rotation is counterclockwise. Because $W \neq 0$ for a complete circuit, $\vec{F}$ is not conservative.

89. (a)

"6–12" Potential

(b) The minimum value is about -0.0107 eV, occurring at a separation of approximately 0.380 nm. Because the function is concave upward at this separation, this separation is one of stable equilibrium, although very shallow.

(c) -6.69×10^{-12} N; 7.49×10^{-11} N

Chapter 7

1. (a)
3. (a) False
   (b) False
5. As she starts pedaling, chemical energy inside her body is converted into kinetic energy as the bike picks up speed. As she rides it up the hill, chemical energy is converted into gravitational potential energy. While freewheeling down the hill, potential energy is converted to kinetic energy, and while braking to a stop, kinetic energy is converted into thermal energy (a more random form of kinetic energy) by the frictional forces acting on the bike.
7. (d)
9. No. From the work–kinetic energy theorem, no total work is being done on the rock, as its kinetic energy is constant. However, the rod must exert a tangential force on the rock to keep the speed constant. The effect of this force is to cancel the component of the force of gravity that is tangential to the trajectory of the rock.
11. 33.6 s
13. 3.04×10^{19} J/y; $\approx 6\%$
15. 1.10×10^6 L/s
17. (c)
19. 3.89 m
21. 5.05 m
23. 25.6°
25. $U = \dfrac{[mg(\sin\theta + \mu_s \cos\theta)]^2}{2k}$
27. $6mg$
29. (c)
31. $6mg$
33. (a) 31.0 m
    (b) -31.7 J
    (c) 33.7 m/s

35. (a) 151 m
    (b) 45.3 m/s
37. (a) $\frac{5}{2}mgL$
    (b) $6mg$
39. (a) 20.2°
    (b) 6.39 m/s
41. $v = L\sqrt{2\dfrac{g}{L}(1 - \cos\theta) + \dfrac{k}{m}\left(\sqrt{\frac{13}{4} - 3\cos\theta} - \frac{1}{2}\right)^2}$
43. (a) 94.2 kJ
    (b) The energy required to do this work comes from chemical energy stored in the body.
    (c) 471 kJ
45. (a) 104 J
    (b) 70.2 J
    (c) 33.8 J
    (d) 2.91 m/s
47. (a) 7.67 m/s
    (b) 58.9 J
    (c) 0.333
49. (a) $W_f = (13.7\text{ N})y$
    (b) $E_{mech} = -(13.7\text{ N})y$
    (c) 1.98 m/s
51. 0.875 m; 2.49 m/s
53. (a) 9.00×10^{13} J
    (b) $\$2.5 \times 10^6$
    (c) 28,400 y
55. 1.88×10^{-28} kg
57. 3.56×10^{14} reactions
59. 0.782 MeV
61. (a) 3.16 kg
    (b) 8.04×10^9 kg
63. (b)
65. 57.6 MJ
67. (a) 0.208
    (b) 3.45 MJ
69. (a) From the FBD we can see that the forces acting on the box are the normal force exerted by the inclined plane, kinetic friction force, and the gravitational force (the weight of the box) exerted by the earth.
    (b) 0.451 m
    (c) 1.33 J
    (d) 2.52 m/s
71. 11.3 kW; -6.77 kW
73. (a) 1.60 kJ
    (b) 619 J
    (c) 23.4 m/s
75. (a) 147 J
    (b) The energy is transferred to the girder from its surroundings, which are warmer than the girder. As the temperature of the girder rises, the atoms in the girder vibrate with a greater average kinetic energy, leading to a larger average separation, which causes the girder's expansion.

77. (a)

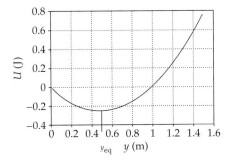

(b) $F = -ky + mg$

(c) $y_{max} = \dfrac{2mg}{k}$

(d) $y_{eq} = \dfrac{mg}{k}$

(e) $W_f = \dfrac{m^2g^2}{2k}$

79. (a) 17.3 m

(b) 4.91 kN

(c) 4.91 m/s²

(d) 13.4 kN, upward

(e) 5.46 kN; 63.9°

(f) 1.44 kN

81. (a) 491 N; 981 N

(b) 9.82 kW; 29.4 kW

(c) 8.85°

(d) 6.36 km/L

83. (a) 17.4 MJ

(b) 1.39×10^{10} J

(c) 9.73×10^9 J

(d) 1.59 MW

87. (a) $v = \sqrt{\dfrac{2mgY}{M + m}}$

(b) $v = \sqrt{\dfrac{2mgY}{M + m}}$

89. $D = 2\sqrt{HL(1 - \cos\theta)}$

91. (a) $K_{max} = mgh + \dfrac{m^2g^2}{2k}$

(b) $x_{max} = \dfrac{mg}{k} + \sqrt{\dfrac{m^2g^2}{k^2} + \dfrac{2mgh}{k}}$

(c) $x = \dfrac{mg}{k} + \sqrt{\dfrac{2m^2g^2}{k^2} + \dfrac{4mgh}{k}}$

93. (a)

(b) 5.39 kJ

Chapter 8

1. A doughnut.

3. (b)

5. No. Consider a 1-kg block with a speed of 1 m/s and a 2-kg block with a speed of 0.707 m/s. The blocks have equal kinetic energies but momenta of magnitude 1 kg·m/s and 1.414 kg·m/s, respectively.

7. $\vec{p}_{recoil} = \vec{p}_{rifle} = -\vec{p}_{bullet}$ or $\vec{p}_{rifle} + \vec{p}_{bullet} = 0$

9. Conservation of momentum requires only that the net external force acting on the system be zero. It does not require the presence of a medium such as air.

11. Think of someone pushing a box across a floor. Her push on the box is equal but opposite to the push of the box on her, but the action and reaction forces act on *different objects*. You can only add forces when they act on the same object.

13. The problem is that the comic situations violate the conservation of momentum! To move forward requires pushing something backward, which Superman doesn't appear to be doing when flying around. In a similar manner, if Superman picks up a train and throws it at Lex Luthor, he (Superman) ought to be tossed backward at a pretty high speed to satisfy the conservation of momentum.

15. The friction of the tire against the road causes the car to slow down. This is rather subtle, as the tire is in contact with the ground without slipping at all times, so as you push on the brakes harder, the force of static friction of the road against the tires must increase. Also, of course, the brakes heat up, and not the tires.

17. Assume that the ball travels at 80 mi/h = 35 m/s. The ball stops in a distance of about 1 cm, so the distance traveled is about 2 cm at an average speed of about 18 m/s. The collision time is $\dfrac{0.02 \text{ m}}{18 \text{ m/s}} \approx 1$ ms.

19. (a) False

(b) True

(c) True

21. (a) The loss of kinetic energy is the same in both cases.

(b) The percentage loss is greatest for the case in which the two objects have oppositely directed velocities of magnitude $\frac{1}{2}v$.

23. (b) is correct because all of 1's kinetic energy is transferred to 2 when $m_2 = m_1$.

25. The water is changing direction when it rounds the corner in the nozzle. Therefore, the nozzle must exert a force on the stream of water to change its direction, and, from Newton's 3rd law, the water exerts an equal but opposite force on the nozzle.

27. No. $\vec{F}_{ext,net} = d\vec{p}/dt$ defines the relationship between the net force acting on a system and the rate at which its momentum changes. The net external force acting on the pendulum bob is the sum of the force of gravity and the tension in the string and these forces do not add to zero.

29. Think of the stream of air molecules hitting the sail. Imagine that they bounce off the sail elastically–their net change in momentum is then roughly twice the change in momentum that they experienced going through the fan. Another way of looking at it: initially, the air is at rest, but after passing through the fan and bouncing off the sail, it is moving backward; therefore, the boat must exert a net force on the air pushing it backward, and there must be a force on the boat pushing it forward.

31. (a) 2.33 s
    (b) 6.74 m/s

33. (0.233 m, 0)

35. (2.00 m, 1.40 m)

37. (1.50 m, 1.36 m)

41. $z_{cm} = \frac{1}{2}R$

43. $\vec{v}_{cm} = (3\ \text{m/s})\hat{i} - (1.5\ \text{m/s})\hat{j}$

45. $\vec{a}_{cm} = (2.4\ \text{m/s}^2)\hat{i}$

47. (a) $F_n = (m_p + m_b)g$
    (b) $F_n = (m_p + 2m_b)g$
    (c) $F_n = m_p g$

49. (a) $F_n = (m_p + m_b)g$
    (b) $F_n = m_p g + m_b g\left(1 + \sqrt{1 + \dfrac{2kh}{m_b g}}\right)$

51. $\vec{v}_{10} = (4\ \text{m/s})\hat{i}$

53. $\vec{v}' = 2v\hat{i} - v\hat{j}$

55. $-\sqrt{\dfrac{gh}{3}}\,\hat{i}$

57. (a) 43.5 J
    (b) $\vec{v}_{cm} = (1.50\ \text{m/s})\hat{i}$
    (c) $\vec{v}_{1,rel} = (3.50\ \text{m/s})\hat{i}$ and $\vec{v}_{2,rel} = (-3.50\ \text{m/s})\hat{i}$
    (d) 36.8 J
    (e) $K_{cm} = 6.75\ \text{J} = K - K_{rel}$

59. (a) 10.8 N·s
    (b) 1.34 kN

61. 1.81 MN·s; 10.602 MN

63. 230 N

65. (a) $\vec{I} = (1.08\ \text{N·s})\hat{i}$ (directed into wall)
    (b) 360 N, into wall
    (c) 0.480 N·s, away from wall
    (d) 3.84 N, away from wall

67. (a) 20.0 m/s
    (b) 20% of the initial kinetic energy is transformed into thermal energy, sound, and the deformation of metal.

69. (a) −2.00 m/s
    (b) The collision was inelastic.

71. (a) $\vec{v}_{cm} = (23.1\ \text{m/s})\hat{i}$
    (b) −254 m/s

73. (a) 5.00 m/s
    (b) 0.250 m
    (c) $v_{1f} = 0; v_{2f} = 7.00\ \text{m/s}$

75. (a) $0.200v_0$
    (b) $0.400v_0$

77. 450 m/s

81. $h = \dfrac{v^2}{8g}\left(\dfrac{m_1}{m_2}\right)^2$

83. 0.0529

85. 1.50×10^6 m/s

87. (a) $\vec{v}_1 = (312\ \text{m/s})\hat{i} + (66.6\ \text{m/s})\hat{j}$
    (b) 5.61 km
    (c) 35.8 kJ

89. 0.913

91. (a) 20% of its mechanical energy is lost.
    (b) 0.894

93. (a) 1.70 m/s
    (b) 0.833

95. (a) 60°
    (b) 2.50 m/s; 4.33 m/s

97. (a) 1.00 m/s; 1.73 m/s
    (b) The collision was elastic.

99. $v_1 = 8.66\ \text{m/s}; v_2 = 5.00\ \text{m/s}$

101. In an elastic collision
$$K_i = K_f = \frac{p_1^2}{2}\left[\frac{m_1^2 + 6m_1 m_2 + m_2^2}{m_1^2 m_2 + m_1 m_2^2}\right] = \frac{p_1'^2}{2}\left[\frac{m_1^2 + 6m_1 m_2 + m_2^2}{m_1^2 m_2 + m_1 m_2^2}\right]$$
If $p_1' = +p_1$, the particles do not collide.

103. (a) $\vec{v}_{cm} = 0$
    (b) $\vec{u}_3 = (-5\ \text{m/s})\hat{i}; \vec{u}_5 = (3\ \text{m/s})\hat{i}$
    (c) $\vec{u}_3' = (5\ \text{m/s})\hat{i}; u_5' = 0.75\ \text{m/s}$
    (d) $\vec{v}_3' = (5\ \text{m/s})\hat{i}; \vec{v}_5' = (-3\ \text{m/s})\hat{i}$
    (e) 60.0 J; 60.0 J

105. (a) 360 kN
    (b) 120 s
    (c) 1.72 km/s

107. (a) $\tau_0 = 1 + \dfrac{a_0}{g}$
    (b) $v_f = gI_{sp}\left(\ln\dfrac{m_0}{m_f} - \dfrac{1}{\tau_0}\left(1 - \dfrac{m_f}{m_0}\right)\right)$
    (c)
    (d) 28.1

109. 0.192 m/s; 31.3 mJ; 12.0 mJ

111. 0.462 m/s

113. (a) $\vec{p} = -(1.10 \times 10^5\ \text{kg·km/h})\hat{i} + (1.05 \times 10^5\ \text{kg·km/h})\hat{j}$
    (b) 43.4 km/h; 46.3° west of north

115. (a) 6.26 m/s
    (b) 20.0 m

117. 3.72 m

119. (a) The velocity of the basketball will be equal in magnitude but opposite in direction to the velocity of the baseball.
    (b) $v_{1f} = 0$
    (c) $v_{2f} = 2v$

121. (a) 29.6 km/s
    (b) 8.10; The energy comes from an immeasurably small slowing of Saturn.

123. 3.00×10^5 m/s

125. (a) 0.600 m/s²
    (b) 960 N

127. No. The driver was traveling at 23.3 km/h.

129. 8.85 kg

131. $\frac{1}{14}r$

133. (a) $0.716^N E_0$
    (b) 55

135. (a) $y_{cm} = \dfrac{v^2}{2L}t^2$

(b) $a_{cm} = \dfrac{v^2}{L}$

(c) $F = \dfrac{vt}{L}\left(\dfrac{v}{gt} + 1\right)Mg$

137. $v_{2f} = \left(\dfrac{m_b}{m_2 + m_b}\right)\left[1 + \dfrac{m_1}{m_1 + m_b}\right]v;$

$v_{1f} = -\dfrac{m_2 m_b(2m_1 + m_b)}{(m_1 + m_b)^2\,(m_2 + m_b)}v$

139. -0.960 m/s^2

141. $v = (1.70\ \mathrm{m}^{1/2}/\mathrm{s})\sqrt{L}$

Chapter 9

1. Because r is greater for the point on the rim, it moves the greater distance. Both turn through the same angle. Because r is greater for the point on the rim, it has the greater speed. Both have the same angular velocity. Both have zero tangential acceleration. Both have zero angular acceleration. Because r is greater for the point on the rim, it has the greater centripetal acceleration.

3. (c)

5. (d)

7. No. A net torque is required to *change* the rotational state of an object. A net torque may decrease the angular speed of an object. All we can say for sure is that a net torque will *change* the angular speed of an object.

9. (b)

11. (b)

13. For a given applied force, this increases the torque about the hinges of the door, which increases the door's angular acceleration, leading to the door being opened more quickly. It is clear that putting the knob far from the hinges means that the door can be opened with less effort (force). However, it also means that the hand on the knob must move through the greatest distance to open the door, so it may not be the quickest way to open the door. Also, if the knob were at the center of the door, you would have to walk around the door after opening it, assuming the door is opening toward you.

15. (b)

17. (b)

19. (a)

21. True. If the sphere is slipping, then there is kinetic friction that dissipates the mechanical energy of the sphere.

23. 10.3%

25. 6.42

27. (a) 15.6 rad/s

(b) 46.8 rad

(c) 7.45 rev

(d) 73.0 m/s^2

29. (a) 40.0 rad/s

(b) 0.960 m/s^2; 192 m/s^2

31. (a) 0.589 rad/s^2

(b) 4.71 rad

33. (d)

35. 1.04 rad/s; 9.92 rev/min

37. (a) 1.87 N·m

(b) 124 rad/s^2

(c) 620 rad/s

39. (a) $g\sin\theta$

(b) Because the line-of-action of the tension passes through the pendulum's pivot point, its lever arm is zero and it causes no torque.

(c) $g\sin\theta$

41. (a) $d\tau_f = \dfrac{2\mu_k Mg}{R^2}r^2 dr$

(b) $\tau_f = \frac{2}{3}MR\mu_k g$

(c) $\Delta t = \dfrac{3R\omega}{4\mu_k g}$

43. 60.0 kg·m^2

45. (a) 28.0 kg·m^2

(b) 28.0 kg·m^2

47. 2.60 kg·m^2

49. (b) $I_{cm} = \frac{1}{12}m(a^2 + b^2)$

51. 5.41×10^{-47} kg·m^2

55. $I = \frac{3}{10}MR^2$

57. $I_x = 3M\left(\dfrac{H^2}{5} + \dfrac{R^2}{20}\right)$

59. (a) 84.6 mJ

(b) 347 rev/min

63. (a) 19.6 kN

(b) 5.89 kN·m

(c) 0.267 rad/s

(d) 1.57 kN

65. (a) 3.62 rad/s

(b) 3.62 rad/s

67. Unless M, the mass of the ladder, is zero, $v_r > v_f$. It is better to let go and fall to the ground.

69. 3.11 m/s^2; $T_1 = 12.5$ N; $T_2 = 13.4$ N

71. 8.23 m/s

73. (a) $a = \dfrac{g}{1 + \dfrac{2M}{5m}}$

(b) $T = \dfrac{2mMg}{5m + 2M}$

75. (a) 72.0 kg

(b) 1.37 rad/s^2; $T_1 = 294$ N; $T_2 = 746$ N

77. (a) $a = \dfrac{g\sin\theta}{1 + \dfrac{m_1}{2m_2}}$

(b) $T = \dfrac{\frac{1}{2}m_1 g\sin\theta}{1 + \dfrac{m_1}{2m_2}}$

(c) $E = m_2 gh$

(d) $E_{bottom} = m_2 gh$

(e) $v = \sqrt{\dfrac{2gh}{1 + \dfrac{m_1}{2m_2}}}$

(f) For $\theta = 0$: $a = T = 0$

For $\theta = 90°$: $a = \dfrac{g}{1 + \dfrac{m_1}{2m_2}}$, $T = \frac{1}{2}m_1 a$, and $v = \sqrt{\dfrac{2gh}{1 + \dfrac{m_1}{2m_2}}}$

For $m_1 = 0$: $a = g\sin\theta$, $T = 0$, and $v = \sqrt{2gh}$

79. 0.0864 m/s^2; 3.14 m/s

81. 0.192 m/s^2; 0.962 N

83. 1.13 kJ

85. 45.9 m

87. $19.5°$

89. (a) $a = \frac{2}{3}g \sin \theta$

    (b) $f_s = \frac{1}{3}mg \sin \theta$

    (c) $\theta_{max} = \tan^{-1}(3\mu_s)$

91. $v' = \sqrt{\frac{4}{3}}v$

93. 223 J

97. (a) $\alpha = \dfrac{2F}{R(M + 3m)}$; counterclockwise

    (b) $a_c = \dfrac{F}{M + 3m}$

    (c) $a_{CB} = -\dfrac{2F}{M + 3m}$

99. (a) 0.400 rad/s^2; 0.200 rad/s^2

    (b) 4.00 N

101. (a) $s_1 = \dfrac{12}{49}\dfrac{v_0^2}{\mu_k g}$, $t_1 = \dfrac{2}{7}\dfrac{v_0}{\mu_k g}$, and $v_1 = \dfrac{5}{7}v_0$

    (b) $5/7$

    (c) 26.6 m; 3.88 s; 5.71 m/s

103. $v = \dfrac{2r\omega_0}{7}$

105. (a) 360 kN

    (b) 120 s

    (c) 1.72 km/s

107. (a) $v = 1.57 v_0$

    (b) $\Delta t = \dfrac{4}{7}\dfrac{v_0}{\mu_k g}$

    (c) $\Delta x = 0.735\dfrac{v_0^2}{\mu_k g}$

111. $I = 2mR^2$

113. 0.134 m

115. (a) 7.36 m/s^2

    (b) 14.7 m/s^2

    (c) 2.43 m/s

117. (a) 780 kJ

    (b) 90.3 N·m; 150 N

    (c) 1380 rev

119. (a) 15.0 m

    (b) 15.4 rad/s

121. (a) S^2

    (b) S^3

    (c) S^5

123. (a) $\omega = \sqrt{\dfrac{4g}{3r}}$

    (b) $F = \frac{7}{3}Mg$

125. (a) $v = \sqrt{\dfrac{2MgD \sin \theta}{M + \dfrac{I}{r^2}}}$

    (b) $f_s = \dfrac{Mg \sin \theta}{1 + \dfrac{R}{r}}$

127. (a) 14.7 m/s^2

    (b) 66.7 cm

129. 41.7 J

131. The solid line on the graph shown below shows the position y of the bucket when it is in free fall and the dashed line shows y under the conditions modeled in this problem.

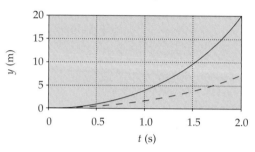

133. (a) 25.7 N

    (b) 3.21 kg

    (c) 1.10 m/s^2

Chapter 10

1. (a) True

    (b) True

    (c) True

3. $90°$

5. (a) Doubling $\vec{p}$ doubles $\vec{L}$.

    (b) Doubling $\vec{r}$ doubles $\vec{L}$.

7. False

9. (e)

11. It is easier to crawl radially outward. In fact, a radially inward force is required just to prevent you from sliding outward.

13. The hardboiled egg is solid inside, so everything rotates with a uniform velocity. By contrast, it is difficult to get the viscous fluid inside a raw egg to start rotating; however, once it is rotating, stopping the shell will not stop the motion of the interior fluid, and, for this reason, the egg may start rotating again after momentarily stopping.

15. (b)

17. (b)

19. (a) The lifting of the nose of the plane rotates the angular momentum vector upward. It veers to the right in response to the torque associated with the lifting of the nose.

    (b) The angular momentum vector is rotated to the right when the plane turns to the right. In turning to the right, the torque points down. The nose will move downward.

21. (b)

23. The center of mass of the rod-and-putty system moves in a straight line, and the system rotates about its center of mass.

25. 4.17 rev/s

27. (a) $2.40 \times 10^{-8} \text{ kg·m}^2/\text{s}$

    (b) 5.22×10^{52}; 2.29×10^{26}

    (c) The quantization of angular momentum is not noticed in macroscopic physics because no experiment can differentiate between $\ell = 2 \times 10^{26}$ and $\ell = 2 \times 10^{26} + 1$.

29. (a) 0.331

  (b) Because experimentally $C < 2/5 = 0.4$, the mass density must be greater near the center of the earth.

31. 10.1 rad/s

33. $\vec{\tau} = FR\hat{k}$

35. (a) $24\hat{k}$

  (b) $-24\hat{j}$

  (c) $-5\hat{k}$

39. $\vec{B} = 4\hat{j} + 3\hat{k}$

45. (a) 54.0 kg·m²/s

  (b) ω increases as the particle approaches the point and decreases as it recedes.

47. (a) 1.33×10^{-5} kg·m²/s

  (b) 1.33×10^{-5} kg·m²/s

  (c) 1.33×10^{-5} kg·m²/s

  (d) 8.83×10^{-5} kg·m²/s; -6.17×10^{-5} kg·m²/s

49. (a) 4.00 N·m

  (b) $(0.192 \text{ rad/s}^2)t$

51. (a) $\tau_{\text{net}} = Rg(m_2 \sin\theta - m_1)$

  (b) $L = vR\left(\dfrac{I}{R^2} + m_1 + m_2\right)$

  (c) $a = \dfrac{g(m_2 \sin\theta - m_1)}{\dfrac{I}{R_2} + m_1 + m_2}$

55. (a) 5.00 rev/s

  (b) 622 J

  (c) The energy comes from your internal energy.

57. 9.67 mm/s

59. (a) $L_0 = r_0 m v_0$

  (b) $K_0 = \frac{1}{2}mv_0^2$

  (c) $T = F_c = m\dfrac{v_0^2}{r_0}$; $W = -\dfrac{3}{2}mv_0^2$

61. 54.7°

63. (a) 3.46×10^{-47} kg·m²

  (b) 1.99 meV; 5.98 meV; 12.0 meV

65. 82.5 m/s

67. $v_{\text{cm}} = \dfrac{mv}{M + m}$; $\omega = \dfrac{mMvd}{\frac{1}{12}ML^2(M + m) + Mmd^2}$

69. $v = \sqrt{\dfrac{(0.5M + 0.8m)(\frac{1}{3}Md^2 + 0.64md^2)g}{0.32dm^2}}$

71. (a) $v_{\text{cm}} = \dfrac{K}{M}$

  (b) $\dfrac{4K}{M}$

  (c) $-\dfrac{2K}{M}$

  (d) $x = \frac{1}{6}\ell$

73. 0.349

75. 12 rad/s; 10.8 J

77. (a) 18.1 J·s

  (b) 0.414 rad/s

  (c) 15.2 s

  (d) 0.0791 J·s

79. (a) $\vec{L} = -(47.7 \text{ kg·m}^2/\text{s})\hat{k}$

  (b) $\vec{\tau} = (15.9 \text{ N·m})\hat{k}$

81. (a) 243 J·s

  (b) 306 J

83. (a) No, L decreases.

  (b) Its kinetic energy is constant.

  (c) v_0 (The kinetic energy remains constant.)

85. Yes.

87. $v_r = \dfrac{\ell\omega}{2L}\sqrt{(L^2 - \ell^2)}$

91. (a) 0.228 rad/s

  (b) 0.192 rad/s

93. 4.47×10^{22} N·m

95. 12.5 rad/s

97. (a) 26.5 rad/s

  (b) $\vec{L} = (0.303 \text{ kg·m}^2/\text{s})e^{(1.41\text{s}^{-1})t}$

Chapter R

1. The friend in the car.

3. Yes. If two events occur at the same time *and* place in one reference frame they occur at the same time *and* place in all reference frames. (Any pair of events that occur at the same time *and* at the same place in one reference frame are called a space-time coincidence.)

5. (a)

7. $1 + (8.61 \times 10^{-11})$

9. 6.00 ns

11. (a) 6.63 m

  (b) 12.6 m

13. (a) 599 m

  (b) 13.4 km

15. (a) 129 y

  (b) 87.6 y

17. (a) 0.600 m

  (b) 2.50 ns

19. 0.800c

21. (a) $4.50 \times 10^{-10}\%$

  (b) 0.142 ms

23. 25.0 min; 25.0 min

25. 60.0 min

27. 0.400c; event B can precede event A provided $v > 0.400c$

29. (a) 11.3 y

  (b) 40.0 y

31. (a) 1.005

  (b) 1.155

  (c) 1.667

  (d) 7.089

33. (a) $0.155E_0$

  (b) $1.29E_0$

  (c) $6.09E_0$

35. 2.97 GeV

39. (b) 0.866c

  (c) 0.999c

41. (a) 0.794%

    (b) 68.7%

43. (a) 0.943

45. (a) 617 eV

    (b) 79.6 eV

    (c) 7.96 eV

47. (a) 0.745

    (b) 5.00 ft

    (c) No. In Keisha's rest frame, the back end of the ladder will clear the door before the front end hits the wall of the shed, while in Ernie's rest frame, the front end will hit the wall of the shed while the back end has yet to reach the door.

Chapter 11

1. (a) False

    (b) True

3. (d)

5. (a)

7. The gravitational field is proportional to the mass within the sphere of radius r *and* inversely proportional to the square of r, i.e., proportional to $r^3/r^2 = r$.

9. (d)

11. $1.08 \times 10^{11} M_s$

13. (a) 2.78 h

    (b) 19.3×10^{42} kg·m²/s; 7.85×10^{42} kg·m²/s; 0.703%

    (c) 4.80×10^{-4} rad/s

15. 84.0 y

17. (a) 1.59×10^{11} m

    (b) 2.71×10^{10} m; 2.91×10^{11} m

19. (a) 90°

    (b) 0.731 AU

21. (a) 1.90×10^{27} kg

    (b) 0.282 m/s²; 0.0356 m/s²

23. (a) 8.18×10^4 s

    (b) 1.22×10^9 m

25. 1.99×10^{30} kg

27. $10w$, where w is your weight on earth.

29. 2.27×10^4 m/s

31. 1.43

33. (a) 7.37 m

    (b) 0.0319 mm

35. 0.605

37. (a) 2.27 kg

    (b) It is the *inertial* mass of m_2.

39. 10^9 m

41. $W = \dfrac{GMm_0}{R}$

43. 6.94 km/s

45. (a) $\vec{F}_{\text{outside}} = -\dfrac{GMm_0}{r^2}\hat{r}$

    (b) $U(r) = -\dfrac{GMm_0}{r}$; $U(R) = -\dfrac{GMm_0}{R}$

    (d) $U(r) = U(R) = -\dfrac{GMm_0}{R}$

(e)

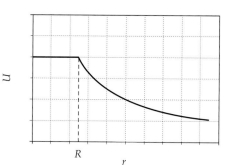

47. 2.38 km/s

49. 19.4 km/s

51. (a) 62.7 MJ

    (b) 17.4 kW·h

    (c) $139

53. (a) 7.31 h

    (b) 1.04 GJ

    (c) 8.72×10^{12} J·s

55. 11.1 GJ

57. $\vec{g} = (4 \text{ N/kg})\hat{i}$

59. (a) $\vec{g} = \dfrac{Gm}{L^2}\hat{i} + \dfrac{Gm}{L^2}\hat{j}$

    (b) $g = \sqrt{2}\,\dfrac{Gm}{L^2}$

61. (a) $\vec{g} = (-1.67 \times 10^{-11} \text{ N/kg})\hat{i}$

    (b) $\vec{g} = (-8.34 \times 10^{-12} \text{ N/kg})\hat{i}$

    (c) 2.48 m

63. (a) $M = \frac{1}{2}CL^2$

    (b) $\vec{g} = \dfrac{2GM}{L^2}\left[\ln\!\left(\dfrac{x_0}{x_0 - L}\right) - \left(\dfrac{L}{x_0 - L}\right)\right]\hat{i}$

65. (a) 0

    (b) 0

    (c) 3.20×10^{-9} N/kg

67. $g_1 = g_2$

69. (a) $F = \dfrac{Gm(M_1 + M_2)}{9a^2}$

    (b) $F = \dfrac{GmM_1}{3.61a^2}$

    (c) 0

71. (a) $F_g = \dfrac{mg}{R}r$

    (b) $F_N = \left(\dfrac{mg}{R} - m\omega^2\right)r$

    (c) The change in mass between you and the center of the earth as you move away from the center is more important than the rotational effect.

73. $g(x) = G\left(\dfrac{4\pi\rho_0 R^3}{3}\right)\left[\dfrac{1}{x^2} - \dfrac{1}{8(x - \frac{1}{2}R)^2}\right]$

75. $\omega = \sqrt{\dfrac{4\pi\rho_0 G}{3}}$

77. 0.104 mm/s

79. (a) $\vec{F} = -\dfrac{GMm}{d^2}\left[1 - \dfrac{\frac{d^3}{4}}{\left\{d^2 + \dfrac{R^2}{4}\right\}^{3/2}}\right]\hat{i}$

    (b) $\vec{F}(R) = -0.821\dfrac{GMm}{R^2}\hat{i}$

81. 249 y

83. (a) $W = GM_E m\left(\dfrac{1}{r_1} - \dfrac{1}{r_2}\right)$

   (b) $W = mgR_E^2\left(\dfrac{1}{R_E} - \dfrac{1}{R_E + h}\right)$

85. 8.96×10^7 m

87. 1.70 Mm

91. $v = 1.64\sqrt{\dfrac{GM}{a}}$

93. For $r < R_1$, $g(r) = 0$; For $r > R_2$, $g(r) = \dfrac{GM}{r^2}$;

   For $R_1 < r < R_2$, $g(r) = \dfrac{GM(r^3 - R_1^3)}{r^2(R_2^3 - R_1^3)}$

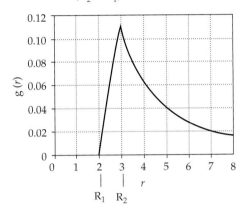

95. $g = \dfrac{2G\lambda}{r}$

97. (b) $U = -\dfrac{GMm_0}{L}\ln\left(\dfrac{x_0 + L/2}{x_0 - L/2}\right)$

99. 33.5 pN

101. (a) The gravitational force is greater on the lower robot, so if it were not for the cable its acceleration would be greater than that of the upper robot, and they would separate. In opposing this separation the cable is stressed.

   (b) 220 km

Chapter 12

1. (a) False

   (b) True

   (c) True

   (d) False

3. No. The definition of the center of gravity does not require that there be any material at its location.

5. This technique works because the center of mass must be directly under the balance point. Hence the intersection of the two lines must be at the center of mass.

7. (b)

9. (c)

11. The tensile strengths of stone and concrete are at least an order of magnitude lower than their compressive strengths, so you want to build compressive structures to match their properties.

13. (b) 200 N/m

15. 318 N

17. (b) Taking long strides requires a larger coefficient of static friction because θ is then large.

   (c) If μ_s is small, that is, there is ice on the surface, θ must be small to avoid slipping.

19.

21. $(x_{cg}, y_{cg}) = \left(\dfrac{\frac{1}{2}a^2b - \pi aR^2 + \pi R^3}{ab - \pi R^2}, \dfrac{1}{2}b\right)$

23. 692 N; 90°; 2.54 kN; No block is required to prevent the mast from moving.

25. 0.728 m

27. $F_2 = \dfrac{1}{2}W$; $F_1 = \dfrac{\sqrt{3}}{2}W$

29. (a) 5.00 m

   (b) 4.87 m

31. $\vec{F}_1 = \dfrac{Mg\sqrt{h(2R - h)}}{h - R}\hat{i} + Mg\hat{j}$

33. (a) $\vec{F} = (30.0\,\text{N})\hat{i} + (30.0\,\text{N})\hat{j}$

   (b) $\vec{F} = (35.0\,\text{N})\hat{i} + (45.0\,\text{N})\hat{j}$

35. (a) $F_n = Mg - F\sqrt{\dfrac{2R - h}{h}}$

   (b) $F_{c,h} = F$

   (c) $F_{c,v} = F\sqrt{\dfrac{2R - h}{h}}$

37. (a) 6.87 N

   (b) 1.65 N·m

   (c) −8.26 N; 15.1 N

39. 636 N; 21.5°

41. (a) 70.7 N

   (b) 1.77 m

   (c) 3.54 m

   (d) 497 N

43. $\tau_{net} = (69.3\,\text{N})b - (40.0\,\text{N})a$

45. $D = \frac{1}{2}(\sqrt{3}b - a)$

47. $y = \dfrac{35.7\,\text{m} - 30.4x}{3.57\,\text{m} - (294\text{m}^{-1})x}$

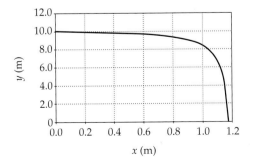

49. $h = \mu_s L \tan \theta \sin \theta$

51. $\mu_s = \dfrac{2h}{L \tan \theta \sin \theta}$

53. 59.0°

55. (a) 41.6 N

    (b) 0.136%

57. 5.01°

61. (a) 1.82×10^6 N/m²

    (b) 6.62 mJ

63. 0.686

65. It will not support the elevator.

69. $F_L = 117$ N; $F_R = 333$ N

71. $w_1 = 1.50$ N; $w_2 = 7.00$ N; $w_3 = 3.50$ N

73. 0.148

75. $\mu_s < 0.500$

77. $\mu_s = \frac{1}{2}(\cot \theta - 1)$

79. (a) 147 N

    (b) 3.62 m

81. The block will tip before it slides.

83. $\mu_s < 0.500$

85. (a) The stick remains balanced as long as the center of mass is between the two fingers. For a balanced stick the normal force exerted by the finger nearest the center of mass is greater than that exerted by the other finger. Consequently, a larger static-frictional force can be exerted by the finger closer to the center of mass, which means the slipping occurs at the other finger.

    (b) The finger farthest from the center of mass will slide inward until the normal force it exerts on the stick is sufficiently large to produce a kinetic-frictional force exceeding the maximum static-frictional force exerted by the other finger. At that point the finger that was not sliding begins to slide, the finger that was sliding stops sliding, and the process is reversed. When one finger is slipping the other is not.

87. (a) 23.0 m/s

    (b) 29.1 m/s

89. (c) $\ell_5 = 1.142$ m; $\ell_{10} = 1.464$ m; $\ell_{100} = 2.594$ m

    (d) Increasing N in the spreadsheet solution suggests that the sum of the individual offsets continues to grow as N increases without bound. The series is, in fact, divergent and the stack of bricks has no maximum offset or length.

91. 566 N

93. $F_n = 2mg$; $F = \dfrac{mg}{\cos \theta}$; $F_W = mg\dfrac{R - r}{\sqrt{R(2r - R)}}$

Chapter 13

1. (e)

3. (c)

5. Nothing. The fish is in neutral buoyancy, so the upward acceleration of the fish is balanced by the downward acceleration of the displaced water.

7. (b)

9. It blows over the ball, reducing the pressure above the ball to below atmospheric pressure.

11. False

13. The buoyant force acting on the ice cubes equals the weight of the water they displace (i.e., $B = w_f = \rho_f V_{fg}$). When the ice melts, the volume of water displaced by the ice cubes will occupy the space previously occupied by the submerged part of the ice cubes. Therefore the water level remains constant.

15. Because the pressure increases with depth, the object will be compressed and its density will increase. Thus it will sink to the bottom.

17. The drawing shows the beaker and a strip within the water. As is readily established by a simple demonstration, the surface of the water is not level while the beaker is accelerated, showing that there is a pressure gradient. That pressure gradient results in a net force on the small element shown in the figure.

19. From Bernoulli's principle, the opening above which the air flows faster will be at a lower pressure than the other one, which will cause a circulation of air in the tunnel from opening 1 toward opening 2. It has been shown that enough air will circulate inside the tunnel even with the slightest breeze outside.

21. 0.673 kg

23. 103 kg

25. 29.8 inHg

27. 230 N

29. 198 atm

31. (a) 14.8 kN

    (b) 0.339 kg

33. 0.453 m

35. $F = \dfrac{\rho g a^3}{8}$

37. 4.36 N

39. (a) 11.1×10^3 kg/m³

    (b) lead

41. 800 kg/m³; 1.11

43. 250 kg/m³

45. 3.89 kg

47. 2.46×10^7 kg

49. 491 kN

51. (a) 9.28 cm/s

    (b) 0.331 cm

    (c) 8.31 cm, in reasonable agreement with everyday experience.

53. (a) 12.0 m/s

    (b) 133 kPa

    (c) The volume flow rates are equal.

55. (a) 4.58 L/min

    (b) 763 cm²

57. 144 kPa

59. (a) 21.2 kg/s

    (b) 636 kg·m/s

    (c) 899 kg·m/s; 899 N

61. (a) $x = 2\sqrt{h(H - h)}$

    (b) $h = \frac{1}{2}H \pm \frac{1}{2}\sqrt{H^2 - x^2}$

63. (b) $P_{top} = P_{atm} - \rho g d$

65. 1.43 mm

67. 93.4 mi/h; Since most major league pitchers can throw a fastball in the low-to-mid-90s, this drag crisis may very well play a role in the game.

69. 0.0137; 0.0115

71. The net force is zero. Neglecting the thickness of the table, the atmospheric pressure is the same above and below the surface of the table.

73. 1061 kg/m^3

75. 65.7%

77. If you are floating, the density (or specific gravity) of the liquid in which you are floating is immaterial as you are in translational equilibrium under the influence of your weight and the buoyant force on your body. Thus the buoyant force on your body is your weight in both (a) and (b).

79. $V = \dfrac{m}{0.96\rho_{\mathrm{W}}}$

81. 11.8 cm

83. 1 m is a reasonable diameter for the pipeline.

85. $h_{\mathrm{A}} = 12.6$ m; $h_{\mathrm{B}} = 9.78$ m

87. (a) 64.6%

    (b) 10.7 kN

    (c) 17.9 m/s^2

89. 3.31×10^{-3} mmHg or 3.31 μmHg

91. 1.37

93. (a) 70.0 m^3

    (b) 7.47 m/s^2

95. (c) 0.126 km^{-1}

97. (a) 33.9 kN

    (b) 39.8 kN; 36.1 kN

99. (c) $h = \left(\sqrt{H} - \dfrac{A_2}{2A_1} \sqrt{2g}\, t \right)^2$

    (d) 1 h 46 min

INDEX

Physical Constants[†]

Atomic mass constant	$m_u = \frac{1}{12}m(^{12}C)$	$1\,u = 1.660\,538\,73(13) \times 10^{-27}\,kg$
Avogadro's number	N_A	$6.022\,141\,99(47) \times 10^{23}\,particles/mol$
Boltzmann constant	$k = R/N_A$	$1.380\,6503(24) \times 10^{-23}\,J/K$ $8.617\,342(15) \times 10^{-5}\,eV/K$
Bohr magneton	$m_B = e\hbar/(2m_e)$	$9.274\,008\,99(37) \times 10^{-24}\,J/T =$ $5.788\,381\,749(43) \times 10^{-5}\,eV/T$
Coulomb constant	$k = 1/(4\pi\epsilon_0)$	$8.987\,551\,788\ldots \times 10^9\,N \cdot m^2/C^2$
Compton wavelength	$\lambda_C = h/(m_e c)$	$2.426\,310\,215(18) \times 10^{-12}\,m$
Fundamental charge	e	$1.602\,176\,462(63) \times 10^{-19}\,C$
Gas constant	R	$8.314\,472(15)\,J/(mol \cdot K) =$ $1.987\,2065(36)\,cal/(mol \cdot K) =$ $8.205\,746(15) \times 10^{-2}\,L \cdot atm/(mol \cdot K)$
Gravitational constant	G	$6.673(10) \times 10^{-11}\,N \cdot m^2/kg^2$
Mass of electron	m_e	$9.109\,381\,88(72) \times 10^{-31}\,kg =$ $0.510\,998\,902(21)\,MeV/c^2$
Mass of proton	m_p	$1.672\,621\,58(13) \times 10^{-27}\,kg =$ $938.271\,998(38)\,MeV/c^2$
Mass of neutron	m_n	$1.674\,927\,16(13) \times 10^{-27}\,kg =$ $939.565\,330(38)\,MeV/c^2$
Permittivity of free space	ϵ_0	$8.854\,187\,817\ldots \times 10^{-12}\,C^2/(N \cdot m^2)$
Permeability of free space	μ_0	$4\pi \times 10^{-7}\,N/A^2$
Planck's constant	h	$6.626\,068\,76(52) \times 10^{-34}\,J \cdot s =$ $4.135\,667\,27(16) \times 10^{-15}\,eV \cdot s$
	$\hbar = h/(2\pi)$	$1.054\,571\,596(82) \times 10^{-34}\,J \cdot s =$ $6.582\,118\,89(26) \times 10^{-16}\,eV \cdot s$
Speed of light	c	$2.997\,924\,58 \times 10^8\,m/s$
Stefan-Boltzmann constant	σ	$5.670\,400(40) \times 10^{-8}\,W/(m^2 \cdot K^4)$

[†] The values for these and other constants can be found in Appendix B as well as on the Internet at http://physics.nist.gov/cuu/Constants/index.html. The numbers in parentheses represent the uncertainties in the last two digits. (For example, 2.044 43(13) stands for 2.044 43 ± 0.000 13.) Values without uncertainties are exact. Values with ellipses are exact (like the number $\pi = 3.1415\ldots$).

Derivatives and Definite Integrals

$$\frac{d}{dx}\sin ax = a\cos ax$$

$$\int_0^\infty e^{-ax}\,dx = \frac{1}{a}$$

$$\int_0^\infty x^2 e^{-ax^2}\,dx = \frac{1}{4}\sqrt{\frac{\pi}{a^3}}$$

$$\frac{d}{dx}\cos ax = -a\sin ax$$

$$\int_0^\infty e^{-ax^2}\,dx = \frac{1}{2}\sqrt{\frac{\pi}{a}}$$

$$\int_0^\infty x^3 e^{-ax^2}\,dx = \frac{4}{a^2}$$

$$\frac{d}{dx}e^{ax} = ae^{ax}$$

$$\int_0^\infty xe^{-ax^2}\,dx = \frac{2}{a}$$

$$\int_0^\infty x^4 e^{-ax^2}\,dx = \frac{3}{8}\sqrt{\frac{\pi}{a^5}}$$

The a in the six integrals is a positive constant.

Vector Products

$$\vec{A} \cdot \vec{B} = AB\cos\theta \qquad \vec{A} \times \vec{B} = AB\sin\theta\,\hat{n} \quad (\hat{n}\text{ obtained using right-hand rule})$$